ISNM
International Series of
Numerical Mathematics
Vol.119

Edited by
K.-H. Hoffmann, München
H. D. Mittelmann, Tempe

Approximation and Computation:

A Festschrift in Honor of Walter Gautschi
Proceedings of the Purdue Conference, December 2-5, 1993

Edited by

R. V. M. Zahar

Birkhäuser
Boston • Basel • Berlin

R.V.M. Zahar
Département d'informatique et de
    recherche opérationnelle
Université de Montréal
C.P. 6128, Succursale "Centre-Ville"
Montréal, Québec
Canada H3C 3J7

Library of Congress Cataloging In-Publication Data

Approximation and computation : a festschrift in honor of Walter
    Gautschi : proceedings of the Purdue conference, December 2-5, 1993
    / edited by R.V.M. Zahar.
        p.   cm. -- (International series of numerical mathematics ;
    vol. 119)
        Includes bibliographical references and indexes.
        ISBN 0-8176-3753-2 (acid-free)
        1. Approximation theory--Congresses. 2. Orthogonal polynomials-
    -Congresses. 3. Numerical integration--Congresses. 4. Functions,
    Special--Congresses.   I. Gautschi , Walter. II. Zahar, R. V. M.
    (Ramsay Vincent Michel), 1941-   . III. Series: International
    series of numerical mathematics ;   v. 119.
    QA221.A634 1994   94-44243
    511'.4--dc20   CIP

Printed on acid-free paper
© Birkhäuser Boston 1994   *Birkhäuser*

ISBN 0-8176-3753-2
ISBN 3-7643-3753-2
Text prepared by editor in TEX.
Printed and bound by Quinn-Woodbine, Woodbine, NJ.
Printed in the U.S.A.

9 8 7 6 5 4 3 2 1

CONTENTS

Walter Gautschi

International Series of Numerical Mathematics, Vol. 119, ©1994 Birkhäuser

FOREWORD*

John R. Rice[†]

Walter Gautschi was born in Basel, Switzerland on December 11, 1927. He went to school there, finishing his education with a Ph.D. in mathematics from the University of Basel under the direction of A.M. Ostrowski. In 1954, he went to Rome for a year as a Research Fellow at the National Institute for Applications of Computation. He then came to the United States for appointments in three laboratories: Harvard Computation Lab in 1955, National Bureau of Standards in 1956, and Oak Ridge National Lab in 1959. During this period he married Erika who had been a neighbor in Basel. They now have four children: Thomas, Theresa, Doris and Caroline.

Ten years after his Ph.D. Walter had established himself as a leading numerical analyst and he joined academia in 1963 as Professor of Mathematics and Computer Science at Purdue. He must have found it comfortable as he has stayed here for 30 years. He did take a couple of years off, once to go to Munich and once to go to Madison, Wisconsin. He must have found Italy comfortable too; he went there about eight times for the summer and he is going again next summer.

Walter has had an outstanding career in research, having published over 100 journal papers *and in four languages*! His linguistic virtuosity is further seen from his two book translations from German to English: *Introduction to Numerical Analysis* by Stoer and Bulirsch, and *Lectures on Numerical Mathematics* by Rutishauser. He also wrote many conference papers. But what I find truly amazing are his other scholarly works: over 80 book reviews, over 400 reviews for Mathematical Reviews, and countless referee reports. I looked over Walter's vita to see where his research interests have led him. He started out studying methods for ordinary differential equations and he wrote ten papers in the field with the last in the 1980's. The "BIG THREE" of his research interests are numerical quadrature, special functions and orthogonal polynomials; in each area he has written over 30 papers. Linear algebra has always been an important tool in Walter's work and about a dozen of his papers can be classified as "pure" linear algebra. Then there are about 25 other papers in other areas which demonstrate his wide breadth of interests.

Walter has excelled equally in teaching. He is consistently rated very highly by the students in spite of also being quite demanding. He has had seven Ph.D. students, concentrating on the Z's. His first student was Ray Zahar (who organized

---

*Opening Remarks at Walter Gautschi's Conference, December 2, 1993
†Department of Computer Sciences, Purdue University, West Lafayette, IN 47907-1398, USA

the program of this conference so admirably) in 1968 and his most recent is Minda Zhang in 1993. The others are J.L. Phillips, L.A. Anderson, Ja-Chen Lin, S.E. Notaris, and Shikang Li.

Walter's superb efforts in reviewing and refereeing show that he has a very strong commitment to his profession *far beyond the call of duty*. This shows up in other ways. He has a total of 76 years of service on the editorial boards of journals. This includes the last 9 years as Managing Editor (or Editor-in-Chief) of Mathematics of Computation. This is perhaps the premier journal in its field, one that handles about 300 papers a year. That means it averages more than one paper submitted every working day of the year. And every one of these papers is considered in some way by the Managing Editor. Finally, Walter has written four very special articles: historical or memorial papers for E.B. Christoffel, Luigi Gatteschi, Yudell L. Luke and Alexander M. Ostrowski.

This is an extraordinary achievement in scholarship. I am pleased to see such an outstanding set of speakers who have come to Purdue to help us celebrate Walter's sixty-fifth birthday, and to honor him and his achievements.

International Series of Numerical Mathematics, Vol. 119, ©1994 Birkhäuser

INTRODUCTION

R.V.M. Zahar*

The sixty-fifth birthday of Walter Gautschi provided an opportune moment for an international symposium in his honor, to recognize his many contributions to mathematics and computer sciences. Conceived by John Rice and sponsored by Purdue University, the conference took place in West Lafayette from December 2 to 5, 1993, and was organized around the four main themes representing Professor Gautschi's principal research interests: Approximation, Orthogonal Polynomials, Quadrature and Special Functions. Thirty-eight speakers — colleagues, co-authors, research collaborators or doctoral students of Professor Gautschi — were invited to present articles at the conference, their lectures providing an approximately equal representation of the four disciplines. Five invited speakers, Germund Dahlquist, Philip Davis, Luigi Gatteschi, Werner Rheinboldt and Stephan Ruscheweyh, were unable to present their talks because of illness or other commitments, although Professors Dahlquist, Gatteschi and Ruscheweyh subsequently contributed articles to these proceedings. Thus, the final program contained thirty-three technical lectures, ten of which were plenary sessions.

Approximately eighty scientists attended the conference, and for some sessions — in particular, Walter's presentation of his entertaining and informative *Reflections and Recollections* — that number was complemented by many visitors and friends, as well as the family of the honoree. A surprise visit by Paul Erdös provided one of the highlights of the conference week. The ambiance at the symposium was extremely collegial, due no doubt to the common academic interests and the personal friendships shared by the participants.

Several friends and associates gave generously of their time and energy for the organization of the conference, and I am particularly thankful to them. At l'Université de Montréal, François Jalbert contributed valued technical advice in the planning and construction of the program, and Bernard Derval provided unlimited systems support. Mark Ocker and Jean Jackson of Purdue University accepted the responsibility for the detailed operation of the technical sessions and social activities, at the same time establishing an excellent rapport with all the conference participants. Four of Professor Gautschi's colleagues require our thanks as well. John Rice obtained funding for the conference, motivated its organization with minimal intervention, and hosted the final event with friendly efficiency. Gene Golub added his support and information during the invitational phase, and played

*Département d'informatique et de recherche opérationnelle, Université de Montréal, C.P. 6128, Succ. "Centre-Ville", Montréal, Québec, Canada H3C 3J7; zahar@iro.umontreal.ca

his inimitable rôle as master of ceremonies at the conference dinner. Frank Olver was distinguished by his quiet and reassuring behind-the-scenes guidance, as well as his intimate tribute to the honoree. Finally, the vitality of the conference owed a great deal to the indomitable humor and personal courtesy of Richard Varga.

These proceedings contain thirty-three articles from the invited speakers of the conference — Richard Askey, Gene Golub and Ed Saff not finding it possible to submit the papers they presented — and five additional articles kindly contributed by the following authors: Brezinski and Van Iseghem, De Villiers and Rohwer, Ehrich, Gupta and Masson, and Kunkle. It is a pleasure to acknowledge the speakers (and many others, too numerous to mention) for their enthusiastic best wishes to Walter Gautschi, and the authors for their supreme efforts in producing such excellent papers. In this volume, the co-authors of each article are listed alphabetically, and the papers themselves are presented in alphabetical order based on the first-named authors. All articles were anonymously refereed, the great majority of them more than once. I am deeply indebted to those referees for their expert, timely and unselfish contributions, especially since most of them were neither participants at the conference nor authors in these proceedings. And I most gratefully acknowledge the diligent assistance of Richard Simard, who accepted the tasks of formatting the papers and of molding the various versions of TeX and other text editors to construct a cogent and uniformly-presented book.

A brief discussion of each of the thirty-eight papers in this volume is given below. Although many of the articles are multidisciplinary, with subject boundaries that cannot be clearly delimited, they have been grouped under the four main themes of the conference, for convenient reference.

Approximation

In the first five listed articles of this group, Gautschi's work on specific subjects is employed, expanded or generalized.

Li studies extended Lagrange-Kronrod interpolation for a continuous function on $[-1, 1]$, where the nodes are chosen as the zeros of orthogonal and Stieltjes polynomials with respect to Chebyshev weight functions. He proves that the interpolants converge in the mean for Chebyshev functions of the first, third and fourth kind.

Mastroianni also considers extended Lagrange interpolation on $[-1, 1]$, where the nodes are the zeros of two types of orthogonal polynomials with (possibly) different weight functions, as well as selected points near the endpoints of the interval. Necessary and sufficient conditions for convergence of the interpolants in Jacobi-weighted L_p spaces are proved.

Flury and Neuenschwander present algorithms for simultaneously reducing several positive definite symmetric matrices to almost diagonal form by similarity transformations, a technique that has applications to statistical models in principal component and canonical correlation analysis.

Dahlquist presents an algorithm for the Fast Fourier Transform of functions which may otherwise be slowly convergent, such as those which are smooth on a logarithmic scale, with particular emphasis on obtaining good accuracy near the origin.

Cash and Zahar review the algorithms of Gautschi and Olver, as well as several modifications and generalizations of those algorithms, for approximating subdominant solutions of recurrence relations and difference systems, and show that they can all be expressed by specific triangular decompositions of the matrix in a particular boundary value system.

Polynomial interpolation in $\mathbb{R}^d$ is considered in three articles.

de Boor describes a method of interpolation for a function defined on a finite subset of $\mathbb{R}^d$, for which the resulting linear system has an infinite number of solutions. He shows that a natural choice for the polynomial interpolating subspace F, which satisfies degree-reduction, scale-invariance and orthogonality, can be computed by a segmented form of Gaussian elimination. It is proved that the resulting F is independent of the elimination process.

Cheney and Lei consider the approximation of a continuous function f on $\mathbb{R}^d$ by quasi-interpolants: infinite linear combinations of base functions, with coefficients that are values of f on some lattice of nodes in $\mathbb{R}^d$. Choices of both nodes and base functions are discussed, with particular emphasis on scattered nodes, and the properties of the interpolants are given.

Kunkle uses quasi-interpolants to strengthen and generalize the proof of Favard's theorem on univariate extensions.

Polynomial interpolation in the complex plane is discussed in two articles, although from completely different standpoints.

Brück, Sharma and Varga consider the difference between the interpolating polynomial approximation and the truncated power series of an analytic function defined on a compact subset E of the complex plane. They show that previous results on equiconvergence can be generalized and extended when the nodes of the Lagrange interpolation polynomial are chosen as zeros of the Faber polynomial with respect to E.

Greiner and Ruscheweyh analyze the quality of polynomial approximations to conformal maps f in the complex plane. They investigate bounds for the largest number ρ such that the image $f(\rho D)$, where D is the unit disk, is included in the image $p_n(D)$ of some univalent polynomial interpolating f at the origin.

Additional problems of approximation are considered in the following articles.

De Villiers and Rohwer present a detailed and complete analysis of quadratic nodal spline interpolation on the real line, and give sharp bounds for the norm of the approximation operator as measured by the Lebesgue constant.

Micchelli and Xu present a method, based on a refinement equation and the two operations of translation and scaling, for constructing smooth wavelets: that is, continuous orthogonal bases on $[0, 1]$ with any desired degree of smoothness.

Rivlin considers polynomials $p(x)$ bounded by 1 at k equally spaced points of the real line, and gives upper bounds for the maximum norm of p on the span of the k points, and for the absolute values of the coefficients of p.

Calvetti, Reichel and Zhang give a detailed description of an adaptive semi-iterative algorithm, based on a modification of Richardson iteration, for the solution of inconsistent semidefinite linear systems. Particular emphasis is placed on the choice of iteration parameters as reciprocal values of Leja points.

Orthogonal Polynomials

Zhang describes collaborative work with Gautschi on a modified Chebyshev algorithm for computing a sequence of orthogonal polynomials of Sobolev type, and analyzes the sensitivity of the associated modified moments.

Brezinski and Van Iseghem define vector orthogonal polynomials of negative integer dimension, and develop a QD-like algorithm for their algebraic expressions.

Gohberg and Landau give a concise proof of the Gohberg-Semençul formula which expresses the inverse of a Toeplitz matrix in terms of two closely related Toeplitz matrices, and describe the formula's relevance in stochastic processes and other applications.

Gupta and Masson examine a polynomial recurrence for the Askey-Wilson polynomial, and give explicit solutions in terms of hypergeometric functions as well as expressions for the associated continued fractions and weight functions.

Gutknecht and Gragg present an extended survey analyzing several techniques for computing formal orthogonal polynomials by solving recurrence formulae with a look-ahead strategy.

Iserles develops a technique, combining Taylor and Dirichlet expansions, for transforming the almost Mathieu equation and its generalizations to three-term recurrences with coefficients that are almost periodic.

Quadrature

The first four articles in this group describe results based on quadrature rules to which Gautschi has made significant contributions.

Notaris gives a state-of-the-art review of Gauss-Kronrod quadrature on real intervals, with special emphasis on the error term, and on desirable properties such as the interlacing of nodes, the inclusion of the nodes in the interval of integration, and the positivity of weight functions.

Milovanović constructs weighted Gauss-Christoffel quadrature rules on $(0, \infty)$, and reduces the expressions for slowly convergent series to integrals involving modified weights that are motivated by the Einstein and Fermi functions.

Ehrich investigates the remainder terms of extended Gaussian quadrature formulae for functions of bounded variation, and shows that the Gauss-Kronrod formula is asymptotically optimal among all formulae with the interlacing property.

Stenger describes several transformations in the complex plane, such as the sinc approximation, and demonstrates how the analysis of numerical quantities (in particular the remainder terms of quadrature formulae) can be aided or simplified by considering the transformed problems.

Integration on the real line is considered in three articles.

Laurie investigates real integrals and quadrature rules based on the expansion of the integral in series of polynomials that are orthogonal with respect to an inner product induced by the rules. He gives numerical results which demonstrate that a nonlinear rule is a promising candidate of the given type.

Lyness considers real integrals with a singularity in the form of an arbitrary monomial term at the origin, and develops the algebraic expression obtained by the trapezoidal rule in a generalized Euler-Maclaurin expansion.

Mastroianni and Monegato derive new estimates for the remainder terms of Gauss-Laguerre and Gauss-Hermite quadrature rules for integrals over infinite intervals.

Aspects of quadrature in the complex plane are discussed in the following two articles.

Jones and Waadeland analyze Szegő quadrature on the unit circle in the complex plane, with nodes as zeros of Szegő polynomials. Convergence as the number of nodes approaches infinity is proved, and upper bounds for the remainder term are given.

Korevaar presents a survey of recent work involving Chebyshev-type quadrature on a compact subset of the complex plane when the number of nodes is large, and discusses applications to potential theory.

Special Functions

Some of Gautschi's contributions to special functions are considered in the first two articles of this group.

Ismail and Muldoon present several theorems on the complete monotonicity of expressions involving gamma and q-gamma functions — for the main purpose of deriving inequalities of those special functions — and develop some bounds in the complex plane.

Temme considers series expansions for the incomplete gamma function with complex arguments, and examines several asymptotic representations that are suitable for numerical computation.

Various aspects of special functions are presented in the following articles.

de Branges introduces a class of weight functions for the upper-half of the complex plane, including a discussion of the related Hilbert spaces of entire functions, and demonstrates that such weight functions can be constructed from the gamma function.

Butcher considers two functional equations which arise in the parallelization of diagonally-implicit multistage methods for ordinary differential equations, and shows how the solutions and their Padé approximations are related to the stability of the numerical methods.

Gatteschi develops asymptotic formulae for the zeros of the Jacobi polynomials, mainly in terms of Bessel functions, and presents inequalities that the zeros satisfy.

Letessier, Valent and Wimp discuss a method for generating differential equations satisfied by combinations of derivatives of known differential equations, and apply the technique to obtain specific differential equations satisfied by hypergeometric functions.

Ng and Wong construct a uniformly valid asymptotic approximation for a damped linear oscillator, by a logical development of known asymptotic techniques of increasing complexity.

Olver analyzes the asymptotic behavior of the generalized exponential integral for large absolute values of the argument and the order, and demonstrates that increased accuracy in asymptotic expansions can be obtained by expanding the remainder term in series of generalized exponential integrals.

Pasquini formulates a method for finding the zeros of polynomial solutions to second order linear homogeneous differential equations by direct conversion to a nonlinear algebraic system with the same zeros. The method is compared to the more traditional one of calculating the eigenvalues of an associated Jacobi matrix, and is shown to be numerically superior in the case of Bessel polynomials.

International Series of Numerical Mathematics, Vol. 119, ©1994 Birkhäuser

Reflections and Recollections

Walter Gautschi[1]

0   I really had no intention to give a talk at this meeting, and it was only after persistent persuasion on the part of Ray Zahar that I agreed to compromise: I am not going to present a technical lecture but rather an informal personal talk reflecting on my career, especially on my student days in Basel, my early attempts of doing mathematics, and the various, often accidental, happenings in my life that led me to get interested in one area or another. In retrospect, I find it indeed rather remarkable that some major lines of research which I pursued during good parts of my life had their origin in what appeared at the time to be insignificant chance events. If there are researchers — and undoubtedly there are — who at an early stage of their lives set their major research goals, conceive grand designs to achieve them, and then spend the rest of their lives putting piece after piece together to complete their program, I am not one of them! Much of my work has been the result of small outside stimuli, and in this sense I am afraid that my career must be characterized as being somewhat ill-conditioned.

Another thought why I decided to give this talk is this: when one reads papers, one often has the feeling of being handed down a highly polished endproduct, all nicely wrapped up, with no hints being provided as to what motivated the work, what were the difficulties that had to be overcome, and what wrong tacks were taken along the way, leading nowhere. While I am not advocating that this kind of information be put into each paper, it would be revealing, nevertheless, if scientists with a substantial record of research to their credit would take time out near the end of their careers to write a personal account of their work and explain the external and internal circumstances that motivated, influenced, and helped them along in their pursuits. Such essays, indeed, could well become valuable source material for future historians.

1   Needless to say, I was born approximately 65 years ago (on December 11, 1927, to be precise) in Basel, along with my twin brother, Werner. From the way our parents named us, it would appear that I was born first — "a" comes before "e". Not only were we twins, but we were actually identical twins. Just how identical we were can be seen from an early photograph taken in the kindergarten swimming pool. (It is the only photograph in my possession that features both of us together.) While it may seem difficult to speculate as to who is who in this picture, there is actually an easy answer. Identical twins are known to exhibit

[1]Department of Computer Sciences, Purdue University, West Lafayette, IN 47907-1398, USA

traits of symmetry; in our case, one of us had his hair parted on the left, the other on the right. If you look at me, you will note that my hairline is on my left. This means in Fig. 1 that I must be the fellow behind, looking away from the camera.

Figure 1: Werner and Walter in kindergarten

It is true that there still remained the problem (for outsiders) of remembering who of us had the hairline on the left and who on the right. Interestingly, nature (or perhaps our parents) provided an answer for that too: one of our names has an "l" in the middle, which stands for "left", while the other has an "r", which stands for "right". And miraculously, this works in English as well as in German! Our likeness continued unabatedly into adult age, as can be seen from Fig. 2 below.

As identical twins we had of course many common interests: we both loved music and studied the piano, often playing four-hand pieces together, and we both eventually became mathematicians. There were also differences. Unfortunately, at the early age of five, my brother became seriously ill with pneumonia and angina. I remember well how pneumonia in those days was dreaded as a deadly disease, and it was indeed extremely fortunate that Werner eventually recovered, after many months in the Children's Hospital in Basel, thanks in large measure to his dedicated physician — then a professor of pediatrics at the University of Basel. Yet his heart was seriously damaged as a result of the illness. This meant that he was not allowed to engage in any strenuous physical activities and had to abstain from sports of any kind. Turning this handicap into an asset, Werner developed a passion for reading, and he became a very studious and scholarly type of person, quite in contrast to my own inclinations, which were more outdoors-oriented. I loved to play soccer out on the street, or on the nearby playground, often until late at night and at the expense of doing homework for school. Inevitably, my grades, especially in Latin and Greek history — we both attended the Humanistic Gymnasium (HG) in Basel — began to slide and deteriorated to a point where, after the first year of Gymnasium, I was advised to change over to the Scientific Gymnasium (Mathematisch-Naturwissenschaftliches Gymnasium — MNG for short) since I did reasonably well in mathematics, physics, and modern languages. I made the change, but in the process lost one year of school. From this point on, Werner, who had no difficulties whatsoever, either in his humanistic or scientific studies, was always one year ahead of me: he graduated with the Matura, began his studies of mathematics at the University of Basel, became an assistant to Professor Ostrowski, and got his Ph.D. degree, always one year before me. He

Figure 2: Werner and Walter in their early thirties

wrote a thesis in linear algebra, which in part dealt with the asymptotic behavior of powers of matrices and in part with the theory of matrix norms. The two parts were published respectively in the *Duke Mathematical Journal* and the *Compositio Mathematica*. At the suggestion of Ostrowski, he went to Princeton to join the group around John von Neumann and Herman Goldstine, to learn about, and participate in, their work on computing eigenvalues of matrices. He soon became interested, however, in statistics and probability, which also flourished at Princeton, and after the year was up moved to Berkeley, where he joined the schools of Jerzy Neyman and David Blackwell. There, for two years, he deepened his knowledge of statistics to a point where he could actively start doing research in this field. After a brief instructorship at Ohio State University, he began his academic career in earnest at Indiana University, but returned to Ohio State two years later in 1959. It was there, soon after he arrived, that his heart gave out and he died of a massive heart attack.[2] A life full of promise came abruptly to an end, and I am tempted to say about Werner — at the risk of sounding presumptuous — what Newton said about Roger Cotes, who also died at an early age: "Had he lived, we would have known something."

2 This may be an appropriate moment to pay tribute to our teachers. The University of Basel has always been (and still is) a relatively small university, but one that has a long and distinguished history going back to 1460. Among the world-class mathematicians who taught at the university were Jakob Bernoulli (from 1687 to 1705), his younger brother Johann Bernoulli, who after Jakob's death succeeded him in his chair, and Johann's sons Daniel and Johann II. Teleki, a Hungarian count who visited Basel, wrote in his diary about Daniel Bernoulli[3]: "This great man had barely a few lads in his audience, and even these, it seemed and he himself said, were very immature." And it was not much different when I started taking classes at the university. There were only two full professors in mathematics, Andreas Speiser and Alexander Ostrowski, outstanding scholars as in Bernoulli's time, and a few lads like myself sitting in their lectures. Both were excellent teachers, each in his own way, Speiser emphasizing the beauty of mathematics, Ostrowski its rigor.

Figure 3: Andreas Speiser, Alexander Ostrowski

Speiser came from a well-established family in Basel, but studied in Göttingen under Hermann Minkowski. He is perhaps best known for his classic book on finite

---

[2]For obituaries, see A. Ostrowski,"Werner Gautschi 1927–1959", *Verh. Naturf. Ges. Basel*, **71**:314–316, 1960; J.R. Blum, "Werner Gautschi, 1927–1959", *Ann. Math. Statist.*, **31**:557, 1960.

[3]E. Bonjour, *Die Universität Basel: von den Anfängen bis zur Gegenwart 1460–1960*. Helbing & Lichtenhahn, Basel, 1960, p. 312.

groups — "still the most beautiful introduction into group theory" according to B.L. van der Waerden — published in Springer's "yellow" series. He was also deeply interested in the philosophy and history of mathematics, and wrote widely on the subject, and in the interplay between mathematics and the arts as manifested, for example, in ornaments. The fact that he was also an excellent pianist made him even more endearing to me, and I remember well the excitement of playing with him — on two grand pianos stored away in the warehouse of the local music store — Mozart's beautiful D Major Sonata for two pianos. Speiser made his most important contribution to the history of mathematics as the general director of a committee overseeing the publication of Euler's collected works. Thirty-seven volumes were published under his leadership, of which he edited eleven. Speiser's views on philosophy, often at odds with contemporary philosophers, are vividly described by Fleckenstein in an obituary[4] where one also finds a list of Speiser's writings and a brief introduction by van der Waerden to his mathematical work.

Ostrowski, on the other hand, was born in Kiev, where already at the age of 18 he studied mathematics privately with Dimitrii Alexandrovich Grave — a student of Chebyshev — and wrote a long paper (in Russian) on Galois fields. He left Kiev shortly before the revolution and went, via Marburg, to Göttingen, where he wrote his thesis under David Hilbert and Edmund Landau. He became an assistant to Felix Klein and while still in Göttingen he was entrusted with the task of editing (with R. Fricke) the first volume of Klein's collected papers. Already famous for his work in algebra and complex function theory, he came to Basel in 1927, where he remained until his retirement in 1958 except for occasional visits to the United States and Canada. In Basel he continued to produce works of great depth and craftsmanship in almost every discipline of mathematics. His own collected works are now available in six volumes[5], and there are several published appreciations of his work, and obituaries[6].

While still a student, I served my mathematical apprenticeship as an assistant to Ostrowski for two years. This was quite a memorable experience. One of the principal duties of Ostrowski's assistants indeed was to literally assist him during his research. Ostrowski had the peculiar habit of "talking" mathematics; when he was doing research — usually in the afternoons — he had to verbalize the whole thought process, even when he was alone in his office. He would stand at the blackboard — his assistant sitting at the table in front of him — and he would spell

---

[4]J.O. Fleckenstein and B.L. van der Waerden, "Zum Gedenken an Andreas Speiser", *Elem. Math.*, **26**:97–102, 1971.

[5]Alexander Ostrowski, "Collected Mathematical Papers", Vols. 1–6, Birkhäuser, Basel, 1983–1985.

[6]W. Gautschi, "To Alexander M. Ostrowski on his ninetieth birthday", *Linear Algebra Appl.*, **52/53**:iii–vi, 1983; P. Lancaster, "Alexander M. Ostrowski, 1893–1986", *Aequationes Math.*, **33**:121–122, 1987; M. Eichler, "Alexander Ostrowski — Über sein Leben und Werk", *Acta Arith.*, **51**:295–298, 1988; R. Jeltsch-Fricker, "In memoriam: Alexander M. Ostrowski (1893 bis 1986)", *Elem. Math.*, **43**:33–38, 1988. [English translation in *Leningrad Math. J.*, **2**:199–203, 1991]; D.K. Faddeev, "On R. Jeltsch-Fricker's paper 'In memoriam: Alexander M. Ostrowski (1893–1986)' ", *Leningrad Math. J.*, **2**:205–206, 1991

out exactly what was going through his mind, why something may or may not be true, how to go about proving it, and if entangled in contradictions, where he might have gone astray. Now it was precisely the assistant's responsibility to prevent this last calamity from occurring! He had to interrupt Ostrowski immediately, as soon as he made a mistake. An awesome responsibility, considering Ostrowski's lightning speed of reasoning, which was legendary. Once things began to crystallize, the time came to "fix" it: he would dictate in short key phrases what he had learned, and the assistant had to write it down on a piece of paper. The next afternoon, Ostrowski would take it from there and continue with his research until it was time again to fix further progress. This could go on for several weeks; but then, when everything finally fell into place, and he had all the key elements in front of him, dutifully recorded by the assistant, Ostrowski was ready to dictate the whole paper. This could be in any of four languages — he was fluent in German, French, English and Italian. Happily, I was familiar with all these languages, having switched from a Latin- and Greek-based Gymnasium to one teaching modern languages. (Russian would have been a problem, but fortunately Ostrowski quit long ago publishing in his native language.) At this point, the assistant became a secretary and had to type the paper on a very old typewriter, but one that had three levels, the top one for mathematical symbols. When galley proofs came back from the printer — and they arrived with great frequency — it was yet another of the assistant's duties to help proofread them. Naturally, all this was a great learning experience for me.

One semester, while I was still his assistant, Ostrowski went on leave to visit the Numerical Analysis group at the Bureau of Standards, then located in Los Angeles. In his place came Rolf Nevanlinna from Zurich, and I was given the pleasant task of assisting him during his visits to Basel. The seminar he gave on geometric function theory still stands out in my mind as one of the highlights of my educational experience.

3   But when did I first receive formal training in Numerical Mathematics? There was a lot of it in Ostrowski's research, as he was just then developing his convergence theory of relaxation methods. Yet, Ostrowski never taught a course in Numerical Analysis. I took my first course — somewhat pompously titled "Scientific Computation" (see Fig. 4) — in my third year at the university. The instructor was Eduard Batschelet, then a Privatdozent at the university and a full-time teacher of mathematics and physics at the HG, the school I flunked out of. He himself had been a student of Ostrowski, and graduated with an interesting thesis on algebraic equations, which analyzed the maximum spread of all the roots when the degree of the polynomial is fixed and all (complex) coefficients have fixed moduli but varying phases. He succeeded in sharpening earlier results of Ostrowski on this problem. His "Habilitationsschrift", however, dealt with finite difference methods for solving elliptic equations with mixed boundary conditions, and he was probably the first to give a satisfactory treatment of curved boundaries for such problems. So Batschelet certainly was the natural person to teach such a course, but not surprisingly — the first electronic computers had just appeared a few

years earlier — the course was taught in the old tradition of Runge and Willers[7].
I mention this course, since it sparked my early interest in ODEs. It was in this
course where we learned about Grammel's graphical method for solving ordinary

Figure 4: Excerpt from Walter's "Kollegienbuch"

differential equations. The basic idea is very simple: if one plots the reciprocal
of a function y of x on a polar axis, with x being the angle, then the endpoint
P will trace a curve C as x varies. If one does the same with the derivative y',
but on an axis turned around by a right angle, one gets another curve C' traced
by the endpoint P' (see Fig. 5). It is elementary to observe that the straight line
connecting P' with P is always tangent to C. Now solving a differential equation
basically means: one knows y' and wants y. Thus, knowing P', the tangential
property above tells us the direction in which P has to move, and we can take
a small step, possibly with suitable refinements as indicated in the figure on the
right, to advance P, hence $1/y$, hence y. The nice thing about this method is that it
can be applied (in principle) to differential equations of arbitrary order by simply
plotting the reciprocals of the successive derivatives on polar axes successively
rotated by right angles and drawing tangents from the highest derivative down to
next lower ones.

This simple method was used by Grammel in a variety of engineering applications. As I learned later from Kamke's monumental Handbook (which devotes

---

[7]Later on, Batschelet became interested in Biostatistics, moved to the United States and became a professor at Catholic University in Washington. He eventually returned to Switzerland
to assume a professsorship at the University of Zurich. There, he died unexpectedly, soon after
retirement. For obituaries, see: F. Halberg, "In memoriam: Edward Batschelet, Chronobiomath-
ematician", *Chronobiologia*, **7**:148–151, 1980; R. Ineichen, "Professor Dr. Eduard Batschelet, 6.
April 1914 bis 3. Oktober 1979", *Elem. Math.*, **35**:105–107, 1980.

a considerable amount of space to this method) and from other sources, Grammel (1889–1964) was a professor at the Technical University of Stuttgart and a prominent researcher in Technical Mechanics. He wrote several books, one on

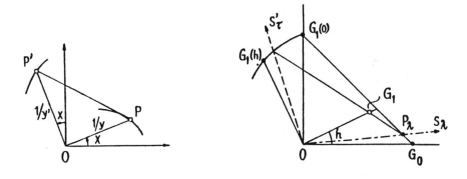

Figure 5: Grammel's method

Hydrodynamics, another on the theory of the spinning top, and his best-known, together with C.B. Biezeno, on Technical Dynamics, a massive 1000-page treatise which he jokingly referred to as a "textbook of mechanics for adults"! He has received many distinctions, among them an Honorary Doctoral Degree from the E.T.H. in Zurich.

When I first learned about this method, I was puzzled by the fact that it used reciprocals of functions rather than functions themselves. It didn't take me long to discover, however, that a similar method could be devised that works directly with the functions. The only difference is that the straight line $\overline{P'P}$ is now not tangential, but normal to the curve C. Thus, instead of advancing tangentially, one advances at a right angle. I told Batschelet about this, and he liked the idea. He in turn must have told Ostrowski, because Ostrowski asked me to give a talk on this "new" method at the annual meeting of GAMM (the older German sister society of SIAM) which in 1951 happened to convene in Freiburg, not far from Basel. That was a rather frightening proposition to me — the young unexperienced graduate student I was — but I had no choice: Ostrowski insisted on it! What had made this experience even more frightening was to discover, when I gave my talk, that the famous Professor Grammel was sitting right in the front row! My presentation must have been awfully flustered, because after the talk Grammel got up and made some comments which, although I can't remember the substance of them, made me feel put in my place. Nevertheless, I wrote up my talk and produced my first published paper [9][8]. On a happier note, at the same meeting I met for the first time Heinz Rutishauser — then a rising star at the E.T.H. in Zurich — who gave talks on the solution of some PDE problems by relaxation on the old Zuse machine Z4, which was then running at the E.T.H., and on his pioneering designs for constructing compilers for high-level programming languages.

4 About this time, I had to start thinking about writing a thesis. I went to see Ostrowski to ask whether he would have an interesting thesis topic for me. I was rather taken aback when he suggested that I might as well continue with this graphical method of mine and expand it into a Ph.D. thesis. I knew that the future was not in graphical methods, but in digital numerical methods, and the thought of spending a lot of time and energy on something that seemed old-fashioned and unrewarding distressed me greatly. But there was no point arguing with Ostrowski: once he had made up his mind, that was it! Fortunately, he left me considerable leeway as to the details of how, and in what direction, the "expansion" should proceed. My decision was to subject the method to a detailed analysis of the error — both local and global — and I labored on it for many months. It was a labor of convenience, not of love! I knew that in order to graduate, one had to write a thesis, and that's what I did. In fairness, however, I must admit that the work also led to some purely mathematical questions regarding systems of difference equations, and I submitted some relevant results to the *Commentarii Mathematici Helvetici*, where they appeared in due time [12]. The rest of the thesis was published in the *Verhandlungen der Naturforschenden Gesellschaft Basel* [10, 11], a journal of the local society of natural sciences.

As I was writing my thesis, I engaged in a considerable amount of "moonlighting". I looked around to see what others had done on the general problem of error analysis for ODEs. A paper that made a particular impression on me was one by R. von Mises, who in 1930 gave what I believe was the first analysis of the propagation of error in the Adams method. He did it only for a single differential equation, however, and I felt it would be worthwhile to extend his ideas to systems of ODEs, especially since my thesis had to deal also with higher-order differential equations. Another exciting event was the appearance of Collatz's book on numerical methods for differential equations. It couldn't have come at a better time! I immediately devoured it from cover to cover, and I felt like I found the Promised Land. All I needed now was a gentle push into a specific direction where I could do meaningful original research. Such a push came in the form of two papers, one of which from no other than Ludwig Bieberbach. (Little did I know that much later I should have had a hand in helping to prove his famous conjecture.) Bieberbach wrote many textbooks — on analytic geometry, algebra, function theory, differential equations, and others — and I knew them all and liked them, because they were written briskly and always to the point. In his book on differential equations, he had a short section on numerical methods, in particular

---

[8]Numbers in brackets refer to the list of publications at the end of the article.

the Runge-Kutta method, for which he stated an upper bound for the truncation error, which, he said, can be obtained " . . . durch einige Rechnung". It was only in 1951, however, that the complete proof — this time for general systems of differential equations — appeared in print, in the then newly established *Zeitschrift für Angewandte Mathematik und Physik* (*ZAMP*), published in Zurich. (How this paper came about, given Bieberbach's peculiar situation after World War II, I do not really know, but I suspect that Ostrowski had something to do with it.) I studied this paper very carefully, because I had just read another paper, by R. Zurmühl, in which a Runge-Kutta-type method was proposed for a single nth-order differential equation, which greatly interested me (perhaps unduly so, in retrospect) because the method had a truncation error of the spectacular order $n + 3$ for the solution itself, and of orders one less for each successive derivative up to the $(n - 2)$nd, for which (and also for the $(n - 1)$st) the order was the usual 5. (It was Rutishauser who later correctly pointed out that the global error was nevertheless of order 4 in all derivatives, just as in the classical Runge-Kutta method.) So I plunged into the formidable task of applying Bieberbach's techniques to Zurmühl's method to obtain truncation error bounds for each derivative. It was quite a tour-de-force, but I prevailed and submitted the results (without proofs) to *ZAMP*, the same journal that published Bieberbach's work.

In another article that was more closely related to my thesis, I proved that the two choices — Grammel's and my own — of plotting curves were the only natural ones possible in the context of solving ODEs. I thought the proof I gave, based as it was on a partial differential equation, was rather pretty. I sent that paper also to *ZAMP*, but it met with a peculiar fate: it seemed to have disappeared into a black hole! For several years I didn't hear anything about it, and — timid and shy as I was — did not dare to inquire. Many years later, when I was working at the Bureau of Standards, Hans Ziegler, a professor of mechanics at the E.T.H., came to visit us, and we had a friendly chat over a cup of coffee. To my great surprise, he asked me whatever happened to that paper of mine submitted to *ZAMP*, which he thought (having been an editor of *ZAMP*) was a nice little piece of work, but he had never seen it in print. The only thing I could tell him was that I didn't have the foggiest idea myself, and, in fact, had long forgotten about it!

5 Soon after I had passed my Ph.D. exam, I received the happy news that I had been awarded a two-year fellowship from a private Swiss foundation. To fill the time until the summer of 1954, when the term of the fellowship started, I was given a one-semester assistantship at the E.T.H. under Professor Saxer. If nothing else, it gave me a first taste of mass education, as I had to conduct the calculus exercise classes in front of several hundred students. More rewarding was Professor Stiefel's seminar on numerical PDEs, which I regularly attended and which brought me in closer contact with Stiefel, Rutishauser and some of their young collaborators.

The fellowship I was to embark on was admirably generous in its conception and purpose: It was to give young people an opportunity to go out into the world and enlarge their horizon, not only technically, but also — and perhaps more importantly — culturally and socially. As I was to spend the first of my two years in Rome, at Mauro Picone's Institute of Computational Mathematics[9], what place could be better than this to expand one's horizon? Here was a city where history is solidified all over the place, from pre-Christian civilization to the grandeur of the Roman empire, from the refined architectures of the Baroque and Renaissance to the more recent expressions of grandiosity. Even a lifetime wouldn't be enough to absorb this enormous richness of cultural heritage, let alone one measly year. Fortunately, the Institute was open only during the morning hours (most people there quit working at 2 p.m.) so every afternoon I was free to roam around the city and explore its art treasures, which is exactly what I did. Although mathematically I learned a great deal by attending courses of Picone on the calculus of variations and of Luigi Fantappiè on his symbolic calculus, and by studying Fichera's book

Figure 6: Picone's farewell lecture prior to his retirement

on linear operators — my first exposure to functional analysis, I began to feel a bit guilty, midway through the year, for not having done any mathematics myself. I was relieved, therefore, when one morning Picone summoned me to his office (he had a battery of push bottons in his office activating bells, which he frequently used to call in his collaborators and staff). Knowing that I sat in on his calculus of

---

[9]Picone was probably the first to conceive the idea of creating a computing laboratory, following his experience in calculating ballistic trajectories during World War I, and he founded one as early as 1927 in Naples. Subsequently, in 1932, it became one of the Institutes of the National Research Council (C.N.R.) in Rome, called "Istituto Nazionale per le Applicazioni del Calcolo" (I.N.A.C.), of which he remained the director until 1960. It still exists today, and is now named after him.

variations course, he showed me a recent paper in which he derived very general necessary conditions for extremal curves to hold, not only in the interior of domains but also on the boundary. He confessed that he could derive these conditions only under the assumption of C^1 regularity, but felt that they should hold under the sole assumption of continuity. Why didn't I take a look at it and see whether I could come up with a proof?

I was happy that finally I had some serious mathematics to think about. It must have been pure luck, however, that it occurred to me right away — the same evening, as I recall — how the desired proof can be crafted via an appropriate sequence of transformations of variables. The argument was surprisingly elementary, and I showed it to Picone the next day. His reaction, rather unexpectedly, was one of exuberance (Picone came originally from Sicily and had all the temperamental attributes of an Italian from the South!); he was extremely profuse in his praise, and was sincerely delighted, without any feelings of envy, that I succeeded where he had failed. I was deeply moved by this outpouring of unselfish encouragement and enthusiasm, and it certainly was something I badly needed to boost my self-confidence. There were times, indeed, when I wondered about my ability to do research. I now felt that, perhaps, I had a chance after all!

Picone, shrewd as he was, asked me to write up the results for publication, not in an Italian journal, but rather in German in the Swiss journal *Commentarii Mathematici Helvetici* [15]. In this way, not only my proof, but also his own new ideas in the calculus of variations, were given a chance to become more widely known outside of Italy. He also suggested that I try to generalize the results to extremal problems in two dimensions. This turned out to be rather more difficult, but eventually I got that done also and wrote it up as a note for the Accademia dei Lincei [14]. At about the same time I learned that my Runge-Kutta paper had been accepted for *ZAMP* [13]. This was a very happy moment in my life: there I was, all of a sudden with three papers ready to go into print! I decided that was enough for mathematics, and I resumed my explorations of the treasures of Rome.

The year in Rome ended all too soon, and it became time to make the big move to the United States. Here, at Aiken's Computation Laboratory at Harvard, I was to spend my second postdoctoral year and get some hands-on experience with electronic computers. There was ample opportunity to do that, as I was assigned a project in physical chemistry that required the computation of extensive numerical tables. I had access during nights to Mark III, a massive electronic computer with magnetic drum storage and lots of vacuum tubes which had a tendency to blow out every so often. The programming (in machine language!) therefore had to be done with great care, allowing for frequent checks to make sure that everything was progressing in good order. But once the program and the computer worked, it was wonderful for me to watch the great speed with which the computations were done. I hadn't seen anything like it before. It also provided added incentive to learn more about numerical analysis, which I did by studying the books of Scarborough, Householder, Milne, and Hildebrand, some of the few textbooks then available. I also sat in on a course of Garrett Birkhoff on heat conduction, and attended his seminar on lattice theory, to keep alive my curiosity in applied and pure mathematics, but I did little original research except for a few attempts — all unsuccessful — to make further progress on the calculus of variations problem that I had worked on in Rome.

6 Once the year in Cambridge drew to a close, the more serious side of life — earning a living — needed attention. Fortunately, this was a time when computers sprang up everywhere, and people to man them were in great demand. So the climate for suitable employment was quite favorable. In addition, Ostrowski had already established contacts with the Bureau of Standards (now the "National Institute of Standards and Technology"), and it was through his recommendations that I was offered a job at the Bureau in Washington, D.C. There were some technical difficulties in hiring me, however, which had to do with the fact that I did not have U.S. citizenship and — more seriously — was a citizen of a country that was not a member of NATO! This made it impossible to employ me directly at the Bureau, but I was given a position of adjunct professor at American University, which allowed me to work under a contract with the Bureau of Standards. My office, however, was on the Bureau grounds, and I paid my respects to the university by offering evening classes in Numerical Analysis.

One of the major projects then in progress at the Bureau was the preparation of the *Handbook of Mathematical Functions*, under the direction of Milton Abramowitz. Most of the chapters had already been assigned, but there were still a few in need of an author. One was the chapter on the error function and related functions, and Milton asked me whether I would be interested in writing it. I knew very little about special functions and nothing about the error function, but I was eager to learn, and I accepted. Having had a solid training in classical analysis, I felt I was up to the job. I began to study all treatises available on special functions, particularly on the class of confluent hypergeometric functions, of which the error function was a special member, and I systematically went through the research literature. I kept a detailed diary in which I collected all the information I could find, systematizing it and rechecking everything. Sometimes this led to original research as, for example, in connection with certain known inequalities for the error function, which I managed to generalize to the incomplete gamma function [16]. This in turn, in a limit case, sharpened an interesting inequality of Eberhard Hopf for the exponential integral, and in another special case gave rise to new inequalities for the gamma function. Curiously, it was only the latter that caught the attention of the public and eventually generated a literature of its own. Another spin-off of my work on the *Handbook*, which I never bothered to publish, was an application of numerical ODE methods to the construction of contour maps for analytic functions. The particular map I needed was for the complex error function. Eventually, after many rounds of revisions and polishing, the chapter got written up in its final form [4]. The result must have pleased Milton, because he asked me to help writing another chapter [3], the one on the exponential integral, which seemed to have run into problems materializing.

During this work on the *Handbook* I became acquainted with J.C.P. Miller's ideas of generating sequences of Bessel functions by backward recursion. This was an idea that greatly fascinated Milton Abramowitz, and I remember him giving a seminar on the subject. His excitement and enthusiasm for this method, and for anything else relating to special functions, were infectious indeed, and I began to take an active interest in this method, all the more so as I could see applications for the generation of repeated integrals of the error function. This started an interest of mine that continued well beyond my tenure at the Bureau, and I will say more about it later.

There was another event at the Bureau that had considerable impact on my professional development: the Bureau received a grant from NSF to conduct a training program in Numerical Analysis for senior university faculty, which under the direction of John Todd was intended, in his own words, " . . . to attract mature mathematicians into an area of vital importance which had been largely neglected". The program ran quite successfully for a year, and was repeated two years later under the direction of Phil Davis. It was at that time that I was asked to give the course on numerical ODEs, which Henry Antosiewicz gave the first time around, but did not want to repeat. Since I had some acquaintance with ODE methods from my student days in Basel, and was eager to deepen it further, I was more than happy to accept. This gave me a welcome opportunity not only to review the older literature on the subject, but also to study the exciting new developments that had occurred just a year or so before when Germund Dahlquist published his convergence and stability theory for general multistep methods. I worked hard to incorporate these new results, along with other important advances due to Hull and Luxemburg, for example. After the program had again successfully run its course, it was decided to assemble all lectures into a book entitled *Survey of Numerical Analysis*. It was edited by John Todd, and I joined forces with Antosiewicz to consolidate and expand our lectures into a survey article for this book [2]. The book appeared in 1962, and our article could well have made some impression were it not for the fact that the same year Henrici's book on ODEs came out, which overshadowed everything!

It was in Washington also where I made repeated attempts at learning Russian. (Remember, it was the time of the sputniks!) I finally succeeded on my third or fourth try, having set aside a few hours every evening to master the Cyrillic alphabet and Russian grammar, and to develop a Russian vocabulary in mathematics. My efforts eventually paid off as I gained access to the original Russian works in approximation theory, whose clarity of exposition I learned to appreciate and admire. It was through these studies of the Russian literature, for example, that I got the idea of employing trigonometric polynomials rather than the usual algebraic ones to develop ODE methods for oscillatory differential equations [21].

At about that time, the contract arrangements between American University and the Bureau of Standards came under scrutiny by Congressional committees, which made the Administration at the Bureau rather jittery, and it was eventually decided to dissolve all these contracts. I was told to look around for another job! It so happened that Alston Householder, who frequently visited the Bureau, became aware of this problem, and on one of these visits invited me to come to Oak Ridge for an interview. I gladly accepted; but not knowing what else to talk about, I gave a lecture on my work on Picone's necessary conditions in the calculus of variations. Rather surprisingly, I found at least one person in the audience — Wallace Givens — who expressed considerable interest in this topic. (I learned only later that as an assistant to Oswald Veblen at Princeton, Givens developed a lasting interest in projective geometry.) I liked the congenial atmosphere that I found prevailing in Householder's "Mathematics Panel", and so, when the offer arrived, I accepted without hesitation.

7 While the few years in Washington were rather hectic, life in Tennessee turned out to be more leisurely, and I had time to think through and sort out the many ideas that were generated during my work at the Bureau. This goes in particular for the idea of backward recursion, which still intrigued me, because I felt there was more to it than met the eye. It gradually dawned on me, and I then saw it confirmed in Satz 2.46C of the second volume of Perron's book, which appeared just about this time, that there are connections with a minimality concept for solutions of difference equations and continued fractions. Perron gave no reference for this theorem, but I had the feeling that it must be much older. In trying to hunt down its origin, I went through Nörlund's 1924 book on the difference calculus, but could not find anything other than lots of references to Pincherle's work which all looked suspicious. So the next logical thing to do was to search through both volumes of Pincherle's selected works. And there it was! In a long and seemingly unrelated paper on the hypergeometric function, I found the theorem I was looking for, namely the equivalence of the existence of minimal solutions (Pincherle called them "integrali distinti") for three-term recurrence relations and the convergence of a certain associated continued fraction. Miller's algorithm, therefore, could now be formulated in terms of continued fractions (which at the same time rid it of potential overflow), and the method no longer appeared as an isolated "trick", but instead firmly grounded on, and inserted in, the main stream of classical analysis. It took several years, though, to write this all up and publish it [31], but from then on, Pincherle's result was resurrected from obscurity and began to enjoy the attention of many researchers who were inspired by its simplicity and usefulness.

I was also very much interested in generalizing this theory to difference equations of arbitrary order, and in fact lectured on this in 1963 at one of the University of Michigan Engineering Summer Conferences that Bob Bartels used to organize, and during which Wilkinson's pioneering work on matrix computation was developed. I never got around, however, to work this out completely, and finally, at Purdue, handed the matter over to my first Ph.D. student — Ray Zahar — who did an admirable job with it.

Naturally, I could not spend all my time on problems I was interested in myself and had obligations to the Laboratory to help whenever help was needed. On occasion, this could bring me in contact with scientists of the greatest stature, such as Eugene Wigner, for example, who was a regular consultant at the Laboratory. One day he called on Householder to provide him with computational assistance. Alston thought that I was the most likely person who could help, and set up a meeting between Wigner and myself. It was one of those encounters that one treasures for the rest of one's life! He was an extremely gentle and unassuming person, always ready to praise the efforts of others and belittle his own. Wigner was then interested in random Hermitian matrices and the distribution of their eigenvalues. This led him to consider certain quantities involving Hermite polynomials, which he wanted computed and plotted. There was really no difficulty for me to carry out these computations, and I provided Wigner with the desired results and plots within a few days. Seeing how extremely grateful he was for this small service, and seeing it acknowledged in a paper he later wrote[10], was rather touching. The experience reminded me very much of the similar one with Picone many years earlier, and I resolved there and then that, should I ever have gifted students of my own — and I was fortunate to have had some later in my life — I would extend to them the same generous support and encouragement that I received from such great men as Picone and Wigner.

There was another encounter that would become of great significance for my future work. It was a chemist, this time, who came to my office with one of those integrals chemists run into, that seem to resist accurate computation. In this case it had the form

$$\int_{-1}^{1} f(x)w(x)dx, \quad w(x) = \frac{4}{\pi^2 + \left[\ln\frac{1+x}{1-x}\right]^2},$$

where f was some complicated but smooth function. Although the "weight function" w vanishes at the endpoints, it has logarithmic singularities there, which must have been the cause of the numerical difficulties. Now I had studied Hildebrand well enough to know that the answer here was Gaussian quadrature: find the orthogonal polynomials that belong to the weight function w, and then use their zeros as quadrature nodes and the Christoffel numbers as the weights. It was also fairly obvious from Hildebrand's text how the required polynomials could be successively generated in terms of the moments of w (only much later did I learn that Stieltjes already knew about this method), which in turn are computable in terms of special functions. So it looked to me as if this were a routine exercise, to be done and over with in a matter of days. Little did I know what lay ahead of me! It took indeed a few days to write the necessary programs, but the results they produced were horrendous, in spite of careful checking and rechecking. The first few polynomials I got looked reasonable enough, but as the degree went up (to produce more accurate Gauss formulae), they completely fell apart. When one gets complex Gaussian nodes, one knows that something is seriously amiss! I had to

---

[10]E.P. Wigner, "Distribution laws for the roots of a random Hermitean matrix". In *Statistical Theories of Spectra*, pages 446–461. Academic Press, New York, 1965. C.E. Porter, ed.

admit to the chemist that, for the moment, I couldn't really suggest anything better than to try standard rules of integration, like the trapezoidal rule or Simpson's rule.

Yet, this embarassing failure got under my skin, and I was determined to regain my honor. The first thing that needed to be done was to find out what it was that caused the failure. I suspected it was a case of ill-conditioning, and I set out to analyze the sensitivity of the problem. This led me naturally into confluent Vandermonde matrices and their conditioning, and I began to study not only confluent, but also ordinary, and later generalized, Vandermonde matrices in order to determine their condition numbers. To nobody's surprise, they grew exponentially with the order of the matrix! These problems occupied me — off and on — for many years to come [22, 24, 52, 53, 62, 78, 97, 129], and indeed, similar questions of conditioning for other problems also captured my interest [36, 45, 47, 64, 89, 7]. The more important issue at hand, however, was to find better methods for generating orthogonal polynomials. This became a task that required a great deal of effort and has kept me busy ever since. The irony of it all is that, had I been more knowledgeable at the time, I would have realized that the polynomials I needed were nothing but the numerator Legendre polynomials of order 1, whose recursion coefficients are simply those of the Legendre polynomials shifted in their indices by 1. Thus, to compute their zeros and get the required Gauss formulae would have been absolutely trivial. It sometimes pays to be ignorant!

8 This brings me to the last chapter of my career — my tenure at Purdue — and, with its more than 30 years, by far the longest one. My narrative, of necessity, must pick up speed, since otherwise I would have to keep you here for several more hours. Actually, most of the ingredients of my research have already been put in place: the seeds have been planted, and it was now a matter of nurturing them and seeing them to fruition. Still, there are a few events worth recounting, as they have given new twists to my work or incentives for further consolidation.

Early on, when Givens was Director of the Applied Mathematics Division at Argonne, I had the privilege of consulting there regularly. I enjoyed the companionship with people like Jim Cody and Henry Thacher, from whom I greatly benefitted in my work of developing software for special functions. (At the time it was all in Algol, which I had picked up in Oak Ridge, but eventually I became converted to Fortran.) Particularly enjoyable were the few summers I spent at Argonne, in the stimulating company of the Todds and of Wilkinson. There was also a group of astrophysicists at Argonne who were interested in Doppler broadening phenomena and therefore in fast methods of computing the Voigt function, which is nothing but the error function in the complex domain. In the late 1960s they organized a one-day workshop to which I was also invited. Having had nothing ready to talk about, I hurriedly put something together on the computation of the complex error function [38, 41]. Curiously, these became two of my papers most frequently cited (by physicists)!

For one reason or another (the circumstances of which I can't recall) I developed an intense interest in oscillatory integrals and in effective methods for computing them. I set out to collect a vast amount of literature, and in the process decided to start work on a survey paper on the subject. I was just about then going to Munich on a sabbatical, and therefore there was ample time to work on the project. It transpired, however, that I became engrossed in a problem of computing Fourier coefficients, which occupied me for the whole year. The result was a long paper on the theory of attenuation factors — the only paper I wrote on Fourier analysis [43]. I did however give a survey talk on the numerical integration of oscillating functions at Catholic University of Leuven, but never got around to writing it up.

On my return from Munich, I was to become the host of Professor Hiroki Yanagiwara from Japan, who himself was to spend a sabbatical with me at Purdue. Although there were some serious problems of communication, because of language barriers, I understood what he was up to: some new quadrature rules with equal coefficients — first studied by Chebyshev — which were to be constructed on the basis of some optimality criterion. I became actively involved in this problem, and fascinated by it, because of the interesting mathematical questions it raised. That was the beginning of a long interest in Chebyshev-type quadrature rules, and I was happy to collaborate on this not only with Yanagiwara [48], but later also with Giovanni Monegato [57], who spent two years with me as a postdoctoral student.

But by far the most significant event came in the Spring of 1979, when Paul Butzer in Aachen wrote me a desperate letter asking me whether I could fill in for Phil Davis, who had to withdraw "for inescapable obligations" from his earlier commitment to present one of the festival lectures (on Gaussian quadrature) at the International Christoffel Symposium. This was to be held in the Fall of the same year in Aachen to honor Christoffel on his 150th anniversary of birth. I wrote back that, yes, I would be willing to accept but that " . . . in arriving at my decision, I had to weigh the prospects of participating at the conference honoring Christoffel, which I find very exciting, against the obligations I have already committed myself to for this summer, which include heavy teaching and proofreading galleys and page proofs of a book", and then asked Butzer for patience in my delivery of the manuscript. The "teaching" I was referring to was the only summer course I ever taught at Purdue, and the "book" was my translation of Stoer and Bulirsch [138]. Needless to say, this was a most frantic summer, but I managed to pull it off and produced the paper in time [6]. Even though it was a lot of work, I benefitted immensely from it, as I got to know the history not only of Gaussian quadrature, but also of orthogonal polynomials in general. Indeed, there was some more "archaeological" work to be done in connection with an algorithm that generates the recursion coefficients of orthogonal polynomials directly from the moments. This time I had to go through Chebyshev's collected works to find it hidden away in a paper on the discrete least squares problem. It is true that this algorithm, like the one of Stieltjes, suffers from the same ill-conditioning that led to my debacle in Oak Ridge, but modified versions of both are now known, and they are presently the pillars of constructive methods for orthogonal polynomials [76, 115]. It was from this point on that I began to concentrate almost exclusively on studying orthogonal polynomials, both their numerical construction and applications to various problems of approximation.

Last, but not least, I am fortunate to have received many stimulations from colleagues and friends, either on their visits to Purdue or on other occasions, which greatly enriched my work. I would like to particularly single out: Gradimir Milovanović, who introduced me to quadrature problems involving Einstein and Fermi functions and their relevance to the summation of slowly convergent series [84], and also contributed significantly to our work on moment-preserving spline approximation [91, 92] and orthogonal polynomials on the semicircle [86, 88, 94]; Richard Varga, who collaborated on joint papers on the remainder of Gaussian quadrature rules for analytic functions [80, 101]; Bernhard Flury, a statistician, who in his work on principal component analysis needed help on the interesting problem of simultaneously reducing several symmetric positive definite matrices to nearly diagonal form by an orthogonal similarity transformation [87]; Stephan Ruscheweyh who helped solve a difficult spectral problem related to Vandermonde matrices [102]; and most memorably, Louis de Branges for getting me involved in his marvelous proof of the Bieberbach conjecture[11] [146]. And it was only last year

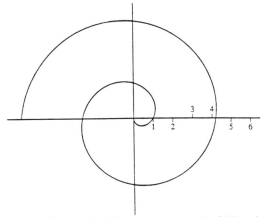

Figure 7: The analytic spiral of Theodorus

that Phil Davis introduced me to the world of spirals and the fascinating problem of finding an analytic interpolant (see Fig. 7) to the classical discrete spiral of Theodorus[12] [8]. I am deeply grateful to all these friends, and to many others who showed an interest in my work and helped me along in one way or another.

[11]L. de Branges, A proof of the Bieberbach conjecture, *Acta Math.*, **154**:137–152, 1985; L. de Branges, Das mathematische Erbe von Ludwig Bieberbach (1886–1982), *Nieuw Arch. Wisk.*, **9**(no. 3):366–370, 1991.

[12]Philip J. Davis, *Spirals: From Theodorus to Chaos*. A K Peters, Wellesley, Massachusetts, 1993.

PUBLICATIONS OF WALTER GAUTSCHI

Books

1. (with H. Bavinck and G.M. Willems) Colloquium Approximatietheorie. MC Syllabus **14**, Mathematisch Centrum Amsterdam, 1971.

Book Chapters

2. (with H.A. Antosiewicz) Numerical methods in ordinary differential equations. Ch. 9 in *Survey of Numerical Analysis*, pages 314–346. McGraw-Hill, New York, 1962. J. Todd, ed.

3. (with W.F. Cahill) Exponential integral and related functions. Ch. 5 in *Handbook of Mathematical Functions*, Nat. Bur. Standards Appl. Math. Ser., **55**:227–251, 1964. M. Abramowitz and I.A. Stegun, eds. [Russian translation by V.A. Ditkin and L.N. Karmazina in *Spravočnik po special'nym funkciyam*, Nauka, 55–79, Moscow, 1979.]

4. Error function and Fresnel integrals. Ch. 7 in *Handbook of Mathematical Functions*, Nat. Bur. Standards Appl. Math. Ser., **55**:295–329, 1964. M. Abramowitz and I.A. Stegun, eds. [Russian translation by V.A. Ditkin and L.N. Karmazina in *Spravočnik po special'nym funkciyam*, Nauka, 119–152, Moscow, 1979.]

5. Computational methods in special functions — a survey. In *Theory and Applications of Special Functions*, pages 1–98. Academic Press, New York, 1975. R. Askey, ed.

6. A survey of Gauss-Christoffel quadrature formulae. In *E.B. Christoffel; The Influence of his Work in Mathematics and the Physical Sciences*, pages 72–147. Birkhäuser, Basel, 1981. P.L. Butzer and F. Fehér, eds.

7. Questions of numerical condition related to polynomials. In *Symposium on Recent Advances in Numerical Analysis*, pages 45–72. Academic Press, New York, 1978. C. de Boor and G.H. Golub, eds. [Revised and reprinted in *Studies in Numerical Analysis*, Studies in Mathematics, Vol. **24**, pages 140–177. The Mathematical Association of America, 1984. G.H. Golub, ed.]

8. The spiral of Theodorus, special functions, and numerical analysis. Supplement A in P.J. Davis, *Spirals: From Theodorus to Chaos*, pages 67–87. A K Peters, Boston, 1993.

Refereed Journals

9. Ein Analogon zu Grammels Methode der graphischen Integration gewöhnlicher Differentialgleichungen. *Z. Angew. Math. Mech.*, **31**:242–243, 1951.

10. Fehlerabschätzungen für die graphischen Integrationsverfahren von Grammel und Meissner-Ludwig. *Verh. Naturforsch. Ges. Basel*, **64**:401–435, 1953.

11. Über die zeichnerischen Ungenauigkeiten und die zweckmässige Bemessung der Schrittlänge beim graphischen Integrationsverfahren von Meissner-Ludwig. *Verh. Naturforsch. Ges. Basel*, **65**:49–66, 1954.

12. Über eine Klasse von linearen Systemen mit konstanten Koeffizienten. *Comment. Math. Helv.*, **28**:186–196, 1954.

13. Über den Fehler des Runge-Kutta-Verfahrens für die numerische Integration gewöhnlicher Differentialgleichungen n-ter Ordnung. *Z. Angew. Math. Phys.*, **6**:456–461, 1955.

14. Una estensione agli integrali doppi di una condizione di Picone necessaria per un estremo. *Atti Accad. Naz. Lincei. Rend. Cl. Sci. Fis. Mat. Nat.*, (8) **20**:283–289, 1956.

15. Bemerkung zu einer notwendigen Bedingung von Picone in der Variationsrechnung. *Comment. Math. Helv.*, **31**:1–4, 1956.

16. Some elementary inequalities relating to the gamma and incomplete gamma function. *J. Math. and Phys.*, **38**:77–81, 1959.

17. Exponential integral $\int_1^\infty e^{-xt} t^{-n} dt$ for large values of n. *J. Res. Nat. Bur. Standards*, **62**:123–125, 1959.

18. Note on bivariate linear interpolation for analytic functions. *Math. Tables Aids Comput.*, **13**:91–96, 1959.

19. Recursive computation of certain integrals. *J. Assoc. Comput. Mach.*, **8**:21–40, 1961.

20. Recursive computation of the repeated integrals of the error function. *Math. Comp.*, **15**:227–232, 1961.

21. Numerical integration of ordinary differential equations based on trigonometric polynomials. *Numer. Math.*, **3**:381–397, 1961.

22. On inverses of Vandermonde and confluent Vandermonde matrices. *Numer. Math.*, **4**:117–123, 1962.

23. Diffusion functions for small argument. *SIAM Rev.*, **4**:227–229, 1962.

24. On inverses of Vandermonde and confluent Vandermonde matrices II. *Numer. Math.*, **5**:425–430, 1963.

25. Algorithm 221 — Gamma function, and Algorithm 222 — Incomplete beta function ratios. *Comm. ACM*, **7**:143–144, 1964. Certification of Algorithm 222. *ibid.*, 244.

26. Algorithm 236 — Bessel functions of the first kind. *Comm. ACM*, **7**:479–480, 1964. Certification of Algorithm 236. *ibid.*, **8**:105–106, 1965.

27. Algorithm 259 — Legendre functions for arguments larger than one. *Comm. ACM*, **8**:488–492, 1965.

28. Computation of successive derivatives of $f(z)/z$. *Math. Comp.*, **20**:209–214, 1966.

29. Algorithm 282 — Derivatives of e^x/x, $\cos(x)/x$, and $\sin(x)/x$. *Comm. ACM*, **9**:272, 1966.

30. Algorithm 292 — Regular Coulomb wave functions. *Comm. ACM*, **9**:793–795, 1966.

31. Computational aspects of three-term recurrence relations. *SIAM Rev.*, **9**:24–82, 1967.

32. Numerical quadrature in the presence of a singularity. *SIAM J. Numer. Anal.*, **4**:357–362, 1967.

33. Construction of Gauss-Christoffel quadrature formulas. *Math. Comp.*, **22**:251–270, 1968.

34. Algorithm 331 — Gaussian quadrature formulas. *Comm. ACM*, **11**:432–436, 1968.

35. Remark on Algorithm 292. *Comm. ACM*, **12**:280, 1969.

36. On the condition of a matrix arising in the numerical inversion of the Laplace transform. *Math. Comp.*, **23**:109–118, 1969.

37. An application of minimal solutions of three-term recurrences to Coulomb wave functions. *Aequationes Math.*, **2**:171–176, 1969; abstract, *ibid.*, **1**:208, 1968.

38. Algorithm 363 — Complex error function. *Comm. ACM*, **12**:635, 1969.

39. (with B.J. Klein) Recursive computation of certain derivatives — a study of error propagation. *Comm. ACM*, **13**:7–9, 1970.

40. (with B.J. Klein) Remark on Algorithm 282. *Comm. ACM*, **13**:53–54, 1970.

41. Efficient computation of the complex error function. *SIAM J. Numer. Anal.*, **7**:187–198, 1970.

42. On the construction of Gaussian quadrature rules from modified moments. *Math. Comp.*, **24**:245–260, 1970.

43. Attenuation factors in practical Fourier analysis. *Numer. Math.*, **18**:373–400, 1972.

44. Zur Numerik rekurrenter Relationen. *Computing*, **9**:107–126, 1972. [English translation in Aerospace Research Laboratories, Report ARL 73-0005, February 1973.]

45. The condition of orthogonal polynomials. *Math. Comp.*, **26**:923–924, 1972.

46. Algorithm 471 — Exponential integrals. *Comm. ACM*, **16**:761–763, 1973.

47. On the condition of algebraic equations. *Numer. Math.*, **21**:405–424, 1973.

48. (with H. Yanagiwara) On Chebyshev-type quadratures. *Math. Comp.*, **28**:125–134, 1974.

49. A harmonic mean inequality for the gamma function. *SIAM J. Math. Anal.*, **5**:278–281, 1974.

50. Some mean value inequalities for the gamma function. *SIAM J. Math. Anal.*, **5**:282–292, 1974.

51. Nonexistence of Chebyshev-type quadratures on infinite intervals. *Math. Comp.*, **29**:93–99, 1975.

52. Norm estimates for inverses of Vandermonde matrices. *Numer. Math.*, **23**:337–347, 1975.

53. Optimally conditioned Vandermonde matrices. *Numer. Math.*, **24**:1–12, 1975.

54. (with L.A. Anderson) Optimal weighted Chebyshev-type quadrature formulas. *Calcolo*, **12**:211–248, 1975.

55. Stime dell'errore globale nei metodi "one-step" per equazioni differenziali ordinarie. *Rend. Mat.*, (2) **8**:601–617, 1975.

56. Comportement asymptotique des coefficients dans les formules d'intégration d'Adams, de Störmer et de Cowell. *C. R. Acad. Sci. Paris Ser. A*, **283**:787–788, 1976.

57. (with G. Monegato) On optimal Chebyshev-type quadratures. *Numer. Math.*, **28**:59–67, 1977.

58. Evaluation of the repeated integrals of the coerror function. *ACM Trans. Math. Software*, **3**:240–252, 1977.

59. Algorithm 521 — Repeated integrals of the coerror function. *ACM Trans. Math. Software*, **3**:301–302, 1977.

60. Anomalous convergence of a continued fraction for ratios of Kummer functions. *Math. Comp.*, **31**:994–999, 1977.

61. Qualche contributo recente sul problema di Chebyshev nella teoria dell'integrazione numerica. *Rend. Sem. Mat. Univ. e Politec. Torino*, **35**:39–44, 1976–77.

62. On inverses of Vandermonde and confluent Vandermonde matrices III. *Numer. Math.*, **29**:445–450, 1978.

63. (with J. Slavik) On the computation of modified Bessel function ratios. *Math. Comp.*, **32**:865–875, 1978.

64. The condition of polynomials in power form. *Math. Comp.*, **33**:343–352, 1979.

65. On the preceding paper "A Legendre polynomial integral" by James L. Blue. *Math. Comp.*, **33**:742–743, 1979.

66. A computational procedure for incomplete gamma functions. *ACM Trans. Math. Software*, **5**:466–481, 1979.

67. Algorithm 542 — Incomplete gamma functions. *ACM Trans Math. Software*, **5**:482–489, 1979.

68. Un procedimento di calcolo per le funzioni gamma incomplete. *Rend. Sem. Mat. Univ. e Politec. Torino*, **37**:1–9, 1979.

69. Families of algebraic test equations. *Calcolo*, **16**:383–398, 1979.

70. (with F. Costabile) Stime per difetto per gli zeri più grandi dei polinomi ortogonali. *Boll. Un. Mat. Ital.*, (5) **17A**:516–522, 1980.

71. (with M. Montrone) Metodi multistep con minimo coefficiente dell'errore globale. *Calcolo*, **17**:67–75, 1980.

72. Minimal solutions of three-term recurrence relations and orthogonal polynomials. *Math. Comp.*, **36**:547–554, 1981.

73. (with R.E. Lynch) Error behavior in optimal relaxation methods. *Z. Angew. Math. Phys.*, **33**:24–35, 1982.

74. A note on the successive remainders of the exponential series. *Elem. Math.*, **37**:46–49, 1982.

75. Polynomials orthogonal with respect to the reciprocal gamma function. *BIT*, **22**:387–389, 1982.

76. On generating orthogonal polynomials. *SIAM J. Sci. Statist. Comput.*, **3**:289–317, 1982.

77. On the convergence behavior of continued fractions with real elements. *Math. Comp.*, **40**:337–342, 1983.

78. The condition of Vandermonde-like matrices involving orthogonal polynomials. *Linear Algebra Appl.*, **52/53**:293–300, 1983.

79. How and how not to check Gaussian quadrature formulae. *BIT*, **23**:209–216, 1983.

80. (with R.S. Varga) Error bounds for Gaussian quadrature of analytic functions. *SIAM J. Numer. Anal.*, **20**:1170–1186, 1983.

81. On Padé approximants associated with Hamburger series. *Calcolo*, **20**:111–127, 1983.

82. Discrete approximations to spherically symmetric distributions. *Numer. Math.*, **44**:53–60, 1984.

83. On some orthogonal polynomials of interest in theoretical chemistry. *BIT*, **24**:473–483, 1984.

84. (with G.V. Milovanović) Gaussian quadrature involving Einstein and Fermi functions with an application to summation of series. *Math. Comp.*, **44**:177–190, 1985. Supplement. *ibid.*, S1–S11.

85. Orthogonal polynomials — constructive theory and applications. *J. Comput. Appl. Math.*, **12/13**:61–76, 1985.

86. (with G.V. Milovanović) Polynomials orthogonal on the semicircle. *Rend. Sem. Mat. Univ. e Politec. Torino*, **Special issue**:179–185, July 1985.

87. (with B. Flury) An algorithm for simultaneous orthogonal transformation of several positive definite matrices to nearly diagonal form. *SIAM J. Sci. Statist. Comput.*, **7**:169–184, 1986.

88. (with G.V. Milovanović) Polynomials orthogonal on the semicircle. *J. Approx. Theory*, **46**:230–250, 1986.

89. On the sensitivity of orthogonal polynomials to perturbations in the moments. *Numer. Math.*, **48**:369–382, 1986.

90. (with F. Caliò and E. Marchetti) On computing Gauss-Kronrod quadrature formulae. *Math. Comp.*, **47**:639–650, 1986.

91. (with G.V. Milovanović) Spline approximations to spherically symmetric distributions. *Numer. Math.*, **49**:111–121, 1986.

92. (with M. Frontini and G.V. Milovanović) Moment-preserving spline approximation on finite intervals. *Numer. Math.*, **50**:503–518, 1987.

93. (with J. Wimp) Computing the Hilbert transform of a Jacobi weight function. *BIT*, **27**:203–215, 1987.

94. (with H.J. Landau and G.V. Milovanović) Polynomials orthogonal on the semicircle II. *Constructive Approx.*, **3**:389–404, 1987.

95. (with M.A. Kovačević and G.V. Milovanović) The numerical evaluation of singular integrals with coth-kernel. *BIT*, **27**:389–402, 1987.

96. (with S.E. Notaris) An algebraic study of Gauss-Kronrod quadrature formulae for Jacobi weight functions. *Math. Comp.*, **51**:231–248, 1988.

97. (with G. Inglese) Lower bounds for the condition number of Vandermonde matrices. *Numer. Math.*, **52**:241–250, 1988.

98. (with T.J. Rivlin) A family of Gauss-Kronrod quadrature formulae. *Math. Comp.*, **51**:749–754, 1988.

99. On the zeros of polynomials orthogonal on the semicircle. *SIAM J. Math. Anal.*, **20**:738–743, 1989.

100. (with S.E. Notaris) Gauss-Kronrod quadrature formulae for weight functions of Bernstein-Szegö type. *J. Comput. Appl. Math.*, **25**:199–224, 1989.

101. (with E. Tychopoulos and R.S. Varga) A note on the contour integral representation of the remainder term for a Gauss-Chebyshev quadrature rule. *SIAM J. Numer. Anal.*, **27**:219–224, 1990.

102. (with A. Córdova and S. Ruscheweyh) Vandermonde matrices on the circle: spectral properties and conditioning. *Numer. Math.*, **57**:577–591, 1990.

103. (with Shikang Li) The remainder term for analytic functions of Gauss-Radau and Gauss-Lobatto quadrature rules with multiple end points. *J. Comput. Appl. Math.*, **33**:315–329, 1990.

104. On the remainder term for analytic functions of Gauss-Lobatto and Gauss-Radau quadratures. *Rocky Mountain J. Math.*, **21**:209–226, 1991.

105. A class of slowly convergent series and their summation by Gaussian quadrature. *Math. Comp.*, **57**:309–324, 1991.

106. On certain slowly convergent series occurring in plate contact problems. *Math. Comp.*, **57**:325–338, 1991.

107. (with Shikang Li) Gauss-Radau and Gauss-Lobatto quadratures with double end points. *J. Comput. Appl. Math.*, **34**:343–360, 1991.

108. On the paper "A continued fraction approximation of the modified Bessel function $I_1(t)$" by P.R. Parthasarathy and N. Balakrishnan. *Appl. Math. Letters*, **4**:47–51, 1991.

109. Quadrature formulae on half-infinite intervals. *BIT*, **31**:438–446, 1991.

110. Spline approximation and quadrature formulae. *Atti Sem. Mat. Fis. Univ. Modena*, **40**:169–182, 1992.

111. On mean convergence of extended Lagrange interpolation. *J. Comput. Appl. Math.*, **43**:19–35, 1992.

112. (with Shikang Li) A set of orthogonal polynomials induced by a given orthogonal polynomial. *Aequationes Math.*, **46**:174–198, 1993.

113. Is the recurrence relation for orthogonal polynomials always stable? *BIT*, **33**:277–284, 1993.

114. On the computation of generalized Fermi-Dirac and Bose-Einstein integrals. *Comput. Phys. Comm.*, **74**:233–238, 1993.

115. Algorithm 726 — ORTHPOL: A package of routines for generating orthogonal polynomials and Gauss-type quadrature rules. *ACM Trans. Math. Software*, **20**:21–62, 1994.

Conference Proceedings

116. Instability of linear second-order difference equations. In *Proc. IFIP Congress 62*, page 207. North-Holland, Amsterdam, 1963. C.M. Popplewell, ed.

117. Computation of transcendental functions by recurrence relations. In *Proc. IFIP Congress 65*, Vol. 2, pages 485–486. Spartan Books, Washington, D.C., 1966. W.A. Kalenich, ed.

118. Advances in Chebyshev quadrature. In *Numerical Analysis*, Lecture Notes Math., Vol. 506, pages 100–121. Springer, Berlin, 1976. G.A. Watson, ed.

119. On generating Gaussian quadrature rules. In *Numerische Integration*, Internat. Ser. Numer. Math., Vol. 45, pages 147–154. Birkhäuser, Basel, 1979. G. Hämmerlin, ed.

120. An algorithmic implementation of the generalized Christoffel theorem. In *Numerical Integration*, Internat. Ser. Numer. Math., Vol. 57, pages 89–106. Birkhäuser, Basel, 1982. G. Hämmerlin, ed.

121. (with G.V. Milovanović) On a class of complex polynomials having all zeros in a half circle. In *Numerical Methods and Approximation Theory*, pages 49–53. Faculty of Electronic Engineering, Univ. Niš, Niš, 1984. G.V. Milovanović, ed.

122. Some new applications of orthogonal polynomials. In *Polynômes Orthogonaux et Applications*, Lecture Notes Math., Vol. 1171, pages 63–73. Springer, Berlin, 1985. C. Brezinski, A. Draux, A.P. Magnus, P. Maroni and A. Ronveaux, eds.

123. Gauss-Kronrod quadrature — a survey. In *Numerical Methods and Approximation Theory III*, pages 39–66. Faculty of Electronic Engineering, Univ. Niš, Niš, 1988. G.V. Milovanović, ed.

124. (with S.E. Notaris) Newton's method and Gauss-Kronrod quadrature. In *Numerical Integration III*, Internat. Ser. Numer. Math., Vol. 85, pages 60–71. Birkhäuser, Basel, 1988. H. Brass and G. Hämmerlin, eds.

125. Orthogonality — conventional and unconventional — in numerical analysis. In *Computation and Control*, Progress in Systems and Control Theory, Vol. 1, pages 63–95. Birkhäuser, Boston, 1989. K. Bowers and J. Lund, eds.

126. Some applications and numerical methods for orthogonal polynomials. In *Numerical Analysis and Mathematical Modelling*, Banach Center Publications, Vol. 24, pages 7–19. PWN Polish Scientific Publishers, Warsaw, 1990. A. Wakulicz, ed.

127. Orthogonal polynomials on the semicircle. In *Numerical Analysis and Mathematical Modelling*, Banach Center Publications, Vol. 24, pages 21–27. PWN Polish Scientific Publishers, Warsaw, 1990. A. Wakulicz, ed.

128. Computational aspects of orthogonal polynomials. In *Orthogonal Polynomials: Theory and Practice*, NATO ASI Series, Series C: Mathematical and Physical Sciences, Vol. 294, pages 181–216. Kluwer, Dordrecht, 1990. Paul Nevai, ed.

129. How (un)stable are Vandermonde systems? In *Asymptotic and Computational Analysis*, Lecture Notes in Pure and Appl. Math., Vol. 124, pages 193–210. Dekker, New York, 1990. R. Wong, ed.

130. Computational problems and applications of orthogonal polynomials. In *Orthogonal Polynomials and Their Applications*, IMACS Annals Comput. Appl. Math., Vol. 9, pages 61–71. Baltzer, Basel, 1991. C. Brezinski, L. Gori and A. Ronveaux, eds.

131. Remainder estimates for analytic functions. In *Numerical Integration — Recent Developments, Software and Applications*, NATO ASI Series, Series C:

Mathematical and Physical Sciences, Vol. 357, pages 133–145. Kluwer, Dordrecht, 1992. T.O. Espelid and A. Ganz, eds.

132. Gauss-type quadrature rules for rational functions. In *Numerical Integration IV*, Internat. Ser. Numer. Math., Vol. 112, pages 111–130. Birkhäuser, Basel, 1993. H. Brass and G. Hämmerlin, eds.

133. Summation of slowly convergent series. In *Numerical Analysis and Mathematical Modelling*, Banach Center Publications, Vol. 29, pages 7–18. PWN Polish Scientific Publishers, Warsaw, 1994. A. Wakulicz, ed.

134. Applications and computation of orthogonal polynomials. In *Proc. SANUM Conference 1992*, Durban, S. Africa. To appear.

135. Applications and computation of orthogonal polynomials (Extended abstract). In *Proc. Conference on Advances in Computational Mathematics*, New Delhi. To appear. H.P. Dikshit and C. Micchelli, eds.

136. Luigi Gatteschi's work on special functions and numerical analysis. *Proc. Internat. Joint Symposium on Special Functions and Artificial Intelligence*, Torino. G. Allasia, ed. To appear.

Proceedings Edited

137. Mathematics of Computation 1943–1993: A Half-Century of Computational Mathematics. In *Proc. Sympos. Appl. Math.*, American Math. Soc., Providence, RI. To appear.

Translations

138. (with R. Bartels and C. Witzgall) "Introduction to Numerical Analysis" by J. Stoer and R. Bulirsch, translated from German, Springer, New York, 1980. [2d ed., Texts in Appl. Math., Vol. 12, Springer, New York, 1993.]

139. "Lectures on Numerical Mathematics" by H. Rutishauser, annotated translation from German, Birkhäuser, Boston, 1990.

Other

140. Ninety-seven book reviews published in *Math. Comp.*, and some 450 reviews of technical articles and books in *Math. Reviews*.

141. (with F. Malmborg) Calculations related to the improved free volume theory of liquids. Harvard Computation Laboratory, Problem Report 100, pages 1–41, 1956.

142. On error reducing multi-step methods (abstract). *Notices Amer. Math. Soc.*, 10:95, 1963.

143. Recognition of Christoffel's work on quadrature during and after his lifetime. In *E.B. Christoffel; The Influence of his Work in Mathematics and the Physical Sciences*, pages 724–727. Birkhäuser, Basel, 1981. P.L. Butzer and F. Fehér, eds.

144. To Alexander M. Ostrowski on his ninetieth birthday. *Linear Algebra Appl.*, **52/53**:xi–xiv, 1983.

145. (with Jet Wimp) In memoriam YUDELL L. LUKE June 26, 1918 – May 6, 1983. *Math. Comp.*, 43:349–352, 1984.

146. Reminiscences of my involvement in de Branges's proof of the Bieberbach conjecture. In *The Bieberbach Conjecture*, Proc. Symp. on the Occasion of the Proof, Math. Surveys Monographs, no. 21, pages 205–211. American Math. Society, Providence, RI, 1986. A. Baernstein II, D. Drasin, P. Duren and A. Marden, eds.

147. A.M. Ostrowski (1893 – 1986). *SIAM Newsletter*, **20**:2,19, January 1987.

148. A conjectured inequality for Hermite interpolation at the zeros of Jacobi polynomials, Problem 87-7. *SIAM Rev.*, **29**:297–298, 1987.

149. (with S.E. Notaris) Problem 6. In *Numerical Integration IV*, Internat. Ser. Numer. Math., Vol. 112, pages 379–380. Birkhäuser, Basel, 1993. H. Brass and G. Hämmerlin, eds.

150. Reflections and recollections. In *Approximation and Computation*, Internat. Ser. Numer. Math. Birkhäuser, Basel, 1994. R.V.M. Zahar, ed.

Ph.D. STUDENTS

R.V.M. Zahar	(1968)
J.L. Phillips	(1969)
L.A. Anderson	(1974)
Ja-Chen Lin	(1988)
S.E. Notaris	(1988)
Shikang Li	(1992)
Minda Zhang	(1993)

POSTGRADUATE STUDENTS

H. Yanagiwara, Japan (1971-72)
G. Monegato, Italy (1974-76)

International Series of Numerical Mathematics, Vol. 119, ©1994 Birkhäuser

GAUSS ELIMINATION BY SEGMENTS AND MULTIVARIATE POLYNOMIAL INTERPOLATION

C. de Boor*

Dedicated to Walter Gautschi on the occasion of his 65th Birthday

Abstract. The construction of a polynomial interpolant to data given at finite pointsets in $\mathbb{R}^d$ (or, most generally, to data specified by finitely many linear functionals) is considered, with special emphasis on the linear system to be solved. Gauss elimination by segments (i.e., by groups of columns rather than by columns) is proposed as a reasonable means for obtaining a description of all solutions and for seeking out solutions with 'good' properties. A particular scheme, due to Amos Ron and the author, for choosing a particular polynomial interpolating space in dependence on the given data points, is seen to be singled out by requirements of degree-reduction, dilation-invariance, and a certain orthogonality requirement. The close connection, between this particular construction of a polynomial interpolant and the construction of an H-basis for the ideal of all polynomials which vanish at the given data points, is also discussed.

1   INTRODUCTION

Polynomial interpolation in d variables has long been studied from the following point of view: one is given a polynomial space, F, and seeks to characterize the point sets $\Theta \subset \mathbb{R}^d$ for which the pair (Θ, F) is **correct** in the sense that F contains, for each g defined at least on Θ, exactly one element which agrees with g on Θ. Further, the space F is either the space Π_k of all polynomials on $\mathbb{R}^d$ of degree $\leq k$, or is, more generally, a D-invariant space spanned by monomials. See R. A. Lorentz [9] for an up-to-date accounting of the many interesting efforts in this direction.

In contrast, de Boor and Ron [5] starts with an arbitrary finite point set $\Theta \subset \mathbb{R}^d$ and proposes a particular choice, denoted Π_Θ, from among the many polynomial spaces F for which (Θ, F) is correct. To be explicit (an *explanation* can be found at the end of Section 7), the definition of Π_Θ makes use of the **least** term $f_\downarrow$ of a function f, which, by definition, is the first, or lowest-degree, nontrivial term in the expansion of f into a sum $f = f_0 + f_1 + f_2 + \ldots$ in which f_j is a homogeneous polynomial of degree j, all j. With this, Π_Θ is defined as the linear span

$$\Pi_\Theta := \operatorname{span}\{f_\downarrow : f \in \exp_\Theta\}$$

*Computer Sciences Department, 1210 W. Dayton St., Madison WI 53706-1685, USA; deboor@cs.wisc.edu of least terms of elements in the linear span $\exp_\Theta$ of exponentials $e_\theta : x \mapsto \exp(\theta \cdot x)$, $\theta \in \Theta$. Because of this, we have come to call the resulting interpolant

$$P_\Theta g$$

to a given function g, defined at least on Θ, the **least interpolant to** g.

The connection between the least interpolant and Gauss elimination (applied to the appropriate Vandermonde matrix) was already explored in de Boor [2] and de Boor and Ron [7]. However, this earlier discussion left the (wrong) impression that a particular variant of Gauss elimination was needed in order to obtain the least interpolant. The present paper corrects this impression, in the process of examining the question of what might single out Π_Θ from among all polynomial spaces F for which (Θ, F) is correct. To be sure, de Boor and Ron [8] already contains such a discussion, but not at all from the point of view of elimination. As in [8], the discussion here is actually carried out in the most general situation possible, namely of interpolation to data of the form Λg, with Λ an arbitrary linear map to $\mathbb{R}^n$ from the space Π of all polynomials on $\mathbb{R}^d$. Since such 'data maps' are of finite rank, it would be trivial to extend them to functions on $\mathbb{R}^d$ other than polynomials.

In the discussion, two properties of Π_Θ are seen to play a particularly important role: (i) P_Θ is **degree-reducing**, i.e.,

$$\deg P_\Theta g \leq \deg g, \quad \forall g \in \Pi;$$

and (ii) Π_Θ is **dilation-invariant**, i.e., for any $r > 0$ and any $p \in \Pi_\Theta$, also $p(r \cdot) \in \Pi_\Theta$. The latter property is equivalent to the fact that Π_Θ **stratifies**, i.e., Π_Θ is the direct sum $\Pi_\Theta = \bigoplus_k \Pi^0_{\Theta,k}$ of its homogeneous subspaces

$$\Pi^0_{\Theta,k} := \Pi_\Theta \cap \Pi^0_k$$

(with Π^0_k the space of all *homogeneous* polynomials of exact degree k (including the zero polynomial)). Also, the most intriguing property of Π_Θ (see de Boor and Ron [8]), namely that

$$\Pi_\Theta = \bigcap_{p|_\Theta = 0} \ker p_\uparrow(D),$$

is looked at anew. Here, $p_\uparrow$ is the **leading** term of the polynomial p, i.e., the last, or highest-degree, nontrivial term in the expansion $p = p_0 + p_1 + p_2 + \ldots$ of p with p_j homogeneous of degree j, all j.

In addition, it is shown how to make use of the construction of Π_Θ by elimination to construct an H-basis for the polynomial ideal of all polynomials which vanish on Θ. This responds to questions raised by algebraists during talks on the algorithm given in [2] and [7].

2

The paper highlights *Gauss elimination by segments*. In this generalization of Gauss elimination, the matrix A to be factored is somehow **segmented**, i.e., is given in the form
$$A = [A_0, A_1, \ldots],$$
with each **segment**, A_j, comprising zero or more consecutive columns of A. (Here and throughout, we use MATLAB notation.) Correspondingly, elimination is to proceed, not column by column, but segment by segment. In the case of polynomial interpolation, the segmentation naturally arises by grouping monomials of the same degree together and because there is no natural ordering for the monomials of the same degree. Because of this particular application, we start the indexing of the segments of A with 0.

By definition, **Gauss elimination by segments** applied to $A = [A_0, A_1, \ldots]$ produces a factorization
$$[A_0, A_1, \ldots,] = M[R_0, R_1, \ldots]$$
with M invertible and
$$R := [R_0, R_1, \ldots]$$
a **segmented row-echelon form**, the segmentation corresponding to that of A. This means that R is block upper triangular, $R = (R_{ij})$ say with $R_j = [R_{0j}; R_{1j}; \ldots]$ having exactly as many columns as A_j does, with each diagonal block R_{jj} onto. (A casual search of the literature failed to turn up earlier occurrences of this concept of a segmented row echelon form. However, in hindsight, the notion is so natural that it is bound to have been used before.)

Except for trivial cases, a segmented matrix has infinitely many such factorizations. Nevertheless, all such factorizations have certain properties in common. For our purposes, the most important property is the fact that, for each j, the row-space of R_{jj} depends only on A (and the segmentation). In particular, it is independent of the details of the elimination process which led to R. We take the trouble to prove this since, as we also show, the row-space of R_{jj} provides all the information needed to construct the space $\Pi_{\Theta,j}^0 = \Pi_\Theta \cap \Pi_j^0$ of homogeneous polynomials of degree j in Π_Θ. We also identify the numerical procedure proposed in [2] and [7] as a particularly stable way for obtaining such a segmented row-echelon form.

For simplicity, the paper deals with *real*-valued functions only. All the results are true if matrix transpose is replaced by conjugate transpose and, correspondingly, certain quantities by their complex conjugate.

Finally, the paper is also meant to illustrate the perhaps debatable point that it is possible to write about polynomial interpolation in several variables without covering entire pages with formulæ.

The paper had its start in a discussion I had with Nira Dyn, in August of 1992, concerning possible alternatives to the algorithm in [7], and subsequent discussions, in February of 1993, concerning possible alternatives to the least interpolant. In particular, the use of column maps below is my response to Nira Dyn's very direct and useful way of describing the least interpolant.

3

2 POLYNOMIALS

A polynomial in d variables is customarily written in the form

$$p(x) = \sum_\alpha x^\alpha c(\alpha), \tag{2.1}$$

with

$$x^\alpha := x(1)^{\alpha(1)} \cdots x(d)^{\alpha(d)}, \quad x = (x(1), \ldots, x(d)) \in \mathbb{R}^d$$

and the sum taken over the set

$$\mathbb{Z}_+^d = \{\alpha \in \mathbb{R}^d : \alpha(j) \in \mathbb{Z}_+, \text{ all } j\}$$

of multi-indices. Further, the coefficient sequence c has finite support.

Since it is important to distinguish between the polynomial p and its value $p(x)$ at the point x, a notation for the power map $x \mapsto x^\alpha$ is needed. There being no standard notation around, we'll use $()^\alpha$ for it. Here is the formal definition:

$$()^\alpha : \mathbb{R}^d \to \mathbb{R} : x \mapsto x^\alpha.$$

In these terms, the polynomial p of (2.1) can be written

$$p = \sum_\alpha ()^\alpha c(\alpha).$$

As indicated already by the unusual positioning of the coefficients, this formula for p can be viewed as the result of applying the 'matrix'

$$X := [()^\alpha : \alpha \in \mathbb{Z}_+^d]$$

with 'columns' $()^\alpha$ to the 'vector' c in the standard way, namely by multiplying the 'column' with index α with the entry of c with index α and summing. Thus,

$$p = Xc.$$

This notation (taken from de Boor [3]) turns out to be very convenient in what is to follow.

Formally, we think of X as a linear map, defined on the space

$$\text{dom } X := (\mathbb{Z}_+^d \to \mathbb{R})_0 := \{c : \mathbb{Z}_+^d \to \mathbb{R} : \# \text{supp } c < \infty\}$$

of finitely supported real-valued sequences indexed by $\mathbb{Z}_+^d$, and mapping into the space $\mathbb{R}^d \to \mathbb{R}$ of real-valued functions of d arguments.

Note that X is 1-1 and its range is Π, the space of polynomials in d indeterminates, interpreted here as functions on $\mathbb{R}^d$. Any linear subspace F of Π

4

of dimension n is the range of a map XW, with W a 1-1 linear map from $\mathbb{R}^n$ into dom X, hence $V := XW$ (or, in more traditional language, the sequence of columns of V) is a basis for F. To be sure, both W and V are linear maps on $\mathbb{R}^n$, hence are, in the language of [3], **column maps**. This language is meant to stress the fact that any linear map B, from n-dimensional coordinate space $\mathbb{F}^n$ to some linear space S over the scalar-field $\mathbb{F}$, is necessarily of the form

$$B = [b_1, \ldots, b_n] : \mathbb{F}^n \to S : c \mapsto \sum_j b_j c(j),$$

with

$$b_j := B\mathbf{i}_j$$

the j**th column of** B and

$$\mathbf{i}_j := (\underbrace{0, \ldots, 0}_{j-1 \text{ zeros}}, 1, 0, \ldots).$$

We denote by

$$\#B$$

the number of columns in the column map B. If the target, S, of the column map $B \in L(\mathbb{F}^n, S)$ is itself a coordinate space, $S = \mathbb{F}^m$ say, hence B is (naturally identified with) an $m \times n$-matrix, then the columns of B as a column map are, indeed, its columns as a matrix.

For any linear map A, whether a column map or not, we denote by

$$\operatorname{ran} A, \quad \ker A$$

its **range** (or set of values) respectively its **kernel** or nullspace.

3   INTERPOLATION

As discussed in the Introduction, we are interested in interpolation at a given finite point set $\Theta \subset \mathbb{R}^d$, of cardinality

$$n := \#\Theta,$$

say. Define

$$\Lambda : \Pi \to \mathbb{R}^n : g \mapsto g|_\Theta := (g(\theta) : \theta \in \Theta).$$

Then, $f \in \Pi$ interpolates to g at Θ iff f solves the equation

$$\Lambda ? = \Lambda g. \tag{3.1}$$

We have introduced the map Λ here, not only for notational convenience, but also in order to stress the fact that most of the discussion to follow is valid for any onto linear map $\Lambda : \Pi \to \mathbb{R}^n$. We call any linear map $\Lambda : \Pi \to \mathbb{R}^n$ a **data map** and note that it is necessarily of the form $\Lambda : g \mapsto (\lambda_i g : i = 1, \ldots, n)$ for some linear functionals λ_i on Π, which we call the **rows** of Λ, for obvious reasons. Given an onto data map Λ, we say that (Λ, F) is **correct** (or, F **is correct for** Λ) if (3.1) has exactly one solution in F for every $g \in \Pi$. In that case, we denote the unique interpolant in F to $g \in \Pi$ by

$$P_F g.$$

Here, for completeness, are certain known facts about interpolation.

Lemma 3.2 *Let F be an n-dimensional polynomial space, let $V \in L(\mathbb{R}^n, \Pi)$ be a basis for F, and let $\Lambda \in L(\Pi, \mathbb{R}^n)$ be onto. Then, the following are equivalent:*
  (i) (Λ, F) is correct.
  (ii) $\Lambda|_F$ is invertible.
  (iii) Λ is 1-1 on F.
  (iv) ΛV is invertible.

  Further, if any one of these conditions holds, then

$$P_F = V(\Lambda V)^{-1} \Lambda.$$

Proof: Since Λ is onto, i.e., its range is all of $\mathbb{R}^n$, correctness of (Λ, F) is equivalent to having the linear map $F \to \mathbb{R}^n : f \mapsto \Lambda f$ be invertible, and, since both domain and target of this map have the same dimension, n, this is equivalent to having this map be 1-1. Finally, since V is a basis for F, the linear map $V|^F : \mathbb{R}^n \to F : a \mapsto Va$ is invertible. Since $\Lambda V = \Lambda|_F V|^F$, it follows that $\Lambda|_F$ is invertible iff ΛV is invertible. ∎

In order to investigate what spaces F are correct for Λ, we look at all possible polynomial interpolants to a given g, i.e., at all possible solutions of (3.1). Since X is 1-1, the solutions of (3.1) are in 1-1 correspondence with solutions of the equation

$$\Lambda X? = \Lambda g.$$

This equation is a system of n linear algebraic equations, albeit in infinitely many unknowns, namely the power coefficients $c(\alpha)$, $\alpha \in \mathbb{Z}_+^d$, of the interpolant. Nevertheless, we may apply Gauss elimination to determine all solutions.

4 GAUSS ELIMINATION

Gauss elimination (with row pivoting) produces a factorization

$$\Lambda X = MR,$$

with M invertible, and R 'in row-echelon form', i.e., 'right-triangular' in the following sense. Each row other than the first is either entirely zero, or else its left-most nonzero entry is strictly to the right of the left-most nonzero entry in the preceding row. The left-most nonzero entry (if any) in a row is called the **pivot element** for that row, and the row is called the **pivot row for** the unknown associated with the column in which the pivot element occurs. An unknown is called **bound** or **free** depending on whether or not it has a pivot row. It is a standard result that an unknown is bound if and only if its column is not in the linear span of the columns to the left of it.

Since ΛX is onto (X being onto Π and Λ being onto), every row of R will be a pivot row. Let $(\alpha_1, \ldots, \alpha_n)$ be the corresponding sequence of indices of bound unknowns, and set $F = \operatorname{ran} V$ with

$$V := [()^{\alpha_1}, \ldots, ()^{\alpha_n}].$$

Then ΛV is invertible, hence (Λ, F) is correct.

This construction is quite arbitrary. As should have been clear from the language used, Gauss elimination depends crucially on the *ordering* of the columns of ΛX, i.e., on the ordering of the columns of X. On the other hand, for $d > 1$, there is no natural ordering of the columns of X, i.e., of the monomials. The best we can do, offhand, is to order the monomials by degree. Here, to be sure,

$$\deg Xc = \max\{|\alpha| := \|\alpha\|_1 = \sum_i \alpha(i) : c(\alpha) \neq 0\}.$$

Equivalently, $\deg p = \min\{k : p \in \Pi_k\}$, with

$$\Pi_k := \operatorname{ran} X_{\leq k},$$

where

$$X_{\leq k} := [X_0, \ldots, X_k]$$

and

$$X_j := [()^{\alpha} : |\alpha| = j].$$

Assume that X is **graded**, i.e., segmented by degree, i.e.,

$$X = [X_0, X_1, \ldots].$$

This has the following happy consequence.

Result 4.1 ([7; (2.5)]). *If X is graded, then polynomial interpolation from the space F spanned by the monomials corresponding to the bound columns of ΛX is degree-reducing, i.e.,*

$$\deg P_F g \leq \deg g \qquad \forall g \in \Pi.$$

Proof: Recall, as we did earlier, that an unknown is bound if and only if its column is not a linear combination of preceding columns. In particular, any column is in the linear span of the bound columns not to the right of it. This implies that, if $\deg g = k$ and therefore $\Lambda g \in \operatorname{ran} \Lambda X_{\leq k}$, then Λg is necessarily in the linear span of the bound columns in $\Lambda X_{\leq k}$. Since $f := P_F g$ is the unique element in F with $\Lambda f = \Lambda g$, it follows that $P_F g$ is already in

$$F_k := F \cap \Pi_k. \qquad \blacksquare$$

In particular, for $d = 1$ and $\Lambda = |_{\Theta}$, we recover in this way the standard choice for F, namely $\Pi_{<n}$. However, for $d > 1$, F still depends on the ordering of the columns within each X_k and there is no natural ordering for them. It is for this reason that we now consider Gauss elimination by segments.

5   GAUSS ELIMINATION BY SEGMENTS

To recall from the Introduction, Gauss elimination by segments applied to the segmented matrix $A = [A_0, A_1, \ldots]$ produces a factorization

$$[A_0, A_1, \ldots] = M[R_0, R_1, \ldots]$$

with M invertible and

$$R := [R_0, R_1, \ldots]$$

a segmented row-echelon form, the segmentation corresponding to that of A. This means that R is block upper triangular, $R = (R_{ij})$, say, with

$$R_j = [R_{0j}; R_{1j}; \ldots] = [U_{jj}; R_{jj}; 0] := \begin{bmatrix} U_{jj} \\ R_{jj} \\ 0 \end{bmatrix},$$

$\#R_j = \#A_j$, and with each diagonal block R_{jj} onto. (The semicolon is used here as in MATLAB.)

While, except for trivial cases, a segmented matrix has infinitely many such factorizations, some properties of such a factorization depend only on A and the particular segmentation used. Here are the basic facts.

Lemma 5.1 *Let A, B, C, M, S, Q, T be matrices with*

$$A = [B, C] = M \begin{bmatrix} S & Q \\ 0 & T \end{bmatrix},$$

M invertible, $\#B = \#S$, and S onto. Then
  (i) The transpose S' of S is (i.e., the rows of S form) a basis for the row-space of B, i.e., for $\operatorname{ran} B'$.
  (ii) The row-space of T, i.e., the space $\operatorname{ran} T'$, depends only on A and $\#T = \#A - \#B$. Explicitly,

$$\operatorname{ran} T' = \{ y \in \mathbb{R}^{\#T} : (0, y) \in \operatorname{ran} A' \}.$$

  (iii) For all $c \perp \operatorname{ran} T'$, there exists b so that $(b, c) \in \ker A$.

Proof: (i): Since $\#B = \#S$, we have $B = MS$, with M invertible, hence $\operatorname{ran} B' = \operatorname{ran} S'M' = \operatorname{ran} S'$.

(ii): Since M is invertible, $\operatorname{ran} A' = \operatorname{ran} \begin{bmatrix} S' & 0 \\ Q' & T' \end{bmatrix}$. In particular, for any $\#T$-sequence y and any $\#S$-sequence x, $(x, y) \in \operatorname{ran} A'$ if and only if

$$(x, y) = (S'z_1, Q'z_1 + T'z_2)$$

for some sequence $z = (z_1, z_2)$. In particular, since S is onto and therefore S' is 1-1, such x is zero iff $z_1 = 0$. In other words, $(0, y) \in \operatorname{ran} A'$ if and only if $y = T'z_2$ for some z_2.

(iii): If $c \perp \operatorname{ran} T'$ then $c \in \ker T$, therefore, for any $\#S$-vector b, $A(b, c) = M(Sb + Qc, 0)$. It follows that $(b, c) \in \ker A$ if and only if $Sb = -Qc$ and, since S is onto by assumption, such a choice for b is always possible. ∎

Corollary 5.2 *Let*

$$\Lambda X = MR$$

with M invertible and $R = (R_{ij})$ a segmented row-echelon form, segmented corresponding to the segmentation $X = [X_0, X_1, \ldots]$ of X. Then,

(i) for any k, R'_{kk} is a basis for

$$\{y \in \mathbb{R}^{r_k} : (0, y) \in \operatorname{ran}(\Lambda X_{\leq k})'\},$$

with $r_k := \#R_{kk}$. In particular, $\operatorname{ran} R'_{kk}$ depends only on Λ.

(ii) for any k, $\#R'_{kk} = \operatorname{rank} R_{kk} = \operatorname{rank} \Lambda X_{\leq k} - \operatorname{rank} \Lambda X_{<k}$.

Proof: (i): This follows from (ii) of Lemma 5.1, with the choices $B := \Lambda X_{<k}$, $C := \Lambda X_k$, $S := (R_{ij} : 0 \leq i, j < k)$, $Q := [R_{0k}; \ldots; R_{k-1,k}]$, and $T := R_{kk}$.

(ii): Since M is invertible,

$$\operatorname{rank} \Lambda X_{\leq k} = \operatorname{rank} R_{\leq k},$$

and the latter number equals $\sum_{j \leq k} \#R'_{jj}$ since $R_{\leq k} = (R_{ij} : i = 0, 1, \ldots; j = 0, \ldots, k)$ is block upper triangular with each diagonal block onto. In particular, each R_{jj} has full row rank, hence $\#R'_{jj} = \operatorname{rank} R_{jj}$, all j. ∎

We briefly discuss the construction of a segmented row-echelon form for a given segmented matrix $A = [A_0, A_1, \ldots]$.

It follows from Lemma 5.1 that, in step 0 of Gauss elimination by segments applied to the segmented matrix $A = [A_0, A_1, \ldots]$, one determines, by some means, some basis R'_{00} for $\operatorname{ran} A'_0$. While this could, of course, be done by Gauss elimination, with the nontrivial rows of the resulting row-echelon-form for A_0 providing the desired basis, i.e., providing the rows for the matrix R_{00}, more stable procedures come to mind. For example, one could construct R'_{00} as an orthogonal basis, or even an orthonormal basis, for the row-space of A_0. The favored numerical procedure for this would be to construct the QR factorization for A'_0, preferably using

9

Householder reflections and column interchanges, and use the resulting orthonormal basis for ran A'_0.

Since R'_{00} is a basis for ran A'_0, there is a unique N_0 so that $A_0 = N_0 R_{00}$. In fact, N_0, or some left inverse for N_0, is usually found during the construction of R_{00}. In any case, since the rows of R_{00} are a basis for the row-space of A_0, N_0 is necessarily 1-1, hence can be extended to some invertible matrix $M_0 = [N_0, \ldots]$. This gives the factorization

$$A = [A_0, A_1, A_2, \cdots] = M_0 \begin{bmatrix} R_{00} & & Q & \\ 0 & T_1 & T_2 & \cdots \end{bmatrix}.$$

Subsequent steps concentrate on the segmented matrix $T := [T_1, T_2, \cdots]$.

This discussion of Gauss elimination by segments was only given in order to indicate the many different ways available for constructing a factorization $A = MR$ of a segmented matrix $A = [A_0, A_1, \ldots]$ into an invertible M and a correspondingly segmented row-echelon form. The *existence* of such a factorization was never in doubt, as any factorization $A = MR$ obtained by Gauss elimination with partial pivoting is seen to be of the desired form after R is appropriately blocked.

6   A "NATURAL" CHOICE FOR F

We have already seen that a correct F can be so chosen that P_F is degree-reducing. The following result characterizes all such choices F.

Result 6.1   ([8]). *If (Λ, F) is correct, then, P_F is degree-reducing if and only if each $F_k = F \cap \Pi_k$ is as large as possible.*

We give a simple proof of this in a moment, for completeness. But first we explore what limits if any might actually be imposed on the dimension of F_k. Since correctness of (Λ, F) implies that Λ is 1-1 on F, Λ must be 1-1 on every F_k, hence, for every k,

$$\dim F_k = \dim \Lambda(F_k) \leq \dim \Lambda X_{\leq k} = \operatorname{rank} \Lambda X_{\leq k}.$$

Moreover, equality is possible here for every k; it is achieved by the subspace F spanned by the monomials corresponding to bound columns in $\Lambda[X_0, X_1, \ldots]$. This proves the following.

Corollary 6.2   *Assume that (Λ, F) is correct. Then, P_F is degree-reducing if and only if*

$$\dim F_k = \operatorname{rank} \Lambda X_{\leq k}, \qquad k = 0, 1, 2, \ldots . \tag{6.3}$$

In other words, if $V = [V_0, V_1, \ldots]$ is a **graded** basis for F, i.e., $V_{\leq k}$ is a basis for F_k for every k, then P_F cannot be degree-reducing unless

$$\#V_{\leq k} = \operatorname{rank} \Lambda X_{\leq k}, \qquad k = 0, 1, 2, \ldots . \tag{6.4}$$

In terms of any segmented row-echelon form $R = (R_{ij})$ for $\Lambda[X_0, X_1, \ldots]$, (6.4) is, by (ii) of Corollary 5.2, equivalent to

$$\#V_k = \#R'_{kk}, \qquad k = 0, 1, 2, \ldots .$$

Note that consideration of a graded basis for F imposes no restriction since such a basis can always be constructed inductively, starting with $k = -1$: With a basis $V_{\leq k}$ for F_k already in hand, a requisite V_{k+1} is obtained by completing $V_{\leq k}$ to a basis for F_{k+1}. For the proof of Result 6.1, we require the evident fact that we are free to include in V_{k+1} any one element from $F_{k+1} \backslash F_k$.

Proof of Result 6.1: If P_F is degree-reducing and G is any polynomial space correct for Λ, then P_F is 1-1 on G (since, by Lemma 3.2, $P_F = V(\Lambda V)^{-1}\Lambda$ with both V and $(\Lambda V)^{-1}$ 1-1, and also Λ is 1-1 on G), therefore, for any k, $\dim G_k = \dim P_F(G_k) \leq \dim F_k$, since $P_F(G_k) \subseteq P_F(\Pi_k) \subseteq \Pi_k \cap F = F_k$ (using the fact that $P_F(\Pi_k) \subseteq \Pi_k$ since P_F is degree-reducing).

Conversely, if P_F fails to be degree-reducing, then there exists g with $k := \deg g < \deg P_F g =: k'$. Since $P_F g \in F_{k'} \backslash F_{k'-1}$, we may (by the last sentence before this proof) include it in a graded basis V for F. Let U be the column map obtained from V by replacing $P_F g$ by g. Then $\Lambda V = \Lambda U$, hence also ΛU is invertible and therefore also $G := \operatorname{ran} U$ is correct for Λ. However, $\dim G_k > \dim F_k$. ∎

We are now ready to discuss use of an appropriately segmented row-echelon form

$$R = (R_{ij}) = M^{-1}\Lambda[X_0, X_1, \cdots]$$

for ΛX in checking whether a polynomial space F satisfying (6.3) is correct for interpolation at Λ. If V is any basis for F, then (Λ, F) is correct iff the matrix ΛV is invertible. With $V =: XW$, this is equivalent to having RW invertible (since M is invertible by assumption). Now assume that $V = [V_0, V_1, \ldots]$ is a graded basis for F. Let

$$V_k =: XW_k, \qquad k = 0, 1, \ldots .$$

Then $V_j = \sum_{i \leq j} X_i W_{ij}$, all i, j, i.e., the column map W, from $\mathbb{R}^n$ into $\operatorname{dom} X$, given by $V =: XW$, is 1-1 and block upper triangular, with W_{ij} having $\#X_i$ rows and $\#V_j$ columns. Since (6.3) holds, $\#V_j = \#R'_{jj}$, all j. Hence, the matrix RW is block upper triangular, with *square* diagonal blocks. Therefore, RW is invertible if and only if $R_{kk}W_{kk}$ is invertible for each k. We have proved the following:

Proposition 6.5 *Let MR be any factorization for $\Lambda[X_0, X_1, \ldots]$ with M invertible and $R = (R_{ij})$ a correspondingly segmented row-echelon form. Assume that the polynomial space F satisfies (6.3), and let $V = XW$ be a graded basis for F, hence $W = (W_{ij})$ is block upper triangular, with W_{ij} having $\#X_i$ rows and $\#R'_{jj}$ columns. Then (Λ, F) is correct if and only if $R_{kk}W_{kk}$ is invertible for every k.*

Note that this condition only involves the diagonal blocks of W. This suggests that we choose all off-diagonal blocks of W to be zero, thus keeping the structure of F simple. This is equivalent to the requirement that

$$F = \bigoplus_k X_k W_{kk},$$

which, in turn, is equivalent to having F be scale-invariant.

Note further that the condition of the Proposition can always be met, since, by construction, each R_{kk} is onto, hence has right inverses.

If R_{kk} is square, then invertibility of $R_{kk} W_{kk}$ implies that W_{kk} is invertible, hence

$$\operatorname{ran} X_k W_{kk} = \operatorname{ran} X_k = \Pi_k^0.$$

In other words, there is then no real choice; regardless of how W_{kk} is chosen, F contains all homogeneous polynomials of degree k.

If R_{kk} is not square, then the invertibility of $R_{kk} W_{kk}$ does not completely determine the kth homogeneous component of F. How is one to choose from among the infinitely many possibilities?

The choice

$$W_{kk} = R'_{kk}, \qquad k = 0, 1, 2, \ldots \tag{6.6}$$

suggests itself, as it guarantees that $R_{kk} W_{kk}$ is invertible (since R_{kk} is onto). Moreover, if R'_{kk} was chosen as an orthonormal basis, then the diagonal block $R_{kk} W_{kk}$ becomes an identity matrix, thus facilitating the calculation (by backsubstitution) of the coefficient vector c of the resulting interpolant Xc from F to the given g. Indeed, the interpolant is of the form XWa, hence a solves the linear system

$$\Lambda X W? = \Lambda g,$$

and therefore solves the equivalent linear system

$$RW? = M^{-1} \Lambda g. \tag{6.7}$$

From a, the coefficient vector c in the power form Xc for the interpolant is obtained as

$$c = Wa.$$

The resulting interpolation scheme P_F has additional nice properties. However, we don't bother to derive them here since P_F has a major flaw: even if $\Lambda : g \mapsto g|_\Theta$, P_F depends on the coordinate system chosen in $\mathbb{R}^d$, i.e., it fails to have the desirable property that

$$P_F T g = T P_F g$$

for all $g \in \Pi$ and all 'reasonable' changes of variables, T.

12

7 EXAMPLE

Suppose that $\Lambda : g \mapsto g|_\Theta$ with $\Theta = \{s_1\theta, \ldots, s_n\theta\}$ for some nontrivial θ and some (scalar) n-set $\{s_1, \ldots, s_n\}$. Then, for each k,

$$\Lambda X_k = [s_1^k, \ldots, s_n^k]' \, [\theta^\alpha : |\alpha| = k].$$

This shows that each ΛX_k has rank 1. Therefore, regardless of how the segmented row-echelon form $R = (R_{ij})$ for $\Lambda[X_0, X_1, \ldots]$ is obtained, each R_{kk} has just one row, and this row is a scalar multiple of the vector $(\theta^\alpha : |\alpha| = k)$. Consequently, $V_k = X_k W_{kk}$ has just one column, namely the polynomial

$$p_k := \sum_{|\alpha|=k} ()^\alpha \theta^\alpha.$$

This is unfortunate, for the following reason. If $\theta = \mathbf{i}_i$ for some i, then F consists of all polynomials of degree $< n$ in the ith indeterminate only, and that is good. However, if θ has more than one nonzero entry, then F usually differs from the natural choice in that case, namely the space of all polynomials of degree $< n$ which are constant in any direction perpendicular to θ. For F to be that space, we would need to choose each p_k as some scalar multiple of

$$\sum_{|\alpha|=k} ()^\alpha \theta^\alpha / \alpha!,$$

e.g., as the polynomial

$$x \mapsto (\theta \cdot x)^k.$$

This means that, instead of $W_{kk} = R'_{kk}$, we should choose

$$W_{kk} = \Omega_k^{-1} R'_{kk}, \tag{7.1}$$

with

$$\Omega_k := \operatorname{diag}(\alpha! : |\alpha| = k). \tag{7.2}$$

The space F resulting from this choice is the **least choice**, i.e., the space

$$\Pi_\Lambda = \bigoplus_k \operatorname{ran} X_k \Omega_k^{-1} R'_{kk} =: \operatorname{ran}[V_0, V_1, V_2, \ldots] \tag{7.3}$$

of de Boor and Ron [8]. In particular, it is the space Π_Θ of [5] in case $\Lambda = \cdot|_\Theta$, i.e., $\Lambda : g \mapsto g|_\Theta$. Correspondingly, we denote by

$$P_\Lambda, \quad \text{resp.} \quad P_\Theta$$

the resulting interpolation projectors.

To be sure, [8] arrived at this choice in a completely different way, as the linear span of all the least terms of elements of a certain space of formal power series. To make the connection between (7.3) and the definition of Π_Λ in [8], let v_i be one of the columns of $[V_0, V_1, V_2, \ldots]$. Then v_i is of the form

$$v_i = \sum_{|\alpha|=k} ()^\alpha/\alpha!\, R(i,\alpha)$$

for some k and, for that k, $R(i,\alpha) = 0$ for all $|\alpha| < k$. This implies that $v_i = f_{i\downarrow}$ (see the Introduction for the definition of $f_\downarrow$), with f_i the formal power series

$$f_i := \sum_\alpha ()^\alpha/\alpha!\, R(i,\alpha).$$

Further, $R = M^{-1}\Lambda X$, for some (invertible) M. Consequently, the f_i form a basis for the space G spanned by the formal power series

$$g_j := \sum_\alpha ()^\alpha/\alpha!\, \lambda_j ()^\alpha, \quad j = 1,\ldots,n,$$

with λ_j the jth row of the data map Λ. Further, if g is any element of G with $\deg g_\downarrow = k$, then, by Lemma 5.1(ii), $g_\downarrow$ is necessarily in the range of $X_k \Omega_k^{-1} R'_{kk}$. So, altogether, this shows that the right-hand side of (7.3) is, indeed, the linear span of all the leasts of elements of G, hence equal to Π_Λ as defined in [8].

Taking, in particular, $\Lambda = |_\Theta$, hence $\lambda_j = \delta_\theta$ (i.e., point evaluation at θ) for some $\theta \in \Theta$, we get

$$g_j = \sum_\alpha ()^\alpha/\alpha!\, \theta^\alpha = \sum_\alpha (\cdot\theta)^\alpha/\alpha! = e_\theta,$$

thus showing that, for this case, (7.3) does, indeed, give the space spanned by all the leasts of linear combinations of exponentials e_θ with $\theta \in \Theta$.

It has been pointed out (by Andreas Felgenhauer of TU Dresden, after a talk on this material) that the simple choice (6.6) actually provides (7.1) if X_k were changed to

$$X_k := [x \mapsto \prod_i x(\beta(i)) : \beta \in \{1,\ldots,d\}^k],$$

i.e., if the monomials were treated as if the indeterminates did not commute, with a corresponding change of Ω_k, to the scalar matrix $k!$.

8 THE LEAST CHOICE

The least choice, (7.3), not only provides the 'right' space in case Θ lies on a straight line, it has a rather impressive list of properties which are detailed in

14

[5,7,8]. Of these, the following is perhaps the most unusual. Its statement uses the notation $p_\uparrow$, for the *leading* term of the polynomial p, as introduced in the Introduction. Further, if $p = Xc$, then

$$p(D) := \sum_\alpha D^\alpha c(\alpha),$$

with

$$D^\alpha = D_1^{\alpha(1)} \cdots D_d^{\alpha(d)},$$

and D_i the derivative with respect to the ith argument.

Result 8.1 ([7]).

$$\Pi_\Theta = \bigcap_{p|_\Theta = 0} \ker p_\uparrow(D).$$

A proof is provided in [8], as part of a more general argument. A direct proof can be found in [4].

It is informative to consider this result in the light of elimination, as a further motivation for the choice (7.1) for the W_{kk}. For this, we continue to denote by MR any factorization of $\Lambda[X_0, X_1, \ldots]$ into an invertible M and a correspondingly segmented row-echelon form.

Lemma 8.2 Let $p \in \Pi_k \backslash \Pi_{<k}$ with $p_\uparrow =: X_k c$. Then, there exists $q \in \Pi_{<k}$ with $\Lambda(p - q) = 0$ if and only if $c \perp \operatorname{ran} R'_{kk}$.

Proof: Suppose that $c \perp \operatorname{ran} R'_{kk}$. Then $R_{kk}c = 0$, hence, by (iii) of Lemma 5.1, there exists b so that $(b, c) \in \ker R_{\leq k}$. This implies that $(b, c) \in \ker \Lambda X_{\leq k}$, i.e., $p := X_{\leq k}(b, c)$ is in $\ker \Lambda$.

Conversely, if p is of (exact) degree k and in $\ker \Lambda$, then $p = X_{\leq k}(b, c)$ for some b, c with $p_\uparrow = X_k c$, and $0 = \Lambda p = \Lambda X_{\leq k}(b, c) = M R_{\leq k}(b, c)$, hence, since M is invertible, also $R_{\leq k}(b, c) = 0$, in particular $R_{kk}c = 0$, i.e., $c \perp \operatorname{ran} R'_{kk}$. ∎

This lemma characterizes $\operatorname{ran} R'_{kk}$ as the orthogonal complement of the set

$$\{c : \exists\{p \in \ker \Lambda\}\ p_\uparrow = X_k c\}$$

of leading coefficients of polynomials of exact degree k in the kernel of Λ. Since

$$\Pi^0_{\Lambda,k} := \Pi_\Lambda \cap \Pi^0_k = X_k \Omega_k^{-1} R'_{kk},$$

this provides the following characterization of the kth homogeneous component of Π_Λ, i.e., of $\Pi^0_{\Lambda,k}$.

Proposition 8.3

$$\Pi^0_{\Lambda,k} = \Pi^0_k \cap \bigcap \{\ker p_\uparrow(D) : p \in \Pi_k \backslash \Pi_{<k}, \Lambda p = 0\}.$$

15

Proof: Assume that q is a *homogeneous* polynomial of degree k, i.e., $q =: X_k c$, and consider the corresponding differential operator $q(D)$. Then $\Pi_{<k} \subset \ker q(D)$ trivially. Hence, for any $g \in \Pi_k \backslash \Pi_{<k}$, setting $g_\uparrow =: X_k a$, we have

$$q(D)g = q(D)(g_\uparrow) = \sum_{|\alpha|=k} c(\alpha)\alpha! a(\alpha).$$

In particular, with Ω_k as given in (7.2),

$$q(D)g = 0 \quad \Longleftrightarrow \quad c \perp \Omega_k a.$$

If now g is an element of Π_Λ, then $g_\uparrow = X_k \Omega_k^{-1} R'_{kk} b$ for some b, therefore

$$\Omega_k a = R'_{kk} b.$$

Consequently, $\Pi_\Lambda \cap \Pi_k \subset \ker q(D)$ if and only if $c \perp \operatorname{ran} R'_{kk}$, i.e., if and only if $R_{kk} c = 0$, i.e., by Lemma 8.2, if and only if q is the leading term of some polynomial in $\ker \Lambda$. ∎

Note that, for the case $\Lambda = |_\Theta$, Result 8.1 is much stronger than Proposition 8.3. For, in this special case, not only is $\Pi^0_{\Theta,k}$ annihilated by $p_\uparrow(D)$ in case $p \in \Pi_k \backslash \Pi_{<k}$ and $p|_\Theta = 0$, but *all of* Π_Θ is annihilated by such $p_\uparrow(D)$. In particular,

$$p_\uparrow(D) X_j \Omega_j^{-1} R'_{jj} = 0$$

for any j.

The reason for this much stronger result in the case $\Lambda = |_\Theta$ is that, in this case, $\ker \Lambda$ is a polynomial ideal and, consequently, Π_Λ is D-invariant, i.e., $f \in \Pi_\Lambda$ and $\alpha \in \mathbb{Z}_+^d$ implies $D^\alpha f$. See [8] for details.

9 CONSTRUCTION OF AN H-BASIS FOR A POLYNOMIAL IDEAL WITH FINITE VARIETY

Customarily, a polynomial ideal $\mathcal{I}$ is specified by some finite generating set for it, i.e., by describing it as

$$\mathcal{I} = \sum_{g \in G} \Pi g$$

for some finite $G \subset \Pi$. Given this description, it is nontrivial to determine whether or not a given $f \in \Pi$ belongs to $\mathcal{I}$, except when G is an **H-basis** for $\mathcal{I}$, meaning that, for every k,

$$\mathcal{I}_k := \mathcal{I} \cap \Pi_k = \sum_{g \in G} \Pi_{k-\deg g} g.$$

Such a generating set is also called a **Macaulay basis** or a **canonical basis** for $\mathcal{I}$.

Indeed, if G is an H-basis for $\mathcal{I}$, then, with

$$\mathcal{I}^0 := \{p_\uparrow : p \in \mathcal{I}\}$$

the **associated homogeneous ideal**, we have

$$\mathcal{I}_k^0 := \mathcal{I}^0 \cap \Pi_k^0 = \sum_{g \in G} \Pi_{k-\deg g}^0 g_\uparrow, \quad \text{all } k.$$

Thus, $f \in \mathcal{I}$ if and only if

$$f_\uparrow = \sum_{g \in G} p_g g_\uparrow$$

for some $p_g \in \Pi_{k-\deg g}^0$, and in addition, for these p_g,

$$f - \sum_{g \in G} p_g g \in \mathcal{I}.$$

Since the last condition involves a polynomial of degree $< \deg f$, this leads to a terminating recursive check.

If an ideal is not given in terms of an H-basis, then it is, in general, nontrivial to construct an H-basis, except when the direct summands $\mathcal{I}_k^0$ are known in the sense that bases are known for them.

In that situation, one can construct an H-basis G for $\mathcal{I}$ by constructing

$$G_k := G \cap \Pi_k$$

inductively, i.e., for $k = 0, 1, 2, \ldots$, so that

$$\mathcal{I}_k = \sum_{g \in G_k} \Pi_{k-\deg g} g, \tag{9.1}$$

as follows.

Assume that we already have G_{k-1} in hand (as is surely the case initially for any k with $\mathcal{I}_{k-1}^0 = \{0\}$, e.g., for $k = 0$). We claim that the choice

$$G_k := G_{k-1} \cup B,$$

with $B \subset \mathcal{I}_k$ so that $[B_\uparrow]$ is a basis for a linear subspace in $\mathcal{I}_k^0$ complementary to

$$\sum_{g \in G_{k-1}} \Pi_{k-\deg g}^0 g_\uparrow,$$

does the job, i.e., satisfies (9.1).

Indeed, if $p \in \mathcal{I}_k$, then $p_\uparrow \in \mathcal{I}_k^0$, hence

$$p_\uparrow = [B_\uparrow]a + \sum_{g \in G_{k-1}} p_g g_\uparrow$$

for some $a \in \mathbf{C}^B$ and some $p_g \in \Pi_{k-\deg g}^0$, all $g \in G_{k-1}$. However, then

$$p - [B]a - \sum_{g \in G_{k-1}} p_g g \quad \in \quad \mathcal{I}_{k-1} = \sum_{g \in G_{k-1}} \Pi_{k-1-\deg g} g,$$

therefore

$$p \in \sum_{g \in B} \Pi_0 g + \sum_{g \in G_{k-1}} \Pi_{k-\deg g} g,$$

which was to be shown.

Note that the H-basis constructed is minimal in the sense that any proper subset of it would generate a proper subideal of $\mathcal{I}$.

An appropriate B can be constructed by: (i) starting with C for which $[C_\uparrow]$ is a basis for $\mathcal{I}_{k-1}^0$ (as constructed in the preceding step), (ii) selecting from the columns of $[()^{1_j} c : c \in C, j = 1, \ldots, d]$ a column map $\tilde{C}$ which is maximal with respect to having $[\tilde{C}_\uparrow]$ 1-1, (iii) extending this to a column map $[\tilde{C}, B]$ with ($B \subset \mathcal{I}$ and) $[\tilde{C}_\uparrow, B_\uparrow]$ a basis for $\mathcal{I}_k^0$.

Of course, this requires that one have in hand, for each k, a $B_k \subset \mathcal{I}$ for which $[B_{k\uparrow}]$ is a basis for $\mathcal{I}_k^0$, and it is in this sense that one needs to "know" the $\mathcal{I}_k^0$. This information is easy to derive in case $\mathcal{I}$ is an ideal with finite codimension or, what is the same, with finite variety.

It is well-known (see, e.g., de Boor and Ron [6] for a detailed retelling of the relevant facts) that a polynomial ideal has finite codimension exactly when it is of the form $\ker \Lambda$, with $\Lambda' =: [\lambda_1, \ldots, \lambda_n]$ a basis for a subspace of Π' of the form

$$\sum_{\theta \in \Theta} \delta_\theta P_\theta(D) = \{\sum_{\theta \in \Theta} \delta_\theta p_\theta(D) : p_\theta \in P_\theta\} \tag{9.2}$$

with Θ, the variety of the ideal, a *finite* subset of $\mathbf{C}^d$, $\delta_\theta : p \mapsto p(\theta)$ the linear functional of evaluation at θ, and each P_θ a D-invariant finite-dimensional linear subspace of Π. To put it differently, a correct polynomial interpolation scheme with data map $\Lambda \in L(\Pi, \mathbb{R}^n)$ is ideal (in the sense of Birkhoff [1]) if and only if it is **Hermite interpolation**, i.e., its set of interpolation conditions is of the form (9.2).

Of course, if $\mathcal{I} = \ker \Lambda$ for some explicitly known map $\Lambda \in L(\Pi, \mathbb{R}^n)$, then membership in $\mathcal{I}$ of a given $f \in \Pi$ is trivially testable: simply compute Λf. On the other hand, an H-basis for such an ideal is likely to have its uses in the construction of error formulæ for the associated polynomial interpolation schemes.

In any case, if $\mathcal{I} = \ker \Lambda$ for some onto $\Lambda \in L(\Pi, \mathbb{R}^n)$, then

$$\mathcal{I} = X \ker R,$$

with

$$\Lambda X = MR$$

any particular factorization of ΛX with a 1-1 M.

In particular, let $R = (R_{ij})$ be a segmented row-echelon form, segmented corresponding to the segmentation $X = [X_0, X_1, \ldots]$. By Lemma 8.2, $p \in \Pi_k^0$ is in $\mathcal{I}_k^0$ iff $c := X_k^{-1}p \in \ker R_{kk}$. Consequently, any basis K_k for $\ker R_{kk}$ provides a basis, $X_k K_k$, for $\mathcal{I}_k^0$. Since $R = (R_{ij})$ is in segmented row-echelon form, each R_{jj} is onto while $R_{ij} = 0$ for $i > j$; therefore we can find $K_0, \ldots, K_{k-1}$ so that $\sum_{j<k} R_{ij} K_j = -R_{ik} K_k$ for all $i < k$, while this holds trivially for all $i \geq k$. This implies that $B_k := \sum_{j \leq k} X_j K_j$ is a column map into $\mathcal{I}$, with $B_{k\uparrow}$ a basis for $\mathcal{I}_k^0$.

If the rows of R_{kk} are, as in [7], constructed as an orthonormal basis (of the row space of a certain matrix and with respect to some inner product), then a convenient choice for K_k is the completion $[R'_{kk}, K_k]$ of R'_{kk} to an orthonormal basis for all of $\mathbb{R}^{\#R_{kk}}$.

10   THE CONSTRUCTION OF RULES

The **least rule for** $\mu \in \Pi'$ **from** $\Lambda \in L(\Pi, \mathbb{R}^n)$ is, by definition, the linear functional μP_Λ. Since $P_\Lambda = V(\Lambda V)^{-1}\Lambda$ (with V some basis for Π_Λ; see Lemma 3.2), we can also write the least rule explicitly in terms of the rows λ_i of Λ as

$$\mu P_\Lambda = a * \Lambda = \sum_i a(i)\lambda_i, \qquad (10.1)$$

with

$$a := \mu V(\Lambda V)^{-1}.$$

Since our basis $V = [V_0, V_1, \ldots] =: XW$ for Π_Λ has the simple segments

$$V_k = X_k \Omega_k^{-1} R'_{kk}, \quad k = 0, 1, 2, \ldots,$$

(see (7.3)) it is not hard to apply the inverse of the matrix $(\Lambda V)' = (\Lambda XW)' = (MRW)'$ to the vector μXW. In fact, the resulting work can be carried out right along with the construction of the factorization if one is willing to record M^{-1} in some convenient form, as we now show.

The idea is to apply the program described in [7], not just to the matrix ΛX but to the 'augmented' matrix $[\Lambda; \mu]X$. (The choice of the word 'augmented' here is quite deliberate; it stresses the fact that, what is about to be described, is nothing but the dual of the standard procedure of applying Gauss elimination to the augmented matrix $[A, b]$ in order to compute a solution of $A? = b$.) For this to work, this additional, last row must never be used as pivot row. To recall, the program in [7] constructs each R'_{kk} as an orthogonal basis for the row space of a certain working array, call it B_k, whose rows consist of the X_k-part of each row not yet used as pivot row. In other words, the columns of B_k, i.e., the entries in each row of B_k, are naturally indexed by α with $|\alpha| = k$. Orthogonality is with respect to the inner product

$$\langle b, c \rangle_k := b \cdot \Omega_k^{-1} c. \tag{10.2}$$

The construction consists of making all rows in B_k not yet used as pivot rows orthogonal to those being used as pivot rows. Without that additional last row, work on B_k ceases once all rows not yet used as pivot rows are zero. With that additional row, work on this segment still ceases when all rows not used as pivot rows are zero, *except* for that additional row. This may leave that additional row nonzero, but it is certain to leave it orthogonal to the rows of the resulting R_{kk} with respect to the inner product (10.2).

At the end, we obtain the factorization

$$\begin{bmatrix} \Lambda \\ \mu \end{bmatrix} X = \begin{bmatrix} M & 0 \\ m & 1 \end{bmatrix} \begin{bmatrix} R \\ u \end{bmatrix},$$

with $\Lambda X = MR$ the earlier factorization, and with u the segmented sequence $(u_0, u_1, \ldots)$ such that

$$u_k \cdot \Omega_k^{-1} R'_{kk} = 0$$

for every nonempty R_{kk}. One verifies that

$$\begin{bmatrix} M & 0 \\ m & 1 \end{bmatrix}^{-1} = \begin{bmatrix} M^{-1} & 0 \\ -mM^{-1} & 1 \end{bmatrix}.$$

(Here, m is a $1 \times (m-1)$-matrix rather than a sequence, hence I write mM^{-1} rather than the (undefined) $m \cdot M^{-1}$.) This implies that

$$(-mM^{-1}\Lambda + \mu)V = (-mM^{-1}\Lambda + \mu)XW = \sum_k u_k \cdot \Omega_k^{-1} R'_{kk} = 0.$$

Consequently, $mM^{-1}\Lambda V = \mu V$, hence the desired coefficient vector a for (10.1) equals the one row of mM^{-1}.

11   A SECOND LOOK AT GAUSS ELIMINATION BY SEGMENTS

Here is a second look at Gauss elimination by segments, from the point of view of the data map $\Lambda \in L(\Pi, \mathbb{R}^n)$. This discussion is applicable to any (segmented) matrix $A = [A_0, A_1, \ldots]$ since an arbitrary matrix $A \in \mathbb{F}^{n \times J}$ can be thought of as such a product ΛX of a row map $\Lambda \in L(S, \mathbb{F}^n)$ with a column map $X \in L(\mathbb{F}^J, S)$ in at least two ways:

(1) $S = \mathbb{F}^n$ and $\Lambda = \text{id}$, hence $X = A = [A(:, j) : j \in J]$.

(2) $S = \mathbb{F}^J$ and $V = \text{id}$, hence $\Lambda = A \in L(S, \mathbb{F}^n)$.

At the same time, it is to be hoped that elimination by segments will be found of help in other applications where it is more natural to do elimination by certain segments rather than by columns. For example, a process requiring a certain amount of column pivoting might be a good prospect.

For this reason, in this section, $X = [X_0, X_1, \ldots]$ is any segmented column map, from $\mathbb{F}_0^J := (J \to \mathbb{F})_0$ into some linear space S, with J some set, and Λ is any onto data map or **row map** on S, and the matrix of interest is

$$A := \Lambda X = [\Lambda X_0, \Lambda X_1, \ldots] =: [A_0, A_1, \ldots],$$

as before.

We have already made use of the following fact: with $\lambda_1, \ldots, \lambda_n$ the rows of the data map $\Lambda \in L(S, \mathbb{F}^n)$, the dual map $\Lambda' \in L(\mathbb{F}^n, S')$ for Λ is of the form

$$\Lambda' = [\lambda_1, \ldots, \lambda_n] : \mathbb{F}^n \to S' : a \mapsto \sum_i a(i)\lambda_i.$$

Since Λ is onto by assumption, Λ' is 1-1, hence a basis for its range.

For each k, let Λ_k be a data map whose rows, when restricted to $S_k := \operatorname{ran} X_{\leq k}$, provide a basis for the linear space

$$L_k := \{\lambda|_{S_k} : \lambda \in L, \ \lambda X_{<k} = 0\}.$$

Equivalently, Λ_k is any onto data map with its $\dim L_k$ rows taken from $\operatorname{ran} \Lambda'$, for which $\Lambda_k X_{<k} = 0$ while $\Lambda_k X_k$ is onto.

However this is actually done numerically, we will have constructed a basis

$$[\Lambda_k' : k = 1, 2, \ldots]$$

for $\operatorname{ran} \Lambda'$. As Λ' is 1-1, hence a basis for its range, this means that we have, in effect, constructed an invertible matrix $M' \in \mathbb{R}^{m \times m}$ so that

$$\Lambda' = [\Lambda_k' : k = 0, 1, 2, \ldots]M'.$$

Thus,

$$\Lambda X = A = MR,$$

with

$$R := [\Lambda_0; \Lambda_1; \Lambda_2; \ldots][X_0, X_1, X_2, \ldots] =: (R_{ij} : i, j = 0, 1, 2, \ldots)$$

a block upper triangular matrix since

$$R_{ij} = \Lambda_i X_j$$

is trivial for $i > j$.

Consider now, in particular, the diagonal blocks,

$$R_{kk} = \Lambda_k X_k.$$

If λ is any linear functional in ran Λ' which vanishes on $\Pi_{<k}$, then there is exactly one coefficient sequence c so that $\lambda = \Lambda'_k c$ on Π_k. In other words, the rows of R_{kk} provide a basis for the linear subspace $\{(\lambda()^\alpha : |\alpha| = k) : \lambda \in \text{ran } \Lambda', \lambda X_{<k} = 0\}$. In particular, R'_{kk} is 1-1, and its range is independent of the particular choice of the map Λ'_k. This recovers (i) of Corollary 5.2 in this more general setting.

Acknowledgements. I acknowledge with pleasure and gratitude various extended discussions I had with Nira Dyn concerning the material of this paper, particularly during a visit to Israel, in February 1993, funded by the US-Israeli Binational Science Foundation.

The work was supported by the National Science Foundation under Grant No. DMS-9224748, by the United States Army under Contract DAAL03-90-G-0090, by the US-Israel Binational Science Foundation under Grant No. 90-00220, and by the Alexander von Humboldt Stiftung.

REFERENCES

[1] Birkhoff G. The algebra of multivariate interpolation. In *Constructive approaches to mathematical models*, pages 345–363. Academic Press, New York, 1979. C.V. Coffman and G.J. Fix, eds.

[2] de Boor C. Polynomial interpolation in several variables. In *Proceedings of the Conference honoring Samuel D. Conte*, pages xxx–xxx. Plenum Press, New York, 199x. R. DeMillo and J.R. Rice, eds.

[3] de Boor C. An alternative approach to (the teaching of) rank, basis and dimension. *Linear Algebra Appl.*, 146:221–229, 1991.

[4] de Boor C. On the error in multivariate polynomial interpolation. *Applied Numerical Mathematics*, 10:297–305, 1992.

[5] de Boor C., Ron A. On multivariate polynomial interpolation. *Constr. Approx.*, 6:287–302, 1990.

[6] de Boor C., Ron A. On polynomial ideals of finite codimension with applications to box spline theory. *J. Math. Anal. Appl.*, 158:168–193, 1991.

[7] de Boor C., Ron A. Computational aspects of polynomial interpolation in several variables. *Math. Comp.*, 58:705–727, 1992.

[8] de Boor C., Ron A. The least solution for the polynomial interpolation problem. *Math. Z.*, 210:347–378, 1992.

[9] Lorentz R.A. Multivariate Birkhoff Interpolation. *Lecture Notes in Mathematics*, v. 1516, Springer-Verlag, Heidelberg, 1992.

International Series of Numerical Mathematics, Vol. 119, ©1994 Birkhäuser

QUANTIZATION OF THE GAMMA FUNCTION[*]

Louis de Branges[†]

To Walter Gautschi on the occasion of his sixty-fifth birthday

Abstract. A class of weight functions for the upper half-plane is introduced which admit a generalization of Fourier analysis. A weighted Hardy space for the upper half-plane is considered, as well as related Hilbert spaces of entire functions. Both the Hardy space and the spaces of entire functions admit dissipative imaginary shifts. The dissipative property implies that the weight function has an analytic extension to a half-plane which contains the upper half-plane and that the ratio of the shifted function to the given function has nonnegative real part in the half-plane. These conditions are satisfied when the weight function is a constant, in which case the Hilbert spaces of entire functions are the Paley-Wiener spaces of Fourier analysis. The simplest nontrivial examples of such weight functions are constructed from the gamma function. The associated Hilbert spaces of entire functions appear in the theory of the Hankel transformation. The spaces have been called Sonine spaces by James and Virginia Rovnyak because they are related to properties of the Hankel transformation, which were discovered in 1880 by N. Sonine. His contributions are often taken as motivation for the Riemann hypothesis because entire functions appear which are related to zeta functions and for which an analogue of the Riemann hypothesis is true. This feature persists in all the Hilbert spaces of entire functions associated with the present generalization of the gamma function. The generalization is seen as a variation on the classical theme of quantization in special function theory.

1 QUANTIZATION

A weight function is a function $W(z)$ which is analytic and without zeros in the upper half-plane. The weighted Hardy space $\mathcal{F}(W)$ is the set of functions $F(z)$ such that the ratio $F(z)/W(z)$ is analytic and of bounded type in the upper half-plane, has square integrable boundary values on the real axis, and satisfies the inequality

$$\log|F(x+iy)/W(x+iy)| \leq \frac{y}{\pi} \int_{-\infty}^{+\infty} \frac{\log|F(t)/W(t)|dt}{(t-x)^2+y^2}$$

for $y > 0$. The space is a Hilbert space in the norm

$$\| F \|_{\mathcal{F}(W)}^2 = \int_{-\infty}^{+\infty} |F(t)/W(t)|^2 dt.$$

---

[*]Research supported by the National Science Foundation
[†]Department of Mathematics, Purdue University, West Lafayette, IN 47907-1395, USA

A continuous linear functional is defined in the space, taking $F(z)$ into $F(w)$, when w is in the upper half-plane. The reproducing kernel of the linear functional is the element

$$\frac{W(z)\bar{W}(w)}{2\pi i(\bar{w}-z)}$$

of the space.

Weight functions are now considered which are well related to imaginary shifts [5]. Let κ be a given positive number. Every element of the space $\mathcal{F}(W)$ is assumed to be of the form $F(z)+F(z+i\kappa)$ for an element $F(z)$ of the space such that $F(z+i\kappa)$ belongs to the space, and the real part of

$$\langle F(t+i\kappa),\ F(t)\rangle_{\mathcal{F}(W)}$$

is assumed to be nonnegative. An equivalent condition is that $W(z)$ has an analytic extension to the half-plane $i\bar{z}-iz>-\kappa$ and that the real part of

$$W(z)/W(z+i\kappa)$$

is nonnegative in the half-plane.

A stronger hypothesis on the weight function is made which results in a quantization of the theory of the gamma function. A quantum q is a given number, $0<q<1$. The given weight function is said to be a quantum gamma function with quantum q if the previous positivity condition is satisfied for every positive number κ such that

$$q\le\exp(-2\pi\kappa).$$

The theory of quantum gamma functions is an application of the theory [3] of Hilbert spaces whose elements are entire functions and which have these properties:

(H1) Whenever $F(z)$ belongs to the space and has a nonreal zero w, the function $F(z)(z-\bar{w})/(z-w)$ belongs to the space and has the same norm as $F(z)$;

(H2) For every nonreal number w the linear functional defined on the space, taking $F(z)$ into $F(w)$, is continuous;

(H3) The function $F^*(z)=\bar{F}(\bar{z})$ belongs to the space whenever $F(z)$ belongs to the space and it always has the same norm as $F(z)$.

The theory of these spaces is related to the theory of entire functions $E(z)$ which satisfy the inequality

$$|E(x-iy)|<|E(x+iy)|$$

when y is positive. If $E(z)$ is any such function, write

$$E(z)=A(z)-iB(z),$$

where $A(z)$ and $B(z)$ are entire functions which are real for real z, and

$$K(w, z) = [B(z)\bar{A}(w) - A(z)\bar{B}(w)]/[\pi(z - \bar{w})].$$

Let $\mathcal{H}(E)$ be the set of entire functions $F(z)$ such that the integral

$$\| F \|^2 = \int_{-\infty}^{+\infty} |F(t)/E(t)|^2 dt$$

is finite and such that the inequality

$$|F(w)|^2 \leq \| F \|^2 K(w, w)$$

holds for all complex numbers w. Then $\mathcal{H}(E)$ is a Hilbert space whose elements are entire functions and which satisfies the axioms (H1), (H2), and (H3). For every complex number w the kernel $K(w, z)$ belongs to the space as a function of z, and the identity

$$F(w) = \langle F(t), K(w, t)\rangle$$

holds for every element $F(z)$ of the space. A Hilbert space whose elements are entire functions, which satisfies the axioms (H1), (H2), and (H3), and which contains a nonzero element is isometrically equal to a space $\mathcal{H}(E)$.

An entire function $E(z)$ is said to be of Pólya class if it has no zeros in the upper half-plane, if it satisfies the inequality

$$|E(x - iy)| \leq |E(x + iy)|$$

when y is positive, and if the modulus of $E(x + iy)$ is a nondecreasing function of positive y for every real number x. A space $\mathcal{H}(E)$ then exists if $E(z)$ and $E^*(z)$ are linearly independent. The Pólya class is stable under bounded-type perturbations: An entire function $E(z)$ is of Pólya class if it has no zeros in the upper half-plane, if it satisfies the above inequality when y is positive, and if an entire function $F(z)$ of Pólya class exists such that $F(z)/E(z)$ is of bounded type in the upper half-plane.

Spaces $\mathcal{H}(E)$ are now considered which are well related to imaginary shifts. Let κ be a given positive number. Every element of the space $\mathcal{H}(E)$ is assumed to be of the form $F(z) + F(z + i\kappa)$ for an element $F(z)$ of the space such that $F(z + i\kappa)$ belongs to the space, and the real part of

$$\langle F(t + i\kappa), F(t)\rangle_{\mathcal{H}(E)}$$

is assumed to be nonnegative. Let q be a given quantum. A space $\mathcal{H}(E)$ is said to be a quantum Sonine space of quantum q if the positivity condition is satisfied for every positive number κ such that

$$q \leq \exp(-2\pi\kappa).$$

25

If $W(z)$ is a quantum gamma function of quantum q, then related quantum Sonine spaces of quantum q exist. Each such space $\mathcal{H}(E)$ is contained contractively in the space $\mathcal{F}(W_\tau)$,

$$W_\tau(z) = W(z) \exp(-i\tau z)$$

for a real number τ, and the inclusion is isometric on the domain of multiplication by z in $\mathcal{H}(E)$. The function $E(z)$ is then of Pólya class. A partial ordering of quantum Sonine spaces associated with $W(z)$ is defined: The space $\mathcal{H}(E(a))$ is less than or equal to the space $\mathcal{H}(E(b))$ if it is contained contractively in $\mathcal{H}(E(b))$ and if the inclusion is isometric on the domain of multiplication by z in $\mathcal{H}(E(a))$. The quantum Sonine spaces which are associated with $W(z)$ are totally ordered. Every quantum Sonine space which is associated with $W(z)$ is the least upper bound of smaller quantum Sonine spaces which are associated with $W(z)$. Every quantum Sonine space which is associated with $W(z)$ is the greatest lower bound of bigger quantum Sonine spaces which are associated with $W(z)$.

The union of the quantum Sonine spaces which are associated with $W(z)$ is dense in the Hilbert space of equivalence classes of measurable functions $F(x)$ of real x such that the integral

$$\| F \|^2 = \int_{-\infty}^{+\infty} |F(t)/W(t)|^2 dt$$

is finite. The intersection of the quantum Sonine spaces which are associated with $W(z)$ contains no nonconstant element.

Assume that $W(z)$ is a quantum gamma function and that $E(z)$ is the defining function of a quantum Sonine space which is associated with $W(z)$. Then the ratio

$$E(z)/W(z)$$

is analytic and of bounded type in the upper half-plane. The mean type of the ratio is called the mean type of the quantum Sonine space associated with $W(z)$. Every real number which is greater than the mean type of a quantum Sonine space associated with $W(z)$ is the mean type of a quantum Sonine space associated with $W(z)$. A fundamental problem is to determine the greatest lower bound of mean types of quantum Sonine spaces which are associated with $W(z)$.

The problem is easily solved when the constant function one almost belongs to some, and hence every, quantum Sonine space which is associated with $W(z)$. This means that

$$[F(z) - F(w)]/(z - w)$$

belongs to the space whenever $F(z)$ belongs to the space for some, and hence every, complex number w. Then $W(z)$ is a function which is analytic and of bounded type in the upper half-plane. If $-\tau$ is the mean type of $W(z)$ in the upper half-plane, then τ is the greatest lower bound of mean types of quantum Sonine spaces which are associated with $W(z)$.

If $W(z)$ is a given quantum gamma function, then a quantum gamma function $W_\epsilon(z)$ of the same quantum exists for every positive number ϵ such that

$$W_\epsilon(z)W_\epsilon^*(z) = W(z)W^*(z) + \epsilon^2$$

for $-\kappa < i\bar{z} - iz < \kappa$, where κ is any positive number such that

$$q \leq \exp(-2\pi\kappa).$$

The choice of the function can be made so that

$$W(z) = \lim W_\epsilon(z)$$

uniformly on compact subsets of the half-plane $-\kappa < i\bar{z} - iz$ for any such positive number κ. The constant function one almost belongs to the quantum Sonine spaces associated with $W_\epsilon(z)$ for every positive number ϵ. For every positive number ϵ define $-\tau_\epsilon$ to be the mean type of $W_\epsilon(z)$ in the upper half-plane. Then

$$\tau = \liminf \tau_\epsilon$$

is greater than or equal to the greatest lower bound of mean types of quantum Sonine spaces which are associated with $W(z)$.

These results are related to those of Arne Beurling and Paul Malliavin [1].

REFERENCES

[1] Beurling A., Malliavin P. On Fourier transforms of measures with compact support. *Acta Math.*, **107**:291–309, 1962.

[2] de Branges L. Self-reciprocal functions. *J. Math. Anal. Appl.*, **9**:433–457, 1964.

[3] de Branges L. *Espaces Hilbertiens de Fonctions Entières.* Masson, Paris, 1972.

[4] de Branges L. The Riemann hypothesis for Hilbert spaces of entire functions. *Bull. Amer. Math. Soc.*, **15**:1–17, 1986.

[5] de Branges L. The convergence of Euler products. *J. Funct. Anal.*, **107**:122–210, 1992.

[6] de Branges L. A conjecture which implies the Riemann hypothesis. *J. Funct. Anal.*, **121**:117–184, 1994.

[7] Li X.J. *The Riemann hypothesis for polynomials orthogonal on the unit circle.* Ph.D. dissertation, Purdue University, 1993. *Math. Nachr.* To appear.

[8] Rovnyak J., Rovnyak V. Self-reciprocal functions for the Hankel transformation of integer order. *Duke Math. J.*, **34**:771–785, 1967.

[9] Rovnyak J., Rovnyak V. Sonine spaces of entire functions. *J. Math. Anal. Appl.*, **27**:68–100, 1969.

[10] Rovnyak V.G. *Self-reciprocal functions*. Ph.D. dissertation, Yale University, 1965.

[11] Rovnyak V.G. Self-reciprocal functions. *Duke Math. J.*, **33**:363–378, 1966.

International Series of Numerical Mathematics, Vol. 119, ©1994 Birkhäuser

Vector Orthogonal Polynomials of Dimension $-d$

C. Brezinski[*]   J. Van Iseghem[†]

Dedicated to Professor Walter Gautschi in honor of his 65th birthday

Abstract. Vector orthogonal polynomials of dimension $-d$ where d is a nonzero positive integer are defined. They are proved to satisfy a recurrence relation with $d + 2$ terms. A Shohat–Favard type theorem and a QD like algorithm are given.
Keywords: *Orthogonal polynomials, biorthogonality, recurrence relations.*

1   INTRODUCTION AND DEFINITIONS

Let $c_{i,j}$ be given complex numbers for $i, j = 0, 1, \ldots$. We define the linear functionals $L_0, L_1, \ldots$ on the space of complex polynomials by

$$L_i\left(x^j\right) = c_{i,j}.$$

The formal biorthogonal polynomials with respect to the family L_i [2] are the polynomials satisfying the biorthogonality conditions

$$L_p\left(x^j P_k^{(i,j)}(x)\right) = 0 \quad \text{for} \quad p = i, \ldots, i + k - 1.$$

They are given by

$$P_k^{(i,j)}(x) = \lambda \begin{vmatrix} c_{i,j} & \cdots & c_{i,j+k} \\ \vdots & & \vdots \\ c_{i+k-1,j} & \cdots & c_{i+k-1,j+k} \\ 1 & \cdots & x^k \end{vmatrix},$$

where λ is an arbitrary nonzero constant which depends on k, i, j. The polynomials have the exact degree k if and only if

$$D_k^{(i,j)} = \begin{vmatrix} c_{i,j} & \cdots & c_{i,j+k-1} \\ \vdots & & \vdots \\ c_{i+k-1,j} & \cdots & c_{i+k-1,j+k-1} \end{vmatrix} \neq 0.$$

AMS(MOS) Classification numbers: 42C05.

[*]Laboratoire d'Analyse Numérique et d'Optimisation, UFR IEEA-M3, Université des Sciences et Technologies de Lille, 59655–Villeneuve d'Ascq cedex – France; brezinsk@omega.univ-lille1.fr

[†]U.F.R. de Mathématiques Pures et Appliquées, Université des Sciences et Technologies de Lille, 59655–Villeneuve d'Ascq cedex – France

In the sequel, we shall assume that this condition holds for all k, i, j and we shall consider the case where the polynomials $P_k^{(i,j)}$ are monic, which corresponds to the choice $\lambda = 1/D_k^{(i,j)}$.

Vector orthogonal polynomials of dimension $d \in \mathbb{N}$ were introduced by Van Iseghem [8]. They correspond to the case where the linear functionals L_i are related by

$$L_i\left(x^{j+1}\right) = L_{i+d}\left(x^j\right),$$

that is

$$L_i\left(x^{j+n}\right) = L_{i+nd}\left(x^j\right). \tag{1.1}$$

When $d = 1$, the usual formal orthogonal polynomials are recovered [1].

Polynomials of dimension $d = -1$ were considered in [5]. They generalize the usual orthogonal polynomials on the unit circle and have applications in Laurent-Padé and two-point Padé approximation [6]. They are obtained by taking $d = -1$ in the relations (1.1) defining the vector orthogonal polynomials, that is by assuming that the linear functionals L_i are related by

$$L_i\left(x^j\right) = L_{i+1}\left(x^{j+1}\right)$$

or $c_{i,j} = c_{i+1,j+1}$. It was proved that, in the definite case (that is when $\forall k, D_k^{(0,0)} \neq 0$) these polynomials satisfy a three–term recurrence relationship of the form (setting for simplicity P_k for $P_k^{(0,0)}$)

$$P_{k+1}(x) = (x + B_{k+1})\, P_k(x) - C_{k+1}\, x\, P_{k-1}(x)$$

with $P_{-1}(x) = 0, P_0(x) = 1$ and

$$\begin{aligned} C_{k+1} &= L_0\left(xP_k\right)/L_0\left(xP_{k-1}\right) \\ B_{k+1} &= C_{k+1}L_{k-1}\left(P_{k-1}\right)/L_k\left(P_k\right). \end{aligned}$$

The non-definite case, which occurs when some of the determinants $D_k^{(0,0)}$ vanish, was treated by Draux [7] by an indirect approach involving the whole table of the usual formal orthogonal polynomials of dimension $d = 1$, and by a direct one in [3]. Formal orthogonal polynomials on a curve were considered in [4]. It was proved that the orthogonal polynomials of dimension $d \geq 1$ or $d \leq -1$ correspond to orthogonality on a curve consisting of discrete points located on the unit circle.

In this paper, we shall consider the case where d is negative or, in other words, the case $-d$ with $d \geq 1$, which corresponds to linear functionals related by

$$L_i(x^{j+1}) = L_{i-d}(x^j)$$

that is $\quad L_i(x^j) = L_{i+d}(x^{j+1}) = L_{i+nd}(x^{j+n}). \tag{1.2}$

In that case, we shall speak of vector orthogonal polynomials of dimension $-d$. We shall make use of new notations which are simplified with respect to the general case. Because of the pseudo–periodicity of the moments we have for $i = sd$, $c_{sd,j} = c_{0,j-s}$. So, the determinants in the expression defining $P_k^{(sd,j)}$ only depend on the difference $j - s$. Thus, after replacing $j - s$ by $-s$ with $s \in \mathbb{Z}$, we shall make use of the notation P_r^s instead of $P_r^{(sd,0)}$ and D_k^s instead of $D_k^{(sd,0)}$. Then we have

$$L_p(P_k^s) = 0, \quad p = sd, \ldots, sd + k - 1,$$

$$P_k^s(x) = P_k^{(sd,0)}(x) = \lambda \begin{vmatrix} c_{0,-s} & \cdots & c_{0,-s+k} \\ \vdots & & \vdots \\ c_{k-1,-s} & \cdots & c_{k-1,-s+k} \\ 1 & \cdots & x^k \end{vmatrix},$$

$$1/\lambda = D_k^s = \begin{vmatrix} c_{0,-s} & \cdots & c_{0,-s+k-1} \\ \vdots & & \vdots \\ c_{k-1,-s} & \cdots & c_{k-1,-s+k-1} \end{vmatrix}.$$

Denoting by $D_k^s(x)$ the numerator of P_k^s, we see that

$$P_k^s(x) = \frac{D_k^s(x)}{D_k^s}, \quad P_k^s(0) = \frac{D_k^{s-1}}{D_k^s}, \quad L_{k+sd}(P_k^s) = \frac{D_{k+1}^s}{D_k^s}.$$

The functionals L_p can be defined for negative values of the index p, since, by the periodicity for all p and k, $L_p(x^i) = c_{p,i} = c_{p+kd,i+k}$.

We shall assume that the polynomials are orthogonal of the exact dimension $-d$, that is that the relation (1.2) does not hold with $-d + 1$ or $-d - 1$.

These polynomials could possibly have applications in the simultaneous Padé approximation of several Laurent series and in simultaneous two-point Padé approximation. Such applications remain to be studied.

2   THE RECURRENCE RELATION

We shall now prove that, for s fixed, the orthogonal polynomials $(P_k^s)_{k \geq 0}$, of dimension $-d$ satisfy a recurrence relation with $d + 2$ terms as in the case of the dimension d [8]. For sake of simplicity, everything is written for $s = 0$, and P_k^s is denoted by P_k. We have:

Theorem 1   If $\forall k \geq 0$, $P_k(0) \neq 0$ and $L_k(P_k) \neq 0$, then the monic vector orthogonal polynomials P_{k+1} satisfy a relation of the form

$$P_{k+1}(x) = (x + b_k)P_k(x) - x \sum_{i=1}^{d} a_i P_{k-i}(x), \quad k = 0, 1, \ldots$$

with $P_0(x) = 1$, and $P_i(x) = 0$ for $i < 0$. The coefficients $a_0, \ldots, a_d$ and b_k (which depend on k) are solutions of the system

$$\sum_{j=1}^{d} a_j L_i(x P_{k-j}) = L_i(x P_k), \quad i = 0, \ldots, d - 1$$

31

$$b_k L_k(P_k) = a_d L_{k-d}(P_{k-d}).$$

If $k < d$, the same recurrence relationship holds. The coefficients $a_{k+1}, \ldots, a_d$ are zero. The coefficients $a_0, \cdots, a_k$ and b_k (which depend on k) are solutions of the system

$$\sum_{j=1}^{k} a_j L_i(x P_{k-j}) = L_i(x P_k), \quad i = 0, \ldots, k-1$$

$$b_k L_k(P_k) = \sum_{j=1}^{k} a_j L_k(x P_{k-j}) - L_k(x P_k).$$

Proof: Since $P_k(0)$ has been assumed to be different from zero, it is possible to consider the polynomial $\bar{P}_k$ normalized by the condition $\bar{P}_k(0) = 1$. So $\bar{P}_k = A_k P_k$, $\bar{P}_{k+1} - \bar{P}_k$ is zero for $x = 0$ and has the exact degree $k + 1$. Let us write it as

$$(\bar{P}_{k+1} - \bar{P}_k)(x) = x \sum_{i=0}^{k} \lambda_i \bar{P}_i(x)$$

$$\forall p, \quad L_p(\bar{P}_{k+1} - \bar{P}_k) = L_{p-d} \left(\sum_{i=0}^{k} \lambda_i \bar{P}_i \right).$$

We suppose first that $k \geq d$. Since $L_p(\bar{P}_{k+1} - \bar{P}_k) = 0$ for $p = d, \cdots, k-1$, we obtain a triangular system with respect to $\lambda_0, \cdots, \lambda_{k-d-1}$, whose diagonal terms are $L_{p-d}(\bar{P}_{p-d}), p = d, \ldots, k-1$. Because these quantities are different from zero by assumption, the existence of the following relation is proved:

$$\bar{P}_{k+1}(x) = (\lambda_k x + 1)\bar{P}_k(x) + x \sum_{i=k-d}^{k-1} \lambda_i \bar{P}_i(x).$$

Similarly, for $p = k$, it follows that

$$-L_k(\bar{P}_k) = \lambda_{k-d} L_{k-d}(\bar{P}_{k-d})$$

which proves that λ_{k-d} is nonzero, or equivalently that the relation cannot be shortened. Let us now come back to the monic polynomials P_k and show how to compute the coefficients of the relation

$$P_{k+1}(x) = (x + b_k)P_k(x) - x \sum_{i=1}^{d} a_i P_{k-i}(x).$$

Now $a_0, \ldots, a_d$ and b_k are known to exist and to be unique. They are given by imposing that $L_i(P_{k+1}) = 0$ for $i = 0, \cdots, d-1$ and for $i = k$, that is

$$\sum_{j=1}^{d} a_j L_i(x P_{k-j}) = L_i(x P_k) \quad i = 0, \ldots, d-1$$

$$b_k L_k(P_k) = a_d L_{k-d}(P_{k-d}).$$

When $k < d$, the proof is obtained by expanding $P_{k+1} - xP_k$ on the basis $\{P_n\}$. Similar relations hold for the coefficients. ∎

Remark 1 It must be noticed that, if the dimension is not $-d$ but $-d-1$, then the orthogonality relations used in the proof of the theorem are no longer true from $i = d$ but only from $i = d + 1$. On the other hand, if the polynomials are of dimension $-d+1$ instead of d, then in the last equation ($p = d - 1$) of the system defining the a_i's, all the terms $L_{d-1}(xP_i) = L_0(xP_i)$ are zero, and then $a_d = 0$, which implies that $L_k(P_k) = 0$.

Because $P_{k+1}(0) = b_k$, the equation defining b_k shows the link between the non–degenerate character of the family and the nonzero value of the polynomials at zero.

3   A SHOHAT–FAVARD THEOREM

We shall now prove the reciprocal of the recurrence relation, namely that if a family of polynomials satisfies a recurrence relation of the form given in the preceding theorem, then it is a family of orthogonal polynomials of dimension $-d$ whose moments $L_i(x^j)$, such that $L_i(x^j) = L_{i+d}(x^{j+1})$, can be calculated.

Let us assume that we are given a family $\{P_k\}$ of monic polynomials such that the recurrence relation of the Theorem 1 holds. First of all, let us remark that the family of linear functionals $\{L_i\}$ is defined apart from a multiplying factor. Indeed, if $L_i(P_k) = 0$ for $i = 0, \ldots, k - 1$, then it also holds if each functional L_i is replaced by $L'_i = \mu_i L_i$ where μ_i is an arbitrary nonzero constant.

For $k = 0$, we have

$$P_1(x) = x + b_0 \text{ and } L_0(P_1) = L_0(x) + b_0 L_0(1) = 0.$$

By the remark above, $L_0(1)$ or $L_0(x)$ can be arbitrarily chosen (to be a nonzero value since, otherwise, we shall have $\forall j$, $L_0(x^j) = 0$) and the preceding relation gives the other quantity. Thus $L_0(1)$ and $L_0(x)$ are known.

For $k = 1$, we have $P_2(x) = (x+b_1)P_1(x) - a_1 x P_0(x)$. The condition $L_0(P_2) = 0$ allows us to obtain the value of $L_0(x^2)$ since $L_0(1)$ and $L_0(x)$ are already known.

We have

$$L_1(P_2) = L_1(xP_1) + b_1 L_1(P_1) - a_1 L_1(xP_0) = 0.$$

By the remark above, among the quantities $L_1(1), L_1(x)$ and $L_1(x^2)$, two of them can be arbitrarily chosen (to be a nonzero value) and the preceding relation determines the third one. Thus, these three quantities are known. And so on.

For $k = d - 1$, we have $P_d(x) = (x + b_{d-1})P_{d-1}(x) - a_1 x P_{d-2}(x) - \cdots - a_{d-1} x P_0(x)$. The condition $L_0(P_d) = 0$ allows us to obtain the value of $L_0(x^d)$ since $L_0(1), \ldots, L_0(x^{d-1})$ are already known. And so on until the condition $L_{d-2}(P_d) = 0$ which gives the value of $L_{d-2}(x^d)$ since $L_{d-2}(1), \ldots, L_{d-2}(x^{d-1})$ are already known.

We have

$$L_{d-1}(P_d) = L_{d-1}(xP_{d-1}) + b_{d-1}L_{d-1}(P_{d-1}) - \sum_{i=1}^{d-1} a_i L_{d-1}(xP_{d-i-1}) = 0.$$

By the remark above, among the quantities $L_{d-1}(1), \ldots, L_{d-1}(x^d)$, d of them can be arbitrarily chosen (to be nonzero values) and the preceding relation determines the $(d+1)$th one. Thus, these $d+1$ quantities are known.

For $k = d$, we have $P_{d+1}(x) = (x + b_d)P_d(x) - a_1 xP_{d-1}(x) - \cdots - a_d xP_0(x)$. The condition $L_0(P_{d+1})$ allows us to obtain the value of $L_0(x^{d+1})$ since $L_0(1), \ldots,$ $L_0(x^d)$ are known. And so on until the condition $L_{d-1}(P_{d+1}) = 0$ which gives the value of $L_{d-1}(x^{d+1})$ since $L_{d-1}(1), \ldots, L_{d-1}(x^d)$ are already known.

We have

$$L_d(P_{d+1}) = L_d(xP_d) + b_d L_d(P_d) - \sum_{i=1}^{d} a_i L_d(xP_{d-i}) = 0.$$

But $L_d(xP_d) = L_0(P_d) = 0$ by the conditions on the functionals.

Similarly $L_d(xP_{d-1}) = L_0(P_{d-1}) = 0, \ldots, L_d(xP_1) = L_0(P_1) = 0$ and $L_d(xP_0) = L_0(P_0) \neq 0$. Thus

$$L_d(P_{d+1}) = b_d L_d(P_d) - a_d L_0(P_0).$$

Setting $P_d(x) = \beta_0 + \cdots + \beta_{d-1}x^{d-1} + x^d$, we have

$$
\begin{aligned}
L_d(P_d) &= \beta_0 L_d(1) + \beta_1 L_d(x) + \cdots + \beta_{d-1}L_d(x^{d-1}) + L_d(x^d) \\
&= \beta_0 L_d(1) + \beta_1 L_0(1) + \cdots + \beta_{d-1}L_0(x^{d-2}) + L_0(x^{d-1}).
\end{aligned}
$$

Thus, in the relation $L_d(P_{d+1}) = 0$, all the quantities are known except $L_d(1)$ which can now be computed if b_d and β_0 are both different from zero. Let us remark that the condition $b_d = 0$ is equivalent to $P_{d+1}(0) = 0$ and that $\beta_0 = 0$ is equivalent to $P_d(0) = 0$. Moreover $L_d(x) = L_0(1), \ldots, L_d(x^{d+2}) = L_0(x^{d+1})$ are also known.

And so on for any greater value of k. Thus, we have proved:

Theorem 2 *Let $\{P_k\}$ be a family of monic polynomials such that $\forall k$, $P_k(0) \neq 0$, and which satisfy the recurrence relation*

$$P_{k+1}(x) = (x + b_k) - x \sum_{i=1}^{d} a_i P_{k-i}(x) \text{ for } k = 0, 1, \ldots$$

with $P_0(x) = 1$ and P_i identically zero for $i < 0$.

These polynomials are vector orthogonal polynomials of dimension $-d$ with recpect to linear functionals L_i whose moments $L_i(x^j)$ can be computed.

Remark 2 As we saw before, when $k = 0$, there is one degree of freedom in the values of the moments since $L_0(1)$ or $L_0(x)$ can be arbitrarily chosen. When $k = 1$, there are 2 degrees of freedom, and so on. When $k = d - 1$, there are d degrees of freedom. So, the vector space of the solutions has dimension $(d!)$, a result similar to the result which holds for the vector orthogonal polynomials of dimension d.

If we define the polynomials $\widetilde{P}_k$ by $\widetilde{P}_k(x) = x^k P_k(x^{-1})$, then these polynomials satisfy

$$\widetilde{P}_{k+1}(x) = (b_k x + 1) - \sum_{j=1}^{d} a_j x^j \widetilde{P}_{k-j}(x)$$

which shows, by Theorem 2, that the polynomials $\widetilde{P}_k$ are vector orthogonal polynomials of dimension $-d$ only in the case $d = 1$ [3].

4   ADJACENT FAMILIES AND A QD LIKE ALGORITHM

Let us now come back to the definition of the biorthogonal polynomials $P_k^{(sd,0)} = P_k^s$, and study the link between two consecutive diagonals. A QD like algorithm, as in the case of vector orthogonality of dimension d [8], will be obtained.

In the sequel, all the determinants D_k^s will be assumed to be nonzero, so the biorthogonal polynomials P_k^s exist, are of the exact degree k, and the monic ones are unique for any k and s. Moreover $P_k^s(0) \neq 0$. The polynomials $\bar{P}_k^s$, normalized to be 1 at zero will also be considered and we shall set $\bar{P}_k^s = A_k^s P_k^s$.

The polynomials are displayed in a two-dimensional array where k is the index of a column (increasing from the left to the right) and s is the index of a diagonal (increasing from the bottom to the top). That is, we have

$$\begin{matrix} \ddots & \vdots & \vdots & \vdots \\ \cdots & P_k^s & P_{k+1}^{s+1} & \cdots \\ \cdots & P_k^{s-1} & P_{k+1}^s & \cdots \\ \vdots & \vdots & \vdots & \ddots \end{matrix}$$

By analogy with the classical QD algorithm, the relations of Theorem 3 will be called the relations (q). We shall indicate the position, in the preceding table, of the polynomials which are known by a $\bullet$, and by a $*$ the position of the polynomial which is computed by the relation under consideration.

Theorem 3 *We have*

$$(q) \begin{cases} P_{k+1}^s(x) = x P_k^{s-1}(x) - q_{k+1}^s P_k^s(x) \quad k \geq 0 \\[2mm] q_{k+1}^s = \dfrac{D_{k+1}^{s-1} D_k^s}{D_k^{s-1} D_{k+1}^s} = \dfrac{L_{(s-1)d+k}(P_k^{s-1})}{L_{sd+k}(P_k^s)}. \end{cases}$$

$$\begin{matrix} & & \bullet \\ & \bullet & * \end{matrix}$$

35

Proof: We first apply the Sylvester identity to the numerator $D_{k+1}^s(x)$ of P_{k+1}^s, where the row k+1 has been taken as the first row.

Renumbering the rows in the classical order, the previous equality is

$$D_{k+1}^s(x)D_k^{s-1} = xD_k^{s-1}(x)D_{k+1}^s - D_k^s(x)D_{k+1}^{s-1}.$$

Dividing by $D_k^{s-1}D_{k+1}^s$ and taking in account that $P_k^s(x) = D_k^s(x)/D_k^s$, it follows that

$$P_{k+1}^s(x) = xP_k^{s-1}(x) - q_{k+1}^s P_k^s(x)$$

with the given expression for q_{k+1}^s. ∎

By analogy with the classical QD algorithm, we shall now find the relation (e), but, as we shall see, it can be given in two different forms $(e1)$ and (e).

Theorem 4 *The relations $(e1)$ and (e), in the table of biorthogonal polynomials of dimension $(-d)$, are the following:*

$$(e1) \begin{cases} P_k^{s+1}(x) = P_k^s(x) + \displaystyle\sum_{i=k-d}^{k-1} \varepsilon_{k,i}^s P_i^{s+1}(x) \quad k \geq d \\[4mm] \varepsilon_{k,k-d}^s = -\dfrac{L_{sd+k}(P_k^s)}{L_{sd+k}(P_{k-d}^{s+1})} \end{cases}$$

$$(e) \begin{cases} \dfrac{A_k^{s+1}}{A_k^s} P_k^{s+1}(x) = P_k^s(x) + x\displaystyle\sum_{i=k-d}^{k-1} e_{k,i}^s P_i^s(x) \quad k \geq d \\[4mm] e_{k,k-d}^s = \dfrac{L_{sd+k}(P_k^s)}{L_{sd+k-d}(P_{k-d}^s)}. \end{cases}$$

Proof: The proof is based on the biorthogonality property of the polynomials. P_k^{s+1} is monic and of degree k, and so it can be expanded on the basis $\{P_i^{s+1}\}_{i\geq 0}$

$$P_k^{s+1}(x) = P_k^s(x) + \sum_{i=0}^{k-1} \lambda_i P_i^{s+1}(x).$$

From the orthogonality with respect to the functional L_p, it follows that

$$\begin{aligned} L_p(P_k^s) &= 0, \qquad p = sd, \ldots, sd + k - 1 \\ L_p(P_k^{s+1}) &= 0, \qquad p = (s+1)d, \ldots, (s+1)d + k - 1. \end{aligned}$$

Writing these equations for $p = (s+1)d, \cdots, sd + k - 1$, $k > d$, we obtain

$$\begin{cases} p = (s+1)d & 0 = \lambda_0 L_{(s+1)d}(P_0^{s+1}) \\ \cdots\cdots\cdots\cdots & \cdots\cdots\cdots\cdots\cdots\cdots\cdots\cdots\cdots\cdots\cdots\cdots\cdots \\ p = sd + k - 1 & 0 = \lambda_0 L_{sd+k-1}(P_0^{s+1}) + \cdots + \lambda_{k-d-1} L_{sd+k-1}(P_{k-d-1}^{s+1}). \end{cases}$$

Since the system is triangular with a non-zero diagonal, $\lambda_0, \ldots, \lambda_{k-d-1}$ are zero and the result follows.

Writing down the equations for the following values of p, a triangular system is obtained with unknowns $\lambda_{k-d}, \ldots, \lambda_{k-1}$. The coefficient of smallest index is

$$\lambda_{k-d} L_{sd+k}(P_{k-d}^{s+1}) = -L_{sd+k}(P_k^s)$$

and so it is nonzero in the non–degenerate case, i.e. the relation cannot be shortened. Thus we obtain the first formula (e1)

$$P_k^{s+1}(x) - P_k^s(x) = \sum_{i=k-d}^{k-1} \varepsilon_{k,i}^s P_i^{s+1}(x).$$

This relation can also be written for the diagonal s. For convenience, let us now use the polynomials $\bar{P}_k^s$ normalized by $\bar{P}_k^s(0) = 1$. We have, by expansion on the basis $\{\bar{P}_i^s\}_i$

$$\bar{P}_k^{s+1}(x) - \bar{P}_k^s(x) = x \sum_{i=0}^{k-1} \lambda_i' \bar{P}_i^s(x).$$

The same proof leads to the following expressions

$$\begin{cases} \bar{P}_k^{(s+1)}(x) - \bar{P}_k^s(x) = x \sum_{i=k-d}^{k-1} \bar{a}_{k,i}^s \bar{P}_i^s(x) \quad k \geq d \\ \\ \bar{a}_{k,k-d}^s = -\dfrac{L_{sd+k}(\bar{P}_k^s)}{L_{sd+k-d}(\bar{P}_{k-d}^s)}. \end{cases}$$

Writing these expressions for the monic polynomials P_k^s, $\bar{P}_k^s = (D_k^s/D_k^{s-1})P_k^s = A_k^s P_k^s$, we obtain the second relation (e)

$$(e) \begin{cases} \dfrac{A_k^{s+1}}{A_k^s} P_k^{s+1}(x) = P_k^s(x) + x \sum_{i=k-d}^{k-1} e_{k,i}^s P_i^s(x) \quad k \geq d \\ \\ e_{k,k-d}^s = \dfrac{L_{sd+k}(P_k^s)}{L_{sd+k-d}(P_{k-d}^s)}. \end{cases}$$

In all cases, the relations cannot be shortened if all the determinants D_k^s are nonzero. ∎

Let us now show how the quantities $\varepsilon_{k,i}^s$ can be expressed in terms of the quantities $e_{k,i}^s$. P_k^s is expressed using (e1) in terms of the ε's. Then each term $x P_i^s$ is expressed in terms of the diagonal $s + 1$ by using the relation (q); it finally follows that

$$
\begin{cases}
\dfrac{A_k^{s+1}}{A_k^s} - e_{k,k-1}^s = 1 \\[2ex]
\varepsilon_{k,i}^s = e_{k,i-1}^s + q_{i+1}^{s+1} e_{k,i}^s & i = k-d+1, \ldots, k-1 \\[2ex]
\varepsilon_{k,k-d}^s = e_{k,k-d}^s q_{k-d+1}^{s+1}.
\end{cases}
$$

Now, as in the classical case, two different proofs of the diagonal relation will give the QD algorithm.

Let us use (q) and then $(e1)$

$$
\begin{aligned}
P_{k+1}^s(x) &= x P_k^{s-1}(x) - q_{k+1}^s P_k^s(x) \\
&= x\left(P_k^s(x) - \sum_{i=k-d}^{k-1} \varepsilon_{k,i}^{s-1} P_i^s(x) \right) - q_{k+1}^s P_k^s(x) \\
&= (x - q_{k+1}^s) P_k^s(x) - x \sum_{i=k-d}^{k-1} \varepsilon_{k,i}^{s-1} P_i^s(x) \\
&= (x - q_{k+1}^s) P_k^s(x) - x \sum_{i=k-d+1}^{k-1} (e_{k,i-1}^{s-1} + q_{i+1}^s e_{k,i}^{s-1}) P_i^s(x) \\
&\qquad\qquad - x e_{k,k-d}^{s-1} q_{k-d+1}^s P_{k-d}^s(x).
\end{aligned}
$$

Similarly, using (e) and then (q), we get

$$
\begin{aligned}
P_{k+1}^s(x) &= \frac{A_{k+1}^{s+1}}{A_{k+1}^s} P_{k+1}^{s+1}(x) - x \sum_{i=k-d+1}^{k} e_{k+1,i}^s P_i^s(x) \\
&= \frac{A_{k+1}^{s+1}}{A_{k+1}^s} \left(x P_k^s(x) - q_{k+1}^{s+1} P_k^{s+1}(x) \right) - x \sum_{i=k-d+1}^{k} e_{k+1,i}^s P_i^s(x) \\
P_{k+1}^s(x) &= \left(\left(\frac{A_{k+1}^{s+1}}{A_{k+1}^s} - e_{k+1,k}^s \right) x - q_{k+1}^{s+1} \frac{A_{k+1}^{s+1}}{A_{k+1}^s} \frac{A_k^s}{A_k^{s+1}} \right) P_k^s(x) \\
&\quad - x q_{k+1}^{s+1} \frac{A_{k+1}^{s+1}}{A_{k+1}^s} \frac{A_k^s}{A_k^{s+1}} \sum_{i=k-d}^{k-1} e_{k,i}^s P_i^s(x) - x \sum_{i=k-d+1}^{k-1} e_{k+1,i}^s P_i^s(x).
\end{aligned}
$$

Finally, identifying the two expressions, we get first

$$
q_{k+1}^{s+1} \frac{A_{k+1}^{s+1}}{A_{k+1}^s} \frac{A_k^s}{A_k^{s+1}} = q_{k+1}^s, \qquad \frac{A_{k+1}^{s+1}}{A_{k+1}^s} - e_{k+1,k}^s = 1,
$$

from which it follows that

$$e^s_{k+1,k} + 1 = \prod_{i=1}^{k+1} \frac{q^s_i}{q^{s+1}_i}.$$

This last relation together with the remaining part of the identification, gives the QD algorithm

$$(QD) \begin{cases} e^s_{k+1,k} + 1 = \displaystyle\prod_{i=1}^{k+1} \frac{q^s_i}{q^{s+1}_i} \\[2mm] e^s_{k,i} q^s_{k+1} + e^s_{k+1,i} = e^{s-1}_{k,i-1} + q^s_{i+1} e^{s-1}_{k,i}, \quad i = k - d + 1, \ldots, k - 1 \\[2mm] e^s_{k,k-d} q^s_{k+1} = e^{s-1}_{k,k-d} q^s_{k-d+1}. \end{cases}$$

The first d columns $(q^s_i)_s$, $i = 1, \ldots, d$ are the initializations of the algorithm. They are given by the formula of Theorem 3 (namely $q^s_1 = c_{0,-s+1}/c_{0,-s}$), $e^s_{k,i} = 0$, $i < 0$. If the d columns $q_{n+1}, \ldots, q_{n+d}$, and also the column $(E^s_{n+d-1})_s$ where $E^s_k = (e^s_{k,i})_i$, $i = k - d, \ldots, k$, are known, then $e^s_{n+d,n+d-1}$ is computed by the first relation, $e^s_{n+d,i}$ for $i = n, \ldots, n + d - 1$ is computed by the second relation, and finally q^s_{n+d+1} is obtained from the third relation.

REFERENCES

[1] Brezinski C. *Padé-Type Approximation and General Orthogonal Polynomials.* ISNM 50, Birkhäuser, Basel, 1980.

[2] Brezinski C. *Biorthogonality and its Applications to Numerical Analysis.* Marcel Dekker, New York, 1991.

[3] Brezinski C. A unified approach to various orthogonalities. *Ann. Fac. Sci. Toulouse,* sér.3, vol.1, fasc.3, 277–292, 1992.

[4] Brezinski C. Formal orthogonality on an algebraic curve. *Annals of Numer. Math.* To appear.

[5] Brezinski C., Redivo Zaglia M. Orthogonal polynomials of dimension -1 in the non-definite case. *Rend. Mat. Roma,* ser. VII, **13**, 1993. To appear.

[6] Bultheel A. *Laurent Series and their Padé Approximations.* Birkhäuser, Basel, 1987.

[7] Draux A. *Polynômes Orthogonaux Formels–Applications.* LNM 974, Springer Verlag, Berlin, 1987.

[8] Van Iseghem J. Vector orthogonal relations. Vector qd–algorithm. *J. Comput. Appl. Math.,* **19**:141–150, 1987.

International Series of Numerical Mathematics, Vol. 119, ©1994 Birkhäuser

AN EXTENSION OF A RESULT OF RIVLIN ON WALSH EQUICONVERGENCE (FABER NODES)

R. Brück[*]   A. Sharma[†]   R.S. Varga[‡]

Dedicated to Walter Gautschi on the occasion of his 65th birthday

Abstract.   We continue our investigations of generalizations of Walsh's equiconvergence theorem. The setting is a compact set E of the complex plane, whose complement is simply connected in the extended complex plane, and the Faber polynomials associated with E. Here, we study equiconvergence phenomena for differences of interpolating polynomials, defined by Lagrange (and Hermite) interpolants in zeros of associated Faber polynomials.

1 INTRODUCTION

We begin with the well-known Walsh equiconvergence theorem [10, p. 153]. With $D_\tau := \{z \in \mathbb{C} : |z| < \tau\}$ and with any R satisfying $1 < R < \infty$, let A_R denote the set of all functions f which are analytic in D_R, but not in $\bar{D}_R$. Then, the Walsh equiconvergence theorem, simply stated, asserts that if $f(z) = \sum_{k=0}^{\infty} a_k z^k$ is in A_R, then

$$\lim_{n \to \infty} [L_n(z; f) - S_n(z; f)] = 0, \qquad z \in D_{R^2}, \qquad (1.1)$$

the convergence being uniform and geometric on every disk $\bar{D}_\mu$ with $\mu < R^2$, where $L_n(z; f)$ denotes the Lagrange interpolant to f in the $(n + 1)^{st}$ roots of unity and where $S_n(z; f) := \sum_{k=0}^{n} a_k z^k$. (Since the convergence to zero in (1.1) takes place in the domain D_{R^2} which is *larger* than the domain D_R of analyticity of f, the result of (1.1) is said to exhibit *overconvergence*.) Rivlin [9] extended (1.1) by replacing $L_n(z; f)$ by polynomials $P_{m,n}(z; f)$ which best approximate f, in the ℓ_2-sense over all polynomials of degree n, in the $(m + 1)^{st}$ roots of unity, where, for a fixed positive integer q (i.e., $q \in \mathbb{N}$), $m := q(n + 1) - 1$ for all $n \in \mathbb{N}$. He showed that

$$\lim_{n \to \infty} [P_{m,n}(z; f) - S_n(z; f)] = 0, \qquad z \in D_{R^{1+q}}, \qquad (1.2)$$

the convergence being uniform and geometric on every disk $\bar{D}_\mu$ with $\mu < R^{1+q}$.

[*]Mathematisches Institut, Justus-Liebig-Universität Giessen, Arndtstrasse 2, D-35392 Giessen, Germany

[†]Department of Mathematics, University of Alberta, Edmonton, Alberta, Canada T6G 2G1

[‡]Institute for Computational Mathematics, Kent State University, Kent, Ohio 44242-0001, USA

In a recent paper [2], we studied the problem of Walsh equiconvergence for domains more general than disks. More precisely, if $\mathbb{C}_\infty$ denotes the extended complex plane, let E be a compact subset (not a point) of the complex plane whose complement, $\mathbb{C}_\infty \backslash E$, is simply connected. By the Riemann mapping theorem, there exists a conformal map ψ of $\{w \in \mathbb{C} : |w| > 1\}$ onto $\mathbb{C}_\infty \backslash E$, where the mapping is normalized at infinity by $\psi(\infty) = \infty$ and $c := \psi'(\infty) > 0$. (The quantity c is called the *capacity* of E.) For $1 < R < \infty$, let $C_R := \{z = \psi(w) : |w| = R\}$ be an outer level curve of E, and let A_R now denote the class of functions which are analytic in $G_R := \mathrm{Int} C_R$, but not in $\bar{G}_R$. If F_k denotes the kth Faber polynomial associated with E and if

$$f(z) = \sum_{k=0}^{\infty} a_k F_k(z) \tag{1.3}$$

is the Faber expansion of f with respect to E, we set

$$S_n(z; f) := \sum_{k=0}^{n} a_k F_k(z). \tag{1.4}$$

As in [2], if we assume that the boundary ∂E of E is a Jordan curve, then the conformal map ψ can be extended to a homeomorphism of $\{w \in \mathbb{C} : |w| \geq 1\}$ onto $\mathbb{C} \backslash \mathrm{Int} E$, so that we may define, for each $m \in \mathbb{N}$, the $m+1$ points $\{z_{k,m}\}_{k=0}^{m}$, where

$$z_{k,m} := \psi(w_{k,m}), \qquad w_{k,m} := \exp\left(\frac{2\pi i k}{m+1}\right), \qquad k = 0, \ldots, m. \tag{1.5}$$

The points $z_{k,m}$ are called the $(m+1)^{st}$ *Fejér nodes* with respect to E. Following Pommerenke [8], we call ∂E an r_0-*analytic curve* $(0 \leq r_0 < 1)$ if the conformal map ψ admits a univalent continuation to $\{w \in \mathbb{C} : |w| > r_0\}$. For $f \in A_R$, let (1.3) be the Faber expansion of f with respect to E, and let $L_m(z; f)$ denote the Lagrange interpolant of f in the Fejér nodes (1.5). If $L_m(z; f) = \sum_{k=0}^{m} b_k F_k(z)$ is the Faber expansion of $L_m(z; f)$, set

$$S_n\left(z; L_m(\cdot; f)\right) := \sum_{k=0}^{n} b_k F_k(z). \tag{1.6}$$

Our extension in [2] of Rivlin's theorem [9] studied the region of equiconvergence of the difference $S_n\left(z; L_m(\cdot; f)\right) - \sum_{k=0}^{n} a_k F_k(z)$.

If we set

$$S_{m,n,j}(z; f) := \sum_{k=0}^{n} a_{k+j(m+1)} F_k(z), \qquad j \in \mathbb{N}_0, \tag{1.7}$$

then for any integer $\ell \in \mathbb{N}$, we considered the difference

$$\Delta_{m,n,\ell}(z; f) := S_n\left(z; L_m(\cdot; f)\right) - \sum_{j=0}^{\ell-1} S_{m,n,j}(z; f). \tag{1.8}$$

The following theorem was proved in [2].

Theorem A [2] *Let ∂E be an r_0-analytic curve for some $r_0 \in [0,1)$, let $f \in A_R$, let $m = q(n+1) - 1$ for a fixed $q \in \mathbb{N}$, and let $\Delta_{m,n,\ell}(z; f)$ be given by (1.8). Then,*

$$\lim_{n \to \infty} \Delta_{m,n,\ell}(z; f) = 0, \qquad z \in G_\lambda, \tag{1.9}$$

the convergence being uniform and geometric on every subset $\bar{G}_\mu$ for $1 \leq \mu < \lambda$, where

$$\lambda := \min\{R^{1+\ell q}; R/r_0^q; R^q/r_0^{q-1}\}, \tag{1.10}$$

with $0^k := 0$ for any nonnegative integer k and $1/0 := \infty$.

Remark. From (1.10), we see that if $q = 1$ and if $0 < r_0 < 1$, then $\lambda = R$, and also that $\lambda \to R$ for arbitrary q as $r_0 \to 1$. Thus, Theorem A gives no overconvergence in these cases. Indeed, the first author has shown in [1] that there is no overconvergence in the case $q = \ell = 1$ and $r_0 > 0$.

In the special case of $E = D_1$ where $r_0 = 0$, (1.10) reduces to $\lambda = R^{1+\ell q}$. For $\ell = 1$, this gives the result of Rivlin [9, Theorem 1] and for $q = 1$, this gives a result of Cavaretta, Sharma and Varga [3, Theorem 1]. If $q \geq \ell + 1$ and $r_0 \leq \frac{1}{R^\ell}$, then (1.10) gives $\lambda = R^{1+\ell q}$, which means that we have the same λ as in the case when E is chosen to be the closed unit disk $\bar{D}_1$ (where $r_0 = 0$).

The object of this paper is to investigate the situation where the nodes of the Lagrange interpolant $L_m(z; f)$ are the zeros of the Faber polynomial $F_{m+1}(z)$, rather than the Fejér nodes of (1.5). In §2, we list some properties of Faber polynomials and state Theorem 1 and outline its proof. Section 3 deals with operators analogous to those of Theorem A, but based on Hermite interpolation using Fejér nodes and Faber nodes, respectively. It consists of a statement of Theorem B (which was given without proof in [2]) and the statement and proof of its analogue using Faber nodes. In §4, we give Theorem C (given without proof in [2]) and prove an analogous result (Theorem 3) using Faber nodes. Sections 5 and 6 are devoted to the proofs of Theorems B and C, respectively.

2 FABER NODES

In this section, we establish an analogue of Theorem A, where we replace the $m+1$ Fejér nodes of (1.5) by the $m + 1$ zeros of the $(m + 1)^{st}$ Faber polynomial with respect to E. We call these zeros *Faber nodes*. It is well known [5, p. 584], for an arbitrary compact set E (not a point) for which $\mathbb{C}_\infty \backslash E$ is simply connected, that the associated Faber polynomials $\{F_n(z)\}_{n \geq 0}$ satisfy

$$\lim_{n \to \infty} |F_n(\psi(w))|^{1/n} = |w|, \qquad |w| > 1, \tag{2.1}$$

uniformly on every closed subset of $\{w \in \mathbb{C} : |w| > 1\}$. For any fixed $R > 1$, (2.1) implies that the zeros of $F_n(z)$ all lie in G_R for any n sufficiently large, which further implies that all accumulation points of the zeros of $\{F_n(z)\}_{n \geq 1}$ must lie in E. Thus, $L_n^*(z; f)$, defined to be the Lagrange interpolant of f in the zeros of the Faber polynomial $F_{n+1}(z)$, is then well defined for all large n. If ∂E is r_0-analytic, then (2.1) holds for $|w| > r_0$ and thus, the Faber nodes all lie in the interior of E for every n sufficiently large. It is known [7] that if E is convex, but not a line segment, then all Faber nodes lie in the interior of E. In the case when E is the line segment $[-1, 1]$, it is well known that the Faber polynomials for E coincide with the classical Chebyshev polynomials of the first kind.

Let $L_m^*(z; f)$ denote the Lagrange interpolant to $f \in A_R$ in the $m + 1$ Faber nodes, i.e., the zeros of the Faber polynomial $F_{m+1}(z)$ of degree $m + 1$. Set

$$\Delta_{m,n,\ell}^*(z; f) := S_n\left(z; L_m^*(\cdot; f)\right) - \sum_{j=0}^{\ell-1} S_{m,n,j}(z; f), \qquad \ell \in \mathbb{N}, \qquad (2.2)$$

where $S_{m,n,j}(z; f)$ is given by (1.7) and where $S_n\left(z; L_m^*(\cdot; f)\right)$ is the Faber expansion of $L_m^*(z; f)$ up to degree n. We now establish:

Theorem 1 *Let $f \in A_R$, let $m = q(n + 1) - 1$ for a fixed $q \in \mathbb{N}$, and let $\Delta_{m,n,\ell}^*(z; f)$ be given by (2.2), where $L_m^*(z; f)$ is the Lagrange interpolant of f in the Faber nodes (i.e., the zeros of $F_{m+1}(z)$ with respect to a compact set E). Then,*

$$\lim_{n \to \infty} \Delta_{m,n,\ell}^*(z; f) = 0, \qquad z \in G_{R^q}, \qquad (2.3)$$

the convergence being uniform and geometric on every subset $\bar{G}_\mu$ for $1 \leq \mu < R^q$.

Proof: We proceed along the lines of the proof of Theorem A of [2]. From the well-known Hermite interpolation formula, we have, for any $r' \in (1, R)$ and any $z \in \mathbb{C}$, that

$$L_m^*(z; f) = \frac{1}{2\pi i} \int_{|\zeta|=r'} f\left(\psi(\zeta)\right) \frac{\psi'(\zeta)}{\psi(\zeta) - z} \cdot \frac{w_m\left(\psi(\zeta)\right) - w_m(z)}{w_m\left(\psi(\zeta)\right)} d\zeta, \qquad (2.4)$$

where $w_m(z) := c^{m+1} F_{m+1}(z)$. From [5, eq. (5.2)], the Faber coefficients a_k of f have the integral representation of

$$a_k = \frac{1}{2\pi i} \int_{|\zeta|=r'} \frac{f\left(\psi(\zeta)\right)}{\zeta^{k+1}} d\zeta, \qquad k = 0, 1, 2, \ldots . \qquad (2.5)$$

Then, from (1.4) and (2.5), we have

$$S_n(z; f) = \frac{1}{2\pi i} \int_{|\zeta|=r'} f\left(\psi(\zeta)\right) \sum_{k=0}^{n} \frac{F_k(z)}{\zeta^{k+1}} d\zeta, \qquad z \in \mathbb{C}. \qquad (2.6)$$

44

From (2.4) and (2.6), we obtain

$$S_n\left(z; L_m^*(\cdot; f)\right)$$

$$= \frac{1}{2\pi i} \int\limits_{|\zeta|=r'} f\left(\psi(\zeta)\right) \left(\frac{1}{2\pi i} \int\limits_{|t|=r} \frac{\psi'(\zeta)}{\psi(\zeta) - \psi(t)} \cdot \frac{w_m(\psi(\zeta)) - w_m(\psi(t))}{w_m(\psi(\zeta))} \sum_{k=0}^{n} \frac{F_k(z)}{t^{k+1}}\, dt\right) d\zeta,$$

$$\tag{2.7}$$

where we choose r and r' such that $1 < r < r' < R$. Furthermore, from (1.7) and (2.5) we easily derive that

$$\sum_{j=0}^{\ell-1} S_{m,n,j}(z; f) = \frac{1}{2\pi i} \int\limits_{|\zeta|=r'} f\left(\psi(\zeta)\right) \frac{\zeta^{\ell(m+1)} - 1}{\zeta^{(\ell-1)(m+1)}(\zeta^{m+1} - 1)} \sum_{k=0}^{n} \frac{F_k(z)}{\zeta^{k+1}}\, d\zeta, \quad z \in \mathbb{C}.$$

$$\tag{2.8}$$

But, as a consequence of the residue theorem (applied in $|t| > r$), it follows that

$$\frac{1}{2\pi i} \int\limits_{|t|=r} \frac{\psi'(\zeta)}{\psi(\zeta) - \psi(t)} \frac{dt}{t} = \frac{1}{\zeta}, \qquad |\zeta| > r,$$

which allows us to express (2.8) as

$$\sum_{j=0}^{\ell-1} S_{m,n,j}(z; f) = \frac{1}{2\pi i} \int\limits_{|\zeta|=r'} f\left(\psi(\zeta)\right) \times$$

$$\tag{2.9}$$

$$\left(\frac{1}{2\pi i} \int\limits_{|t|=r} \frac{\psi'(\zeta)}{\psi(\zeta) - \psi(t)} \cdot \frac{\zeta}{t} \cdot \frac{\zeta^{\ell(m+1)} - 1}{\zeta^{(\ell-1)(m+1)}(\zeta^{m+1} - 1)} \sum_{k=0}^{n} \frac{F_k(z)}{\zeta^{k+1}}\, dt\right) d\zeta.$$

From (2.7) and (2.9), we thus have an integral representation for $\Delta_{m,n,\ell}^*(z; f)$ of (2.2) as the difference of two double integrals.

For the Faber polynomials F_{n+1}, it is known from [5, eq. (2.7)] that $F_0(z) = 1$ and

$$F_n\left(\psi(w)\right) = w^n + \sum_{\nu=1}^{\infty} \alpha_{n,\nu} w^{-\nu}, \qquad |w| > 1,\ n \in \mathbb{N}, \tag{2.10}$$

uniformly on closed subsets of $\{w \in \mathbb{C} : |w| > 1\}$, where, from [5, eq. (4.9)],

$$|\alpha_{n,k}| \leq \sqrt{\frac{n}{k}}, \qquad n, k \in \mathbb{N}. \tag{2.11}$$

From (2.11), it readily follows that

$$\left|\sum_{\nu=1}^{\infty} \alpha_{n,\nu} w^{-\nu}\right| \leq \frac{\sqrt{n}}{|w| - 1}, \qquad |w| > 1,\ n \in \mathbb{N}. \tag{2.12}$$

Next, as previously, we set

$$w_n(z) := c^{n+1} F_{n+1}(z).$$

From (2.10) and (2.12), it is easy to deduce that, for any t and ζ with $1 < |t| < |\zeta|$,

$$\frac{w_m\left(\psi(\zeta)\right) - w_m\left(\psi(t)\right)}{w_m\left(\psi(\zeta)\right)} = \frac{\zeta^{m+1} - t^{m+1}}{\zeta^{m+1}} \left(1 + O(1)\frac{\sqrt{m}}{|\zeta|^m}\right), \qquad m \to \infty, \quad (2.13)$$

uniformly on closed subsets of $\{\zeta \in \mathbb{C} : |\zeta| > 1\} \times \{t \in \mathbb{C} : |t| > 1\}$. Choosing ρ such that $1 < \rho < r < r' < R$ and using (2.7), (2.9), (2.10) and (2.13), we obtain (with $z = \psi(w)$)

$$\Delta^*_{m,n,\ell}(z;f) =: \sum_{j=1}^{4} \Delta^{*(j)}_{m,n,\ell}(z;f)$$

$$= \frac{1}{2\pi i} \int_{|\zeta|=r'} f\left(\psi(\zeta)\right) \left(\frac{1}{2\pi i} \int_{|t|=r} \frac{\psi'(\zeta)}{\psi(\zeta) - \psi(t)} \sum_{j=1}^{4} K^{(j)}_{m,n,\ell}(w,\zeta,t)\, dt\right) d\zeta,$$

where

$$K^{(1)}_{m,n,\ell}(w,\zeta,t) := \frac{\zeta^{m+1} - t^{m+1}}{\zeta^{m+1}} \left(1 + O\left(\frac{1}{\rho^m}\right)\right) \sum_{k=1}^{n} \sum_{\nu=1}^{\infty} \alpha_{k,\nu} w^{-\nu} t^{-k-1}$$

$$- \frac{\zeta}{t} \cdot \frac{\zeta^{\ell(m+1)} - 1}{\zeta^{(\ell-1)(m+1)}(\zeta^{m+1} - 1)} \sum_{k=1}^{n} \sum_{\nu=1}^{\infty} \alpha_{k,\nu} w^{-\nu} \zeta^{-k-1}, \tag{2.14}$$

$$K^{(2)}_{m,n,\ell}(w,\zeta,t) := \frac{t^{n+1} - w^{n+1}}{(t-w)t^{n+1}} - \frac{\zeta}{t} \cdot \frac{\zeta^{n+1} - w^{n+1}}{(\zeta-w)\zeta^{n+1}} \cdot \frac{\zeta^{\ell(m+1)} - 1}{\zeta^{(\ell-1)(m+1)}(\zeta^{m+1} - 1)}, \tag{2.15}$$

$$K^{(3)}_{m,n,\ell}(w,\zeta,t) := O\left(\frac{1}{\rho^m}\right) \frac{\zeta^{m+1} - t^{m+1}}{\zeta^{m+1}} \cdot \frac{t^{n+1} - w^{n+1}}{(t-w)t^{n+1}}, \tag{2.16}$$

and

$$K^{(4)}_{m,n,\ell}(w,\zeta,t) := -\frac{t^{m+1}}{\zeta^{m+1}} \cdot \frac{t^{n+1} - w^{n+1}}{(t-w)t^{n+1}}. \tag{2.17}$$

(i) Because $1 < |t| < |\zeta|$, then letting n tend to infinity gives

$$K_1(w,\zeta,t) := \lim_{n\to\infty} K^{(1)}_{m,n,\ell}(w,\zeta,t)$$

$$= \sum_{k=1}^{\infty} \sum_{\nu=1}^{\infty} \alpha_{k,\nu} w^{-\nu} t^{-k-1} - \frac{\zeta}{t} \sum_{k=1}^{\infty} \sum_{\nu=1}^{\infty} \alpha_{k,\nu} w^{-\nu} \zeta^{-k-1},$$

where the two double series on the right side are convergent, uniformly on closed subsets of $\{w \in \mathbb{C} : |w| > 1\} \times \{t \in \mathbb{C} : |t| > 1\}$. The residue theorem now implies that

$$\frac{1}{2\pi i} \int\limits_{|t|=r} \frac{\psi'(\zeta)}{\psi(\zeta) - \psi(t)} K_1(w, \zeta, t)\, dt = 0,$$

for all $\zeta \in \mathbb{C}$ with $|\zeta| > r > 1$. Thus, $\lim\limits_{n\to\infty} \Delta_{m,n,\ell}^{*(1)}(z; f) = 0$, locally uniformly on $\mathbb{C}$.

(ii) Again using the residue theorem, we see that for $|w| > r'$,

$$\frac{1}{2\pi i} \int\limits_{|t|=r} \frac{\psi'(\zeta)}{\psi(\zeta) - \psi(t)} K_{m,n,\ell}^{(2)}(w, \zeta, t)\, dt = K_{m,n,\ell}^{(2)}(w, \zeta, \zeta)$$

$$= \frac{\zeta^{n+1} - w^{n+1}}{(\zeta - w)\zeta^{n+1}} \cdot \frac{1 - \zeta^{(\ell-1)(m+1)}}{\zeta^{(\ell-1)(m+1)}(\zeta^{m+1} - 1)} = \begin{cases} 0, & \ell = 1, \\ O(1)\left(\dfrac{|w|}{(r')^{1+q}}\right)^n, & \ell \geq 2. \end{cases}$$

But, as r' is any number satisfying $1 < r' < R$, it follows that $\lim\limits_{n\to\infty} \Delta_{m,n,\ell}^{*(2)}(z; f) = 0$, uniformly on $\bar{G}_\mu$ for every $1 \leq \mu < R^{1+q}$ when $\ell \geq 2$.

(iii) Also, it is obvious from the expression for $K_{m,n,\ell}^{(3)}(w, \zeta, t)$ that for $|w| > r'$,

$$K_{m,n,\ell}^{(3)}(w, \zeta, t) = O(1)\left(\frac{|w|}{r\rho^q}\right)^n,$$

so that $\lim\limits_{n\to\infty} \Delta_{m,n,\ell}^{*(3)}(z; f) = 0$, uniformly on $\bar{G}_\mu$ for $1 \leq \mu < R^{1+q}$.

(iv) Finally, in order to estimate $\Delta_{m,n,\ell}^{*(4)}$, consider

$$F_{m,n}(w, \zeta) := \frac{1}{2\pi i} \int\limits_{|t|=r} \frac{\psi'(\zeta)}{\psi(\zeta) - \psi(t)} K_{m,n,\ell}^{(4)}(w, \zeta, t)\, dt.$$

Since, from [5, eq. (2.9)],

$$\frac{\psi'(\zeta)}{\psi(\zeta) - \psi(t)} = \frac{1}{\zeta - t} + \sum_{\nu=1}^{\infty} \sum_{k=1}^{\infty} \alpha_{\nu,k} t^{-k} \zeta^{-\nu-1} \tag{2.18}$$

holds uniformly on closed subsets of $\{t \in \mathbb{C} : |t| > 1\} \times \{\zeta \in \mathbb{C} : |\zeta| > 1\}$, where the coefficients $\alpha_{\nu,k}$ are defined by (2.10), it follows from (2.17) that

$$F_{m,n}(w, \zeta)$$
$$= -\frac{1}{2\pi i} \int\limits_{|t|=r} \frac{1}{\zeta^{m+2}} \left\{ \sum_{k=0}^{\infty} \sum_{j=0}^{n} \zeta^{-k} w^j t^{k+m-j} + \sum_{\nu=1}^{\infty} \sum_{k=1}^{\infty} \sum_{j=0}^{n} \alpha_{\nu,k} \zeta^{-\nu} w^j t^{m-k-j} \right\} dt.$$

Since $k + m - j = k + q(n + 1) - 1 - j \geq 0$, the integral of the double sum above vanishes and the triple sum gives a contribution to the integral only when $k = m + 1 - j$, so that

$$F_{m,n}(w, \zeta) = -\frac{1}{\zeta^{m+2}} \sum_{\nu=1}^{\infty} \sum_{j=0}^{n} \alpha_{\nu, m+1-j} w^j \zeta^{-\nu}. \tag{2.19}$$

Again using (2.11), we obtain

$$F_{m,n}(w, \zeta) = O(1) \frac{1}{(r')^m} \sum_{\nu=1}^{\infty} \sum_{j=0}^{n} \sqrt{\frac{\nu}{m+1-j}} \, |w|^j (r')^{-\nu}$$

$$= O(1) \frac{1}{(r')^m} \sum_{\nu=1}^{\infty} \frac{\sqrt{\nu}}{(r')^\nu} \sum_{j=0}^{n} |w|^j = O(1) \left(\frac{|w|}{(r')^q} \right)^n,$$

which implies that $\lim_{n \to \infty} \Delta_{m,n,\ell}^{*(4)}(z; f) = 0$, uniformly on $\bar{G}_\mu$ for every $1 \leq \mu < R^q$. On combining the above results of (i)-(iv), we have the desired result of (2.3) of Theorem 1. ∎

Remarks. (1) We do not know if $\lambda = R^q$ is best possible in (2.3). However, we can improve our result if ∂E is r_0-analytic. In this case, we have from [5, eq. (4.2)] that $\alpha_{n,k} = O(\beta^{n+k})$, for every β such that $r_0 < \beta < 1$. Then (2.12) can be improved by the bound

$$\left| \sum_{\nu=1}^{\infty} \alpha_{n,\nu} w^{-\nu} \right| = O(\beta^n), \qquad |w| > \beta, \; n \in \mathbb{N}, \tag{2.20}$$

which leads, for $\beta < |t| < |\zeta|$, to

$$\frac{w_m\left(\psi(\zeta)\right) - w_m\left(\psi(t)\right)}{w_m\left(\psi(\zeta)\right)} = \frac{\zeta^{m+1} - t^{m+1}}{\zeta^{m+1}} \left(1 + O\left(\left(\frac{\beta}{|\zeta|} \right)^m \right) \right), \qquad m \to \infty. \tag{2.21}$$

An examination of the proof of Theorem 1 then shows that the estimates of $\Delta_{m,n,\ell}^{*(1)}(z; f)$ and $\Delta_{m,n,\ell}^{*(2)}(z; f)$ remain unchanged. But from (2.16) we now have, because of (2.21), that

$$K_{m,n,\ell}^{(3)}(w, \zeta, t) = O(1) \left(\frac{|w| \beta^q}{r(r')^q} \right)^n,$$

so that $\lim_{n \to \infty} \Delta_{m,n,\ell}^{*(3)}(z; f) = 0$, uniformly on $\bar{G}_\mu$ for every $1 \leq \mu < R(R/r_0)^q$.

Finally, from (2.19) and $|\zeta| = r'$, we have

$$F_{m,n}(w,\zeta) = O(1)\frac{1}{(r')^m}\sum_{\nu=1}^{\infty}\sum_{j=0}^{n}\beta^{\nu+m-j}|w|^j(r')^{-\nu}$$

$$= O(1)\left(\frac{\beta}{r'}\right)^m\sum_{\nu=1}^{\infty}\left(\frac{\beta}{r'}\right)^\nu\sum_{j=0}^{n}\left(\frac{|w|}{\beta}\right)^j = O(1)\left(\frac{|w|\beta^{q-1}}{(r')^q}\right)^n.$$

Thus, $\lim_{n\to\infty}\Delta_{m,n,\ell}^{*(4)}(z;f) = 0$, uniformly on $\bar{G}_\mu$ for every $1 \le \mu < r_0(R/r_0)^q$. Combining all the above, we have

$$\lim_{n\to\infty}\Delta_{m,n,\ell}^*(z;f) = 0, \qquad z \in G_\lambda,$$

where

$$\lambda := \begin{cases} r_0(R/r_0)^q, & \ell = 1, \\ \min\left\{r_0(R/r_0)^q; R^{q+1}\right\}, & \ell \ge 2. \end{cases} \tag{2.22}$$

(2) A further improvement may be achieved in the case of an ellipse E_δ (where $\delta > 1$ and where ∂E_δ is the image of the circle $\{w \in \mathbb{C} : |w| = \delta\}$ under the map $w \to \frac{1}{2}(w + \frac{1}{w})$, and is therefore r_0-analytic with $r_0 = 1/\delta$). Then, we have $F_n(\psi(w)) = w^n + \frac{1}{\delta^{2n}w^n}$, and an examination of the proof of Theorem 1 shows that in this case

$$\lambda = \begin{cases} r_0(R/r_0)^{2q-1}, & \ell = 1, \\ \min\{R(R/r_0)^{2q-1}; R^{q+1}\}, & \ell \ge 2. \end{cases} \tag{2.23}$$

(3) The previous remark also applies to the case of the segment $E = [-1,1]$ (where $\delta = 1$) and gives

$$\lambda = \begin{cases} R^{2q-1}, & \ell = 1, \\ R, & q = 1, \\ R^{q+1}, & \ell \ge 2. \end{cases} \tag{2.24}$$

For $\ell = 1$, this is Theorem 2 of Rivlin [9].

(4) The main reason for the different results in Theorem A and Theorem 1 (in the case when ∂E is r_0-analytic) may be explained as follows. If E is the closed unit disk D_1, then the Fejér nodes coincide with the roots of unity while the Faber nodes are all zero, so that $L_n^*(z;f) \equiv S_n(z;f)$.

3   HERMITE INTERPOLATION

In [2], we stated without proof a result analogous to Theorem A, after replacing Lagrange interpolation with Hermite interpolation in Fejér nodes. For $s \in \mathbb{N}$ and $f \in A_R$, we denote by $H_{s(m+1)-1}(z;f)$ the Hermite interpolation polynomial to

$f, f', \ldots, f^{(s-1)}$ in the $(m+1)^{st}$ Fejér nodes on a compact set E. Then for $p, q \in \mathbb{N}$ with $sq \geq p$, and $m = q(n+1) - 1$, we considered, as in [2], the operator

$$\Delta_{m,n}^{p,s}(z; f) := S_{p(n+1)-1}\left(z; H_{s(m+1)-1}(\cdot, f)\right) - S_{p(n+1)-1}(z; f). \tag{3.1}$$

With the above notation and the notation that $\lfloor t \rfloor$ denotes the integral part of the real number t, we state the following corrected form of:

Theorem B [2] Let ∂E be an r_0-analytic curve for some $r_0 \in [0,1)$, let $f \in A_R$, let $m = q(n+1) - 1$ for a fixed $q \in \mathbb{N}$, let $s, p \in \mathbb{N}$ be such that $sq \geq p$, and let $\Delta_{m,n}^{p,s}(z; f)$ be given by (3.1). Then,

$$\lim_{n \to \infty} \Delta_{m,n}^{p,s}(z; f) = 0, \qquad z \in G_\lambda, \tag{3.2}$$

the convergence being uniform and geometric on every subset $\bar{G}_\mu$ for $1 \leq \mu < \lambda$, where

$$\lambda := \begin{cases} \min\left\{R/r_0^{q/p}; R^{sq/p}/r_0^{(q/p)-1}; R^\sigma\right\}, & \text{if } q \geq p, \\ \min\left\{R/r_0^{q/p}; R^{sq/p}; R^\sigma\right\}, & \text{if } q < p, \end{cases} \tag{3.3}$$

where, if $p = tq + \tau$ with $t \in \mathbb{N}_0$ and $0 \leq \tau < q$, then $\sigma := (qs + \tau)/p$ if $0 < \tau < p$ and $\sigma := q(s+1)/p$ if $\tau = 0$.

Here, we shall consider the Hermite interpolant to $f, f', \ldots, f^{(s-1)}$ in the $(m+1)^{st}$ Faber nodes. We denote it by $H_{s(m+1)-1}^*(z; f)$. We set

$$\Delta_{m,n}^{*p,s}(z; f) := S_{p(n+1)-1}\left(z; H_{s(m+1)-1}^*(\cdot; f)\right) - S_{p(n+1)-1}(z; f), \tag{3.4}$$

where $S_{p(n+1)-1}(z; f)$ denotes the $(p(n+1)-1)^{st}$ section of the Faber expansion of f.

We next establish:

Theorem 2 Let $f \in A_R$, let $m = q(n+1) - 1$ for a fixed $q \in \mathbb{N}$, let $s, p \in \mathbb{N}$ be such that $sq \geq p$, and let $\Delta_{m,n}^{*p,s}(z; f)$ be given by (3.4) for the Faber nodes with respect to a compact set E. Then,

$$\lim_{n \to \infty} \Delta_{m,n}^{*p,s}(z; f) = 0, \qquad z \in G_\lambda, \tag{3.5}$$

the convergence being uniform and geometric on every subset $\bar{G}_\mu$ for $1 \leq \mu < \lambda$, where

$$\lambda := \min\{R^{1+q/p}; R^{sq/p}\}. \tag{3.6}$$

Proof: In analogy to (2.7) and (2.6), it is easy to see that

$$S_{p(n+1)-1}(z; H_{s(m+1)-1}^*(\cdot, f)) = \frac{1}{2\pi i} \int\limits_{|\zeta|=r'} f(\psi(\zeta))$$

$$\times \left(\frac{1}{2\pi i} \int\limits_{|t|=r} \frac{\psi'(\zeta)}{\psi(\zeta) - \psi(t)} \cdot \frac{w_m^s(\psi(\zeta)) - w_m^s(\psi(t))}{w_m^s(\psi(\zeta))} \sum_{k=0}^{p(n+1)-1} \frac{F_k(z)}{t^{k+1}} \, dt \right) d\zeta, \tag{3.7}$$

50

and

$$S_{p(n+1)-1}(z; f) = \frac{1}{2\pi i} \int\limits_{|\zeta|=r'} f\left(\psi(\zeta)\right) \left(\frac{1}{2\pi i} \int\limits_{|t|=r} \frac{\psi'(\zeta)}{\psi(\zeta) - \psi(t)} \cdot \frac{\zeta}{t}^{p(n+1)-1} \sum_{k=0} \frac{F_k(z)}{\zeta^{k+1}} \, dt \right) d\zeta,$$

(3.8)

where $1 < r < r' < R$.

Writing $w_m(z) = c^{m+1} F_{m+1}(z)$, we use (2.10) and (2.12) to show that

$$\frac{w_m^s\left(\psi(\zeta)\right) - w_m^s\left(\psi(t)\right)}{w_m^s\left(\psi(\zeta)\right)} = \frac{\zeta^{s(m+1)} - t^{s(m+1)}}{\zeta^{s(m+1)}} \left(1 + O(1)\frac{\sqrt{m}}{|\zeta|^m} \right), \qquad (3.9)$$

uniformly on closed subsets of $\{\zeta \in \mathbb{C} : |\zeta| > 1\} \times \{t \in \mathbb{C} : |t| > 1\}$.

Choosing ρ such that $1 < \rho < r < r' < R$ and putting (3.7), (3.8), (3.9) and (2.10) together, we obtain (with $z = \psi(w)$)

$$\Delta_{m,n}^{*p,s}(z; f) = \sum_{j=1}^{4} \Delta_{m,n,j}^{*p,s}(z; f), \qquad (3.10)$$

where

$$\Delta_{m,n,j}^{*p,s}(z; f) := \frac{1}{2\pi i} \int\limits_{|\zeta|=r'} f\left(\psi(\zeta)\right) \left(\frac{1}{2\pi i} \int\limits_{|t|=r} \frac{\psi'(\zeta)}{\psi(\zeta) - \psi(t)} K_{m,n,j}(w, \zeta, t) \, dt \right) d\zeta,$$

$$j = 1, 2, 3, 4.$$

The kernels $K_{m,n,j}(w, \zeta, t)$ $(j = 1, 2, 3, 4)$ are given explicitly as follows:

$$K_{m,n,1}(w, \zeta, t) := \frac{\zeta^{s(m+1)} - t^{s(m+1)}}{\zeta^{s(m+1)}} \left(1 + O\left(\frac{1}{\rho^m}\right) \right) \sum_{k=1}^{p(n+1)-1} \sum_{\nu=1}^{\infty} \frac{\alpha_{k,\nu}}{w^\nu t^{k+1}}$$

$$-\frac{\zeta}{t} \sum_{k=1}^{p(n+1)-1} \sum_{\nu=1}^{\infty} \frac{\alpha_{k,\nu}}{w^\nu \zeta^{k+1}},$$

(3.11)

$$K_{m,n,2}(w, \zeta, t) := \frac{t^{p(n+1)} - w^{p(n+1)}}{(t-w)t^{p(n+1)}} - \frac{\zeta}{t} \cdot \frac{\zeta^{p(n+1)} - w^{p(n+1)}}{(\zeta - w)\zeta^{p(n+1)}}, \qquad (3.12)$$

$$K_{m,n,3}(w, \zeta, t) := O\left(\frac{1}{\rho^m}\right) \frac{\zeta^{s(m+1)} - t^{s(m+1)}}{\zeta^{s(m+1)}} \cdot \frac{t^{p(n+1)} - w^{p(n+1)}}{(t-w)t^{p(n+1)}}, \qquad (3.13)$$

51

and

$$K_{m,n,4}(w,\zeta,t) := -\frac{t^{s(m+1)}}{\zeta^{s(m+1)}} \cdot \frac{t^{p(n+1)} - w^{p(n+1)}}{(t-w)t^{p(n+1)}}. \tag{3.14}$$

As in the proof of Theorem 1, it can be shown that $\lim_{n\to\infty} \Delta^{*p,s}_{m,n,1}(z;f) = 0$, locally uniformly on $\mathbb{C}$. Using the residue theorem, we can see, exactly as in the proof of Theorem 1, that $\Delta^{*p,s}_{m,n,2}(z;f) \equiv 0$. Moreover it is obvious that for $|w| > r'$,

$$K_{m,n,3}(w,\zeta,t) = O(1)\left(\frac{|w|^p}{\rho^q r^p}\right)^n, \tag{3.15}$$

so that $\lim_{n\to\infty} \Delta^{*p,s}_{m,n,3}(z;f) = 0$, uniformly on $\bar{G}_\mu$ for $1 \le \mu < R^{1+(q/p)}$.

Finally, in order to estimate $\Delta^{*p,s}_{m,n,4}(z;f)$, we consider

$$F_{m,n}(w,\zeta) := \frac{1}{2\pi i} \int_{|t|=r} \frac{\psi'(\zeta)}{\psi(\zeta) - \psi(t)} K_{m,n,4}(w,\zeta,t)\, dt, \tag{3.16}$$

and recall from (2.18) that

$$\frac{\psi'(\zeta)}{\psi(\zeta) - \psi(t)} = \frac{1}{\zeta - t} + \sum_{\nu=1}^{\infty}\sum_{k=1}^{\infty} \alpha_{\nu,k} t^{-k} \zeta^{-\nu-1}. \tag{3.17}$$

The integrand in (3.16), when expanded in powers of t and on using (3.17), becomes

$$-\frac{1}{\zeta^{s(m+1)+1}}\left[\sum_{k=0}^{\infty}\sum_{j=0}^{p(n+1)-1} \zeta^{-k} w^j t^{k+s(m+1)-j-1} + \right.$$

$$\left. \sum_{\nu=1}^{\infty}\sum_{k=1}^{\infty}\sum_{j=0}^{p(n+1)-1} \alpha_{\nu,k}\zeta^{-\nu} w^j t^{-k+s(m+1)-j-1} \right].$$

Since $k + s(m+1) - j - 1 = k + sq(n+1) - j - 1 \ge 0$, the integral of the double sum vanishes and only the triple sum gives a contribution to the integral $F_{m,n}(w,\zeta)$ when $k = s(m+1) - j$. Thus, we have

$$F_{m,n}(w,\zeta) = -\frac{1}{\zeta^{s(m+1)+1}}\sum_{\nu=1}^{\infty}\sum_{j=0}^{p(n+1)-1} \alpha_{\nu,s(m+1)-j} w^j \zeta^{-\nu}. \tag{3.18}$$

Again using the inequality of (2.11), we obtain

$$F_{m,n}(w,\zeta) = O(1)\frac{1}{(r')^{sm}}\sum_{\nu=1}^{\infty}\sum_{j=0}^{p(n+1)-1} \sqrt{\frac{\nu}{s(m+1)-j}}\, |w|^j (r')^{-\nu}$$

$$= O(1)\frac{1}{(r')^{sm}}\sum_{\nu=1}^{\infty}\frac{\sqrt{\nu}}{(r')^\nu}\sum_{j=0}^{p(n+1)-1} |w|^j = O(1)\left(\frac{|w|^p}{(r')^{sq}}\right)^n,$$

which implies that $\lim_{n\to\infty} \Delta^{*p,s}_{m,n,4}(z;f) = 0$, uniformly on $\bar{G}_\mu$ for every $1 \leq \mu < R^{sq/p}$. Combining the above results then gives the desired result of (3.6) of Theorem 2. ∎

Remarks. (1) We do not know whether λ in (3.6) is best possible. However, if ∂E is r_0-analytic, we can improve our result. In this case we have for every β such that $r_0 < \beta < 1$,

$$\frac{w^s_m(\psi(\zeta)) - w^s_m(\psi(t))}{w^s_m(\psi(\zeta))} = \frac{\zeta^{s(m+1)} - t^{s(m+1)}}{\zeta^{s(m+1)}} \left(1 + O\left(\left(\frac{\beta}{|\zeta|}\right)^m\right)\right).$$

Then, an examination of the above proof shows that in this case the estimates of $\Delta^{*p,s}_{m,n,1}(z;f)$ and $\Delta^{*p,s}_{m,n,2}(z;f)$ remain unchanged. But

$$K_{m,n,3}(w,\zeta,t) = O(1)\left(\frac{\beta^q |w|^p}{r^{q+p}}\right),$$

so that $\lim_{n\to\infty} \Delta^{*p,s}_{m,n,3}(z;f) = 0$, uniformly on $\bar{G}_\mu$ for every $1 \leq \mu < R(R/r_0)^{q/p}$.

Finally, from (3.18) (since now $\alpha_{n,k} = O(\beta^{n+k})$), we obtain

$$F_{m,n}(w,\zeta) = O(1)\frac{1}{(r')^{sm}} \sum_{\nu=1}^{\infty} \sum_{j=0}^{p(n+1)-1} \beta^{\nu+s(m+1)-j} |w|^j (r')^{-\nu}$$

$$= O(1)\left(\frac{\beta}{r'}\right)^{sm} \sum_{\nu=1}^{\infty} \left(\frac{\beta}{r'}\right)^{\nu} \sum_{j=0}^{p(n+1)-1} \left(\frac{|w|}{\beta}\right)^j = O(1)\left(\frac{|w|^p \beta^{sq-p}}{(r')^{sq}}\right)^n,$$

and this implies that $\lim_{n\to\infty} \Delta^{*p,s}_{m,n,4}(z;f) = 0$, uniformly on $\bar{G}_\mu$ for every $1 \leq \mu < r_0(R/r_0)^{sq/p}$. Therefore, when the boundary curve ∂E is an r_0-analytic curve,

$$\lim_{n\to\infty} \Delta^{*p,s}_{m,n}(z;f) = 0, \qquad z \in G_\lambda,$$

where

$$\lambda := \min\left\{r_0\left(\frac{R}{r_0}\right)^{sq/p}; R\left(\frac{R}{r_0}\right)^{q/p}\right\}.$$

(2) A further improvement may be achieved in the case of an ellipse E_δ (where $\delta > 1$ and where ∂E_δ is r_0-analytic with $r_0 = 1/\delta$). An examination of the proof of Theorem 2 shows that in this case,

$$\lambda := \min\left\{r_0\left(\frac{R}{r_0}\right)^{(2sq/p)-1}; R\left(\frac{R}{r_0}\right)^{2q/p}\right\}.$$

This also applies to the case of a segment $E = [-1,1]$ (where $\delta = 1$) and gives

$$\lambda := \min\left\{R^{(2sq/p)-1}; R^{1+(2q/p)}\right\}.$$

4   MIXED HERMITE AND LAGRANGE INTERPOLATION

We next consider the case where Faber sections of Hermite interpolants are compared with Faber sections of Lagrange interpolants. This is analogous to Theorem C of [2] which was stated without proof. More precisely, for $s, p \in \mathbb{N}$ with $s \geq \max\{p; 2\}$, we set

$$D_{p,s,n}(z; f) := S_{p(n+1)-1}\left(z; \{H_{s(n+1)-1}(\cdot; f) - L_{s(n+1)-1}(\cdot; f)\}\right), \qquad (4.1)$$

where $H_{s(n+1)-1}(z; f)$ and $L_{s(n+1)-1}(z; f)$ are Hermite and Lagrange interpolants in Fejér nodes with respect to a compact set E. The following theorem was announced in [2].

Theorem C [2] *Let ∂E be an r_0-analytic curve for some $r_0 \in [0, 1)$, let $f \in A_R$, let $s, p \in \mathbb{N}$ be such that $s \geq \max\{p, 2\}$, and let $D_{p,s,n}(z; f)$ be given by (4.1). Then,*

$$\lim_{n \to \infty} D_{p,s,n}(z; f) = 0, \qquad z \in G_\lambda, \qquad (4.2)$$

the convergence being uniform and geometric on every subset $\bar{G}_\mu$ for $1 \leq \mu < \lambda$, where

$$\lambda := \left\{ \begin{array}{ll} \left. \begin{array}{ll} R^{s+2} & \text{for } p = 1 \text{ and } s \text{ even} \\ R^{(s+1)/p}; & \text{otherwise} \end{array} \right\} & \text{if } r_0 = 0, \\ \min\{R/r_0^{1/p}; R^{s/p}\}, & \text{if } r_0 > 0. \end{array} \right. \qquad (4.3)$$

We shall next prove an analogue of Theorem C, using Hermite interpolation and Lagrange interpolation in Faber nodes. Let $H^*_{s(n+1)-1}(z; f)$ be the Hermite interpolant to f in the zeros of $(F_{n+1}(z))^s$, and let $L^*_{s(n+1)-1}(z; f)$ denote the Lagrange interpolant to f in the zeros of $F_{s(n+1)}(z)$. As in earlier sections, $S_{p(n+1)-1}(z; f)$ denotes the Faber section of degree $p(n + 1) - 1$ of the Faber expansion of f. Set

$$D^*_{p,s,n}(z; f) := S_{p(n+1)-1}\left(z; \{H^*_{s(n+1)-1}(\cdot; f) - L^*_{s(n+1)-1}(\cdot; f)\}\right). \qquad (4.4)$$

We next establish:

Theorem 3 *Let $f \in A_R$, let $s, p \in \mathbb{N}$ be such that $s \geq \max\{p, 2\}$, and let $D^*_{p,s,n}(z; f)$ be given by (4.4). Then,*

$$\lim_{n \to \infty} D^*_{p,s,n}(z; f) = 0, \qquad z \in G_\lambda, \qquad (4.5)$$

the convergence being uniform and geometric on every subset $\bar{G}_\mu$ for $1 \leq \mu < \lambda$, where

$$\lambda := R^{1+(1/p)}. \qquad (4.6)$$

Proof: It is easy to verify that the following integral representation holds:

$$D^*_{p,s,n}(z;f)$$

$$= \frac{1}{2\pi i} \int\limits_{|\zeta|=r'} f(\psi(\zeta)) \left(\frac{1}{2\pi i} \int\limits_{|t|=r} \frac{\psi'(\zeta)}{\psi(\zeta) - \psi(t)} K_{s,n}(\zeta,t) \sum_{k=0}^{p(n+1)-1} \frac{F_k(z)}{t^{k+1}} \, dt \right) d\zeta,$$

(4.7)

where $K_{s,n}(\zeta,t)$ is the difference between the kernels of Hermite and Lagrange interpolants and where $1 < r < r' < R$. Then,

$$K_{s,n}(\zeta,t) := \frac{w_n^s(\psi(\zeta)) - w_n^s(\psi(t))}{w_n^s(\psi(\zeta))} - \frac{w_{s(n+1)-1}(\psi(\zeta)) - w_{s(n+1)-1}(\psi(t))}{w_{s(n+1)-1}(\psi(\zeta))}$$

$$= \frac{w_{s(n+1)-1}(\psi(t))}{w_{s(n+1)-1}(\psi(\zeta))} - \frac{w_n^s(\psi(t))}{w_n^s(\psi(\zeta))}.$$

(4.8)

Using (2.10) and (2.12), we then have

$$K_{s,n}(\zeta,t) = \frac{t^{s(n+1)}}{\zeta^{s(n+1)}} O\left(\frac{1}{\rho^{n+1}} \right),$$

uniformly on closed subsets of $\{\zeta \in \mathbb{C} : |\zeta| > 1\} \times \{t \in \mathbb{C} : |t| > 1\}$, where ρ is chosen such that $1 < \rho < r < r' < R$. Moreover (with $z = \psi(w)$),

$$K_{s,n}(\zeta,t) \sum_{k=0}^{p(n+1)-1} \frac{F_k(z)}{t^{k+1}} = O\left(\frac{1}{\rho^{n+1}} \right) \frac{t^{s(n+1)}}{\zeta^{s(n+1)}} \cdot \frac{t^{p(n+1)} - w^{p(n+1)}}{(t-w)t^{p(n+1)}}$$

$$+ O\left(\frac{1}{\rho^{n+1}} \right) \frac{t^{s(n+1)}}{\zeta^{s(n+1)}} \sum_{k=1}^{p(n+1)-1} \sum_{\nu=1}^{\infty} \alpha_{k,\nu} w^{-\nu} t^{-k-1}.$$

(4.9)

Since $|t| < |\zeta|$ and because of (2.11), the second term on the right side of (4.9) is bounded above by

$$O\left(\frac{1}{\rho^{n+1}} \right) \left(\frac{|t|}{|\zeta|} \right)^{s(n+1)} \cdot \sum_{k=1}^{p(n+1)-1} \sqrt{k}|t|^{-k-1} \cdot \sum_{\nu=1}^{\infty} \frac{1}{\sqrt{\nu}} |w|^{-\nu},$$

and this tends to zero as $n \to \infty$. The first term on the right of (4.9) can be estimated by

$$O(1) \left(\frac{|w|^p}{\rho r^p} \right)^n,$$

which gives the desired result of (4.6) of Theorem 3. ∎

55

Remarks. (1) We do not know whether λ of (4.6) is best possible. However, we are able to improve our result if ∂E is r_0-analytic. In this case, we have

$$\lim_{n\to\infty} D^*_{p,s,n}(z;f) = 0, \qquad z \in G_\lambda,$$

with $\lambda := R(R/r_0)^{1/p}$.

(2) A further improvement may be achieved in the case of an ellipse E_δ (where $\delta > 1$ and where ∂E_δ is r_0-analytic with $r_0 = 1/\delta$). An examination of the above proof shows that in this case λ is given by

$$\lambda := R(R/r_0)^{2/p}.$$

This also applies to the case of the segment $E = [-1,1]$ (where $\delta = 1$) and gives $\lambda := R^{1+2/p}$.

5   PROOF OF THEOREM B

Since the proof of Theorem B was not given in [2], we outline it briefly here. Observe that the formula for $\Delta^{p,s}_{m,n}(z;f)$ remains the same as the difference of (3.7) and (3.8). We have to keep in mind that $w_n(\psi(t))$, based on Fejér nodes, satisfies (from Lemma 3.1 in [4] where ∂E is a r_0-analytic curve, with $r_0 \in [0,1)$)

$$w_m(\psi(w)) = c^{m+1}(w^{m+1} - 1)(1 + O(\beta^m)), \qquad (5.1)$$

for any β such that $r_0 < \beta < 1$, uniformly on closed subsets of $\{w \in \mathbb{C} : |w| > r_0\}$, which gives

$$\frac{w_m(\psi(\zeta)) - w_m(\psi(t))}{w_m(\psi(\zeta))} = \frac{\zeta^{m+1} - t^{m+1}}{\zeta^{m+1} - 1}(1 + O(\beta^m)), \qquad (5.2)$$

and

$$\frac{w^s_m(\psi(\zeta)) - w^s_m(\psi(t))}{w^s_m(\psi(\zeta))} = \frac{(\zeta^{m+1} - 1)^s - (t^{m+1} - 1)^s}{(\zeta^{m+1} - 1)^s}(1 + O(\beta^m)), \qquad (5.3)$$

uniformly on closed subsets of $\{\zeta \in \mathbb{C} : |\zeta| > 1\} \times \{t \in \mathbb{C} : |t| > 1\}$. Here again, we write

$$\Delta^{p,s}_{m,n}(z;f) = \sum_{j=1}^4 \Delta_{m,n,j}(z;f),$$

where for $j = 1,2,3,4$ we have

$$\Delta_{m,n,j}(z;f) = \frac{1}{2\pi i}\int_{|\zeta|=r'} f(\psi(\zeta))\left(\frac{1}{2\pi i}\int_{|t|=r}\frac{\psi'(\zeta)}{\psi(\zeta) - \psi(t)}K_{m,n,j}(w,\zeta,t)\,dt\right)d\zeta.$$

$$(5.4)$$

The kernels $K_{m,n,j}(w, \zeta, t)$ are defined as follows:

$$K_{m,n,1}(w, \zeta, t) := \frac{(\zeta^{m+1} - 1)^s - (t^{m+1} - 1)^s}{(\zeta^{m+1} - 1)^s} (1 + O(\beta^m)) \sum_{k=1}^{p(n+1)-1} \sum_{\nu=1}^{\infty} \frac{\alpha_{k,\nu}}{w^\nu t^{k+1}}$$

$$- \frac{\zeta}{t} \sum_{k=1}^{p(n+1)-1} \sum_{\nu=1}^{\infty} \frac{\alpha_{k,\nu}}{w^\nu \zeta^{\nu+1}},$$

$$\text{(5.5)}$$

$$K_{m,n,2}(w, \zeta, t) := \frac{t^{p(n+1)} - w^{p(n+1)}}{(t - w)t^{p(n+1)}} - \frac{\zeta}{t} \cdot \frac{\zeta^{p(n+1)} - w^{p(n+1)}}{(\zeta - w)\zeta^{p(n+1)}}, \quad \text{(5.6)}$$

$$K_{m,n,3}(w, \zeta, t) := O(\beta^m)\frac{(\zeta^{m+1} - 1)^s - (t^{m+1} - 1)^s}{(\zeta^{m+1} - 1)^s} \cdot \frac{t^{p(n+1)} - w^{p(n+1)}}{(t - w)t^{p(n+1)}}, \quad \text{(5.7)}$$

and

$$K_{m,n,4}(w, \zeta, t) := -\frac{(t^{m+1} - 1)^s}{(\zeta^{m+1} - 1)^s} \cdot \frac{t^{p(n+1)} - w^{p(n+1)}}{(t - w)t^{p(n+1)}}. \quad \text{(5.8)}$$

The estimate of $\Delta_{m,n,1}(z; f)$ is obtained as in the proof of Theorem 2, and we obtain $\lim\limits_{n \to \infty} \Delta_{m,n,1}(z; f) = 0$, locally uniformly on $\mathbb{C}$. Further, using the residue theorem, we obtain

$$\frac{1}{2\pi i} \int_{|t|=r} \frac{\psi'(\zeta)}{\psi(\zeta) - \psi(t)} K_{m,n,2}(w, \zeta, t)dt = K_{m,n,2}(w, \zeta, \zeta) = 0$$

for all w, ζ, which yields $\lim\limits_{n \to \infty} \Delta_{m,n,2}(z; f) = 0$ for all $z \in \mathbb{C}$. For the estimate of $\Delta_{m,n,3}(z; f)$, we examine the kernel $K_{m,n,3}(w, \zeta, t)$ and we see that for $|w| > r$,

$$K_{m,n,3}(w, \zeta, t) = O(1) \left(\frac{|w|^p \beta^q}{r^p} \right)^n,$$

and thus $\lim\limits_{n \to \infty} \Delta_{m,n,3}(z; f) = 0$, uniformly on $\bar{G}_\mu$ for $1 \leq \mu < R/r_0^{q/p}$. To find the estimate of $\Delta_{m,n,4}(z; f)$, we use (2.18) to obtain

$$\frac{1}{2\pi i} \int_{|t|=r} \frac{\psi'(\zeta)}{\psi(\zeta) - \psi(t)} K_{m,n,4}(w, \zeta, t)\, dt = I_1(w, \zeta) + I_2(w, \zeta), \quad \text{(5.9)}$$

where

$$I_1(w, \zeta) := \frac{1}{2\pi i} \int_{|t|=r} \frac{1}{\zeta - t} K_{m,n,4}(w, \zeta, t)dt,$$

and

$$I_2(w,\zeta) := \frac{1}{2\pi i} \int\limits_{|t|=r} A(\zeta,t) K_{m,n,4}(w,\zeta,t)\,dt,$$

with $A(\zeta,t) := \sum\limits_{\nu=1}^{\infty}\sum\limits_{k=1}^{\infty} \alpha_{\nu,k} t^{-k}\zeta^{-\nu-1}$. Expanding the integrand in $I_1(w,\zeta)$ in powers of t, we obtain

$$\frac{1}{\zeta - t} K_{m,n,4}(w,\zeta,t) = \frac{-1}{(\zeta^{m+1}-1)^s} \sum_{k=0}^{\infty}\sum_{j=0}^{s}\sum_{\nu=0}^{p(n+1)-1} \binom{s}{j} \frac{(-1)^{s-j}w^\nu}{\zeta^{k+1}} t^{k+j(m+1)-\nu-1},$$

and its integral, with respect to t, will vanish except when $k = \nu - j(m+1) \geq 0$.

On setting $k = \nu - j(m+1) \geq 0$ in the above display, it follows, as $s \geq p/q$ and $m+1 = q(n+1)$ from the hypotheses of Theorem B, that the above summation index j satisfies $0 \leq j \leq \min\left\{\frac{p(n+1)-1}{q(n+1)} \,;\, s\right\} = \frac{p}{q} - \frac{1}{q(n+1)}$. Thus,

$$I_1(w,\zeta) = \frac{-1}{\zeta(\zeta^{m+1}-1)^s} \sum_{j=0}^{\lfloor \frac{p}{q} - \frac{1}{q(n+1)}\rfloor} \binom{s}{j}(-1)^{s-j}\zeta^{jq(n+1)} \sum_{\nu=jq(n+1)}^{p(n+1)-1}\left(\frac{w}{\zeta}\right)^\nu.$$

On writing $p = qt + \tau$, where t is a nonnegative integer and where $0 \leq \tau < q$, it follows that

$$I_1(w,\zeta) = \begin{cases} O(1)\left(\dfrac{|w|^p}{(r')^{sq+\tau}}\right)^n & \text{if } 0 < \tau < q, \\[4mm] O(1)\left(\dfrac{|w|^p}{(r')^{q(s+1)}}\right)^n & \text{if } 0 = \tau, \end{cases} \tag{5.10}$$

which implies that $\lim\limits_{n\to\infty} \Delta_{m,n,4}^{(1)}(z;f) = 0$ uniformly on $\bar{G}_\mu$ for every $1 \leq \mu < R^\lambda$, where

$$\lambda := \begin{cases} \dfrac{qs+\tau}{p} & \text{if } 0 < \tau < q, \\[4mm] \dfrac{q(s+1)}{p} & \text{if } 0 = \tau. \end{cases}$$

In the special case when $p < q$, then $p = tq + \tau$ implies $t = 0$ and $\tau = p > 0$, so that the first display in (5.10) applies with $\tau = p$.

In order to estimate $I_2(w,\zeta)$, we again expand the integrand in powers of t and obtain

$$A(\zeta,t) K_{m,n,4}(w,\zeta,t)$$

$$= \frac{-1}{(\zeta^{m+1}-1)^s} \sum_{\mu=1}^{\infty}\sum_{k=1}^{\infty}\sum_{j=0}^{s}\sum_{\nu=0}^{p(n+1)-1} \binom{s}{j}(-1)^{s-j}\alpha_{\mu,k}\zeta^{-\mu-1}w^\nu t^{-k+j(m+1)-\nu-1}.$$

Then, all the integrals vanish except when $k = j(m+1) - \nu \geq 1$. Then, as $s \geq p/q$,

$$I_2(w, \zeta) = \frac{-1}{(\zeta^{m+1} - 1)^s} \sum_{\mu=1}^{\infty} \sum_{\nu=0}^{p(n+1)-1} \sum_{\substack{j=1 \\ jq(n+1) \geq \nu+1}}^{s} \binom{s}{j} \frac{(-1)^{s-j} \alpha_{\mu, j(m+1)-\nu} w^{\nu}}{\zeta^{\mu+1}},$$

and using the known fact [5, eq. (4.2)] that $\alpha_{\mu,k} = O(1)\beta^{\mu+k}$, we obtain

$$I_2(w, \zeta) = O(1) \frac{1}{(r')^{sqn}} \sum_{\substack{\nu=0}}^{p(n+1)-1} \sum_{\substack{j=1 \\ jq(n+1) \geq \nu+1}}^{s} \beta^{jq(n+1)} \left(\frac{|w|}{\beta}\right)^{\nu}$$

$$= O(1) \frac{1}{(r')^{sqn}} \sum_{\ell=0}^{p-1} \sum_{\nu=0}^{n} \sum_{\substack{j=1 \\ jq(n+1) \geq \nu+1+\ell(n+1)}}^{s} \beta^{jq(n+1)} \left(\frac{|w|}{\beta}\right)^{\nu+\ell(n+1)}$$

$$= O(1) \frac{1}{(r')^{sqn}} \beta^n \sum_{\ell=0}^{p-1} |w|^{\ell(n+1)} \sum_{\nu=0}^{n} \left(\frac{|w|}{\beta}\right)^{\nu} = O(1) \left(\frac{|w|^p}{(r')^{sq}}\right)^n,$$

$$\tag{5.11}$$

since $jq - \ell \geq \frac{\nu+1}{n+1}$ which implies $jq - \ell \geq 1$. Therefore, we obtain

$$\Delta_{m,n,4}^{(2)}(z; f) := \frac{1}{2\pi i} \int_{|\zeta|=r'} f(\psi(\zeta)) I_2(w, \zeta) d\zeta \to 0 \qquad \text{as } n \to \infty,$$

uniformly on $\bar{G}_\mu$ for every $1 \leq \mu < R^{sq/p}$.

In the special case $q \geq p$, we have $jq(n+1) \geq jp(n+1) \geq \nu + 1 + \ell(n+1)$ for all $j = 1, \ldots, p$, $\nu = 0, \ldots, n$, and $\ell = 0, \ldots, p-1$ so that

$$I_2(w, \zeta) = O(1) \left(\frac{|w|^p \beta^{q-p}}{(r')^{sq}}\right)^n,$$

which shows that $\lim_{n \to \infty} \Delta_{m,n,4}^{(2)}(z; f) = 0$, uniformly on $\bar{G}_\mu$ for every $1 \leq \mu < R^{sq/p}/r_0^{-1+q/p}$. Combining the above results then gives the desired result (3.3) of Theorem B. ∎

Remarks. (1) For arbitrary p, q, s, as $r_0 \to 1$, we obtain $\lambda \to R$. In the case $s = p$ and $r_0 > 0$, we obtain

$$\lambda = \begin{cases} \min\{R^{1+q}; R/r_0^{q/p}; R^q/r_0^{-1+q/p}\}, & \text{if } q \geq p, \\ \min\{R^q; R/r_0^{q/p}\}, & \text{if } q < p. \end{cases} \tag{5.12}$$

If $r_0 = 0$, i.e., $E = \bar{D}_1$, we have $\lambda = R^{1+sq/p}$ for $q \geq p$ and $\lambda = R^{(1+sq)/p}$ for $q < p$. If in addition $s = p$ and $q = 1$, Theorem B gives a special case of Theorem 3 of [3].

(2) We do not know whether λ of (3.3) is best possible. However, we are able to improve our result if E is the ellipse E_δ. An examination of the proof then yields that Theorem B holds with

$$\lambda = \begin{cases} \min\{R^{1+sq/p}; R^{1+q/p}/r_0^{2q/p}; R^{-1+(s+1)q/p}/r_0^{2(-1+q/p)}\}, & q \geq p, \\ \min\{R^{1+q/p}/r_0^{2q/p}; R^{sq/p}\}, & q < p. \end{cases} \tag{5.13}$$

(3) The previous remark also applies when $\delta = 1$, i.e., $E = [-1,1]$, and we obtain

$$\lambda = \begin{cases} \min\{R^{1+q/p}; R^{(s+1)q/p-1}\}, & q \geq p, \\ \min\{R^{1+q/p}; R^{sq/p}\}, & q < p. \end{cases} \tag{5.14}$$

6 PROOF OF THEOREM C

As in the proof of Theorem 3 (Sec. 4), the form of the integral representation of the operator $D_{p,s,n}(z;f)$ is given by (4.7) where the kernel $K_{s,n}(\zeta,t)$ is given by (4.8). Since $w_n\left(\psi(t)\right)$ is based on Fejér nodes, we have

$$K_{s,n}(\zeta,t) = \frac{t^{s(n+1)} - 1}{\zeta^{s(n+1)} - 1}\left(1 + O(\beta^{sn})\right) - \frac{(t^{n+1}-1)^s}{(\zeta^{n+1}-1)^s}\left(1 + O(\beta^n)\right)$$

$$= K_{s,n}^{(1)}(\zeta,t) + K_{s,n}^{(2)}(\zeta,t), \tag{6.1}$$

where

$$K_{s,n}^{(1)}(\zeta,t) := \frac{t^{s(n+1)} - 1}{\zeta^{s(n+1)} - 1} - \frac{(t^{n+1}-1)^s}{(\zeta^{n+1}-1)^s} \tag{6.2}$$

and

$$K_{s,n}^{(2)}(\zeta,t) := \frac{t^{s(n+1)} - 1}{\zeta^{s(n+1)} - 1}O(\beta^{sn}) - \frac{(t^{n+1}-1)^s}{(\zeta^{n+1}-1)^s}O(\beta^n). \tag{6.3}$$

Since ∂E is an r_0-analytic curve, then $1 > \beta > r_0$ and with $z = \psi(w)$ for $|w| > 1$, we have

$$\sum_{k=0}^{p(n+1)-1} \frac{F_k(z)}{t^{k+1}} = \frac{t^{p(n+1)} - w^{p(n+1)}}{(t-w)t^{p(n+1)}} + \sum_{k=1}^{p(n+1)-1}\sum_{\nu=1}^{\infty} \alpha_{k,\nu}w^{-\nu}t^{-k-1}$$

$$=: A_{p,n,1}(w,t) + A_{p,n,2}(w,t).$$

Since

$$\lim_{n\to\infty} A_{p,n,2}(w,t) = \sum_{k=1}^{\infty}\sum_{\nu=1}^{\infty} \alpha_{k,\nu}w^{-\nu}t^{-k-1},$$

uniformly on closed subsets of $\{w \in \mathbb{C} : |w| > 1\} \times \{t \in \mathbb{C} : |t| > 1\}$, we obtain

$$\lim_{n\to\infty} K_{s,n}^{(j)}(\zeta,t)A_{p,n,2}(w,t) = 0 \qquad (j = 1,2), \tag{6.4}$$

uniformly for $|\zeta| > |t| \geq \delta$, $|w| \geq \delta$ for any $\delta > 1$. It is therefore enough to consider

$$
\begin{cases}
K_{p,s,n}^{(1)}(w,\zeta,t) := K_{s,n}^{(2)}(\zeta,t)A_{p,n,1}(w,t), \\
K_{p,s,n}^{(2)}(w,\zeta,t) := K_{s,n}^{(1)}(\zeta,t)A_{p,n,1}(w,t).
\end{cases}
\tag{6.5}
$$

We now set

$$
D_{p,s,n}^{(j)}(z;f) := \frac{1}{2\pi i} \int\limits_{|\zeta|=r'} f\left(\psi(\zeta)\right) \left(\frac{1}{2\pi i} \int\limits_{|t|=r} \frac{\psi'(\zeta)}{\psi(\zeta) - \psi(t)} K_{p,s,n}^{(j)}(w,\zeta,t)\, dt \right) d\zeta.
\tag{6.6}
$$

Estimate of $D_{p,s,n}^{(1)}(z;f)$

From (6.3), we see that $K_{s,n}^{(2)}(\zeta,t) = O(\beta^n)$, uniformly for $|\zeta| > |t| \geq \delta > 1$, so that for $|w| > r$, we obtain

$$
K_{p,s,n}^{(1)}(w,\zeta,t) = O(1) \left(\frac{\beta|w|^p}{r^p} \right)^n,
$$

and thus $\lim\limits_{n\to\infty} D_{p,s,n}^{(1)}(z;f) = 0$, uniformly on $\bar{G}_\mu$ for every $1 \leq \mu < R/r_0^{1/p}$. Note that if $r_0 = 0$, then $D_{p,s,n}^{(1)}(z;f) \equiv 0$.

Estimate of $D_{p,s,n}^{(2)}(z;f)$

Since

$$
\frac{\psi'(\zeta)}{\psi(\zeta) - \psi(t)} = \frac{1}{\zeta - t} + \sum_{\mu=1}^{\infty} \sum_{k=1}^{\infty} \alpha_{\mu,k} t^{-k} \zeta^{-\mu-1},
\tag{6.7}
$$

uniformly on closed subsets of $\{t \in \mathbb{C} : |t| > 1\} \times \{\zeta \in \mathbb{C} : |\zeta| > 1\}$ where $\alpha_{\mu,k} = O(\beta^{\mu+k})$, we can write

$$
\frac{\psi'(\zeta)}{\psi(\zeta) - \psi(t)} K_{p,s,n}^{(2)}(w,\zeta,t) = \sum_{j=1}^{3} B_{p,s,n}^{(j)}(w,\zeta,t),
$$

where we have set

$$
\begin{cases}
B_{p,s,n}^{(1)}(w,\zeta,t) := \dfrac{t^{s(n+1)} - 1}{\zeta^{s(n+1)} - 1} \cdot \dfrac{t^{p(n+1)} - w^{p(n+1)}}{(\zeta - t)(t - w)t^{p(n+1)}}, \\[4mm]
B_{p,s,n}^{(2)}(w,\zeta,t) := -\dfrac{(t^{n+1} - 1)^s}{(\zeta^{n+1} - 1)^s} \cdot \dfrac{t^{p(n+1)} - w^{p(n+1)}}{(\zeta - t)(t - w)t^{p(n+1)}}, \\[4mm]
B_{p,s,n}^{(3)}(w,\zeta,t) := K_{p,s,n}^{(2)}(w,\zeta,t) \displaystyle\sum_{\mu=1}^{\infty} \sum_{k=1}^{\infty} \alpha_{\mu,k} t^{-k} \zeta^{-\mu-1}.
\end{cases}
\tag{6.8}
$$

61

We now evaluate

$$I_j(w,\zeta) := \frac{1}{2\pi i} \int\limits_{|t|=r} B_{p,s,n}^{(j)}(w,\zeta,t)dt \qquad (j=1,2,3).$$

In order to find $I_1(w,\zeta)$, we note that $(\zeta^{s(n+1)}-1)B_{p,s,n}^{(1)}(w,\zeta,t)$ when expanded in powers of t yields

$$\sum_{k=0}^{\infty}\sum_{\nu=0}^{p(n+1)-1} \zeta^{-k-1}w^\nu t^{k+s(n+1)-\nu-1} - \sum_{k=0}^{\infty}\sum_{\nu=0}^{p(n+1)-1} \zeta^{-k-1}w^\nu t^{k-\nu-1}.$$

In the first sum, $k+s(n+1)-\nu-1 \geq 0$ and so its integral vanishes. The integral over the second sum gives a contribution only when $k=\nu$ and so

$$I_1(w,\zeta) = -\sum_{\nu=0}^{p(n+1)-1} \zeta^{-\nu-1}w^\nu/(\zeta^{(s(n+1)}-1)$$

$$= -\frac{\zeta^{p(n+1)} - w^{p(n+1)}}{(\zeta-w)(\zeta^{s(n+1)}-1)\zeta^{p(n+1)}}. \tag{6.9}$$

The integrand in $I_2(w,\zeta)$, after multiplying by $(\zeta^{n+1}-1)^s$, has the following expansion in powers of t :

$$-\sum_{k=0}^{\infty}\sum_{j=0}^{s}\sum_{\nu=0}^{p(n+1)-1} \binom{s}{j}(-1)^{s-j}\zeta^{-k-1}w^\nu t^{k+j(n+1)-\nu-1}.$$

On integrating with respect to t on the circle $|t| = r$, all the integrals vanish except when $k = \nu - j(n+1) \geq 0$ (then $\nu \geq j(n+1)$ and thus $j \leq p-1$) and therefore

$$I_2(w,\zeta) = \frac{-1}{(\zeta^{n+1}-1)^s} \sum_{j=0}^{p-1}\sum_{\nu=j(n+1)}^{p(n+1)-1} \binom{s}{j}(-1)^{s-j}w^\nu \zeta^{j(n+1)-\nu-1}$$

$$= -\frac{w^{p(n+1)}\sum_{j=0}^{p-1}\binom{s}{j}(-1)^{s-j}\zeta^{j(n+1)} - \zeta^{p(n+1)}\sum_{j=0}^{p-1}\binom{s}{j}(-1)^{s-j}w^{j(n+1)}}{(\zeta-w)(\zeta^{n+1}-1)^s\zeta^{p(n+1)}}. \tag{6.10}$$

From (6.9) and (6.10), we obtain after a slight rearrangement

$$I_1(w,\zeta) + I_2(w,\zeta) = \frac{w^{p(n+1)} - \zeta^{p(n+1)}}{(\zeta - w)\zeta^{p(n+1)}} \cdot \frac{(\zeta^{n+1} - 1)^s - (-1)^s(\zeta^{s(n+1)} - 1)}{(\zeta^{s(n+1)} - 1)(\zeta^{n+1} - 1)^s}$$

$$- \frac{w^{p(n+1)} \sum_{j=1}^{p-1} \binom{s}{j}(-1)^{s-j}\zeta^{j(n+1)} - \zeta^{p(n+1)} \sum_{j=1}^{p-1} \binom{s}{j}(-1)^{s-j}w^{j(n+1)}}{(\zeta - w)(\zeta^{n+1} - 1)^s\zeta^{p(n+1)}}$$

$$=: J_1(w,\zeta) + J_2(w,\zeta).$$

Since

$$(\zeta^{n+1}-1)^s-(-1)^s(\zeta^{s(n+1)}-1) = \begin{cases} 2\zeta^{s(n+1)}\left(1 + O\left(\dfrac{1}{\zeta^{n+1}}\right)\right), & \text{if } s \text{ is odd,} \\[2mm] -s\zeta^{(s-1)(n+1)}\left(1 + O\left(\dfrac{1}{\zeta^{n+1}}\right)\right), & \text{if } s \text{ is even,} \end{cases}$$

it follows that

$$J_1(w,\zeta) = \begin{cases} O(1)\left(\dfrac{|w|^p}{(r')^{p+s}}\right)^n, & s \text{ is odd,} \\[3mm] O(1)\left(\dfrac{|w|^p}{(r')^{p+s+1}}\right)^n, & s \text{ is even.} \end{cases} \tag{6.11}$$

Furthermore, $J_2(w,\zeta) \equiv 0$ for $p = 1$ and for $p > 1$,

$$J_2(w,\zeta) = O(1)\left(\frac{|w|^p}{(r')^{s+1}}\right)^n. \tag{6.12}$$

Therefore combining (6.11) and (6.12), we obtain

$$\frac{1}{2\pi i} \int\limits_{|\zeta|=r'} f(\psi(\zeta))\left[I_1(w,\zeta) + I_2(w,\zeta)\right] d\zeta \to 0, \qquad \text{as } n \to \infty,$$

uniformly on $\bar{G}_\mu$ for every $1 \le \mu < \lambda$, where

$$\lambda = \begin{cases} R^{s+2}, & \text{for } p = 1 \text{ and } s \text{ even,} \\ R^{(s+1)/p}, & \text{otherwise.} \end{cases}$$

This proves the theorem for $r_0 = 0$ because then $I_3(w,\zeta) \equiv 0$.

Estimate of $I_3(w, \zeta)$

In order to estimate $I_3(w, \zeta)$, we expand $(\zeta^{s(n+1)} - 1)B^{(3)}_{p,s,n}(w, \zeta, t)$ in powers of t. Then we have

$$(\zeta^{s(n+1)} - 1)B^{(3)}_{p,s,n}(w, \zeta, t) = S_1 + S_2 + S_3,$$

where

$$
\begin{cases}
S_1 := \displaystyle\sum_{\mu=1}^{\infty} \sum_{k=1}^{\infty} \sum_{\nu=0}^{p(n+1)-1} \alpha_{\mu,k} \zeta^{-\mu-1} w^\nu t^{-k+s(n+1)-\nu-1}, \\[4mm]
S_2 := -\displaystyle\sum_{\mu=1}^{\infty} \sum_{k=1}^{\infty} \sum_{\nu=0}^{p(n+1)-1} \alpha_{\mu,k} \zeta^{-\mu-1} w^\nu t^{-k-\nu-1}, \\[4mm]
S_3 := -\left(1 + O\!\left(\dfrac{1}{\zeta^{n+1}}\right)\right) \displaystyle\sum_{\mu=1}^{\infty} \sum_{k=1}^{\infty} \sum_{j=0}^{s} \sum_{\nu=0}^{p(n+1)-1} \binom{s}{j} \dfrac{(-1)^{s-j} \alpha_{\mu,k} w^\nu}{\zeta^{\mu+1}} t^{-k+j(n+1)-\nu-1}.
\end{cases}
$$

$$(6.13)$$

Since the power of t in the sum S_2 is $-k-\nu-1 \leq -2$, we have no contribution from the integral of S_2. Term by term integration of the first term S_1 is non-zero only when $k = s(n+1) - \nu$ and so its contribution is

$$S_{11} = \frac{1}{\zeta^{s(n+1)} - 1} \sum_{\mu=1}^{\infty} \sum_{\nu=0}^{p(n+1)-1} \alpha_{\mu,s(n+1)-\nu} \zeta^{-\mu-1} w^\nu.$$

Similarly, the integration of S_3 with respect to t vanishes except when $k = j(n+1) - \nu$ (note that then $j(n+1) \geq \nu+1$ and then $j \neq 0$). This yields

$$S_{31} = \left(1 + O\!\left(\frac{1}{\zeta^{n+1}}\right)\right) \frac{1}{\zeta^{s(n+1)} - 1} \sum_{\mu=1}^{\infty} \sum_{\nu=0}^{p(n+1)-1}$$

$$\sum_{\substack{j=1 \\ j(n+1) \geq \nu+1}}^{s} \binom{s}{j} (-1)^{s-j} \alpha_{\mu,j(n+1)-\nu} \zeta^{-\mu-1} w^\nu.$$

Since $\alpha_{\mu,k} = O(1)\beta^{\mu+k}$, we obtain

$$S_{11} = O(1) \frac{1}{(r')^{sn}} \sum_{\mu=1}^{\infty} \left(\frac{\beta}{r'}\right)^\mu \beta^{sn} \sum_{\nu=0}^{p(n+1)-1} \left(\frac{w}{\beta}\right)^\nu = O(1) \left(\frac{\beta^{s-p}|w|^p}{(r')^s}\right)^n,$$

and

$$S_{31} = O(1)\frac{1}{(r')^{sn}} \sum_{\mu=1}^{\infty} \left(\frac{\beta}{r'}\right)^{\mu} \sum_{\substack{j=1 \\ j(n+1)\geq\nu+1}}^{s} \sum_{\nu=0}^{p(n+1)-1} \beta^{j(n+1)} \left(\frac{w}{\beta}\right)^{\nu}$$

$$= O(1)\frac{1}{(r')^{sn}} \sum_{\nu=0}^{p(n+1)-1} \beta^{\nu+1} \left(\frac{w}{\beta}\right)^{\nu} = O(1)\left(\frac{|w|^p}{(r')^s}\right)^n.$$

It therefore follows that

$$\frac{1}{2\pi i} \int_{|\zeta|=r'} f(\psi(\zeta)) I_3(w,\zeta) d\zeta \to 0, \qquad \text{as } n \to \infty,$$

uniformly on $\bar{G}_\mu$ for every $1 \leq \mu < R^{s/p} \min\{1/r_0^{-1+s/p}, 1\} = R^{s/p}$. Combining the above results then gives the desired result of (4.3) of Theorem C. ∎

Remarks. (1) For $s = p$ and $r_0 > 0$, we have from (4.3) that $\lambda = R$, so that Theorem C gives no overconvergence. Also $\lambda \to R$ as $r_0 \to 1$ for arbitrary p and s. If $0 < r_0 < \frac{1}{R^{s-p}}$, then $\lambda = R^{s/p}$.

(2) If $r_0 = 0$, then λ from (4.3) is best possible as can be seen by the example $f(z) = 1/(R-z)$, but we do not know if this is the case when $r_0 > 0$. However, we are able to improve our result if E is the ellipse E_δ. An examination of the proof shows that Theorem C holds with

$$\lambda = \min\{R^{1+1/p}/r_0^{2/p}; R^{s/p}\}.$$

(3) The above remark also applies when $\delta = 1$, i.e., $E = [-1, 1]$ and we obtain

$$\lambda = \begin{cases} R & \text{for } s = p, \\ R^{1+1/p} & \text{for } s > p. \end{cases}$$

REFERENCES

[1] Brück R. On the failure of Walsh's equiconvergence theorem for Jordan domains. *Analysis*, **13**:229–234, 1993.

[2] Brück R., Sharma A., Varga R.S. An extension of a result of Rivlin on Walsh equiconvergence. In *Advances in Computational Mathematics*: New Delhi, India, 1993, pages 225–234. World Scientific Publishing Co. Ptl. Ltd., Singapore, 1994. H.P. Dikshit and C.A. Micchelli, eds.

[3] Cavaretta A.S. Jr., Sharma A., Varga R.S. Interpolation in the roots of unity: An extension of a theorem of J. L. Walsh. *Resultate Math.*, **3**:155–191, 1980.

[4] Curtiss J.H. Convergence of complex Lagrange interpolation polynomials on the locus of the interpolation points. *Duke Math. J.*, **32**:187–204, 1965.

[5] Curtiss J.H. Faber polynomials and the Faber series. *Amer. Math. Monthly*, **78**:577–596, 1971.

[6] Gaier D. *Vorlesungen über Approximation im Komplexen*. Birkhäuser Verlag, Basel, 1980.

[7] Kövari T., Pommerenke Ch. On Faber polynomials and Faber expansions. *Math. Z.*, **99**:193–206, 1967.

[8] Pommerenke Ch. Über die Verteilung der Fekete-Punkte. *Math. Ann.*, **168**:111–127, 1967.

[9] Rivlin T.J. On Walsh equiconvergence. *J. Approx. Theory*, **36**:334–345, 1982.

[10] Walsh J.L. *Interpolation and Approximation by Rational Functions in the Complex Domain*, 5th Ed. American Math. Soc., Providence, R.I., 1969.

International Series of Numerical Mathematics, Vol. 119, ©1994 Birkhäuser

The Parallel Solution of Ordinary Differential Equations and some Special Functions[*]

J. C. Butcher[†]

Dedicated to Walter Gautschi on the occasion of his 65th birthday

Abstract. This paper concerns the functions defined by the functional equations

$$\exp(z) = \phi\left(\frac{z}{\exp(z)}\right),$$

$$(1-z)\exp(tz) = \psi\left(\frac{z}{(1-z)\exp(tz)}\right)$$

and their relationship to the numerical solution of ordinary differential equations in a parallel environment. It will be shown that ϕ and its Padé approximations are related to stability questions for type 3 DIMSIM methods and that ψ and its Padé approximations have a similar relationship to type 4 DIMSIMs. An introduction to DIMSIM methods will be presented as a special class of general linear methods. The particular features which identify DIMSIMs from the much wider class of general linear methods are the natural meaning of the quantities being approximated, high stage-order and ease of implementation in a sequential or parallel environment. In particular type 3 and 4 DIMSIMs are intended for parallel implementation and for non-stiff and stiff differential equation systems respectively. Particular methods will be derived from approximations to ϕ and ψ selected for their stability properties.

1 INTRODUCTION

In this paper we discuss the functions defined by the functional equations

$$\exp(z) = \phi\left(\frac{z}{\exp(z)}\right), \tag{1.1}$$

$$(1-z)\exp(tz) = \psi\left(\frac{z}{(1-z)\exp(tz)}\right), \tag{1.2}$$

and their relationship to the numerical solution of ordinary differential equations in a parallel environment. Traditional methods do not adapt easily to parallelism (across the method) because there is no scope for calculating stage values simultaneously in single-stage methods such as linear multistep. On the other hand, the order of Runge-Kutta methods is limited to the number of sequential stages

---

[*]This work was supported by the New Zealand Foundation for Research, Science and Technology
[†]Department of Mathematics, The University of Auckland, New Zealand

[5]. In the search for methods where stage-parallelism makes sense, it is natural to look next at the wider class of general linear methods and the special family within this class known as "diagonally-implicit multistage integration methods" or DIMSIMs. The particular methods we consider for parallel computation are known as type 3 and type 4 DIMSIMs (type 3 for non-stiff problems and type 4 for stiff problems). Section 2 is devoted to an introduction to these methods. In §3 we discuss the function ϕ and, in particular, its Padé approximations. The immediate superdiagonal of the Padé table is suggested as a fruitful place to look for useful methods because of the relationship between these approximations and the stability polynomial of a related type 3 DIMSIM. Section 4 is devoted to the introduction of a particular DIMSIM based on a member of this superdiagonal. Section 5 is concerned with the function ψ and its Padé approximations. In the same way that ϕ is related to stability questions for type 3 methods, intended for non-stiff problems, ψ is related to stability questions for type 4 methods, for stiff problems. Finally, in §6 we will discuss a particular third order A-stable method based on a member of the second row of the Padé table for ψ.

2 AN INTRODUCTION TO DIMSIMS

A general linear method for the numerical solution of ordinary differential equations can be characterized by a partitioned $(s + r) \times (s + r)$ matrix

$$\begin{bmatrix} A & U \\ B & V \end{bmatrix}.$$

These methods were introduced in [1] as a framework for analyzing existing methods and identifying possible new and useful methods. Because the class of general linear methods is very large, the subclass of "DIMSIMs" has recently been proposed [2]. In this paper we will make the assumptions inherent in this special case.

Suppose that at the start of step number n, which is supposed to advance the numerical approximation from x_{n-1} to $x_n = x_{n-1} + h$, r vectors of information concerning the solution at or near x_{n-1} are available. Denote this collection of vectors by

$$y^{[n-1]} = \begin{bmatrix} y_1^{[n-1]} \\ y_2^{[n-1]} \\ \vdots \\ y_r^{[n-1]} \end{bmatrix}$$

and denote by $y^{[n]}$ the corresponding collection of quantities computed in the step.

Let the vectors consisting of the s internal stages computed within the step and the derivatives evaluated at these stages be denoted by

$$Y = \begin{bmatrix} Y_1 \\ Y_2 \\ \vdots \\ Y_s \end{bmatrix}, \quad F = \begin{bmatrix} F_1 \\ F_2 \\ \vdots \\ F_s \end{bmatrix} = \begin{bmatrix} f(Y_1) \\ f(Y_2) \\ \vdots \\ f(Y_s) \end{bmatrix},$$

where the differential equation whose solution is being approximated is the autonomous N dimensional system

$$y'(x) = f(y(x)).$$

With this terminology and the use of Kronecker products of the form

$$A \otimes I_N = \begin{bmatrix} a_{11}I_N & a_{12}I_N & \cdots & a_{1s}I_N \\ a_{21}I_N & a_{22}I_N & \cdots & a_{2s}I_N \\ \vdots & \vdots & & \vdots \\ a_{s1}I_N & a_{s2}I_N & \cdots & a_{ss}I_N \end{bmatrix},$$

the values of Y and $y^{[n]}$ are defined as solutions, assuming they exist, of the algebraic equations

$$\begin{bmatrix} Y \\ y^{[n]} \end{bmatrix} = \begin{bmatrix} A \otimes I_N & U \otimes I_N \\ B \otimes I_N & V \otimes I_N \end{bmatrix} \begin{bmatrix} hF \\ y^{[n-1]} \end{bmatrix}.$$

The characteristic features of diagonally implicit multistage integration methods are:

(i) the form of the matrix A: it should be of lower triangular form with equal entries on the diagonal elements;

(ii) the nature of the quantities to be approximated by $y_i^{[n]}$, $i = 1, 2, \ldots, r$: it should have a value given by a weighted Taylor series

$$\sum_{j=0}^{p} h^j \alpha_{ij} y^{(j)}(x_n) + O(h^{p+1}),$$

where p is the order of the method;

(iii) the stage order q should be close to p: for $q \geq p - 1$,

$$Y_i = y(x_{n-1} + hc_i) + O(h^{q+1}),$$

for $i = 1, 2, \ldots, s$; and

(iv) V has low rank.

69

Amongst the still large family of DIMSIMs, we will confine our attention in this paper to methods in which $A = 0$, and to those in which A is a diagonal matrix with positive diagonals. These are known as type 3 and type 4 DIMSIMs respectively. They are selected for special consideration because the stages can be evaluated in parallel. Type 3 methods are intended for the solution of non-stiff and type 4 for stiff problems.

3 THE FUNCTION ϕ

It is a simple matter, by comparing coefficients, to verify that ϕ and ϕ^{-1} have the following series expansions

$$\phi(Z) = 1 + Z + \frac{3}{2}Z^2 + \frac{8}{3}Z^3 + \frac{125}{24}Z^4 + \frac{54}{5}Z^5 + \frac{16807}{720}Z^6 + O\left(Z^7\right),$$

$$\phi(Z)^{-1} = 1 - Z - \frac{1}{2}Z^2 - \frac{2}{3}Z^3 - \frac{9}{8}Z^4 - \frac{32}{15}Z^5 - \frac{625}{144}Z^6 + O\left(Z^7\right),$$

suggesting the following result.

Theorem 3.1 *For $|Z| < e^{-1}$, $\phi(Z)$ and $\phi(Z)^{-1}$ are given by*

$$\phi(Z) = 1 + \sum_{k=1}^{\infty} \frac{(k+1)^{k-1}}{k!} Z^k$$

and

$$\phi(Z)^{-1} = 1 - Z - \sum_{k=2}^{\infty} \frac{(k-1)^{k-1}}{k!} Z^k.$$

This result is proved in [3], making use of the Lagrange expansion [6] for the solution to the functional equation $z = Z\exp(z)$, where we require expansions for both z/Z and $\exp(-z)$.

The application of this function is to the numerical solution of non-stiff ordinary differential equation systems where all the stages of the method are to be evaluated in parallel. The methods in question are type 3 DIMSIMs and are characterized by a matrix of coefficients of the form

$$\begin{bmatrix} 0 & I \\ B & V \end{bmatrix},$$

where $p = q = r = s$, with p the order of the method, q the stage order, r the number of vectors passed from step to step, and s the number of stages. The stability polynomial for such a method takes the form

$$\begin{aligned} P\left(w, z\right) &= \det\left(wI - V - zB\right) \\ &= w^r - w^{r-1}\left(\alpha_0 - z\beta_1\right) - w^{r-2}\left(z\alpha_1 - z^2\beta_2\right) - \cdots - \left(z^{r-1}\alpha_{r-1} - z^r\beta_r\right) \end{aligned} \quad (3.1)$$

if V is assumed to be of rank 1. Because of (3.1) the order of the method is r, $P\left(\exp(z),z\right)=O\left(z^{r+1}\right)$ and this implies that

$$\exp(z)=\frac{\alpha_0+\alpha_1(z\exp(-z))+\cdots+\alpha_{r-1}(z\exp(-z))^{r-1}}{1+\beta_1(z\exp(-z))+\cdots+\beta_r(z\exp(-z))^r}+O\left(z^{r+1}\right),$$

so that

$$\frac{\alpha_0+\alpha_1 Z+\cdots+\alpha_{r-1}Z^{r-1}}{1+\beta_1 Z+\cdots+\beta_r Z^r}$$

is a rational approximation of order r to ϕ defined by (1.1). To simplify the numerical method as much as possible and to lower the dependence of each step on remote steps in the past, the choice of Padé approximations close to the diagonal are suggested. For example, in the case $p=q=5$, the $(3,2)$ Padé approximation is found to be

$$\frac{1-\frac{228}{85}Z+\frac{451}{340}Z^2}{1-\frac{313}{85}Z+\frac{1193}{340}Z^2-\frac{133}{204}Z^3}. \tag{3.2}$$

4   A SPECIAL TYPE 3 METHOD

Using the approach discussed in [2], a DIMSIM can be derived with stability determined by the rational approximation (3.2). It is characterized by the matrices

$$B=\begin{bmatrix} \frac{109}{5100} & -\frac{554}{1275} & -\frac{468}{425} & \frac{874}{1275} & -\frac{601}{5100} \\ -\frac{307}{61200} & \frac{1409}{30600} & -\frac{1193}{2550} & \frac{24121}{30600} & -\frac{6277}{61200} \\ \frac{157}{15300} & -\frac{217}{3825} & \frac{211}{1275} & -\frac{1177}{3825} & -\frac{1973}{15300} \\ -\frac{329}{20400} & \frac{923}{10200} & -\frac{171}{850} & \frac{6013}{10200} & \frac{4481}{20400} \\ \frac{5087}{15300} & -\frac{6422}{3825} & \frac{4376}{1275} & -\frac{12482}{3825} & \frac{43757}{15300} \end{bmatrix},$$

$$V=\begin{bmatrix} 0 & \frac{1}{255} & \frac{7}{170} & \frac{10}{17} & \frac{11}{30} \\ 0 & \frac{1}{255} & \frac{7}{170} & \frac{10}{17} & \frac{11}{30} \\ 0 & \frac{1}{255} & \frac{7}{170} & \frac{10}{17} & \frac{11}{30} \\ 0 & \frac{1}{255} & \frac{7}{170} & \frac{10}{17} & \frac{11}{30} \\ 0 & \frac{1}{255} & \frac{7}{170} & \frac{10}{17} & \frac{11}{30} \end{bmatrix}.$$

This method, which has abscissae at $[-4,-3,-2,-1,0]$, can be interpreted as a scheme in which step values are repeatedly reevaluated as in a predictor-corrector scheme, but where the reevaluations are carried out in parallel with predictions and initial corrections in later steps. A preliminary analysis of the new method suggests that it does take advantage of parallelism to improve on the performance available from a traditional Adams-Bashforth-Moulton PECE scheme. Its error constant is only $-13489/122400\approx0.110204$ compared with $-3/160$ for the Moulton corrector of the same order. Because the PECE method carries out two function evaluations per step, the Moulton error constant should be multiplied by 32 for a fair comparison. This gives the value -0.6.

Error estimates for this method are very convenient because in step number n, the values imported from the previous steps are approximations to $y(x_n - 4h)$, $y(x_n - 3h)$, $y(x_n - 2h)$, $y(x_n - h)$ and $y(x_n)$. The last four of these are equal to the first four of the approximations exported from the current step for the subsequent step. A detailed analysis shows that

$$
\begin{aligned}
y_1^{[n]} - y_2^{[n-1]} &= -\frac{601}{5100} y^{(6)}(x_n) h^6 + O\left(h^7\right), \\
y_2^{[n]} - y_3^{[n-1]} &= -\frac{6277}{61200} y^{(6)}(x_n) h^6 + O\left(h^7\right), \\
y_3^{[n]} - y_4^{[n-1]} &= -\frac{1973}{15300} y^{(6)}(x_n) h^6 + O\left(h^7\right), \\
y_4^{[n]} - y_5^{[n-1]} &= \frac{4481}{20400} y^{(6)}(x_n) h^6 + O\left(h^7\right),
\end{aligned}
$$

and any of these may be made use of to estimate the local truncation error in the step of $-13489/122400 y^{(6)}(x_n) h^6 + O\left(h^7\right)$.

Interpolation for dense output, or for the computation of retarded values in a functional differential equation, is also straightforward. One method suggested for this, is to compute a Nordsieck vector evaluated at x_n. These approximations are given by

$$
\bar{y}^{[n]} = \overline{B} h F + \overline{V} y^{[n-1]} \approx
\begin{bmatrix}
y(x_n) \\
h y'(x_n) \\
h^2 y''(x_n) \\
\vdots \\
h^5 y^{(5)}(x_n)
\end{bmatrix},
$$

where

$$
\overline{B} =
\begin{bmatrix}
-\frac{329}{20400} & \frac{923}{10200} & -\frac{171}{850} & \frac{6013}{10200} & \frac{4481}{20400} \\
0 & 0 & 0 & 0 & 1 \\
\frac{1}{4} & -\frac{4}{3} & 3 & -4 & \frac{25}{12} \\
\frac{11}{12} & -\frac{14}{3} & \frac{19}{2} & -\frac{26}{3} & \frac{35}{12} \\
\frac{3}{2} & -7 & 12 & -9 & \frac{5}{2} \\
1 & -4 & 6 & -4 & 1
\end{bmatrix},
\quad
\overline{V} =
\begin{bmatrix}
0 & \frac{1}{255} & \frac{7}{170} & \frac{10}{17} & \frac{11}{30} \\
0 & 0 & 0 & 0 & 0 \\
0 & 0 & 0 & 0 & 0 \\
0 & 0 & 0 & 0 & 0 \\
0 & 0 & 0 & 0 & 0 \\
0 & 0 & 0 & 0 & 0
\end{bmatrix}.
$$

In addition to using the Nordsieck vector for interpolation purposes, it is also possible to use these approximations to adjust for change of stepsize. Suppose the step is to be changed from h to ρh. Using the standard Nordsieck approach, we multiply $\bar{y}^{[n]}$ by the matrix

$$
D = \operatorname{diag}\left(1, \rho, \rho^2, \ldots, \rho^5\right).
$$

The vector $y^{[n]}$, appropriate to the altered stepsize is then formed from the new $\overline{y}^{[n]}$ by multiplying by a matrix W with (i,j) element equal to

$$w_{ij} = \frac{(i-4)^{j-1}}{(j-1)!}.$$

That is,

$$W = \begin{bmatrix} 1 & -3 & \frac{9}{2} & -\frac{9}{2} & \frac{27}{8} & -\frac{81}{40} \\ 1 & -2 & 2 & -\frac{4}{3} & \frac{2}{3} & -\frac{4}{15} \\ 1 & -1 & \frac{1}{2} & -\frac{1}{6} & \frac{1}{24} & -\frac{1}{120} \\ 1 & 0 & 0 & 0 & 0 & 0 \\ 1 & 1 & \frac{1}{2} & \frac{1}{6} & \frac{1}{24} & \frac{1}{120} \end{bmatrix}.$$

Thus

$$y^{[n]} = WD(\rho)\overline{y}^{[n]},$$

where it can be seen that for $\rho = 1$, the same value of $y^{[n]}$ is found as would have been given by

$$y^{[n]} = BhF + Vy^{[n-1]},$$

because $B = W\overline{B}$ and $V = W\overline{V}$.

An alternative procedure, of evaluating the Nordsieck vector at $x_n + h$, is not proposed because it would use an approximation to $y(x)$ at a more advanced value of x than has been checked by an error estimate involving $hf(y(x))$. Such a procedure would not be robust against possible discontinuous behavior.

5   THE FUNCTION ψ

From a comparison of the first few terms of the Taylor expansions of ψ defined by (1.2) and ψ^{-1}, the general form of the terms can be ascertained. They are given in the following.

Theorem 5.1

$$\psi(Z) = 1 + \sum_{k=1}^{\infty} (-1)^{k+1} \frac{L'_{k+1}((k+1)t)}{k+1} Z^k,$$

$$\psi(Z)^{-1} = 1 + (1-t)Z + \sum_{k=2}^{\infty} (-1)^{k+1} t \frac{L'_k((k-1)t)}{k} Z^k.$$

The proof of this result will be given in a later paper [4]. As for theorem 3.1, it makes use of the Lagrange expansion [6] but in this case for the functional equation $z = Z(1-z)\exp(tz)$, where the expansions for z/Z and $\exp(-tz)/(1-z)$ are required. In [4] a special study will be made of the top row of the Padé table for which the associated methods have remarkable stability (for quite high values of r, they are A-stable and strongly stable at ∞, for a suitably chosen value of t).

73

However, in this paper we focus our attention on the second row of the Padé table The (2,3) Padé approximation to ψ is given by

$$\frac{1 - \frac{2}{3}Z - \frac{1}{2}Z^2}{1 - \frac{2}{3}Z} \tag{5.1}$$

and this leads to the stability polynomial

$$
\begin{aligned}
P(w, z) &= \det(w(1 - z)I - (1 - z)V - zB) \\
&= w^3(1 - z)^3 - (1 + \frac{2}{3}z)w^2(1 - z)^2 + \frac{2}{3}zw(1 - z) + \frac{1}{2}z^2.
\end{aligned}
$$

A-stability of a method with this stability function is equivalent to the statement that there do not exist complex numbers w and z such that (i) $P(w, z) = 0$, (ii) $\operatorname{Re} z \leq 0$ and (iii) $|w| > 1$. By the maximum modulus principle, we can replace (ii) by $\operatorname{Re} z = 0$.

To prove that the method would actually be A-stable, we can use a recursive argument based on the Schur criterion for a polynomial having all its zeros in the closed unit disk. This criterion is based on the observation that if the polynomial

$$g_n(w) = \alpha_0 w^n + \alpha_1 w^{n-1} + \cdots + \alpha_n$$

has the property that $|\alpha_0| > |\alpha_n|$ then all zeros of g_n are in the open unit disk if and only if the same is true for the polynomial

$$g_{n-1}(w) = \beta_0 w^{n-1} + \beta_1 w^{n-2} + \cdots + \beta_{n-1},$$

where

$$[\beta_0, \beta_1, \ldots, \beta_{n-1}] = \bar{\alpha}_0[\alpha_0, \alpha_1, \ldots, \alpha_{n-1}] - \alpha_n[\bar{\alpha}_n, \bar{\alpha}_{n-1}, \ldots, \bar{\alpha}_1].$$

Using this criterion leads to a sequence of polynomials in w and y of increasingly low order in w until the following condition is arrived at. For all $y > 0$,

$$
\begin{aligned}
y^4(4 \;&+\; 12y^2 + 11y^4 + 4y^6)(6192 + 44624y^2 + 129540y^4 \\
&+\; 194244y^6 + 162299y^8 + 77244y^{10} + 20880y^{12} + 2880y^{14}) > 0.
\end{aligned}
$$

This is a sufficient condition for A-stability, because if any of the intermediate Schur polynomials had changed sign for some $y > 0$, the same would have been true for this poynomial.

6 AN A-STABLE TYPE 4 METHOD

After some manipulation a type 4 DIMSIM can be derived with the stability given by the Padé approximation (5.1).

$$B = \begin{bmatrix} -\frac{29}{36} & \frac{19}{9} & -\frac{23}{36} \\ -\frac{8}{9} & \frac{34}{9} & -\frac{11}{9} \\ -\frac{53}{36} & \frac{49}{9} & -\frac{47}{36} \end{bmatrix}, \qquad V = \begin{bmatrix} -\frac{1}{2} & \frac{8}{3} & -\frac{7}{6} \\ -\frac{1}{2} & \frac{8}{3} & -\frac{7}{6} \\ -\frac{1}{2} & \frac{8}{3} & -\frac{7}{6} \end{bmatrix}.$$

The abscissae for this method are $-2, -1, 0$. Just as for the type 3 method discussed in §4, error estimates are possible by comparing the first two stage values with the second and third stage values found in the previous step. Write $Y_1^{[n]}$, $Y_2^{[n]}$ and $Y_3^{[n]}$ as the stage values computed in step number n. It is then found that

$$Y_1^{[n]} - Y_2^{[n-1]} = -\frac{23}{36} y^{(4)}(x_n) h^4 + O\left(h^5\right),$$

$$Y_2^{[n]} - Y_3^{[n-1]} = -\frac{11}{9} y^{(4)}(x_n) h^4 + O\left(h^5\right),$$

and either of these may be made use of to estimate the local truncation error in the step of $-\frac{43}{72} y^{(4)}(x_n) h^4 + O\left(h^5\right)$.

As for the type 3 method introduced in §4, we can compute a Nordsieck vector

$$\overline{y}^{[n]} = \overline{B} h F + \overline{V} y^{[n-1]} \approx \begin{bmatrix} y(x_n) \\ h y'(x_n) \\ h^2 y''(x_n) \\ h^3 y^{(3)}(x_n) \end{bmatrix},$$

where

$$\overline{B} = \begin{bmatrix} -\frac{8}{9} & \frac{34}{9} & -\frac{2}{9} \\ 0 & 0 & 1 \\ \frac{1}{2} & -2 & \frac{3}{2} \\ 1 & -2 & 1 \end{bmatrix}, \qquad \overline{V} = \begin{bmatrix} -\frac{1}{2} & \frac{8}{3} & -\frac{7}{6} \\ 0 & 0 & 0 \\ 0 & 0 & 0 \\ 0 & 0 & 0 \end{bmatrix}.$$

The output vector $y^{[n]}$ can be recovered using $y^{[n]} = W \overline{y}^{[n]}$ for fixed stepsize, where W is now given by

$$W = \begin{bmatrix} 1 & -2 & \frac{3}{2} & -\frac{2}{3} \\ 1 & -1 & 0 & 0 \\ 1 & 0 & -\frac{1}{2} & -\frac{1}{3} \end{bmatrix}.$$

For variable stepsize we compute $y^{[n]} = W D(\rho) \overline{y}^{[n]}$, where $D = \operatorname{diag}\left(1, \rho, \rho^2, \rho^3\right)$.

Although a full stability analysis for variable stepsize is not yet available, it is at least possible to analyze stability at 0 and ∞. Let

$$R(\rho, z) = W D(\rho) \left(V + \frac{z}{1-z} B \right)$$

be the magnification matrix for the standard linear test problem assuming that a stepsize ratio of ρ has been applied after the step. What would be desirable would be to find $a < 1$ and $b > 1$, with a as small as possible and b as large as possible, and a constant C such that for any integer N, any sequence of N steps with all stepsize ratios in $[a, b]$ would yield a product of all matrices $R(\rho, 0)$ to be bounded in norm by C with a similar restriction on products of the form $R(\rho, \infty)$. Unfortunately, the best $[a, b]$ interval that can be found has quite a small width and this suggests that this method, with the step change mechanism we have described, will not respond very stably to fluctuations in stepsize. Undoubtedly, alternative step change procedures could be considered for this or similar methods which would not suffer this disadvantage to the same extent. Work on this question is continuing.

REFERENCES

[1] Butcher J.C. On the convergence of numerical solutions to ordinary differential equations. *Math. Comp.*, **20**:1–10, 1966.

[2] Butcher J.C. Diagonally-implicit multi-stage integration methods. *Appl. Numer. Math.*, **11**:347–363, 1993.

[3] Butcher J.C. General linear methods for the parallel solution of ordinary differential equations. *World Sci. Ser. Appl. Anal.*, **2**:99–111, 1993.

[4] Butcher J.C. General linear methods for the parallel solution of stiff ordinary differential equations. In preparation.

[5] Iserles A., Nørsett S.P. On the theory of parallel Runge-Kutta methods. *IMA J. Numer. Anal.*, **10**:463–488, 1990.

[6] Pólya G., Szegö G. *Aufgaben und Lehrsätze aus der Analysis I.* 1925. Revised and enlarged English translation: Problems and Theorems in Analysis, I, Springer, Berlin, 1972.

International Series of Numerical Mathematics, Vol. 119, ©1994 Birkhäuser

An Adaptive Semi-iterative Method for Symmetric Semidefinite Linear Systems

D. Calvetti*   L. Reichel†   Q. Zhang‡

Dedicated to Walter Gautschi on the occasion of his 65th birthday

Abstract. The development of semi-iterative methods for the solution of large linear nonsingular systems has received considerable attention. However, these methods generally cannot be applied to the solution of singular linear systems. We present a new adaptive semi-iterative method tailored for the solution of large sparse symmetric semidefinite linear systems. This method is a modification of Richardson iteration and requires the determination of relaxation parameters. We want to choose relaxation parameters that yield rapid convergence, and this requires knowledge of an interval $[a, b]$ on the real axis that contains most of the nonvanishing eigenvalues of the matrix. Such an interval is determined during the iterations by computing certain modified moments. Computed examples show that our adaptive iterative method typically requires a smaller number of iterations and much fewer inner product evaluations than an appropriate modification of the conjugate gradient algorithm of Hestenes and Stiefel. This makes our scheme particularly attractive to use on certain parallel computers on which the communication required for inner product evaluations constitutes a bottleneck.

1 INTRODUCTION

Large linear systems of equations

$$A\boldsymbol{x} = \boldsymbol{b}, \qquad A \in \mathbb{R}^{n \times n}, \qquad \boldsymbol{x}, \boldsymbol{b} \in \mathbb{R}^n, \tag{1.1}$$

with a symmetric and positive semidefinite matrix arise in several applications. For instance, consider the solution of the least-squares problem

$$\min_{\boldsymbol{y} \in \mathbb{R}^m} \|C\boldsymbol{y} - \boldsymbol{d}\|, \tag{1.2}$$

where $\boldsymbol{d} \in \mathbb{R}^n$ and $C \in \mathbb{R}^{n \times m}$, $n > m$, is a matrix of rank strictly smaller than m. The normal equations associated with (1.2) give rise to a symmetric semidefinite linear system. For another example, consider the underdetermined system

$$C^T \boldsymbol{z} = \boldsymbol{b}. \tag{1.3}$$

*Department of Pure and Applied Mathematics, Stevens Institute of Technology, Hoboken, NJ 07030. E-mail: calvetti@mathrs1.stevens-tech.edu

†Department of Mathematics and Computer Science, Kent State University, Kent, OH 44242. Research supported in part by NSF grant DMS-9205531. E-mail: reichel@mcs.kent.edu

‡Department of Mathematics and Computer Science, Kent State University, Kent, OH 44242. Research supported in part by NSF grant DMS-9205531. E-mail: qzhang@mcs.kent.edu

It can be solved by first computing the solution x^* of the symmetric semidefinite, possibly inconsistent, linear system $C^T C x = b$. The solution $z^* \in \mathbb{R}^m$ of (1.3) can then be determined from $z^* = C x^*$. Other applications of symmetric semidefinite linear systems arise in the modeling of Markov chains and the solution of boundary value problems for partial differential equations with Neumann or periodic boundary conditions; see, e.g., [2, 14, 16, 18].

Semi-iterative methods designed for the solution of positive definite linear systems of equations are not well suited for the solution of semidefinite linear systems. This can be seen as follows: Let x_0 be an initial approximate solution of (1.1). The iterates x_k generated by a semi-iterative method can be written as

$$x_k = p_k(A)x_0 + q_{k-1}(A)b, \tag{1.4}$$

where p_k is a polynomial of degree k defined by the method such that $p_k(0) = 1$, and

$$q_{k-1}(\lambda) := \lambda^{-1}(1 - p_k(\lambda)). \tag{1.5}$$

Define the residual vectors

$$r_k := b - A x_k, \qquad k \geq 0.$$

It follows from (1.4) and (1.5) that

$$r_k = p_k(A) r_0, \qquad k \geq 0. \tag{1.6}$$

Because of the relation (1.6) the polynomials p_k are often referred to as residual polynomials; the polynomials q_{k-1} given by (1.5) are referred to as auxiliary polynomials.

Let $\mathcal{R}(A)$ and $\mathcal{N}(A)$ denote the range and null space of A, respectively. Since A is symmetric, $\mathcal{N}(A) = \mathcal{R}(A)^\perp$. Decompose the right-hand side of (1.1) as follows:

$$b = b_\mathcal{R} + b_\mathcal{N}, \qquad b_\mathcal{R} \in \mathcal{R}(A), \quad b_\mathcal{N} \in \mathcal{N}(A).$$

The consistent system

$$A x = b_\mathcal{R} \tag{1.7}$$

has many solutions when A is singular, and we denote by x^* the unique solution that satisfies $x^* - x_0 \in \mathcal{R}(A)$. In particular, if A is singular, then the solution x^* depends on the choice of initial vector x_0. Note that x^* is a least-squares solution of (1.1), i.e.,

$$\|A x^* - b\| = \min_{x \in \mathbb{R}^n} \|A x - b\| = \|b_\mathcal{N}\|. \tag{1.8}$$

Throughout this paper $\| \cdot \|$ denotes the Euclidean vector norm on $\mathbb{R}^n$ or the induced matrix norm on $\mathbb{R}^{n \times n}$. If $x_0 \in \mathcal{R}(A)$, then $x^* \in \mathcal{R}(A)$, and therefore x^* is the unique solution of (1.7) and (1.8) of minimal norm. This solution is generally of interest in applications, and it can be obtained, e.g., by letting $x_0 := 0$.

In view of (1.4) the error

$$e_k := x^* - x_k \tag{1.9}$$

satisfies

$$
\begin{aligned}
e_k &= x^* - p_k(A)x_0 - q_{k-1}(A)(Ax^* + b_\mathcal{N}) \\
&= p_k(A)e_0 - q_{k-1}(0)b_\mathcal{N}
\end{aligned}
$$

and, in particular,

$$\|e_k\|^2 = \|p_k(A)e_0\|^2 + \|q_{k-1}(0)b_\mathcal{N}\|^2. \tag{1.10}$$

It follows from (1.10) that if $b_\mathcal{N} \neq 0$, i.e., if the system (1.1) is inconsistent, then x_k converges to the solution x^* as k increases for any choice of x_0 if and only if the the polynomials p_k and q_{k-1} satisfy

$$\lim_{k \to \infty} p_k(A)v = 0, \qquad \forall\, v \in \mathcal{R}(A), \tag{1.11}$$

and

$$\lim_{k \to \infty} q_{k-1}(0) = 0. \tag{1.12}$$

For future reference, we note that it follows from (1.5) that

$$q_{k-1}(0) = -p_k'(0). \tag{1.13}$$

Semi-iterative methods designed for the solution of nonsingular linear systems are based on the use of residual polynomials that satisfy (1.11), but not (1.12); see, e.g., [8, 19]. It follows from (1.10) that application of these methods to the solution of singular inconsistent linear systems yields iterates x_k that do not converge as k increases. Computed examples showing that the conjugate gradient method does not yield convergent iterates when applied to inconsistent positive semidefinite linear systems are presented in [14].

Semidefinite consistent linear systems can be solved by iterative methods that violate (1.12) provided that the influence on the computed solution of roundoff errors can be ignored. In view of that the effects of roundoff errors often cannot be neglected, also semidefinite consistent systems should generally be solved by iterative methods that satisfy both (1.11) and (1.12).

The problems encountered when trying to apply iterative methods designed for the solution of nonsingular linear systems to the solution of semidefinite systems has spurred the development of parameter dependent semi-iterative methods that satisfy both conditions (1.11) and (1.12), and therefore are well suited for the solution of semidefinite linear systems; see, e.g., [7, 9, 13, 20]. These methods require that a region in the complex plane, e.g., an interval, that contains most of the nonvanishing eigenvalues of A, be explicitly known in order to determine suitable relaxation parameters.

The present paper describes an adaptive semi-iterative method that does not require a priori knowledge of the spectrum. The method is a modification of Richardson iteration and can be written as

$$x_{k+1} \quad := \quad x_k - \tau_k \delta_{k-1} A\tilde{r}_{k-1}, \qquad k = 1, 2, \dots , \tag{1.14}$$

$$x_1 \quad := \quad x_0 := 0, \tag{1.15}$$

where

$$\tilde{r}_{k-1} \quad := \quad \tilde{r}_{k-2} - \delta_{k-2} A\tilde{r}_{k-2}, \qquad k = 2, 3, \dots , \tag{1.16}$$

$$\tilde{r}_0 \quad := \quad r_0. \tag{1.17}$$

We refer to the $\delta_{k-1} \in \mathbb{R}$ as relaxation parameters, and to the $\tilde{r}_{k-1}$ as pseudoresidual vectors. They are the residual vectors of a Richardson iteration method closely related to the iterations (1.14). The $\tilde{r}_{k-1}$ generally do not converge to zero as k increases. The τ_k are auxiliary parameters that satisfy

$$\tau_k := -\sum_{j=0}^{k-1} \delta_j, \qquad k = 1, 2, \dots . \tag{1.18}$$

The residual polynomials associated with the method (1.14)-(1.15) are given by

$$p_{k+1}(\lambda) \quad = \quad (1 - \tau_k \lambda) \prod_{j=0}^{k-1} (1 - \delta_j \lambda), \qquad k = 1, 2, \dots , \tag{1.19}$$

$$p_1(\lambda) \quad = \quad p_0(\lambda) = 1. \tag{1.20}$$

It is easy to verify that $p'_k(0) = 0$ for all $k \geq 0$, which in view of (1.13) secures that the condition (1.12) is satisfied. We wish to determine the relaxation parameters δ_{k-1} so that the error vectors e_k given by (1.9) converge rapidly to zero, or equivalently, so that the convergence in (1.11) is rapid. Our adaptive iterative method chooses the relaxation parameters δ_{k-1} to be reciprocal values of Leja points (defined below) for intervals $[a_j, b_j]$ on the positive real axis. We want these intervals to contain (most of) the nonvanishing eigenvalues of A. The intervals are determined during the iterations in the following manner: we compute certain modified moments during the iterations and use the modified moments and relaxation parameters as input to the modified Chebyshev algorithm. This algorithm yields a small $m \times m$ symmetric tridiagonal matrix $\hat{T}_m$, whose eigenvalues approximate some eigenvalues of A. Associated with $2m$ modified moments is an m-point Gaussian quadrature rule. The nodes of this rule are given by the eigenvalues of $\hat{T}_m$ and the weights are given by the squares of the first components of the normalized eigenvectors of $\hat{T}_m$. The extreme eigenvalues of $\hat{T}_m$ whose associated Gaussian weights are not "tiny" are used to update the interval $[a_j, b_j]$.

Eigenvalues of $\hat{T}_m$ associated with tiny weights are ignored, because these eigenvalues are highly sensitive to perturbations in the modified moments, e.g., caused by round-off errors.

Modified moments were first used by Golub and Kent [11] in the context of an adaptive iterative method; they apply modified moments to determine a suitable interval for the Chebyshev iteration method. An application of modified moments to adaptive Richardson iteration is described in [3]. These methods as well as the one of the present paper rely on the modified Chebyshev algorithm. Numerical properties of the modified Chebyshev algorithm have been studied extensively by Gautschi [10].

Section 2 discusses the convergence of our iterative scheme under the assumption that the interval $[a_j, b_j]$ contains the nonvanishing eigenvalues of A. The computations of modified moments and of tridiagonal matrices $\hat{T}_m$ that yield estimates of eigenvalues of the matrix A are discussed in §3. An algorithm for our adaptive modified Richardson iteration method is presented in §4, and §5 contains computed examples that compare our algorithm with a modification of the conjugate gradient algorithm by Hestenes and Stiefel. Concluding remarks can be found in §6.

2 CONVERGENCE

Let $\mathbb{I} := [a, b]$ be an interval on the positive real axis that contains all nonvanishing eigenvalues of A. In this section we define Leja points and apply their properties to show that the iterative method (1.14)-(1.15), with the relaxation parameters δ_{k-1} chosen to be reciprocal values of Leja points for the interval $\mathbb{I}$, yields iterates x_{k+1} that converge with an asymptotically optimal rate of convergence with respect to $\mathbb{I}$. Leja points for fairly general compact sets in the complex plane $\mathbb{C}$ were introduced by Edrei [6] and Leja [15]. Leja points for the interval $\mathbb{I} = [a, b]$ are defined as follows: Introduce the weight function $\omega(\lambda) = \lambda$, and let

$$z_0 := b. \tag{2.1}$$

Choose the point z_k so that

$$\prod_{j=0}^{k-1} |z_k - z_j| \omega(z_k) = \max_{z \in \mathbb{I}} \prod_{j=0}^{k-1} |z - z_j| \omega(z), \qquad z_k \in \mathbb{I}, \qquad k = 1, 2, \ldots . \tag{2.2}$$

Equation (2.2) might not determine the point z_k uniquely. We call any sequence of points $z_0, z_1, \ldots$, which satisfy (2.1)-(2.2) a sequence of Leja points for $\mathbb{I}$, and we briefly refer to the points z_k in such a sequence as Leja points for $\mathbb{I}$. Edrei [6] and Leja [15] studied these points when the weight function in (2.2) is $\omega(\lambda) = 1$. The asymptotic properties of the Leja points is the same for the weight functions $\omega(\lambda) = \lambda$ and $\omega(\lambda) = 1$. Computed examples in §6 show the former weight function to yield a better choice of the first few relaxation parameters $\delta_k := 1/z_k$.

It follows from results in [15] that the Leja points for $[a, b]$ are uniformly distributed on $[a, b]$ with respect to the density function

$$d\sigma(\lambda) := \frac{1}{\pi}(\lambda - a)^{-1/2}(b - \lambda)^{-1/2}. \tag{2.3}$$

We note in passing that the zeros of Chebyshev polynomials for the interval $[a, b]$ are also uniformly distributed on $[a, b]$ with respect to the density function (2.3).

We now consider the convergence of the iterative method (1.14)-(1.15) with parameters $\delta_k := 1/z_k$, when the z_k are Leja points for $\mathbb{I} = [a, b]$. First we introduce some notation. Let $\boldsymbol{\delta} := \{\delta_k\}_{k=0}^{\infty}$ and define the asymptotic convergence factor

$$\kappa(\boldsymbol{\delta}) := \varlimsup_{k \to \infty} \sup_{e_0 \in \mathcal{R}(A)} \left(\frac{\|e_k\|}{\|e_0\|} \right)^{1/k}, \tag{2.4}$$

where the error vectors e_k are given by (1.9). Introduce the spectral factorization

$$A = Q\Lambda Q^T, \tag{2.5}$$

where

$$
\begin{aligned}
\Lambda &= \operatorname{diag}[\lambda_1, \lambda_2, \dots \lambda_n], & 0 = \lambda_1 = \dots = \lambda_{\mu-1} < \lambda_\mu \leq \lambda_{\mu+1} \leq \dots \leq \lambda_n, \\
Q &= [q_1, q_2, \dots, q_n], & Q^T Q = I.
\end{aligned}
$$

It follows from (1.10), and the fact that $q_{k-1}(0) = 0$ for $k \geq 1$, that

$$\|e_k\| = \|p_k(A)e_0\| = \|p_k(\Lambda)Q^T e_0\|.$$

Substituting this expression into (2.4) yields

$$\kappa(\boldsymbol{\delta}) = \varlimsup_{k \to \infty} \max_{\lambda \in \lambda_+(A)} |p_k(\lambda)|^{1/k},$$

where $\lambda_+(A)$ denotes the set of nonvanishing eigenvalues of A. In general, $\lambda_+(A)$ is not explicitly known, but assume that an interval $\mathbb{I} := [a, b]$ which contains $\lambda_+(A)$ is explicitly known, and introduce the asymptotic convergence factor with respect to $\mathbb{I}$,

$$\kappa(\boldsymbol{\delta}, \mathbb{I}) := \varlimsup_{k \to \infty} \max_{\lambda \in \mathbb{I}} |p_k(\lambda)|^{1/k} \geq \kappa(\boldsymbol{\delta}).$$

The following lemma shows that $\kappa(\boldsymbol{\delta}, \mathbb{I})$ is minimal when the relaxation parameters δ_k are chosen as reciprocal values of Leja points for $\mathbb{I}$. In the proof of the lemma we use the polynomials

$$\tilde{p}_k(\lambda) := \prod_{j=0}^{k-1}(1 - \delta_j \lambda), \qquad k = 1, 2, \dots, \tag{2.6}$$

$$\tilde{p}_0(\lambda) := 1, \tag{2.7}$$

which we refer to as pseudoresidual polynomials because they satisfy

$$\tilde{r}_k = \tilde{p}_k(A)\tilde{r}_0. \tag{2.8}$$

Lemma 2.1 (Convergence) *Let $\lambda_+(A) \subset \mathbb{I} = [a, b]$ with $0 < a < b < \infty$. Assume that the iteration parameters δ_k are reciprocal values of Leja points z_k for $\mathbb{I}$. Then*

$$\kappa(\boldsymbol{\delta}, \mathbb{I}) = \inf_{\boldsymbol{\eta}} \kappa(\boldsymbol{\eta}, \mathbb{I}) = \rho(a, b) := \frac{\mu}{1 + \sqrt{1 - \mu^2}}, \tag{2.9}$$

where $\mu = \frac{b-a}{b+a}$. In particular, if $\mathbb{I}'$ is a subinterval of $\mathbb{I}$, then

$$\inf_{\boldsymbol{\eta}} \kappa(\boldsymbol{\eta}, \mathbb{I}') \leq \inf_{\boldsymbol{\eta}} \kappa(\boldsymbol{\eta}, \mathbb{I}). \tag{2.10}$$

Proof: It follows from

$$p_k(\lambda) = (1 - \tau_{k-1}\lambda)\tilde{p}_{k-1}(\lambda), \qquad k \geq 2,$$

that

$$\varlimsup_{k\to\infty} \max_{\lambda\in\mathbb{I}} |p_k(\lambda)|^{1/k} \leq \varlimsup_{k\to\infty} \max_{\lambda\in\mathbb{I}} |\tilde{p}_k(\lambda)|^{1/k}. \tag{2.11}$$

Conversely,

$$\min_{\boldsymbol{\delta}} \max_{\lambda\in\mathbb{I}} |p_k(\lambda)| \geq \min_{\boldsymbol{\delta}} \max_{\lambda\in\mathbb{I}} |\tilde{p}_k(\lambda)|.$$

Thus,

$$\varlimsup_{k\to\infty} \max_{\lambda\in\mathbb{I}} |p_k(\lambda)|^{1/k} = \varlimsup_{k\to\infty} \max_{\lambda\in\mathbb{I}} |\tilde{p}_k(\lambda)|^{1/k}. \tag{2.12}$$

The pseudoresidual polynomials $\tilde{p}_k$ are residual polynomials for Richardson iteration and it is well known that

$$\varlimsup_{k\to\infty} \max_{\lambda\in\mathbb{I}} |\tilde{p}_k(\lambda)|^{1/k} = \rho(a, b),$$

see, e.g., [3, 8, 17] for details. Inequality (2.10) follows from the observation that $\frac{\partial\rho}{\partial a} < 0$ and $\frac{\partial\rho}{\partial b} > 0$. $\blacksquare$

The smallest interval allowed in Lemma 2.1 is $[a, b] = [\lambda_\mu, \lambda_n]$. In view of (2.10), we call this interval the asymptotically optimal interval for A.

3 MODIFIED MOMENTS

In this section we define modified moments and described how they can be used to determine estimates of the extreme eigenvalues of A while computing the iterates x_k by (1.14)-(1.15). Our formulas are analogous with those derived by Golub and Kent [11] for obtaining eigenvalue estimates while computing iterates by the Chebyshev iteration method. Let A have spectral factorization (2.5) and let $\tilde{p}_k$ be the pseudoresidual polynomials defined by (2.6)-(2.7). Express $\tilde{r}_0$, given by (1.17), in the basis of eigenvectors

$$\tilde{r}_0 = \sum_{j=1}^{n} w_j q_j. \tag{3.1}$$

Then it follows from (2.5) that

$$\tilde{r}_k = \sum_{j=1}^{n} w_j \tilde{p}_k(\lambda_j) q_j.$$

Introduce the density function

$$d\sigma_n := \sum_{j=\mu}^{n} w_j^2 \lambda_j \delta(\lambda - \lambda_j), \tag{3.2}$$

where the λ_j are the eigenvalues of A, the w_j are the coefficients defined by (3.1) and $\delta(\lambda)$ is the Dirac δ-function. We refer to $d\sigma_n$ as the spectral density function for the restriction of A to $\mathcal{R}(A)\backslash\{0\}$ associated with $\tilde{r}_0$. Define the inner product

$$(\tilde{r}_k, \tilde{r}_\ell) := \tilde{r}_k^T A \tilde{r}_\ell = \sum_{j=\mu}^{n} w_j^2 \lambda_j \tilde{p}_k(\lambda_j) \tilde{p}_\ell(\lambda_j) = \int_{\mathbb{R}} \tilde{p}_k(\lambda) \tilde{p}_\ell(\lambda) d\sigma_n.$$

From the vectors $\tilde{r}_k$ and $\tilde{r}_0$ one can compute modified moments for $d\sigma_n$ with respect to the pseudoresidual polynomials $\tilde{p}_k$ by

$$\nu_k := \int_{\mathbb{R}} \tilde{p}_k(\lambda) d\sigma_n = (\tilde{r}_k, \tilde{r}_0) \tag{3.3}$$

without explicitly knowing $d\sigma_n$. Assume that we have computed the $2m$ modified moments $\{\nu_k\}_{k=0}^{2m-1}$. We can apply the modified Chebyshev algorithm to determine recursion coefficients α_k and $\beta_k \geq 0$ for the first m monic orthogonal polynomials with respect to $d\sigma_n$ from the modified moment $\{\nu_k\}_{k=0}^{2m-1}$ and the relaxation parameters $\{\delta_k\}_{k=0}^{2m-1}$, which also are recursion coefficients for the pseudoresidual polynomials (2.6)

$$\tilde{p}_k(\lambda) = (1 - \delta_{k-1}\lambda)\tilde{p}_{k-1}(\lambda), \qquad k \geq 1,$$

with $\tilde{p}_0$ given by (2.7). The modified Chebyshev algorithm is described, e.g., in [10, 11]. In the computed examples of §5, we have used the version given in [3, Algorithm 3.1]. We assume that m is small enough so that no invariant subspace is found. Then the modified Chebyshev algorithm [3, Algorithm 3.1] yields the tridiagonal matrix

$$T_m := \begin{bmatrix} \alpha_0 & 1 & & & & \\ \beta_1 & \alpha_1 & 1 & & & \\ & \beta_2 & \alpha_2 & 1 & & \\ & & & \ddots & & \\ & & & \ddots & \ddots & 1 \\ & & & & \beta_{m-1} & \alpha_{m-1} \end{bmatrix}, \tag{3.4}$$

whose entries are the recursion coefficients for the monic orthogonal polynomials with respect to the measure $d\sigma_n$. A diagonal similarity transformation of T_m yields a symmetric tridiagonal matrix which we denote by $\hat{T}_m$.

Lemma 3.1 *The symmetric tridiagonal $m \times m$ matrix $\hat{T}_m$ can also be determined by the Lanczos process with initial vector $A^{1/2}\tilde{r}_0$. Therefore the eigenvalues $\hat{\lambda}_1 \leq \hat{\lambda}_2 \leq \ldots \leq \hat{\lambda}_m$ of $\hat{T}_m$ satisfy*

$$\lambda_\mu \leq \hat{\lambda}_1, \qquad \hat{\lambda}_m \leq \lambda_n. \tag{3.5}$$

Proof: Let $\pi_0, \pi_1, \ldots$, be orthonormal polynomials with respect to the inner product

$$< f, g >:= \int_{\mathbb{R}} f(\lambda)g(\lambda)d\sigma_n \tag{3.6}$$

and introduce the $n \times m$ Vandermonde-like matrix

$$P := \begin{bmatrix} \pi_0(\lambda_1) & \pi_1(\lambda_1) & \ldots & \pi_{m-1}(\lambda_1) \\ \cdot & \cdot & & \cdot \\ \cdot & \cdot & & \cdot \\ \cdot & \cdot & & \cdot \\ \pi_0(\lambda_n) & \pi_1(\lambda_n) & \ldots & \pi_{m-1}(\lambda_n) \end{bmatrix},$$

as well as the vector

$$\boldsymbol{\pi}_m := [\pi_m(\lambda_1), \pi_m(\lambda_2), \ldots, \pi_m(\lambda_n)]^T.$$

Then, for some constant c_m,

$$P\hat{T}_m = \Lambda P + c_m \boldsymbol{\pi}_m e_m^T, \tag{3.7}$$

where $e_j := [0, \ldots, 0, 1, 0, \ldots, 0]^T$ denotes the jth axis vector in $\mathbb{R}^m$. Define the diagonal matrix

$$D := \mathrm{diag}(\lambda_1^{1/2}w_1, \lambda_2^{1/2}w_2, \ldots, \lambda_n^{1/2}w_n).$$

It follows from (3.7) that

$$DP\hat{T}_m = \Lambda DP + c_m D\boldsymbol{\pi}_m e_m^T$$

and therefore

$$QDP\hat{T}_m = Q\Lambda Q^T QDP + c_m QD\boldsymbol{\pi}_m e_m^T. \tag{3.8}$$

We have

$$QDPe_j = \boldsymbol{\pi}_{j-1}(A)A^{1/2}\tilde{r}_0, \qquad 1 \leq j \leq m,$$

and due to the orthonormality of the polynomials π_{j-1} with respect to the inner product (3.6), the matrix $V_m := QDP$ satisfies $V_m^T V_m = I_m$, where I_m is the

$m \times m$ identity matrix. Moreover, the columns of V_m form a basis of the Krylov subspace

$$\mathcal{K}_m(A, A^{1/2}\tilde{r}_0) := \text{span}\{A^{1/2}\tilde{r}_0, A^{3/2}\tilde{r}_0, \dots, A^{m-1/2}\tilde{r}_0\}.$$

Equality (3.8) can be written as

$$V_m \hat{T}_m = A V_m + c_m v_{m+1} e_m^T,$$

which shows that the matrix $\hat{T}_m$ can be determined by applying the Lanczos process to A with initial vector $A^{1/2}\tilde{r}_0$. Finally,

$$\hat{\lambda}_1 = \min_{x \in \mathcal{K}_m(A, A^{1/2}\tilde{r}_0)} \frac{x^T A x}{x^T x} \geq \lambda_\mu, \qquad \hat{\lambda}_m = \max_{x \in \mathcal{K}_m(A, A^{1/2}\tilde{r}_0)} \frac{x^T A x}{x^T x} \leq \lambda_n$$

shows (3.5). ∎

The eigenvalues $\{\hat{\lambda}_j\}_{j=1}^m$ of $\hat{T}_m$ are the nodes, and the squares of the first components of the eigenvectors of length $\nu_0^{1/2}$ yield the weights $\hat{w}_j^2$, of an m-point Gaussian quadrature rule associated with the density function (3.2),

$$G_m f := \sum_{j=1}^m f(\hat{\lambda}_j) \hat{w}_j^2. \tag{3.9}$$

We will use both the nodes $\hat{\lambda}_j$ and the weights $\hat{w}_j^2$ to determine intervals $[a_j, b_j]$ that contain most of the nonvanishing eigenvalues of A. Details are presented in the next section. The nodes and weights of the quadrature rule (3.9) can be computed conveniently from $\hat{T}_m$ by the Golub-Welsch [12] algorithm.

4 MODIFIED RICHARDSON ITERATION

In this section we present an algorithm for our adaptive iterative method and we discuss some computational aspects. We begin with describing how the relaxation parameters are computed.

4.1 Determination of relaxation parameters

Throughout this paper we will assume that

$$[a_0, b_0] \subset [\lambda_\mu, \lambda_n]. \tag{4.1}$$

For many matrices the choice $a_0 = b_0 = \frac{1}{n}\text{trace}(A)$ gives an interval that satisfies this requirement. The assumption (4.1) is not essential for our scheme, but it simplifies the algorithm. It is quite straightforward to modify the Algorithm 4.2

so that it determines an initial interval that satisfies (4.1). Assuming that (4.1) holds, we show how to determine new intervals such that

$$[a_l, b_l] \subset [\lambda_\mu, \lambda_n], \qquad l \geq 0. \tag{4.2}$$

Thus, assume that we know an interval $[a_j, b_j]$ that satisfies (4.2), and assume further that we already have carried out q iterations by the scheme (1.14) with relaxation parameters $\delta_0, \delta_1, \ldots, \delta_{q-1}$. We then would like to determine new relaxation parameters δ_k, $k \geq q$, as reciprocal values of Leja points for the set $\mathbb{I} = [a_j, b_j]$ in the presence of the points $z_k = 1/\delta_k$ for $0 \leq k < q$. In view of that

$$\prod_{j=0}^{k-1} (\lambda - z_j) = \tilde{p}_k(\lambda) \prod_{j=0}^{k-1} (-z_j),$$

Leja points for $\mathbb{I} = [a_j, b_j]$ can be determined by maximizing $|\tilde{p}_k(\lambda)|$ over $\mathbb{I}$. This maximum is used in our criterion for deciding when to update the interval $\mathbb{I}$, see below, and is part of the output of the following algorithm.

Algorithm 4.1 (Computation of relaxation parameters)
Input: a_j, b_j, ℓ, q, $\{z_k\}_{k=0}^{q-1}$ $(z_k = 1/\delta_k)$;
Output: δ_q, $\mu_q := \max_{a_j \leq \lambda \leq b_j} |\tilde{p}_q(\lambda)|$;
if $q = 0$ then
$\qquad z_0 := b_j;\ \delta_0 := 1/z_0;\ \mu_0 := 1;$
else
$\qquad$ *Determine $z_q \in [a_j, b_j]$, so that*

$$|\tilde{p}_q(z_q)|\omega(z_q) = \max_{z \in [a_j, b_j]} |\tilde{p}_q(z)|\omega(z);$$

$\qquad \delta_q := 1/z_q;\ \mu_q := |\tilde{p}_q(z_q)|;$
endif ;

The points $z_0, z_1, \ldots, z_{q-1}$ serve as memory of previous iterations when the new relaxation parameter δ_q is determined. The presence of this memory has the effect that relaxation parameters δ_q, determined after the interval $\mathbb{I} = [a_j, b_j]$ has been increased, are distributed so that eigenvector components in the residual error that have not been damped before will be damped more heavily than other eigenvector components until the the points $z_0, z_1, \ldots, z_q$, for some $q \geq k$, are distributed roughly according to the density function (2.3).

4.2 Updating the interval

In §3 we showed how to compute modified moments associated with the pseudoresidual polynomials $\tilde{p}_k$ during the iterations, and how these modified moments and the relaxation parameters can be used to determine a symmetric tridiagonal matrix $\hat{T}_m$. Assume for the moment that the eigenvalues $\hat{\lambda}_1 \leq \hat{\lambda}_2 \leq \ldots \leq \hat{\lambda}_m$ and Gaussian weights $\hat{w}_j^2$ associated with $\hat{T}_m$ have been computed. We will now discuss their application to updating the interval $[a_j, b_j]$. Thus, we would like to determine a new interval $[a_{j+1}, b_{j+1}]$ that gives faster convergence than $[a_j, b_j]$ and satisfies (4.2). It was shown in [3, Lemma 4.2] that Gaussian nodes $\hat{\lambda}_j$ associated with tiny Gaussian weights $\hat{w}_j^2$ are sensitive to perturbations. This result is independent of the size of the modified moment ν_0, which can be considered a scaling factor. We therefore may assume that $\nu_0 = 1$ and that the Gaussian weights are normalized to satisfy

$$\sum_{j=1}^{m} \hat{w}_j^2 = 1. \tag{4.3}$$

Computational experience from the solution of numerous problems confirms that when the Gaussian weight $\hat{w}_1^2$ associated with $\hat{\lambda}_1$ is tiny then, indeed, the computed value of $\hat{\lambda}_1$ may be contaminated by a large error and should not be used to update the interval. Similarly, when $\hat{w}_m^2$ is tiny, the computed value of $\hat{\lambda}_m$ should not be used to update the interval. This suggests the updating scheme

$$\text{if } \hat{w}_1^2 \geq \epsilon_w \text{ then } a_{j+1} \quad := \quad \min\{a_j, \hat{\lambda}_1\} \text{ else } a_{j+1} := a_j, \tag{4.4}$$

$$\text{if } \hat{w}_m^2 \geq \epsilon_w \text{ then } b_{j+1} \quad := \quad \max\{b_j, \hat{\lambda}_m\} \text{ else } b_{j+1} := b_j. \tag{4.5}$$

In view of (3.5) the updating formulas (4.4) and (4.5) give an interval $[a_{j+1}, b_{j+1}]$ that satisfies (4.2). The choice of ϵ_w is not very critical. We have found that, when the rate of convergence can be increased significantly by letting a_{j+1} be smaller than a_j, the Gaussian weight $\hat{w}_1^2$ typically is quite large. Similarly, when the rate of convergence can be increased significantly by letting b_{j+1} be larger than b_j, the weight $\hat{w}_m^2$ typically is large. In the computed examples we used $\epsilon_w = 0.1/m$. Our computational experience suggests that for reasons of numerical stability m should be chosen fairly small, e.g., $m = 5$.

4.3 Computation of modified moments

Formula (3.3) shows how to compute one modified moment in each iteration. It was shown in [3] how to determine two modified moments in each iteration by a Richardson iteration method. The results in [3] carry over to the iterative method of the present paper, and make it possible to update the intervals $[a_j, b_j]$ more frequently than if only one modified moment is computed in each iteration. The ability to update the intervals often is particularly important in the beginning of the iterations, when many of the eigenvalues of A might lie far outside the initial interval $[a_0, b_0]$ used to determine the first relaxation parameters. In iterations when modified moments are evaluated, Algorithm 4.2 below determines two modified moments from each pseudoresidual vector $\tilde{r}_k$.

We now turn to the question of when to compute modified moments. We would like to avoid determining modified moments very often, because their evaluation requires the computation of inner products, and this computation can be a bottleneck on some modern computers. We also note that the computation of the Euclidean vector norm requires the evaluation of an inner product. In order to keep the number of inner product computations small, we only evaluate the relative change $\|\Delta x_k\|/\|x_k\|$ in our stopping criterion of Algorithm 4.2 every m iterations. Here and below $\Delta x_k := x_k - x_{k-1}$.

Assume that the relaxation parameters are generated by Algorithm 4.1 as reciprocal values of Leja points for $[a_j, b_j]$. It follows from (2.8) that, for all $k \geq 0$,

$$\frac{\tilde{r}_k^T A \tilde{r}_k}{\tilde{r}_0^T A \tilde{r}_0} \leq \max_{\lambda \in \lambda_+(A)} |\tilde{p}_k(\lambda)|^2.$$

Assume that for some value of $k > 0$ the inequality

$$\frac{\tilde{r}_k^T A \tilde{r}_k}{\tilde{r}_0^T A \tilde{r}_0} > \max_{\lambda \in [a_j, b_j]} |\tilde{p}_k(\lambda)|^2 \tag{4.6}$$

holds. Then, clearly, $[a_j, b_j] \subsetneq [\lambda_\mu, \lambda_n]$, and it may be appropriate to increase the interval used to determine relaxation parameters. We use this updating criterion in Algorithm 4.2. We remark that the maximum on the right-hand side of (4.6) is evaluated when determining Leja point z_k for $[a_j, b_j]$ by Algorithm 4.1. Note that the criterion (4.6) for updating the interval is independent of the expected asymptotic rate of convergence of the iterative method. Updating criteria that depend on the expected asymptotic rate of convergence, obtained by Lemma 2.1, are more difficult to use, because it is not easy to judge whether a finite number of iterates determined are indeed the first few elements of a sequence that converges with a desired rate.

Computational experience indicates that it is not worthwhile to try to update the intervals $[a_j, b_j]$ too frequently, because then we often obtain that $[a_{j+1}, b_{j+1}] = [a_j, b_j]$. In fact, $[a_j, b_j] \subsetneq [a_{j+1}, b_{j+1}]$ only if the residual vectors determined when using Leja points in the interval $[a_j, b_j]$ are sufficiently rich in eigenvector components associated with eigenvalues of A outside $[a_j, b_j]$. We try to achieve this by keeping each interval $[a_j, b_j]$, $j \geq 1$, for at least $2m$ iterations.

4.2 Algorithm for modified Richardson iteration

The following algorithm is used in the computed examples of §5. For sake of simplicity we use the same notation as in previous sections. It is easy to see that only very few of the vectors used by the algorithm have to be stored simultaneously. Note that the inner products required in order to update the interval $[a_j, b_j]$ may already have been computed when evaluating modified moments. In our count of inner products in §5 we assume that the algorithm is implemented so as to avoid the computation of unnecessary inner products.

Algorithm 4.2 (Modified Richardson iteration)
Input: $A \in \mathbb{R}^{n \times n}$, $b \in \mathbb{R}^n$, $\epsilon, a_0, b_0 \in \mathbb{R}$, $m \in \mathbb{N}$;

Output: *Approximate solution* $x_k \in \mathbb{R}^n$;
$\tilde{r}_0 := b$; $x_1 := 0$; $s_0 := A\tilde{r}_0$;
$j := 0$; $k := 1$; $t_0 := 0$; $\epsilon_w := 0.1/m$;
while $\|\Delta x_k\|/\|x_k\| \geq \epsilon$ **do**
    if $\tilde{r}_{k-1}^T s_{k-1}/\tilde{r}_0^T s_0 \leq \max_{a_j \leq \lambda \leq b_j} |\tilde{p}_{k-1}(\lambda)|^2$ **then**
        for $l := 1, 2, \ldots, m$ **do**
            Compute δ_{k-1} *by Algorithm 4.1 for the interval* $[a_j, b_j]$;
            $t_k := t_{k-1} - \delta_{k-1}$; $\Delta x_{k+1} := t_k \delta_{k-1} s_{k-1}$;
            $x_{k+1} := x_k - \Delta x_{k+1}$; $\tilde{r}_k := \tilde{r}_{k-1} - \delta_{k-1} s_{k-1}$;
            $s_k := A\tilde{r}_k$; $k := k + 1$;
        end l;
    else
        $\tilde{\nu}_0 := \tilde{r}_{k-1}^T s_{k-1}$;
        for $l := 1, 2, \ldots, m$ **do**
            Compute δ_{k-1} *by Algorithm 4.1 for the interval* $[a_j, b_j]$;
            $\tilde{\delta}_{2l-2} := \delta_{k-1}$; $\tilde{\delta}_{2l-1} := \delta_{k-1}$;
            $t_k := t_{k-1} - \delta_{k-1}$; $\Delta x_{k+1} := t_k \delta_{k-1} s_{k-1}$;
            $x_{k+1} := x_k - \Delta x_{k+1}$; $\tilde{r}_k := \tilde{r}_{k-1} - \delta_{k-1} s_{k-1}$;
            $\tilde{\nu}_{2l-1} := \tilde{r}_k^T s_{k-1}$; $s_k := A\tilde{r}_k$;
            $\tilde{\nu}_{2l} := \tilde{r}_k^T s_k$; $k := k + 1$;
        end l;
        if $\|\Delta x_k\|/\|x_k\| \geq \epsilon$ **then**
            Use the modified moments $\{\tilde{\nu}_l\}_{l=0}^{2m-1}$ *and the parameters*
            $\{\tilde{\delta}_l\}_{l=0}^{2m-1}$ *as input to the modified Chebyshev algorithm*
            (see, e.g., [3, Algorithm 3.1]) to determine the tridiagonal
            matrix T_m *of order* m *given by (3.4)*;
            Symmetrize T_m *to obtain* $\hat{T}_m$;
            Compute the Gaussian quadrature rule associated with $\hat{T}_m$
            by the Golub-Welsch algorithm with weights normalized
            according to (4.3);
            Determine new interval endpoints a_{j+1} *and* b_{j+1} *by (4.4)*
            and (4.5);
            $j := j + 1$;
            for $l := 1, 2, \ldots, m$ **do**
                Compute δ_{k-1} *by Algorithm 4.1 for the interval* $[a_j, b_j]$;
                $t_k := t_{k-1} - \delta_{k-1}$; $\Delta x_{k+1} := t_k \delta_{k-1} s_{k-1}$;
                $x_{k+1} := x_k - \Delta x_{k+1}$; $\tilde{r}_k := \tilde{r}_{k-1} - \delta_{k-1} s_{k-1}$;
                $s_k := A\tilde{r}_k$; $k := k + 1$;
            end l;
        endif;
    endif;
endwhile;

5   COMPUTED EXAMPLES

In this section we describe the results of some numerical examples illustrating the performance of our adaptive modified Richardson algorithm for the solution of semidefinite inconsistent linear systems of equations. The computer programs used were all written in FORTRAN 77 and the numerical experiments were carried out on an IBM RISC 6000/550 workstation in double precision arithmetic, i.e., with approximately 15 significant decimal digits. Our test problems use singular matrices with well-known properties and illustrate some of the features of our adaptive modified Richardson method.

The linear systems of equations in the first two examples are obtained by discretizing the Poisson equation

$$-\Delta u = f \tag{5.1}$$

on the unit square $\Omega := \{(x, y) : 0 < x, y < 1\}$ with Neumann boundary condition $\frac{\partial}{\partial n} u = 0$. The right-hand side function in (5.1) is $f(x, y) = (x - \frac{1}{2})(y - \frac{1}{2})$. We discretize (5.1) by an $N \times N$-point grid, including the boundary points, and we approximate the Laplacian Δ by the standard five-point stencil at the interior points. In order to generate a linear system with a symmetric matrix, the normal derivative is approximated by a first order divided difference. This yields a system of $n = N^2$ linear equations for the N^2 unknowns $\{u_{i,j}\}_{i,j=1}^{N}$ which approximate the solution of the boundary value problem at the grid points (ih, jh), where $h := \frac{1}{N-1}$. Scaling each equation by h^2 gives us a linear system of equations (1.1) with an $n \times n$ symmetric positive semi-definite matrix which contains the vector $e := [1, \ldots, 1]^T$ in its null space.

We solve these linear systems of equations by our adaptive modified Richardson scheme (Algorithm 4.2). This scheme is denoted "MR" (for modified Richardson) in the tables below. In all computed examples we let $m := 5$ and $\epsilon_w := 0.02$. In addition to the number of iterations, we report the number of inner product computations required, where we also count each evaluation of the norm of a vector as an inner product evaluation. The number of times Algorithm 4.2 determines a tridiagonal matrix $\hat{T}_m$ in order to update the interval used for the computation of relaxation parameters is shown in the tables in the columns labeled "updates". In all examples for all algorithms, we let $x_0 := 0$.

For comparison, we also solve the linear systems by the conjugate gradient (CG) method. We use the ORTHODIR implementation of the CG method described by Ashby et al. [1], modified to use the initial search direction Ar_0. It can be shown that the residual polynomials p_k and auxiliary polynomials q_{k-1} associated with this implementation of the CG method satisfy both (1.11) and (1.12); see [4] for details. In this CG implementation, the search directions are generated by a three-term recurrence relation, and therefore each iteration requires the computation of three inner products and one matrix-vector multiplication. The same stopping criterion is used as in Algorithm 4.2, i.e., every $m = 5$ iterations we evaluate $\|\Delta x_k\|/\|x_k\|$ in order to determine whether this quotient is smaller

91

Table 5.1: Poisson equation with Neumann boundary condition; $n = 3600$

iterative method	initial interval	final interval	updates	iterations	inner products
CG				130	441
MR	[4.0,4.0]	[.0058,7.974]	3	125	103
MR	[.01,7.9]	[.0056,7.993]	3	110	94

Table 5.2: Poisson equation with Neumann boundary condition; $n = 6400$

iterative method	initial interval	final interval	updates	iterations	inner products
CG				165	560
MR	[4.0,4.0]	[.0032,7.978]	3	165	127
MR	[.01,7.9]	[.0032,7.993]	3	110	94
MR	[.006,7.9]	[.0032,7.994]	3	145	115

than the value of the parameter ϵ. This stopping criterion adds two inner product evaluations every five iterations to our count of inner product computations. This algorithm is denoted by "CG" in the tables.

An initial interval $[a_0, b_0]$ is required in order to start the iterations in the MR algorithm. For many linear systems of equations, a small interval containing $\frac{1}{n}\text{trace}(A)$ is a good choice of initial interval. Numerical experiments show that using modified moments it is possible to correct a "bad" choice of initial interval at the cost of a few additional iterations.

Example 5.1 We solve the Poisson equation (5.1) on a uniform grid with 60×60 grid points, i.e, the order of the matrix A is $n = 3600$. The iterations are terminated as soon as the ratio of the Euclidean norm of the correction to the approximate solution vector and the approximate solution vector satisfies $\|\Delta x_k\|/\|x_k\| < 5 \cdot 10^{-5}$. The results of numerical experiments are described in Table 5.1. Two different initial intervals were selected for the MR method. We remark that the MR method for this problem requires fewer iterations, and substantially fewer inner products, than the CG method even when the initial interval $[a_0, b_0]$ is very small.

Example 5.2 In this example we solve the Poisson equation (5.1) on a uniform grid with 80×80 grid points, i.e., we solve a symmetric linear system of equations with a matrix of order $n = 6400$. The stopping criterion and the initial intervals used for Algorithm 4.2 are the same as in Example 5.1. Numerical results in Table 5.2 show that a "bad" choice of initial interval for the MR method can be corrected at the cost of a few additional iterations.

The last four examples consider singular systems of linear equations with 4000×4000 diagonal matrices obtained from the matrix $A := \text{diag}(1, 2, \ldots, 4000)$ by setting some of its diagonal entries equal to zero. The right-hand side vector in

Table 5.3: $A = \text{diag}(0, \ldots, 0, 11, \ldots, 4000); \; n = 4000$

iterative method	initial interval	final interval	updates	iterations	inner products
CG				95	322
MR	[1000,3000]	[34.66,3999]	5	90	100
MR	[40,3900]	[40,3996]	2	80	67

Table 5.4: $A = \text{diag}(0, \ldots, 0, 11, \ldots, 4000); \; n = 4000$

iterative method	initial interval	final interval	updates	iterations	inner products
MR	[1000,3000]	[12.79,3999]	6	180	163
MR	[40,3900]	[12.90,4000]	5	180	154

these examples is $b := [1, 2, \ldots, 4000]^T$. Thus, the systems are inconsistent.

Example 5.3 We solve a linear system of equation with a 4000×4000 diagonal matrix $A := \text{diag}(0, \ldots, 0, 11, \ldots, 4000)$. The results shown in Table 5.3 compare the CG method with the adaptive modified Richardson method for different choices of initial interval. Table 5.3 shows that for $[a_0, b_0] = [1000, 3000]$ the number of iterations for the MR method is approximately the same as for the CG method, but the number of inner product evaluations required is much smaller. The table also shows that other choices of initial interval can reduce the iteration count for the MR method. Note that the final intervals determined for both initial intervals are considerably smaller than the asymptotically optimal interval $[11, 4000]$.

Example 5.4 In this example we consider the system of linear equations of Example 5.3 and we solve it by the modified Richardson algorithm using the weight function $w(z) = 1$ to determine the Leja points. It is clear from comparing the results of Tables 5.3 and 5.4 that the number of iterations required by the MR method to achieve convergence when we use $w(z) = 1$ is about twice the number of iterations necessary when using $w(z) = z$. The difference in the iteration count depends on that the final interval determined with the weight function $w(z) = 1$ is larger than when $w(z) := z$ is used.

Example 5.5 We solve a linear system of equation with a 4000×4000 diagonal matrix, $A := \text{diag}(1, \ldots, 3990, 0, \ldots, 0)$. The results presented in Table 5.5 indicate that the number of iterations required by our adaptive semi-iterative algorithm is much smaller than the number of steps needed by the CG algorithm, even if we choose a "bad" initial interval for the MR algorithm. We remark that for this example the MR algorithm with initial interval $[a_0, b_0] = [50, 3990]$ achieves convergence in about one third of the number of iterations required by the CG method.

Table 5.5: $A = \text{diag}(1, \ldots, 3990, 0, \ldots, 0); \; n = 4000$

iterative method	initial interval	final interval	updates	iterations	inner products
CG				235	798
MR	[1000,3000]	[33.46,3989]	5	90	100
MR	[500,3000]	[17.33,3989]	4	110	103
MR	[40,3990]	[40,3985]	2	80	67

Table 5.6: $A = \text{diag}(1, \ldots, 1990, 0, \ldots, 0, 2001, \ldots, 4000); \; n = 4000$

iterative method	initial interval	final interval	updates	iterations	inner products
CG				240	769
MR	[1000,3000]	[33.74,3999]	5	95	103
MR	[100,3000]	[17.12,4000]	4	115	106
MR	[40,3900]	[40,3996]	2	80	67

Example 5.6 In this example we solve a linear system of equations with the 4000×4000 matrix $A := \text{diag}(1, \ldots, 1990, 0, \ldots, 0, 2001, \ldots, 4000)$. Table 5.6 shows that the MR algorithm requires fewer iterations than the CG method for a variety of initial intervals. The table also suggests that a judicious choice of initial interval can reduce the number of iterations required.

6 CONCLUSION

We have described an adaptive semi-iterative scheme for the solution of inconsistent positive semidefinite linear system of equations. The iterations parameters for this scheme are reciprocal values of Leja points for an interval containing (most of) the positive part of the spectrum. This interval is determined during the iterations by interpreting inner products of certain vectors computed during the iterations as modified moments with respects to certain polynomials. We compared our adaptive semi-iterative scheme, which is a modification of Richardson iteration, with an ORTHODIR implementation of the conjugate gradient algorithm suitable for the solution of inconsistent positive semidefinite linear systems of equations. Numerical experiments indicate that the modified Richardson algorithm can achieve convergence in much fewer steps than the conjugate gradient algorithm. In addition, the small number of inner product evaluations required by modified Richardson iteration makes this scheme attractive for implementation on parallel computers on which inner product computations are relatively slow due to the communication they require.

REFERENCES

[1] Ashby S.F., Manteuffel T.A., Saylor P.E. A taxonomy for conjugate gradient methods. *SIAM J. Numer. Anal.*, **27**:1542–1568, 1990.

[2] Barker G.P., Yang S.-J. Semi-iterative and iterative methods for singular M-matrices. *SIAM J. Matrix Anal. Appl.*, **9**:168–180, 1988.

[3] Calvetti D., Reichel L. Adaptive Richardson iteration based on Leja points. *Math. Comp.* To appear.

[4] Calvetti D., Reichel L., Zhang Q. Conjugate gradient algorithms for symmetric inconsistent linear systems. In *Proceedings of the Lanczos Centenary Conference*. SIAM, Philadelphia. To appear. M. Chu, R. Plemmons, D. Brown and D. Ellison, eds.

[5] Dax A. The convergence of linear stationary iterative processes for solving singular unstructured systems of linear equations. *SIAM Review*, **32**:611–635, 1990.

[6] Edrei A. Sur les déterminants récurrents et les singularités d'une fonction donnée par son développement de Taylor. *Composito Math.*, **7**:20–88, 1939.

[7] Eiermann M., Marek I., Niethammer W. On the solution of singular linear systems of algebraic equations by semi-iterative methods. *Numer. Math.*, **53**:265–283, 1988.

[8] Eiermann M., Niethammer W., Varga R.S. A study of semi-iterative methods for nonsymmetric systems of linear equations. *Numer. Math.*, **47**:505–533, 1985.

[9] Eiermann M., Reichel L. On the application of orthogonal polynomials to the iterative solution of singular linear systems of equations. In *Vector and Parallel Computing*, pages 285–297. Ellis Horwood, Chichester, 1989. J. Dongarra, I. Duff, P. Gaffney and S. McKee, eds.

[10] Gautschi W. On generating orthogonal polynomials. *SIAM J. Sci. Stat. Comput.*, **3**:289–317, 1982.

[11] Golub G.H., Kent M.D. Estimates of eigenvalues for iterative methods. *Math. Comp.*, **53**:619–626, 1989.

[12] Golub G.H., Welsch J.H. Calculation of Gauss quadrature rules. *Math. Comp.*, **23**:221–230, 1969.

[13] Hanke M., Hochbruck M. A Chebyshev-like semi-iteration for inconsistent linear systems. *Elec. Trans. Numer. Anal.*, **1**:89–103, 1993.

95

[14] Kaaschieter E.F. Preconditioned conjugate gradients for solving singular systems. *J. Comput. Appl. Math.*, **24**:265–275, 1988.

[15] Leja F. Sur certaines suites liées aux ensembles plans et leur application à la représentation conforme. *Ann. Polon. Math.*, **4**:8–13, 1957.

[16] Lewis J.G., Rehm R.G. The numerical solution of a nonseparable elliptic partial differential equation by preconditioned conjugate gradients. *J. Res. Nat. Bureau Standards*, **85**:367–390, 1980.

[17] Reichel L. The application of Leja points to Richardson iteration and polynomial preconditioning. *Linear Algebra Appl.*, **154-156**:389–414, 1991.

[18] Schreiber R. Generalized iterative methods for semidefinite linear systems. *Linear Algebra Appl.*, **62**:219–230, 1984.

[19] Varga R.S. *Matrix Iterative Analysis*. Prentice-Hall, Englewood Cliffs, NJ, 1962.

[20] Woźniakowski H. Numerical stability of the Chebyshev method for the solution of large linear systems. *Numer. Math.*, **28**:191–209, 1977.

International Series of Numerical Mathematics, Vol. 119, ©1994 Birkhäuser

A UNIFIED APPROACH TO RECURRENCE ALGORITHMS

J.R. Cash[*]   R.V.M. Zahar[†]

*Dedicated with respect and gratitude, to Walter Gautschi
on the occasion of his sixty-fifth birthday*

Abstract. We show that the known methods for the computation of subdominant solutions of linear difference equations, either scalar recurrences or difference systems, are based on the replacement of an initial value problem by a system of linear equations expressed in boundary value form. This fact allows the development of a single uniform analysis for the convergence of the methods, and for the condition of the problem. Moreover, it is demonstrated that each of the resulting recurrence algorithms is mathematically equivalent to a type of triangular decomposition of the linear algebraic system, all algorithms being a form of either LU or UL factorization.

1 INTRODUCTION: THREE-TERM RECURRENCE RELATIONS

During the past fifty years, several dozen algorithms have been formulated for the solution of linear pth order difference equations, to avoid the instability of straightforward recursion when the desired solution is asymptotically subdominant with respect to at least one of the complementary solutions. Despite the apparent diversity of these algorithms, each of them can in fact be derived as the result of a two-stage mathematical process. First, the dimension $p - k$ of the space of solutions which dominate the desired solution is determined, and a boundary value problem with k initial conditions and $p - k$ final conditions (often arbitrary zeros) is formulated. And secondly, the resulting linear algebraic system is solved by a variant of the well known numerical methods for linear systems, such as Gaussian elimination.

The simplest case of instability in forward recurrence occurs in the computation of a solution $f(n)$, $n = 1, 2, ..., E$ of the second order scalar homogeneous difference equation (the *three-term recurrence relation*)

$$a_2(n)y(n+2) + a_1(n)y(n+1) + a_0(n)y(n) = 0, \qquad n = 0, 1, 2, \ldots \qquad (1.1)$$

[*]Department of Mathematics, Imperial College of Science, Technology and Medicine, 180 Queen's Gate, London SW7 2BZ, England; j.cash@ic.ac.uk

[†]Département d'informatique et de recherche opérationnelle, Université de Montréal, C.P. 6128, Succ. Centre-Ville, Montréal, Québec, H3C 3J7, Canada; zahar@iro.umontreal.ca

when $f(n)$ is *minimal* in the sense that there exists another linearly independent solution $g(n)$ such that

$$\lim_{n \to \infty} \frac{f(n)}{g(n)} = 0.$$

Indeed, a relative error of order ϵ in either $f(0)$ or $f(1)$ will induce a relative error proportional to $\epsilon g(n)/f(n)$ in the computed values of $f(n)$ and thus the error is unbounded as $n \to \infty$. The first algorithm which addressed this problem of instability and which employed recurrence in a nontrivial way, was formulated by J.C.P. Miller specifically for the computation of such minimal solutions.

Although Miller originally designed his technique for the computation of Taylor series solutions to linear ordinary differential equations (c. 1936), it was first published in the introduction to the British Association Tables of Bessel functions [3]. *Miller's backward recurrence algorithm*, as it has become known, can be described for those homogeneous equations (1.1) in which $a_0(n) \neq 0$ for all n, as follows. Let $N \geq E$, and let $z^{[N]}(n)$ be computed by:

$$v^{[N]}(N) = 1, \qquad v^{[N]}(N+1) = 0$$

$$v^{[N]}(n) = -\frac{1}{a_0(n)} \left[a_1(n) v^{[N]}(n+1) + a_2(n) v^{[N]}(n+2) \right], \quad n = N-1, N-2, \ldots, 0$$

$$z^{[N]}(n) = \frac{f(0)}{v^{[N]}(0)} v^{[N]}(n), \qquad n = 0, 1, \ldots, E. \tag{1.2}$$

Then $z^{[N]}(n)$ is an estimate of $f(n)$. Scheme (1.2) simply corresponds to the arbitrary choice of consecutive values 1 and 0 for N large enough, backward recurrence of (1.1) for n decreasing to 0, and a final normalization so that the single initial condition $z^{[N]}(0) = f(0)$ is satisfied. The most eloquent commentary on Miller's algorithm can be found in the article of Walter Gautschi [6] where, amongst a great many other results and applications, it is proved that for fixed n,

$$\lim_{N \to \infty} z^{[N]}(n) = f(n) \tag{1.3}$$

if and only if f is a minimal solution of (1.1) satisfying $f(0) \neq 0$. The remarkable fact that in the limit, a solution of a second-order linear difference equation can be computed from only one given initial value is due to the minimality of f which, as Gautschi notes, restricts f to a one-dimensional subspace of the solution space.

In parallel with the development of Miller's algorithm, an alternative method for computing a minimal solution, presented once again in the context of Bessel's equation, was discussed by L. Fox [5] at the suggestion of Miller. The initial condition $y(0) = f(0)$ is specified as in Miller's algorithm, and for large N the value $y(N+1) = 0$ is taken arbitrarily. A solution of the homogeneous equation (1.1) which is subject to these two *boundary conditions*, thus satisfies the linear algebraic system

$$y(0) = f(0)$$

$$a_2(n)y(n+2) + a_1(n)y(n+1) + a_0(n)y(n) = 0 \qquad n = 0, 1, \ldots, N-1 \qquad (1.4)$$

$$y(N+1) = 0$$

which has a tridiagonal matrix. Although Fox suggested that system (1.4) be solved by relaxation, a more direct method is to introduce the temporary value 1 in the penultimate component $y(N)$, solve the system backwards, and then use the first equation to scale the resulting vector. But the latter method is precisely that given by Miller's algorithm (1.2) with $E = N$. A third and more obvious method is to solve system (1.4) by Gaussian elimination, as was discussed by F.W.J. Olver in the article [12] which also extended Miller's technique to non-minimal solutions of inhomogeneous equations (but which applies equally well to minimal solutions) and which presented an automatic numerical method for determining a value of N for which convergence is achieved.

In [12], the first contribution was to show that Miller's algorithm could be regarded as a method for solving the tridiagonal system (1.4) by backwards elimination. In order to demonstrate this equivalence, it was supposed that Gaussian elimination (without pivoting) could be applied to the equations (1.4) in reverse order, or in other words, in the direction of decreasing n. Although Olver expressed his analysis directly in terms of difference equations, we choose to introduce the method as a type of matrix factorization in order to have a common framework for comparison with other techniques. Indeed, we express (1.4) in matrix form as $\mathbf{Ay=b}$, or

$$
\begin{bmatrix}
1 \\
a_0(0)\ a_1(0)\ a_2(0) \\
\quad a_0(1)\ a_1(1) \quad a_2(1) \\
\quad\quad \ddots \\
\quad\quad a_0(N{-}2)\ a_1(N{-}2)\ a_2(N{-}2) \\
\quad\quad\quad a_0(N{-}1)\ a_1(N{-}1)\ a_2(N{-}1) \\
\quad\quad\quad\quad 1
\end{bmatrix}
\begin{bmatrix}
y(0) \\ y(1) \\ y(2) \\ \vdots \\ y(N{-}1) \\ y(N) \\ y(N{+}1)
\end{bmatrix}
=
\begin{bmatrix}
f(0) \\ 0 \\ 0 \\ \vdots \\ 0 \\ 0 \\ 0
\end{bmatrix}
$$

$$(1.5)$$

and when $\mathbf{A}^{-1}$ exists, we consider the general decomposition $\mathbf{A=UL}$, where $\mathbf{U}$ and $\mathbf{L}$ are of the form

$$
\mathbf{U} =
\begin{bmatrix}
1 \\
\quad u_1(0)\ u_2(0) \\
\quad\quad u_1(1)\ u_2(1) \\
\quad\quad\quad \ddots \\
\quad\quad\quad u_1(N{-}2)\ u_2(N{-}2) \\
\quad\quad\quad\quad u_1(N{-}1)\ u_2(N{-}1) \\
\quad\quad\quad\quad\quad 1
\end{bmatrix}
\qquad (1.6a)
$$

$$\mathbf{L} = \begin{bmatrix} 1 & & & & & & \\ \ell_0(0) & \ell_1(0) & & & & & \\ & \ell_0(1) & \ell_1(1) & & & & \\ & & & \ddots & & & \\ & & & & \ell_0(N-2) & \ell_1(N-2) & \\ & & & & & \ell_0(N-1) & \ell_1(N-1) \\ & & & & & & 1 \end{bmatrix}. \quad (1.6b)$$

(For simplicity of notation, we drop the superscript $[N]$.) In order for the decomposition $\mathbf{A} = \mathbf{UL}$ to be unique, it is necessary to add N conditions on the $4N$ elements of $\mathbf{U}$ and $\mathbf{L}$, because $\mathbf{A}$ contains only $3N$ variables. Olver defined $\ell_1(N) = 1$ and took the extra linear conditions to be

$$\ell_0(n-1) = -\ell_1(n), \qquad n = 1, 2, \ldots, N,$$

so that, on equating $\mathbf{A}$ with the product $\mathbf{UL}$, the elements of $\mathbf{L}$ and $\mathbf{U}$ satisfy

$$\ell_1(N) = 1, \qquad \ell_1(N-1) = \frac{-a_1(N-1)}{a_0(N-1)}$$

$$\ell_1(n) = \frac{-a_1(n)\ell_1(n+1) - a_2(n)\ell_1(n+2)}{a_0(n)}, \qquad n = N-2, N-3, \ldots, 0 \quad (1.7a)$$

and

$$u_1(n) = \frac{-a_0(n)}{\ell_1(n+1)}, \qquad u_2(n) = \frac{a_2(n)}{\ell_1(n+1)}, \qquad n = N-1, N-2, \ldots, 0 \quad (1.7b)$$

provided that $a_0(n) \neq 0$ and $\ell_1(n) \neq 0$ for all $n \leq N - 1$. The linear system $\mathbf{ULy}=\mathbf{b}$ can now be solved in the usual way by putting $\mathbf{Ly}=\mathbf{x}$, so that $\mathbf{Ux}=\mathbf{b}$. Solving first the upper triangular system $\mathbf{Ux}=\mathbf{b}$, it follows that $\mathbf{x}=\mathbf{b}$. Then the lower triangular system $\mathbf{Ly}=\mathbf{b}$ has the solution $\mathbf{y}$ given by

$$\begin{aligned} y(0) &= f(0) \\ y(n) &= \frac{\ell_1(n)}{\ell_1(n-1)} y(n-1), \qquad n = 1, \ldots, N \\ y(N+1) &= 0. \end{aligned} \quad (1.8)$$

An examination of (1.7a) reveals that the sequence $\ell_1(n)$ is a solution of the homogeneous equation (1.1) with the two end conditions $\ell_1(N) = 1, \ell_1(N+1) = 0$. It follows immediately that the calculation of $\ell_1(n)$ for $n = N - 1, N - 2, \ldots, 0$ corresponds exactly to the backward recurrence calculation of $v^{[N]}(n)$ in Miller's algorithm (1.2). Moreover, by expanding (1.8) we have $y(n) = [\ell_1(n)/\ell_1(0)]f(0)$, which is identical to the last equation of (1.2). Therefore, the forward substitution stage of the above UL decomposition corresponds exactly to the final scaling of Miller's algorithm. Thus, for the general boundary value system (1.4), we have the result that this particular form of reverse Gaussian elimination is equivalent to solving the linear difference equations (1.7a) and (1.8). Although we have given explicit expressions (1.7b) for the elements of $\mathbf{U}$, those expressions are not actually required, as usual in UL decomposition. Indeed, because of the special form of the right side in (1.5), the elements of $\mathbf{U}$ need not even be calculated.

We assume throughout this article that $E \ll N$, which is the typical case. Thus Miller's algorithm and its UL equivalent require about $4N$ arithmetic floating point operations. However, N operations can be saved in this algorithm and others if it is assumed that (1.1) is divided symbolically by $a_2(n) \neq 0$, so that the coefficient of $y(n+2)$ is 1. In order to permit a general discussion of matrix decomposition, we do not take $a_2(n) \equiv 1$ in this article.

In the computation of $v^{[N]}(n)$ by (1.2) (equivalently $\ell_1(n)$ by (1.7a)) numerical overflow can often occur as n decreases. In [6], Gautschi suggests that one direct method of avoiding this difficulty is to solve for the ratios $r(n) \equiv v^{[N]}(n+1)/v^{[N]}(n)$ instead of the $v^{[N]}(n)$, since it must be assumed in any case that $v^{[N]}(n) \neq 0$. After division of the three-term recurrence in (1.2) by $v^{[N]}(n)$, it follows that the sequence $y(n)$ can be computed by the algorithm

$$
\begin{aligned}
r(N) &= 0 \\
r(n) &= -\frac{a_0(n)}{a_1(n) + a_2(n)r(n+1)}, && n = N-1, \ldots, 0, \\
y(0) &= f(0) \\
y(n) &= r(n-1)y(n-1), && n = 1, \ldots, E.
\end{aligned}
\tag{1.9}
$$

(The intimate connection between the values $r(n)$ and a truncated continued fraction defined by the $\{a_i(n)\}$ is used to great analytical advantage in [6] to prove asymptotic convergence of $z^{[N]}(n)$.) Although the backward recurrence in (1.9) is nonlinear, the entire algorithm can also be expressed as a form of Gaussian elimination of (1.5), the basic elimination step of which is also nonlinear. Indeed, for $n = 0, 1, \ldots, N-1$, let

$$
u_1(n) = a_1(n) + a_2(n)r(n+1), \qquad u_2(n) = a_2(n).
$$

Then the $r(n)$ calculated by (1.9) satisfy the equation $\mathbf{A=UL}$, where $\mathbf{UL}$ is given by

$$
\begin{bmatrix}
1 & & & & \\
u_1(0)\,u_2(0) & & & \\
& u_1(1)\,u_2(1) & & \\
& & \ddots & \\
& & & u_1(N-1)\,u_2(N-1) \\
& & & & 1
\end{bmatrix}
\begin{bmatrix}
1 & & & & \\
-r(0) & 1 & & & \\
& -r(1)\,1 & & \\
& & \ddots & \\
& & & -r(N-1)\,1 & \\
& & & & 1
\end{bmatrix},
$$

as can be verified by direct multiplication. It follows, therefore, that algorithm (1.9) is mathematically equivalent to UL decomposition, with $\mathbf{L}$ unit diagonal, followed by the solution of $\mathbf{Ux=b}$ (giving the trivial solution $\mathbf{x=b}$) and $\mathbf{Ly=x}$. The complexity of the algorithm is $3N$, a decrease of N in comparison with Miller's algorithm.

Instead of solving (1.5) by UL decomposition, Olver [12] presented an algorithm based on LU decomposition, for the principal reason that an estimate of the truncation error could be obtained automatically during the forward elimination phase of Gaussian decomposition. Using the same terminology for $\mathbf{L}$ and $\mathbf{U}$ as in (1.5), Olver showed that when the N additional conditions for $\mathbf{U}$ are taken to be $u_2(0) = -1$, and

$$u_1(n) = -u_2(n+1), \qquad n = 0, 1, ..., N - 2, \tag{1.10}$$

then $u_2(1) = -a_1(0)/a_2(0)$ and

$$u_2(n+1) = -\frac{a_1(n)u_2(n) + a_0(n)u_2(n-1)}{a_2(n)}, \qquad n = 2, 3, \ldots, N-2, \tag{1.11}$$

$$\ell_0(n) = \frac{-a_0(n)}{u_2(n)}, \qquad \ell_1(n) = \frac{a_2(n)}{u_2(n)}, \qquad n = 0, 1, ..., N-1.$$

Thus $u_2(n+1)$ is a solution of (1.1) and is computed by forward recurrence. The algorithm in [12] is then equivalent to the solution of $\mathbf{Lx=b}$ and $\mathbf{Uy=x}$; that is, the recurrence (1.11) followed by

$$
\begin{aligned}
x(0) \quad &= \quad f(0) \\
x(n+1) \quad &= \quad \frac{a_0(n)}{a_2(n)}, \qquad && n = 0, 1, ..., N-1, \\
y(N+1) \quad &= \quad 0 \\
y(n) \quad &= \quad \frac{u_2(n)y(n+1) - x(n)}{u_2(n+1)}, \qquad && n = N, N-1, ..., 0.
\end{aligned}
\tag{1.12}
$$

This algorithm for calculating $y(n), n = 0, 1, ..., E$ has complexity $8N$, but the extra work enables the estimation of the truncation error. In addition, the algorithm of (1.11) and (1.12) is easily extended to the computation of intermediate (non-minimal) solutions, for which Olver originally designed it [12].

Shortly after the publications of Gautschi [6] and Olver [12], their algorithms were applied to scalar linear recurrence relations of arbitrary order in [16] and by Olver [11] where convergence of the solutions was discussed. Olver, in particular, suggested that by regarding Olver's algorithm as a form of LU decomposition, stability could be analyzed conveniently. For minimal solutions of three-term recurrences, a combined form of the algorithms of Miller and Olver was presented in [13]. A pivoted form of Olver's algorithm for second order inhomogeneous equations, which also permitted the estimation of the truncation error, was given in [4], and different forms of LU decomposition for the same equations were considered in [14], where their complexities were discussed. Gautschi's algorithm was generalized to second order inhomogeneous recurrences by Aggarwal and Burgmeier [1]. A detailed analysis of LU decomposition for equations of arbitrary order was later given by Lozier [9], and block LU decomposition of first order difference systems of arbitrary dimension was discussed in [18] and further analyzed in [20]. The LU decomposition method of [14] was presented in the book of Wimp [15], along with a survey of known recurrence algorithms, as well as several pertinent and interesting applications.

In §2, we present the decompositions associated with scalar recurrence algorithms, and in §3 we generalize those decompositions to two algorithms for difference systems. In the final section of this article, we show how the analysis of convergence and conditioning for difference systems is aided by the uniform representation of recurrence algorithms, and we list some recent results.

2 SCALAR RECURRENCES

We consider algorithms for the solution of the pth order scalar difference equation, homogeneous or inhomogeneous, of the form

$$\sum_{i=0}^{p} a_i(n)y(n+i) = b(n), \qquad n = 0, 1, 2, \ldots \tag{2.1}$$

where the $\{a_i(n)\}$ and $\{b(n)\}$ are given sequences of complex numbers, and we suppose that the desired solution $f(n)$ is required for $n = 0, 1, ..., E$. In this section, we discuss the basic algorithms for (2.1), showing again that they can be regarded as matrix factorization methods for the specific boundary value systems.

In formulating the appropriate linear system, it is usually assumed that f is dominated by $p - k$ complementary solutions, is not dominated by k complementary solutions, and that the first k initial values of $f(n)$ are known. Several quite different definitions of dominance are given in the literature, an intuitive one being that a solution f_i dominates f if

$$\liminf_{n\to\infty} \frac{f(n)}{f_i(n)} = 0. \tag{2.2}$$

For the current section, we adopt definition (2.2), postponing a more precise discussion of dominance until §4. We then consider the solution of the boundary value system

$$
\begin{aligned}
y(n) &= f(n), & n &= 0, 1, \ldots, k-1 \\
\sum_{i=0}^{p} a_i(n)y(n+i) &= b(n), & n &= 0, 1, \ldots, N-1 \\
y(n) &= \tilde{f}(n), & n &= N+k, \ldots, N+p-1
\end{aligned}
\tag{2.3}
$$

103

where the values $\tilde{f}(n), n = N + k, \ldots, N + p - 1$ are either estimates of the corresponding values of f or arbitrary zeros. Writing system (2.3) in matrix form as $\mathbf{Ay=b}$, we have

$$
\mathbf{A} = \begin{bmatrix}
1 & & & & & & & & \\
& \ddots & & & & & & & \\
& & 1 & & & & & & \\
a_0(0) & \cdots & a_{k-1}(0) & a_k(0) & \cdots & & a_p(0) & & \\
& & & & \ddots & & & & \\
& & a_0(N-1) & \cdots & a_k(N-1) & a_{k+1}(N-1) & \cdots & a_p(N-1) \\
& & & & & 1 & & & \\
& & & & & & \ddots & & \\
& & & & & & & 1
\end{bmatrix}
$$
(2.4)

$$\mathbf{y} = [y(0), \ldots, y(k-1), y(k), \ldots, y(N+k-1), y(N+k), \ldots, y(N+p-1)]^T,$$

$$\mathbf{b} = \left[f(0), \ldots, f(k-1), b(0), \ldots, b(N-1), \tilde{f}(N+k), \ldots, \tilde{f}(N+p-1)\right]^T.$$

We consider either UL or LU factorizations, where in both cases

$$
\mathbf{U} = \begin{bmatrix}
1 & & & & & & & & \\
& \ddots & & & & & & & \\
& & 1 & & & & & & \\
& & u_k(0) & u_{k+1}(0) & \cdots & & u_p(0) & & \\
& & & \ddots & & & & & \\
& & & u_k(N-1) & u_{k+1}(N-1) & \cdots & u_p(N-1) & \\
& & & & 1 & & & & \\
& & & & & \ddots & & & \\
& & & & & & 1
\end{bmatrix}
$$
(2.5a)

$$
\mathbf{L} = \begin{bmatrix}
1 & & & & & & & & \\
& \ddots & & & & & & & \\
& & 1 & & & & & & \\
\ell_0(0) & \cdots & \ell_{k-1}(0) & \ell_k(0) & & & & & \\
& & & & \ddots & & & & \\
& & \ell_0(N-1) & \ell_1(N-1) & \cdots & \ell_k(N-1) & & \\
& & & & & 1 & & & \\
& & & & & & \ddots & & \\
& & & & & & & 1
\end{bmatrix}
$$
(2.5b)

Thus, each of the matrices $\mathbf{A}$, $\mathbf{U}$ and $\mathbf{L}$ is square, of dimension $N + p$. In order to distinguish between the total matrices and sub-matrices with some initial or final rows absent, we will use the notation $\mathbf{A}[i\!:\!j]$ to stand for the $(j - i + 1) \times (N + p)$ matrix formed from the consecutive rows i to j of $\mathbf{A} \equiv \mathbf{A}[1 : N+p]$. We use a compatible notation for the vectors $\mathbf{b}$ and $\mathbf{y}$.

In the remainder of this section, we present the matrices $\mathbf{U}$ and $\mathbf{L}$ for the known scalar algorithms. As may be noted by the three decompositions given in §1, the derivation of the formulae for a specific algorithm is far from a trivial task, and often requires a good deal of guesswork. Once the elements of the triangular matrices are presented, however, it is a simple matter to verify those formulae by direct multiplication, and so we generally omit that algebra.

2.1 Minimal solutions of homogeneous equations

We consider first the computation of a *minimal* solution (one which is dominated by $p - 1$ complementary solutions) of the homogeneous form of (2.1). The natural extension of Miller's algorithm (1.2) to pth order equations is given by

$$
\begin{aligned}
v^{[N]}(N) &= 1; \quad v^{[N]}(n) = 0, \quad n = N + 1, \ldots, N + p - 1 \\
v^{[N]}(n) &= -\frac{1}{a_0(n)} \sum_{i=1}^{p} a_i(n) v^{[N]}(n+i), \quad n = N-1, N-2, \ldots, 0 \\
z^{[N]}(n) &= \frac{f(0)}{v^{[N]}(0)} v^{[N]}(n), \quad n = 0, 1, \ldots, E.
\end{aligned} \tag{2.6}
$$

Taking $k = 1$ in (2.4) and defining as in §1, $\ell_1(N) = 1$ and

$$
\ell_0(n-1) = -\ell_1(n), \quad n = 1, 2, \ldots, N,
$$

it follows from the multiplication $\mathbf{A} = \mathbf{U}\mathbf{L}$ that $\ell_1(n)$ satisfies

$$
\ell_1(N) = 1; \quad \ell_1(n) = 0, \quad n = N + 1, \ldots, N + p - 1
$$

$$
\ell_1(n) = -\frac{1}{a_0(n)} \sum_{i=1}^{p} a_i(n) \ell_1(n+i), \quad n = N-1, \ldots, 0, \tag{2.7}
$$

corresponding to the backward recurrence for $v^{[N]}(n)$ in (2.6). For $n = 0, 1, \ldots,$ $N - 1$, the elements of $\mathbf{U}$ satisfy $u_1(n) = -a_0(n)/\ell_1(n+1)$ and for $i = 2, \ldots, p$,

$$
u_i(n) = \frac{u_{i-1}(n)\ell_1(n) - a_i(n)}{\ell_1(n+i)} \equiv -\frac{\displaystyle\sum_{s=0}^{i-1} a_s(n-1)\ell_1(n+s-1)}{\ell_1(n+i-1)\ell_1(n+i)}
$$

provided that $\ell_1(n) \neq 0$ for all n. Neither the explicit expression nor the calculation of the $u_i(n)$ is actually required, however, because the solution of $\mathbf{U}\mathbf{L}\mathbf{y} = \mathbf{b}$ reduces to $\mathbf{L}\mathbf{y} = \mathbf{b}$ or

$$
y(n) = \frac{\ell_1(n)}{\ell_1(0)} f(0), \quad n = 0, 1, \ldots, N - 1,
$$

which, when added to (2.7), completes the equivalence between (2.6) and the UL decomposition. When $p \ll N$ as well as $E \ll N$, both of which we assume throughout, the complexity of Miller's algorithm and its UL equivalent is $2pN$.

We note that, although it seems necessary to take the final conditions $v^{[N]}(n) = 0$, $n = N + 1, \ldots, N + p - 1$ in Miller's algorithm (2.6), the same does not hold for the UL decomposition. Indeed, estimates $\tilde{f}(n)$, $n = N + 1, \ldots, N + p - 1$, of the solution can simply be included on the right side of (2.3). In the latter case, however, the solution of $\mathbf{Ux=b}$ does not reduce to $\mathbf{x=b}$, and so an extra backward recurrence for $\mathbf{x}$ needs to be computed (requiring $2(p - 1)N$ operations) and the values of $\mathbf{x}$ must be added to the right side of the last equation in (2.6). In this event, the complexity is increased to $2(2p - 1)N$, or almost twice that of Miller's algorithm.

At the suggestion of Gautschi, his algorithm for three-term recurrences was generalized to homogeneous equations of arbitrary order in [16], as follows. We define $r(n) = y(n+1)/y(n)$ and divide (2.1) by $y(n)$, so that

$$\sum_{i=0}^{p} a_i(n) \prod_{j=0}^{i-1} r(n+j) = 0. \qquad (2.8)$$

If we formally take $r(n) = 0$ for $n = N, N + 1, \ldots, N + p - 2$, and solve for $r(n)$ in (2.8) backwards for $n = N - 1, N - 2, \ldots, 0$ by a type of Horner scheme, the following algorithm is obtained.

$$
\begin{aligned}
r(N) &= r(N{+}1) = \cdots = r(N{+}p{-}2) = 0 \\
s_p(n) &= a_p(n) \\
s_i(n) &= a_i(n) + r(n{+}i)s_{i+1}(n), \quad i = p{-}1, p{-}2, \ldots, 1 \\
r(n) &= -\frac{a_0(n)}{s_1(n)} \\
y(0) &= f(0) \\
y(n) &= r(n{-}1)y(n{-}1), \qquad n = 1, 2, \ldots, E.
\end{aligned}
\left. \phantom{\begin{aligned}r\\s\\s\\r\\y\\y\end{aligned}} \right\} n = N{-}1, \ldots, 0
\qquad (2.9)
$$

The values $s_i(n)$ and $r(n)$ are the elements in the matrix factorization $\mathbf{A=UL}$, where $k = 1$ in (2.4), and where in (2.5),

$$
\begin{aligned}
u_i(n) &= s_i(n), & i &= 1, \ldots, p - 1, \\
\ell_0(n) &= -r(n), & \ell_1(n) &= 1,
\end{aligned}
\qquad (2.10)
$$

for all $n = 0, 1, \ldots, N - 1$, as can be verified after substitution of (2.10) into (2.5) and a little algebra. Algorithm (2.9) is then equivalent to the solution of $\mathbf{Ux=b}$, giving $\mathbf{x=b}$, and then the solution of $\mathbf{Ly=b}$, the calculation of the $u_i(n)$ being necessary only for the construction of $\mathbf{L}$. The complexity of (2.9) is $(2p - 1)N$, the formulation in terms of the $r(n)$ effectively saving N operations in comparison with Miller's algorithm.

For minimal solutions of homogeneous recurrences, the one initial condition $y(0) = f(0)$ is often replaced by the more general linear condition

$$\sum_{m=0}^{\infty} \lambda(m)f(m) = S, \tag{2.11}$$

where $\{\lambda(m)\}$ and S are known, in which case algorithm (2.9) can be modified as follows [6], [12], [16], [7]. Replacing the upper limit in (2.11) by $N + p - 1$, $f(m)$ by $y(m)$, and then dividing by $y(0)$ gives the truncated equation

$$\sum_{m=0}^{N+p-1} \lambda(m) \prod_{j=0}^{m-1} r(j) = \frac{S}{y(0)}, \tag{2.12}$$

where by definition, $r(N) = r(N+1) = \cdots = r(N+p-2) = 0$. If we denote the left side of (2.12) by $t(0)$ and define $t(N+1) = \cdots = t(N+p-1) = 0$, then $t(0)$ can be calculated by the Horner scheme starting with $t(N+1) = 0$, and

$$t(n) = \lambda(n) + r(n)t(n+1), \qquad n = N, \ldots, 0. \tag{2.13}$$

Algorithm (2.9) is then modified to be

$$
\left.
\begin{aligned}
r(N) &= r(N+1) = \cdots = r(N+p-2) = 0 \\
t(N) &= \lambda(N) \\
s_p(n) &= a_p(n) \\
s_i(n) &= a_i(n) + r(n+i)s_{i+1}(n), \quad i = p-1, p-2, \ldots, 1 \\
r(n) &= -\frac{a_0(n)}{s_1(n)} \\
t(n) &= \lambda(n) + r(n)t(n+1) \\
y(0) &= S/t(0) \\
y(n) &= r(n-1)y(n-1), \qquad n = 1, 2, \ldots, E.
\end{aligned}
\right\}
\begin{aligned}
\\ \\ \\ \\ n = N-1, \ldots, 0 \\ \\ \\
\end{aligned}
\tag{2.14}
$$

To show how the latter algorithm relates to the solution of a linear system, we note first that the defining system (2.4) must be modified so that its first row (the only initial condition) is replaced by the truncated form of (2.11). Letting $A[2:N+p]$ be the sub-matrix containing the last $N + p - 1$ rows of the matrix A in (2.4), and defining $\lambda = [\lambda(0), \lambda(1), \ldots, \lambda(N+p-1)]$, the new system can be written as

$$\begin{bmatrix} \lambda \\ A[2:N+p] \end{bmatrix} y = \begin{bmatrix} S \\ 0 \end{bmatrix}. \tag{2.15}$$

It is straightforward to verify that algorithm (2.14) is equivalent to the solution of (2.15) by UL decomposition, with L exactly as in (2.10), but with U given by

$$\begin{bmatrix} t \\ U[2:N+p] \end{bmatrix},$$

where $\mathbf{t} = [t(0), t(1), \ldots, t(N+p-1)]$ and where the $u_i(n)$ are specified in (2.10). The complexity of (2.14) is $(2p + 1)N$, the extra $2N$ computations over that of (2.9) being required for the calculation of $\mathbf{t}$.

Algorithms (2.9) and (2.14) are particularly advantageous when estimates for $r(n)$, $n = N, \ldots, N + p - 2$, are known, because it is only necessary to substitute those estimates in place of the former zeros and to calculate $t(n)$, $n = N + p - 1, \ldots, N$. In the UL form, this change simply corresponds to adding the data $\ell_0(n) \equiv -r(n), n = N, \ldots, N + p - 2$, to the sub-diagonals of rows $N + 2$ to $N + p$ in both $\mathbf{A}$ and $\mathbf{L}$. But the solution of $\mathbf{Ux=b}$ still reduces to $\mathbf{x=b}$, unlike the UL form of Miller's algorithm. Thus the complexity of each algorithm (2.9) and (2.14) is independent of the final values taken for $r(n)$.

For minimal solutions of three-term recurrences, a modification of Olver's algorithm was given in [13] to overcome the numerical difficulties encountered when the values $u(n)$ of (1.11) are small for some initial region $0 \leq n \leq M$ and are therefore subject to numerical cancellation. We describe the algorithm for inhomogeneous equations with $p = 2$, and for the general linear condition (2.11). The value M is chosen so that $\mathbf{A}[M+2 : N+2]$ is diagonally dominant, and the system

$$\left[\begin{array}{c} \mathbf{e} \\ \mathbf{A}[M+2:N+2] \end{array} \right] \left[\mathbf{v}[M+1:N+2] \right] = \left[\begin{array}{c} 1 \\ \mathbf{0} \end{array} \right], \qquad (2.16)$$

where $\mathbf{e} = [1, 0, \ldots, 0]$, is solved by LU decomposition (which permits the estimation of the truncation error). Miller's algorithm (1.2) with $N = M$ is then applied, but with the conditions given by

$$v^{[N]}(M) = 1, \qquad v^{[N]}(M+1) = v(M+1).$$

Finally, $y(n)$ is calculated by the normalization

$$y(n) = \frac{S}{\sum\limits_{m=0}^{N+1} \lambda(m) v^{[N]}(m)} v^{[N]}(n), \qquad n = 0, 1, \ldots, E, \qquad (2.17)$$

where $v^{[N]}(n) = v(n), n = M+2, \ldots, N+1$. As described, therefore, this algorithm is equivalent to an LU decomposition of rows $M+2$ to $N+2$ of (2.4), followed by a UL decomposition of rows 1 to $M + 2$. When the computation of the denominator in (2.17) is included, the complexity of the algorithm is $10N - 4M$.

Both mathematically and algorithmically, this method can be rephrased as a combination of the algorithms of Olver and Gautschi, as we now show. Because the right hand vector of (2.4) is defined by $\mathbf{b}[M+2:N+2] = \mathbf{0}$, the values $\mathbf{v}(n)$ calculated by (2.16) satisfy

$$r(n) \equiv \frac{y(n+1)}{y(n)} = \frac{v(n+1)}{v(n)}, \qquad n = M, \ldots, N, \qquad (2.18)$$

where **y** is the solution of (2.4). Therefore, once the solution **v** in the LU decomposition of (2.16) is determined, the ratios (2.18) can be calculated, and the vector **y** of (2.4) satisfies

$$-r(n)y(n) + y(n+1) = 0, \qquad n = M, \ldots, N. \tag{2.19}$$

Replacing the last $N - M + 1$ rows of **A** with (2.19) gives the system **By**=**b**, or

$$\begin{bmatrix} \lambda \\ \mathbf{A}[2:M{+}1] \\ \mathbf{L}[M{+}2:N{+}2] \end{bmatrix} \mathbf{y} = \begin{bmatrix} S \\ 0 \\ 0 \end{bmatrix} \tag{2.20}$$

where λ is defined in (2.15), and where we have used the notation $\mathbf{L}[M{+}2:N{+}2]$ to denote that rows $M + 2$ to $N + 2$ of **B** contain $-r(n)$ on the sub-diagonal, and the values 1 on the diagonal. The structure of **B** is therefore analogous to that of (2.15), but with given values of $r(n)$, $n = M, \ldots, N$, and so algorithm (2.14) can be applied directly to (2.20). The complexity of this modification is $10N - 5M$, with a saving of M arithmetic operations, and (2.20) has the additional advantage that estimates of $r(n)$, $n = N + 1, \ldots, N + p - 2$ can also be included without any increase in complexity.

The algorithm of [13] clearly can be employed to compute minimal solutions of scalar equations of arbitrary order p. Indeed, the value $N+2$ in (2.16) and (2.20) needs only to be replaced by $N + p$ to effect the generalization. Moreover, it is easy to adapt the algorithm for applications in other circumstances. For instance, instead of employing Miller's algorithm for the initial region $n = 0, \ldots, M$, forward recurrence could be used if it is known that the desired solution is dominant in that region. A method based on the last technique was formulated for the third and fourth order recurrences arising in the series calculation of the eigenvectors and eigenvalues of differential, multiple eigenvalue problems in [2].

2.2 Intermediate solutions of general equations

When $1 < k < p$ in (2.3), the natural generalization of Olver's algorithm described in [16],[11] and analyzed extensively in [9] has become the standard in the field. Except for the computation of the truncation error estimate, it is well known that the method is equivalent to the solution of (2.4) by a type of LU decomposition. For second order recurrences, it was noted by Van der Cruyssen [14] that if **L** is taken to be unit lower triangular instead of employing the normalization given in (1.10), then the complexity of the algorithm is reduced. Since there seems to be no reason for not taking **L** to be unit lower triangular in the case of pth order scalar recurrences, we shall therefore refer to Olver's algorithm as that traditional method of LU decomposition (i.e. without pivoting) and consequent solution of (2.4), which has a complexity of $[2(k + 1)(p - k) + 3k + 1]N$ when $f(0)$ is given.

For second order inhomogeneous recurrences, it was shown in [4] that the estimation of the truncation error in Olver's algorithm could be maintained even

109

if pivoting was employed. It may sometimes be desirable to employ pivoting for pth order equations as well, producing a PLU decomposition, where $\mathbf{P}$ is a permutation matrix, instead of an LU decomposition. Because row pivoting has the effect of changing the upper bandwith of $\mathbf{U}$ from $p - k$ to p, the complexity of the pivoted form of Olver's algorithm can increase to a possible $[2(k + 1)p + 3k + 1]N$.

Algorithms which initially seem unrelated to LU decomposition can be shown to be equivalent. Such is the case with the method for second order inhomogeneous recurrences presented by Aggarwal and Burgmeier [1], which we now describe. The authors consider the case $a_2(n) \equiv 1$, and derive the equations for their method by first seeking two sequences $\{\kappa(n)\}$ and $\{\epsilon(n)\}$ such that

$$y(n+1) - \kappa(n+1)y(n) = \epsilon(n+1), \qquad n = 0, 1, 2, \ldots . \tag{2.21}$$

Multiplying the latter equation by $\omega(n+1)$, where $\{\omega(n)\}$ is a sequence to be determined, and subtracting from (2.21) with n replaced by $n + 1$, gives

$$y(n+2) - \Big[\kappa(n+2) + \omega(n+1)\Big]y(n+1) + \kappa(n+1)\omega(n+1)y(n)$$

$$= \epsilon(n+2) - \omega(n+1)\epsilon(n+1).$$

Equating coefficients in the last equation with those of (2.1) gives

$$\omega(n+1) = \frac{a_0(n)}{\kappa(n+1)} \tag{2.22a}$$

$$\kappa(n+2) = -a_1(n) - \omega(n+1), \tag{2.22b}$$

$$\epsilon(n+2) = b(n) + \omega(n+1)\epsilon(n+1). \tag{2.22c}$$

Equations (2.22) and (2.21) are then used to define the basic algorithms of [1], which employ either forward or backward recurrence, depending on the dominance of the desired solution.

The scheme defined in (2.22) is none other than LU decomposition with $\mathbf{L}$ unit diagonal, and is thus equivalent to that in [14]. Indeed, if in (2.5) with $k = 1$ we define

$$\ell_0(n) = -\omega(n+1), \qquad u_1(n) = -\kappa(n+2), \qquad u_2(n) = 1, \qquad n = 0, 1, \ldots, N-1,$$

and take $\kappa(1) = -1$, then (2.22a) and (2.22b) follow after the multiplication of $\mathbf{L}$ and $\mathbf{U}$. Moreover, specifying $x(n) = \epsilon(n+1)$, expression (2.22c) becomes equivalent to the solution of $\mathbf{Lx=b}$. Lastly, (2.21) is equivalent to $\mathbf{Uy=x}$.

The connection between triangular decomposition and the algorithms for scalar equations serves as a motivation for the construction of algorithms for difference systems.

3 LINEAR DIFFERENCE SYSTEMS

From the standpoint of LU decomposition, linear systems formed by sequences of scalar equations of arbitrary order were discussed by Lozier [9], and the corresponding algorithms were analyzed in detail (including brief comments on stability). Only two algorithms, however, have been formulated specifically for subdominant solutions of pth order difference systems of the type

$$\mathbf{y}(n) = \mathbf{A}(n)\mathbf{y}(n-1) + \mathbf{b}(n), \qquad n = 1, 2, 3, \ldots, \tag{3.1}$$

and we now present those methods. To facilitate the discussion, we use the notations $\mathbf{y}_1(n)$ and $\mathbf{y}_2(n)$ to represent, respectively, the first k and the last $p - k$ components of the vector $\mathbf{y}(n)$. Similarly, we partition $p \times p$ matrices $\mathbf{A}$ as

$$\mathbf{A} = \begin{bmatrix} \mathbf{A}_{11} & \mathbf{A}_{12} \\ \mathbf{A}_{21} & \mathbf{A}_{22} \end{bmatrix}, \tag{3.2}$$

where $\mathbf{A}_{11}$ is $k \times k$, and the other $\mathbf{A}_{ij}$ are of consistent dimensions. For the present section, we say that a complementary solution $\mathbf{f}_i$ of (3.1) dominates the desired particular solution $\mathbf{f}$ if, in the L_∞ norm,

$$\liminf_{n \to \infty} \frac{\|\mathbf{f}(n)\|}{\|\mathbf{f}_i(n)\|} = 0, \tag{3.3}$$

deferring a slightly more thorough discussion of dominance until §4. We suppose that $\mathbf{f}$ is dominated by $p - k$ complementary solutions, is not dominated by k complementary solutions, and that $\mathbf{f}_1(0)$ is known. Thus, we consider the solution of the boundary value problem

$$\begin{aligned} \mathbf{y}_1(0) &= \mathbf{f}_1(0) \\ \mathbf{y}(n) &= \mathbf{A}(n)\mathbf{y}(n-1) + \mathbf{b}(n), \qquad n = 1, 2, \ldots, N \\ \mathbf{y}_2(N) &= \tilde{\mathbf{f}}_2(N) \end{aligned} \tag{3.4}$$

where $\tilde{\mathbf{f}}_2(N)$ is an estimate of $\mathbf{f}_2(N)$, or arbitrarily $\mathbf{0}$. Representing (3.4) by $\mathbf{A}\mathbf{y} = \mathbf{b}$, we have

$$\mathbf{A} = \begin{bmatrix} \mathbf{I}_k & & & & & \\ -\mathbf{A}(1) & \mathbf{I} & & & & \\ & -\mathbf{A}(2) & \mathbf{I} & & & \\ & & & \ddots & & \\ & & & -\mathbf{A}(N) & \mathbf{I} & \\ & & & & & \mathbf{I}_{p-k} \end{bmatrix} \tag{3.5}$$

where $\mathbf{I}_k, \mathbf{I}_{p-k}$ are identity matrices of dimensions k and $p - k$ respectively.

3.1 Orthogonal decomposition

We discuss first the algorithm of Mattheij [10] which, although given for homogeneous equations, can be generalized to inhomogeneous systems (3.1) as follows.

One constructs a sequence $\mathbf{Q}(0), \mathbf{Q}(1), \mathbf{Q}(2), \ldots$ of pth order unitary matrices $(\mathbf{Q}^*(n)\mathbf{Q}(n) = \mathbf{I})$ such that

$$\mathbf{A}(n)\mathbf{Q}(n-1) = \mathbf{Q}(n)\mathbf{L}(n), \qquad n = 1, 2, \ldots, \tag{3.6}$$

where each $\mathbf{L}(n)$ is block lower-triangular of the form

$$\mathbf{L}(n) = \left[\begin{array}{cc} \mathbf{L}_{11}(n) & \mathbf{0} \\ \mathbf{L}_{21}(n) & \mathbf{L}_{22}(n) \end{array} \right]$$

and $\mathbf{L}_{11}(n)$, $\mathbf{L}_{22}(n)$ are lower triangular. (Because it was assumed in [10] that $\mathbf{f}_2(0)$ was given, an upper-triangular matrix was taken instead of $\mathbf{L}(n)$. We use a lower-triangular matrix merely to conform to our previous terminology.) Since (3.6) implies that the original difference system (3.1) can be re-written as

$$
\begin{aligned}
\mathbf{Q}^*(n)\mathbf{y}(n) &= \mathbf{Q}^*(n)\mathbf{A}(n)\mathbf{Q}(n-1)\mathbf{Q}^*(n-1)\mathbf{y}(n-1) + \mathbf{Q}^*(n)\mathbf{b}(n) \\
&= \mathbf{L}(n)\mathbf{Q}^*(n-1)\mathbf{y}(n-1) + \mathbf{Q}^*(n)\mathbf{b}(n), \qquad n = 1, 2, \ldots,
\end{aligned}
$$

the method requires the initial change of dependent variable

$$\mathbf{z}(n) = \mathbf{Q}^*(n)\mathbf{y}(n), \qquad \mathbf{c}(n) = \mathbf{Q}^*(n)\mathbf{b}(n). \tag{3.7}$$

Then if $\mathbf{z}_1(0)$ can be found, the equation

$$\mathbf{z}_1(n) = \mathbf{L}_{11}(n)\mathbf{z}_1(n-1) + \mathbf{c}_1(n), \qquad n = 1, 2, \ldots, N \tag{3.8a}$$

can be solved in the forward direction for $\mathbf{z}_1(1), \mathbf{z}_1(2), \ldots, \mathbf{z}_1(N)$. Finally,

$$\mathbf{z}_2(n) = \mathbf{L}_{21}(n)\mathbf{z}_1(n-1) + \mathbf{L}_{22}(n)\mathbf{z}_2(n-1) + \mathbf{c}_2(n), \qquad n = N, N-1, \ldots, 0 \tag{3.8b}$$

is solved in the backward direction for $\mathbf{z}_2(N-1), \ldots, \mathbf{z}_2(0)$, starting with the value $\mathbf{z}_2(N) = \tilde{\mathbf{f}}_2(N)$.

Mattheij's method can also be described as a type of decomposition for the full matrix $\mathbf{A}$ of the linear system (3.5), as follows. Block diagonal matrices $\mathbf{Q}$ and $\bar{\mathbf{Q}}$,

$$\mathbf{Q} = \mathrm{diag}\left[\mathbf{Q}(0), \mathbf{Q}(1), \ldots, \mathbf{Q}(N)\right],$$

$$\bar{\mathbf{Q}} = \mathrm{diag}\left[\mathbf{I}_k, \mathbf{Q}^*(1), \mathbf{Q}^*(2), \ldots, \mathbf{Q}^*(N), \mathbf{I}_{p-k}\right],$$

are found so that $\bar{\mathbf{Q}}\mathbf{A}\mathbf{Q} = \bar{\mathbf{L}}$, where

$$\bar{\mathbf{L}} = \left[\begin{array}{ccccccc} \mathbf{Q}_{11}(0) & \mathbf{Q}_{12}(0) & & & & & \\ -\mathbf{L}_{11}(1) & \mathbf{0} & \mathbf{I}_k & \mathbf{0} & & & \\ -\mathbf{L}_{21}(1) & -\mathbf{L}_{22}(1) & \mathbf{0} & \mathbf{I}_{p-k} & & & \\ & & & \ddots & & & \\ & & & & -\mathbf{L}_{11}(N) & \mathbf{0} & \mathbf{I}_k & \mathbf{0} \\ & & & & -\mathbf{L}_{21}(N) & -\mathbf{L}_{22}(N) & \mathbf{0} & \mathbf{I}_{p-k} \\ & & & & & & \mathbf{Q}_{21}(N) & \mathbf{Q}_{22}(N) \end{array} \right]. \tag{3.9}$$

The system $\mathbf{Ay=b}$ is thereby converted to $\bar{\mathbf{Q}}A\mathbf{QQ}^*\mathbf{y} = \bar{\mathbf{Q}}\mathbf{b}$, or $\bar{\mathbf{L}}\mathbf{z} = \mathbf{c}$, with $\mathbf{z} = \mathbf{Q}^*\mathbf{y}$ and $\mathbf{c} = \bar{\mathbf{Q}}\mathbf{b}$. Thus, except for its first and last rows, $\bar{\mathbf{L}}\mathbf{z}= \mathbf{c}$ represents a difference system of the form (3.1), but because each $\mathbf{L}(n)$ is block lower-triangular, the vector components are decoupled as signified in equations (3.8).

In order to start the recurrences in (3.8), suitable methods for computing $\mathbf{z}_1(0)$ and $\mathbf{z}_2(N)$ must be found. Since $\mathbf{z}(0) = \mathbf{Q}^*(0)\mathbf{y}(0)$, a convenient $\mathbf{Q}(0)$ must be chosen so that $\mathbf{z}_1(0)$ can be calculated. The choice of $\mathbf{Q}(0)$ would appear to be crucial to the success of the method, however, for once it is chosen, the sequence $\mathbf{Q}(0), \mathbf{Q}(1), \ldots, \mathbf{Q}(N)$ is uniquely defined by (3.6). Although several techniques for choosing $\mathbf{Q}(0)$ are discussed in [10], we mention only that it is necessary in (3.9) to have $\mathbf{Q}_{12}(0) = \mathbf{0}$ for the explicit computation of $\mathbf{z}_1(0)$ and the successful start of the forward recurrence (3.8a). The most convenient choice of $\mathbf{z}_2(N)$ is zero as usual. The numerical stability of the entire scheme remains to be analyzed.

We note that although a lower triangular matrix is employed in (3.6), the method does *not* correspond to an LU decomposition of the entire matrix $\mathbf{A}$ of (3.5). If the QL decomposition in (3.6) is performed by Householder's method, then the total number of floating point operations of the method described by (3.7) and (3.8) is $[\frac{4}{3}p^3 + 3p^2]N$, and is independent of k.

3.2 Block LU decomposition

We now describe a method [18] which is based on block LU decomposition of the matrix $\mathbf{A}$ in (3.5), and which also permits the estimation of the truncation error to be computed during the forward elimination [20]. The method was developed as a generalization of a matrix form of Miller's algorithm for solving (3.1) when it is homogeneous and when $\mathbf{f}$ is minimal [17].

We rewrite system (3.4) as

$$\mathbf{Ay} = \mathbf{b}, \tag{3.10}$$

where $\mathbf{A}$ is the matrix

$$
\begin{bmatrix}
\mathbf{I}_k & & & & & & & \\
-\mathbf{A}_{21}(1) & -\mathbf{A}_{22}(1) & \mathbf{0} & \mathbf{I}_{p-k} & & & & \\
-\mathbf{A}_{11}(1) & -\mathbf{A}_{12}(1) & \mathbf{I}_k & \mathbf{0} & & & & \\
& & & -\mathbf{A}_{21}(2) & -\mathbf{A}_{22}(2) & \mathbf{0} & \mathbf{I}_{p-k} & \\
& & & -\mathbf{A}_{11}(2) & -\mathbf{A}_{12}(2) & \mathbf{I}_k & \mathbf{0} & \\
& & & & \ddots & & \ddots & \\
& & & & & -\mathbf{A}_{21}(N) & -\mathbf{A}_{22}(N) & \mathbf{0} & \mathbf{I}_{p-k} \\
& & & & & -\mathbf{A}_{11}(N) & -\mathbf{A}_{12}(N) & \mathbf{I}_k & \mathbf{0} \\
& & & & & & & & \mathbf{I}_{p-k}
\end{bmatrix}
$$

and

$$\mathbf{y} = [\mathbf{y}_1(0), \mathbf{y}_2(0), \mathbf{y}_1(1), \mathbf{y}_2(1), \mathbf{y}_1(2), \ldots, \mathbf{y}_2(N-1), \mathbf{y}_1(N), \mathbf{y}_2(N)]^T,$$

$$\mathbf{b} = \left[\mathbf{f}_1(0), \mathbf{b}_2(1), \mathbf{b}_1(1), \mathbf{b}_2(2), \mathbf{b}_1(2), \ldots, \mathbf{b}_2(N), \mathbf{b}_1(N), \tilde{\mathbf{f}}_2(N)\right]^T.$$

113

For notational convenience, the blocks of each $\mathbf{A}(n)$ have been interchanged in order for the square matrix $-\mathbf{A}_{22}(n)$ to appear on the diagonal. Taking advantage of the special structure of $\mathbf{A}$, we obtain by direct LU decomposition the block form given by

$$
\mathbf{L} =
\begin{bmatrix}
\mathbf{I}_k & & & & & & & \\
-\mathbf{A}_{21}(1) & \mathbf{L}_{22}(1) & & & & & & \\
-\mathbf{A}_{11}(1) & \mathbf{L}_{12}(1) & \mathbf{I}_k & & & & & \\
& & -\mathbf{A}_{21}(2) & \mathbf{L}_{22}(2) & & & & \\
& & -\mathbf{A}_{11}(2) & \mathbf{L}_{12}(2) & \mathbf{I}_k & & & \\
& & & & & \ddots & & \\
& & & & & -\mathbf{A}_{21}(N) & \mathbf{L}_{22}(N) & \\
& & & & & -\mathbf{A}_{11}(N) & \mathbf{L}_{12}(N) & \mathbf{I}_k \\
& & & & & & & \mathbf{I}_{p-k}
\end{bmatrix}
$$

$$(3.11a)$$

$$
\mathbf{U} =
\begin{bmatrix}
\mathbf{I}_k & & & & & & & \\
& -\mathbf{U}_{22}(1) & \mathbf{0} & \mathbf{W}_{22}(1) & & & & \\
& & \mathbf{I}_k & -\mathbf{V}_{12}(1) & & & & \\
& & & -\mathbf{U}_{22}(2) & \mathbf{0} & \mathbf{W}_{22}(2) & & \\
& & & & \mathbf{I}_k & -\mathbf{V}_{12}(2) & & \\
& & & & & \ddots & & \\
& & & & & -\mathbf{U}_{22}(N) & \mathbf{0} & \mathbf{W}_{22}(N) \\
& & & & & & \mathbf{I}_k & -\mathbf{V}_{12}(N) \\
& & & & & & & \mathbf{I}_{p-k}
\end{bmatrix},
$$

$$(3.11b)$$

where the $\mathbf{L}_{22}(n)$ are unit lower triangular and the $\mathbf{U}_{22}(n)$ are upper triangular. The individual blocks of (3.11) satisfy the recurrences defined by $\mathbf{V}_{12}(0) = \mathbf{0}$, and for $n = 1, 2, \ldots, N$

$$\mathbf{L}_{22}(n)\mathbf{U}_{22}(n) = \mathbf{A}_{21}(n)\mathbf{V}_{12}(n-1) + \mathbf{A}_{22}(n) \qquad (3.12a)$$

$$\mathbf{L}_{22}(n)\mathbf{W}_{22}(n) = \mathbf{I}_{p-k} \qquad (3.12b)$$

$$\mathbf{L}_{12}(n)\mathbf{U}_{22}(n) = \mathbf{A}_{11}(n)\mathbf{V}_{12}(n-1) + \mathbf{A}_{12}(n) \qquad (3.12c)$$

$$\mathbf{V}_{12}(n) = \mathbf{L}_{12}(n)\mathbf{W}_{22}(n). \qquad (3.12d)$$

The first equation (3.12a) signifies that the LU decomposition $\mathbf{L}_{22}(n)\mathbf{U}_{22}(n)$ of its right side is formed. Then (3.12b) gives an expression for $\mathbf{W}_{22}(n)$, (3.12c) for $\mathbf{L}_{12}(n)$, and finally (3.12d) for $\mathbf{V}_{12}(n)$. Of course, all these matrices are obtained automatically during the normal row-by-row LU decomposition of (3.10), expressions (3.12) serving only to emphasize the relations between them.

The solution of (3.10) then reduces to $\mathbf{Lx} = \mathbf{b}$, or, for $n = 1, 2, \ldots, N$,

$$\mathbf{L}_{22}(n)\mathbf{x}_2(n-1) = \mathbf{A}_{21}(n)\mathbf{x}_1(n-1) + \mathbf{b}_2(n),$$

$$\mathbf{x}_1(n) = \mathbf{A}_{11}(n)\mathbf{x}_1(n-1) - \mathbf{L}_{12}(n)\mathbf{x}_2(n-1) + \mathbf{b}_1(n),$$

followed by the solution of $\mathbf{U}\mathbf{y} = \mathbf{x}$, or $\mathbf{y}_2(N) = \tilde{\mathbf{f}}_2(N)$ and for $n = N, N-1, \ldots, 1,$

$$\mathbf{y}_1(n) = \mathbf{V}_{12}(n)\mathbf{y}_2(n) + \mathbf{b}_2(n),$$

$$\mathbf{U}_{22}(n)\mathbf{y}_2(n-1) = \mathbf{W}_{22}(n)\mathbf{y}_2(n) - \mathbf{x}_2(n-1),$$

and $\mathbf{y}_1(0) = \mathbf{f}_1(0)$.

Although pivoting seems to be a pleasant luxury for scalar equations, it would be absolutely foolhardy *not* to employ pivoting with difference systems. Indeed, the entire success of the LU decomposition depends on the nonsingularity of the matrices on the right hand side of (3.12a), and the method will fail if $\mathbf{A}_{22}(1)$ is singular for instance, which can easily occur in practice. We use a convenient form of pivoting, which causes no fill-in of the matrix of (3.10) and which preserves the structures of (3.11) up to a permutation, a method suggested by Lam [8]. Starting with the second row of blocks in (3.10), $p - k$ steps of row pivoting are performed, followed by k steps of column pivoting, alternating in this way until the last block of order $p - k$ is encountered, where row pivoting is employed. Elimination by rows is used throughout. If we denote the permutation matrices introduced by row pivoting as $\mathbf{P}(n)$ and those introduced by column pivoting as $\mathbf{Q}(n)$, then the pivoted form of the algorithm produces the decomposition $\mathbf{A} = \mathbf{PLUQ}$, where $\mathbf{P}(N+1)$ is of order $p - k$, $\mathbf{Q}(0) = \mathbf{I}$ and

$$\mathbf{P} = \mathbf{diag}\,[\mathbf{I}_k, \mathbf{P}(1), \ldots, \mathbf{P}(N), \mathbf{P}(N+1)], \qquad \mathbf{Q} = \mathbf{diag}\,[\mathbf{Q}(0), \mathbf{Q}(1), \ldots, \mathbf{Q}(N)].$$

In terms of difference equations, the pivoting method has the effect of transforming (3.1) to

$$\mathbf{P}^*(n)\mathbf{Q}(n)\mathbf{y}(n) = \mathbf{P}^*(n)\,[\mathbf{Q}(n)\mathbf{A}(n)\mathbf{Q}^*(n-1)]\,\mathbf{Q}(n-1)\mathbf{y}(n-1) + \mathbf{P}^*(n)\mathbf{Q}(n)\mathbf{b}(n), \tag{3.13}$$

which, after the transformations $\bar{\mathbf{y}}(n) = \mathbf{Q}(n)\mathbf{y}(n)$, $\bar{\mathbf{b}}(n) = \mathbf{Q}(n)\mathbf{b}(n)$ and $\bar{\mathbf{A}}(n) = \mathbf{Q}(n)\mathbf{A}(n)\mathbf{Q}^*(n-1)$, and dropping the $\mathbf{P}^*(n)$, becomes

$$\bar{\mathbf{y}}(n) = \bar{\mathbf{A}}(n)\bar{\mathbf{y}}(n-1) + \bar{\mathbf{b}}(n),$$

an equation of the exact same form as (3.1). Consequently, the complexity of the pivoted algorithm is identical to that of the algorithm without pivoting, the only additional work in the pivoted form being the storing of the permutations.

After some long algebra, the number of arithmetic operations for this system form of the LU algorithm can be calculated as

$$\left[\frac{2}{3}q^3 + (k+1)q^2 + \left(2pk + 5k + \frac{4}{3}\right)q + 2pk\right]N,$$

where $q \equiv p - k$. Consequently, in comparison with the orthogonal decomposition algorithm, the LU scheme requires about $1/2$ the number of operations when k is near 0 (backward recurrence), decreasing to a factor of about $1/[\frac{2}{3}p + \frac{3}{2}]$ when k is approximately equal to p (forward recurrence).

4 CONVERGENCE AND CONDITIONING

Because of the well known method of casting pth order scalar recurrence relations in the system form (1.3), each of the methods presented in the previous sections, whether for scalar difference equations or for linear difference systems, homogeneous or inhomogeneous, can be regarded as a technique for solving the boundary value problem (3.4). If, for the desired solution f, we could specify $\mathbf{f}_2(N)$, then convergence of the solution to (3.4), when it exists, would not have to be discussed at all. The value of $\mathbf{f}_2(N)$ is seldom known, however, so it must either be estimated by an asymptotic formula or be taken to be zero arbitrarily. Thus, the solution of (3.4), which we denote by $\mathbf{y}^{[N]}(n)$ for $n = 0, 1, \ldots, N$, is an estimate of $\mathbf{f}(n)$, and the question of $\mathbf{y}^{[N]}(n)$ converging in the sense that

$$\lim_{N \to \infty} \mathbf{y}^{[N]}(n) = \mathbf{f}(n) \tag{4.1}$$

becomes relevant. Indeed, it seems natural to define convergence as follows.

Definition 4.1 *The solution* $\mathbf{y}^{[N]}(n)$ *of* (3.4) **converges** *to* $\mathbf{f}(n)$ *if and only if, for N sufficiently large, the solution vectors* $\mathbf{y}^{[N]}(n)$, $n = 0, 1, \ldots, N$, *of* (3.4) *exist and are unique, and if* (4.1) *holds for each fixed n.*

Necessary and sufficient conditions for the convergence of the solution of the inhomogeneous boundary value system (3.4) are given in [18], [19] and can be described as follows. We recall that if $\mathbf{f}_1(n), \ldots, \mathbf{f}_p(n)$ form a *fundamental system* of linearly independent solutions to the homogeneous form of equation (3.1), then the matrix $\mathbf{F}(n) = [\mathbf{f}_1(n), \mathbf{f}_2(n), \ldots, \mathbf{f}_p(n)]$ with the solution $\mathbf{f}_i(n)$ as its i-th column, is called a *fundamental matrix*. The following result is proved in [19].

Theorem 4.2 *Let f be the desired solution of the difference system* (3.1). *Then the solution* $\mathbf{y}^{[N]}(n)$ *of* (3.4) *converges to* $\mathbf{f}(n)$ *if and only if* (3.1) *possesses a fundamental matrix* $\mathbf{F}$ *such that*

(a) $\mathbf{F}_{12}(0) = 0$ *and* $\mathbf{F}_{22}^{-1}(N)$ *exists for all N sufficiently large, and*

(b) $\lim\limits_{N \to \infty} \mathbf{F}_{22}^{-1}(N) \left[\tilde{\mathbf{f}}_2(N) - \mathbf{f}_2(N) \right] = 0.$

In other words, $\mathbf{F}(0)$ must be reducible, dividing the complementary solution space into two subspaces containing k and $p-k$ solutions respectively, in such a way that $\mathbf{F}_{22}(N)$ *dominates* the solution $\mathbf{f}_2(N)$ in the sense of the equation given in (b).

Even if a floating point estimate $\tilde{\mathbf{f}}_2(N)$ for $\mathbf{f}_2(N)$ were known to within a unit rounding error ϵ, each component $\tilde{f}_{2j}(N) - f_{2j}(N)$ of the difference $\tilde{\mathbf{f}}_2(N) - \mathbf{f}_2(N)$

would be of the form $\epsilon_j f_{2j}(N)$, where $|\epsilon_j| \leq \epsilon$, and so the presence of $\tilde{\mathbf{f}}_2(N)$ in **(b)** is not particularly relevant. Moreover, as we saw in (3.13), the type of pivoting used for the LU algorithm of §3 has the effect of permuting the components of **f**. Thus, it would be more realistic to consider the permuted solution $\bar{\mathbf{f}}(N) = \mathbf{Q}(N)\mathbf{f}(N)$ and the fundamental matrix $\bar{\mathbf{F}}(n) = \mathbf{Q}(N)\mathbf{F}(N)$ of the difference equation

$$\bar{\mathbf{y}}(n) = [\mathbf{Q}(n)\mathbf{A}(n)\mathbf{Q}^*(n-1)]\,\bar{\mathbf{y}}(n-1). \tag{4.2}$$

On the basis of Theorem 4.2, these ideas lead to the following.

Definition 4.3 *Let* $\{\mathbf{Q}(n)\}$ *be a sequence of permutation matrices, and let* $\bar{\mathbf{F}}(n)$ *be a fundamental matrix for (4.2) such that* $\bar{\mathbf{F}}_{12}(0) = 0$. *Then the subspace of solutions* $\{\mathbf{f}_{k+1}, \mathbf{f}_{k+2}, \ldots, \mathbf{f}_p\}$ *is said to* **dominate** **f** *if and only if*

$$\lim_{N \to \infty} \bar{\mathbf{F}}_{22}^{-1}(N)\bar{\mathbf{f}}_2(N) = 0 \tag{4.3}$$

for all possible $\{\mathbf{Q}(n)\}$.

Evidently, it follows that $\{\mathbf{f}_{k+1}, \ldots, \mathbf{f}_p\}$ dominates **f** if and only if (4.3) holds for all constant **Q**, independent of n. Thus, the concepts of both convergence and dominance can be related by the necessary and sufficient conditions of Theorem 4.2. The value k in (3.4) should be chosen, therefore, in such a way that $\{\mathbf{f}_{k+1}, \mathbf{f}_{k+2}, \ldots, \mathbf{f}_p\}$ is the largest subspace which dominates **f**. The hypotheses of the above theorem also play a role in the study of the condition of (3.4).

The conditioning of difference systems has received little attention in the published literature. We now present briefly some recent results [21], which show how the solution of (3.4) is affected by perturbations in all the data. The context of our discussion is as follows. We suppose that a unique solution **f** of (3.4) exists (Theorem 4.2) and we remark that if $\mathbf{f}(M) = 0$ for some M, $0 \leq M \leq N$, then the recurrence theoretically becomes decoupled at M because $\mathbf{f}(M+1) = \mathbf{b}(M+1)$, and therefore the difference systems for $0 \leq n < M$ and $M < n \leq N$ can be analyzed separately. Thus, we assume without loss of generality that $\mathbf{f}(n) \neq 0$ for $n = 0, 1, \ldots, N$. We also assume that the matrices $\mathbf{A}(n)$ are nonsingular for all n, which assures that each $\mathbf{y}(n-1)$ is uniquely defined in terms of each $\mathbf{y}(n)$, and vice versa.

In order to examine its condition, we suppose that (3.4) is perturbed to a problem of the form

$$\hat{\mathbf{y}}_1(0) = [\mathbf{f}_1(0) + \eta(0)] \tag{4.4a}$$

$$\hat{\mathbf{y}}(n) = [\mathbf{A}(n) + \mathbf{E}(n)]\hat{\mathbf{y}}(n-1) + [\mathbf{b}(n) + \mathbf{e}(n)], \qquad i = 1, 2, \ldots, N \tag{4.4b}$$

$$\hat{\mathbf{y}}(N) = [\tilde{\mathbf{f}}_2(N) + \eta(N)], \tag{4.4c}$$

where each of the relative perturbations is bounded by the parameter ϵ, in the sense that

$$\|\eta(0)\| \leq \|\mathbf{f}_1(0)\|\epsilon, \quad \|\mathbf{E}(n)\| \leq \|\mathbf{A}(n)\|\epsilon, \quad \|\mathbf{e}(n)\| \leq \|\mathbf{b}(n)\|\epsilon, \quad \eta(N)\| \leq \|\tilde{\mathbf{f}}_2(N)\|\epsilon, \tag{4.5}$$

and where the L_∞ vector norm and its subordinate matrix norm are employed. When (4.4) possesses a unique solution, we denote it by $\hat{\mathbf{f}}$ and we measure the condition of problem (3.4) by the *error function*

$$r(n) \equiv \sup_{(4.5)} \frac{\|\hat{\mathbf{f}}(n) - \mathbf{f}(n)\|}{\|\mathbf{f}(n)\|}, \qquad (4.6)$$

the subscript (4.5) signifying that the supremum is taken subject to all conditions in (4.5). The *relative condition number* is defined to be $c(n) \equiv \lim_{\epsilon \to 0} r(n)/\epsilon$, or in other words, the factor of ϵ in a convergent perturbation ϵ-series for $r(n)$, when such a series exists. We say that problem (3.4) is *well-conditioned* if and only if $\sup_{\{0 \le n \le N\}} c(n)$ is bounded by a specified constant which, of course, depends on the context.

Before presenting a theorem on the condition of (3.4), we introduce another notation. For fundamental matrices $\mathbf{F}$, we use the partition $\mathbf{F}(n) = [\mathbf{F}_1(n)\ \mathbf{F}_2(n)]$, where $\mathbf{F}_1(n)$ is of dimension $p \times k$ and $\mathbf{F}_2(n)$ is $p \times (p - k)$. If the matrix has an inverse sign, we write

$$\mathbf{F}^{-1}(n) = \left[\begin{array}{c} [\mathbf{F}^{-1}(n)]_1 \\ [\mathbf{F}^{-1}(n)]_2 \end{array} \right],$$

where $[\mathbf{F}^{-1}(n)]_1$ is of dimension $k \times p$ and $[\mathbf{F}^{-1}(n)]_2$ is $(p - k) \times p$. Therefore, it follows that

$$\mathbf{F}(n)\mathbf{F}^{-1}(s) = \mathbf{F}_1(n)[\mathbf{F}^{-1}(s)]_1 + \mathbf{F}_2(n)[\mathbf{F}^{-1}(s)]_2,$$

a simple expression which demonstrates that the Green's function for (3.4) can be partitioned as the sum of two separate functions of similar form. It is important to note, however, that when $\mathbf{F}$ is the fundamental matrix for (3.4), each of the two expressions on the right depends on the column order in $\mathbf{F}$, whereas the Green's function on the left is independent of that order (indeed, independent of $\mathbf{F}$).

With these notations, we define the amplification factors (cf. [7])

$$G_1(s,n) \equiv \|\mathbf{F}_1(n)[\mathbf{F}^{-1}(s)]_1\| \frac{\|\mathbf{f}(s)\|}{\|\mathbf{f}(n)\|}, \qquad G_2(s,n) \equiv \|\mathbf{F}_2(n)[\mathbf{F}^{-1}(s)]_2\| \frac{\|\mathbf{f}(s)\|}{\|\mathbf{f}(n)\|}. \qquad (4.7)$$

Since $\|\mathbf{F}_1(n)[\mathbf{F}^{-1}(s)]_1\|$ is a measure of the growth of the solutions $\{\mathbf{f}_1, \mathbf{f}_2, \ldots, \mathbf{f}_k\}$ over the interval s to $n > s$, the factor $G_1(s,n)$ is a measure of that growth divided by the growth of $\mathbf{f}$ over the same interval, and will be small if $\{\mathbf{f}_1, \mathbf{f}_2, \ldots, \mathbf{f}_k\}$ does not dominate $\mathbf{f}$. Similarly, $G_2(s,n)$ will be small, but this time for $n < s$, if $\{\mathbf{f}_{k+1}, \mathbf{f}_{k+2}, \ldots, \mathbf{f}_p\}$ dominates f over the interval n to s. The following theorem [21] demonstrates that the boundary value problem incorporates the best behaviors of both $G_1(s,n)$ and $G_2(s,n)$ as long as the dominance relations are respected.

Theorem 4.4 *Suppose that* $\{\mathbf{f}_{k+1}, \mathbf{f}_{k+2}, \ldots, \mathbf{f}_p\}$ *dominates* $\mathbf{f}$ *(Definition 4.3), and let* N *be large enough so that*

$$\|\bar{\mathbf{F}}_{22}^{-1}(N)\bar{\mathbf{f}}_2(N)\| \le \epsilon.$$

Let $G_1(s, n)$ and $G_2(s, n)$ be calculated by (4.7). Then the relative condition number of (3.4) satisfies

$$c(n) \leq G_1(0, n) + \sum_{s=1}^{n} G_1(s, n)[1 + G_1(s-1, s)]$$

$$+ G_2(N, n) + \sum_{s=n+1}^{N} G_2(s, n)[1 + G_2(s, s-1)]$$

Moreover, if $\mathbf{f}_1(0) \neq \mathbf{0}$ and $\mathbf{f}_2(N) \neq \mathbf{0}$, then $c(n) \geq G_1(0, n) + G_2(N, n)$.

According to this theorem, therefore, $c(n)$ will be small when the boundary value system (3.4) converges, and as long *there exists* a partition $\mathbf{F}(n) = [\mathbf{F}_1(n) \ \mathbf{F}_2(n)]$ such that $\mathbf{F}_1(n)$ does not dominate $\mathbf{f}$, and such that $\mathbf{F}_2(n)$ does dominate $\mathbf{f}$.

Although the analyses of convergence and conditioning benefit from the uniform representation of algorithms for difference equations, single results applying to any linear system (3.4) on which such algorithms are based, the same cannot be said about stability. The important point is that we are not interested in the error of the total vector $\mathbf{y} = [\mathbf{y}^T(0), \mathbf{y}^T(1), \ldots, \mathbf{y}^T(N)]^T$ of (3.4) to which the analysis of Wilkinson can be applied, but in the errors of the individual vectors $\mathbf{y}(n)$, which depend on the detailed operation of the independent algorithms. A great deal of work remains to be done in the area of stability for recurrence algorithms.

REFERENCES

[1] Aggarwal V.B., Burgmeier J.W. An algorithmic decomposition system for second order recurrence relations. *BIT* **18**:241–258, 1978.

[2] Arscott F.M., Taylor P.J., Zahar R.V.M. On the numerical construction of ellipsoidal wave equations. *Math. Comp.*, **40**:367–380, 1983.

[3] British Association for the Advancement of Science. *Bessel Functions, Part I, Mathematical Tables v. 10.*, Cambridge University Press, Cambridge, 1952.

[4] Cash J.R. A note on Olver's algorithm for the solution of second-order linear difference equations. *Math. Comp.*, **35**:767–772, 1980.

[5] Fox L. The solution by relaxation methods of ordinary differential equations. *Proc. Camb. Phil. Soc.*, **45**:50–68, 1949.

[6] Gautschi W. Computational aspects of three-term recurrence relations. *SIAM Review*, **9**:24–83, 1967.

[7] Gautschi W. Zur Numerik rekurrenter Relationen. *Computing*, **9**:107–126, 1972. (English translation in Aerospace Research Laboratories, Report ARL 73-0005, February 1973.)

[8] Lam D.C.L. *Implementation of the box scheme and model analysis of diffusion-correction equations.* Ph.D. dissertation. University of Waterloo, Waterloo, Ontario, 1974.

[9] Lozier D.W. *Numerical solution of linear difference equations.* Ph.D. dissertation. University of Maryland, College Park, Maryland, 1980.

[10] Mattheij R.M.M. Stable computation of dominated solutions of linear recursions. Report 7803, Mathematisch Instituut, Nijmegen, 1978.

[11] Oliver J. The numerical solution of linear recurrence relations. *Numer. Math.*, 11:349–360, 1968.

[12] Olver F.W.J. Numerical solution of second-order linear difference equations. *Jour. Res. Nat. Bur. Standards*, **71B**:111–129, 1967.

[13] Olver F.W.J., Sookne D.J. Note on backward recurrence algorithms. *Math. Comp.*, **26**:941–947, 1972.

[14] Van der Cruyssen P. A reformulation of Olver's algorithm for the numerical solution of second-order linear difference equations. *Numer. Math.*, **32**:159–166, 1979.

[15] Wimp J. *Computation with Recurrence Relations.* Pitman, Boston, London, Melbourne, 1984.

[16] Zahar R.V.M. *Computational algorithms for linear difference equations.* Ph.D. dissertation. Purdue University, Lafayette, Indiana, 1968.

[17] Zahar R.V.M. A mathematical analysis of Miller's algorithm. *Numer. Math.*, **27**:427–447, 1977.

[18] Zahar R.V.M. The computation of subdominant solutions of linear difference systems. Report 434, Département d'IRO, Université de Montréal, 1982.

[19] Zahar R.V.M. Convergence of asymptotic staircase systems. *Cong. Numer.*, **51**:257–264, 1986.

[20] Zahar R.V.M. A generalization of Olver's algorithm for linear difference systems. In *Asymptotic and Computational Analysis*, pages 535–551. Dekker, New York and Basel, 1990. R. Wong, ed.

[21] Zahar R.V.M. The condition of difference systems and of block bidiagonal linear systems. To be published.

International Series of Numerical Mathematics, Vol. 119, ©1994 Birkhäuser

QUASI-INTERPOLATION ON IRREGULAR POINTS

E. W. Cheney*   Junjiang Lei*

Dedicated to Walter Gautschi on the occasion of his 65th birthday

Abstract.   A quasi-interpolant is an operator L having the form

$$Lf = \sum_{i=1}^{\infty} f(y_i)g_i. \tag{1.1}$$

The points y_i are called "nodes"; they are prescribed in $\mathbb{R}^n$. The entities g_i are prescribed functions from $\mathbb{R}^n$ to $\mathbb{R}$. The case of irregularly situated nodes is of particular interest. We investigate the question of how to select the "base functions" g_i in order to obtain favorable estimates of $\|Lf - f\|$.

1   INTRODUCTION

We assume throughout that the nodes y_i are points in $\mathbb{R}^n$, where n is fixed. The function f is defined on the sequence $[y_i]$, or possibly on some larger subset of $\mathbb{R}^n$. The base functions g_i are real valued and defined everywhere in $\mathbb{R}^n$. The terminology "quasi-interpolation" was apparently introduced around 1971 by Birkhoff, de Boor, and Fix. Early references are [15, 6]. However, there are many prior examples of operators having the form of a quasi-interpolant. One has only to recall the Bernstein polynomial operators or the operators introduced by Schoenberg in 1946 [20] as significant instances. Indeed, quasi-interpolation has played an important rôle in approximation theory for many years.

When the set of nodes is the integer lattice $\mathbb{Z}^n$, one can sometimes allow the functions g_i to be shifts of a single basic function. This case has been extensively studied. See [7, 16, 17]. A single basic function can also be used when the nodes are irregular. Buhmann, Dyn and Levin [12] obtained such a quasi-interpolant by using radial basis functions.

Here we take a different approach. One procedure included in our theory is to start with a partition of $\mathbb{R}^n$ into (possibly irregular) cells, then interpolate the function on each cell by a polynomial, and finally smooth the resulting piecewise polynomial by convolution with a smooth kernel.

In all cases, we seek error estimates of the following generic type:

$$\|Lf - f\| \le c|f|_m \delta^m, \tag{1.2}$$

---

* Department of Mathematics, University of Texas at Austin, Austin, TX 78712

where the norm on the left may be a supremum norm on $\mathbb{R}^n$, $|f|_m$ is a Sobolev seminorm involving only m-th order derivatives of f, and δ is the density of the node-set, defined to be the smallest number r such that each closed ball of radius r contains a node. An estimate of the form given in (1.2) will allow the user to have confidence in the quality of the approximation, especially if it is feasible to combine L with the "dilation" (or "scaling") operators S_h defined by

$$(S_h f)(x) = f(hx). \tag{1.3}$$

2   TERMINOLOGY AND NOTATION

Some terminology is described here. We denote by $\Pi_m(\mathbb{R}^n)$ the linear space of all polynomials of degree at most m in n real variables. A basis for this space is

$$\{x \mapsto x^\alpha : \alpha \in \mathbb{Z}_+^n \ , \ |\alpha| \le m\}. \tag{2.1}$$

(Usual multi-index notation is used.) The dimension of $\Pi_m(\mathbb{R}^n)$ is $\binom{n+m}{n}$.

Let $\{x_1, \ldots, x_N\}$ be a finite set of $\mathbb{R}^n$. We say that this set is in *general position* with respect to $\Pi_m(\mathbb{R}^n)$ if

$$N = \dim \Pi_m(\mathbb{R}^n) = \binom{n+m}{n} \tag{2.2}$$

and

$$\sum_{i=1}^N |p(x_i)| > 0 \quad \text{for each nonzero } p \text{ in } \Pi_m(\mathbb{R}^n). \tag{2.3}$$

Corresponding to any point $x \in \mathbb{R}^n$ there is a *point evaluation functional* x^*, whose definition is

$$x^*(f) = f(x). \tag{2.4}$$

With this terminology, we can say that a set $\{x_1, \ldots, x_N\}$ is in general position with respect to $\Pi_m(\mathbb{R}^n)$ if the set $\{x_1^*, \ldots, x_N^*\}$ is a basis for the algebraic dual of $\Pi_m(\mathbb{R}^n)$. Equivalently, if $\{p_1, \ldots, p_N\}$ is a basis for $\Pi_m(\mathbb{R}^n)$, then

$$\det\big(p_i(x_j)\big)_{i,j=1}^N \ne 0. \tag{2.5}$$

A *partition of unity* in $\mathbb{R}^n$ is a sequence of functions $\theta_1, \theta_2, \ldots$ in $L^1(\mathbb{R}^n)$ such that $\sum_{i=1}^\infty \theta_i(x) = 1$ for all $x \in \mathbb{R}^n$.

If $\{x_1, \ldots, x_N\}$ is in general position with respect to $\Pi_m(\mathbb{R}^n)$, then there exist polynomials $\ell_1, \ldots, \ell_N$ in $\Pi_m(\mathbb{R}^n)$ such that $\ell_j(x_i) = \delta_{ij}$. We shall call $\{\ell_1, \ldots, \ell_N\}$ the *Lagrangian basis* of $\Pi_m(\mathbb{R}^n)$ associated with $\{x_1, \ldots, x_N\}$. We have then the familiar Lagrange interpolation process: the polynomial $p = \sum_{j=1}^N c_j \ell_j$ has the property $p(x_i) = c_i$ for $1 \le i \le N$.

Various linear operators are needed. They are

$$(f * g)(x) = \int_{\mathbb{R}^n} f(x - y)g(y)\,dy \qquad (x \in \mathbb{R}^n) \tag{2.6}$$

$$(S_\lambda f)(x) = f(\lambda x) \qquad (\lambda \in \mathbb{R}) \tag{2.7}$$

$$(T_v f)(x) = f(x - y) \qquad (v, x \in \mathbb{R}^n). \tag{2.8}$$

These are called "convolution," "scaling," and "translation."

The Sobolev space $W_{k,\infty}(\mathbb{R}^n)$ contains all functions $f : \mathbb{R}^n \to \mathbb{R}$ whose derivatives $f^{(\alpha)}$ are absolutely continuous when $|\alpha| < k$ and essentially bounded when $|\alpha| = k$. We put

$$|f|_{j,\infty} = \sum_{|\alpha|=j} \|f^{(\alpha)}\|_\infty \text{ and } \|f\|_{k,\infty} = \sum_{j=1}^{k} |f|_{j,\infty}. \tag{2.9}$$

We use $B(x, r) = \{y \in \mathbb{R}^n : |y - x| \le r\}$.

3 GENERAL APPROXIMATION OPERATORS

In this section we discuss approximation by "scaling" or "dilating" a linear operator.

Let E be a normed linear space whose elements are functions from $\mathbb{R}^n$ to $\mathbb{R}$. Let L be a linear operator from E into E. We can interpret Lf as an approximation to the function f, and then seek to understand how good an approximation it is. Thus, estimates of $\|Lf - f\|_\infty$ are called for.

Let S_λ denote a scaling operator defined in Equation (2.7). We assume that $0 < \lambda < \infty$. There are many situations in approximation theory where interest is focused on the family of operators $S_\lambda^{-1} L S_\lambda$. In particular, it is desirable to have estimates of the form

$$\|S_\lambda^{-1} L S_\lambda f - f\|_\infty = O(\lambda^m) \text{ as } \lambda \to 0. \tag{3.1}$$

An example of this conclusion arises in the next result, which is well-known and elementary.

Lemma 3.1 *Let L be an operator on the Sobolev space $W_{m,\infty}(\mathbb{R}^n)$. The following properties of L are equivalent:*

$$\|Lf - f\|_\infty \le c|f|_{m,\infty} \qquad (f \in W_{m,\infty}); \tag{3.2}$$

$$\|S_\lambda^{-1} L S_\lambda f - f\|_\infty \le c|f|_{m,\infty}\lambda^m \qquad (f \in W_{m,\infty}, \; \lambda > 0). \tag{3.3}$$

Proof: We obtain (3.2) by letting $\lambda = 1$ in (3.3). Conversely,

$$\|S_\lambda^{-1} L S_\lambda f - f\|_\infty = \|S_\lambda^{-1}(LS_\lambda - S_\lambda)f\|_\infty$$
$$= \|LS_\lambda f - S_\lambda f\|_\infty \le c|S_\lambda f|_{m,\infty} = c|f|_{m,\infty}\lambda^m. \qquad \blacksquare$$

Lemma 3.1 indicates that in order to have rapid convergence of $S_\lambda^{-1} L S_\lambda f$ to f as $\lambda \to 0$, we should consider operators L that possess error bounds like the one in Inequality (3.2). The next lemma provides a condition on L for this.

Lemma 3.2 *Let L be a linear operator from $C(\mathbb{R}^n)$ into $C(\mathbb{R}^n)$ such that for some $m \in \mathbb{N}$ and $A > 0$,*

$$|(Lf)(x)| \le A \sup_{y \in \mathbb{R}^n} \frac{|f(y)|}{(1 + |x - y|)^m} \qquad (f \in C(\mathbb{R}^n) , \ x \in \mathbb{R}^n). \qquad (3.4)$$

(This inequality is vacuous for any function giving $+\infty$ on the right.) For such an L the following two properties are equivalent:

$$Lp = p \quad \text{for all} \ \ p \in \Pi_{m-1}(\mathbb{R}^n); \qquad (3.5)$$

$$\|Lf - f\|_\infty \le A|f|_{m,\infty} \quad \text{for all} \ \ f \in W_\infty(\mathbb{R}^n). \qquad (3.6)$$

Proof: Obviously (3.6) implies (3.5) because each $p \in \Pi_{m-1}$ satisfies $|p|_{m,\infty} = 0$. To prove the other implication, assume (3.5). Let p be the Taylor polynomial of order $m - 1$ for f at x. Let $r = f - p$. By the remainder formula in Taylor's theorem, we have

$$|(Lf)(x) - f(x)| = |(Lf)(x) - p(x)| = |(L(f - p))(x)| = |(Lf)(x)|$$

$$\le A \sup_{y \in \mathbb{R}^n} \frac{|r(y)|}{(1 + |x - y|)^m} = A \sup_{y \in \mathbb{R}^n} \left| \sum_{|\alpha|=m} \frac{f^\alpha(\xi(y)) \frac{1}{\alpha!}(y - x)^\alpha}{(1 + |y - x|)^m} \right|$$

$$\le A \sup_{y \in \mathbb{R}^n} \sum_{|\alpha|=m} |f^\alpha(\xi(y))| = A|f|_{m,\infty}. \qquad \blacksquare$$

For want of a better term, we call the property expressed in Inequality (3.4) the "local decay property of order m." Combining Lemmas 3.1 and 3.2, we arrive at the equivalence of approximation order with polynomial reproduction. Although this equivalence has long been known for quasi-interpolation (see, for example, [7, 20]), it is of interest to establish this result for a broader class of linear operators. The formal statement follows.

Theorem 3.3 *Let L be a linear operator from $C(\mathbb{R}^n)$ into $C(\mathbb{R}^n)$ that has the local decay property of order m, (3.4). Then the following two statements are equivalent:*

(a) $Lp = p$ *for all* $p \in \Pi_{m-1}(\mathbb{R}^n)$;

(b) $\|S_\lambda^{-1} L S_\lambda f - f\| = O(\lambda^m)$ *for* $f \in W_{m,\infty}$.

A different hypothesis, termed "uniformly local," has been used by de Boor in [4]. He pointed out that if a quasi-interpolant is uniformly local and satisfies (a) above, then it satisfies (b). An operator L is defined to be uniformly local if there exist positive constants B and r such that for $f \in C(\mathbb{R}^n)$ and $x \in \mathbb{R}^n$,

$$|(Lf)(x)| \le B \sup_{|y-x| \le r} |f(y)|. \qquad (3.7)$$

Lemma 3.4 *If L is uniformly local then it has the local decay property in Equation (3.4) for each $m = 1, 2, 3, \ldots$.*

Proof: Assume the property in Inequality (3.7). Then

$$|(Lf)(x)| \le B \sup_{|x-y|\le r} |f(y)| = B \sup_{|x-y|\le r} \frac{|f(y)|(1+|x-y|)^m}{(1+|x-y|)^m}$$

$$\le B(1+r)^m \sup_{y\in\mathbb{R}^n} \frac{|f(y)|}{(1+|x-y|)^m}. \qquad \blacksquare$$

Our next goal, Theorem 3.6, will indicate some conditions on the general quasi-interpolation operator of Equation (1.1) under which the local decay property will hold.

Lemma 3.5 *If* $[y_1, y_2, \ldots]$ *is a sequence in* $\mathbb{R}^n$ *such that* $2\sigma := \inf_{i\ne j} |y_i - y_j| > 0$, *then for any* $\gamma < -n$,

$$\sum_{i=1}^{\infty} \{1 + |y_i|\}^{\gamma} \le 1 + \left(\frac{3}{\sigma}\right)^n |n + \gamma|^{-1}.$$

Proof: The sequence of open cells $B(y_i, \sigma)$ is disjoint. If an inequality $\nu\sigma \le |y_i| \le (\nu+1)\sigma$ is true for $1 \le i \le k$ and for some fixed $\nu \ge 1$, then

$$\bigcup_{i=1}^{k} B(y_i, \sigma) \subset B\big(0, (\nu+2)\sigma\big) \setminus B\big(0, (\nu-1)\sigma\big).$$

By considering the Lebesgue measures of these cells, we conclude that

$$k\sigma^n \le \big[(\nu+2)\sigma\big]^n - \big[(\nu-1)\sigma\big]^n$$

and therefore $k \le 3^n \nu^{n-1}$, as is easily proved using the binomial theorem. Then

$$\sum_{i=1}^{\infty} \{1 + |y_i|\}^{\gamma} = \sum_{|y_i|<\sigma} \big(1 + |y_i|\big)^{\gamma} + \sum_{\nu=1}^{\infty} \sum_{\nu\sigma\le|y_i|<(\nu+1)\sigma} \big(1 + |y_i|\big)^{\gamma}$$

$$\le 1 + \sum_{\nu=1}^{\infty} 3^n \nu^{n-1}(1 + \sigma\nu)^{\gamma} = 1 + 3^n \sum_{\nu=1}^{\infty} \left(\frac{\nu}{1+\sigma\nu}\right)^{n-1} (1 + \sigma\nu)^{\gamma+n-1}$$

$$\le 1 + 3^n \sigma^{1-n} \int_0^{\infty} (1 + \sigma t)^{\gamma+n-1}\, dt = 1 + (3/\sigma)^n |\gamma + n|^{-1}. \qquad \blacksquare$$

Theorem 3.6 *Refer to Equation (1.1), and assume that*

$$\inf_{i\ne j} |y_i - y_j| > 0, \qquad (3.8)$$

$$\sup_x \sup_i |g_i(x)|\big(1 + |x - y_i|\big)^{\gamma} < \infty \qquad (3.9)$$

for some $\gamma > m+n$. *Then the operator* L *has the local decay property of order* m, *as in Inequality (3.4).*

Proof: Denote the left side of Inequality (3.9) by M. Then

$$|(Lf)(x)| \le \sum_{i=1}^{\infty} |f(y_i)| \, |g_i(x)| \le \sum_{i=1}^{\infty} |f(y_i)| M \left(1 + |x - y_i|\right)^{-\gamma}$$

$$= \sum_{i=1}^{\infty} |f(y_i)| \left(1 + |x - y_i|\right)^{-m} M \left(1 + |x - y_i|\right)^{-\gamma + m}$$

$$\le M \sup_{y \in \mathbb{R}^n} \frac{|f(y)|}{\left(1 + |x - y|\right)^m} \sum_{i=1}^{\infty} \left(1 + |x - y_i|\right)^{-\gamma + m}.$$

The series in the last estimate is finite, by Lemma 3.5. In fact, it has an upper bound independent of x, because Lemma 3.5 can be applied to the set $\{x - y_i : i \in \mathbb{N}\}$. ∎

4   QUASI-INTERPOLATION WITH GROUPED SCATTERED NODES

In this section, it is assumed that a partition of unity $\{\theta_i : i \in \mathbb{N}\}$ has been chosen in $\mathbb{R}^n$. This means that $\theta_i \in L^1(\mathbb{R}^n)$ for all i and that

$$\sum_{i=1}^{\infty} \theta_i(x) = 1 \qquad (x \in \mathbb{R}^n). \tag{4.1}$$

For each i there is given also a set of nodes

$$Y_i = \{y_{i1}, y_{i2}, \ldots, y_{iN}\} \tag{4.2}$$

that is in general position with respect to $\Pi_{m-1}(\mathbb{R}^n)$. Thus, in particular, $N = \dim \Pi_{m-1}(\mathbb{R}^n)$. Two further conditions are imposed on these node sets:

$$d := \sup_i \operatorname{diam}(Y_i) < \infty \tag{4.3}$$

$$\sigma := \inf_{i \ne j} \frac{1}{2} |y_{i1} - y_{j1}| > 0. \tag{4.4}$$

Since each Y_i is in general position with respect to $\Pi_{m-1}(\mathbb{R}^n)$, there exist basic Lagrange polynomials characterized by the conditions

$$\ell_{ij} \in \Pi_{m-1}(\mathbb{R}^n) \qquad (i \in \mathbb{N}, \ 1 \le j \le N) \tag{4.5}$$

$$\ell_{ij}(y_{ik}) = \delta_{jk} \qquad (i \in \mathbb{N}, \ 1 \le j, k \le N). \tag{4.6}$$

We can now define a quasi-interpolant operator M by the equation

$$Mf = \sum_{i=1}^{\infty} \sum_{j=1}^{N} f(y_{ij}) \theta_i \ell_{ij}. \tag{4.7}$$

Lemma 4.1 *If $\gamma < -n$ then for all $x \in \mathbb{R}^n$,*

$$\sum_{i=1}^{\infty}\sum_{j=1}^{N}(1+|x-y_{ij}|)^{\gamma} \leq N\left\{1+\left(\frac{2d}{\sigma}+1\right)^n+\frac{(6/\sigma)^n}{|n+\gamma|}\right\}. \qquad (4.8)$$

Proof: The hypotheses on $\{y_{ij}\}$, as expressed in (4.3) and (4.4), remain true if these points are translated by any fixed vector. Consequently, we may assume $x = 0$ in the inequality to be proved, (4.8). The sum on the left side is broken into two parts for separate analysis. Let

$$S_1 := \sum_{|y_{i1}|\leq 2d}\sum_{j=1}^{N}(1+|y_{ij}|)^{\gamma}. \qquad (4.9)$$

The number of indices i that satisfy $|y_{i1}| \leq 2d$ is at most $(2d/\sigma+1)^n$, as we now shall prove. If, say, $|y_{i1}| \leq 2d$ for $1 \leq i \leq k$ then

$$\bigcup_{i=1}^{k} B(y_{i1},\sigma) \subset B(0, 2d+\sigma). \qquad (4.10)$$

By the definition of σ in (4.4), the open balls $B(y_{i1},\sigma)$ are mutually disjoint. Hence, by considering Lebesgue measures in (4.10), we conclude that

$$k\sigma^n \leq (2d+\sigma)^n$$

or $k \leq (2d/\sigma+1)^n$, as asserted. Thus, from (4.9) we have

$$S_1 \leq \left(\frac{2d}{\sigma}+1\right)^n N. \qquad (4.11)$$

Now define

$$S_2 := \sum_{|y_{i1}|> 2d}\sum_{j=1}^{N}(1+|y_{ij}|)^{\gamma}. \qquad (4.12)$$

For the indices i involved in S_2 we have

$$|y_{i1}| \leq |y_{i1}-y_{ij}|+|y_{ij}| \leq d+|y_{ij}| \leq \frac{1}{2}|y_{i1}|+|y_{ij}|.$$

Hence $|y_{ij}| \geq \frac{1}{2}|y_{i1}|$, and

$$S_2 \leq \sum_{|y_{i1}|> 2d}\sum_{j=1}^{N}(1+\frac{1}{2}|y_{i1}|)^{\gamma} = N\sum_{|y_{i1}|> 2d}(1+\frac{1}{2}|y_{i1}|)^{\gamma}.$$

Applying Lemma 3.5 to the sequence $\{\frac{1}{2}y_{i1}\}$ we obtain

$$S_2 \leq N\left[1+\left(\frac{6}{\sigma}\right)^n|n+\gamma|^{-1}\right]. \qquad (4.13)$$

The estimates (4.11) and (4.13) complete the proof. $\blacksquare$

Theorem 4.2 *If there is a γ in the interval $(-\infty, -n - m)$ such that*

$$|\theta_i(x)\ell_{ij}(x)| \leq c\big(1 + |x - y_{ij}|\big)^{\gamma} \qquad (i \in \mathbb{N} , \ 1 \leq j \leq N , \ x \in \mathbb{R}^n) \qquad (4.14)$$

then the operator M has the local decay property (3.4) and the polynomial reproduction property (3.5).

Proof: From Equation (4.7), we have

$$
\begin{aligned}
|(Mf)(x)| &\leq \sum\sum |f(y_{ij})| c\big(1 + |x - y_{ij}|\big)^{\gamma} \\
&= c\sum\sum \frac{|f(y_{ij})|}{\big(1 + |x - y_{ij}|\big)^m}\big(1 + |x - y_{ij}|\big)^{\gamma + m} \\
&\leq c\sup_y \frac{|f(y)|}{\big(1 + |x - y|\big)^m}\sum\sum\big(1 + |x - y_{ij}|\big)^{\gamma + m}.
\end{aligned}
$$

An application of the preceding lemma gives

$$|(Mf)(x)| \leq A\sup_y \frac{|f(y)|}{\big(1 + |x - y|\big)^m}$$

with

$$A = cN\left[1 + \left(\frac{2d}{\sigma} + 1\right)^n + \left(\frac{6}{\sigma}\right)^n |n + m + \gamma|^{-1}\right].$$

The polynomial reproduction property is immediate: For $p \in \Pi_{m-1}(\mathbb{R}^n)$,

$$Mp = \sum_i \theta_i \sum_j p(y_{ij})\ell_{ij} = \sum_i \theta_i p = p. \qquad \blacksquare$$

Theorem 4.3 *Let H be a function in $L^1(\mathbb{R}^n)$ such that for some $m \in \mathbb{N}$, $\gamma < -m - n$, $c_1 > 0$*

$$|H(x)| \leq c_1\big(1 + |x|\big)^{\gamma} \qquad (4.15)$$

$$\int H(x)\,dx = 1 \ , \ \int H(x)x^{\alpha}\,dx = 0 \ \text{ for } \ 0 < |\alpha| < m. \qquad (4.16)$$

Assume also hypothesis (4.14). Then the operator $Lf = H * Mf$ has the local decay property (3.4) and the polynomial reproduction property (3.5).

Proof: A lemma in [17] shows that if two functions on $\mathbb{R}^n$ decay at a rate s (where $s < -n$) then the same is true of their discrete convolution. The proof applies also to the continuous convolution. According to (4.14), the function $g(x) = (\theta_i\ell_{ij})(x + y_{ij})$ satisfies an inequality like (4.15). The same is therefore true of $H * g$, and this implies that

$$|(H * \theta_i\ell_{ij})(x)| \leq c_3\big(1 + |x - y_{ij}|\big)^{\gamma}.$$

The constant c_3 does not depend on i or j. Now we have

$$|(Lf)(x)| \le \sum_{i=1}^{\infty} \sum_{j=1}^{N} |f(y_{ij})| \, |(H * \theta_i \ell_{ij})(x)|$$

$$\le c_3 \sum_{i=1}^{\infty} \sum_{j=1}^{N} |f(y_{ij})| (1 + |x - y_{ij}|)^{\gamma}.$$

From this point on, the calculation in the preceding proof can be applied.

For the polynomial reproduction property, we note first that (4.16) allows us to conclude that $H * p = p$ for all $p \in \Pi_{m-1}$. Indeed, if $p(x) = x^{\alpha}$, $|\alpha| < m$, then

$$(H * p)(x) = \int H(y)(x - y)^{\alpha} \, dy$$

$$= \int H(y) \sum_{0 \le \beta \le \alpha} \binom{\alpha}{\beta} (-1)^{\beta} y^{\beta} x^{\alpha - \beta} \, dy.$$

By (4.16) the sum in this last line reduces to the sole term corresponding to $\beta = 0$. Then the first equation in (4.16) completes the proof. Finally, we use Theorem 4.2 to get $Lp = H * Mp = H * p = p$, for all $p \in \Pi_{m-1}(\mathbb{R}^n)$. $\blacksquare$

Remark. It is well-known that the moment conditions expressed in Equations (4.16) are equivalent to the following conditions on the Fourier transform $\widehat{H}$ of H:

$$\widehat{H}(0) = 1 \tag{4.17}$$

$$(D^{\alpha}\widehat{H})(0) = 0 \quad \text{for} \ \ 0 < |\alpha| < m. \tag{4.18}$$

5 THE GROUPING OF NODES

In the preceding section, the nodes to be used for quasi-interpolation were presented as a doubly subscripted array $\{y_{ij}\}$ and were assumed to have properties enumerated in (4.2), (4.3), and (4.4).

Now we wish to begin with a sequence $Y = \{y_i\}$ that has not been grouped in the manner of Section 4. We assume only that Y has finite density δ. The number δ has the property that each closed ball of radius δ intersects Y, and it is the smallest such number. The objective is to select a grouped subset of Y that can be used to construct the quasi-interpolants M and L of Section 4. The functions θ_i can be taken to be characteristic functions of a set of cubes that partition $\mathbb{R}^n$ in a regular manner.

Lemma 5.1 *If the set $\{z_1, z_2, \ldots, z_N\}$ is in general position with respect to $\Pi_{m-1}(\mathbb{R}^n)$, then the same is true of $\{\lambda z_1, \lambda z_2, \ldots, \lambda z_N\}$ and $\{z_1 + y, z_2 + y, \ldots, z_N + y\}$ for any $\lambda \in \mathbb{R} \setminus 0$ and any $y \in \mathbb{R}^n$.*

Proof: Select a basis $\{p_1, p_2, \ldots, p_N\}$ for $\Pi_{m-1}(\mathbb{R}^n)$ such that each p_i is a monomial, x^α. Since $(\lambda x)^\alpha = \lambda^{|\alpha|} x^\alpha$, we see that $\det(p_i(\lambda z_j)) = \lambda^M \det(p_i(z_j))$, where $M = \sum\{|\alpha| : |\alpha| < m\}$. Thus $\{\lambda z_1, \lambda z_2, \ldots, \lambda z_N\}$ is in general position if $\lambda \neq 0$. Now define $q_i(x) = p_i(x - y)$. It is easily verified that $\{q_1, q_2, \ldots, q_N\}$ is also a basis for $\Pi_{m-1}(\mathbb{R}^n)$. Since

$$\det\big(q_i(z_j + y)\big) = \det\big(p_i(z_j)\big) \neq 0,$$

we see that $\{z_1 + y, z_2 + y, \ldots, z_N + y\}$ is in general position. ∎

Lemma 5.2 *If* $\{z_1, z_2, \ldots, z_N\}$ *is in general position with respect to* $\Pi_{m-1}(\mathbb{R}^n)$, *then there is an* $\varepsilon > 0$ *such that for any norm* $\| \cdot \|$ *on* $\Pi_{m-1}(\mathbb{R}^n)$,

$$\inf_{\substack{x_1, \ldots, x_N \in \mathbb{R}^n \\ |x_i - z_i| \leq \varepsilon}} \inf_{\substack{p \in \Pi_{m-1}(\mathbb{R}^n) \\ \|p\| = 1}} \max_{1 \leq i \leq N} |p(x_i)| > 0. \tag{5.1}$$

Proof: Let $\{p_1, p_2, \ldots, p_N\}$ be any basis for $\Pi_{m-1}(\mathbb{R}^n)$. Then the determinant of $(p_i(z_j))$ is not zero. by continuity, there is an $\varepsilon > 0$ such that $\det(p_i(x_j)) \neq 0$ whenever $|x_i - z_i| \leq \varepsilon$ for $1 \leq i \leq N$. Let $\{x_i\}$ satisfy this condition. Then $\{x_i\}$ is in general position. Hence if $\|p\| = 1$ then $\max_i |p(x_i)| > 0$. By compactness and continuity we arrive at (5.1). ∎

If the left side of Equation (5.1) is denoted by $1/M$ then we have, by homogeneity,

$$\|p\| \leq M \max_i |p(x_i)| \qquad \big(p \in \Pi_{m-1}(\mathbb{R}^n),\ |x_i - z_i| \leq \epsilon\big). \tag{5.2}$$

Observe that M depends on the norm $\| \cdot \|$, on the set $\{z_i\}$, and on the number ε.

Theorem 5.3 *If* δ, n, *and* m *are given, then there correspond two positive numbers* r *and* M *having the following property. If* Y *is a set of density* δ *in* $\mathbb{R}^n$ *and if* Q *is a cube* $u + [0, r)^n$, *then* $Y \cap Q$ *contains points* $y_1, y_2, \ldots, y_N$ *such that*

$$\sup_{x \in Q} |p(x)| \leq M \max_{1 \leq i \leq N} |p(y_i)| \qquad \big(p \in \Pi_{m-1}(\mathbb{R}^n)\big). \tag{5.3}$$

Proof: Let $\{z_1, z_2, \ldots, z_n\}$ be a set in general position with respect to $\Pi_{m-1}(\mathbb{R}^n)$. Let ε be given by Lemma 5.2. Select r so that $Q^* = v + [0, r)^n \supset \bigcup_{i=1}^N B((\delta/\varepsilon)z_i, \delta)$ for some v. Put $\|p\| = \sup_{x \in Q^*} |p(\varepsilon x/\delta)|$. By Lemma 5.2 there is an M such that for $p \in \Pi_{m-1}(\mathbb{R}^n)$ we have

$$|x_i - z_i| \leq \varepsilon \ \Rightarrow\ \|p\| \leq M \max_i |p(x_i)|. \tag{5.4}$$

Now let $Q = u + [0, r)^n$. Since Y has density δ, we can select points

$$y_i \in Y \cap B\left(\frac{\delta}{\varepsilon} z_i + u - v, \delta\right).$$

130

We quickly verify that $y_i \in Q$ and $\left|\frac{\varepsilon}{\delta}(y_i - i + v) - z_i\right| \leq \varepsilon$. Hence, from (5.4) we have

$$\|p\| = \max_{x \in Q^*} \left| p\left(\frac{\varepsilon x}{\delta}\right) \right| \leq M \max_i \left| p\left(\frac{\varepsilon}{\delta}(y_i - u + v)\right) \right|.$$

Applying this to $p(\frac{\delta}{\varepsilon} \cdot + u - v)$, we obtain (5.3). ∎

As a consequence of Theorem 5.3, if a set Y of finite density is given, we can partition $\mathbb{R}^n$ into cubes $Q_\alpha = r\alpha + [0, r)^n$, $\alpha \in \mathbb{Z}^n$, and in each cube Q_α we can find points $y_{\alpha 1}, y_{\alpha 2}, \ldots, y_{\alpha N}$ of Y in general position. These points can be chosen, moreover, so that the corresponding Lagrangian functions $\ell_{\alpha i}$ will have the property

$$\sup_{\alpha \in \mathbb{Z}^n} \sup_{x \in Q_\alpha} \max_{1 \leq i \leq N} |\ell_{\alpha i}(x)| \leq M.$$

This follows from Inequality (5.3). We are now able to satisfy Inequality (4.14) in Theorem 4.2 by taking θ_α to be the characteristic function of Q_α.

6 THE GENERAL CASE OF SCATTERED NODES

In this section, we begin with a prescribed node set Y in $\mathbb{R}^n$. We assume that Y has finite density, δ, so that each closed ball of radius δ must contain at least one point of Y. By the main result of Section 5, there exists a doubly-subscripted subset of Y,

$$\{y_{\alpha j} : \alpha \in \mathbb{Z}^n, \ 1 \leq j \leq N\} \tag{6.1}$$

such that the associated Lagrange basic polynomials $\ell_{\alpha j}$ obey the inequality

$$\sup_{\alpha \in \mathbb{Z}^n} \max_{1 \leq j \leq N} \sup_{x \in Q_\alpha} |\ell_{\alpha j}(x)| \leq M. \tag{6.2}$$

Recall that $Q_\alpha = r\alpha + [0, r)^n$, and that r and M are positive numbers that emerge upon applying Theorem 5.3.

Next we assume that a partition of unity $\{\theta_\alpha\}$ is available that satisfies

$$\sup_{\alpha \in \mathbb{Z}^n} \sup_{x \in \mathbb{R}^n} |\theta_\alpha(x)| \big(1 + |x - r\alpha|\big)^\beta < \infty \tag{6.3}$$

for some $\beta > n + 2m - 1$.

Lemma 6.1 *For each triple (n, m, r), where $n \in \mathbb{N}$, $m \in \mathbb{N}$, and $r > 0$, there is a constant $c(n, m, r)$ such that for all $p \in \Pi_{m-1}(\mathbb{R}^n)$,*

$$|p(x)| \leq c(n, m, r)\big(1 + |x|\big)^{m-1} \sup_{v \in [0,r)^n} |p(v)|. \tag{6.4}$$

Proof: For $p \in \Pi_{m-1}(\mathbb{R}^n)$, define

$$\|p\|_m = \sup_{x \in \mathbb{R}^n} \frac{|p(x)|}{\left(1 + |x|\right)^{m-1}}. \tag{6.5}$$

131

It is easily verified that $\|p\|_m < \infty$ and that Equation (6.5) defines a norm on $\Pi_{m-1}(\mathbb{R}^n)$. By the equivalence of norms on this finite-dimensional space, there is a constant c such that $\|p\|_m \leq c\|p\|_\infty$, where $\|\ \|_\infty$ is the supremum norm on the cube $[0,r]^n$. This inequality implies (6.4). ∎

Lemma 6.2 *Under the hypotheses (6.2) and (6.3), there exists a constant M_3 such that*

$$|\theta_\alpha(x)\ell_{\alpha j}(x)| \leq M_3\big(1 + |x - r\alpha|\big)^{m-\beta-1}. \tag{6.6}$$

Proof: From (6.2) we have

$$|\ell_{\alpha j}(r\alpha + v)| \leq M \qquad \big(v \in [0,r]^n\big).$$

By Lemma 6.1 this allows us to infer that

$$|\ell_{\alpha j}(x)| \leq c(n,m,r)M\big(1 + |x - r\alpha|\big)^{m-1}. \tag{6.7}$$

If the supremum in (6.3) is denoted by M_2, then

$$|\theta_\alpha(x)| \leq M_2\big(1 + |x - r\alpha|\big)^{-\beta}. \tag{6.8}$$

By combining (6.7) and (6.8) we obtain (6.6), with $M_3 = M_2 c(n,m,r)M$. ∎

Theorem 6.3 *Under the hypotheses (6.2) and (6.3), the operator M defined by*

$$Mf = \sum_{\alpha \in \mathbb{Z}^n} \sum_{j=1}^N f(y_{\alpha j})\theta_\alpha \ell_{\alpha j}$$

has the local decay property of order m, as defined in Inequality (3.4).

Proof: This will be an application of Theorem 4.2. Since $m - \beta - 1 < m - (n + 2m - 1) - 1 = -n - m$, it suffices to prove that, for some c,

$$|\theta_\alpha(x)\ell_{\alpha j}(x)| \leq c\big(1 + |x - y_{\alpha j}|\big)^{m-\beta-1}. \tag{6.9}$$

Lemma 6.2 gives a similar estimate, and the link between (6.6) and (6.9) will be explained here. We note that $y_{\alpha j} - r\alpha \in [0,r)^n$, and consequently $|y_{\alpha j} - r\alpha| \leq r\sqrt{n}$. This leads to

$$1 + |x - y_{\alpha j}| \leq 1 + |x - r\alpha| + |r\alpha - y_{\alpha j}|$$
$$\leq 1 + |x - r\alpha| + rn^{1/2}$$
$$\leq 1 + |x - r\alpha| + rn^{1/2} + rn^{1/2}|1 - r\alpha|$$
$$= (1 + rn^{1/2})\big(1 + |x - r\alpha|\big).$$

Since $m - \beta < 0$, we conclude that

$$\big(1 + |x - y_{\alpha j}|\big)^{m-\beta-1} \geq (1 + rn^{1/2})^{m-\beta-1}\big(1 + |x - r\alpha|\big)^{m-\beta-1}$$

and that

$$\big(1 + |x - r\alpha|\big)^{m-\beta-1} \leq (1 + rn^{1/2})^{\beta-m+1}\big(1 + |x - y_{\alpha j}|\big)^{m-\beta-1}.$$

Using this last inequality with (6.6) gives us

$$|\theta_\alpha(x)\ell_{\alpha j}(x)| \leq M_3(1 + rn^{1/2})^{\beta-m+1}\big(1 + |x - y_{\alpha j}|\big)^{m-\beta-1}.$$

Hence (6.9) is true with $c = M_3(1 + rn^{1/2})^{\beta-m+1}$. ∎

7 TWO OPTIMIZATION PROBLEMS

In Section 5, an algorithm was suggested for selecting nodes in general position from a set having finite density. The algorithm proceeds as follows. Let n and m be given, and let $\{z_1, z_2, \ldots, z_N\}$ be in general position with respect to $\Pi_{m-1}(\mathbb{R}^n)$. Here N is the dimension of $\Pi_{m-1}(\mathbb{R}^n)$. An application of Lemma 5.2 determines an $\varepsilon > 0$. Now if Y is any set of density δ in $\mathbb{R}^n$ and if v is fixed in $\mathbb{R}^n$, we can select

$$y_i \in B\left(v + \frac{\delta}{\varepsilon}z_i, \delta\right) \cap Y \qquad (1 \le i \le N).$$

Here B denotes a closed ball having the given center and radius. The set $\{y_1, y_2, \ldots, y_N\}$ is in general position, as is easily verified, using Lemma 5.1 and 5.2. Moreover, the corresponding Lagrange polynomials ℓ_j satisfy

$$\|T_v \ell_j\| \le M$$

where $\| \ \|$ is any norm, T_v is the translation operator, and M depends only on n, m, δ, ε, and $\| \ \|$. Then number ε, in turn, depends on the set $\{z_1, z_2, \ldots, z_N\}$. Two optimization problems can be recognized here. First, to get a large ε, and second, to get a small M. A careful formulation of the first problem is as follows: find points $z_1, z_2, \ldots, z_N$ constrained by $|z_i| \le 1$ so that the implication

$$\max_{1 \le i \le N} |x_i - z_i| < \varepsilon \implies \{x_1, x_2, \ldots, x_N\} \text{ is in general position}$$

is true for the largest possible ε.

A partial solution of this problem when $m = 2$ and $n = 2$ is given by the next theorem.

Theorem 7.1 *When $m = 2$ and $n = 2$, the maximum value of ε is at least $3/4$.*

Proof: Let

$$z_1 = (0, 1), \quad z_2 = \left(\frac{\sqrt{3}}{2}, \frac{-1}{2}\right) \quad \text{and} \quad z_3 = \left(\frac{-\sqrt{3}}{2}, \frac{-1}{2}\right).$$

These are the vertices of an equilateral triangle, and they lie on the unit circle. If x_1, x_2, x_3 are three points satisfying $|x_i - z_i| < 3/4$, then they are in general position. To prove this, suppose that they are not. Then they are colinear. Suppose that they occur in the order x_2, x_1, x_3 on the line. Since $|x_2 - z_2| < 3/4$, the 2nd coordinates satisfy $|x_{22} - z_{22}| < 3/4$ or $|x_{22} + 1/2| < 3/4$. Hence $x_{22} < 1/4$. Similarly, $x_{32} < 1/4$. Hence $x_{12} < 1/4$ by convexity. But then $|z_1 - x_1| \ge 1 - x_{12} > 3/4$. ∎

133

A related optimization problem is to find points $z_1, z_2, \ldots, z_N$ in the unit ball of $\mathbb{R}^n$ that maximize the determinant $\det(p_i(z_j))$, where $\{p_1, p_2, \ldots, p_N\}$ is a basis for $\Pi_{m-1}(\mathbb{R}^n)$. Such a determinant is called a *Vandermonde* determinant. Although much is known about this problem when $n = 1$, the general case has not been solved. Bos, in [8,9], has studied the problem. In particular, he has addressed the *Fejér question*, which asks (for a given compact set K in $\mathbb{R}^n$) whether there are points $x_1, x_2, \ldots, x_N$ in K such that the corresponding Lagrange functions satisfy

$$\sup_{x \in K} \sum_{i=1}^{N} \ell_i^2(x) = 1. \tag{7.1}$$

In 1932, Fejér proved that when the Vandermonde determinant is maximized for points $x_i \in [-1, 1] = K \subset \mathbb{R}$ the resulting Lagrange polynomials satisfy (7.1). This is true for all m. Bos proved in [9] that for $m = 4$ and $n = 2$ the Fejér question has a negative answer when K is the unit ball. He proved the same negative result for arbitrary $n \geq 3$ and $m \geq 4$ when K is taken to be the unit sphere in $\mathbb{R}^n$.

REFERENCES

[1] Beatson R.K., Light W.A. Quasi-interpolation in the absence of polynomial reproduction. Pre-print, May 1992.

[2] Beatson R.K., Powell M.J.D. Univariate multiquadric approximation: quasi-interpolation to scattered data. Report DAMPT-NA7, Cambridge University, 1990.

[3] de Boor C. Quasi-interpolants and approximation power of multivariate splines. In *Computation of Curves and Surfaces*, pages 313–345. Kluwer, 1990. W. Dahmen, M. Gasca and C.A. Micchelli, eds.

[4] de Boor C. Approximation order without quasi-interpolants. In *Approximation Theory VII*, pages 1–18. Academic Press, New York, 1993. E.W. Cheney, C.K. Chui, and L.L. Schumaker, eds.

[5] de Boor C. The quasi-interpolant as a tool in elementary polynomial spline theory. In *Approximation Theory*, pages 269–276. Academic Press, New York, 1973. G.G. Lorentz, ed.

[6] de Boor C., Fix G. Spline approximation by quasi-interpolants. *J. Approx. Theory*, 8:19–45, 1973.

[7] de Boor C., Jia R.Q. Controlled approximation and a characterization of the local approximation order. *Proc. Amer. Math. Soc.*, 95:547–553, 1985.

[8] Bos L. A characteristic of points in R^2 having Lebesgue function of polynomial growth. *J. Approx. Theory*, 56:316–329, 1989.

[9] Bos L. Some remarks on the Fejér problem for Lagrange interpolation in several variables. *J. Approx. Theory*, **60**:133–140, 1990.

[10] Buhmann M.D. Convergence of univariate quasi-interpolation using multiquadrics. *IMA J. Numer. Anal.*, **8**:365–384, 1988.

[11] Buhmann M.D. On quasi-interpolation with radial basis functions. *J. Approx. Theory*, **72**:103–130, 1993.

[12] Buhmann M.D., Dyn N., Levin D. On quasi-interpolation with radial basis functions on non-regular grids. Submitted to *Constr. Approx.*.

[13] Chui C.K., Diamond H. A natural formulation of quasi-interpolation by multivariate splines. *Proc. Amer. Math. Soc.*, **99**:643–646, 1987.

[14] Chui C.K., Shen X.C., Zhong L. On Lagrange polynomial quasi-interpolation. In *Topics in Polynomials of one and Several Variables and their Applications*, pages 125–142. World Scientific Publishers, Singapore, 1993. T.H. Rassias et al., eds.

[15] Frederickson P.O. Quasi-interpolation, extrapolation, and approximation in the plane. In *Proceedings of the Manitoba Conference on Numerical Mathematics*, pages 159–167. Dept. Comput. Sci., University of Manitoba, 1971.

[16] Jia R.Q., Lei J. Approximation by multi-integer translates of functions having global support. *J. Approx. Theory*, **72**:2–23, 1993.

[17] Light W.A., Cheney E.W. Quasi-interpolation with translates of a function having non-compact support. *Constr. Approx.*, **8**:35–48, 1992.

[18] Rabut C. How to build quasi-interpolants with application to polyharmonic B-splines. In *Curves and Surfaces*. Academic Press, New York, 1991. P.J. Laurent, A. Le Méhauté and L.L. Schumaker, eds.

[19] Rabut C. *B-Splines polyharmoniques cardinales: interpolation, quasi-interpolation, filtrage*. Thesis, Toulouse, 1990.

[20] Schoenberg I.J. Contributions to the problem of approximation of equidistant data by analytic functions. *Quart. Appl. Math.*, **4**:45–99, 112–141, 1946.

[21] Smith P.W., Ward J.D. Quasi-interpolants from spline interpolation operators. *Constr. Approx.*, **6**:97–110, 1990.

135

International Series of Numerical Mathematics, Vol. 119, ©1994 Birkhäuser

A Multigrid FFT Algorithm for Slowly Convergent Inverse Laplace Transforms

Germund Dahlquist[†]

Dedicated to Walter Gautschi on the occasion of his 65th birthday

Abstract. The algorithm is based on an exact relation, due to Cooley, Lewis and Welch, between the Discrete Fourier Transform and the periodic sums, associated with a function and its Fourier Transform in a similar way as in the Poisson summation formula. It makes use of several equidistant grids, with the same number of points covering m different symmetric intervals of length L, 2L, 4L, 8L,..., where it applies FFT and spline interpolation to the midpoints of the grid.

Typically the number of arithmetic operations per computed function value is about twice as large as for the FFT, but the distribution of points is more adequate for many applications, because the union of the grids is, globally, an approximately equidistant grid on a logarithmic scale. Some properties and applications of the algorithm will be discussed.

Novel features emphasized in this article include a new initial condition, where there is some relation to the work of Walter Gautschi on Practical Fourier Analysis, and the use of convolution smoothing to improve the performance on problems with discontinuities in the time domain.

1 INTRODUCTION AND BASIC FORMULAS

Let $f(t)$ be a real function and let $\hat{f}(\omega)$ be its Fourier Transform. We use the following notation,

$$\hat{f}(\omega) = \int_{-\infty}^{\infty} e^{-i\omega t} f(t)dt, \quad f(t) = \frac{1}{2\pi} \int_{-\infty}^{\infty} e^{i\omega t} \hat{f}(\omega)d\omega. \tag{1.1}$$

$\hat{f}(\omega)$ is the given function, computable for any real ω. We want to compute $f(t)$ on a grid, such that it is possible to compute $f(t)$, by interpolation (or extrapolation), to (almost) a prescribed accuracy, for any real t, or, if this is an unreasonable request, for any t on a prescribed interval.

The author is more concerned about good accuracy when $|t|$ is small than most writers on the subject of inverse Laplace Transforms, and $\hat{f}(\omega)$ is therefore computed for larger values of ω than usual. The "special market" for the algorithm to be described, is when the product of the length of the intervals, inside which

---

AMS categories: 65R10, 42A04, 65B20, 65D07.

[†]Department of Computing Sciences, Royal Institute of Technology, S-10044, Stockholm, Sweden

$f(t)$ and $\hat{f}(\omega)$ cannot be neglected, is very large. Fortunately, with our algorithm, the number of function values and the amount of work grow only like the logarithm of this product. The ideal situation is, if $\hat{f}(\omega)$ or $f(t)$ has a smooth graph in a log-log diagram, for large arguments, but the use of the algorithm is not restricted to such cases.

The original form of the algorithm, Dahlquist (1993), is presented here in §2, together with a *new initial condition* that increases the efficiency in important special cases; see also Example 1. (The examples are in §4.) In addition to numerous new comments throughout this article, there are some new features, such as:

- the *delta modification*; see the end of §2 and Example 2. This is a seemingly advantageous alternative to the Gauss modification introduced by Dahlquist, loc.cit., in order to avoid a catastrophic cancellation, that may occur for small values of t, in problems where the above-mentioned product is very large;
- the use of *convolution smoothing* to improve the performance on problems with discontinuities in the time domain; see Example 4.

The algorithm is based on Discrete Fourier Analysis, but in this article we shall emphasize its application to the inversion of a one-sided Laplace Transform, a problem that has important applications to several branches of science and technology, such as:

- probabilty distributions in actuarial mathematics, e.g., Bohman(1960), and queueing problems, e.g., Abate and Whitt (1992);
- electrical power engineering, e.g., Pizarro and Eriksson (1991);
- radionuclide migration, see e.g., Hodgkinson et al. (1984).

Let $s = \sigma + i\omega$ and consider the Laplace Transform of a real function g,

$$\hat{g}(s) = \int_0^\infty e^{-st} g(t) dt = \int_0^\infty e^{-i\omega t} e^{-\sigma t} g(t) dt. \tag{1.2}$$

Assume that this integral converges absolutely. By the Fourier inversion formula,

$$\frac{1}{2\pi} \int_{-\infty}^\infty e^{i\omega t} \hat{g}(\sigma + i\omega) d\omega = \begin{cases} e^{-\sigma t} g(t), & \text{if } t > 0; \\ \frac{1}{2} g(0+), & \text{if } t = 0; \\ 0, & \text{if } t < 0. \end{cases} \tag{1.2'}$$

This is equivalent to the Complex Inversion Formula for the Laplace Transform. Set

$$\hat{f}_\sigma(\omega) = 2\text{Re } \hat{g}(s), \quad f_\sigma(t) = e^{-\sigma|t|} g(|t|). \tag{1.3}$$

Note that *both functions are real and even*. The relation between them can be written as a Cosine Transform, but we prefer to write it in the following form, which is obtained by (1.2'), after a short calculation, for all real t,

$$f_\sigma(t) = \frac{1}{2\pi} \int_{-\infty}^\infty e^{i\omega t} \hat{f}_\sigma(\omega) d\omega, \tag{1.3'}$$

The use of $\hat{f}_\sigma(\omega) = 2\mathrm{Re}\,\hat{g}(s)$, instead of $\hat{g}(s)$ itself, is advantageous, not only due to the shortcuts and memory savings possible in the FFT in the real, symmetric case, but also because the real part decays faster than $\hat{g}(\sigma + i\omega)$, as $\omega \to \infty$, if $g(0) \neq 0$.

For example, if $\hat{g}(s) = 1/(1 + s)$, we have

$$g(t) = e^{-t}, \quad f_0(t) = e^{-|t|}, \quad \hat{g}(i\omega) = (1 + i\omega)^{-1}, \quad \mathrm{Re}\,\hat{g}(i\omega) = (1 + \omega^2)^{-1}.$$

The fact that the discontinuity at $t = 0$ for g — note that $g(t) = 0$ for $t < 0$ — becomes a discontinuity only in the derivative for f_0, is related to this.

It is possible to move the line of integration to the left across *simple poles* of $\hat{g}(s)$, also when one works with the real part only, but the formula differs a little from the usual residue formula; see Dahlquist (1992). Notice that the formulas (1.3), (1.3′), are not valid if σ is less than the convergence abscissa of the integral (1.2), even if the integral in (1.3′) is convergent.

The numerical inversion of Laplace Transforms has a reputation of being a very ill-conditioned problem. It is so, if one has access to data from the real axis only. Gautschi (1969) showed that the condition numbers of matrices that occurred with a method, due to Bellman et al. (1966), which used n equidistant real data points, may increase like 8^n or $(3 + \sqrt{8})^n$, depending on the choice of numerical quadrature formulas.

If the function $\hat{g}(s)$ can be computed for arbitrary complex values, then (1.3) and (1.3′), offer a much better-conditioned formulation of the problem, but also with this formulation it is important that the line of integration is not far to the right since, by (1.3), the reconstruction of $g(t)$ from $f_\sigma(t)$ involves a multiplication of $f_\sigma(t)$ — and its errors — by $e^{\sigma|t|}$. We shall return to the choice of σ in §3.

The reason why we choose a line of integration parallel to the imaginary axis is that we want to take advantage of the techniques of Discrete Fourier Analysis, in particular the fast Fourier Transform that solves the problem simultaneously for all t, so that the amount of arithmetic operations per function value increases only proportionally to the logarithm of the number of function values. Talbot (1979) shows how to modify the contour of integration, in order to make the trapezoidal rule very efficient, in particular if $g(t)$ is wanted at a small number of points, but this is a different setting of the problem.

Let $p, \hat{p}$, be positive constants. Consider two periodic functions,

$$\hat{f}(\omega, \hat{p}) = \sum_{r=-\infty}^{\infty} \hat{f}(\omega + 2r\hat{p}), \quad f(t, p) = \sum_{m=-\infty}^{\infty} f(t + 2mp). \tag{1.4}$$

Note that $\hat{f}(\omega, \hat{p})$ is determined everywhere by its values for $|\omega| \leq \hat{p}$. Similarly, $f(t, p)$ is determined by its values for $|t| \leq p$.

Let the Fourier expansion of $f(t, p)$ be

$$f(t, p) = \sum_{k=-\infty}^{\infty} a_k e^{itk\pi/p}.$$

139

After some algebraic manipulations, which are valid if the expansion in (1.4), which defines $f(t,p)$, is uniformly convergent for $t \in [-p,p]$, it is found that $a_k = \frac{1}{2p}\hat{f}(\frac{k\pi}{p})$. Put $\pi/p = \Delta\omega$. Then the Fourier expansion reads

$$f(t, \frac{\pi}{\Delta\omega}) = \frac{1}{2\pi} \sum_{k=-\infty}^{\infty} \hat{f}(k\Delta\omega)e^{itk\Delta\omega}\Delta\omega. \tag{1.5}$$

This is *Poisson's summation formula*, usually stated for $t = 0$, $\Delta\omega = 1$, hence $p = \pi$. (The derivation indicated above is in fact similar to a classical derivation of the Poisson formula, see e.g. Courant and Hilbert (1931, p.64).) The right-hand side is seen to equal the trapezoidal approximation to the integral

$$f(t) = \frac{1}{2\pi} \int_{-\infty}^{\infty} \hat{f}(\omega)e^{it\omega}d\omega,$$

which is very useful even with a moderately small step size $\Delta\omega$, if f has a compact support, or if $f(x)$ decays rapidly as $|x| \to \infty$, since the error is

$$f(t,p) - f(t) = f(t \pm 2p) + f(t \pm 4p) + f(t \pm 6p) + \ldots, \quad (p = \pi/\Delta\omega). \tag{1.6}$$

This is a well known quantitative expression, see e.g., Fettis (1955), for the fact, that the trapezoidal rule, with a constant step size over an *infinite* interval, is often very much more than second order accurate. Note that, if $|t| < p$, the arguments of f on the right-hand side are all greater than $p = \pi/\Delta\omega$. If $\Delta\omega$ is small enough, the truncation error can be neglected. If $\hat{f}(\omega) \to 0$ slowly, as $\omega \to \infty$, and if $\Delta\omega$ is small, a very large number of terms are needed in the expansion (1.5).

If $f(t)$ is real and even, then the same is true for $\hat{f}(\omega)$, and the Fourier expansion is the real part of $a_0 + 2\sum_{k=1}^{\infty} a_k z^k$, $z = e^{it\Delta\omega}$. Several writers recommend various techniques of convergence acceleration, such as Euler's transformation, the epsilon algorithm or Gustafson's Chebyshev acceleration, see e.g., Abate and Whitt (1992), Gustafson (1990). Both have extensive bibliographies. Such techniques become, however, less useful when $t\Delta\omega$ is small; the expansion is then "slowly alternating".

Among other things, this dilemma led me to consider the use of several related grids, both in the time domain and the frequency domain. Another reason was a reflexion over the standard practice to estimate the aliasing effect by the comparison of results from two grids. Wouldn't it be possible to use more than two grids, and wouldn't it be possible to use the results not only for the estimation of the error, but also for improving the result? The algorithm to be described is a kind of affirmative answer to these questions.

In the time domain we denote by $G(p)$ a grid of $N + 1$ equidistant points of the interval $[-p,p]$, $-p + 2kp/N$, $k = 0, 1, 2, \ldots, N$. We consider m such grids $G(p_j)$, where $p_j = 2p_{j-1}$, $j = 0, 1, \ldots, m - 1$. N is the same for all grids.

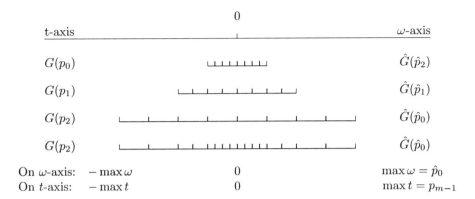

Figure 0: The structure of the grids used in the algorithm.

Since the largest frequency that can be resolved by a grid with spacing $\Delta t = 2p/N$ is

$$\hat{p} = \frac{\pi}{\Delta t} = \frac{\pi N}{2p}, \quad \text{i.e.,} \quad p\hat{p} = \frac{1}{2}\pi N, \tag{1.7}$$

it is natural to use m grids $\hat{G}(\hat{p}_j)$ in the frequency domain, with $\hat{p}_j = \frac{1}{2}\pi N/p_j$, $j = 0, 1, \ldots, m-1$. Note that $\hat{p}_j = \hat{p}_{j-1}$. Set

$$\hat{G} = \bigcup_{j=0}^{m-1} \hat{G}(\hat{p}_j), \quad G = \bigcup_{j=0}^{m-1} G(p_j), \tag{1.8}$$

In Figure 0, $m = 3, N = 8$. More typical values are $m = 20, N = 128$. Note that the sets $\hat{G}$ and G are similar in structure to the set of floating point numbers. Locally they are equidistant, but globally they are more like being equidistant on a logarithmic scale. We shall deal with problems where G and $\hat{G}$ are adequate grids for the interpolation of the functions f and $\hat{f}$. The local time scale of the function $f(t)$ should increase about linearly with t, and analogously for its Fourier Transform $\hat{f}$.

The number of values needed of the real and even function $\hat{f}$ equals half of the number of points in $\hat{G}$, i.e. $\frac{1}{4}N(m+1)+1$, and the same number is also useful as an approximate measure of the amount of work needed in the algorithm that will be described below. With a terminology borrowed from music, we set oct $= \log_2(\max\omega/\min\Delta\omega)$, where oct is an abbreviation for the number of octaves. The number of octaves depends on the function $\hat{f}$ and on the tolerance set for the error of computation. (One can reduce oct by restricting the range in the time domain to $t > \text{lowt}$ (say).) Note that, by (1.7), we also have

$$\text{oct} = \log_2(\max\omega/\min\Delta\omega) = \log_2(\max t/\min\Delta t) = \log_2 N + m - 2, \tag{1.9}$$

where m is the number of grids. Also note that $\texttt{oct} = \log_2(\max\omega\,\max t/\pi)$, and "the special market" for our algorithm, as described at the beginning of this paper, can therefore be expressed as problems with a large number of octaves. The disadvantage of the usual FFT for such a problem is that it uses a constant step size.

We see that, for a given number of octaves, it is much more expensive to increase $\log_2 N$ by 1 and decrease m by 1 than to do it the other way around. (The number of function values is multiplied by, respectively, $2m/(m+1)$ and $(m+2)/(2m+2)$.) It turns out that it is often advantageous to use a large value of m, while N is to be kept as small as possible with the given tolerance.

The basis of our algorithm is the remarkable formula (1.11) below, due to Cooley, Lewis and Welch, (1967,Theorem 1) that is easily derived from Poisson's Summation Formula, restricted to the grid $G(p)$. For convenient reference we collect the notations introduced above, and add some immediate consequences of them.

$$\Delta\omega = 2\hat{p}/N = \pi/p, \quad p\hat{p} = \frac{1}{2}\pi N, \quad \Delta t = 2p/N = \pi/\hat{p}, \quad \Delta\omega\Delta t = 2\pi/N. \quad (1.10)$$

Put $t = j\Delta t$, $j = 0, 1, 2, \ldots, N-1$, into the Poisson Formula (1.5). Then

$$f(j\Delta t, p) = \frac{\Delta\omega}{2\pi} \sum_{k=-\infty}^{\infty} \hat{f}(k\Delta\omega)e^{jk2\pi i/N} = \frac{\Delta\omega}{2\pi} \sum_{k=0}^{N-1} \sum_{r=-\infty}^{\infty} \hat{f}(k\Delta\omega + 2r\hat{p})e^{jk2\pi i/N}$$

since, if k is increased by rN, then the exponent is unchanged modulo $2\pi i$, and $k\Delta\omega$ is increased by $rN\Delta\omega = 2r\hat{p}$. Hence

$$f(j\Delta t, p) = \frac{\Delta\omega}{2\pi} \sum_{k=0}^{N-1} \hat{f}(k\Delta\omega, \hat{p})e^{jk2\pi i/N}. \quad (1.11)$$

This shows that *the Inverse Discrete Fourier Transform (IDFT) with N terms expresses an exact relation between the periodic functions $f(t, p)$ and $\hat{f}(\omega, \hat{p})$, for any $p, \hat{p}$, such that $p\hat{p} = \frac{1}{2}\pi N$.* Note that *it relates the coarse grids $\hat{G}(\hat{p})$ in the frequency domain with the fine grids $G(p)$ in the time domain,* and vice versa.

The (direct) Discrete Fourier Transform (DFT) then reads, (since $\Delta\omega\Delta t/2\pi = 1/N$),

$$\hat{f}(k\Delta\omega, \hat{p}) = \Delta t \sum_{j=0}^{N-1} f(j\Delta t, \hat{p})e^{-jk2\pi i/N}. \quad (1.11')$$

The rearrangement of terms in this derivation is most easily legitimated under the assumption that the sum in (1.5) is absolutely convergent, see Dahlquist (1993). We have, however, had good experience with the algorithm described in the next section, which is based on (1.11), also when the sum is conditionally convergent only; see Example 3.

142

2   DESCRIPTION OF THE ALGORITHM

The following simple observation is very important to us:

$$f(t,p) = \sum_{j=-\infty}^{\infty} f(t + 4jp) + \sum_{j=-\infty}^{\infty} f(t - 2p + 4jp),$$

i.e.

$$f(t, p) = f(t, 2p) + f(t - 2p, 2p). \tag{2.1}$$

The corresponding relation for $\hat{f}$ reads,

$$\hat{f}(\omega, \hat{p}) = \hat{f}(\omega, 2\hat{p}) + \hat{f}(\omega - 2\hat{p}, 2\hat{p}). \tag{2.1'}$$

Our algorithm uses these formulas as recurrence relations. It proceeds from coarser to finer grids, first in the frequency domain, then in the time domain, after the Inverse Discrete Fourier Transform has made the transition between the domains.

If the $4\hat{p}$-periodic function $\hat{f}(\cdot, 2\hat{p})$ is known on $\hat{G}(2\hat{p})$, then $\hat{f}(\omega, \hat{p})$ is directly determined for $\omega \in \hat{G}(2\hat{p})$, but some interpolation is necessary for $\omega \in \hat{G}(\hat{p}) \backslash \hat{G}(2\hat{p})$. This has to be done with some care, since it would be much too restrictive to assume that interpolation can be applied to the first term on the right-hand side of (2.1'). So, we subtract $\hat{f}(\omega)$ from both sides of (2.1'),

$$\hat{f}(\omega, \hat{p}) - \hat{f}(\omega) = \left(\hat{f}(\omega, 2\hat{p}) - \hat{f}(\omega)\right) + \hat{f}(\omega - 2\hat{p}, 2\hat{p}). \tag{2.2}$$

It follows from the definition of the $2\hat{p}$-periodic function $\hat{f}(\cdot, \hat{p})$ that *if* $|\omega| \le \hat{p}$, *then the left-hand side of (2.2) depends only on values* $\hat{f}(y)$ *with* $|y| > \hat{p}$*. The same is true also for each of the two terms on the right-hand side.* This makes a great difference for the class of functions which we are dealing with.

The first part of the algorithm is as follows. Start on the coarsest grid, and let the initial condition be, $\hat{f}(\omega, \hat{p}_0) = \hat{f}(\omega)$, $\omega \in G(\hat{p}_0)$.

Then for $\hat{p} = \hat{p}_1, \hat{p}_2, \ldots, \hat{p}_{m-1}$, apply (2.2) for $\omega \in \hat{G}(2\hat{p})$. Then use cubic spline interpolation with a "not a knot condition", De Boor (1978), in order to obtain $f(\omega, \hat{p}) - f(\omega)$, $\omega \in \hat{G}(\hat{p}) \backslash \hat{G}(2\hat{p})$.

Then the IDFT yields $f(t,p)$, $p = \frac{1}{2}\pi N/\hat{p}$, $t \in G(p)$, for $p = p_0, p_1, \ldots, p_{m-1}$.

So, the input of *the second part* of the algorithm is, $f(t,p)$, $t \in G(p)$, $p = p_{m-1}, p_{m-2}, \ldots, p_0$, and the output will be $f(t)$, $t \in G = \bigcup G(p_j)$. Again, start on the coarsest grid, with the initial condition, $f(t) = f(t, p_{m-1})$, $t \in G(p_{m-1})$.

Then, for $p = p_{m-1}, p_{m-2}, \ldots, p_0$, $f(t, p) - f(t)$ is computed by means of the analog of (2.2) that reads,

$$f(t,p) - f(t) = (f(t, 2p) - f(t)) + f(t - 2p, 2p), \quad t \in G(2p). \tag{2.3}$$

As in the first part, cubic spline interpolation is used to obtain $f(t,p) - f(t)$, $t \in G(p)\backslash G(2p)$. Finally, $f(t)$, $t \in G(p)$ is, of course, obtained by $f(t) = f(t,p) - \big(f(t,p) - f(t)\big)$.

One can show that, apart from rounding errors, $\hat{f}(\omega, \hat{p})$, and hence $f(t,p)$ too, are exact results, if $\hat{f}$ is replaced by *a piecewise cubic spline interpolant* $\hat{s}$, with the following properties:

- $\hat{s}(\omega) = 0$ for $\omega > p_0$.
- For $j = 1, 2, \ldots, m-1$, $\hat{s}(\omega)$ is, for $\hat{p}_j \leq \omega \leq 2\hat{p}_j$, a cubic spline determined by interpolation of $\hat{f}(w)$ at $\omega = \hat{p}_j + 4k\hat{p}_j/N$, $k = 0, 1, \ldots, \frac{1}{4}N$, with "not a knot condition" at $k = 1$ and $k = \frac{1}{4}N - 1$. Similarly for $-2\hat{p}_j < \omega < -\hat{p}_j$.

A similar interpretation is possible for the second part, if f is replaced by a piecewise spline interpolant in the time domain. The effect of this replacement seems hard to estimate in advance.

The basic assumption of the method is that these approximations are satisfactory, both in the frequency and in the time domain. The question of how to choose the parameters of the algorithm in order to achieve this, will be dealt with in §3.

The interpolation plays an important role in this algorithm. Note, e.g., that for $\omega \in \hat{G}_{m-1}$, the computation of $\hat{f}(\omega, \hat{p}_{m-1})$ to desired accuracy, directly from (1.4), would require the summation of all terms for which $|\omega + r\hat{p}_{m-1}| < \hat{p}_0$. This means about 2^m terms for each ω. In total this would require $\frac{1}{2}N2^m$ function values, while the algorithm, thanks to its use of interpolation, needs only $\frac{1}{4}N(m+1)$ function values. The number of arithmetic operations is a modest multiple of this, including the solution of tridiagonal systems for the spline interpolation and the Fast Fourier Transforms where, strictly speaking, the factor "$m + 1$" also should contain a term that is a modest multiple of $\log_2 N$.

The above description may be called *the original version* of the algorithm. We shall now describe some *modifications* that in some cases greatly increase its numerical stability or its efficiency.

There may be a *risk of catastrophic cancellation* at the computation of $f(t)$ for small values of t. In the second part of the algorithm, $f(t)$ is obtained as the difference between $f(t,p)$ and another term. By the definition of $f(t,p)$,

$$f(t,p) = \sum_{j=-\infty}^{\infty} f(t + 2jp) \approx \frac{1}{2p} \int_{-\infty}^{\infty} f(t+x)dx = \frac{\hat{f}(0)}{2p}.$$

The approximation is valid if p is small. We note that $|t| \leq p$. We see that there is numerical cancellation if $|f(p)| \ll |\hat{f}(0)|/(2p)$, i.e. for small values of p, if $\hat{f}(0) \neq 0$. We also see, however, that we can reduce this error considerably, if we make $\hat{f}(0) = 0$.

Dahlquist (1993) suggested the so-called *Gauss modification* of the algorithm, where $\hat{f}(\omega) - \hat{f}(0) \exp(-\omega^2/(2k^2))$ is substituted for $\hat{f}(\omega)$, if the origin is a maximum for $\hat{f}$, and k is an estimate of $-\hat{f}''(0) > 0$. At the exit, the exact inverse

Fourier Transform of the Gauss term, i.e., $\hat{f}(0) \exp(-t^2 k^2/2)k/\sqrt{2\pi}$, is added to $f(t)$.

Now a simpler alternative has been tested that is called the *delta modification*, where simply $\hat{f}(0)$ is set equal to zero, with no other changes of $\hat{f}(\omega)$. At the exit, $\hat{f}(0)/(2 \max t)$ is added to $f(t)$.

The explanation of this rule is as follows. The delta modification means that $\hat{f}(0)\hat{\phi}(\omega)$ is subtracted from $\hat{f}(\omega)$ on $\hat{G}$, where $\hat{\phi}(\omega)$ is a discrete delta function, i.e., $\hat{\phi}(0) = 1$, while $\hat{\phi}(\omega) = 0$ at all other points on $\hat{G}$. It follows that $\hat{\phi}(\omega, \hat{p})$ is a discrete delta function on $\hat{G}(\hat{p})$, with a periodic continuation outside $\hat{G}(\hat{p})$. By (1.11), its IDFT is $\phi(t,p) = \Delta\omega/(2\pi) = 1/(2p)$, $t \in G(p)$, $p\hat{p} = \frac{1}{2}\pi N$. Substitute ϕ for f in the recurrence relation (2.3), and consider $\phi(u,q) = 1/(2q)$ as given. It is easily verified, by induction, that $\phi(t) = 1/(2p_{m-1}) = 1/\max t$, $t \in G$, is the unique solution of this recurrence relation with the initial condition $\phi(t) = \phi(t, p_{m-1}) = 1/(2p_{m-1})$, $t \in G(p_{m-1})$.

Note that the discrete delta function $\hat{\phi}(\omega)$, can be interpreted as a discretization of a Dirac pulse $\min \Delta\omega\delta(\omega)$, whose inverse Fourier Transform is the constant $\min \Delta\omega/(2\pi) = 1/(2 \max t)$, in exact agreement with the the algorithm.

The delta modification has practically the same effect on Example 2, as the Gauss modification had on the same example in Dahlquist (1993). We do not yet know which is the best. Since the delta modification is based on an exact property of the discrete delta function, we prefer it, at present. The fact that the response of the algorithm to the Gauss function is not exactly equal to the inverse Fourier transform of the Gauss function can have a visible effect on the error curve for large values of t, when $N \leq 16$, i.e. cases where a modification is not at all needed.

Note that these modifications also reduce the loss of digital information due to outshifting in the computation of $\hat{f}(0, \hat{p})$, when $|\hat{f}(\pm\hat{p})| \ll |\hat{f}(0)|$. In this respect the Gauss modification has the advantage of being useful sometimes also for $\hat{f}(\omega, \hat{p})$, $\omega \neq 0$.

Next, we consider a modification concerning *a new initial condition for the first part of the algorithm*. We want to find something better than $\hat{f}(\omega, \hat{p}_0) = \hat{f}(\omega)$, $\omega \in \hat{G}(\hat{p}_0)$. For this recent work I have had inspiration from the work of Walter Gautschi (1972) on attenuation factors in practical Fourier analysis, although from a converse point of view. Suppose that

$$\hat{f}(y) \approx \alpha|y|^{-\beta}, \text{ for } |y| > \hat{p}, \tag{2.4}$$

with a relative error less than ε. If $|\omega| < \hat{p}$, this can be applied to $y = |\omega + 2r\hat{p}|$, since $|\omega + 2r\hat{p}| > \hat{p}$ for $r = \pm 1, \pm 2, \ldots$. By the addition of these approximate equalities for $r = \pm 1, \pm 2, \ldots$,

$$\hat{f}(\omega, \hat{p}) - \hat{f}(\omega) \approx \sum_{r=-\infty}^{\infty} \alpha|\omega + 2r\hat{p}|^{-\beta} - \alpha|\omega|^{-\beta} = \alpha|2\hat{p}|^{-\beta}\left(\sum_{r=-\infty}^{\infty} |z + r|^{-\beta} - |z|^{-\beta}\right),$$

where we have set $z = \omega/(2\hat{p})$. Again the relative error is less than ε.

Convenient formulas for such sums are given by Gautschi (1972), when β is an even number. For $\beta = 2$ we obtain $\hat{f}(\omega,\hat{p}) - \hat{f}(\omega) \approx \alpha|2\hat{p}|^{-2}\left((\pi/\sin\pi z)^2 - (1/z)^2\right)$, and finally, by (2.4),

$$\hat{f}(\omega,\hat{p}) - \hat{f}(\omega) \approx \hat{f}(\hat{p})\frac{\pi^2}{4}\left(\frac{1}{\sin^2\pi z} - \frac{1}{(\pi z)^2}\right), \quad \beta = 2, \quad \hat{p} = \hat{p}_0, \tag{2.5}$$

with a relative error less than 2ε. If $\omega > 0$, $\omega \in \hat{G}(\hat{p})$, then $z = 0, \frac{1}{N}, \frac{2}{N}, \ldots \frac{1}{2}$, and the multiplier of $\hat{f}(\hat{p})$ in (2.5) increases smoothly from $\frac{1}{12}\pi^2 \approx 0.8$ to $\frac{1}{4}\pi^2 - 1 \approx 1.5$.

The case $\beta = 2$ is particularly interesting, since an asymptotic expansion of the form

$$\hat{f}_\sigma(\omega) = 2\int_0^\infty f_\sigma(t)e^{-i\omega t}dt \sim 2f'_\sigma(0)\omega^{-2} + 2f'''_\sigma(0)\omega^{-4} + \ldots, \quad (\omega \gg 1), \tag{2.6}$$

can be obtained by repeated integration by parts. It is valid, e.g., if f_σ and its derivatives are absolutely integrable on $(0,\infty)$. Since $g(t) = 0$ for $t < 0$, f_σ has a singularity at $t = 0$ that is entirely responsible for the slow convergence towards zero of $\hat{f}_\sigma(\omega)$.

Roughly speaking, the new initial condition takes care of the first term of this expansion. The exploratory part of the program, see §3, determines $\hat{p}_0$ so that the effect of the *second* term only is negligible for $\omega \geq \hat{p}_0$. With the old initial condition, $\hat{p}_0$ is to be chosen so that the effect of the *first* term of the relevant expansion of $\hat{f}(\omega)$ is negligible.

The new initial condition yields a smaller value of $\hat{p}_0$, hence the number of function values is reduced, sometimes by more than 70%. See Example 1 of §4. That the savings do not become even larger, is due to the principal advantage of the multigrid structure that the number of function values, in the original version, varies linearly with $\log\hat{p}_0$. The old initial condition is still used, when the asymptotic behavior of $\hat{f}(\omega)$ is of a different type.

Finally, *the termination criterion of the second part* can also be modified. In some situations the program stops at $p = p_r$ for some $r > 0$. For example, if $f(t)$ or some low derivative has discontinuities for some positive values t, the user can give a value to p_r, in order to avoid getting such points in an interval, where interpolation is used. There are also cases, where a positive value of r is chosen by the program, e.g. if it detects memory shortage. See also §3. This may give a reduced accuracy for $t < p_r$ only.

3 HOW TO CHOOSE THE PARAMETERS OF THE ALGORITHM

The algorithm contains a number of parameters, tol, acct, lowt, given by the user, and σ, $\hat{p}_0 = \max\omega = \pi/\min\Delta t$, $\min\Delta\omega = \pi/\max t$, m, N, r, normally computed by an exploratory program. Dahlquist (1993) describes such a program where the computer becomes acquainted with $\hat{g}$ through its values on a very coarse grid with a very large number of octaves, four points per octave. This grid is referred to as "the very coarse grid". The program uses this, with some assistance of the user, to choose values of the parameters of the method, in order to hopefully satisfy the tolerance tol for the absolute error in $f_\sigma(t)$.

These parameter values are to be considered as useful suggestions for a first run that should be followed by a faster second run. In the second run, $\log_2(N)$ and $\log_2(\max\omega)$ are decreased by 1. Equation (1.9) is still valid. The second run requires no new function values.

The level of the error of the first run can be estimated by $1/15$ of the difference between the results of the two runs, assuming that the spline interpolation errors, which are $O(N^{-4})$, dominate the errors due to the finite values of $\max\omega$ and $\max t$. The latter parameters are in fact chosen with some margin, since it is relatively inexpensive to increase m (instead of N), as mentioned in §1. So this assumption is rather trustworthy, except possibly for very small values of t. It is informative to plot the difference between the two runs versus t. It is, like the true error, a rather ragged curve. See §4, where both curves are plotted in some cases. If the difference between the two computations is too large, which does not often happen, a new run can be done with a larger value of N, or some other change.

The program first choses σ, i.e., the location of the line of integration. This must be done before the values are computed on the coarse grid. It was mentioned above that, for the inversion of a Laplace Transform, our algorithm will be applied to $\hat{f}_\sigma(\omega) = 2\mathrm{Re}\,\hat{g}(\sigma + i\omega)$, and the result is $g(|t|) = f_\sigma(t)\exp(\sigma|t|)$, where $f_\sigma(t)$ is the inverse Fourier Transform of $\hat{f}_\sigma(\omega)$. It is important to realize that *not only $f_\sigma(t)$, but also its errors are multiplied by $e^{\sigma|t|}$*.

It is reasonable to choose $\sigma = 0$, if $\hat{g}(s)$ is analytic on and to the right of the imaginary axis, and if the user is satisfied by using the same tolerance for the absolute error of $g(t)$ for all t. If $\hat{g}$ has a singularity on the imaginary axis, the user must provide a value for a parameter named acct, the default value of which is ∞. It stands for *the length of an interval on the t-axis, along which it is acceptable for the error tolerance of $g(t)$ to be increased by a factor of 10*. Then the program chooses $\sigma = \ln 10/\text{acct}$. The quantity acct is, for example, to be chosen so that some simple asymptotic approximation is sufficiently accurate for (say) $t > 3\text{acct}$. If one wants smaller absolute errors for large values of t, it is possible to choose acct < 0, hence $\sigma < 0$, provided that $\hat{g}(s)$ has appropriate analyticity properties. Cooley, Lewis and Welch (1970, §4) has an interesting discussion of the the choice of σ and N for the single-grid case, in order to obtain a balance between aliasing and rounding errors.

Next, the program investigates the values of $\hat{g}(\sigma + i\omega)$, mainly its real part, for large and small values of $|\omega|$, in order to choose, respectively, $\hat{p}_0 = \max\omega$ and $\min\Delta\omega$, and hence also $\max t$ and oct. The determination of $\min\Delta\omega$, if $|\hat{f}(\omega)|$ has its maximum at the origin, has been discussed at some length by Dahlquist (1993), so we omit this. If the maximum is not at the origin, see e.g., Example 1, the user is, at present, asked to suggest a value of $\max t$ that is then converted by the program to min $\Delta\omega = \pi/\max t$. (This can certainly be made more automatic.)

There is more to be said about the investigations at the upper end. We assume that $g(s)$ is analytic for $\mathrm{Re}\, s \geq \sigma$, and that

$$\hat{g}(\sigma + i\omega) \to 0, \text{ as } \omega \to \infty, \quad \int_{-\infty}^{\infty} |\hat{g}^{(k)}(\sigma + i\omega)| d\omega < \infty, \quad k = 1, 2, \dots. \quad (3.1)$$

Set $s = \sigma + i\omega$, $\hat{f}(\omega) = 2\mathrm{Re}\hat{g}(\sigma + i\omega)$. First it is inspected if $s\hat{g}(s)$ converges to some limit c reasonably fast as $\omega \to \infty$; a lemma of Doetsch, see e.g., Dahlquist (1993), requires "as fast as $s^{-\delta}$ for some $\delta > 0$". If so, *a useful conclusion is that* $g(0+) = c$.

Next, a relation of the form $\hat{f}(\omega) \approx \alpha\omega^{-\beta}$ is established, first locally, near the upper end of the very coarse grid. If $0 < \beta \leq 1$ the Fourier integral is conditionally convergent, and the user is asked to provide a strictly positive value of the parameter named `lowt`, by which he states that the error is allowed to exceed `tol` when $t < \text{lowt}$ (unless he has already done so). Then, it is inspected if the approximation holds over several octaves, with the same α, β. If so, the case is a candidate for the new type of initial condition which has, at the time of writing, been tried for $\beta = 2$ only.

For the *old initial condition*, $\hat{p}_0 = \max\omega$ is to be chosen to make an estimate of $\frac{2}{\pi}|\int_{\hat{p}_0}^{\infty} \hat{f}_\sigma(\omega) \cos\omega t d\omega|$ less than ε, where ε is the part of `tol` allowed for this error source. Note that t is a multiple of $\min \Delta t = \pi/\hat{p}_0$. Hence $\sin t\hat{p}_0 = 0$. By integration by parts, we obtain, under the assumptions (3.1), with a remainder that we omit,

$$\frac{2}{\pi} \int_{\hat{p}_0}^{\infty} \hat{f}_\sigma(\omega) \cos\omega t d\omega = \frac{2}{\pi} \cos(t\hat{p}_0)\left(t^{-2}\hat{f}'(\hat{p}_0) - t^{-4}\hat{f}'''(\hat{p}_0) + \dots\right). \quad (3.2)$$

First, suppose that the user allows the error to exceed `tol` for $t < \text{lowt}$. (For example, Davies and Martin (1979) use de facto `lowt` $= 0.5$ in their comparative tests—a nice way to make life easy.) As stated in §1, $\hat{p}_0$ is of the form $2^\gamma\pi$, where γ is an integer. If we consider only the first term of (3.2), it follows that γ should then be the smallest integer such that

$$f'(2^{\gamma'}\pi) < \frac{1}{2}\pi(\text{lowt})^2\varepsilon, \quad \forall\gamma' \geq \gamma, \quad \hat{p}_0 = 2^\gamma\pi. \quad (3.3)$$

If $\hat{p}_0 > \pi/\text{lowt}$, the program choses $r = \max(r_0, \log_2(\hat{p}_0\text{lowt}/\pi))$, where $r_0 \geq 0$ is the smallest value of r that can be accepted for other reasons. This means that even though the values $f(t, p_j)$ are needed for $j \geq r$ only, $\hat{f}(\omega, \hat{p}_j)$ are computed also for $j < r$, in order that $\hat{f}(\omega, \hat{p}_j)$ should become accurate enough for $j \geq r$.

Now assume that $\beta > 1$. If no positive value of `lowt` has been provided, then we set $r = 0$, and put `lowt` $= \pi/\hat{p}_0 = 2^{-\gamma'}$ in (3.3), and seek the smallest integer γ, which satisfies this modified version of (3.3). This value of $\hat{p}_0$ is accepted unless another estimate of $\hat{p}_0$ that is obtained from the inequality $\frac{2}{\pi} \int_{\hat{p}_0}^{\infty} |\hat{f}(\omega)| d\omega < \varepsilon$, is smaller. For this estimation, the program employs numerical integration together with extrapolation of $\hat{f}(\omega)$ outside the very coarse grid. For details, see Dahlquist (1993).

With the *new initial condition*, at present only for $\beta = 2$, the program computes, by means of values on the very coarse grid, an approximation $\hat{f}_\sigma(\omega) \approx \alpha_1 \omega^{-2} + \alpha_2 \omega^{-4}$, where a few percent accuracy is enough for α_2. Then $\hat{p}_0$ is determined so that $\frac{2}{\pi} \int_{\hat{p}_0}^{\infty} \alpha_2 \omega^{-4} d\omega < \varepsilon$. (Some details about accuracy checks are omitted.) Note that, by (2.6), $g'(0+) = f'_\sigma(0+) + \sigma f_\sigma(0+) = \frac{1}{2}\alpha_1 + \sigma g(0+)$. The analysis on the coarse grid yields a rather accurate estimate of α_1. It was mentioned above how to estimate $g(0+)$.

There are two more parameters, N and m, which are linked by (1.9), according to which $\log_2 N + m - 2 = \text{oct}$. They are chosen by means of a rough estimate of the local interpolation errors on the very coarse grid of the frequency domain, together with the fact that the spline interpolation error on finer grids is then approximately proportional to N^{-4}. It is, of course, doubtful to let the very coarse grid be the starting point of an application of asymptotics, but the issue is just to determine N that is restricted to be a power of 2.

These three error sources and the rounding error are allowed to contribute with 10% of `tol` each. This leaves 60% of the tolerance to the interpolation error in the frequency domain, which we are unable to estimate in advance, but it is also approximately proportional to N^{-4}. The check of this error source is implicitly done only by the second run described above. Cases where `tol` is considerably exceeded, have been very rare in our numerical experiments, except for very small values of t, and examples that obviously violate the assumptions of the algorithm, e.g., if $f(t)$ or $f'(t)$ has discontinuities; see Example 4.

4 NUMERICAL EXAMPLES

Four rather different examples are chosen in order to illuminate some features of the method. Error curves, i.e. the absolute value of the difference, *divided by the tolerance*, between the computed and the exact solution, as a function of t, are shown in logarithmic diagrams. The error curve is usually a ragged curve. It is easily distinguished from a smoother curve in the same diagram, that shows the computed solution itself (not divided by the tolerance). Some of the diagrams also contains a darker ragged curve, usually located above the error curve. It shows the difference, also divided by `tol`, of the results of two runs with different value of N, as described in §3. By and large, the examples confirm the not quite convincing considerations according to which a smoothed upper envelope of the difference curve provides, after division by 15, a rather reliable estimate of the level of the error.

`maxerr` and `rmserr` are, respectively, the maximal absolute error and the root mean square error, both computed on the time interval $[\text{lowt}, \min(\max t, \text{acct})]$.

149

errat0 is the absolute value of the error at $t = 0$.

The experiments were carried out on personal computers with the Math Coprocessor Intel 80487, with 53 bits mantissa. The program consists of a few mfiles, written in the interpretative language Matlab. We have no experience of using this algorithm with lower precision, but the experience from the use of tol $= 10^{-12}$ with this high precision, indicate that one rarely looses more than 2 or 3 decimal digits, due to rounding errors.

After the introduction of the Gauss and delta modifications, the rounding errors have been a less severe limitation for the accuracy achieved than the shortage of memory, with a DOS implementation on a PC. It is, however, very important that the function values $\hat{g}(s)$ be computed with a small *relative* error, also when $|s|$ is very large, since the comparatively few computed function values, fval $= \frac{1}{4}(m + 1)N$, represent a usually much larger number of terms, 2^m, in the truncated sum that defines $\hat{f}(\omega, p)$, so the average "weight" is high for an individual computed value. It is therefore always necessary to *critically examine the formulas that define* $g(s)$, and sometimes necessary to rewrite them, in order to avoid the loss of numerical information; e.g., a subexpression like $\sqrt{s+1} - \sqrt{s}$ should be replaced by $1/(\sqrt{s+1} + \sqrt{s})$.

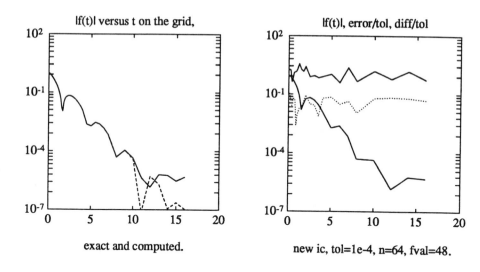

Figure 1a: The second experiment of Example 1, tol $= 10^{-4}$.
Left: $|f(t)|$ versus t. The solid and dashed curves are interpolated through grid values of, respectively, the exact and the computed solution.
Right: The absolute error (lower curve), and the difference between the results of two runs (upper curve) versus t, both divided by tol.

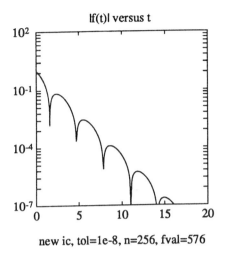

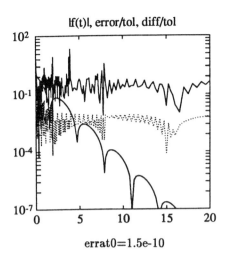

new ic, tol=1e-8, n=256, fval=576

errat0=1.5e-10

Figure 1b: The fourth experiment of Example 1, ($\texttt{tol} = 10^{-8}$).
Left: $|f(t)|$ versus t. Right: The absolute error (lower curve), and the difference between the results of two runs (upper curve) versus t, both divided by $\texttt{tol}$.

Example 1 Let $\hat{g}(s) = (s+1)/((s+1)^2+1)$, $\sigma = 0$. Then $g(t) = e^{-t}\cos t$. Since the maximum of $\operatorname{Re}\hat{g}(s)$ is not at the origin, the user is, in this example, asked to give the value of $\max t$, so that the computer can compute $\min \delta\omega$.

Six experiments are reported here, in order to illustrate the difference between the old and the new initial condition, over a wide range of tolerances.

Init.cond.	tol	max t	$\hat{p}_0$	oct	m	N	fval	maxerr	errat0
old*	10^{-4}	16	201	10	7	32	64	$1.2\,10^{-5}$	$9\,10^{-6}$
new	10^{-4}	16	50	8	5	32	48	$1.2\,10^{-5}$	$2\,10^{-5}$
old	10^{-8}	32	$3.4\,10^9$	35	29	256	1920	$1.6\,10^{-8}$	$5\,10^{-11}$
new	10^{-8}	32	$1.6\,10^3$	14	8	256	576	$3.2\,10^{-9}$	$1\,10^{-10}$
new	10^{-12}	32	$5\,10^4$	18	10	1024	2816	$1.3\,10^{-11}$	$1\,10^{-13}$
new*	10^{-12}	32	$5\,10^4$	19	10	2048	5632	$7.7\,10^{-13}$	$9\,10^{-16}$

The asterisk on the first line of this table indicates that the user gave the value $\texttt{lowt} = 1/16$ (always a power of 2). $\texttt{lowt} = 0$ in the other five cases. The asterisk on the last line indicates that $N = 2048$ was given manually, since the maximum error had become too large (on the previous line) with the automatically chosen $N = 1024$. The rms error is, however, $1.4\,10^{-12}$, with $N = 1024$.

Example 2 Let $\hat{g}(s) = \frac{1}{2}\sqrt{\pi}s^{-3/2}$. Then $g(t) = t^{1/2}$. Note that $\hat{g}(s)$ is singular at the origin, and slowly convergent at ∞. Correspondingly, $g(t)$ is unbounded, and

151

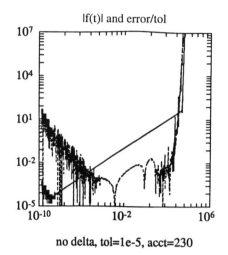
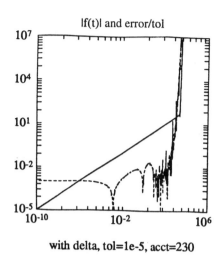

no delta, tol=1e-5, acct=230   with delta, tol=1e-5, acct=230

Figure 2: Example 2, without and with delta modification, lowt = 0, acct = 230. The solid curve is the computed solution; the ragged curve shows error/tol.

$g'(0+)$ is infinite. We ran three experiments, in order to to illustrate in particular "the boundary layer", the effect of the delta modification and the effect of the use of a positive lowt. We chose, in the first two experiments, tol $= 10^{-5}$ and acct $= 100 \ln 10 = 230$, hence $\sigma = 0.01$. Some numerical data are given here:

delta	lowt	tol	fval	acct	N	m	$\hat{p}_0$	maxerr	errat0
1	0	10^{-5}	1536	230	128	47	$3.5\,10^{12}$	$1.6\,10^{-6}$	$4.2\,10^{-7}$
0	0	10^{-5}	1536	230	128	47	$3.5\,10^{12}$	$4.7\,10^{-4}$	$4.2\,10^{-7}$
0	2^{-7}	10^{-3}	136	60	32	16	$1.6\,10^{3}$	$1.2\,10^{-4}$	$2.0\,10^{-2}$

When lowt > 0, $r = 2$; in the other cases $r = 0$. We see in Fig. 2 that the modification causes a strong improvement for $t < 10^{-5}$. We also see that the error increases strongly for $t > 600$, but up to this point the error is less than the tolerance; note that acct $= 230$. (In a real application, one can, of course, cut off "the boundary layer" from the diagram.)

Example 3 Let $\hat{g}(s) = \sqrt{\pi/s}$. Hence $g(t) = \sqrt{1/t}$ is determined by a conditionally convergent integral, and $g(0+)$ is infinite.

An experiment was run with TOL $= 10^{-3}$, since we ran out of memory with our standard tolerance 10^{-5}. A look at Fig. 3 shows however, that outside the boundary layers the accuracy is much better than requested, and that the norms of the *relative error*, denoted maxrel, rmsrel, are most interesting, although, as

152

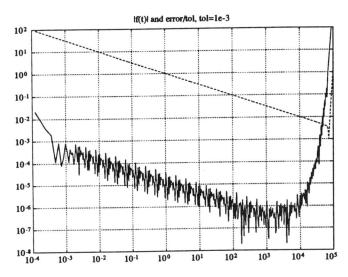

Figure 3: A conditionally convergent case, Example 3,
$\hat{g}(s) = \sqrt{\pi/s}$ lowt $= 10^{-4}$, acct $= 10^4$.

usual, the absolute error was plotted. These norms are measured on the interval $[6.1\,10^{-4}, 3840]$, i.e. the "boundary layers" were discarded. The first three parameters of the following table are chosen by the user. The others are computed by the program.

TOL	lowt	acct	fval	N	m	max t	$\hat{p}_0$	maxrel	rmsrel	f(0)
10^{-3}	10^{-4}	10^4	1216	128	37	$1.3\,10^5$	$1.1\,10^8$	$1.7\,10^{-7}$	$3.9\,10^{-8}$	$1.6\,10^4$

The positive surprise of this example indicates that there may be a better way to analyze the behavior for large ω, in order to distinguish between cases of singular behavior at $t = 0$ only, and cases of bad smoothness at other values of t. This is likely to be improved by the new type of initial conditions, which has, however, not yet been implemented for the asymptotic behavior of $g(s)$ in this example.

Example 4 Some examples, where $g(t)$ has a step discontinuity, have been studied. We applied convolution smoothing, where $\hat{f}_\sigma(\omega)$ is multiplied by $\exp(-(a_2\omega)^2)$. This results, approximately, in the convolution of $f_\sigma(t)$ with $c\exp(-(t/(2a_2))^2)$. The first case is

$$\hat{g}(s) = \frac{1 - e^{-s}}{s}, \quad g(t) = 1 - H(t - 1),$$

where H is the Heaviside unit step function. When $N = 512$, $p_r = 2$, i.e., $\Delta t = 1/128$, the results obtained in Fig. 4a, with $a_2 = 3/512$ show a good compromise

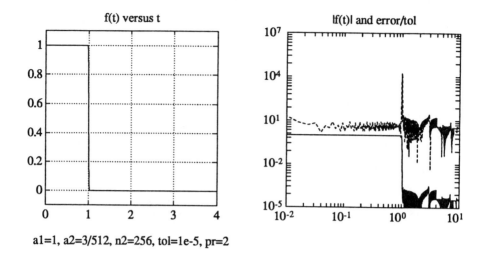

al=1, a2=3/512, n2=256, tol=1e-5, pr=2

Figure 4a: Reconstruction of a rectangular pulse with convolution smoothing; $a_2 = 3/512$, $\Delta t = 1/128$, tol $= 10^{-5}$, $(n_2 = N/2)$.

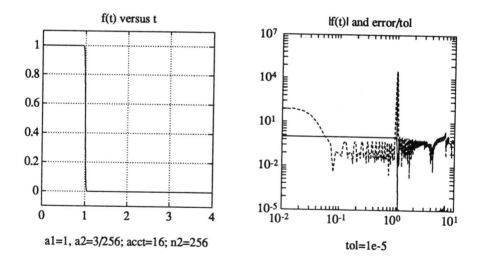

al=1, a2=3/256; acct=16; n2=256   tol=1e-5

Figure 4b: Reconstruction of a rectangular pulse with more smoothing; $a_2 = 3/256$, $\Delta t = 1/128$, tol $= 10^{-5}$, $(n_2 = N/2)$.

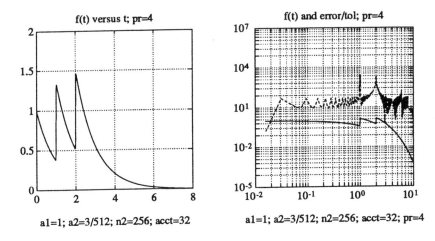

f(t) versus t; pr=4

al=1; a2=3/512; n2=256; acct=32

f(t) and error/tol; pr=4

al=1; a2=3/512; n2=256; acct=32; pr=4

Figure 4c: Reconstruction of $g(t)$ according to (4.1), with convolution smoothing, $a_2 = 3/512$, $\Delta t = 1/64$, tol $= 10^{-5}$.

between the case with $a_2 = 0$ that had visible oscillations, and the case shown in Fig. 4b, $a_2 = 3/256$, with better smoothing, but the discontinuity is seen to be less steep.

We also performed experiments with

$$\hat{g}(s) = \frac{1 + e^{-a_1 s} + e^{-2s}}{1 + s}, \quad g(t) = e^{-t} + e^{a_1 - t} H(t - a_1) + e^{2-t} H(t - 2). \quad (4.1)$$

The result shown in Fig. 4c, with $a_1 = 1$, $a_2 = 3/512$, $N = 512$, $p_r = 4$, i.e., $\Delta t = 1/64$, is considerably better than the results without convolution smoothing reported by Dahlquist (1993). The other parameters are $p_0 = 804$, $m = 7$, $r = 2$, fval $= 1024$, max $t = 64$, and the average absolute value of the error is $3\,10^{-2}$.

Dahlquist (1993) discusses how cases like this should be handled, if possible, with shift operations in the time domain, in order to avoid a function in the frequency domain with oscillations that are not well resolved by the grid, for large values of ω. This example violates the assumptions of the algorithm, both in the frequency domain and the time domain. In view of this, the results shown here are better than expected.

REFERENCES

[1] Abate J., Whitt W. The Fourier-series method for inverting transforms of probability distributions. *Queueing Systems*, 10:5–88, 1992.

[2] Bellman R., Kalaba R., Lockett J. *Numerical Inversion of the Laplace Transform.* Amer. Elsevier, New York, 1966.

155

[3] Bohman H. A method to calculate the distribution function when the characteristic function is known. *Ark. Mat.*, **4**:99–157, 1960.

[4] Cooley J., Lewis P., Welch P. Application of the Fast Fourier Transform to computation of Fourier integrals, Fourier series, and convolution integrals. *IEEE Trans.*, AU-**15**:79–84, 1967.

[5] Cooley J., Lewis P., Welch P. The Fast Fourier Transform: programming considerations in the calculation of sine, cosine and Laplace transforms. *J. Sound and Vibration*, **12**:315–337, 1970.

[6] Courant R., Hilbert D. *Methoden der Mathematischen Physik, vol.1*. J.Springer, Berlin, page 64, 1931.

[7] Dahlquist G. On an inversion formula for Laplace transforms that uses the real part only. Report TRTITA-NA-9213, NADA, Royal Inst. Techn., Stockholm, 1992.

[8] Dahlquist G. A "multigrid" extension of the FFT for the numerical inversion of Fourier and Laplace transforms. *BIT*, **33**:85–112, 1993.

[9] Davies B., Martin B. Numerical inversion of the Laplace transform: a survey and comparison of methods. *J. Comp. Phys.*, **33**:1–32, 1979.

[10] DeBoor C. *A Practical Guide to Splines*. Springer, New York, 1978.

[11] Dubner H., Abate J. Numerical inversion of Laplace transforms by relating them to the finite Fourier cosine transform. *JACM*, **15**:115–123, 1968.

[12] Fettis H.E. Numerical calculation of certain definite integrals by Poisson's summation formula. *MTAC*, **9**:85–92, 1955.

[13] Gautschi W. On the condition of a matrix arising in the numerical inversion of the Laplace transform. *Math. Comp.*, **23**:109–118, 1969.

[14] Gautschi W. Attenuation factors in practical Fourier analysis. *Numer. Math.*, **18**:373–400, 1972.

[15] Gustafson S-Å Computing inverse Laplace transforms using convergence acceleration. In *Proc. of the Second Conference on Computing and Control*, Aug. 1990. Birkhäuser, Zürich, 1990. K.L. Bowers and J. Lund, eds.

[16] Hodgkinson D.F., Lever D.A., England T.H Mathematical modelling of radionuclide migration through fractured rock using numerical inversion of Laplace transforms. *Ann. Nucl. Energy*, **11**:111, 1984.

[17] Pizarro M., Eriksson R. Modelling of the ground mode of transmission lines in time domain simulations. In *7th Int. Symp. on High Voltage Engineering*, Aug. 1991, Dresden.

[18] Talbot A. The accurate numerical inversion of Laplace transforms. *J. Inst. Maths. Applics.*, **23**:97–120, 1979.

International Series of Numerical Mathematics, Vol. 119, ©1994 Birkhäuser

SHARP BOUNDS FOR THE LEBESGUE CONSTANT IN QUADRATIC NODAL SPLINE INTERPOLATION

J.M. De Villiers[*]   C.H. Rohwer[†]

To Walter Gautschi on his sixty-fifth birthday

Abstract. It is known that, in the construction of polynomial spline approximation operators, the two properties of locality and interpolation at all of the knots are incompatible for quadratic and higher order splines. In previous work, the authors employed the concept of additional (or secondary) knots to explicitly construct a nodal spline approximation operator W which possesses, for arbitrary order m, the properties of locality, interpolation at particular (primary) knots, as well as exactness on the polynomial class of order m. In addition, estimates were obtained which implied a rather crude upper bound, depending only on m and the local primary mesh ratio, for the Lebesgue constant $||W||$. The principal aim of this paper is to explicitly calculate, in the quadratic case $m = 3$, sharp upper and lower bounds for $||W||$. The exact value $||W|| = 1.25$ in the case of equidistant primary knots is then immediately derivable. Some implications of the results, as well as an application in quadrature, are pointed out and briefly discussed.

1 INTRODUCTION

Suppose that $[a, b]$ is a finite interval on the real line $\mathbb{R}$, and for a given integer $m \geq 2$, let the integer n satisfy $n \geq m - 1$. Let the points

$$a = x_0 < x_1 < x_2 < \ldots < x_{(m-1)n} = b$$

be given, and define

$$\xi_{m,i} := x_{(m-1)i}, \quad i = 0, 1, \ldots, n.$$

The two partitions $\triangle_{m,n}$ and $\Pi_{m,n}$ of $[a, b]$ given by

$$\triangle_{m,n} := \{x_i; i = 0, 1, \ldots, (m-1)n\}, \quad \Pi_{m,n} := \{\xi_{m,i}; \ i = 0, 1, \ldots, n\},$$

then clearly satisfy $\Pi_{m,n} \subset \triangle_{m,n}$. We shall refer to the points of $\Pi_{m,n}$ and $\triangle_{m,n} \backslash \Pi_{m,n}$, respectively, as the *primary* and *secondary points* (or *knots*) corresponding to the partition $\triangle_{m,n}$.

---

[*]Department of Mathematics, University of Stellenbosch, 7600 Stellenbosch, South Africa; jmdv@sunvax.sun.ac.za

[†]Department of Mathematics, University of Stellenbosch, 7600 Stellenbosch, South Africa; chr@maties.sun.ac.za

We write $\mathbb{P}^m$ for the set of polynomials of order m (degree $\leq m - 1$), and denote by $S_{m,n}$ the set of polynomial splines of order m and with simple knots at the points $x_i, i = 1, \ldots, (m-1)n - 1$, so that $S_{m,n} \subset C^{m-2}[a, b]$. We shall write $B[a, b]$ for the set of real-valued functions on $[a, b]$, and employ the usual symbol δ_{ij} for the Kronecker delta.

In the papers [5, 6], De Villiers and Rohwer constructed a *local* spline approximation operator $W : B[a, b] \to S_{m,n}$ with the properties of *interpolation at the primary points* $\Pi_{m,n}$:

$$(Wf)(\xi_{m,i}) = f(\xi_{m,i}), \quad i = 0, 1, \ldots, n, \quad f \in B[a, b], \tag{1.1}$$

as well as *optimal order polynomial reproduction*:

$$Wp = p, \quad p \in \mathbb{P}^m. \tag{1.2}$$

Noting that, in the linear case $m = 2$, Wf is trivially given by the piecewise linear interpolant of f, *we shall henceforth assume that $m \geq 3$*. For these values of m it follows from the results in [3, 5, 6] that the defining formula for Wf on $[a, b]$ can be written in the form

$$(Wf)(x) = \sum_{i=p_j}^{q_j} f(\xi_{m,i})w_{m,i}(x), \quad x \in [\xi_{m,j}, \xi_{m,j+1}], \quad j = 0, 1, \ldots, n - 1, \tag{1.3}$$

with

$$p_j := \max\{0, j - i_1 + 1\}, \quad q_j := \min\{n, j + i_0\}, \tag{1.4}$$

$$i_0 := \begin{cases} \frac{1}{2}(m+1) & m \text{ odd,} \\ \frac{1}{2}m + 1 & m \text{ even,} \end{cases} \quad i_1 := (m+1) - i_0, \tag{1.5}$$

and where the relevant values on $[a, b]$ of the functions $w_{m,i}$ can be calculated from the formulas

$w_{m,i}(x)$

$$= \begin{cases} \displaystyle\prod_{k=0,k\neq i}^{m-1} \frac{x - \xi_{m,k}}{\xi_{m,i} - \xi_{m,k}}, & x \in [a, \xi_{m,i_1-1}], \quad (i \leq m - 1), \\[3ex] \displaystyle\sum_{r=0}^{m-2} \sum_{j=j_0}^{j_1} \alpha_{i,r,j} N_{(m-1)(i+j)+r}^m(x), & x \in [\xi_{m,i_1-1}, \xi_{m,n-i_0+1}], \quad (n \geq m), \\[3ex] \displaystyle\prod_{k=0,k\neq n-i}^{m-1} \frac{x - \xi_{m,n-k}}{\xi_{m,i} - \xi_{m,n-k}}, & x \in [\xi_{m,n-i_0+1}, b], \quad (i \geq n - (m-1)), \end{cases} \tag{1.6}$$

with

$$j_0 := \max\{-i_0, i_1 - 2 - i\}, \quad j_1 := \min\{-i_0 + (m-1), n - i_0 - i\}. \tag{1.7}$$

158

The B-spline series in (1.6) is written in terms of the set $\{N_k^m; k = (m-1)(i_1-2),$ $(m-1)(i_1-2)+1,\ldots,(m-1)(n-i_0+1)-1\} \subset S_{m,n}$ of normalized B-splines as defined (see [2, p. 108]) by

$$N_k^m(x) := (x_{k+m} - x_k)[x_k,\ldots,x_{k+m}](\cdot - x)_+^{m-1},$$

whereas the B-spline coefficients $\alpha_{i,r,j}$ can be calculated directly from the explicit formula

$$\alpha_{i,r,j} = \frac{1}{(m-1)!} \sum_{\substack{1 \leq \nu_k \leq m-1 \\ k \in M_j; \nu_k \text{distinct}}} \prod_{k=-i_0; k \neq j}^{-i_0+(m-1)} \frac{\xi_{m,i+j-k} - x_{(m-1)(i+j)+\nu_k+r}}{\xi_{m,i+j-k} - \xi_{m,i}}, \quad (1.8)$$

with M_j denoting the integer set

$$M_j := \{-i_0, -i_0+1, \ldots, -i_0+(m-1)\}\backslash\{j\}, \quad j = j_0, j_0+1, \ldots, j_1. \quad (1.9)$$

As also noted in [6], we see from (1.3) and (1.6) that the spline interpolant Wf coincides, on the intervals $[a, \xi_{m,i_1-1}]$ and $[\xi_{m,n-i_0+1}, b]$, with the Lagrange polynomial interpolants of f with respect to, respectively, the point sets $\{\xi_{m,i}; i = 0, 1, \ldots, m-1\}$ and $\{\xi_{m,n-i}; i = 0, 1, \ldots, m-1\}$. Thus, for $n = m-1$, the operator W is identical to the Lagrange interpolation operator, whereas for $n \geq m$, it is a smooth extension thereof, having recalled also the spline continuity property $Wf \in C^{m-2}[a,b]$. Hence the operator W can be interpreted as a nodal spline generalization of the Lagrange interpolation operator, where exactness on $I\!\!P^m$ has been preserved, and where W is *local* in the sense that, for a fixed $j \in \{0, 1, \ldots, n-1\}$, the value of Wf at a point $x \in (\xi_{m,j}, \xi_{m,j+1})$ depends on the values of f only at $(m+1)$ or fewer neighboring primary knots.

In order to analyze the approximation properties of W further, we next define, following [8, p. 26], the *Lebesgue constant*

$$\|W\| := \sup\{\|Wf\| ; f \in C[a,b], \|f\| \leq 1\}, \quad (1.10)$$

where we have employed the notation $\|g\| := \max_{a \leq x \leq b} |g(x)|$ for the maximum norm of a function $g \in C[a,b]$. It can then be shown from (1.3) and (1.10) that

$$\|W\| = \max_{0 \leq j \leq n-1} \max_{\xi_{m,j} \leq x \leq \xi_{m,j+1}} \sum_{i=p_j}^{q_j} |w_{m,i}(x)|. \quad (1.11)$$

Now define, for a given primary partition $\Pi_{m,n}$ of $[a,b]$, the *local primary mesh ratio* $R_{m,n}$ by

$$R_{m,n} := \max_{1 \leq i,j \leq n; |i-j|=1} \frac{|\xi_{m,i} - \xi_{m,i-1}|}{|\xi_{m,j} - \xi_{m,j-1}|}. \quad (1.12)$$

Note in particular that $R_{m,n} \geq 1$, with $R_{m,n} = 1$ if and only if $\Pi_{m,n}$ is an equidistant partition.

Recalling from [3, Theorem 5.1] the uniform bound

$$|w_{m,i}(x)| \leq \left[\sum_{\lambda=1}^{m-1} (R_{m,n})^{\lambda} \right]^{m-1}, \quad a \leq x \leq b, \quad i = 0, 1, \ldots, n, \qquad (1.13)$$

it is now an easy matter to derive, using also from (1.11), (1.4) and (1.5), the upper bound

$$\|W\| \leq (m+1) \left[\sum_{\lambda=1}^{m-1} (R_{m,n})^{\lambda} \right]^{m-1}. \qquad (1.14)$$

Next, we appeal to standard theory to estimate the interpolation error $\|f - Wf\|$, for $f \in C^k[a,b], k = 0, 1, \ldots, m$. After defining the *primary mesh norm* $H_{m,n}$ by

$$H_{m,n} := \max_{1 \leq i \leq n} |\xi_{m,i} - \xi_{m,i-1}|, \qquad (1.15)$$

and the *modulus of continuity* $\omega(f; \delta)$ of a function $f \in C[a,b]$ by

$$\omega(f; \delta) := \max_{|x-y| \leq \delta; \, x,y \in [a,b]} |f(x) - f(y)|,$$

we can employ well-known techniques of error estimation (see e.g. [2, Chapter 12], [8, Section 3.3]) to derive the estimates

$$\|f - Wf\| \leq \begin{cases} \|W\| \, \omega(f; \, mH_{m,n}), \quad f \in C[a,b], \\[2mm] (1 + \|W\|) \, c_{m,k} H_{m,n}^k \, \|f^{(k)}\|, \quad f \in C^k[a,b], \ k = 1, 2, \ldots, m, \end{cases} \qquad (1.16)$$

with the positive numbers $c_{m,k}$ bounded by

$$c_{m,k} \leq \begin{cases} \dfrac{(\pi m)^k (m-k)!}{2^{2k} m!}, \quad k = 1, 2, \ldots, m-1, \\[4mm] \dfrac{m^m}{2^{2m-1} m!}, \quad k = m. \end{cases} \qquad (1.17)$$

Here the two upper bounds in (1.17) are essentially due to two Jackson-type estimates for polynomials as given in, respectively, [1, p. 147, Jackson Theorem V] and [2, p. 31, equation (10)].

It is clear from (1.14) and (1.16) that, for a sequence $\Pi_{m,n}, n = m-1, m, \ldots,$ of primary partitions satisfying $H_{m,n} \to 0, n \to \infty$, as well as the condition that the sequence $R_{m,n}, n = m-1, m, \ldots,$ is uniformly bounded with respect to n, the error $\|f - Wf\|$ exhibits optimal order of convergence.

160

A notable feature of the estimates (1.14) and (1.16) are their independence of the placement of the secondary knots. This fact could be exploited for specific purposes, for example by letting the secondary knots merge with their primary neighbors, at the possible cost of an order of continuity, but deriving known, and possibly new, interpolation operators in the process.

The upper bound (1.14) on $||W||$, which holds for arbitrary order m, is a rather crude estimate. We proceed to show, in Section 2 below, that *sharp* upper and lower bounds can be obtained in the quadratic case $m = 3$ by means of explicit calculations; in particular, in the case of equidistant primary knots, for which $R_{3,n} = 1$, the estimate $||W|| \leq 16$, as obtained from (1.14), can be replaced by the exact value $||W|| = 1.25$, as noted in Corollary 2.2 below.

2 SHARP BOUNDS FOR THE QUADRATIC CASE

We consider in this section specifically the *quadratic case* $m = 3$, so that $n \geq 2$, and perform explicit calculations in order to derive sharp bounds for $||W||$, which could then in turn be inserted into the bounds (1.16).

We shall in this section drop the subscript $m(= 3)$ from $w_{3,i}, \xi_{3,i}, H_{3,n}, R_{3,n}$, and write, respectively, w_i, ξ_i, H_n, R_n.

Now recall from [6, pp. 207-208] that a direct (but lengthy) calculation for the case $m = 3$ by means of the formulas (1.3), ..., (1.9) yields the results

$$(Wf)(x) = \sum_{i=p_j}^{q_j} f(\xi_i) w_i(x), \quad x \in [\xi_j, \xi_{j+1}], \quad j = 0, 1, \ldots, n-1, \quad (2.1)$$

with

$$p_j := \max\{0, j-1\} \quad, \quad q_j := \min\{n, j+2\}, \quad (2.2)$$

where the relevant values on $[a, b]$ of the functions w_i can be calculated from the formulas

$$w_i(x) = \begin{cases} \prod_{k=0,k\neq i}^{2} \dfrac{x - \xi_k}{\xi_i - \xi_k}, & x \in [a, \xi_1], \quad (i \leq 2) \\[2mm] s_i(x), & x \in [\xi_1, \xi_{n-1}], \quad (n \geq 3), \\[2mm] \prod_{k=0,k\neq n-i}^{2} \dfrac{x - \xi_{n-k}}{\xi_i - \xi_{n-k}}, & x \in [\xi_{n-1}, b], \quad (i \geq n-2), \end{cases} \quad (2.3)$$

and where, in the notation

$$t_j := x_{2i+j}, \quad j = -4, -3, \ldots, 4, \quad (2.4)$$

the *quadratic nodal spline* s_i satisfies

$$s_i(x) = \begin{cases} s(x;\, t_{-2}, t_0, t_1, t_2, t_3, t_4), & x \in [\xi_i, \xi_{i+2}], \\[2mm] s(-x;\, -t_2, -t_0, -t_{-1}, -t_{-2}, -t_{-3}, -t_{-4}), & x \in [\xi_{i-2}, \xi_i], \end{cases} \quad (2.5)$$

161

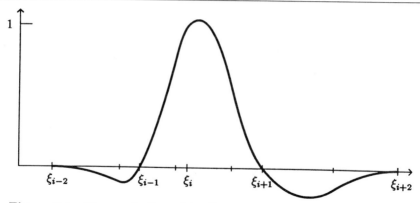

Figure 2.1: The quadratic nodal spline s_i on a randomly chosen partition.

with

$$
s(x;\, t_{-2}, t_0, t_1, t_2, t_3, t_4) = \begin{cases} 1 + (x - t_0)[A - C(x - t_0)], & x \in [t_0, t_1], \\[2mm] (t_2 - x)[B + D(t_2 - x)], & x \in [t_1, t_2], \\[2mm] (x - t_2)[-B + E(x - t_2)], & x \in [t_2, t_3], \\[2mm] -F(t_4 - x)^2, & x \in [t_3, t_4], \end{cases} \tag{2.6}
$$

in terms of the coefficients $A, B, \ldots, F$ defined by the formulas

$$
\left.
\begin{aligned}
A &= \frac{(t_2 - t_0) - (t_0 - t_{-2})}{(t_2 - t_0)(t_0 - t_{-2})}, & B &= \frac{t_4 - t_2}{(t_4 - t_0)(t_2 - t_0)}, \\[2mm]
C &= \frac{(t_4 - t_0)(t_1 - t_0) + (t_2 - t_0)(t_4 - t_0) + (t_0 - t_{-2})(t_2 - t_1)}{2(t_4 - t_0)(t_2 - t_0)(t_0 - t_{-2})(t_1 - t_0)}, \\[2mm]
D &= \frac{(t_4 - t_0)(t_1 - t_0) + (t_2 - t_0)(t_0 - t_{-2}) + (t_0 - t_{-2})(t_2 - t_1)}{2(t_4 - t_0)(t_2 - t_0)(t_0 - t_{-2})(t_2 - t_1)}, \\[2mm]
E &= \frac{(t_3 - t_2) + (t_4 - t_2)}{2(t_4 - t_0)(t_2 - t_0)(t_3 - t_2)}, & F &= \frac{t_3 - t_2}{2(t_4 - t_0)(t_2 - t_0)(t_4 - t_3)}.
\end{aligned}
\right\} \tag{2.7}
$$

Note in particular from (2.7) that the coefficients B, C, D, E and F are strictly positive, so that we can easily deduce the shape and sign properties of the quadratic nodal spline s_i from (2.5) and (2.6). An illustrating example is drawn in Fig. 2.1.

The explicit expressions above now enable us to derive the following sharp upper and lower bounds for the quadratic Lebesgue constant $\|W\|$.

Theorem 2.1 *In the quadratic case* $m = 3$, *and for* $n \geq 2$,

$$1.25 \leq ||W|| \leq 1 + \frac{R_n^2}{2(1 + R_n)}. \tag{2.8}$$

Proof: With the definition

$$w(x) := \sum_{i=p_j}^{q_j} |w_i(x)|, \quad x \in [\xi_j, \xi_{j+1}], \quad j = 0, 1, \ldots, n-1, \tag{2.9}$$

it will clearly suffice, according to (1.11), to show that

$$1.25 \leq \max_{a \leq x \leq b} w(x) \leq 1 + \frac{R_n^2}{2(1 + R_n)}. \tag{2.10}$$

Our method of proof will consist of maximizing $w(x)$ first, for $n \geq 3$, over the interior interval $[\xi_1, \xi_{n-1}]$, and then, for $n \geq 2$, over the endpoint intervals $[a, \xi_1]$ and $[\xi_{n-1}, b]$, after which the resulting maxima are bounded from above and below.
(a) To maximize $w(x)$ on $[\xi_1, \xi_{n-1}]$ for $n \geq 3$, we fix the index $j \in \{1, 2, \ldots,$ $n-2\}$ and let $x \in [\xi_j, \xi_{j+1}]$. But then, from (2.2) and (2.3), we have $p_j = j - 1$, $q_j = j + 2$ and $w_i(x) = s_i(x)$, whence, in (2.9),

$$w(x) = \sum_{i=j-1}^{j+2} |s_i(x)| = -s_{j-1}(x) + s_j(x) + s_{j+1}(x) - s_{j+2}(x)$$
$$= 1 - 2[s_{j-1}(x) + s_{j+2}(x)], \tag{2.11}$$

having also deduced the signs of $s_i(x)$ on $[\xi_j, \xi_{j+1}]$ from (2.5), (2.6) and (2.7) (cf. the example drawn in Fig. 2.2), and then exploiting the fact that

$$\sum_{i=j-1}^{j+2} s_i(x) = 1, \quad x \in [\xi_1, \xi_{n-1}],$$

as is clear from (2.3), (1.3) and (1.2). Now substitute the formulas for $s_{j-1}(x)$ and $s_{j+2}(x)$, as obtained from (2.5), (2.6) and (2.7), into (2.11). We find, after some algebra, the expressions

$$w(x) = \begin{cases} 1 + (x - \xi_j)[K(x - \xi_j) + L], & x \in [\xi_j, x_{2j+1}], \\ 1 + (\xi_{j+1} - x)[M(\xi_{j+1} - x) + N], & x \in [x_{2j+1}, \xi_{j+1}], \end{cases} \tag{2.12}$$

where, in the notation $\alpha := \xi_j - \xi_{j-1}$, $\beta := \xi_{j+1} - \xi_j$, $\gamma := \xi_{j+2} - \xi_{j+1}$, $\delta := x_{2j+1} - \xi_j$, $\tag{2.13}$

we have the values

163

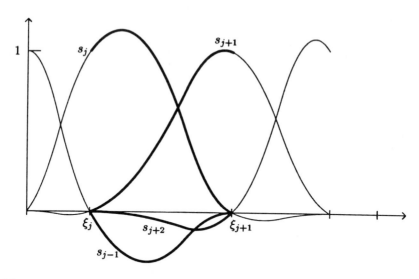

Figure 2.2: The quadratic nodal splines $\{s_i, i = j - 1, \ldots, j + 2\}$ on $[\xi_j, \xi_{j+1}]$.

$$
\left.
\begin{aligned}
K &= \frac{\beta[\alpha(\alpha + \beta) - \gamma(\beta + \gamma)] - \delta[\alpha(\alpha + \beta) + \gamma(\beta + \gamma)]}{\alpha\gamma\delta(\alpha + \beta)(\beta + \gamma)}, \\[2mm]
L &= \frac{2\beta}{\alpha(\alpha + \beta)}, \\[2mm]
M &= \frac{\beta[\gamma(\beta + \gamma) - \alpha(\alpha + \beta)] - (\beta - \delta)[\alpha(\alpha + \beta) + \gamma(\beta + \gamma)]}{\alpha\gamma(\beta - \delta)(\alpha + \beta)(\beta + \gamma)}, \\[2mm]
N &= \frac{2\beta}{\gamma(\beta + \gamma)}.
\end{aligned}
\right\}
\tag{2.14}
$$

It is easily seen from (2.11) that, since $s_{j-1}, s_{j+2} \in S_{3,n} \subset C^1[a, b]$, the functions w and w' are continuous on $[\xi_j, \xi_{j+1}]$. Moreover, using (2.12), (2.13) and (2.14), we calculate the values

$$
w(\xi_j) = 1 = w(\xi_{j+1}),
$$

$$
w'(\xi_j) = L > 0, \qquad w'(\xi_{j+1}) = -N < 0,
$$

and

$$
w'(x_{2j+1}) = 2\frac{\alpha(\alpha + \beta)(\beta - \delta) - \gamma(\beta + \gamma)\delta}{\alpha\gamma(\alpha + \beta)(\beta + \gamma)}.
$$

Hence the sign of the derivative $w'(x_{2j+1})$ depends on the position of the secondary knot x_{2j+1} in the primary interval $[\xi_j, \xi_{j+1}]$ in the sense that

$$w'(x_{2j+1}) \begin{cases} > 0 & , & 0 < \delta < \delta_0, \\ = 0 & , & \delta = \delta_0, \\ < 0 & , & \delta_0 < \delta < \beta, \end{cases}$$

with

$$\delta_0 := \left[\frac{\alpha(\alpha + \beta)}{\alpha(\alpha + \beta) + \gamma(\beta + \gamma)} \right] \beta,$$

having noted also from the definition (2.13) that $0 < \delta < \beta$.

Since the function w clearly attains its maximum over the interval $[\xi_j, \xi_{j+1}]$ on $(\xi_j, x_{2j+1}]$ or (x_{2j+1}, ξ_{j+1}) according to whether, respectively, $w'(x_{2j+1}) \leq 0$ or $w'(x_{2j+1}) > 0$, a straight forward calculus analysis then yields the maxima

$$\max_{\xi_j \leq x \leq \xi_{j+1}} w(x)$$

$$= \begin{cases} 1 + \dfrac{\beta^2 \alpha(\alpha + \beta)}{\gamma(\beta + \gamma)[\alpha(\alpha + \beta) + \gamma(\beta + \gamma) + (\beta - \delta)^{-1}\beta(\alpha - \gamma)(\alpha + \beta + \gamma)]} \\ \qquad\qquad\qquad\qquad\qquad\qquad\qquad \text{for} \quad 0 < \delta \leq \delta_0, \\[2mm] 1 + \dfrac{\beta^2 \gamma(\beta + \gamma)}{\alpha(\alpha + \beta)[\alpha(\alpha + \beta) + \gamma(\beta + \gamma) + \delta^{-1}\beta(\gamma - \alpha)(\alpha + \beta + \gamma)]} \\ \qquad\qquad\qquad\qquad\qquad\qquad\qquad \text{for} \quad \delta_0 < \delta < \beta. \end{cases} \qquad (2.15)$$

Next, treating the two cases $\alpha \leq \gamma$ and $\alpha > \gamma$ in (2.15) separately, we derive the uniform bound

$$\max_{\xi_j \leq x \leq \xi_{j+1}} w(x) \leq 1 + \frac{1}{2} \max\left\{ \frac{\beta^2}{\alpha(\alpha + \beta)}, \frac{\beta^2}{\gamma(\beta + \gamma)} \right\}, \quad 0 < \delta < \beta. \qquad (2.16)$$

(b) To obtain, for $n \geq 2$, a bound similar to (2.16) in the endpoint intervals, we first suppose that $x \in [a, \xi_1]$, for which, from (2.9), (2.3) and (1.4), (1.5), we have

$$w(x) = \sum_{i=0}^{2} |\ell_i(x)| = \ell_0(x) + \ell_1(x) - \ell_2(x),$$

with $\{\ell_i, i = 0, 1, 2\}$ denoting the Lagrange fundamental polynomials defined as in the top line of (2.3) by

$$\ell_i(x) := \prod_{k=0, k\neq i}^{2} \frac{x - \xi_k}{\xi_i - \xi_k}, \quad i = 0, 1, 2.$$

Exploiting the fact that $\sum_{i=0}^{2} \ell_i(x) = 1$, $x \in \mathbb{R}$, we get, in the notation (2.13) with $j = 0$,

$$w(x) = 1 - 2\,\ell_2(x) = 1 + \frac{2}{\gamma(\beta + \gamma)}(x - \xi_0)(\xi_1 - x),$$

and thus

$$\max_{a \leq x \leq \xi_1} w(x) = 1 + \frac{1}{2}\,\frac{\beta^2}{\gamma(\beta + \gamma)}. \qquad (2.17)$$

A similar procedure applied to $[\xi_{n-1}, b]$ yields the analogous maximum

$$\max_{\xi_{n-1} \leq x \leq b} w(x) = 1 + \frac{1}{2}\,\frac{\beta^2}{\alpha(\alpha + \beta)}, \qquad (2.18)$$

in the notation (2.13) with $j = n - 1$.

(c) The desired upper bound in (2.10) now follows from the maxima (2.16), (2.17) and (2.18), together with the bounds, derived from the definition (1.12),

$$\frac{\beta^2}{\alpha(\alpha + \beta)} = \left[\frac{\alpha}{\beta}\left(\frac{\alpha}{\beta} + 1\right)\right]^{-1} \leq \left[\frac{1}{R_n}\left(\frac{1}{R_n} + 1\right)\right]^{-1} = \frac{R_n^2}{1 + R_n},$$

and, similarly,

$$\frac{\beta^2}{\gamma(\beta + \gamma)} \leq \frac{R_n^2}{1 + R_n}.$$

(d) To prove the lower bound in (2.10), we first observe from (2.15), after having treated the two cases $\alpha \leq \gamma$ and $\alpha > \gamma$ again separately, that, for $n \geq 3$,

$$\max_{\xi_j \leq x \leq \xi_{j+1}} w(x) \geq 1 + \frac{1}{2}\min\left\{\frac{\beta^2}{\alpha(\alpha + \beta)}, \frac{\beta^2}{\gamma(\beta + \gamma)}\right\}, \quad 0 < \delta < \beta.$$

Recalling also, for $n \geq 2$, the maxima (2.17) and (2.18), we deduce that there exists at least one index $j \in \{0, 1, \ldots, n - 2\}$ such that the maxima of $w(x)$ over successive intervals $[\xi_j, \xi_{j+1}]$ and $[\xi_{j+1}, \xi_{j+2}]$ are bounded below by

$$\left.\begin{array}{l} \displaystyle\max_{\xi_j \leq x \leq \xi_{j+1}} w(x) \geq 1 + \frac{1}{2}\,\frac{\beta^2}{\gamma(\beta + \gamma)}, \\[2mm] \displaystyle\max_{\xi_{j+1} \leq x \leq \xi_{j+2}} w(x) \geq 1 + \frac{1}{2}\,\frac{\gamma^2}{\beta(\beta + \gamma)}, \end{array}\right\}$$

and thus

$$\max_{\xi_j \leq x \leq \xi_{j+2}} w(x) \geq 1 + \frac{1}{2}\max\left\{\frac{r^2}{r + 1}, \frac{1}{r(r + 1)}\right\}, \qquad (2.19)$$

where we have introduced the notation $r := \beta/\gamma$. Moreover, since the definition (1.12) clearly implies the bounds $1/R_n \leq r \leq R_n$, we can apply a routine calculus procedure to derive the result

$$\max\left\{\frac{r^2}{r+1}, \ \frac{1}{r(r+1)}\right\} \geq \frac{1}{2}, \quad \frac{1}{R_n} \leq r \leq R_n. \tag{2.20}$$

The desired lower bound in (2.10) now follows by combining (2.19) and (2.20). ∎

Corollary 2.2 *Suppose the primary partition $\Pi_{3,n}$ is equidistant. Then*

$$\|W\| = 1.25.$$

Proof: Set $R_n = 1$ in (2.8). ∎

Remarks. (a) The bounds in (2.8) are *sharp* in the sense that, for any given value of $n \geq 2$, we can always construct a primary partition $\Pi_{3,n}$ such that these bounds are actually attained. To demonstrate this fact, we choose an arbitrary positive number $R \geq 1$, and define the primary partition $\Pi_{3,n}$ by

$$\xi_0 = a, \qquad \xi_i = a + \frac{(b-a)[R+(i-1)]}{R+(n-1)}, \quad i = 1, 2, \ldots, n, \tag{2.21}$$

so that clearly $R_n = R$. Also, $\beta = R\gamma$ in (2.17), whereas $\alpha = \beta = \gamma$ in (2.16) and (2.18), so that the upper bound in (2.10) is actually attained in the endpoint interval $[a, \xi_1]$, and thus $\|W\| = 1 + \frac{1}{2}R_n^2/(1+R_n)$ for the specific primary partition (2.21). The sharpness of the lower bound ($= 1.25$) in (2.8) is shown by choosing $R = 1$ in the argument above, as has already (indirectly) been noted in Corollary 2.2.

(b) Suppose $\{\Pi_{3,n}; n = 2, 3, \ldots\}$ is a sequence of primary partitions for which, like the example given by (2.21), the corresponding Lebesgue constants $\|W\| = 1 + \frac{1}{2}R_n^2/(1 + R_n), n = 2, 3, \ldots$, and suppose $R_n \to \infty, n \to \infty$, so that also $\|W\| \to \infty, n \to \infty$. The *theorem of uniform boundedness* [7, p. 203] then implies the existence of a function $f \in C[a, b]$ such that the associated error sequence $\|f - Wf\|$ diverges for $n \to \infty$. This fact emphasizes again the importance of the uniform boundedness of the sequence $R_n, n = 2, 3, \ldots$, as a sufficient condition for the convergence properties (1.16) for $H_n \to 0$ of the nodal spline interpolation operator W.

(c) Note that, for equidistant $\Pi_{3,n}$, Corollary 2.2 implies that W is a generalization of the Lagrange polynomial interpolation operator with the property that the Lebesgue constant $\|W\|$ remains equal to the value 1.25 for all $n \geq 2$.

(d) In the case of odd order m, and equidistant knots $x_i, i = 0, 1, \ldots,$ $(m-1)n$, it has been shown [3, Theorem 4.1] that the quadrature rule

$$Q[f] := \int_a^b (Wf)(x)dx$$

is precisely the Gregory rule of even order $(m+1)/2$. Hence, in the quadratic case $m = 3$, the error estimates

$$
\|f - Wf\| \leq
\begin{cases}
\frac{5}{4}\omega(f; 3H_n), & f \in C[a, b], \\[2mm]
\frac{9\pi}{16} H_n\|f'\|, & f \in C'[a, b], \\[2mm]
\frac{27\pi^2}{128} H_n^2\|f''\|, & f \in C^2[a, b], \\[2mm]
\frac{81}{256} H_n^3\|f'''\|, & f \in C^3[a, b],
\end{cases}
$$

as obtained from (1.16) and Corollary 2.2, yield reasonably sharp order constants for the quadrature error $\left(:= \int_a^b f(x)dx - Q[f] \right)$ corresponding to the Gregory rule of order two, also known as the Lacroix rule.

REFERENCES

[1] Cheney E.W. *Introduction to Approximation Theory*. McGraw-Hill, New York, 1966.

[2] de Boor C. *A Practical Guide to Splines*. Springer-Verlag, Berlin, New York, 1978.

[3] De Villiers J.M. A convergence result in nodal spline interpolation. *J. Approx. Theory*, **74**:266–279, 1993.

[4] De Villiers J.M. A nodal spline interpolant for the Gregory rule of even order. *Numer. Math.*, **66**:123–137, 1993.

[5] De Villiers J.M., Rohwer C.H. Optimal local spline interpolants. *J. Comput. Appl. Math*, **18**:107–119, 1987.

[6] De Villiers J.M., Rohwer C.H. A nodal spline generalization of the Lagrange interpolant. In *Progress in Approximation Theory*, pages 201–211. Academic Press, San Diego, 1991. P. Nevai and A. Pinkus, eds.

[7] Powell M.J.D. *Approximation Theory and Methods*. Cambridge University Press, Cambridge, 1981.

[8] Nürnberger G. *Approximation by Spline Functions*. Springer-Verlag, Berlin, Heidelberg, 1989.

International Series of Numerical Mathematics, Vol. 119, ©1994 Birkhäuser

On the Error of Extended Gaussian Quadrature Formulae for Functions of Bounded Variation

Sven Ehrich*

Dedicated to Professor Walter Gautschi on the occasion of his 65th birthday

Abstract. We investigate the remainder R_{2n+1} of (minimum node) extended Gaussian quadrature formulae Q_{2n+1} by means of the constants $\varrho_V(R_{2n+1})$, which are best possible in the error bound

$$|R_{2n+1}[f]| \le \varrho_V(R_{2n+1}) \operatorname{Var}(f)$$

for all functions f of bounded variation $\operatorname{Var}(f)$. For the most often used Gauss–Kronrod extensions Q_{2n+1}^{GK}, we prove that there holds

$$\lim_{n \to \infty} (2n+1) \varrho_V(R_{2n+1}^{GK}) = \frac{\pi}{2}.$$

As a consequence, we obtain that, among all extended Gaussian formulae whose additional nodes interlace with the Gaussian ones, the Gauss–Kronrod formula Q_{2n+1}^{GK} is asymptotically optimal with respect to ϱ_V.

1 INTRODUCTION AND STATEMENT OF THE MAIN RESULT

A quadrature formula Q_n of degree $\deg(R_n) = s \ge 0$ is a linear functional defined by (see e.g. Brass [1])

$$Q_n[f] = \sum_{\nu=1}^{n} a_{\nu,n} f(x_{\nu,n}), \qquad -\infty < x_{1,n} < \cdots < x_{n,n} < \infty, \qquad (1.1)$$

$$\int_{-1}^{1} f(x)\, dx = Q_n[f] + R_n[f], \qquad R_n[p_\mu] \begin{cases} = 0, & \mu = 0, \ldots, s, \\ \ne 0, & \mu = s+1, \end{cases}$$

where $p_\mu(x) = x^\mu$. A quadrature formula is called positive if the weights $a_{\nu,n}$, $\nu = 1, \ldots, n$, are nonnegative, and Q_n is called interpolatory if

$$\deg(R_n) \ge n - 1. \qquad (1.2)$$

For two quadrature formulae Q_n and Q_m, the former is called an extension of the latter if $n \ge m$ and $Q_n - Q_m$ is of the form $D[f] = \sum_{\nu=1}^{n} b_{\nu,n} f(x_{\nu,n})$.

It is well known that the Gaussian formula $Q_n^G[f] = \sum_{\nu=1}^{n} a_{\nu,n}^G f(x_{\nu,n}^G)$ can be defined by $\deg(R_n^G) = 2n - 1$. Therefore, the minimum number of additional nodes

---

*Institut für Mathematik, Universität Hildesheim, D–31141 Hildesheim, Germany

for a nontrivial interpolatory extension of Q_n^G is $n + 1$, and since our primary interest will be in interpolatory extensions, in the following we will call a quadrature formulae Q_{2n+1} a (minimum node) extended Gaussian formula if Q_{2n+1} is of the form

$$Q_{2n+1}[f] = \sum_{\nu=1}^{n} A_{\nu,n} f(x_{\nu,n}^G) + \sum_{\mu=1}^{n+1} B_{\mu,n+1} f(\xi_{\mu,n+1}). \tag{1.3}$$

An extended Gaussian formula Q_{2n+1} is said to have the interlacing property, if there holds

$$-\infty < \xi_{1,n+1} < x_{1,n}^G < \xi_{2,n+1} < x_{2,n}^G < \cdots < x_{n,n}^G < \xi_{n+1,n+1} < \infty. \tag{1.4}$$

Furthermore, the Gauss–Kronrod formula Q_{2n+1}^{GK} is defined by

$$Q_{2n+1}^{GK}[f] = \sum_{\nu=1}^{n} A_{\nu,n}^{GK} f(x_{\nu,n}^G) + \sum_{\mu=1}^{n+1} B_{\mu,n+1}^{GK} f(\xi_{\mu,n+1}^K), \qquad \deg(R_{2n+1}^{GK}) \geq 3n + 1. \tag{1.5}$$

It follows from a result of Szegö [18] that real Gauss–Kronrod formulae exist uniquely, and that they have the interlacing property (1.4), while all nodes are symmetrically distributed inside the integration interval. Monegato [13, 14] proved the positivity of Q_{2n+1}^{GK}.

In 1988, Gautschi [10] provided a complete survey of the considerable literature that had followed the original work of Kronrod in 1964 [11, 12]. Also in this survey, many of the important open problems that are connected with Gauss–Kronrod quadrature were stated. A systematic investigation of the remainder functional R_{2n+1}^{GK}, which at that time was an open question (see Monegato [15, p. 155]), has only recently begun for smooth, i.e., analytic or k-times differentiable functions (see also the article of Notaris in this volume [16]). Very little is known, however, about the behavior of R_{2n+1}^{GK} for low-order continuous functions. This question, on the other hand, is of no less practical interest, since numerical quadrature routines are often expected to work also for "non-smooth" functions.

Following Gautschi [9, p. 115], we take as a convenient means of measuring the quadrature error for functions of a given continuity class, the error constants that originate from the well-known Peano representation of the remainder functional, since these constants allow a comparison of quadrature formulae on a common basis (see [9, p. 115]). For extended quadrature formulae, such a comparison is useful in at least two respects: first, to justify the basic assumption that the extended formula gives a better approximation than the original formula, and second, to compare different extended formulae, e.g., formulae using the same amount of functional data.

In this paper, we investigate the error constants $\varrho_V(R_n)$ which are best possible in the inequality

$$|R_n[f]| \leq \varrho_V(R_n)\mathrm{Var}(f) \tag{1.6}$$

for all functions f of bounded variation $\mathrm{Var}(f)$. It is well known (see e.g. Brass [1, p. 42], Gautschi [9, p. 115]) that $\varrho_V(R_n)$ exists if Q_n is of type (1.1), and that

$$\varrho_V(R_n) = \sup_{x \in [-1,1]} |K_1(R_n, x)|, \qquad (1.7)$$

where $K_1(R_n, \cdot)$ is the first Peano kernel of R_n defined by

$$K_1(R_n, x) = -(x+1)_+ + \sum_{\nu=1}^{n} a_{\nu,n}(x - x_{\nu,n})_+^0, \qquad (1.8)$$

$$u_+^0 = \begin{cases} 1, & u > 0, \\ \frac{1}{2}, & u = 0, \\ 0, & u < 0, \end{cases} \qquad u_+ = \begin{cases} u, & u > 0, \\ 0, & u \le 0. \end{cases} \qquad (1.9)$$

In view of (1.6), the constants $\varrho_V(R_n)$ are of interest if not much is known on the smoothness of the function f, and in particular if the existence of bounded derivatives is unknown.

The following result describes the general situation for extended Gaussian quadrature formulae which need not necessarily be positive nor interpolatory.

Proposition 1.1 Let $n \in \mathbb{N}$. If Q_{2n+1} is an extended Gaussian formula which has the interlacing property (1.4), then sharp lower bounds for $\varrho_V(R_{2n+1})$ are given by

$$\varrho_V(R_{2n+1}) \ge \frac{1}{2} |x_{n/2,n}^G| \qquad \text{for even } n, \qquad (1.10)$$

$$\varrho_V(R_{2n+1}) \ge \frac{1}{4} |x_{(n-1)/2,n}^G| \qquad \text{for odd } n. \qquad (1.11)$$

Proof: For the proof of Proposition 1.1, we observe that the first Peano kernel defined in (1.8) has constant derivative -1 between two successive nodes of Q_{2n+1}. Therefore, between the two successive nodes $x_{\nu,n}^G$ and $x_{\nu+1,n}^G$ of the Gaussian formula Q_n^G, the supremum of the first Peano kernel of an extended Gaussian formula is minimized if the additional node in $(x_{\nu,n}^G, x_{\nu+1,n}^G)$ is chosen as the midpoint $(x_{\nu,n}^G + x_{\nu+1,n}^G)/2$, and the zeros of the Peano kernels are chosen as $x_{\nu,n}^G + (x_{\nu+1,n}^G - x_{\nu,n}^G)/4$ and $x_{\nu,n}^G + 3(x_{\nu+1,n}^G - x_{\nu,n}^G)/4$. The value of the supremum in $(x_{\nu,n}^G, x_{\nu+1,n}^G)$ is then $(x_{\nu+1,n}^G - x_{\nu,n}^G)/4$. In view of the well known convexity of the Gaussian nodes (see e.g. [19, Theorem 6.3.3]), this quantity is bounded by $(x_{n/2+1,n}^G - x_{n/2,n}^G)/4 = \frac{1}{2}|x_{n/2,n}^G|$ for even n, and by $\frac{1}{4}|x_{(n-1)/2,n}^G|$ for odd n. On the other hand, these values are attained by the extension with the additional points $\xi_{1,n+1} := -\xi_{n+1,n+1} := x_{1,n}^G/3 - 2/3$, $\xi_{\nu,n+1} := (x_{\nu-1,n}^G + x_{\nu,n}^G)/2$, $\nu = 2, \ldots, n$, and with associated weights $A_{1,n} := A_{n,n} := x_{1,n}^G/12 + x_{2,n}^G/4 + 1/3$, $B_{1,n+1} := B_{n+1,n+1} := 2x_{1,n}^G/3 + 2/3$, $A_{\nu,n} = (x_{\nu+1,n}^G - x_{\nu-1,n}^G)/4$, $\nu = 2, \ldots, n-1$, $B_{\mu,n+1} = (x_{\mu,n}^G - x_{\mu-1,n}^G)/2$, $\mu = 2, \ldots, n$. ∎

Remark 1.2 By a result of Gatteschi [8], it follows from (1.11) that for $n \geq 3$

$$\varrho_V(R_{2n+1}) \geq \frac{1}{4} \sin\left\{\frac{2\pi}{2n+1}\left(1 - \frac{1}{2(2n+1)^2}\right)\right\}. \tag{1.12}$$

Using general results of Förster und Petras [7], we immediately obtain an upper bound.

Proposition 1.3 *Let $n \in \mathbb{N}$. If the extended Gaussian formula Q_{2n+1} is positive and satisfies the interpolatory condition (1.2), then*

$$\varrho_V(R_{2n+1}) \leq \varrho_V(R_n^G)\left\{1 - \sin\frac{\pi}{2n+1}\right\}^{-1}. \tag{1.13}$$

Remark 1.4 A slightly sharper bound than the one in Proposition 1.3 can be obtained with the help of [7, Eq. (1.12)] if Q_{2n+1} is also assumed to be symmetric.

Remark 1.5 Förster and Petras [7] derived exact values of $\varrho_V(R_n^G)$, from which they obtained the following bounds [7, Remark 1]:

$$\frac{\pi}{2n+1}\left(1 - \frac{1}{2(2n+1)^2}\right) < \varrho_V(R_n^G) < \frac{\pi}{2n+1} \qquad \text{for odd } n, \tag{1.14}$$

$$\sin\left\{\frac{\pi}{2n+1}\left(1 - \frac{3}{4(2n+1)^2}\right)\right\} < \varrho_V(R_n^G) < \sin\frac{\pi}{2n+1} \qquad \text{for even } n. \tag{1.15}$$

We obtain from Proposition 1.1 and Proposition 1.3 that for the Gauss–Kronrod formula Q_{2n+1}^{GK} there holds

$$\frac{\pi}{2} \leq \limsup_{n\to\infty} (2n+1)\,\varrho_V(R_{2n+1}^{GK}) \leq \pi. \tag{1.16}$$

Using a further result of Förster and Petras [7, Theorem 1.3], we obtain the still sharper upper bound

$$\limsup_{n\to\infty} (2n+1)\,\varrho_V(R_{2n+1}^{GK}) \leq \frac{2}{3}\pi, \tag{1.17}$$

and, to our knowledge, no sharper result exists until now. Our main result is:

Theorem 1.6 *Let $n \in \mathbb{N}$, and let R_{2n+1}^{GK} be the remainder of the Gauss–Kronrod formula defined in (1.5). Then*

$$\lim_{n\to\infty} (2n+1)\,\varrho_V(R_{2n+1}^{GK}) = \frac{\pi}{2}. \tag{1.18}$$

We obtain the following corollary as a consequence of Theorem 1.6 and Proposition 1.1.

Corollary 1.7 *Let $n \in \mathbb{N}$. Among all extended Gaussian formulae Q_{2n+1} that have the interlacing property (1.4), the Gauss–Kronrod quadrature formula Q_{2n+1}^{GK} is asymptotically optimal in the class of functions of bounded variation, i.e.,*

$$\limsup_{n \to \infty} \frac{\varrho_V(R_{2n+1}^{GK})}{\varrho_V(R_{2n+1})} \leq 1. \tag{1.19}$$

Using Theorem 1.6 and Remark 1.5, we state a second corollary that might be particularly useful for numerical quadrature routines that use Gauss and Gauss–Kronrod formulae, since it gives rise to a practical error estimate for the wide class of functions of bounded variation.

Corollary 1.8 *Let $n \in \mathbb{N}$, and let R_n^G and R_{2n+1}^{GK} be the remainders of the Gaussian formula Q_n^G and the Gauss–Kronrod formula Q_{2n+1}^{GK}, respectively. Then*

$$\lim_{n \to \infty} \frac{\varrho_V(R_{2n+1}^{GK})}{\varrho_V(R_n^G)} = \frac{1}{2}. \tag{1.20}$$

Remark 1.9 Similar results as in Corollary 1.8, but for the classes $C^s[-1, 1]$, were proved by Brass and Förster [2] and in [4]. More precisely, for the well-known Peano constants $c_s(R_n)$, which are best possible in the inequality

$$|R_n[f]| \leq c_s(R_n) \sup_{x \in [-1,1]} |f^{(s)}(x)|, \tag{1.21}$$

for all $f \in C^s[-1, 1]$, it is proved in [2, §5] that

$$\frac{c_{2n}(R_{2n+1}^{GK})}{c_{2n}(R_n^G)} \leq \text{const } \sqrt[4]{n} \left(\frac{16}{25\sqrt{5}} \right)^n. \tag{1.22}$$

For every fixed $s \in \mathbb{N}$, it is proved in [4] that

$$\lim_{n \to \infty} \frac{c_s(R_{2n+1}^{GK})}{c_s(R_n^G)} = 2^{-s}. \tag{1.23}$$

2 PROOF OF THE MAIN RESULT (THEOREM 1.6)

The following Lemma is a direct consequence of [3, Corollary 1 and Eq. (65)].

Lemma 2.1 *Let $\epsilon > 0$ be fixed. Then uniformly for all $-1 + \epsilon \leq x_{\nu-1,n}^G < \xi_{\nu,n+1}^K < x_{\nu,n}^G \leq 1 - \epsilon$*

$$\text{(i)} \quad \xi_{\nu,n+1}^K - x_{\nu-1,n}^G = (1 + o(1)) \frac{\pi}{2n+1} \sin \frac{n - \nu + 3/2}{n + 1/2} \pi,$$

$$\text{(ii)} \quad x_{\nu,n}^G - \xi_{\nu,n+1}^K = (1 + o(1)) \frac{\pi}{2n+1} \sin \frac{n - \nu + 1}{n + 1/2} \pi.$$

Defining the numbers $b_\nu^+(R_n)$ and $b_\nu^-(R_n)$ (see [7, Eq. (3.2)]) by

$$b_\nu^+(R_n) := K_1(R_n, x_{\nu,n}) + a_{\nu,n}/2, \qquad b_\nu^-(R_n) := b_\nu^+(R_n) - a_{\nu,n}, \qquad \nu = 1, \ldots, n, \tag{2.1}$$

we find

$$\varrho_V(R_n) = \max_{\nu=1,\ldots,n} \{|b_\nu^+(R_n)|, |b_\nu^-(R_n)|\}. \tag{2.2}$$

Since Q_{2n+1}^{GK} is symmetric, i.e., $x_{n+1-\nu,n}^G = -x_{\nu,n}^G$ and $\xi_{n+2-\mu,n+1}^K = -\xi_{\mu,n+1}^K$ as well as $A_{n+1-\nu,n}^{GK} = A_{\nu,n}^{GK}$ and $B_{n+2-\mu,n+1}^{GK} = B_{\mu,n+1}^{GK}$, we have

$$K_1(R_{2n+1}^{GK}, x) = -K_1(R_{2n+1}^{GK}, -x), \qquad x \in [-1, 1], \tag{2.3}$$

hence we can restrict our considerations to the interval $[-1, 0]$.

For $0 < \epsilon < 1/2$, let $I_\epsilon = [-1, -1+\epsilon)$, and let $\nu_1 = \nu_1(n)$ and $\nu_2 = \nu_2(n)$ be defined by

$$\nu_1 = \min\{\nu : x_{\nu,n}^G \geq -1+\epsilon\}, \qquad \nu_2 = \min\{\mu : \xi_{\mu,n+1}^K \geq -1+\epsilon\}, \tag{2.4}$$

which implies $0 \leq \nu_2 - \nu_1 \leq 1$.

We first consider the supremum of the first Peano kernel for $x \notin I_\epsilon$. Let

$$B_1(R_{2n+1}^{GK}, x) = \begin{cases} -x - 1, & x \in [-1, \xi_{1,n+1}^K], \\ -x + \dfrac{x_{\nu-1,n}^G + \xi_{\nu,n+1}^K}{2}, & x \in (x_{\nu-1,n}^G, \xi_{\nu,n+1}^K), \\ & \nu = 2, \ldots, n+1, \\ -x + \dfrac{\xi_{\nu,n+1}^K + x_{\nu,n}^G}{2}, & x \in (\xi_{\nu,n+1}^K, x_{\nu,n}^G), \\ & \nu = 1, \ldots, n, \\ -x + 1, & x \in (\xi_{n+1,n+1}^K, 1], \end{cases} \tag{2.5}$$

and let

$$2\, B_1(R_{2n+1}^{GK}, x) = B_1(R_{2n+1}^{GK}, x-0) + B_1(R_{2n+1}^{GK}, x+0) \tag{2.6}$$

if $x \in \{x_{1,n}^G, \ldots, x_{n,n}^G, \xi_{1,n+1}^K, \ldots, \xi_{n+1,n+1}^K\}$. It is proved in [4, Proof of Lemma 3] that

$$\lim_{n\to\infty} (2n+1)\,|K_1(R_{2n+1}^{GK}, x) - B_1(R_{2n+1}^{GK}, x)| = 0 \tag{2.7}$$

uniformly for all $x \notin I_\epsilon$.

If $x \notin I_\epsilon$, then $x \in (x_{\nu_1-1,n}^G, 0]$, and we find

$$\limsup_{n\to\infty} (2n+1) \sup_{x \notin I_\epsilon} |K_1(R_{2n+1}^{GK}, x)|$$

$$= \limsup_{n\to\infty} (2n+1) \sup_{x \notin I_\epsilon} |B_1(R_{2n+1}^{GK}, x)|$$

$$\leq \frac{1}{2} \limsup_{n\to\infty} (2n+1) \max\{\, \xi_{\nu,n+1}^K - x_{\nu-1,n}^G, x_{\nu,n}^G - \xi_{\nu,n+1}^K \tag{2.8}$$

$$: \nu_1 \leq \nu \leq \lfloor (n+2)/2 \rfloor\}$$

174

For $\nu_1 < \nu \leq \lfloor (n+2)/2 \rfloor$, none of the nodes $x_{\nu-1,n}^G$, $\xi_{\nu,n+1}^K$ or $x_{\nu,n}^G$ lies in I_ϵ. Now suppose that, for every $n \in \mathbb{N}$, the maximum in (2.8) for the restricted range $\nu_1 < \nu \leq \lfloor (n+2)/2 \rfloor$ is attained for $\nu^* = \nu^*(n)$. By Lemma 2.1, we can estimate

$$\frac{1}{2} \limsup_{n \to \infty} (2n+1) \max\{\xi_{\nu^*,n+1}^K - x_{\nu^*-1,n}^G, x_{\nu^*,n}^G - \xi_{\nu^*,n+1}^K\} \quad (2.9)$$

$$= \frac{\pi}{2} \limsup_{n \to \infty} \sin \frac{n - \nu^* + 1}{n + 1/2} \pi \leq \frac{\pi}{2}.$$

On the other hand, we obtain the value $\frac{\pi}{2}$ for the special choice $\nu^*(n) = \lfloor (n+2)/2 \rfloor$. For the remaining case $\nu = \nu_1$, we can estimate

$$\max\{\xi_{\nu_1,n+1}^K - x_{\nu_1-1,n}^G, x_{\nu_1,n}^G - \xi_{\nu_1,n+1}^K\} < x_{\nu_1,n}^G - x_{\nu_1-1,n}^G, \quad (2.10)$$

and, by the inequalities (see [7, Lemma 2.1])

$$a_{\nu-1,n}^G < x_{\nu,n}^G - x_{\nu-1,n}^G < a_{\nu,n}^G, \qquad \nu = 1, \ldots, \left\lfloor \frac{n+1}{2} \right\rfloor, \quad (2.11)$$

we obtain that (2.10) is bounded by $a_{\nu_1,n}^G$, and hence by $a_{\nu_2,n}^G$. For sufficiently large n, there holds $x_{\nu_2,n}^G \in I_{2\epsilon}$, and we can use the inequality (see Förster and Petras [6, Corollary 1])

$$a_{\nu,n}^G \leq \frac{\pi}{n + 1/2} \sin \phi_{\nu,n}^G \quad (2.12)$$

in order to obtain the bound

$$(2n+1) a_{\nu_2,n}^G \leq 2\pi \sin 2 \arccos (-1 + 2\epsilon). \quad (2.13)$$

Therefore, with $\epsilon \to 0$ an arbitrarily small upper bound can be found for $\limsup_{n \to \infty} (2n+1) a_{\nu_2,n}^G$, and this shows that the case $\nu = \nu_1$ is of no importance for the maximum in (2.8). Thus, the right hand side in (2.8) is equal to $\frac{\pi}{2}$.

We now consider the case $x \in I_\epsilon$, and we will make use of the following result of Förster [5] about the weights of positive quadrature formulae.

Theorem 2.2 (Förster [5]) *Let $n, r \in \mathbb{N}$, and let $Q_r[f] = \sum_{\nu=1}^{r} a_{\nu,r} f(x_{\nu,r})$, $Q_r \neq Q_n^G$, be a positive quadrature formula satisfying $\deg(R_r) \geq 2n - 1$. Furthermore, let $x_{\nu,r} = -\infty$ for $\nu \leq 0$ and $x_{\nu,r} = \infty$ for $\nu \geq r + 1$, and let the corresponding weights be equal to zero. Then*

$$\sum_{x_{k,n}^G \leq x_{\nu,r} \leq x_{l,n}^G} a_{\nu,r} \leq \sum_{\nu=k}^{l} a_{\nu,n}^G \leq \sum_{x_{k-1,n}^G < x_{\nu,r} < x_{l+1,n}^G} a_{\nu,r}, \quad (2.14)$$

$$0 \leq k \leq l \leq n + 1,$$

where the first inequality sign is valid if and only if $k = 0$ and $l = n + 1$, and the second equality sign is valid if and only if $k = 1$ and $l = n$.

We shall apply Theorem 2.2 with $Q_r = Q_{2n+1}^{GK}$. From the left inequality in Theorem 2.2, it follows that

$$b_{2\nu}^+(R_{2n+1}^{GK}) \leq -1 - x_{\nu,n}^G + \sum_{\rho=1}^{\nu} a_{\rho,n}^G = b_\nu^+(R_n^G), \qquad \nu = 1, \ldots, n, \qquad (2.15)$$

while, by the right inequality in Theorem 2.2, it follows that

$$b_{2\nu}^-(R_{2n+1}^{GK}) \geq b_\nu^-(R_n^G), \qquad \nu = 1, \ldots, n. \qquad (2.16)$$

Similarly, we use the left inequality in Theorem 2.2 for the estimate

$$b_{2\nu-1}^+(R_{2n+1}^{GK}) \leq b_\nu^+(R_n^G) + x_{\nu,n}^G - \xi_{\nu,n+1}^K - A_{\nu,n}^{GK} \leq b_\nu^+(R_n^G) + x_{\nu,n}^G - x_{\nu-1,n}^G, \quad (2.17)$$

and the right inequality in Theorem 2.2 for

$$b_{2\nu-1}^-(R_{2n+1}^{GK}) \geq b_{\nu-1}^-(R_n^G) + x_{\nu-1,n}^G - \xi_{\nu,n+1}^K - A_{\nu-1,n}^{GK} \geq b_{\nu-1}^-(R_n^G) - (x_{\nu,n}^G - x_{\nu-1,n}^G). \qquad (2.18)$$

Now, using (2.11) and the fact that (see [7, Eq. (3.5)])

$$0 < b_\nu^+(R_n^G) < a_{\nu,n}^G, \qquad 0 < -b_\nu^-(R_n^G) < a_{\nu,n}^G, \qquad \nu = 1, \ldots, n, \qquad (2.19)$$

we can further estimate in (2.15)-(2.18) and obtain

$$\max\{|b_{2\nu-1}^+(R_{2n+1}^{GK})|, |b_{2\nu}^+(R_{2n+1}^{GK})|, |b_{2\nu-1}^-(R_{2n+1}^{GK})|, |b_{2\nu}^-(R_{2n+1}^{GK})|\} \leq 2 a_{\nu,n}^G. \quad (2.20)$$

By (2.11), this upper bound is increasing with $\nu \leq \nu_2$, which leads to

$$\sup_{x \in I_\epsilon} |K_1(R_{2n+1}^{GK}, x)| \leq 2 a_{\nu_2,n}^G. \qquad (2.21)$$

Now, the same argumentation as above shows that there holds

$$\limsup_{n \to \infty} (2n + 1) \sup_{x \in I_\epsilon} |K_1(R_{2n+1}^{GK}, x)| = 0, \qquad (2.22)$$

which completes the proof since (2.22) means that the supremum of the first Peano kernel $K_1(R_{2n+1}^{GK})$ asymptotically cannot be attained in I_ϵ. ∎

Acknowledgements. The author gratefully acknowledges the helpful remarks of the two referees. He also wishes to express his appreciation for the hospitality that he enjoyed at the symposium.

REFERENCES

[1] Brass H. *Quadraturverfahren.* Vandenhoeck und Ruprecht, Göttingen 1977.

[2] Brass H., Förster K.–J. On the estimation of linear functionals. *Analysis*, **7**:237–258, 1987.

[3] Ehrich S. Asymptotic properties of Stieltjes polynomials and Gauss–Kronrod quadrature formulae. *Hildesheimer Informatik*–Berichte 7/93. Submitted for publication.

[4] Ehrich S. A note on Peano constants of Gauss–Kronrod quadrature schemes. *Hildesheimer Informatik*–Berichte 17/93. Submitted for publication.

[5] Förster K.–J. A comparison theorem for linear functionals and its application in quadrature. In *Numerical Integration II*. ISNM **57**:66-76, Birkhäuser, Basel 1982. G. Hämmerlin, ed.

[6] Förster K.–J., Petras K. On estimates for the weights in Gaussian quadrature in the ultraspherical case. *Math. Comp.*, **55**:243–264, 1990.

[7] Förster K.–J., Petras K. Error estimates in Gaussian quadrature for functions of bounded variation. *SIAM J. Numer. Anal.*, **28**:880-889, 1991.

[8] Gatteschi L. New inequalities for the zeros of Jacobi polynomials. *SIAM J. Math. Anal.*, **18**:1549–1562, 1987.

[9] Gautschi W. A survey of Gauss–Christoffel quadrature formulae. In *E.B. Christoffel, The Influence of his Work on Mathematics and the Physical Sciences*, pages 72–147. Birkhäuser, Basel 1981. P.L. Butzer and F. Fehér, eds.

[10] Gautschi W. Gauss-Kronrod quadrature–a survey. In *Numerical Methods and Approximation Theory III*, pages 39–66. Niš 1988. G.V. Milovanović, ed.

[11] Kronrod A.S. Integration with control of accuracy (Russian). *Dokl. Akad. Nauk SSSR*, **154**:283–286, 1964.

[12] Kronrod A.S. *Nodes and Weights for Quadrature Formulae. Sixteen Place Tables* (Russian). Izdat "Nauka", Moscow 1964. [English translation: Consultants Bureau, New York, 1965.]

[13] Monegato G. A note on extended Gaussian quadrature rules. *Math. Comp.*, **30**:812–817, 1976.

[14] Monegato G. Positivity of weights of extended Gauss–Legendre quadrature rules. *Math. Comp.*, **32**:243–245, 1978.

[15] Monegato G. Stieltjes polynomials and related quadrature rules. *SIAM Review*, **24**:137–158, 1982.

[16] Notaris S.E. An overview of results on the existence or nonexistence and the error term of Gauss–Kronrod quadrature formulae. This volume.

[17] Piessens R., de Doncker E., Überhuber C., Kahaner D.K. *QUADPACK — A Subroutine Package for Automatic Integration.* Springer Series in Computational Mathematics 1, Berlin 1983.

[18] Szegö G. Über gewisse orthogonale Polynome, die zu einer oszillierenden Belegungsfunktion gehören. *Math. Ann.*, **110**:501–513, 1934. [Collected Papers, R. Askey (ed.), vol. **2**:545–557.]

[19] Szegö G. *Orthogonal Polynomials.* AMS Colloq. Publ. **23**, Providence 1975.

International Series of Numerical Mathematics, Vol. 119, ©1994 Birkhäuser

Simultaneous Diagonalization Algorithms with Applications in Multivariate Statistics

Bernard D. Flury*   Beat E. Neuenschwander†

Dedicated to Walter Gautschi on his 65th birthday

Abstract. The following problem arises from multivariate statistical models in principal component and canonical correlation analysis. Let

$$\mathbf{S} = \begin{bmatrix} \mathbf{S}_{11} & \cdots & \mathbf{S}_{1k} \\ \vdots & \ddots & \vdots \\ \mathbf{S}_{k1} & \cdots & \mathbf{S}_{kk} \end{bmatrix}$$

denote a positive definite symmetric (pds) matrix of dimension $pk \times pk$, partitioned into submatrices $\mathbf{S}_{ij}$ of dimension $p \times p$ each, and suppose we wish to find a nonsingular $p \times p$ matrix $\mathbf{B}$ such that all $\mathbf{B}'\mathbf{S}_{ij}\mathbf{B}$ are "almost diagonal". More precisely, for a partitioned $pk \times pk$ matrix

$$\mathbf{A} = \begin{bmatrix} \mathbf{A}_{11} & \cdots & \mathbf{A}_{1k} \\ \vdots & \ddots & \vdots \\ \mathbf{A}_{k1} & \cdots & \mathbf{A}_{kk} \end{bmatrix}$$

we define the parallel–diagonal operator as

$$\mathrm{pdiag}(\mathbf{A}) = \begin{bmatrix} \mathrm{diag}(\mathbf{A}_{11}) & \cdots & \mathrm{diag}(\mathbf{A}_{1k}) \\ \vdots & \ddots & \vdots \\ \mathrm{diag}(\mathbf{A}_{k1}) & \cdots & \mathrm{diag}(\mathbf{A}_{kk}) \end{bmatrix}$$

and suggest to use $\det\{\mathrm{pdiag}(\mathbf{A})\}/\det(\mathbf{A})$ as a measure of deviation from "parallel–diagonality", provided $\mathbf{A}$ is pds. For a nonsingular $p \times p$ matrix $\mathbf{B}$, we study the function

$$\Phi(\mathbf{B}; \mathbf{S}) = \frac{\det[\mathrm{pdiag}\{(\mathbf{I}_k \otimes \mathbf{B})'\mathbf{S}(\mathbf{I}_k \otimes \mathbf{B})\}]}{\det[(\mathbf{I}_k \otimes \mathbf{B})'\mathbf{S}(\mathbf{I}_k \otimes \mathbf{B})]}.$$

The matrix $\mathbf{B}$ which minimizes Φ is said to transform $\mathbf{S}$ to almost parallel–diagonal form. We give an algorithm for minimizing Φ over $\mathbf{B}$ in (i) the group of orthogonal $p \times p$ matrices, and (ii) the set of nonsingular $p \times p$ matrices such that $\mathrm{diag}(\mathbf{B}'\mathbf{B}) = \mathbf{I}_p$, and study its convergence. Statistical applications of the algorithm occur in maximum likelihood estimation of (i) common principal components for dependent random vectors (Neuenschwander 1994), and (ii) common canonical variates (Neuenschwander and Flury 1994). This work generalizes and extends the FG diagonalization algorithm of Flury and Gautschi (1986).

* Department of Mathematics, Indiana University, Rawles Hall, Bloomington, IN 47405, USA
† Institut für Sozial und Präventivmedizin, Universität Bern, Finkenhubelweg 11, 3012 Bern, Switzerland; work supported by a grant from the Swiss National Science Foundation

1 INTRODUCTION

Suppose $\mathbf{A}_1, \ldots, \mathbf{A}_k$ are real symmetric matrices of dimension $p \times p$; then the $\mathbf{A}_j$ are *simultaneously diagonalizable* by the same orthogonal transformation if there exists an orthogonal $p \times p$ matrix $\mathbf{B}$ such that $\mathbf{B}'\mathbf{A}_i\mathbf{B} = \mathbf{D}_i$ is diagonal for all $i = 1, \ldots, k$. Simultaneous diagonalizability has been used in psychometrics [DeLeeuw and Pruzanky 1978] and in multivariate statistics [Flury 1988], and perhaps in other areas. If the $\mathbf{A}_i$ are observed matrices rather than model parameters, they are typically affected by sampling error or by errors of measurement, and may then be "close" to being simultaneously diagonalizable in some sense. In such cases it is reasonable to find an orthogonal matrix $\mathbf{B}$ such that all $\mathbf{B}'\mathbf{A}_i\mathbf{B}$ are "almost diagonal". Two approaches have appeared in the literature: DeLeeuw and Pruzanky (1978) suggested to find $\mathbf{B}$ such as to minimize $\sum_{i=1}^{k}$ off $(\mathbf{B}'\mathbf{A}_i\mathbf{B})$, where off $(\mathbf{M})$ is the sum of all squared off–diagonal elements of $\mathbf{M}$. For $\mathbf{M}$ positive definite symmetric (pds), Flury and Gautschi (1986) proposed

$$\varphi(\mathbf{M}) = \det(\text{diag}\mathbf{M})/\det \mathbf{M} \qquad (1.1)$$

as a measure of deviation from diagonality. This can be motivated by Hadamard's inequality, which shows that $\varphi(\mathbf{M}) \geq 1$, with equality exactly if $\mathbf{M} = \text{diag } \mathbf{M}$. Hence, for k positive definite symmetric matrices $\mathbf{A}_1, \ldots, \mathbf{A}_k$ and associated positive weights $n_1, \ldots, n_k$, Flury and Gautschi (1986) proposed to minimize the function

$$\Phi(\mathbf{B}) = \Pi_{i=1}^{k}[\varphi(\mathbf{B}'\mathbf{A}_i\mathbf{B})]^{n_i} \qquad (1.2)$$

over the group of orthogonal matrices. The same minimization problem arises in maximum likelihood estimation of k covariance matrices of p–variate multivariate normal distributions under the assumption of common principal components (Flury 1988). Minimization of Φ can be done using a Jacobi–type algorithm (Flury and Gautschi 1986) which is currently available in the software packages IMSL (subroutine KPRIN) and NTSYS (procedure CPCA).

In this paper we present two algorithms that were originally motivated by generalizations and modifications of common principal component analysis (Neuenschwander 1991, 1994; Neuenschwander and Flury 1994). The statistical connection will be outlined in more detail in Example 2 of §4. In the sequel, $\mathbf{A}$ will denote a pds matrix of dimension $pk \times pk$ $(p, k \geq 2)$, partitioned into k^2 submatrices of dimension $p \times p$ as

$$\mathbf{A} = (\mathbf{A}_{ij}) = \begin{bmatrix} \mathbf{A}_{11} & \cdots & \mathbf{A}_{1k} \\ \vdots & \ddots & \vdots \\ \mathbf{A}_{k1} & \cdots & \mathbf{A}_{kk} \end{bmatrix} \qquad (1.3)$$

and we will focus on simultaneous transformation of all $\mathbf{A}_{ij}$ to nearly diagonal form. Throughout the paper we will frequently use the Kronecker matrix product $(\otimes)$, and the commutation matrix $\mathbf{I}_{(n,m)}$. [$\mathbf{I}_{(n,m)}$ is a permutation matrix of dimension $nm \times nm$, such that $\mathbf{I}_{(n,m)}\text{vec}\mathbf{M} = \text{vec}\mathbf{M}'$ for an $n \times m$ matrix $\mathbf{M}$; see Magnus (1988), Graybill (1969), Henderson & Searle (1979)].

180

To motivate our approach, we shall define parallel–diagonality of a partitioned $pk \times pk$ matrix $\mathbf{A}$. It will always be implicity assumed that the partition is as in Eq. (1.3).

Definition 1 $\mathbf{A}$ is *parallel–diagonal* if $\mathbf{A}_{ij} = \text{diag}\mathbf{A}_{ij}$ for all submatrices $\mathbf{A}_{ij}$. The "pdiag" operator transforms $\mathbf{A}$ into parallel–diagonal form, i.e.,

$$\text{pdiag}(\mathbf{A}) = \begin{bmatrix} \text{diag}\mathbf{A}_{11} & \cdots & \text{diag}\mathbf{A}_{1k} \\ \vdots & \ddots & \vdots \\ \text{diag}\mathbf{A}_{k1} & \cdots & \text{diag}\mathbf{A}_{kk} \end{bmatrix}. \tag{1.4}$$

With this notation, we have the following

Lemma 1 *For* $\mathbf{A}$ *pds,* $\det(\text{pdiag}\mathbf{A}) \geq \det \mathbf{A}$, *with equality exactly if* $\mathbf{A}$ *is parallel–diagonal.*

Proof: Let

$$\mathbf{M} = \mathbf{I}_{(p,k)}\mathbf{A}\mathbf{I}_{(k,p)} = \begin{bmatrix} \mathbf{M}_{11} & \cdots & \mathbf{M}_{1p} \\ \vdots & \ddots & \vdots \\ \mathbf{M}_{p1} & \cdots & \mathbf{M}_{pp} \end{bmatrix}.$$

where all $\mathbf{M}_{h\ell}$ have dimension $k \times k$. Since $\mathbf{I}_{(p,k)}$ is orthogonal, it follows that $\det\mathbf{A} = \det\mathbf{M}$. Moreover, $\mathbf{A} = \text{pdiag}\mathbf{A}$ exactly if $\mathbf{M}_{h\ell} = \mathbf{O}$ for all $h \neq \ell$, i.e., if $\mathbf{M}$ is block–diagonal. Then a generalization of Hadamard's inequality (for a proof, see Flury and Neuenschwander 1994) tells us that

$$\det\mathbf{M} \leq \Pi_{h=1}^{p}\det \mathbf{M}_{hh}$$

with equality exactly if $\mathbf{M}$ is block–diagonal. ∎

Hence, we suggest

$$\det (\text{pdiag } \mathbf{A})/\det\mathbf{A}$$

as a measure of deviation of $\mathbf{A}$ from parallel–diagonality (always with p and k considered fixed). This can further be justified by Lemma 2, for which we need more notation. For $\alpha \in I\!R$, let $\mathbf{A}_{ij}^{(\alpha)}$ denote the matrix with the same diagonal elements as $\mathbf{A}_{ij}$, and off–diagonal entries multiplied by α, i.e.,

$$\mathbf{A}_{ij}^{(\alpha)} = \alpha\mathbf{A}_{ij} + (1 - \alpha)\text{diag}\mathbf{A}_{ij}.$$

We will now vary α from 0 to 1 (thus "inflating" pdiag$\mathbf{A}$ to $\mathbf{A}$), and study the determinant as a function of α. The result, summarized in Lemma 2, generalizes Lemma 1 from Flury and Gautschi (1986).

Lemma 2 *For* $\mathbf{A}$ *pds and*

$$\mathbf{A}^{(\alpha)} = \begin{bmatrix} \mathbf{A}_{11}^{(\alpha)} & \cdots & \mathbf{A}_{1k}^{(\alpha)} \\ \vdots & \ddots & \vdots \\ \mathbf{A}_{k1}^{(\alpha)} & \cdots & \mathbf{A}_{kk}^{(\alpha)} \end{bmatrix},$$

the function $d(\alpha) = \det \mathbf{A}^{(\alpha)}$ *is decreasing for* $\alpha \in [0, 1]$. *If* $\mathbf{A}$ *is not parallel–diagonal, then* $d(\alpha)$ *is strictly decreasing.*

Proof: Let $\mathbf{M}_\alpha = \mathbf{I}_{(p,k)} \mathbf{A}^{(\alpha)} \mathbf{I}_{(k,p)}$ and $\mathbf{M} = (\mathbf{M}_{ij}) = \mathbf{I}_{(p,k)} \mathbf{A} \mathbf{I}_{(k,p)}$; then

$$\mathbf{M}_\alpha = \begin{bmatrix} \mathbf{M}_{11} & \alpha\mathbf{M}_{12} & \cdots & \alpha\mathbf{M}_{1p} \\ \alpha\mathbf{M}_{21} & \mathbf{M}_{22} & \cdots & \alpha\mathbf{M}_{2p} \\ \vdots & \vdots & \ddots & \vdots \\ \alpha\mathbf{M}_{p1} & \alpha\mathbf{M}_{p2} & \cdots & \mathbf{M}_{pp} \end{bmatrix}$$

and $\det\mathbf{M}_\alpha = \det\mathbf{A}^{(\alpha)}$. Let $\mathrm{bdiag}(\mathbf{M})$ denote the block–diagonal matrix with the same diagonal blocks of size $k \times k$ as $\mathbf{M}$, i.e.,

$$\mathrm{bdiag}(\mathbf{M}) = \begin{bmatrix} \mathbf{M}_{11} & \mathbf{O} & \cdots & \mathbf{O} \\ \mathbf{O} & \mathbf{M}_{22} & \cdots & \mathbf{O} \\ \vdots & \vdots & \ddots & \vdots \\ \mathbf{O} & \mathbf{O} & \cdots & \mathbf{M}_{pp} \end{bmatrix}. \tag{1.5}$$

Then $\mathrm{bdiag}(\mathbf{M})$ is positive definite, and

$$\det\mathbf{M}_\alpha = \det[\alpha\mathbf{M} + (1-\alpha)\mathrm{bdiag}\mathbf{M}]$$
$$= \det(\mathrm{bdiag}\mathbf{M})\det[\alpha\mathbf{R} + (1-\alpha)\mathbf{I}_{pk}]$$

where $\mathbf{R} = (\mathrm{bdiag}\mathbf{M})^{-1/2}\mathbf{M}(\mathrm{bdiag}\mathbf{M})^{-1/2}$. The matrix $\mathbf{R}$ is pds, with diagonal entries equal to 1. Let $\rho_h > 0$ $(h = 1, \ldots, pk)$ denote the eigenvalues of $\mathbf{R}$, and assume that not all ρ_h are equal to 1. Then

$$d_1(\alpha) := \log\det[\alpha\mathbf{R} + (1-\alpha)\mathbf{I}_{pk}]$$
$$= \sum_{h=1}^{pk} \log[1 + \alpha(\rho_h - 1)].$$

We show now that $d_1(\alpha)$ is monotonically decreasing for $\alpha \in [0,1]$. Let

$$d_2(\alpha) = \partial d_1(\alpha)/\partial\alpha$$
$$= \sum_{h=1}^{pk} \frac{\rho_h - 1}{1 + \alpha(\rho_h - 1)}$$

and notice that all denominators are strictly positive for $\alpha \in [0,1]$ because all ρ_h are positive. Then $d_2(0) = \sum_{h=1}^{pk}(\rho_h - 1) = \mathrm{tr}\mathbf{R} - pk = 0$, and $\partial d_2(\alpha)/\partial\alpha < 0$ for all $\alpha \in [0,1]$. Hence d_2 is strictly decreasing in $[0,1]$, implying together with $d_2(0) = 0$ that d_1 is strictly decreasing. Noticing that $\rho_1 = \cdots = \rho_{pk} = 1$ is equivalent to $\mathbf{R} = \mathbf{I}_{pk}$, which is in turn equivalent to $\mathbf{M} = \mathrm{bdiag}(\mathbf{M})$, completes the proof. ∎

We are now ready to define measures of "parallel–diagonalizability" for a partitioned pds matrix

$$\mathbf{S} = (\mathbf{S}_{ij}) = \begin{bmatrix} \mathbf{S}_{11} & \cdots & \mathbf{S}_{1k} \\ \vdots & \ddots & \vdots \\ \mathbf{S}_{k1} & \cdots & \mathbf{S}_{kk} \end{bmatrix}$$

of dimension $pk \times pk$. Define

$$\varphi(\mathbf{B}; \mathbf{S}) := \frac{\det[\mathrm{pdiag}\{(\mathbf{I}_k \otimes \mathbf{B})'\mathbf{S}(\mathbf{I}_k \otimes \mathbf{B})\}]}{\det[(\mathbf{I}_k \otimes \mathbf{B})'\mathbf{S}(\mathbf{I}_k \otimes \mathbf{B})]} \tag{1.6}$$

for nonsingular $p \times p$ matrices $\mathbf{B}$. Let $\mathcal{O}(p)$ denote the group of orthogonal $p \times p$ matrices, and let

$$\varphi_0(\mathbf{S}) = \min_{\mathbf{B} \in \mathcal{O}(p)} \varphi(\mathbf{B}; \mathbf{S}); \tag{1.7}$$

then φ_0 measures the degree to which $\mathbf{S}$ can be parallel–diagonalized by a single orthogonal matrix $\mathbf{B}$, with $\varphi_0(\mathbf{S}) = 1$ exactly if all $\mathbf{B}'\mathbf{S}_{ij}\mathbf{B}$ are diagonal, and $\varphi_0(\mathbf{S}) > 1$ otherwise. Note that for $\mathbf{B} \in \mathcal{O}(p)$ we could as well just minimize the numerator of (1.6) because the denominator is equal to $\det(\mathbf{S})$, but the definition of φ as in Eq. (1.6) has the advantage of giving a fixed lower bound to φ_0.

Consider next minimization of (1.6) over $\mathbf{B}$ in the group of nonsingular $p \times p$ matrices. Suppose $\mathbf{D} = \mathrm{diag}(d_1, \ldots, d_p)$ is a nonsingular diagonal matrix; then

$$\varphi(\mathbf{BD}; \mathbf{S}) = \frac{\det[\mathrm{pdiag}\{(\mathbf{I}_k \otimes \mathbf{BD})'\mathbf{S}(\mathbf{I}_k \otimes \mathbf{BD})\}]}{\det[(\mathbf{I}_k \otimes \mathbf{BD})'\mathbf{S}(\mathbf{I}_k \otimes \mathbf{BD})]}. \tag{1.8}$$

The denominator in (1.8) is $[\det(\mathbf{I}_k \otimes \mathbf{D})\det(\mathbf{I}_k \otimes \mathbf{B})]^2 \det \mathbf{S} = (\det \mathbf{D})^{2k} \det[(\mathbf{I}_k \otimes \mathbf{B})'\mathbf{S}(\mathbf{I}_k \otimes \mathbf{B})]$. To evaluate the numerator, notice that (with fixed p and k)

$$\mathrm{pdiag}\mathbf{M} = \mathbf{I}_{(k,p)} \cdot \mathrm{bdiag}[\mathbf{I}_{(p,k)}\mathbf{M}\mathbf{I}_{(k,p)}] \cdot \mathbf{I}_{(p,k)}, \tag{1.9}$$

where "bdiag" is the block–diagonal operator as in (1.5). Hence, the numerator in (1.8) is

$$\det[\mathrm{pdiag}\{(\mathbf{I}_k \otimes \mathbf{D})(\mathbf{I}_k \otimes \mathbf{B})'\mathbf{S}(\mathbf{I}_k \otimes \mathbf{B})(\mathbf{I}_k \otimes \mathbf{D})\}] \\ = \det[\mathrm{pdiag}\{(\mathbf{D} \otimes \mathbf{I}_k)\mathbf{I}_{(p,k)}(\mathbf{I}_k \otimes \mathbf{B})'\mathbf{S}(\mathbf{I}_k \otimes \mathbf{B})\mathbf{I}_{(k,p)}(\mathbf{D} \otimes \mathbf{I}_k)\}]. \tag{1.10}$$

But for $\mathbf{D} = \mathrm{diag}(d_1, \ldots, d_p)$, and $\mathbf{M} = (\mathbf{M}_{ij})$, we have

$$\mathrm{bdiag}\{(\mathbf{D} \otimes \mathbf{I}_k)\mathbf{M}(\mathbf{D} \otimes \mathbf{I}_k)\} = \begin{bmatrix} d_1^2\mathbf{M}_{11} & \mathbf{O} & \cdots & \mathbf{O} \\ \mathbf{O} & d_2^2\mathbf{M}_{22} & \cdots & \mathbf{O} \\ \vdots & \vdots & \ddots & \vdots \\ \mathbf{O} & \mathbf{O} & \cdots & d_p^2\mathbf{M}_{pp} \end{bmatrix} \tag{1.11}$$

183

and hence Eq. (1.10) is the same as

$$\left(\Pi_{j=1}^{p}d_{j}^{2}\right)^{k}\det[\mathrm{bdiag}\{\mathbf{I}_{(p,k)}(\mathbf{I}_{k}\otimes\mathbf{B})'\mathbf{S}(\mathbf{I}_{k}\otimes\mathbf{B})\mathbf{I}_{(k,p)}\}]$$
$$=(\det\mathbf{D})^{2k}\det[\mathrm{pdiag}\{(\mathbf{I}_{k}\otimes\mathbf{B})'\mathbf{S}(\mathbf{I}_{k}\otimes\mathbf{B})\}]\,,$$

which finally yields $\varphi(\mathbf{BD};\mathbf{S})=\varphi(\mathbf{B};\mathbf{S})$ for any nonsingular diagonal matrix $\mathbf{D}$. It is therefore reasonable to maximize φ over a restricted class of nonsingular matrices. A convenient constraint (corresponding to choosing $\mathbf{D}=[\mathrm{diag}(\mathbf{B}'\mathbf{B})]^{-1/2}$) is to assume that all columns of $\mathbf{B}$ have unit length. Hence, let

$$\mathcal{A}(p)=\{\mathbf{B};\mathbf{B}\text{ nonsingular, }\mathrm{diag}(\mathbf{B}'\mathbf{B})=\mathbf{I}_{p}\}\,,\tag{1.12}$$

and define

$$\varphi_{1}(\mathbf{S}):=\min_{\mathbf{B}\in\mathcal{A}(p)}\varphi(\mathbf{B};\mathbf{S})\,.\tag{1.13}$$

Let $\mathbf{B}_{0}=\mathbf{B}_{0}(\mathbf{S})$ denote an orthogonal matrix for which $\varphi(\mathbf{B};\mathbf{S})$ takes its minimum in $\mathcal{O}(p)$; then any permutation of the columns of $\mathbf{B}_{0}$ or multiplication of columns by -1 will yield an orthogonal matrix that gives the same minimum. We will typically ignore such "trivial" nonuniqueness. The same remark holds for a matrix $\mathbf{B}_{1}=\mathbf{B}_{1}(\mathbf{S})$ for which $\varphi(\mathbf{B};\mathbf{S})$ takes its minimum in $\mathcal{A}(p)$.

For minimization of φ in $\mathcal{O}(p)$, Neuenschwander (1991, 1994) has shown that the following equation system must hold for $\mathbf{B}_{0}=[\mathbf{b}_{1},\ldots,\mathbf{b}_{p}]$:

$$\mathbf{b}_{m}'\left(\sum_{i=1}^{k}\sum_{j=1}^{k}a_{m\ell,ij}\mathbf{S}_{ij}\right)\mathbf{b}_{\ell}=0\quad\text{for all }m\neq\ell,\qquad 1\leq m,\ell\leq p\,,\tag{1.14}$$

where

$$\mathbf{A}_{m\ell}=(a_{m\ell,ij})=\mathbf{P}_{m}^{-1}-\mathbf{P}_{\ell}^{-1}\qquad(m\neq\ell)\tag{1.15}$$

and

$$\mathbf{P}_{h}=\begin{bmatrix}\mathbf{b}_{h}'\mathbf{S}_{11}\mathbf{b}_{h}&\cdots&\mathbf{b}_{h}'\mathbf{S}_{1k}\mathbf{b}_{h}\\\vdots&\ddots&\vdots\\\mathbf{b}_{h}'\mathbf{S}_{k1}\mathbf{b}_{h}&\cdots&\mathbf{b}_{h}'\mathbf{S}_{kk}\mathbf{b}_{h}\end{bmatrix}\tag{1.16}$$

$$=(\mathbf{I}_{k}\otimes\mathbf{b}_{h})'\mathbf{S}(\mathbf{I}_{k}\otimes\mathbf{b}_{h})\,,\qquad h=1,\ldots,p\,.$$

In §2 we present and analyze an algorithm, called the *orthogonal FG+ algorithm*, which solves the equation system (1.14) – (1.16) under the constraint that $[\mathbf{b}_{1},\ldots,\mathbf{b}_{p}]$ is orthogonal.

For minimization of φ in $\mathcal{A}(p)$, Neuenschwander and Flury (1994) have shown that the following equation system holds for a matrix $\mathbf{B}_{1}=[\mathbf{b}_{1},\ldots,\mathbf{b}_{p}]\in\mathcal{A}(p)$ which minimizes φ:

$$\mathbf{b}_{m}'\left(\sum_{i=1}^{k}\sum_{j=1}^{k}a_{m,ij}\mathbf{S}_{ij}\right)\mathbf{b}_{\ell}=0\qquad\text{for all }m\neq\ell,\qquad 1\leq m,\ell\leq p\,,\tag{1.17}$$

$$\mathbf{b}'_m \left(\sum_{i=1}^{k} \sum_{j=1}^{k} a_{\ell,ij} \mathbf{S}_{ij} \right) \mathbf{b}_\ell = 0 \qquad \text{for all } m \neq \ell, \qquad 1 \leq m, \ell \leq p, \qquad (1.18)$$

where

$$\mathbf{A}_h = (a_{h,ij}) = [(\mathbf{I}_k \otimes \mathbf{b}_h)' \mathbf{S} (\mathbf{I}_k \otimes \mathbf{b}_h)]^{-1}, \qquad h = 1, \ldots, p. \qquad (1.19)$$

An algorithm for solving (1.17) – (1.19) under the constraints $\mathbf{b}'_h \mathbf{b}_h = 1$ (i.e., $\mathbf{B} \in \mathcal{A}(p)$) is presented and discussed in §3; it is called the *nonorthogonal FG^+ algorithm* because $\mathbf{B}$ is not necessarily orthogonal. Both the orthogonal and the nonorthogonal FG^+ algorithms are inspired (as the names suggest) by the FG–diagonalization algorithm of Flury and Gautschi (1986), but require substantial adaptation and are more difficult to analyze.

2   THE ORTHOGONAL FG^+ ALGORITHM

2.1 Introduction

The orthogonal FG^+ algorithm provides a numerical solution to the equation system (1.14) – (1.16). It consists of two nested algorithms, called F^+ and G^+ respectively, both of them iterative. The F^+ algorithm follows a cyclic scheme, choosing an index pair (m, ℓ), and rotating the columns $(\mathbf{b}_m, \mathbf{b}_\ell)$ of the current orthogonal matrix $\mathbf{B}$ such that equation (1.14) is satisfied. The proper rotation angle is provided by the G^+ algorithm, which solves the same equation system (1.14) – (1.16) for the special case $p = 2$. Thus, the orthogonal FG^+ algorithm is a generalization of the cyclic Jacobi algorithm (Golub and Van Loan 1983, pp. 299–300).

The algorithm requires two constants $\varepsilon_F > 0$ and $\varepsilon_G > 0$, which determine convergence. The input consists of a positive definite symmetric matrix $\mathbf{S}$ of dimension $pk \times pk$, partitioned into k^2 submatrices $\mathbf{S}_{ij}$ of dimension $p \times p$, and the numbers p and k. The output consists of an orthogonal $p \times p$ matrix $\mathbf{B} = (\mathbf{b}_1, \ldots, \mathbf{b}_p)$, which (in theory, i.e., if $\varepsilon_F = \varepsilon_G = 0$) solves the equation system (1.14) – (1.16). In both the F^+ and the G^+ algorithms iterations are counted; the counts may be used to stop the algorithm in cases of slow convergence. See the remarks in §2.6.

2.2 The orthogonal F^+ algorithm

The algorithm consists of the following steps.

Step OF_0^+: (initialize) Put $f \leftarrow 0$, and $\mathbf{B} \leftarrow \mathbf{I}_p$.

Step OF_1^+: (update) Put $\mathbf{B}^{(f)} \leftarrow \mathbf{B}$, and $f \leftarrow f + 1$.

Step OF_2^+: (sweep) Repeat steps OF_{21}^+ to OF_{24}^+ in cyclic order for all pairs of indices (m, ℓ), $1 \leq m < \ell \leq p$.

Step OF_{21}^+: Put $\mathbf{H} \leftarrow [\mathbf{b}_m \, , \, \mathbf{b}_\ell]$, where $\mathbf{b}_m$ and $\mathbf{b}_\ell$ are the current mth and ℓth columns of $\mathbf{B}$.

Step OF_{22}^+: Put $\mathbf{T} \leftarrow (\mathbf{I}_k \otimes \mathbf{H})' \mathbf{S} (\mathbf{I}_k \otimes \mathbf{H})$.

Step OF_{23}^+: Call the orthogonal G^+ algorithm with input matrix $\mathbf{T}$ to obtain an orthogonal 2×2 matrix $\mathbf{Q}$.

Step OF_{24}^+: In the matrix $\mathbf{B}$, replace the mth and ℓth columns by $\mathbf{b}_m^*$ and $\mathbf{b}_\ell^*$, respectively, where $[\mathbf{b}_m^*, \mathbf{b}_\ell^*] = \mathbf{HQ}$.

Step OF_3^+: (convergence) If $\|\mathbf{B} - \mathbf{B}^{(f)}\| < \varepsilon_F$ for some matrix norm $\| \cdot \|$, return $\mathbf{B}$ and stop. Else, go to step OF_1^+.

2.3 The orthogonal G^+ algorithm

The input to this algorithm consists of a pds matrix $\mathbf{T}$ of dimension $2k \times 2k$, partitioned into submatrices $\mathbf{T}_{ij}$ of dimension 2×2 as

$$\mathbf{T} = \begin{bmatrix} \mathbf{T}_{11} & \cdots & \mathbf{T}_{1k} \\ \vdots & \ddots & \vdots \\ \mathbf{T}_{k1} & \cdots & \mathbf{T}_{kk} \end{bmatrix} .$$

The algorithm has the following steps.

Step OG_0^+: Put $g \leftarrow 0$, and $\mathbf{Q} \leftarrow \mathbf{I}_2$.

Step OG_1^+: Put $\mathbf{Q}^{(g)} \leftarrow \mathbf{Q}$, and $g \leftarrow g + 1$

Step OG_2^+: Put $\mathbf{M}_\ell \leftarrow (\mathbf{I}_k \otimes \mathbf{q}_\ell)' \mathbf{T} (\mathbf{I}_k \otimes \mathbf{q}_\ell)$, where $\mathbf{q}_\ell = \ell$th column of $\mathbf{Q}$, $\ell = 1, 2$.

Step OG_3^+: Put $\mathbf{A}^* = \begin{bmatrix} a_{11}^* & \cdots & a_{1k}^* \\ \vdots & \ddots & \vdots \\ a_{k1}^* & \cdots & a_{kk}^* \end{bmatrix} \leftarrow \mathbf{M}_1^{-1} - \mathbf{M}_2^{-1}$, and

$\mathbf{T}^* \leftarrow \sum_{i=1}^k \sum_{j=1}^k a_{ij}^* \mathbf{T}_{ij}$

Step OG_4^+: Compute $\mathbf{Q} = \begin{bmatrix} c & -s \\ s & c \end{bmatrix}$ as the orthogonal matrix which diagonalizes $\mathbf{T}^*$, choosing $c = \cos \alpha$ and $s = \sin \alpha$ such that $|\alpha| \leq \pi/4$.

Step OG_5^+: If $\|\mathbf{Q} - \mathbf{Q}^{(g)}\| < \varepsilon_G$ for some matrix norm $\| \cdot \|$, return $\mathbf{Q}$ and stop. Else, go to step OG_1^+.

2.4 Analysis of the orthogonal FG^+ algorithm

We will show that the algorithm does not stop (in theory, i.e., for $\varepsilon_F = 0$) until the columns of the orthogonal matrix $\mathbf{B}$ provide a solution to equations (1.14) to (1.16). Note that the matrix in parentheses of (1.14) is always symmetric although the $\mathbf{S}_{ij}$

are typically not symmetric for $i \neq j$. Symmetry holds because the matrices $\mathbf{A}_{m\ell}$ (1.15) are symmetric. Hence (1.14) looks like a system of equations to determine the eigenvectors of a single symmetric matrix. The difficulty is that $\mathbf{A}_{m\ell}$ is a different matrix for each pair of indices (m, ℓ), and hence the problem is more complicated than just a computation of eigenvectors and eigenvalues.

To show that the algorithm provides a solution to the equation system, consider first the F^+ part. In each sweep (i.e., each sequence of $p(p-1)/2$ pairwise rotations), $\mathbf{B}$ remains unchanged unless at least one of the index pairs (m, ℓ) returns an orthogonal 2×2 matrix $\mathbf{Q}$ which is different from $\mathbf{I}_2$. Hence the orthogonal F^+ algorithm will stop only after all calls to the orthogonal G^+ algorithm in a full sweep have returned $\mathbf{I}_2$.

Consider now the call to G^+ in step OF_{23}^+, for a fixed pair of indices (m, ℓ). Let $\mathbf{H} = [\mathbf{b}_m, \mathbf{b}_\ell]$ consist of the current mth and ℓth columns of $\mathbf{B}$; then the input to the G^+ algorithm is the pds matrix

$$\mathbf{T} = \begin{bmatrix} \mathbf{T}_{11} & \cdots & \mathbf{T}_{1k} \\ \vdots & \ddots & \vdots \\ \mathbf{T}_{k1} & \cdots & \mathbf{T}_{kk} \end{bmatrix} = (\mathbf{I}_k \otimes \mathbf{H})'\mathbf{S}(\mathbf{I}_k \otimes \mathbf{H})$$

of dimension $2k \times 2k$, with $\mathbf{T}_{ij} = \mathbf{H}'\mathbf{S}_{ij}\mathbf{H}$. The orthogonal G^+ algorithm deals with minimization of $\varphi(\mathbf{Q}; \mathbf{T})$ over $\mathbf{Q} \in \mathcal{O}(2)$, i.e., for the special case $p = 2$. The equivalent to equations (1.14) – (1.16) based on $\mathbf{T}$ is

$$\mathbf{q}_1' \left(\sum_{i=1}^{k} \sum_{j=1}^{k} a_{ij}^* \mathbf{T}_{ij} \right) \mathbf{q}_2 = 0, \qquad (2.4.1)$$

where $\mathbf{Q} = [\mathbf{q}_1, \mathbf{q}_2]$ is orthogonal,

$$\mathbf{A}^* = \mathbf{M}_1^{-1} - \mathbf{M}_2^{-1} \qquad (2.4.2)$$

and

$$\mathbf{M}_h = (\mathbf{I}_k \otimes \mathbf{q}_h)'\mathbf{T}(\mathbf{I}_k \otimes \mathbf{q}_h), \qquad h = 1, 2. \qquad (2.4.3)$$

If the current $\mathbf{Q}$ solves the equation system (2.4.1) – (2.4.3), then the next iteration will leave $\mathbf{Q}$ unchanged, and the algorithm stops. Otherwise, iterations continue until (up to the precision determined by ε_G) $\mathbf{Q}$ remains stable. So suppose now that $\mathbf{Q} = [\mathbf{q}_1, \mathbf{q}_2]$ is a solution, which is returned to the F^+ algorithm. In step OF_{24}^+ the orthogonal matrix $\mathbf{B}$ is updated to $\mathbf{B}^*$ (say), which is the same as $\mathbf{B}$ except for the mth and ℓth columns $\mathbf{b}_m^*$ and $\mathbf{b}_\ell^*$, where

$$(\mathbf{b}_m^*, \mathbf{b}_\ell^*) = (\mathbf{b}_m, \mathbf{b}_\ell)\mathbf{Q}$$

or

$$\mathbf{b}_m^* = \mathbf{H}\mathbf{q}_1, \quad \mathbf{b}_\ell^* = \mathbf{H}\mathbf{q}_2. \qquad (2.4.4)$$

187

Now $\mathbf{B}^*$ is still orthogonal because replacing the two columns is a Jacobi rotation. Moreover, $\mathbf{b}_m^*$ and $\mathbf{b}_\ell^*$ satisfy the (m, ℓ)th equation in (1.14). To see this, notice that the matrices $\mathbf{P}_m$ and $\mathbf{P}_\ell$ in (1.16) have the form

$$
\begin{aligned}
\mathbf{P}_m &= (\mathbf{I}_k \otimes \mathbf{b}_m^*)'\mathbf{S}(\mathbf{I}_k \otimes \mathbf{b}_m^*) \\
&= (\mathbf{I}_k \otimes \mathbf{q}_1)'(\mathbf{I}_k \otimes \mathbf{H})'\mathbf{S}(\mathbf{I}_k \otimes \mathbf{H})(\mathbf{I}_k \otimes \mathbf{q}_1) \\
&= (\mathbf{I}_k \otimes \mathbf{q}_1)'\mathbf{T}(\mathbf{I}_k \otimes \mathbf{q}_1) \\
&= \mathbf{M}_1
\end{aligned}
\tag{2.4.5}
$$

and similarly, $\mathbf{P}_\ell = \mathbf{M}_2$. Hence, in Eq. (1.15), $\mathbf{A}_{m\ell} = \mathbf{P}_m^{-1} - \mathbf{P}_\ell^{-1} = \mathbf{M}_1^{-1} - \mathbf{M}_2^{-1}$, and

$$
\mathbf{b}_m^{*'} \left(\sum_{i=1}^{k} \sum_{j=1}^{k} a_{m\ell,ij} \mathbf{S}_{ij} \right) \mathbf{b}_\ell^* = \mathbf{q}_1' \left(\sum_{i=1}^{k} \sum_{j=1}^{k} a_{ij}^* \mathbf{H}' \mathbf{S}_{ij} \mathbf{H} \right) \mathbf{q}_2 = 0
$$

because $\mathbf{T}_{ij} = \mathbf{H}'\mathbf{S}_{ij}\mathbf{H}$.

2.5 Convergence of the orthogonal FG^+ algorithm

From the preceding analysis it is clear that the algorithm stops only if a solution to the equation system (1.14) – (1.16) has been found. However, this solution might correspond to any stationary point of the function φ rather than to a minimum. In this section we show that the algorithm converges indeed to a minimum, if proper precautions are taken.

Let us first see how the value of φ changes after a call to the G^+ algorithm. Suppose the current orthogonal matrix in step F_2^+ is $\mathbf{B} = [\mathbf{b}_1, \ldots, \mathbf{b}_p]$, and the current rotation pair is (m, ℓ). Before the call to the G^+ algorithm, we have

$$
\begin{aligned}
\varphi(\mathbf{B}; \mathbf{S}) &= \det[\mathrm{pdiag}\{(\mathbf{I}_k \otimes \mathbf{B})'\mathbf{S}(\mathbf{I}_k \otimes \mathbf{B})\}]/\det \mathbf{S} \\
&= (\Pi_{h=1}^{p} \det \mathbf{P}_h) / \det \mathbf{S},
\end{aligned}
\tag{2.5.1}
$$

where $\mathbf{P}_h = (\mathbf{I}_k \otimes \mathbf{b}_h)'\mathbf{S}(\mathbf{I}_k \otimes \mathbf{b}_h)$ as in Eq. (1.16). After the call to G^+, $\mathbf{B}$ is replaced by $\mathbf{B}^*$ (with new columns $\mathbf{b}_m^*$ and $\mathbf{b}_h^*$), and the value of φ is

$$
\varphi(\mathbf{B}^*; \mathbf{S}) = \frac{1}{\det \mathbf{S}} (\det \mathbf{P}_m^*)(\det \mathbf{P}_\ell^*) \prod_{\substack{h=1 \\ h \neq m, \ell}}^{p} \det \mathbf{P}_h
$$

with $\mathbf{P}_m^* = (\mathbf{I}_k \otimes \mathbf{b}_m^*)'\mathbf{S}(\mathbf{I}_k \otimes \mathbf{b}_m^*)$, and $\mathbf{P}_\ell^*$ analogous. But from Eq. (2.4.5) we know that $\mathbf{P}_m$ and $\mathbf{P}_\ell$ are the same matrices as $\mathbf{M}_1$ and $\mathbf{M}_2$ upon entry in the G^+ algorithm, and $\mathbf{P}_m^*$ and $\mathbf{P}_\ell^*$ are the same as $\mathbf{M}_1$ and $\mathbf{M}_2$ after the G^+ algorithm has converged. Hence, if we can show that the G^+ algorithm reduces the value of $(\det \mathbf{M}_1)(\det \mathbf{M}_2)$, we can conclude that the sequence of orthogonal matrices $\mathbf{B}^{(f)}$ produced by the F^+ algorithm converges to a minimum, i.e.,

$$
\varphi(\mathbf{B}^{(f+1)}; \mathbf{S}) \leq \varphi(\mathbf{B}^{(f)}; \mathbf{S})
$$

with equality exactly if a minimum has been reached.

Suppose the G^+ algorithm is entered with the pds matrix $\mathbf{T} = (\mathbf{T}_{ij})$ of dimension $2k \times 2k$, and assume the current value of the orthogonal matrix $\mathbf{Q} \in \mathcal{O}(2)$ is $\mathbf{I}_2$ — otherwise, replace $\mathbf{T}$ by $(\mathbf{I}_k \otimes \mathbf{Q})'\mathbf{T}(\mathbf{I}_k \otimes \mathbf{Q})$. Let

$$\mathbf{M} = \begin{bmatrix} \mathbf{M}_1 & \mathbf{R} \\ \mathbf{R}' & \mathbf{M}_2 \end{bmatrix} = \mathbf{I}_{(2,k)}\mathbf{T}\mathbf{I}_{(k,2)} \qquad (2.5.2)$$

and recall that we are interested in the value of $(\det\mathbf{M}_1)(\det\mathbf{M}_2)$. Consider a rotation by an angle $\alpha = \sin^{-1} x$, i.e., an orthogonal matrix

$$\mathbf{Q} = \begin{bmatrix} \sqrt{1-x^2} & -x \\ x & \sqrt{1-x^2} \end{bmatrix} =: [\mathbf{q}_1, \mathbf{q}_2]. \qquad (2.5.3)$$

After the rotation we have

$$\mathbf{T}_x := (\mathbf{I}_k \otimes \mathbf{Q})'\mathbf{T}(\mathbf{I}_k \otimes \mathbf{Q}) \qquad (2.5.4)$$

and the updated matrix $\mathbf{M}_x$ becomes

$$\mathbf{M}_x = \begin{bmatrix} \mathbf{M}_{1x} & \mathbf{R}_x \\ \mathbf{R}'_x & \mathbf{M}_{2x} \end{bmatrix} = \mathbf{I}_{(2,k)}\mathbf{T}_x\mathbf{I}_{(k,2)}, \qquad (2.5.5)$$

where

$$\mathbf{M}_{1x} = (\mathbf{I}_k \otimes \mathbf{q}_1)'\mathbf{T}(\mathbf{I}_k \otimes \mathbf{q}_1),$$
$$\mathbf{M}_{2x} = (\mathbf{I}_k \otimes \mathbf{q}_2)'\mathbf{T}(\mathbf{I}_k \otimes \mathbf{q}_2)$$

and

$$\mathbf{R}_x = (\mathbf{I}_k \otimes \mathbf{q}_1)'\mathbf{T}(\mathbf{I}_k \otimes \mathbf{q}_2).$$

Writing $\mathbf{T}_{ij} = \begin{bmatrix} t_{ij,11} & t_{ij,12} \\ t_{ij,21} & t_{ij,21} \end{bmatrix}$, $(i, j = 1, \ldots, k)$, we have

$$\mathbf{q}'_1\mathbf{T}_{ij}\mathbf{q}_1 = (1 - x^2)t_{ij,11} + x\sqrt{1-x^2}(t_{ij,12} + t_{ij,21}) + x^2 t_{ij,22}$$

and

$$\mathbf{q}'_2\mathbf{T}_{ij}\mathbf{q}_2 = (1 - x^2)t_{ij,22} - x\sqrt{1-x^2}(t_{ij,12} + t_{ij,21}) + x^2 t_{ij,11},$$

and we can therefore write

$$\mathbf{M}_{1x} = \mathbf{M}_1 + x(\mathbf{R} + \mathbf{R}') + \mathbf{O}_k(x^2),$$
$$\mathbf{M}_{2x} = \mathbf{M}_2 - x(\mathbf{R} + \mathbf{R}') + \mathbf{O}_k(x^2), \qquad (2.5.6)$$

where $\mathbf{O}_k(x^2)$ is a $k \times k$ matrix such that $\lim_{x \to 0} \mathbf{O}_k(x^2)/x^2$ exists.

The following lemma is well known but stated here because it is used repeatedly.

Lemma 3 *If* $\mathbf{U}$ *and* $\mathbf{V}$ *are symmetric* $k \times k$ *matrices,* $\mathbf{U}$ *positive definite, then* $\det[\mathbf{U} + \mathbf{V}] = (\det \mathbf{U}) \prod_{h=1}^{k}(1 + \lambda_h)$, *where* λ_h *are the eigenvalues of* $\mathbf{U}^{-1}\mathbf{V}$.

Using Lemma 3, we obtain

$$\det \mathbf{M}_{1x} = \det[\mathbf{M}_1 + x(\mathbf{R} + \mathbf{R}')] + O(x^2)$$

and

$$\det \mathbf{M}_{2x} = \det[\mathbf{M}_2 - x(\mathbf{R} + \mathbf{R}')] + O(x^2).$$

Writing $\lambda_h(h = 1, \ldots, k)$ for the eigenvalues of $\mathbf{M}_1^{-1}(\mathbf{R} + \mathbf{R}')$ and $\tau_h(h = 1, \ldots, k)$ for those of $\mathbf{M}_2^{-1}(\mathbf{R} + \mathbf{R}')$, and using Lemma 3 again, we conclude that

$$
\begin{aligned}
(\det \mathbf{M}_{1x})&(\det \mathbf{M}_{2x}) \\
&= \det[\mathbf{M}_1 + x(\mathbf{R} + \mathbf{R}')] \cdot \det[\mathbf{M}_2 - x(\mathbf{R} + \mathbf{R}')] + O(x^2) \\
&= (\det \mathbf{M}_1)(\det \mathbf{M}_2) \prod_{h=1}^{k}(1 + x\lambda_i)(1 - x\tau_i) + O(x^2) \\
&= (\det \mathbf{M}_1)(\det \mathbf{M}_2)f(x) + O(x^2),
\end{aligned}
\tag{2.5.7}
$$

where

$$
\begin{aligned}
f(x) &= 1 + x \sum_{h=1}^{k}(\lambda_i - \tau_i) \\
&= 1 + x \cdot \mathrm{tr}[(\mathbf{M}_1^{-1} - \mathbf{M}_2^{-1})(\mathbf{R} + \mathbf{R}')] \\
&= 1 + x \cdot \mathrm{tr}[\mathbf{A}^*(\mathbf{R} + \mathbf{R}')].
\end{aligned}
\tag{2.5.8}
$$

Putting $\mathbf{T}^* = \sum_{i=1}^{k} \sum_{j=1}^{k} a_{ij}^* \mathbf{T}_{ij}$ as in step OG_3^+, and noticing that

$$
\mathbf{R} = \begin{bmatrix} t_{11,12} & \cdots & t_{1k,12} \\ \vdots & \ddots & \vdots \\ t_{k1,12} & \cdots & t_{kk,12} \end{bmatrix},
$$

we obtain $\mathrm{tr}[\mathbf{A}^*(\mathbf{R} + \mathbf{R}')] = \sum_{i=1}^{k} \sum_{j=1}^{k} a_{ij}^*(t_{ji,12} + t_{ij,12}) = 2t_{12}^*$, implying that

$$f(x) = 1 + 2t_{12}^* x.\tag{2.5.9}$$

If $t_{12}^* > 0$, then $x = \sin \alpha$ should be chosen negative because a small rotation with $\alpha < 0$ will decrease $(\det \mathbf{M}_{1x})(\det \mathbf{M}_{2x})$. Similarly, if $t_{12}^* < 0$, then x should be chosen positive. It turns out that the G^+ algorithm always chooses the right direction, as we now show.

First notice that $t_{11}^* = \mathrm{tr}(\mathbf{A}^*\mathbf{M}_1)$, $t_{22}^* = \mathrm{tr}(\mathbf{A}^*\mathbf{M}_2)$, and therefore

$$
\begin{aligned}
t_{22}^* - t_{11}^* &= \mathrm{tr}[\mathbf{A}^*(\mathbf{M}_2 - \mathbf{M}_1)] \\
&= \mathrm{tr}[\mathbf{M}_1^{-1}\mathbf{M}_2] + \mathrm{tr}[\mathbf{M}_2^{-1}\mathbf{M}_1] - 2k.
\end{aligned}
$$

Lemma 4 *If* $\mathbf{U}$ *and* $\mathbf{V}$ *are pds matrices of dimension* $k \times k$, *then* $\mathrm{tr}[\mathbf{U}^{-1}\mathbf{V}] + \mathrm{tr}[\mathbf{V}^{-1}\mathbf{U}] \geq 2k$, *with equality exactly if* $\mathbf{U} = \mathbf{V}$.

Proof: Let $\lambda_1, \ldots, \lambda_k (> 0)$ denote the eigenvalues of $\mathbf{U}^{-1}\mathbf{V}$; then

$$\mathrm{tr}[\mathbf{U}^{-1}\mathbf{V}] + \mathrm{tr}[\mathbf{V}^{-1}\mathbf{U}] = \sum_{h=1}^{k} \left(\lambda_h + \frac{1}{\lambda_h} \right) \geq 2k$$

because $h(x) := x + x^{-1}$ has a minimum at $x = 1$. $\blacksquare$

Thus Lemma 4 implies

$$t_{22} - t_{11} \geq 0 \tag{2.5.10}$$

with equality exactly if $\mathbf{M}_1 = \mathbf{M}_2$. The orthogonal matrix $\mathbf{Q} = \begin{bmatrix} c & -s \\ s & c \end{bmatrix}$ (with

$c = \cos\alpha$, $s = \sin\alpha$) which diagonalizes $\mathbf{T}^* = \begin{bmatrix} t_{11}^* & t_{12}^* \\ t_{21}^* & t_{22}^* \end{bmatrix}$ is given by

$$\frac{s}{c} = \begin{cases} \frac{1}{2}(r - \sqrt{4 + r^2}) < 0 & \text{if } r > 0, \\ \frac{1}{2}(r + \sqrt{4 + r^2}) > 0 & \text{if } r < 0, \end{cases} \tag{2.5.11}$$

where $r = (t_{22}^* - t_{11}^*)/t_{12}^*$. By (2.5.10), the sign of r is the same as the sign of t_{12}^*, assuming that we use the convention $|\alpha| \leq \pi/4$. Hence, $s < 0$ exactly if $t_{12}^* > 0$, and $s > 0$ exactly if $t_{12}^* < 0$, which shows that the G^+ algorithm always chooses the right direction of rotation.

Finally, we can obtain an approximate value of s, and an approximation to the improvement achieved in an iteration step of the G^+ algorithm. Since $s = c \cdot (r + \sqrt{4 + r^2})/2$, and

$$\lim_{r \to \infty} r(r - \sqrt{4 + r^2}) = \lim_{r \to -\infty} r(r + \sqrt{4 + r^2}) = -2,$$

we have, approximately, for large $|r|$ (i.e., relatively small $|t_{12}^*|$),

$$s \approx \frac{1}{2}\left(\frac{-2}{r}\right) = -\frac{1}{r} = \frac{t_{12}^*}{t_{11}^* - t_{22}^*} \tag{2.5.12}.$$

Using this approximation for x in the function $f(x)$, we have

$$f(s) = 1 + 2t_{12}^* s \approx 1 - 2(t_{12}^*)^2/(t_{22}^* - t_{11}^*).$$

Hence, provided that $|t_{12}^*|/(t_{22}^* - t_{11}^*)$ is small (which means that we are close to a solution of the equation system), we can expect an iteration step of the G^+ algorithm to change the value of $(\det\mathbf{M}_1)(\det\mathbf{M}_2)$ to

$$(\det \mathbf{M}_{1s})(\det \mathbf{M}_{2s}) \approx (\det \mathbf{M}_1)(\det \mathbf{M}_2)\left(1 - 2\frac{(t_{12}^*)^2}{t_{22}^* - t_{11}^*}\right). \tag{2.5.13}$$

In the examples of §4 we will see that the speed of convergence depends on the input matrix $\mathbf{T}$. Generally speaking, the "closer" the matrices $\mathbf{T}_{ij} + \mathbf{T}_{ij}$ $(i, j = 1, \ldots, k)$ are to being simultaneously diagonalizable, the faster the algorithm will converge. In the ideal case of exact simultaneous diagonalizability of all $\mathbf{T}_{ij} + \mathbf{T}_{ji}$, one single iteration of the G^+ algorithm yields the solution because any linear combination $\sum_{i=1}^{k} \sum_{j=1}^{k} a_{ij}^* \mathbf{T}_{ij}$ (with $a_{ij}^* = a_{ji}^*$) has the same eigenvectors. For the general p–dimensional case, if all $\mathbf{S}_{ij} + \mathbf{S}_{ji}$ have a common set of eigenvectors, then the F^+ algorithm performs exactly like a cyclic Jacobi algorithm, and convergence is quadratic (Golub and VanLoan 1983).

2.6 Remarks.

(1) Some care must be taken to prevent oscillating between different but equivalent versions of the orthogonal matrix. This is done in step OG_4^+ by a proper choice of $\mathbf{Q}$ among all orthogonal 2×2 matrices that diagonalize $\mathbf{T}^*$.

(2) In the G^+ algorithm it is typically not necessary to iterate until convergence has been reached according to ε_G. Particularly during the first sweeps of the F^+ algorithm, doing many iterations on the G–level is superfluous because the same pair of indices (m, ℓ) will be used in later rotations again. Hence, it is reasonable to impose an upper limit $g_{\max}$ and stop the G^+ algorithm when $g > g_{\max}$. We used $g_{\max} = 5$ in our program.

(3) In §2.5 we have seen that the orthogonal G^+ algorithm always chooses the right direction of rotation, and have given an approximation to the decrease in the function value; see Eq. (2.5.13). However, we have not shown that the function value always decreases. It might actually happen that the chosen angle of rotation is too large, leading to an increase in the value of the function. To prevent this, one may evaluate φ in each iteration of the G^+ algorithm, and if there was an increase instead of a decrease, reduce the rotation angle repeatedly to half its value until φ decreases.

Interestingly, it seems that this ugly fix is not necessary in practice: In thousands of examples with randomly generated matrices, convergence was always monotonic, i.e., each iteration of the G^+ algorithm strictly decreased the value of φ, and the direction of the rotation never changed. That is, if the current angle of rotation is α, and the minimum occurs at α^*, then the next iteration appears to *always* update α to a value between the old α and α^*, the only exception being the case of *exact* parallel–diagonalizability, in which one step leads to the exact solution. We conjecture therefore that the orthogonal FG^+ algorithm always converges monotonically, even without the "quick fix" indicated here, but we have not been able to prove it.

(4) The algorithm (as well as the nonorthogonal version to be discussed in §3) has been programmed in *GAUSS*; the program is available from the authors upon request.

3 THE NONORTHOGONAL FG^+ ALGORITHM

3.1 Introduction

The nonorthogonal FG^+ algorithm solves the equation system $(1.17) - (1.19)$ numerically. Its structure is essentially the same as in the orthogonal case. The F^+ algorithm follows a cyclic scheme of updating pairs $(\mathbf{b}_m, \mathbf{b}_\ell)$ of columns of the matrix $\mathbf{B}$ (which is nonsingular but not necessary orthogonal), choosing the new pair $(\mathbf{b}_m^*, \mathbf{b}_\ell^*)$ such that equations (1.17) and (1.18) are satisfied. The transformation for the update is provided by the G^+ algorithm, which solves the same equation system $(1.17) - (1.19)$ for the special case $p = 2$. The G^+ algorithm is itself iterative, and requires the simultaneous diagonalization of two pds matrices of dimension 2×2.

The algorithm requires two constants $\varepsilon_F > 0$ and $\varepsilon_G > 0$ which determine convergence. The input consists of a positive definite symmetric matrix $\mathbf{S}$ of dimension $pk \times pk$, partitioned into k^2 submatrices $\mathbf{S}_{ij}$ of dimension $p \times p$. The output consists of a nonsingular matrix $\mathbf{B} = (\mathbf{b}_1, \ldots, \mathbf{b}_p)$ which solves $(1.17) - (1.19)$, with diag $(\mathbf{B}'\mathbf{B}) = \mathbf{I}_p$.

3.2 The nonorthogonal F^+ algorithm

This algorithm works identically to the orthogonal F^+ algorithm of §2.2, except for steps OF_{23}^+ and OF_{24}^+, which are replaced as follows.

Step NF_{23}^+: Call the nonorthogonal G^+ algorithm with input matrix $\mathbf{T}$, obtain a nonsingular 2×2 matrix $\mathbf{Q} = (\mathbf{q}_1, \mathbf{q}_2)$.

Step NF_{24}^+: In the matrix $\mathbf{B}$, replace the mth and ℓth columns by $\mathbf{b}_m^*$ and $\mathbf{b}_\ell^*$, where $\mathbf{b}_m^* = \mathbf{Hq}_1/(\mathbf{q}_1'\mathbf{H}'\mathbf{Hq}_1)^{1/2}$, $\mathbf{b}_\ell^* = \mathbf{Hq}_2/(\mathbf{q}_2'\mathbf{H}'\mathbf{Hq}_2)^{1/2}$.

3.3 The nonorthogonal G^+ algorithm

Steps OG_3^+ and OG_4^+ in the orthogonal G^+ algorithm of §2.3 are replaced by the following steps:

Step NG_3^+: Put

$$\mathbf{A}_h^* = \begin{bmatrix} a_{h,11}^* & \cdots & a_{h,1k}^* \\ \vdots & \ddots & \vdots \\ a_{h,k1}^* & \cdots & a_{h,kk}^* \end{bmatrix} \leftarrow \mathbf{M}_h^{-1} \qquad (h = 1, 2)$$

and

$$\mathbf{T}_h^* \leftarrow \sum_{i=1}^{k} \sum_{j=1}^{k} a_{h,ij}^* \mathbf{T}_{ij} \qquad (h = 1, 2).$$

193

Step NG_4^+: Find the 2×2 matrix

$$\mathbf{Q} = \begin{bmatrix} c_1 & c_2 \\ s_1 & s_2 \end{bmatrix},$$

where $\mathbf{Q}'\mathbf{T}_h\mathbf{Q}$ is diagonal ($h = 1, 2$), such that $c_1^2 + s_1^2 = c_2^2 + s_2^2 = 1, c_1 > 0$, $c_1 \geq |c_2|, \det\mathbf{Q} > 0$.

All other steps remain unchanged. Note that the conditions imposed on $\mathbf{Q}$ in step 4 are needed to prevent the algorithm from alternating between different but equivalent versions. Standardizing the columns of $\mathbf{Q}$ to unit length is not strictly necessary but helpful to check for convergence.

3.4 Analysis of the nonorthogonal FG^+ algorithm

The analysis parallels that of §2.4 closely. In step NF_{23}^+ the nonorthogonal G^+ algorithm is called with input matrix $\mathbf{T} = (\mathbf{I}_k \otimes \mathbf{H})'\mathbf{S}(\mathbf{I}_k \otimes \mathbf{H})$, where $\mathbf{H} = [\mathbf{b}_m , \mathbf{b}_\ell]$ contains the current mth and ℓth columns of B. The nonorthogonal G^+ algorithm solves equation system (1.17) – (1.19) for the special case $p = 2$, the equations being

$$\mathbf{q}_1' \left(\sum_{i=1}^{k} \sum_{j=1}^{k} a_{1,ij}^* \mathbf{T}_{ij} \right) \mathbf{q}_2 = 0, \qquad (3.4.1)$$

$$\mathbf{q}_1' \left(\sum_{i=1}^{k} \sum_{j=1}^{k} a_{2,ij}^* \mathbf{T}_{ij} \right) \mathbf{q}_2 = 0. \qquad (3.4.2)$$

Here,

$$\mathbf{A}_h^* = (a_{h,ij}^*) = [(\mathbf{I}_k \otimes \mathbf{q}_h)'\mathbf{T}(\mathbf{I}_k \otimes \mathbf{q}_h)]^{-1}, \qquad h = 1, 2, \qquad (3.4.3)$$

where $\mathbf{Q} = [\mathbf{q}_1, \mathbf{q}_2] \in \mathcal{A}(2)$, i.e., $\mathbf{Q}$ is nonsingular, and $\mathrm{diag}(\mathbf{Q}'\mathbf{Q}) = \mathbf{I}_2$. Suppose the current matrix $\mathbf{Q}$ is a solution to (3.4.1) – (3.4.3), and is returned to the F^+ algorithm. In step NF_{24}^+, $\mathbf{b}_m$ and $\mathbf{b}_\ell$ are updated to

$$\mathbf{b}_m^* = c_1 \cdot \mathbf{Hq}_1 , \quad \mathbf{b}_h^* = c_2 \cdot \mathbf{Hq}_2 ,$$

where $c_h = (\mathbf{q}_h'\mathbf{H}'\mathbf{Hq}_h)^{-1/2}$, $h = 1, 2$. The updated matrix $\mathbf{B}^*$ is still nonsingular because $\mathbf{b}_m^*$ and $\mathbf{b}_\ell^*$ span the same subspace as $\mathbf{b}_m$ and $\mathbf{b}_\ell$. Moreover, $\mathbf{b}_m^*$ and $\mathbf{b}_\ell^*$ solve the (m, ℓ)th equations in (1.17) and (1.18). To see this, notice that the matrices $\mathbf{A}_h$ in (1.19) have the form

$$\begin{aligned} \mathbf{A}_m = (a_{m,ij}) &= [(\mathbf{I}_k \otimes \mathbf{b}_m^*)'\mathbf{S}(\mathbf{I}_k \otimes \mathbf{b}_m^*)]^{-1} \\ &= [c_1^2(\mathbf{I}_k \otimes \mathbf{q}_1)'(\mathbf{I}_k \otimes \mathbf{H})'\mathbf{S}(\mathbf{I}_k \otimes \mathbf{H})(\mathbf{I}_k \otimes \mathbf{q}_1)]^{-1} \\ &= \mathbf{A}_1^*/c_1^2 \end{aligned}$$

and similarly, $\mathbf{A}_\ell = \mathbf{A}_2^*/c_2^2$. Hence, in equation (1.17),

$$
\mathbf{b}_m^{*'}\left(\sum_{i=1}^k\sum_{j=1}^k a_{m,ij}\mathbf{S}_{ij}\right)\mathbf{b}_\ell^* = c_1^2\mathbf{q}_1'\left(\sum_{i=1}^k\sum_{j=1}^k a_{1,ij}^*\mathbf{H}'\mathbf{S}_{ij}\mathbf{H}/c_1^2\right)\mathbf{q}_2
$$

$$
= \mathbf{q}_1'\left(\sum_{i=1}^k\sum_{j=1}^k a_{1,ij}^*\mathbf{T}_{ij}\right)\mathbf{q}_2 = 0,
$$

and similarly for equation (1.18).

It remains to show that the nonorthogonal G^+ algorithm provides a solution to equation system (3.4.1) – (3.4.3). Write $\mathbf{T}_1^*$ and $\mathbf{T}_2^*$ for the 2×2 matrices in parentheses of (3.4.1) and (3.4.2), i.e., $\mathbf{T}_h^* = \sum_{i=1}^k\sum_{j=1}^k a_{h,ij}^*\mathbf{T}_{ij}$. We will show below that the $\mathbf{T}_h^*$ are positive definite. Hence, there exists a nonsingular matrix $\mathbf{Q}^*$ such that $\mathbf{Q}^{*'}\mathbf{T}_1^*\mathbf{Q}^* = \mathbf{I}_2$ and $\mathbf{Q}^{*'}\mathbf{T}_2^*\mathbf{Q}^* = \mathrm{diag}(s_1, s_2)$ (Searle 1982, p. 313). Rescaling the two columns of $\mathbf{Q}^*$ to unit length thus gives the matrix $\mathbf{Q}$ as required in step NG_4^+. In other words, step NG_4^+ can always be performed, and the G^+ algorithm will iterate until a solution is found. It remains to show that $\mathbf{T}_1^*$ and $\mathbf{T}_2^*$ are positive definite. The following two lemmas will be needed.

Lemma 5 *Suppose the $k \times k$ matrix $\mathbf{A} = (a_{ij})$ is pds, and $\mathbf{T} = \left(\mathbf{T}_{ij}\right)$ is a pds matrix of dimension $kq \times kq$, where each $\mathbf{T}_{ij}$ has dimension $q \times q$. Then the $kq \times kq$ partitioned matrix*

$$
\mathbf{R} = \begin{bmatrix} a_{11}\mathbf{T}_{11} & \cdots & a_{1k}\mathbf{T}_{1k} \\ \vdots & \ddots & \vdots \\ a_{k1}\mathbf{T}_{k1} & \cdots & a_{kk}\mathbf{T}_{kk} \end{bmatrix} \tag{3.4.4}
$$

is pds.

Proof: Symmetry of (3.4.4) is clear. All entries in $\mathbf{R}$ are also entries of the matrix $\mathbf{A} \otimes \mathbf{T}$, which has dimension $k^2q \times k^2q$ and is positive definite (Magnus and Neudecker 1988, p. 29). In fact, $\mathbf{R}$ can be obtained from $\mathbf{A} \otimes \mathbf{T}$ by retaining only rows and columns $(j-1)(k+1)q+1, \ldots, (j-1)(k+1)q+q$ for $j = 1, \ldots, k$, and deleting the remaining rows and columns. This can be written in the form

$$
\mathbf{R} = \mathbf{J}(\mathbf{A} \otimes \mathbf{T})\mathbf{J}'
$$

where $\mathbf{J}$ is a binary matrix of dimension $kq \times k^2q$, with rank kq. Hence, $\mathbf{R}$ is positive definite. ∎

Incidentally, Lemma 5 proves that the Hadamard product of two pds matrices is pds. See Magnus (1988, §7.8) for similar manipulations as those used in the foregoing proof.

Lemma 6 *Suppose the $kq \times kq$ partitioned matrix*

$$
\mathbf{R} = \begin{bmatrix} \mathbf{R}_{11} & \cdots & \mathbf{R}_{1k} \\ \vdots & \ddots & \vdots \\ \mathbf{R}_{k1} & \cdots & \mathbf{R}_{kk} \end{bmatrix}
$$

195

is pds. Then $\sum_{i=1}^{k}\sum_{j=1}^{k}\mathbf{R}_{ij}$ is pds.

Proof: Let $\mathbf{1}_k \in I\!\!R^k$ denote the vector with 1 in each position. Then, for any $\mathbf{x} \in I\!\!R^q, \mathbf{x} \neq \mathbf{0}$,

$$\mathbf{x}' \left(\sum_{i=1}^{k}\sum_{j=1}^{k}\mathbf{R}_{ij} \right) \mathbf{x} = (\mathbf{1}_k \otimes \mathbf{x})'\mathbf{R}(\mathbf{1}_k \otimes \mathbf{x}) > 0$$

because $\mathbf{R}$ is pds. ∎

Using Lemmas 5 and 6 with $q = 2$ shows that $\mathbf{T}_1^*$ and $\mathbf{T}_2^*$ are both pds. This concludes the analysis of the nonorthogonal FG^+ algorithm. No proof of convergence has been established yet, although there is little doubt that the algorithm always converges, as is suggested by many numerical examples.

4   NUMERICAL EXAMPLES

In this section we present three numerical examples to illustrate how the algorithms work.

Example 1   While experimenting with the FG^+ algorithm, we repeatedly generated random pds matrices in order to find "pathological" cases. This example represents a case with relatively slow convergence. The dimensions are $k = p = 2$, so only the G^+ algorithm is needed. The input matrix is

$$\mathbf{S} = \begin{bmatrix} 2.246 & -2.257 & 2.135 & -1.204 \\ -2.257 & 3.869 & -3.219 & 0.618 \\ 2.135 & -3.219 & 3.175 & -1.264 \\ -1.204 & 0.618 & -1.264 & 3.080 \end{bmatrix}.$$

Writing

$$\mathbf{B} = \mathbf{B}(\alpha_1, \alpha_2) = \begin{bmatrix} \cos\alpha_1 & \cos\alpha_2 \\ \sin\alpha_1 & \sin\alpha_2 \end{bmatrix},$$

we computed values of the function

$$\varphi(\mathbf{B}; \mathbf{S}) = \frac{\det[\mathrm{pdiag}\{(\mathbf{I}_2 \otimes \mathbf{B})'\mathbf{S}(\mathbf{I}_2 \otimes \mathbf{B})\}]}{\det\{(\mathbf{I}_2 \otimes \mathbf{B})'\mathbf{S}(\mathbf{I}_2 \otimes \mathbf{B})\}}$$

for $-\frac{\pi}{2} \leq \alpha_1 \leq \frac{\pi}{2}$ and $0 \leq \alpha_2 \leq \pi$. Figure 1 shows a contour plot of $\varphi^*(\mathbf{B}; \mathbf{S}) = 100 \log \varphi(\mathbf{B}; \mathbf{S})$.

The two minima in Figure 1 are equivalent because they correspond to exchanging α_1 with α_2. The solid straight line correspond to $\alpha_2 = \alpha_1 + \frac{\pi}{2}$, i.e., to $\mathbf{B}$ being orthogonal. Along the line $\alpha_2 = \alpha_1 \pmod{\pi}$ the function φ is not defined; consequently contour lines were not drawn beyond $\varphi^* = 600$.

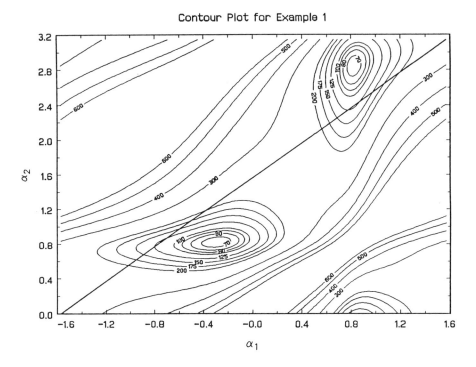

Figure 1: Contour plot of $\varphi^*(\alpha_1, \alpha_2) = 100 \, \log \varphi(\mathbf{B}; \mathbf{S})$ in Example 1.

We ran the nonorthogonal G^+ algorithm with $\varepsilon_G = 10^{-10}$ and initial angles $\alpha_1 = .2$, $\alpha_2 = \alpha_1 + \pi/2$; the reason for using this particular starting point is given later. The following table illustrates how the algorithm performed during the first four steps.

iteration	α_1	α_2	φ^*
0	.20000	1.77080	263.59460
1	.02657	1.18882	180.67337
2	−.22541	.87022	68.66674
3	−.28207	.82865	63.59281
4	−.29039	.82358	63.50503

The algorithm stopped after $g = 8$ iterations, with $\alpha_1 = -.29171$, $\alpha_2 = .82285$, which corresponds to $\mathbf{Q} = \begin{bmatrix} .958 & .680 \\ -.286 & .733 \end{bmatrix}$, and $\varphi^* = 63.50351$.

The orthogonal FG^+ algorithm was run with the same starting values $\alpha_1 = .2$, $\alpha_2 = \alpha_1 + \pi/2$. The following table gives a protocol of the performance of the

197

algorithm.

g	α_1	α_2	φ^*
0	0.20000	1.77080	263.59460
1	0.19811	1.76891	263.59106
2	0.19349	1.76428	263.56998
3	0.18226	1.75305	263.44645
4	0.15549	1.72629	262.75573
5	0.09521	1.66601	259.37455
6	−0.01862	1.55218	247.98626
7	−0.16238	1.40842	229.89591
8	−0.28654	1.28426	213.99752
9	−0.40263	1.16817	196.27278
10	−0.53183	1.03896	167.84282
11	−0.65636	0.91443	134.31930
12	−0.72497	0.84582	122.30881
13	−0.74604	0.82476	121.11160
14	−0.75112	0.81968	121.04155
15	−0.75227	0.81852	121.03791
16	−0.75253	0.81826	121.03772
17	−0.75259	0.81820	121.03771

Iteration 18 did not produce any further changes, and the algorithm stopped. The solution thus found corresponds to

$$\mathbf{Q} = \begin{bmatrix} .730 & .684 \\ .684 & .730 \end{bmatrix}.$$

The algorithm took initially very small steps, but then moved rapidly towards the minimum in iterations 5 to 12. This behavior is explained by the starting point $\alpha_1 = .2$ which is near the maximum of $\varphi(\mathbf{B}(\alpha_1, \alpha_1 + \pi/2))$, as illustrated in Figure 2. As mentioned before, this is a somewhat "pathological" (and therefore, interesting) example, which exhibits extremely slow convergence. Notice also that minimization over $\mathbf{B} \in \mathcal{O}(2)$ corresponds to minimization along the diagonal line in Figure 1, which passes rather far from the minimum reached when α_1 and α_2 vary independently.

The "almost–parallel diagonal" matrices

$$\mathbf{F} = (\mathbf{I}_2 \otimes \mathbf{B})'\mathbf{S}(\mathbf{I}_2 \otimes \mathbf{B})$$

were computed for both the nonorthogonal and the orthogonal solution; they are

$$\mathbf{F} = \begin{bmatrix} 3.62358 & -0.49584 & 3.22781 & 1.04475 \\ -0.49584 & 0.86755 & -0.76414 & -0.88556 \\ 3.22781 & -0.76414 & 3.86346 & 0.77863 \\ 1.04475 & -0.88556 & 0.77863 & 1.86349 \end{bmatrix}$$

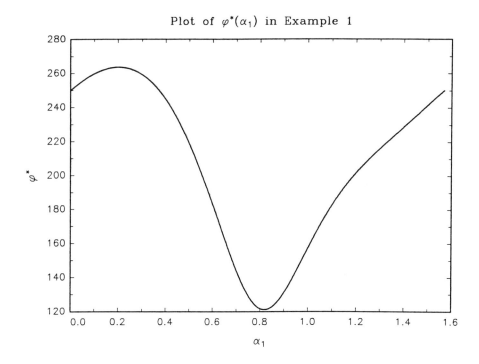

Figure 2: Graph of $\varphi^*(\alpha_1, \alpha_1 + \pi/2)$ in Example 1.

for the nonorthogonal case, and

$$\mathbf{F} = \begin{bmatrix} 5.25646 & -0.95767 & 3.63296 & 1.61943 \\ -0.95767 & 0.85854 & -0.39557 & -0.87996 \\ 3.63296 & -0.39557 & 4.39190 & -0.03544 \\ 1.61943 & -0.87996 & -0.03544 & 1.86310 \end{bmatrix}$$

for the orthogonal case. Neither of them appears to be particularly close to being parallel–diagonal, but the purpose of this example was to illustrate the performance of the G^+ algorithm rather than to give a good application. Note that the slow convergence is typical for situations where the input matrix is far from being parallel–diagonalizable. Indeed, if $\mathbf{S}$ is exactly parallel–diagonalizable, then the G^+ algorithm converges in one iteration.

Example 2 This is a real–data example, the statistical aspects of which are discussed in more detail in Flury and Neuenschwander (1994). The input matrix is

$$S = \begin{bmatrix} 9.247 & 7.972 & 7.274 & 6.894 & 6.110 & 6.252 \\ 7.972 & 11.248 & 7.032 & 9.016 & 6.189 & 8.521 \\ 7.274 & 7.032 & 8.782 & 8.071 & 7.128 & 7.400 \\ 6.894 & 9.016 & 8.071 & 11.082 & 6.663 & 9.862 \\ 6.110 & 6.189 & 7.128 & 6.663 & 8.171 & 8.232 \\ 6.252 & 8.521 & 7.400 & 9.862 & 8.232 & 12.680 \end{bmatrix}$$

with $k = 3$ and $p = 2$. The partition correspond to $k = 3$ instars (stages of development) of water striders, and the size of the submatrices corresponds to the $p = 2$ variables "length of femur" and "length of tibia". The matrix S is actually the covariance matrix of log–measurements obtained on 88 specimens. The data was kindly provided by Dr. C.P. Klingenberg, Department of Entomology, University of Alberta.

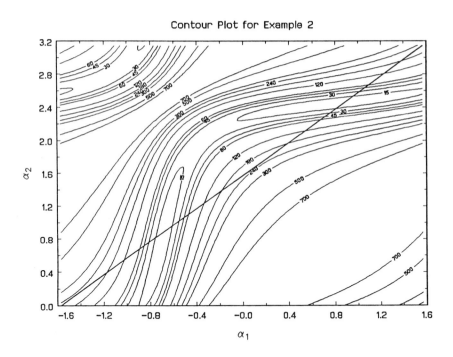

Figure 3: Contour plot of $\varphi^*(\alpha_1, \alpha_2) = 100 \ \log \varphi(B; S)$ in Example 2.

Since $p = 2$, we need to run only the G^+ algorithm again. Both the orthogonal and the nonorthogonal version were run using $\alpha_1 = 0$, $\alpha_2 = \pi/2$ as initial values (i.e., $\mathbf{B} = \mathbf{I}_2$), which gave an initial function value of $\varphi^* = 100 \log \varphi(\mathbf{B}; \mathbf{S}) = 263.595$. With $\varepsilon_G = 10^{-10}$, the nonorthogonal G^+ algorithm converged in 12 iterations to the solution $\alpha_1 = -.70299, \alpha_2 = .75151$, corresponding to $\mathbf{B} = \begin{bmatrix} .763 & .731 \\ -.646 & .683 \end{bmatrix}$ and $\varphi^* = 7.6303$. The orthogonal G^+ algorithm converged in 4 iterations to the solution $\alpha_1 = -.68245, \alpha_2 = .88835$, corresponding to $\mathbf{B} = \begin{bmatrix} .776 & .631 \\ -.631 & .776 \end{bmatrix}$, with $\varphi^* = 7.781$. Figures 3 and 4 illustrate this minimization problem analogously to Figures 1 and 2. Evidently, the straight line in Figure 3 passes near the unconstrained minimum. The contours for $\varphi^* = 15$ in Figure 3 show a rather elongated "valley", which explains why the nonorthogonal algorithm took so many iterations to locate the minimum precisely. In contrast, Figure 4 illustrates why the orthogonal algorithm converges so rapidly, despite the rather bad choice of the starting point.

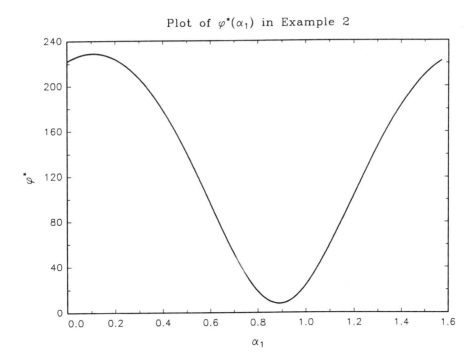

Plot of $\varphi^*(\alpha_1)$ in Example 2

Figure 4: Graph of $\varphi^*(\alpha_1, \alpha_1 + \pi/2)$ in Example 2.

The "almost parallel–diagonal" matrices $\mathbf{F} = (\mathbf{I}_3 \otimes \mathbf{B})'\mathbf{S}(\mathbf{I}_3 \otimes \mathbf{B})$, reported here with only few decimal digits, are

$$F = \begin{bmatrix} 2.22 & 0.58 & 1.13 & 0.34 & 0.98 & -0.02 \\ 0.58 & 18.13 & 0.48 & 15.03 & -0.08 & 13.44 \\ 1.13 & 0.48 & 1.78 & 0.40 & 1.33 & 0.33 \\ 0.34 & 15.03 & 0.40 & 17.91 & -0.40 & 15.42 \\ 0.98 & -0.08 & 1.33 & -0.40 & 1.94 & -0.64 \\ -0.02 & 13.44 & 0.33 & 15.42 & -0.64 & 18.49 \end{bmatrix}$$

for the nonorthogonal case, and

$$F = \begin{bmatrix} 2.24 & 0.65 & 1.15 & 0.50 & 0.98 & 0.12 \\ 0.65 & 18.26 & 0.64 & 15.14 & 0.06 & 13.65 \\ 1.15 & 0.64 & 1.80 & 0.52 & 1.33 & 0.47 \\ 0.50 & 15.14 & 0.52 & 18.07 & -0.27 & 15.66 \\ 0.98 & 0.06 & 1.33 & -0.27 & 1.91 & -0.52 \\ 0.12 & 13.65 & 0.47 & 15.66 & -0.52 & 18.94 \end{bmatrix}$$

for the orthogonal case. Evidently, transforming all submatrices S_{ij} to almost diagonal form is much more successful here than in Example 1. In the statistical context of this example, this meant that a multivariate model with strong constraints on the covariance matrices appears to fit the data well. The particular model used here is called a "common canonical variates" model, in which it is assumed that all $p \times p$ submatices ψ_{ij} of the $pk \times pk$ covariance matrix ψ can be diagonalized simultaneously. If the common diagonalization matrix is orthogonal, then we are in the special case of a common principal component model for dependent random vectors (Neuenschwander 1994).

Example 3 This final example, with $k = 2$ and $p = 4$, serves to illustrate the performance of the F^+ algorithm. The input matrix is

$$S = \begin{bmatrix} 90.4 & -32.8 & -27.2 & 9.6 & 0.0 & -30.0 & -10.0 & 0.0 \\ -32.8 & 109.6 & -9.6 & -27.2 & -30.0 & 0.0 & 0.0 & -10.0 \\ -27.2 & -9.6 & 109.6 & -32.8 & -10.0 & 0.0 & 0.0 & -30.0 \\ 9.6 & -27.2 & -32.8 & 90.4 & 0.0 & -10.0 & -30.0 & 0.0 \\ 0.0 & -30.0 & -10.0 & 0.0 & 109.6 & 30.4 & 12.8 & 7.2 \\ -30.0 & 0.0 & 0.0 & -10.0 & 30.4 & 109.6 & 7.2 & 12.8 \\ -10.0 & 0.0 & 0.0 & -30.0 & 12.8 & 7.2 & 90.4 & 49.6 \\ 0.0 & -10.0 & -30.0 & 0.0 & 7.2 & 12.8 & 49.6 & 90.4 \end{bmatrix},$$

which was constructed such that $S_{11}, S_{12} + S_{21}$, and S_{22} have exactly one common eigenvector $\frac{1}{2}(1, 1, 1, 1)'$. The FG^+ algorithm was run with $\varepsilon_F = \varepsilon_G = 10^{-10}$, and $g_{max} = 5$ (i.e., the G^+ algorithm stops after at most 5 iterations, even if convergence has not been reached). Each sweep of the F^+ algorithm consists of $\binom{p}{2} = 6$ calls to the G^+ algorithm. The initial matrix B was chosen as I_4 for both the orthogonal and the nonorthogonal case. The following table summarizes the

202

performance of the algorithm, where $\varphi = \det[\text{pdiag}\{(\mathbf{I}_2 \otimes \mathbf{B})'\mathbf{S}(\mathbf{I}_2 \otimes \mathbf{B})\}]/\det[(\mathbf{I}_2 \otimes \mathbf{B})'\mathbf{S}(\mathbf{I}_2 \otimes \mathbf{B})]$, and $\bar{g}$ is the average number of iterations of the G^+ algorithm per call.

sweep	orthogonal FG^+		nonorthogonal FG^+	
	$\bar{g}$	$\log \varphi$	$\bar{g}$	$\log \varphi$
0		1.56286		1.56286
1	4.50	.37307	4.17	.34297
2	3.67	.08668	3.17	.02288
3	3.17	.07747	2.17	.01709
4	2.33	.07746	1.50	.01709
5	1.67	.07746	1.00	.01709
6	1.00	.07746		

In the orthogonal case, the minimum was reached at

$$\mathbf{B} = \begin{bmatrix} 0.500 & -0.192 & 0.650 & -0.539 \\ 0.500 & 0.181 & -0.712 & -0.459 \\ 0.500 & 0.687 & 0.217 & 0.480 \\ 0.500 & -0.676 & -0.155 & 0.518 \end{bmatrix}$$

with the "almost–parallel diagonal" matrix $\mathbf{F} = (\mathbf{I}_2 \otimes \mathbf{B})'\mathbf{S}(\mathbf{I}_2 \otimes \mathbf{B})$,

$$\mathbf{F} = \begin{bmatrix} 40.00 & 0.00 & 0.00 & 0.00 & -40.00 & 0.00 & 0.00 & 0.00 \\ 0.00 & 146.86 & -22.63 & 10.35 & 0.00 & 35.09 & -8.63 & -0.07 \\ 0.00 & -22.63 & 120.94 & 15.28 & 0.00 & -8.63 & 24.75 & -2.68 \\ 0.00 & 10.35 & 15.28 & 92.20 & 0.00 & -0.07 & -2.68 & -19.83 \\ -40.00 & 0.00 & 0.00 & 0.00 & 160.00 & 0.00 & 0.00 & 0.00 \\ 0.00 & 35.09 & -8.63 & -0.07 & 0.00 & 40.63 & -4.90 & 1.18 \\ 0.00 & -8.63 & 24.75 & -2.68 & 0.00 & -4.90 & 79.54 & 2.56 \\ 0.00 & -0.07 & -2.68 & -19.83 & 0.00 & 1.18 & 2.56 & 119.83 \end{bmatrix}.$$

In the nonorthogonal case, the minimum was reached at

$$\mathbf{B} = \begin{bmatrix} 0.500 & -0.234 & 0.600 & -0.5851 \\ 0.500 & 0.228 & -0.656 & -0.404 \\ 0.500 & 0.671 & 0.351 & 0.407 \\ 0.500 & -0.665 & -0.2925 & 0.578 \end{bmatrix}$$

with

$$\mathbf{F} = \begin{bmatrix} 40.00 & 0.00 & 0.00 & 0.00 & -40.00 & 0.00 & 0.00 & 0.00 \\ 0.00 & 149.71 & 0.65 & 1.30 & 0.00 & 36.18 & -2.66 & -1.52 \\ 0.00 & 0.65 & 112.66 & 5.40 & 0.00 & -2.66 & 21.73 & -4.45 \\ 0.00 & 1.30 & 5.40 & 88.17 & 0.00 & -1.52 & -4.45 & -18.79 \\ -40.00 & 0.00 & 0.00 & 0.00 & 160.00 & 0.00 & 0.00 & 0.00 \\ 0.00 & 36.18 & -2.66 & -1.52 & 0.00 & 41.45 & -1.47 & -1.10 \\ 0.00 & -2.66 & 21.73 & -4.45 & 0.00 & -1.47 & 75.86 & -5.35 \\ 0.00 & -1.52 & -4.45 & -18.79 & 0.00 & -1.10 & -5.35 & 118.38 \end{bmatrix}.$$

Both algorithms have recovered the common eigenvector. Although constructed, this example is fairly typical for the performance of the FG^+ algorithm in real-data applications arising in statistical situations like the one of Example 2.

5 FINAL REMARKS

The algorithms presented in this paper originated from statistical models in which constraints are imposed on the covariance matrix of a multivariate random vector. Under the assumption of multivariate normality of the data, the sample covariance matrix follows a Wishart distribution, and the problem of finding maximum likelihood estimates under the constraints of common principal components and common canonical variates, respectively, is equivalent to the minimization problems described in §1. A condition for existence of the maximum likelihood estimates is that the sample covariance matrix is positive definite. This condition is satisfied if sampling is from a nonsingular normal distribution and if the sample size exceeds the dimension of the random vector; thus the assumption of positive definiteness, which allowed us to establish a Hadamard type inequality in §1, seems natural.

Yet one might consider simultaneous diagonalization problems in which the matrices are not necessarily pds. In the context of ordinary common principal component analysis, this problem can be addressed with a least squares algorithm proposed by DeLeeuw and Pruzansky (1978); see also the discussion in Flury (1988, §9.8). Using a least squares criterion instead of our function φ of Eq. (1.6) may be feasible. Alternatively, one might consider adding a matrix $\alpha \mathbf{I}_p$ to all matrices $\mathbf{S}_{ii}$ on the diagonal of $\mathbf{S}$, thus replacing $\mathbf{S}$ by $\mathbf{S}^* = \mathbf{S} + \alpha \mathbf{I}_{pk}$, choosing α such that $\mathbf{S}^*$ is positive definite.

The same modification might be useful to make sure that a unique minimum exists, a problem we have ignored altogether in this paper. The examples in Flury and Gautschi (1986) may be used to construct situations in which the function φ has more than one minimum. In such cases the FG^+ algorithm may converge to one of the local minima and fail to find the global one, and multiple starting points may be needed to locate the global minimum. However, as the discussion of conditions for uniqueness in Flury and Gautschi (1986) suggests, multiple minima are unlikely in situations where the statistical models of simultaneous diagonalizability are approximately satisfied. In other words, one would expect multiple minima to be a problem when the matrices are "far from being simultaneously diagonalizable".

Acknowledgement. We would like to thank a referee for his comments on an earlier draft of this paper. This work is dedicated to Walter Gautschi, whose expertise and advice were a rich source of inspiration to the first author at the early stages of simultaneous diagonalization algorithms.

REFERENCES

[1] DeLeeuw J., Pruzansky S. A new computational method for the weighted Euclidean distance model. *Psychometrika*, **43**:479–490, 1978.

[2] Flury B. *Common Principal Components and Related Multivariate Models.* Wiley, New York, 1988.

[3] Flury B., Gautschi W. An algorithm for simultaneous orthogonal transformation of several positive definite symmetric matrices to nearly diagonal form. *SIAM Journal on Scientific and Statistical Computing*, **7**:169–184, 1986.

[4] Flury B., Neuenschwander B.E. Principal component models for patterned covariance matrices, with applications to canonical correlation analysis of several sets of variables. In *Descriptive Multivariate Analysis*, Oxford University Press, 1994. W.J. Krzanowski, ed. In press.

[5] Golub G.H., Van Loan C.F. *Matrix Computations.* The Johns Hopkins University Press, Baltimore, 1983.

[6] Graybill F.A. *Introduction to Matrices with Applications in Statistics.* Wadsworth, Belmont (CA), 1969.

[7] Henderson H.V., Searle S.R. Vec and vech operators for matrices, with some uses in Jacobians and multivariate statistics. *Canadian Journal of Statistics*, **7**:65–81, 1979.

[8] Magnus J.R. *Linear Structures.* Charles Griffin & Co., London, 1988.

[9] Magnus J.R., Neudecker H. *Matrix Differential Calculus with Applications in Statistics and Econometrics.* Wiley, New York, 1988.

[10] Neuenschwander B.E. Common Principal Components for Dependent Random Vectors. Unpublished PhD thesis, University of Bern (Switzerland), Dept. of Statistics, 1991.

[11] Neuenschwander B.E. A common principal component model for dependent random vectors, 1994. Submitted for publication.

[12] Neuenschwander B.E., Flury B. Common canonical variates, 1994. Submitted for publication.

[13] Searle S.R. *Matrix Algebra Useful for Statistics.* Wiley, New York, 1982.

International Series of Numerical Mathematics, Vol. 119, ©1994 Birkhäuser

ON SOME APPROXIMATIONS FOR THE ZEROS
OF JACOBI POLYNOMIALS

Luigi Gatteschi*

Dedicated to Walter Gautschi on his sixty-fifth birthday

Abstract. This paper is concerned with some approximations and inequalities for the zeros of the Jacobi polynomials $P_n^{(\alpha,\beta)}(\cos\vartheta)$ in terms of the zeros of the Bessel function $J_\alpha(x)$. These approximations, obtained by using certain asymptotic expansions of $P_n^{(\alpha,\beta)}(\cos\vartheta)$ as $n \to \infty$, are uniformly valid only for the zeros lying in the interval $0 < \vartheta \leq \pi - \delta$. On the other hand, the inequalities hold for all the zeros.

1 INTRODUCTION

In two previous papers [6,10], as well as in a joint paper with Baratella [2], I have considered some asymptotic approximations for $n \to \infty$ of Jacobi polynomials $P_n^{(\alpha,\beta)}(\cos\vartheta)$ in terms of Bessel functions. Here, the properties of the zeros $\vartheta_{n,k}(\alpha,\beta), k = 1, 2, \ldots, n$, of $P_n^{(\alpha,\beta)}(\cos\vartheta)$, that have been derived or suggested by such approximations, will be summarized and improved. A complete description of earlier results can be found in Szegö [12, Chapter 6].

We shall consider the zeros of the function

$$u_n^{(\alpha,\beta)}(\vartheta) = \left(\sin\frac{\vartheta}{2}\right)^{\alpha+1/2} \left(\cos\frac{\vartheta}{2}\right)^{\beta+1/2} P_n^{(\alpha,\beta)}(\cos\vartheta) \tag{1.1}$$
$$(\alpha, \beta > -1, \quad 0 \leq \vartheta \leq \pi),$$

which satisfies the differential equation

$$\frac{d^2u}{d\vartheta^2} + \left(N^2 + \frac{1/4 - \alpha^2}{4\sin^2\vartheta/2} + \frac{1/4 - \beta^2}{4\cos^2\vartheta/2}\right) u = 0, \tag{1.2}$$

where

$$N = n + \frac{\alpha + \beta + 1}{2}. \tag{1.3}$$

1991 *Mathematics Subject Classification. Primary 33C45, 41A60*
This work was supported by the Consiglio Nazionale delle Ricerche of Italy and by the Ministero dell'Università e della Ricerca Scientifica e Tecnologica of Italy
*Dipartimento di Matematica dell'Università, Torino, Italy

Throughout this paper, we assume the parameters α and β to be bounded. The stronger conditions

$$-\frac{1}{2} \leq \alpha \leq \frac{1}{2}, \quad -\frac{1}{2} \leq \beta \leq \frac{1}{2}, \tag{1.4}$$

may be required when we are dealing with inequalities or explicit bounds for the error terms. It is also convenient to introduce a new parameter ν defined by

$$\nu = \left[N^2 + \frac{1 - \alpha^2 - 3\beta^2}{12} \right]^{1/2}. \tag{1.5}$$

We first recall the inequality (Gatteschi [7])

$$\vartheta_{n,k}(\alpha, \beta) \leq \frac{j_{\alpha,k}}{\nu}, \quad -\frac{1}{2} \leq \alpha, \beta \leq \frac{1}{2}, \tag{1.6}$$

for $k = 1, 2, \ldots, n$, where $j_{\alpha,k}$ is the kth zero of the Bessel function $J_\alpha(x)$. Here the equality sign holds when $\alpha^2 = \beta^2 = 1/4$.

Further, the following more general asymptotic results can be obtained by using a certain Hilb-type asymptotic representation of the Jacobi polynomials (Gatteschi [6]).

Theorem 1.1 (Gatteschi [8]) *Let $\alpha > -1$ and $\beta > -1$. Then, as $n \to \infty$,*

$$\vartheta_{n,k}(\alpha, \beta) = \frac{j_{\alpha,k}}{\nu} \left[1 - \frac{4 - \alpha^2 - 15\beta^2}{720\nu^4} \left(\frac{j_{\alpha,k}^2}{2} + \alpha^2 - 1 \right) \right] + j_{\alpha,k}^5 O(n^{-7}), \tag{1.7}$$

where the bound $O(n^{-7})$ holds uniformly for $k = 1, 2, \ldots, [\gamma n]$, where γ is a fixed constant in $(0, 1)$.

We point out the following consequence:

Corollary 1.2 *We have, as $n \to \infty$,*

$$\vartheta_{n,k}(\alpha, \beta) = \frac{j_{\alpha,k}}{\nu} + j_{\alpha,k}^2 O(n^{-5}), \tag{1.8}$$

uniformly for $k = 1, 2, \ldots, [\gamma n]$, where γ is a fixed constant in $(0, 1)$. Furthermore,

$$\vartheta_{n,k}(\alpha, \beta) = \frac{j_{\alpha,k}}{\nu} + O(n^{-5}), \tag{1.9}$$

if k is fixed and $n \to \infty$.

Next, we recall an asymptotic formula which was obtained independently by Frenzen and Wong [4,5] and Gatteschi [8].

Theorem 1.3 *Let $\alpha > -1$ and $\beta > -1$. Then, as $n \to \infty$,*

$$\vartheta_{n,k}(\alpha,\beta) = \frac{j_{\alpha,k}}{N} - \frac{1}{4N^2}\left[\left(\frac{1}{4} - \alpha^2\right)\left(\frac{2}{t} - \cot\frac{t}{2}\right)\right.$$
$$\left. + \left(\frac{1}{4} - \beta^2\right)\tan\frac{t}{2}\right] + tO(n^{-4}),$$

(1.10)

where

$$t = \frac{j_{\alpha,k}}{N}, \qquad N = n + \frac{\alpha+\beta+1}{2}.$$

The O-term is uniformly bounded for all values of $k = 1, 2, \ldots, [\gamma n]$, where γ is a fixed number in $(0,1)$.

Further information concerning this formula will be given in Section 3. Here we only remark that (1.10) reduces to (1.9) if k is fixed. Indeed, by observing that, as $t \to 0$,

$$\tan\frac{t}{2} = \frac{t}{2} + O(t^3), \qquad \frac{2}{t} - \cot\frac{t}{2} = \frac{t}{6} + O(t^3),$$

from (1.10) we obtain, when k is fixed and $n \to \infty$,

$$\vartheta_{n,k}(\alpha,\beta) = \frac{j_{\alpha,k}}{N}\left(1 - \frac{1 - \alpha^2 - 3\beta^2}{24N^2}\right) + O(n^{-5}),$$

which is the same as (1.9) written in terms of the parameter N.

The close relationship between the result in Theorem 1.1 and the one in Theorem 1.3 is a motivating point of this paper. Analogous relationships will be used in establishing some new results which involve the parameter ν.

2 AN ASYMPTOTIC EXPANSION OF JACOBI POLYNOMIALS

In this section we shall consider an asymptotic representation of the function $u_n^{(\alpha,\beta)}(\vartheta)$. A similar expansion, but involving the parameter N, has been given by Frenzen and Wong [4]. We shall use Olver's differential equation theory [11, Chapter 12], as Elliott [3] did for the asymptotics of $P_n^{(\alpha,\beta)}(z)$ when z is complex.

A very simple one-term representation is given by the following Hilb-type formula (Gatteschi [6])

$$u_n^{(\alpha,\beta)}(\vartheta) \equiv \left(\sin\frac{\vartheta}{2}\right)^{\alpha+1/2}\left(\cos\frac{\vartheta}{2}\right)^{\beta+1/2} P_n^{(\alpha,\beta)}(\cos\vartheta)$$
$$= 2^{-1/2}\nu^{-\alpha}\frac{\Gamma(n+\alpha+1)}{n!}\vartheta^{1/2}J_\alpha(\nu\vartheta) + \eta_0(\alpha,\vartheta), \qquad \alpha > -1,$$

(2.1)

where

$$\eta_0(\alpha,\vartheta) = \begin{cases} \vartheta^{\alpha+9/2}O(n^\alpha) & \text{if} \quad 0 \le \vartheta \le c/n, \\ \vartheta^3 O(n^{-3/2}) & \text{if} \quad c/n \le \vartheta \le \pi - \delta, \end{cases}$$

(2.2)

and c and δ are fixed positive constants.

For convenience, we put

$$A = \frac{1}{4} - \alpha^2, \qquad B = \frac{1}{4} - \beta^2. \tag{2.3}$$

In order to establish a formal series solution of (1.2), we first observe that, as $\vartheta \to 0$,

$$A\left(\frac{1}{4\sin^2 \vartheta/2} - \frac{1}{\vartheta^2}\right) + B\frac{1}{4\cos^2 \vartheta/2} = \frac{1 - \alpha^2 - 3\beta^2}{12} + O(\vartheta^2).$$

Then we write the equation (1.2) in the form

$$\frac{d^2 u}{d\vartheta^2} + \left(\nu^2 + \frac{A}{\vartheta^2}\right) u = f(\vartheta)u, \tag{2.4}$$

where

$$f(\vartheta) = A\left(\frac{1}{\vartheta^2} + \frac{1}{12} - \frac{1}{4\sin^2 \vartheta/2}\right) + B\left(\frac{1}{4} - \frac{1}{4\cos^2 \vartheta/2}\right). \tag{2.5}$$

According to Olver [11], equation (2.4) is formally satisfied by

$$u(\vartheta) = \vartheta^{1/2} J_\alpha(\nu\vartheta) \sum_{s=0}^{\infty} \frac{A_s(\vartheta)}{\nu^{2s}} + \vartheta^{3/2} J_{\alpha+1}(\nu\vartheta) \sum_{s=0}^{\infty} \frac{B_s(\vartheta)}{\nu^{2s+1}}. \tag{2.6}$$

The functions $A_s(\vartheta)$ and $B_s(\vartheta)$ are analytic for $0 \le \vartheta < \pi$ and defined recursively, starting from $A_0(\vartheta) = 1$, by

$$\vartheta B_s(\vartheta) = -\frac{1}{2} A_s'(\vartheta) - \frac{1+2\alpha}{2} \int_0^\vartheta \frac{A_s'(t)}{t} dt + \frac{1}{2} \int_0^\vartheta f(t) A_s(t) dt,$$

$$A_{s+1}(\vartheta) = \frac{1}{2} \vartheta B_s'(\vartheta) - \alpha B_s(\vartheta) - \frac{1}{2} \int \vartheta f(\vartheta) B_s(\vartheta) d\vartheta, \tag{2.7}$$

for $s = 1, 2, \ldots$, where the constant of integration is chosen so that $A_{s+1}(0) = 0$ for $s = 1, 2, \ldots$.

It remains to identify the above function $u(\vartheta)$ with $u_n^{(\alpha,\beta)}(\vartheta)$. This is easily done by using (2.1). We find the representation

$$u_n^{(\alpha,\beta)}(\vartheta) = 2^{-1/2} \nu^{-\alpha} \frac{\Gamma(n+a+1)}{n!} \left[\vartheta^{1/2} J_\alpha(\nu\vartheta) \sum_{s=0}^{m} \frac{A_s(\vartheta)}{\nu^{2s}} \right.$$

$$\left. + \vartheta^{3/2} J_{\alpha+1}(\nu\vartheta) \sum_{s=0}^{m-1} \frac{B_s(\vartheta)}{\nu^{2s+1}} + \varepsilon_m(\alpha, \vartheta) \right]. \tag{2.8}$$

We now set

$$W_{0,\vartheta}(\vartheta B_m) = \exp\{\frac{\lambda(\alpha)}{\nu}V_{0,\vartheta}(\vartheta B_0)\}V_{0,\vartheta}(\vartheta B_m),$$

where $V_{a,b}(\phi)$ denotes the total variation of the function ϕ in the interval (a,b) and $\lambda(\alpha)$ is a bounded function of α. Then the remainder term may be put in the form (see [11, p.444 and 448])

$$\varepsilon_m(\alpha,\vartheta) = \begin{cases} \vartheta^{\alpha+1/2}W_{0,\vartheta}(\vartheta B_m)O(\nu^{-2m-1+\alpha}) & \text{if } 0 \le \vartheta \le c/n, \\ W_{0,\vartheta}(\vartheta B_m)O(\nu^{-2m-3/2}) & \text{if } c/n \le \vartheta \le \pi - \delta. \end{cases} \tag{2.9}$$

The first coefficients in (2.8) are given by

$$A_0(\vartheta) = 1, \qquad \vartheta B_0(\vartheta) = \frac{1}{4}g(\vartheta),$$

$$A_1(\vartheta) = \frac{1}{8}g'(\vartheta) - \frac{1+2\alpha}{8}\frac{g(\vartheta)}{\vartheta} - \frac{1}{32}g^2(\vartheta), \tag{2.10}$$

where

$$g(\vartheta) = A\left(\cot\frac{\vartheta}{2} - \frac{2}{\vartheta} + \frac{\vartheta}{6}\right) - B\left(\tan\frac{\vartheta}{2} - \frac{\vartheta}{2}\right). \tag{2.11}$$

We observe that

$$g(\vartheta) = -\frac{4-\alpha^2-15\beta^2}{360}\vartheta^3 + O(\vartheta^5),$$

hence $\vartheta B_0(\vartheta) = O(\vartheta^3)$, and, from (2.7), $A_s(\vartheta) = O(\vartheta^2), \vartheta B_s(\vartheta) = O(\vartheta)$ for $s = 1, 2, \ldots$. Therefore, the variation $V_{0,\vartheta}(\vartheta B_m)$ is bounded. More precisely, we have

$$V_{0,\vartheta}(\vartheta B_m) = \int_0^\vartheta \left|(tB_m(t))'\right| dt = \begin{cases} O(\vartheta^3) & \text{if } m = 0, \\ O(\vartheta) & \text{if } m = 1, 2, \ldots. \end{cases}$$

Therefore, (2.9) yields

$$\varepsilon_m(\alpha,\vartheta) = \begin{cases} \vartheta^{\alpha+3/2}O(\nu^{-2m-1+\alpha}) & \text{if } 0 < \vartheta \le c/n, \\ \vartheta O(\nu^{-2m-3/2}) & \text{if } c/n \le \vartheta \le \pi - \delta, \end{cases} \tag{2.12}$$

for $m = 1, 2, \ldots$, and

$$\varepsilon_0(\alpha,\vartheta) = \begin{cases} \vartheta^{\alpha+7/2}O(\nu^{-1+\alpha}) & \text{if } 0 < \vartheta \le c/n, \\ \vartheta O(\nu^{-3/2}) & \text{if } c/n \le \vartheta \le \pi - \delta, \end{cases}$$

in the case $m = 0$. Note that this last bound, which corresponds to the remainder term in the Hilb-type formula (2.1), is slightly worse than (2.2).

If we set $m = 1$ in (2.8) we get

$$u_n^{(\alpha,\beta)}(\vartheta) = 2^{-1/2}\nu^{-\alpha}\frac{\Gamma(n+\alpha+1)}{n!}\left[\vartheta^{1/2}J_\alpha(\nu\vartheta)\left(1 + \frac{A_1(\vartheta)}{\nu^2}\right)\right.$$

$$\left. + \vartheta^{3/2}J_{\alpha+1}(\nu\vartheta)\frac{B_0(\vartheta)}{\nu} + \varepsilon_1(\alpha,\vartheta)\right], \tag{2.13}$$

where $B_0(\vartheta), A_1(\vartheta)$ and $\varepsilon_1(\alpha,\vartheta)$ may be evaluated by means of (2.10), (2.11) and (2.12).

The explicit determination of the other coefficients seems to be rather difficult.

3 ASYMPTOTICS AND INEQUALITIES FOR THE ZEROS

In this section we first derive from (2.13) an asymptotic representation of the zeros of $P_n^{(\alpha,\beta)}(\cos\vartheta)$. It is sufficient to consider the equation

$$\left(1 + \frac{A_1(\vartheta)}{\nu^2}\right) J_\alpha(\nu\vartheta) + \frac{\vartheta B_0(\vartheta)}{\nu} J_{\alpha+1}(\nu\vartheta) + \vartheta^{-1/2}\varepsilon_1(\alpha,\vartheta) = 0,$$

that is, from (2.10), (2.11) and (2.12), the equation

$$J_\alpha(\nu\vartheta) - \frac{1}{4\nu}[Aa(\vartheta) + Bb(\vartheta)]J_{\alpha+1}(\nu\vartheta) + \vartheta^{1/2}O(\nu^{-7/2}) = 0, \qquad (3.1)$$

where the functions $a(\vartheta)$ and $b(\vartheta)$ are defined by

$$a(\vartheta) = \frac{2}{\vartheta} - \cot\frac{\vartheta}{2} - \frac{\vartheta}{6}, \qquad b(\vartheta) = \tan\frac{\vartheta}{2} - \frac{\vartheta}{2}. \qquad (3.2)$$

The bound for the error term holds uniformly for $0 < \vartheta \leq \pi - \delta$.

We now put in (3.1)

$$\nu\vartheta = \nu\vartheta_{n,k}(\alpha,\beta) = j_{\alpha,k}(1+h),$$

and take into account that, from Corollary 1.2, $h = O(\nu^{-3})$ when $0 < \vartheta \leq \pi - \delta$. Since

$$J_\alpha(j_{\alpha,k}(1+h)) = hj_{\alpha,k}J'_\alpha(j_{\alpha,k}) + j^2_{\alpha,k}O(\nu^{-6}),$$

and

$$\begin{aligned}
J_{\alpha+1}(j_{\alpha,k}(1+h)) &= J_{\alpha+1}(j_{\alpha,k}) + j_{\alpha,k}O(\nu^{-3}) \\
&= -J'_\alpha(j_{\alpha,k}) + j_{\alpha,k}O(\nu^{-3}),
\end{aligned}$$

we easily obtain

$$hj_{\alpha,k}J'_\alpha(j_{\alpha,k}) + \frac{1}{4\nu}[Aa(j_{\alpha,k}/\nu) + Bb(j_{\alpha,k}/\nu)]J'_\alpha(j_{\alpha,k}) = j^{1/2}_{\alpha,k}O(\nu^{-4}).$$

Since

$$J'_\alpha(j_{\alpha,k}) = j^{-1/2}_{\alpha,k}O(1), \quad \alpha > -1,$$

with $O(1)$ bounded for $k = 1, 2, \ldots$, it follows that

$$h = -\frac{1}{4\nu}[Aa(j_{\alpha,k}/\nu) + Bb(j_{\alpha,k}/\nu)] + j_{\alpha,k}O(\nu^{-4}).$$

Thus, we can state the following ν-analogue of Theorem 1.3.

Theorem 3.1 *Let $\alpha > -1$ and $\beta > -1$. Then, as $n \to \infty$,*

$$\vartheta_{n,k}(\alpha,\beta) = \frac{j_{\alpha,k}}{\nu} - \frac{1}{4\nu^2}\left[A\left(\frac{2}{t} - \cot\frac{t}{2} - \frac{t}{6}\right)\right. \tag{3.3}$$
$$\left. +B\left(\tan\frac{t}{2} - \frac{t}{2}\right)\right] + j_{\alpha,k}O(\nu^{-5}),$$

where

$$t = \frac{j_{\alpha,k}}{\nu}, \quad A = \frac{1}{4} - \alpha^2, \quad B = \frac{1}{4} - \beta^2,$$

$$\nu = \left[\left(n + \frac{\alpha+\beta+1}{2}\right)^2 + \frac{1-\alpha^2-3\beta^2}{12}\right]^{1/2}.$$

The O-term is uniformly bounded for all values of $k = 1, 2, \ldots, [\gamma n]$, where γ is a fixed number in $(0,1)$.

Let $\vartheta_{n,k}^{(1)}$ and $\vartheta_{n,k}^{(2)}$ be the approximations of $\vartheta_{n,k} \equiv \vartheta_{n,k}(\alpha,\beta)$ obtained by omitting the O-terms in (1.10) and (3.3), respectively. We remark that, since $\vartheta_{n,k}^{(1)} = \vartheta_{n,k}^{(2)} + j_{\alpha,k}O(\nu^{-5})$, Theorem 3.1 cannot be considered an improvement of Theorem 1.3. A more exact characterization of these approximations is possible when α and β satisfy the inequalities (1.4). We recall that in this case we have (Gatteschi[9])

$$\vartheta_{n,k}^{(1)} = \frac{j_{\alpha,k}}{N} - \frac{1}{4N^2}\left[A\left(\frac{2}{t} - \cot\frac{t}{2}\right) + B\tan\frac{t}{2}\right] \le \vartheta_{n,k} \le \frac{j_{\alpha,k}}{\nu} \tag{3.4}$$
$$(t = j_{\alpha,k}/N, \quad k = 1, 2, \ldots, n).$$

Here we shall prove that a slightly better inequality holds with the lower bound $\vartheta_{n,k}^{(1)}$ replaced by $\vartheta_{n,k}^{(2)}$.

Theorem 3.2 *Let $-1/2 \le \alpha, \beta \le 1/2$. Then*

$$\vartheta_{n,k}^{(2)} = \frac{j_{\alpha,k}}{\nu} - \frac{1}{4\nu^2}\left[A\left(\frac{2}{t} - \cot\frac{t}{2} - \frac{t}{6}\right) + B\left(\tan\frac{t}{2} - \frac{t}{2}\right)\right] \le \vartheta_{n,k} \le \frac{j_{\alpha,k}}{\nu} \tag{3.5}$$
$$(t = j_{\alpha,k}/\nu, \quad k = 1, 2, \ldots, n).$$

Furthermore, this inequality is better than (3.4).

The proof of (3.5) is based on the well-known Sturm comparison theorem, in the form given by Szegö [12, p.19], applied to the differential equation (1.2), i.e.,

$$\frac{d^2u}{d\vartheta^2} + (N^2 + \frac{A}{4\sin^2\vartheta/2} + \frac{B}{4\cos^2\vartheta/2})u = 0. \tag{3.6}$$

In order to construct a comparison equation, we observe that (3.3) gives, as $n \to \infty$

$$j_{\alpha,k} = \nu\vartheta_{n,k} + \frac{1}{4\nu}\left[A\left(\frac{2}{\vartheta_{n,k}} - \frac{\vartheta_{n,k}}{6} - \cot\frac{\vartheta_{n,k}}{2}\right) + B\left(\tan\frac{\vartheta_{n,k}}{2} - \frac{\vartheta_{n,k}}{2}\right)\right] + O(\frac{1}{n^4}),$$

where the O-term is uniformly bounded with respect to k as in Theorem 3.1. Therefore, we set

$$h(\vartheta) = \nu\vartheta + \frac{1}{4\nu}\left[A\left(\frac{2}{\vartheta} - \frac{\vartheta}{6} - \cot\frac{\vartheta}{2}\right) + B\left(\tan\frac{\vartheta}{2} - \frac{\vartheta}{2}\right)\right], \qquad (3.7)$$

and assume as comparison equation

$$\frac{d^2 z}{d\vartheta^2} + F(\vartheta)z = 0, \qquad (3.8)$$

where

$$F(\vartheta) = \frac{1}{2}\frac{h'''}{h'} - \frac{3}{4}\left(\frac{h''}{h'}\right)^2 + A\left(\frac{h'}{h}\right)^2 + (h')^2,$$

which is satisfied by the function

$$z(\vartheta) = \left(\frac{h}{h'}\right)^{1/2} J_\alpha[h(\vartheta)]. \qquad (3.9)$$

Since the function $h(\vartheta)$ defined by (3.7) is slightly different from the function used in [9], we shall give only a sketch of the proof.

First, we show that

$$F(\vartheta) \geq N^2 + \frac{A}{4\sin^2\vartheta/2} + \frac{B}{4\cos^2\vartheta/2} \qquad (3.10)$$
$$(0 < \vartheta < \pi, \qquad \nu \geq 1).$$

We have

$$h'^2(\vartheta) = N^2 + \frac{A}{4\sin^2\vartheta/2} + \frac{B}{4\cos^2\vartheta/2} - \frac{A}{\vartheta^2} + \frac{1}{16\nu^2}\left(Aa' + Bb'\right)^2.$$

Hence,

$$F(\vartheta) - N^2 - \frac{A}{4\sin^2\vartheta/2} - \frac{B}{4\cos^2\vartheta/2}$$
$$= \frac{1}{2h'}F_1(\vartheta) + \frac{A(h'\vartheta + h)}{h^2\vartheta^2}F_2(\vartheta) + \frac{1}{16\nu^2}\left(Aa' + Bb'\right)^2,$$

where

$$F_1(\vartheta) = h''' - \frac{3}{2}\frac{h''^2}{h'}, \qquad F_2(\vartheta) = h'\vartheta - h,$$

and it suffices to show that $F_1(\vartheta) \geq 0$ and $F_2(\vartheta) \geq 0$.

To this end, we note that $a(\vartheta)$ and $b(\vartheta)$, defined by (3.2), are positive increasing functions for $0 < \vartheta < \pi$, as are all their derivatives. Moreover the following expansions hold [1, p.75]

$$a(\vartheta) = \frac{\vartheta^3}{360} + \frac{\vartheta^5}{15120} + \ldots + \frac{(-1)^{n-1}2B_{2n}}{(2n)!}\vartheta^{2n-1} + \ldots \qquad (|\vartheta| < 2\pi),$$

$$b(\vartheta) = \frac{\vartheta^3}{24} + \frac{\vartheta^5}{240} + \ldots + \frac{(-1)^{n-1}2(2^n - 1)B_{2n}}{(2n)!}\vartheta^{2n-1} + \ldots \qquad (|\vartheta| < \pi),$$

$$(3.11)$$

where B_{2n} is the $2n$-th Bernoulli number.

It is easily seen that $F_2(\vartheta) \geq 0$ for $0 < \vartheta < \pi$. Indeed, we have

$$F_2(\vartheta) = \frac{A}{4\nu}(a'\vartheta - a) + \frac{B}{4\nu}(b'\vartheta - b) \geq 0.$$

To examine $F_1(\vartheta)$, we first consider the interval $0 < \vartheta \leq \pi/2$, where we have

$$Aa''' + Bb''' \geq \frac{A}{60} + \frac{B}{4}, \qquad Aa'' + Bb'' \leq A\left(\frac{32}{\pi^3} - 1\right) + B \leq \frac{8}{\pi^3},$$

thus

$$F_1(\vartheta) = \frac{1}{4\nu}\left[Aa''' + Bb''' - \frac{3}{2}\frac{(Aa'' + Bb'')^2}{4\nu^2 + Aa' + Bb'}\right]$$

$$\geq \frac{1}{4\nu}\left\{\frac{A}{60} + \frac{B}{4} - \frac{3}{\nu^2}\left[A\left(\frac{32}{\pi^3} - 1\right) + B\right]\frac{1}{\pi^3}\right\} \geq 0.$$

Similar considerations may be applied to

$$4\nu(4\nu^2 + Aa' + Bb')F_1(\vartheta)$$

$$= 4\nu^2 Aa''' + A^2\left(a'a''' - \frac{3}{2}a''^2\right) + AB(a'b''' + a'''b')$$

$$+ 4\nu^2 Bb''' + B^2\left(b'b''' - \frac{3}{2}b''^2\right) - 3ABa''b''$$

for the interval $\pi/2 \leq \vartheta \leq \pi$. More precisely, it is straightforward to prove that

$$4\nu^2 Aa''' + A^2\left(a'a''' - \frac{3}{2}a''^2\right) + AB(a'b''' + a'''b') \geq A\left[4\nu^2 a'''\left(\frac{\pi}{2}\right) - \frac{3}{8}a''^2(\pi)\right] \geq 0.$$

Further, observing that

$$b'b''' - \frac{3}{2}b''^2 = -\frac{1}{4}\frac{\tan^2 \vartheta/2}{\cos^2 \vartheta/2},$$

we have

$$4\nu^2 Bb''' + B^2(b'b''' - \frac{3}{2}b''^2) - 3ABa''b'' = B\left[4\nu^2 b''' - \frac{B\tan^2 \vartheta/2}{4\cos^2 \vartheta/2} - 3Aa''b''\right]$$

$$\geq \frac{B}{\cos^2 \vartheta/2}\left[4\nu^2\left(\frac{1}{4} + \frac{3}{4}\tan^2 \vartheta/2\right) - \frac{1}{16}\tan^2 \vartheta/2 - \frac{3}{8}a''(\pi)\tan \vartheta/2\right] \geq 0,$$

which completes the proof of the nonnegativity of $F_1(\vartheta)$ in the interval $\pi/2 \leq \vartheta < \pi$. With both $F_1(\vartheta) \geq 0$ and $F_2(\vartheta) \geq 0$ established, (3.10) is proved.

By applying the Sturm-Szegö theorem, we may compare the positive zeros of the function defined by (3.9) with the zeros of the solution $u_n^{(\alpha,\beta)}(\vartheta)$ of (3.6). Finally, the same arguments as used in [9] prove the inequality (3.5).

The last statement of the theorem is easily verified. Indeed, we have

$$\vartheta_{n,k}^{(2)} = \frac{j_{\alpha,k}}{\nu} - \frac{1}{4\nu^2}\left[Aa(j_{\alpha,k}/\nu) + Bb(j_{\alpha,k}/\nu)\right]$$

$$\geq \frac{j_{\alpha,k}}{N}\left(1 - \frac{1-\alpha^2-3\beta^2}{24N^2}\right) - \frac{1}{4N^2}\left[Aa(j_{\alpha,k}/N) + Bb(j_{\alpha,k}/N)\right] = \vartheta_{n,k}^{(1)}.$$

We remark that the inequality (3.5) holds for all zeros $\vartheta_{n,k}(\alpha,\beta)$ of $P_n^{(\alpha,\beta)}(\cos \vartheta)$. It can be used as an approximation except for the largest values of k. However, the reflection formula

$$P_n^{(\alpha,\beta)}(\cos(\pi - \vartheta)) = (-1)^n P_n^{(\beta,\alpha)}(\cos \vartheta),$$

may be applied. Consequently, for the zeros we have

$$\vartheta_{n,k}(\alpha,\beta) = \pi - \vartheta_{n,n-k+1}(\beta,\alpha), \tag{3.12}$$

which allows us to approximate all the zeros.

In Table 1 the exact values of the zeros $\vartheta_{16,k}(-0.3, 0.4)$, $k = 1, 2, \ldots, 16$, are compared with the upper and lower bounds given by (3.5). Here, use has been made of the relationship (3.12), for $k = 1, 2, \ldots, 8$ and for $k = 9, 10, \ldots, 16$, respectively.

k	Lower bound	Exact value	Upper bound
1	0.11617 69274	0.11617 69304	0.11617 69334
2	0.30464 01233	0.30464 01313	0.30464 02325
3	0.49409 72096	0.49409 72230	0.49409 76820
4	0.68374 70499	0.68374 70697	0.68374 83287
5	0.87346 55912	0.87346 56186	0.87346 83357
6	1.06321 57989	1.06321 58363	1.06322 09382
7	1.25298 26087	1.25298 26596	1.25299 14266
8	1.44275 85069	1.44275 85770	1.44277 27774
9	1.63251 13088	1.63253 91044	1.63253 92339
10	1.82230 35653	1.82232 09419	1.82232 10334
11	2.01209 05942	2.01210 09182	2.01210 09842
12	2.20186 92492	2.20187 49329	2.20187 49809
13	2.39163 32367	2.39163 60190	2.39163 60535
14	2.58136 78944	2.58136 90154	2.58136 90392
15	2.77102 95874	2.77102 98980	2.77102 99129
16	2.96040 90175	2.96040 90481	2.96040 90552

Table 1: Zeros of $P_{16}^{(-0.3,0.4)}(\cos \vartheta)$

In the ultraspherical case $\alpha = \beta$ we have the following corollary.

Corollary 3.3 *Let $-1/2 \leq \alpha \leq 1/2$ and let $\vartheta_{n,k}(\alpha)$ be the nth zero of the ultraspherical polynomial $P_n^{(\alpha,\alpha)}(\cos \vartheta)$. Then,*

$$\frac{j_{\alpha,k}}{\nu} - \frac{1-4\alpha^2}{8\nu^2}\left(\frac{\nu}{j_{\alpha,k}} - \cot\frac{j_{\alpha,k}}{\nu} - \frac{j_{\alpha,k}}{3\nu}\right) \leq \vartheta_{n,k}(\alpha,\alpha) \leq \frac{j_{\alpha,k}}{\nu} \qquad (3.13)$$
$$(k = 1, 2, \ldots, [n/2]),$$

with

$$\nu = \left[(n + \alpha + 1/2)^2 + (1 - 4\alpha^2)/12\right]^{1/2}.$$

REFERENCES

[1] Abramowitz M., Stegun I.A. *Handbook of Mathematical Functions*. Applied Mathematics Series **55**. National Bureau of Standards, Washington, DC, 1954.

[2] Baratella P., Gatteschi L. The bounds for the error term of an asymptotic expansion of Jacobi polynomials. In *Orthogonal Polynomials and their Applications*, Proc. Internat. Symp., Segovia/Spain, 1986. Lect. Notes Math. **1329**:203–221, 1988.

[3] Elliott D. Uniform asymptotic expansions of the Jacobi polynomials and an associated function. *Math. Comp.*, **25**:309–315, 1971.

[4] Frenzen C.L., Wong R. A uniform asymptotic expansions of the Jacobi polynomials with error bounds. *Canad. J. Math.*, **37**:979–1007, 1985.

[5] Frenzen C.L., Wong R. The Lebesgue constants for Jacobi series. *Rend. Semin. Mat. Univ. Politec. Torino*, Fasc. Spec., Special Functions: Theory and Computation, 117–148, 1985.

[6] Gatteschi L. Una nuova rappresentazione asintotica dei polinomi di Jacobi. *Rend. Semin. Mat. Univ. Politec. Torino*, **27**:165–184, 1967–68.

[7] Gatteschi L. Una nuova disuguaglianza per gli zeri dei polinomi di Jacobi. *Atti Accad. Sci. Torino*, Cl. Sci. Fis. Mat. Natur., **103**:259–265, 1968–69.

[8] Gatteschi L. On the zeros of Jacobi polynomials and Bessel functions. *Rend. Semin. Mat. Univ. Politec. Torino*, Fasc. spec., Special Functions: Theory and Computation 149–177, 1985.

[9] Gatteschi L. New inequalities for the zeros of Jacobi polynomials. *SIAM J. Math. Anal.*, **18**:1549–1562, 1987.

[10] Gatteschi L. New error bounds for asymptotic approximations of Jacobi polynomials and their zeros. To appear in *Rend. Mat.*

[11] Olver F.W. *Asymptotics and Special Functions*. Academic Press, New York, 1974.

[12] Szegö G. *Orthogonal Polynomials*. Colloquium Publications, Vol. 23, 4th ed., Amer. Math. Soc., Providence, RI, 1975.

International Series of Numerical Mathematics, Vol. 119, ©1994 Birkhäuser

PREDICTION AND THE INVERSE OF TOEPLITZ MATRICES

I. Gohberg[*]   H. J. Landau[†]

Dedicated to Walter Gautschi on his 65th birthday

Abstract. Toeplitz matrices represent the discrete analogue of convolutions, and the problem of inverting them is often encountered. The inverse of a Toeplitz matrix is no longer Toeplitz. However, thanks to a formula of Gohberg and Semençul, it can be expressed in terms of two closely related triangular Toeplitz matrices. Here we use an analogy with predicting stationary stochastic processes to motivate a simple proof of this formula, as well as of the main facts in the classical trigonometric moment problem.

1 INTRODUCTION

A *Toeplitz* matrix is one whose entries are constant on diagonals: $c_{i,j} = c_{i-j}$, hence the transformation it defines is the discrete analogue of a convolution. Toeplitz matrices therefore serve naturally to model the action of linear digital filters, and the problem of inverting them is often encountered. The inverse of a Toeplitz matrix $\mathbf{C}_k$ is no longer Toeplitz. However, the *Gohberg-Semençul formula* [3, p. 86] expresses it in terms of two closely related triangular Toeplitz matrices. Specifically, for a Hermitian $\mathbf{C}_k$, when the solution $(v_0, \ldots, v_k)$ of

$$\mathbf{C}_k \begin{pmatrix} v_0 \\ \vdots \\ v_k \end{pmatrix} = \begin{pmatrix} 1 \\ 0 \\ \cdot \\ 0 \end{pmatrix}$$

has $v_0 \neq 0$, then

$$\mathbf{C}_k^{-1} = v_0^{-1}(\mathbf{L}_k \mathbf{L}_k^* - \Lambda_k \Lambda_k^*) \,,$$

where $\mathbf{L}_k$ and Λ_k are lower-triangular Toeplitz matrices having as first column $(v_0, \ldots, v_k)$ and $(0, \bar{v}_k, \ldots, \bar{v}_1)$, respectively, and $*$ denotes the adjoint. This expression is particularly useful for repeated application, since the Toeplitz form of $\mathbf{L}_k$ and Λ_k allows the right-hand side to be implemented by fast algorithms. There now exist many different proofs and generalizations of this formula, the arguments often involving calculation. Here we will get a clearer view of it by connecting the

[*]Tel-Aviv University, Tel-Aviv, Israel
[†]AT&T Bell Laboratories, Murray Hill, New Jersey 07974

Toeplitz matrix to a problem of predicting a stationary stochastic process. That the solution $(v_0, \ldots, v_k)$ can be interpreted as a best prediction has long been known [14, 15, 17]; our contribution is to draw further conclusions from this fact by passing to the question of covariance extension. More specifically, the so-called maximum entropy extension of the matrix has a simple inverse, and its kth section can be adjusted to invert the original finite matrix; the two components of the formula represent that section and its adjustment. This approach will, along the way, also yield easily the main facts in the classical trigonometric moment problem.

All these results follow from considering the problem in a suitable orthonormal basis. While prediction motivates this development, none of the arguments actually depend on it.

2   STATIONARY STOCHASTIC PROCESSES

A *stochastic process* (e.g. [17]) is a sequence $\{X_k\}$, $-\infty < k < \infty$, of random variables, together with a family of probability distributions specifying probabilities for the joint occurrence of values for finite subsets of the $\{X_k\}$. These probability distributions in turn determine the *expected value*, denoted by $\mathcal{E}(\cdot)$, for functions of the random variables. We will say that two such functions P and Q are *uncorrelated* if $\mathcal{E}(P\overline{Q}) = 0$. Throughout, we will assume that the mean value $\mathcal{E}(X_k)$ of the variables is zero.

We call the expectations $\mathcal{E}(X_i \overline{X}_j)$ *autocorrelation coefficients* or *covariances*. The process is termed *stationary in the wide sense* if, for all i and j, $\mathcal{E}(X_i \overline{X}_j)$ depends only on $(j - i)$:

$$\mathcal{E}(X_i \overline{X}_j) = c_{j-i} \ . \tag{2.1}$$

Suppose that, for some fixed r,

$$c_k \equiv \mathcal{E}(X_0 \overline{X}_k) \ , \quad 0 \leq k \leq r \ ,$$

is a block of correlation coefficients of such a process; by definition,

$$c_{-k} = \overline{c_k} \ .$$

Denoting by $\mathbf{C}_r$ the Hermitian Toeplitz matrix whose top row consists of $c_0, c_1, \ldots, c_r$, setting $\alpha = (a_0, \ldots, a_r)$, and letting (α, β) be the Euclidean scalar product

$$(\alpha, \beta) \equiv \sum_{k=0}^{r} \alpha_k \overline{\beta}_k \ ,$$

we find by (2.1),

$$(\mathbf{C}_r \alpha, \alpha) = \mathcal{E}\left(\left| \sum_{k=0}^{r} a_k X_k \right|^2 \right) \geq 0 \ . \tag{2.2}$$

Thus if none of the variables $\{X_k\}$ is a linear combination of a finite number of others, which we assume, $\mathbf{C}_r$ is *positive definite*.

3 PREDICTION

When the correlation coefficients $c_0, \ldots, c_r$ are given, we can use them to try to estimate the value of a given X_j from observed values of neighboring X_k. This is *prediction*, which we want to make as good as possible, in the sense of minimizing the average square of the error [17]. The prediction is called *forward* if the observations precede X_j, or *backward* if they follow X_j. Supposing the prediction to be a linear functional of the observations — no loss of generality for Gaussian processes — we seek

$$\min_{c_1, \ldots, c_r} \mathcal{E} \left(\left| X_0 + \sum_{k=1}^{r} c_k X_k \right|^2 \right) \qquad (3.1)$$

to determine residual error in the best backward prediction, which we denote by $X_0 + \sum_{k=1}^{r} e_k X_k$, and similarly

$$\min_{c_1, \ldots, c_r} \mathcal{E} \left(\left| \sum_{k=0}^{r-1} c_k X_k + X_r \right|^2 \right) \qquad (3.2)$$

for the best forward prediction error, $X_r + \sum_{k=0}^{r-1} p_k X_k$. These elements have a simple characterization if we interpret $\mathcal{E}(| \cdot |^2)$ as the square of a Hilbert space norm. Accordingly, for two vectors $\alpha = (\alpha_0, \ldots, \alpha_r)$ and $\beta = (\beta_0, \ldots, \beta_r)$, we define a *new scalar product*

$$[\alpha, \beta] \equiv \mathcal{E} \left(\left(\sum_{i=0}^{r} \alpha_i X_i \right) \overline{\left(\sum_{i=0}^{r} \beta_i X_i \right)} \right) = (\mathbf{C}_r \alpha, \beta) . \qquad (3.3)$$

We can also associate the variable X_k with z^k, $|z| = 1$, hence a vector α with the polynomial $A(z) = \sum_{k=0}^{r} \alpha_k z^k$, and thereby think of (3.3) as defining a Hilbert space scalar product also on (trigonometric) polynomials of degree r. By connecting stochastic processes with complex analysis, this idea has far-reaching consequences. By definition, two vectors, or polynomials, are orthogonal in the above scalar product if the corresponding linear combinations of $\{X_i\}$ are uncorrelated.

The minimization of (3.1) asks for the point in the subspace spanned by $X_1, \ldots, X_r$ that lies closest to X_0 in the norm defined by $[\,,\,]$. It is now obvious geometrically, and can of course be verified by a variation, that the best choice is characterized by having $X_0 + \sum_{k=1}^{r} e_k X_k$ orthogonal to that subspace, that is,

$$\left[X_0 + \sum_{k=1}^{r} e_k X_k, X_j \right] = 0 , \qquad 1 \le j \le r ,$$

or, by (3.3),

$$\mathbf{C}_r(1, e_1, \ldots, e_r) = (\gamma, 0, \ldots, 0) , \qquad (3.4)$$

for some γ, where, as we do henceforth, we have written all vectors in row form for notational convenience. Also for convenience we will, depending on context, use A_r indiscriminately to denote either a polynomial $A_r(z) = \sum_{k=0}^{r} \alpha_k z^k$ or the vector $(\alpha_0, \ldots, \alpha_r)$ of its coefficients. On setting $E_r = (1, e_1, \ldots, e_r)$, we have, from (3.3) and (3.4),

$$\gamma = [E_r, E_r] > 0 , \qquad (3.5)$$

hence γ is the prediction error of (3.1). $E_r(z)$ is known also as the Szegő polynomial of degree r: that of least norm having constant term 1. Analogously, the forward prediction is characterized by

$$\mathbf{C}_r(p_0, \ldots, p_{r-1}, 1) = (0, \ldots, 0, \gamma) .$$

A consequence of the symmetry of Toeplitz matrices is that

$$p_k = \overline{e_{r-k}} ,$$

a fact we can express compactly in terms of the corresponding polynomials $E_r(z)$ and $P_r(z)$, in which $e_0 = p_r = 1$, by $P_r(z) = z^r \overline{E_r(1/z)}$ or, since $1 = |z|^2 = z\overline{z}$, by

$$P_r(z) = z^r \overline{E_r(z)} . \qquad (3.6)$$

4 COVARIANCE EXTENSION

When only a finite sequence of correlations is given, a nondegenerate process is not completely specified, and it is natural to ask for possible continuations of the sequence to $c_{r+1}, c_{r+2}, \ldots$. While any extension which maintains the positive definiteness of the Toeplitz covariance matrices is allowed, the point of view of prediction suggests a particular choice. The guiding principle is that, in forming the best r-step prediction for the process, we have already used all the information contained in the covariances, and so are not entitled, simply by an arbitrary choice of c_{r+1}, to diminish the prediction error from what it had been. Since the prediction now comes from a bigger subspace, it will improve unless it coincides with E_r. To bring this about, we require that E_r, determined by (3.4), also satisfy

$$\mathbf{C}_{r+1}(e_0, \ldots, e_r, 0) = (\gamma, 0, \ldots, 0) . \qquad (4.1)$$

This defines c_{r+1} uniquely by

$$\overline{c}_{r+1} e_0 + \overline{c}_r e_1 + \cdots + \overline{c}_1 e_r = 0 ,$$

the condition needed for (4.1). An important consequence is that $\mathbf{C}_{r+1}$ remains positive definite or, equivalently, that the scalar product (3.3), when extended to

$(r+2)$-vectors — that is, to polynomials of degree $r+1$ — defines a positive norm. For by definition, or explicitly from (4.1), $E_{r+1}(z) = E_r(z)$, hence

$$[E_r(z), zB_r(z)] = [E_{r+1}(z), zB_r(z)] = 0 , \qquad (4.2)$$

with any B_r of degree r. As $E_r(0) = 1$, we can write any polynomial $A_{r+1}(z)$ of degree $r + 1$ in the form

$$A_{r+1}(z) = \alpha E_r(z) + zB_r(z) ,$$

whence by (4.2),

$$[A_{r+1}, A_{r+1}] = |\alpha|^2 [E_r, E_r] + [zB_r(z), zB_r(z)] ,$$

and $[zB_r, zB_r] = [B_r, B_r]$ by stationarity. Thus $[A_{r+1}, A_{r+1}] > 0$ unless $A_{r+1} \equiv 0$. Moreover, by (4.2), $E_r(z)$ is orthogonal to $zE_r(z)$; the forward predictor $P_{r+1}(z) = zP_r(z)$ by (3.6), and is similarly orthogonal to $P_r(z)$.

Continuing the extension by this principle produces a semi-infinite Toeplitz matrix $\mathbf{C}$, defined by a sequence of one-step extensions. It is known as the *maximum-entropy extension* of $\mathbf{C}_r$ because the Gaussian process which can be constructed to have these correlations maximizes the entropy rate

$$\lim_{k \to \infty} -\frac{1}{k} \int \cdots \int p \log p \; dx_1 \cdots dx_k$$

among all processes for which the sequence of covariances begins with $c_0, \ldots, c_r$ [12].

The semi-infinite matrix $\mathbf{C}$ allows the scalar product $[\, , \,]$ of (3.3), originally defined for polynomials of degree r, to be extended to the larger space $\mathcal{H}$ of arbitrary polynomials, together with their limits in the norm $[v, v] \equiv (\mathbf{C}v, v)$.

5 TWO ORTHOGONAL BASES

Because $\mathbf{C}$ is Toeplitz and $\mathbf{C}E_r = \gamma$, the coefficients of $\mathbf{C}z^k E_r$ — that is, of $\mathbf{C}$ applied to a shifted E_r — all vanish beyond z^k; thus $\mathbf{C}z^k E_r$ corresponds to a polynomial with leading term γz^k. By (3.3), it follows that $z^k E_r$ is orthogonal (in the scalar product $[\, , \,]$ of $\mathcal{H}$) to the span of $\{z^j\}$, $j > k$, hence to $z^j E_r(z)$ for $j > k$, and as the order in the scalar product is immaterial, also for $j < k$. We conclude that $\{\gamma^{-1/2} z^k E_r(z)\}$, $k \geq 0$, form an orthonormal set in $\mathcal{H}$. To find the span of that set, we consider the orthogonal complement, i.e., vectors $Q \in \mathcal{H}$ for which

$$0 = [z^k E_r, Q] = (\mathbf{C}z^k E_r, Q) , \qquad k \geq 0 . \qquad (5.1)$$

Supposing q_m to be the first nonvanishing component of Q, the right-hand side, for $k = m$, becomes γq_m, hence (5.1) implies that $Q = 0$. We conclude that the system forms an orthonormal basis for all of $\mathcal{H}$. This is the essential feature of the present extension.

As in (3.6), the best forward prediction of X_{r+k} from $X_0, \ldots, X_{r+k-1}$ is the mirror image of the best backward prediction of X_0 from $X_1, \ldots, X_{r+k}$, and therefore corresponds to $z^k P_r(z)$. Since this polynomial is necessarily orthogonal to $1, z, \ldots, z^{r+k-1}$, it is also orthogonal to $z^j P_r(z)$, $j < k$. Consequently, $\{\gamma^{-1/2} z^k P_r(z)\}$, $k \geq 0$, likewise form an orthonormal set in $\mathcal{H}$. As, in common with $P_r(z)$, all are orthogonal to $1, z, \ldots, z^{r-1}$, the orthogonal complement of their span contains all polynomials of degree $r-1$. If Q is orthogonal to $1, z, \ldots, z^{r-1}$, as well as to all $\{z^k P_r(z)\}$, $k \geq 0$, then since the leading coefficient of $P_r(z)$ does not vanish, Q is likewise orthogonal to z^r, $z^{r+1}, \ldots$, hence also to $\{z^k E_r(z)\}$, $k \geq 0$, but we have seen that this implies $Q = 0$. Consequently $\{\gamma^{-1/2} z^k P_r(z)\}$, $k \geq 0$, form an orthonormal basis for the entire orthogonal complement in $\mathcal{H}$ of the subspace of polynomials of degree $(r-1)$. By the identical argument, the orthogonal complement of polynomials of degree r is spanned by that basis for $k \geq 1$.

6   ORTHONORMAL EXPANSION

Using the basis $\{\gamma^{-1/2} z^k E_r\}$, we have, for any $Q \in \mathcal{H}$,

$$Q = \gamma^{-1} \sum_{k=0}^{\infty} [Q, z^k E_r] z^k E_r = \gamma^{-1} \sum_{k=0}^{\infty} (\mathbf{C}Q, z^k E_r) z^k E_r(z) , \qquad (6.1)$$

the coefficients $a_k \equiv [Q, z^k E_r]$ being square-summable:

$$\gamma^{-1} \sum_{k=0}^{\infty} |a_k|^2 = [Q, Q] .$$

On applying this to $Q = 1 = (1, 0, \ldots) \in \mathcal{H}$, we find

$$1 = \gamma^{-1} \sum_{k=0}^{\infty} a_k z^k E_r(z) = A(z) E_r(z) \qquad (6.2)$$

on $|z| = 1$, with $A(z) = \gamma^{-1} \sum a_k z^k$ and $\gamma^{-1} \sum |a_k|^2 = [1, 1] = c_0$. Since the coefficients $\{a_k\}$ are square-summable, $A(z)$ is square-integrable on $|z| = 1$, and extendible to $|z| < 1$ as an analytic function. By analytic continuation, (6.2) holds also in $|z| \leq 1$. It follows that $A(z) = 1/E_r(z)$ in $|z| \leq 1$, hence $A(z)$ is a nonvanishing rational function in $|z| \leq 1$; it can have no poles on $|z| = 1$, being square-integrable there. This means that also $E_r(z) \neq 0$ in $|z| \leq 1$, so has a bounded reciprocal:

$$|1/E_r(z)| \leq b < \infty , \qquad |z| = 1 . \qquad (6.3)$$

Returning to (6.1), set $\mathbf{C}Q = R$, so that $Q = \mathbf{C}^{-1}R$, obtaining

$$\mathbf{C}^{-1}R = \gamma^{-1} \sum (R, z^k E_r) z^k E_r = \gamma^{-1} \mathbf{L}\mathbf{L}^* R , \qquad (6.4)$$

or

$$\mathbf{C}^{-1} = \gamma^{-1}\mathbf{L}\mathbf{L}^{*} \, ,$$

with $\mathbf{L}$ the semi-infinite lower-triangular Toeplitz matrix, having $E_r = (e_0, \ldots,$ $e_r, 0, 0, \ldots)$ as its first column. $\mathbf{L}$ therefore forms a *banded* matrix. Since $\mathbf{L}$ and $\mathbf{L}^*$ are clearly bounded in ℓ^2, the Hilbert space of vectors with square-summable coefficients, so is $\mathbf{C}^{-1}$.

Now the result of applying $\mathbf{L}$ to a vector $(q_0, q_1, \ldots) \in \ell^2$ corresponds to the product $E_r(z)Q(z)$; consequently, $\mathbf{L}^{-1}$ corresponds to multiplication of $Q(z)$ by $1/E_r(z)$. Hence the norm of $\mathbf{L}^{-1}$ in ℓ^2 is given by

$$\frac{1}{2\pi}\int_{|z|=1}\left|\frac{1}{E_r(z)}Q(z)\right|^2 d\theta \leq \frac{b^2}{2\pi}\int_{|z|=1}|Q(z)|^2 d\theta = b^2\sum_{k=0}^{\infty}|q_k|^2 \, ,$$

the inequality stemming from (6.3). It follows that $\mathbf{L}^{-1}$ is bounded in ℓ^2, as therefore is

$$\mathbf{C} = (\mathbf{C}^{-1})^{-1} = \gamma(\mathbf{L}^{-1})^*\mathbf{L}^{-1} \, . \tag{6.5}$$

By (6.2), $\mathbf{L}^{-1}$ is represented by the semi-infinite lower-triangular Toeplitz matrix whose first column consists of $(a_0, a_1, \ldots)$.

7   THE TRIGONOMETRIC MOMENT PROBLEM

Given complex numbers $c_0, \ldots, c_r$, the trigonometric moment problem asks when they can be expressed as successive Fourier coefficients of a positive measure, that is, in the form

$$c_k = \int_{|z|=1} z^k d\mu(\theta) \, , \quad 0 \leq k \leq r \, , \quad d\mu \geq 0 \, . \tag{7.1}$$

This question has been the source for far-ranging developments in analysis [1, 13]. A necessary condition is the positive definiteness of the Toeplitz matrix $\mathbf{C}_r$, for if (7.1) holds, then

$$(\mathbf{C}_r\alpha, \alpha) = \int_{|z|=1}\left|\sum a_k z^k\right|^2 d\mu(\theta) \geq 0 \, .$$

Sufficiency of this condition is not easily established, but the preceding considerations give an immediate proof. For by (6.5),

$$c_k \equiv (\mathbf{C}z^k, 1) = \gamma(\mathbf{L}^{-1}z^k, \mathbf{L}^{-1}1) = \gamma\int\frac{z^k}{E_r(z)}\frac{1}{\overline{E_r(z)}}\frac{d\theta}{2\pi} \, ,$$

exhibiting $\gamma d\theta/2\pi|E_r(z)|^2$ as a measure that serves in (7.1).

8   THE GOHBERG-SEMENÇUL FORMULA

We can derive an expression for $\mathbf{C}_r^{-1}$ by modifying that for $\mathbf{C}^{-1}$. To this end, suppose that

$$\mathbf{C}_r X_r = Y_r \ ,$$

where $X_r = (x_0, \ldots, x_r)$ and $Y_r = (y_0, \ldots, y_r)$ are vectors of length $r+1$; we desire

$$X_r = \mathbf{C}_r^{-1} Y_r \ .$$

We extend X_r by zero beyond x_r, denoting the result by $(X_r|\,0)$. Then $\mathbf{C}(X_r|\,0)$ will coincide with Y_r in the first $r+1$ components, and will have other components thereafter, that is,

$$\mathbf{C}(X_r|\,0) = (Y_r|\,0) - (0\,|Q)$$

with Q as yet unknown. Applying $\mathbf{C}^{-1}$, we obtain

$$(X_r|\,0) + \mathbf{C}^{-1}(0\,|Q) = \mathbf{C}^{-1}(Y_r|\,0) \ . \tag{8.1}$$

Now $\mathbf{C}^{-1}(0\,|Q)$ is orthogonal in $\mathcal{H}$ to all polynomials R_r of degree r, since

$$[\mathbf{C}^{-1}(0\,|Q), (R_r|\,0)] = ((0\,|Q), (R_r|\,0)) = 0 \ ,$$

hence (8.1) is a decomposition of $\mathbf{C}^{-1}(Y_r|\,0)$ into its components in, and orthogonal to, the space of these polynomials. As we have seen, the latter subspace has $\{z^k P_r\}$, $k \geq 1$, as an orthonormal basis. Since in a Hilbert space the projection of an element onto a subspace is given by expansion of that element in any orthonormal basis for the subspace, we have

$$
\begin{aligned}
\mathbf{C}^{-1}(0\,|Q) &= \gamma^{-1} \sum_{k\geq 1} [\mathbf{C}^{-1}(Y_r|\,0), z^k P_r] z^k P_r \\
&= \gamma^{-1} \sum_{k\geq 1} ((Y_r|\,0), z^k P_r) z^k P_r = \gamma^{-1} \mathbf{\Lambda}\mathbf{\Lambda}^*(Y_r|\,0) \ , \tag{8.2}
\end{aligned}
$$

where $\mathbf{\Lambda}$ is a lower-triangular Toeplitz matrix with first column

$$z P_r(t) = (0, \overline{e}_r, \overline{e}_{r-1}, \ldots, \overline{e}_1, \overline{e}_0, 0, \ldots) \ .$$

On expressing $\mathbf{C}^{-1}(Y_r|\,0)$ by (6.4), and using (8.2), (8.1) becomes

$$X_r = \gamma^{-1}(\mathbf{L}_r \mathbf{L}_r^* - \mathbf{\Lambda}_r \mathbf{\Lambda}_r^*) Y_r \ ,$$

with $\mathbf{L}_r$ and $\mathbf{\Lambda}_r$ the $(r+1) \times (r+1)$ principal minors of $\mathbf{L}$ and $\mathbf{\Lambda}$, respectively. This is the Gohberg-Semençul formula [3], here obtained for positive definite matrices $\mathbf{C}_r$. It remains valid when $\mathbf{C}_r$ is not definite, provided only that the terms on the right-hand side can be defined.

Theorem (Gohberg-Semençul). *Let $\mathbf{C}_r$ be a Hermitian Toeplitz matrix. Suppose that $\mathbf{C}_r$, as well as its principal minor $\mathbf{C}_{r-1}$, are nonsingular, and let $(e_0, \ldots, e_r)$ denote the solution of*

$$\mathbf{C}_r(e_0, \ldots, e_r) = (1, 0, \ldots, 0) \ . \tag{8.3}$$

Then

$$\mathbf{C}_r^{-1} = e_0^{-1}(\mathbf{L}_r \mathbf{L}_r^* - \mathit{\Lambda}_r \mathit{\Lambda}_r^*) \ ,$$

with $\mathbf{L}_r$ and $\mathit{\Lambda}_r$ the $(r+1) \times (r+1)$ lower-triangular Toeplitz matrices having first columns $(e_0, \ldots, e_r)$ and $(0, \bar{e}_r, \ldots, \bar{e}_1)$, respectively.

Proof: We apply an analytic continuation, as in [3, p. 94]. For real α larger than any of the eigenvalues of $-\mathbf{C}_r$, the matrix $(\alpha \mathbf{I} + \mathbf{C}_r)$ is positive definite, hence

$$(\alpha \mathbf{I} + \mathbf{C}_r)^{-1} = e_0(\alpha)^{-1}(\mathbf{L}_n(\alpha)\mathbf{L}_n^*(\alpha) - \mathit{\Lambda}_n(\alpha)\mathit{\Lambda}_n^*(\alpha)) \ , \tag{8.4}$$

where the right-hand matrices are defined in terms of $(e_0(\alpha), \ldots, e_r(\alpha))$, the solution of (8.3) for $\alpha \mathbf{I} + \mathbf{C}_r$; we note the change of normalization from (3.4), which here leads to $[E_r, E_r] = e_0$, in place of (3.5). Now the left-hand side of (8.4) is an analytic (matrix-valued) function of α away from the eigenvalues of $-\mathbf{C}_r$; since $\mathbf{C}_r$ is not singular, $\alpha = 0$ lies in the domain of analyticity. Similarly, the entries $e_j(\alpha)$ on the right-hand side of (8.4) are, by Cramer's rule, quotients of polynomials in α, with denominator $\det(\alpha \mathbf{I} + \mathbf{C}_r)$, which likewise does not vanish at $\alpha = 0$; further, by the hypothesis concerning $\mathbf{C}_{r-1}$, neither does $e_0(\alpha) = \det(\alpha \mathbf{I} + \mathbf{C}_{r-1})/\det(\alpha \mathbf{I} + \mathbf{C}_r)$. Hence the right-hand side is also analytic in the complement of the same eigenvalues, and all its terms are well-defined away from the zeros of $\det(\alpha \mathbf{I} + \mathbf{C}_{r-1})$. We conclude that the identity (8.4) extends by analytic continuation to the entire domain of analyticity, hence in particular to $\alpha = 0$, establishing the theorem. ∎

We can also give a constructive proof. For this purpose, we note that we have used the context of prediction to motivate the choice of $\mathbf{C}$, but that we could have introduced that extension independently, without reference to processes. We can thus replicate the preceding constructions even when $\mathbf{C}$ is not positive, although $[\,,\,]$ no longer has a geometric meaning. Extending $\mathbf{C}_r$ as before, but only until $\mathbf{C}_{2r}$ since $\mathbf{C}$ is now not necessarily bounded, we obtain a decomposition of $\mathbf{C}_{2r}^{-1}(Y_r|0)$ analogous to (8.1), and show that it is unique by virtue of the invertibility of $\mathbf{C}_r$. The expansions (6.4) and (8.2) of the two components remain valid. We omit these details.

9 REMARKS

Toeplitz matrices arise in many areas, and problems involving them have a long history. To our knowledge, N. Levinson was the first to use the structure of a Toeplitz matrix to solve a system of linear equations recursively [14]. The first recursive algorithm for computing the elements of the inverse of a Toeplitz matrix was given by W. F. Trench [15]. Independently, such recursions were also proposed by L. M. Kutikov [10, 11]. The first global formula for the inverse of a Toeplitz matrix appeared in a paper of G. Baxter and I. I. Hirschman, Jr. [2], which was the starting point for the work of I. Gohberg and A. A. Semençul. The first results, valid only under additional restrictions, were published as an appendix by Semençul in the lecture notes [4]; the final form of the Gohberg-Semençul formula appeared in [6] (see also [3]). This global form shows how the structure of the original matrix is reflected in the inverse. The formula can also be deduced from the recursions in [15, 10]; for details, see [9, 16]. There also exist generalizations of it for the block-Toeplitz case [5] as well as for the continuous case [6]. The formulas mentioned above have an interesting interpretation and proofs in the framework of the theory of displacement rank discovered in [8] and developed in a number of subsequent papers; see also [7].

REFERENCES

[1] Akhiezer N.I. *The Classical Moment Problem*. Hafner, New York, 1965.

[2] Baxter G., Hirschman I.I. Jr. An explicit inversion formula for finite-section Wiener-Hopf operators. *Bull. Amer. Math. Soc.*, **70**:820–823, 1964.

[3] Gohberg I.C., Feldman I.A. *Convolution Equations and Projection Methods for their Solution*. Transl. Math. Monographs **41**, Amer. Math. Soc., Providence, R.I., 1974.

[4] Gohberg I.C., Feldman I.A. *Projection methods for solving Wiener-Hopf equations*. Acad. Sci. Moldovian SSR. Inst. Math. with Computing Center, Akad. Nauk Moldov. SSR, Kishiniev, 1967. (Russian.)

[5] Gohberg I., Heinig G. Inversion of finite Toeplitz matrices composed of elements of a noncommutative algebra. *Revue Roumaine de Mathématiques Pures et Appliquées*, **19**:623–663, 1974.

[6] Gohberg I., Semençul A.A. On the inverse of finite Toeplitz matrices and their continuous analogs. *Math. Issled.*, **7**:238–253, 1972. (Russian.)

[7] Heinig G., Rost K. Algebraic methods for Toeplitz-like matrices and operators. In *Operator Theory: Advances and Applications*, **13**, Birkhäuser, Basel, Boston, Berlin, 1984.

[8] Kailath T., Kung S.Y., Morf M. Displacement ranks of matrices and linear equations. *J. Math. Anal. Appl.*, **68**:395–407, 1979; *Bull. Amer. Math. Soc.*, **1**:769–773, 1979.

[9] Kailath T., Viera A., Morf M. Inverses of Toeplitz operators, innovations, and orthogonal polynomials. *SIAM Rev.*, **20**:106–119, 1978.

[10] Kutikov L.M. Inversion of correlation matrices and some problems of self-tuning. *Izv. Akad. Nauk SSSR, Ser. Tech.-Cybernetics*, **5**, 1965. (Russian.)

[11] Kutikov L.M. On the structure of matrices, inverse to correlation matrices for vector random processes. *J. of Computational Math. and Math. Physics*, **7**(4), 1967. (Russian.)

[12] Landau H.J. Maximum entropy and the moment problem. *Bull. Amer. Math. Soc.*, **16**:47–77, 1987.

[13] *Moments in Mathematics*. Proc. Symp. Appl. Math., **37**. Amer. Math. Soc., Providence, R.I., 1987. H.J. Landau, ed.

[14] Levinson N. The Wiener rms error criterion in filter design and prediction. *J. Math. Phys.*, **25**:261–278, 1947.

[15] Trench W.F. An algorithm for the inversion of finite Toeplitz matrices. *J. SIAM*, **12**:515–522, 1964.

[16] Trench W.F. A note on a Toeplitz inversion formula. *Lin. Alg. Appl.*, **129**:55–61, 1990.

[17] Yaglom A.M. *A Introduction to the Theory of Stationary Random Functions*. Prentice-Hall, Englewood Cliffs, NJ, 1962.

International Series of Numerical Mathematics, Vol. 119, ©1994 Birkhäuser

STABLE LOOK-AHEAD VERSIONS OF THE EUCLIDEAN AND CHEBYSHEV ALGORITHMS

William B. Gragg[*]   Martin H. Gutknecht[†]

Dedicated to Walter Gautschi on the occasion of his 65th birthday

Abstract. We first review the basic relations between the regular formal orthogonal polynomials (FOPs) for a sequence of moments (Markov parameters), the nonsingular leading principal submatrices of the moment matrix $\mathbf{M}$ (which is an infinite Hankel matrix), the distinct entries on the main diagonal of the Padé table for the symbol of $\mathbf{M}$ (which is the generating function or z-transform of the moments), the corresponding continued fraction (which is a J-fraction or a P-fraction), and the Euclidean algorithm for power series in ζ^{-1}, which in the generic case is seen to reduce to the Chebyshev algorithm. The underlying recurrences are a special case of the general recurrences that are the basis of the Cabay-Meleshko algorithm which, in contrast to the aforementioned tools, is (weakly) stable. While, in the Toeplitz solver terminology, the Cabay-Meleshko algorithm is of Levinson type, we also outline the corresponding $O(N^2)$ Schur-type algorithm and a related $O(N \log^2 N)$ algorithm. Finally, we sketch three look-ahead strategies of which two are applicable to the $O(N \log^2 N)$ algorithm also.

1 INTRODUCTION

Given a sequence of moments (Markov parameters),

$$\mu_k := \int_{-\infty}^{\infty} \zeta^k d\omega(\zeta) \quad (k = 0, 1, \ldots) \tag{1.1}$$

the Chebyshev algorithm [16, §III], [25], if it does not break down, produces the recurrence coefficients of the formal orthogonal polynomials (FOPs) q_n that belong to these moments and to the underlying measure $d\omega$. These coefficients can also be computed with the unsymmetric Lanczos algorithm for polynomials [51], also known as the Stieltjes procedure [25], [61] where, however, the moments need not be known in advance, but the information for determining the recurrence coefficients is obtained in a more stable way from other integrals with respect to the same measure as one proceeds. The breakdown conditions for both algorithms are the same: they no longer work when the FOPs do not satisfy the classic three-term recurrence. The Chebyshev algorithm has been the method of choice for the

[*]Naval Postgraduate School, Department of Mathematics, Monterey, CA 93943-5000, USA; gragg@math.nps.navy.mil

[†]Interdisciplinary Project Center for Supercomputing, ETH Zurich, ETH-Zentrum, CH-8092 Zurich, Switzerland; mhg@ips.ethz.ch

case where the measure itself is not known, but only the moments. Unfortunately, the problem of computing the measure or the recurrence coefficients of the FOPs is typically very ill conditioned [25], even when we assume a situation where the measure is determined uniquely.

The FOPs are the denominators of the diagonal Padé approximants of the generating function $h(\zeta) := \sum \mu_k \zeta^{-k-1}$ of the given moments. The classic three-term recurrence of the FOPs is therefore also the recurrence of the Padé forms (p_n, q_n), and, indirectly, of the Padé approximants p_n/q_n. It is valid if and only if h can be expanded into a Jacobi fraction (J-fraction). If this is not the case, h can still be expanded into another continued fraction called a P-fraction [53], [54] that was already known to Chebyshev [14]. The underlying recurrence is still a three-term recurrence with one polynomial coefficient, but that one may now be a polynomial of degree greater than 1. The construction of P-fractions can be understood as an application of the Euclidean algorithm to power series in ζ^{-1}. There is also an intimate relationship to the partial realization problem of systems theory where the Euclidean algorithm reappears in the form of the Berlekamp-Massey algorithm [3], [55]. These connections between P-fractions, distinct entries on the diagonal of the Padé table, the Euclidean algorithm, and the partial realization problem are by now well known, see, e.g., [8], [9], [31], [32], [49]. They extend to matrix-valued functions [1], [18], [23], [65], [66]. There is also an analogous connection between the Euclidean algorithm and a recurrence for the distinct entries on an antidiagonal of the Padé table [56]. That recurrence is just a variation of the diagonal one, however. It is a special case of the classical Kronecker algorithm for rational interpolation [50], [67], which, incidentally, was also discovered by Chebyshev before [13], but will not be discussed further here. Finally, there is a close relationship between the recurrence for P-fractions and a nongeneric extension of the Lanczos algorithm [31], [36]. In fact, the latter is equivalent to its polynomial form, which is the generalization of the Stieltjes procedure that was described in [36].

In theory these approaches work without any genericalness assumption. In particular, we need to point out that the Euclidean algorithm for power series is actually an extension of the Chebyshev algorithm to the general situation. It is mathematically equivalent to the nongeneric generalization of the Chebyshev algorithm proposed by Golub and Gutknecht [28], but the details of the two algorithms differ. However, all these algorithms are in general numerically unstable in that there are data that produce in floating-point arithmetic roundoff errors that are arbitrarily much larger than the condition number of the problem would suggest. This instability is due to the fact that some intermediate results can correspond to problems that are arbitrarily badly conditioned. This observation is also of importance for exact arithmetic, since some of these badly conditioned intermediate results can be arbitrarily large numbers requiring an arbitrary number of digits.

The key to stability is to avoid these ill-conditioned intermediate results. Of course, each well-conditioned FOP could be computed stably by Gaussian elimination with pivoting, but for q_N this would require $O(N^3)$ operations. In contrast, the algorithms mentioned above produce in $O(N^2)$ operations a finite subsequence of

232

FOPs. Moreover, the Euclidean algorithm (and hence, the Chebyshev algorithm) can be combined with recursive doubling and fast polynomial multiplication to reduce the complexity to $O(N \log^2 N)$.

The look-ahead Lanczos algorithm for polynomials introduced in [38] (which might be called a look-ahead Stieltjes procedure) generates a sequence of well-conditioned FOPs in $O(N^2)$ operations. It can be understood as a formal block Gram-Schmidt orthogonalization process, and constructs each new well-conditioned FOP by a recurrence that involves the FOPs from the two previous blocks. Its importance lies in the fact that by mapping the polynomials into Krylov space vectors we obtain a generalization of the nonsymmetric Lanczos algorithm that no longer suffers from the so-called serious breakdowns and near-breakdowns [21], [36], [38], [59], [60].

At the same time when the look-ahead Lanczos algorithm for polynomials was developed, Cabay and Meleshko [12], [57] conceived another (weakly) stable procedure for computing well-conditioned Padé approximants along a diagonal of the Padé table (and thus, a series of well-conditioned FOPs). This procedure and a variation of it that might be called a look-ahead Euclidean algorithm (or, look-ahead Chebyshev algorithm) are what we want to discuss here primarily. It still can be combined with recursive doubling and fast polynomial multiplication and performs then in $O(N \log^2 N)$ operations if we assume that the number of successive ill-conditioned FOPs (i.e., the block size) does not grow with N [10]. There are extensions of this approach to vector-valued and matrix-valued Padé approximation and rational interpolation; see, e.g., [2], [10], [11], [33], [34], [64]. However, in these more general settings, stability is still an issue.

Although we start from power series with complex coefficients, most of the results we cite are true for coefficients from an arbitrary field, and many even persist when the coefficients are from a noncommutative ring; see, e.g., Bultheel [7], Fuhrmann [23], Geddes [26], Van Barel and Bultheel [65], [66].

The algorithms mentioned above can all be used as fast solvers for linear systems with an $N \times N$ Hankel matrix $\mathbf{M}_N$, because the FOPs q_n $(n = 0, \ldots, N-1)$ provide us with a symmetric block UDL decomposition of the inverse of this matrix, $\mathbf{M}_N^{-1} = \mathbf{R}_N \mathbf{D}_N^{-1} \mathbf{R}_N^T$, in which the unit upper triangular matrix $\mathbf{R}_N$ contains in its columns the coefficients of the FOPs q_n. The matrix $\mathbf{D}_N$ is diagonal if and only if the classical three-term recurrence of the Jacobi fraction holds. On the other hand, such a diagonal $\mathbf{D}_N$ can be achieved if and only if all leading principal submatrices of $\mathbf{M}_N$ are nonsingular. In general $\mathbf{D}_N$ is block diagonal, but can have arbitrarily badly conditioned diagonal blocks, unless the conditioning of these blocks is kept under control, at the cost of increasing the size of some of these blocks.

Conversely, a number of fast Hankel solvers that have been introduced without reference to Padé approximation or continued fractions can be seen to be variations of some of the procedures mentioned above. We strongly recommend not to miss the Padé connection. Historically, orthogonal polynomials originated from the theory of continued fractions, which is today just part of the Padé theory. Not only the recursions, but many other theoretical results, are best understood

233

from the Padé theory. With this paper we also want to promote this view.

Notation. We denote the vector space of polynomials with complex coefficients by $\mathcal{P}$, and let $\mathcal{P}_m$ be the subspace of polynomials p of exact degree ∂p at most m. Similarly, $\mathcal{L}$ denotes the vector space of formal Laurent series $f(z) = \sum_{k=-\infty}^{\infty} \phi_k z^k$ with complex coefficients. $\mathcal{L}_m^* \subset \mathcal{L}$ is the subspace of elements f with $\phi_k = 0$ for $k > m$. In particular, $\mathcal{L}_0^*$ consists of all formal power series in ζ^{-1}. If $f \in \mathcal{L}_m^*$ and $\phi_m \neq 0$, then we say again that f has exact degree $\partial f = m$. The notation $f(\zeta) = O_-(\zeta^m)$ is equivalent to $f \in \mathcal{L}_m^*$.

2 MOMENTS, FORMAL ORTHOGONAL POLYNOMIALS, AND PADÉ APPROXIMANTS

Given a real or complex sequence of *moments* $\{\mu_k\}_{k=0}^{\infty}$, a linear functional Φ : $\mathcal{P} \to \mathbb{C}$ is determined by the values

$$\Phi(\zeta^k) := \mu_k \quad (k = 0, 1, \ldots) \tag{2.1}$$

it takes on the monomials.

The classical moment problems consist in finding, for real moments μ_k, a nonnegative *measure* $d\omega$ on the positive real axis (*Stieltjes moment problem*) or the real axis (*Hamburger moment problem*) such that

$$\mu_k = \Phi(\zeta^k) = \int_{-\infty}^{\infty} \zeta^k d\omega(\zeta) \quad (k = 0, 1, \ldots). \tag{2.2}$$

When the infinite Hankel *moment matrix* $\mathbf{M} := [\mu_{i+k}]_{i,k=0}^{\infty}$ is positive semidefinite, such a measure exists, and there is a finite (if $\mathbf{M}$ is singular) or an infinite sequence $\{q_n\}$ of *orthogonal polynomials* associated to it. These orthogonal polynomials satisfy $\partial q_n = n$ $(n = 0, 1, \ldots)$ and

$$\Phi(\zeta^k q_n) = 0 \quad \text{if } 0 \leq k < n, \tag{2.3}$$

$$\Phi(\zeta^n q_n) \neq 0. \tag{2.4}$$

This implies that

$$\Phi(q_m q_n) = \delta_{m,n} \delta_n, \tag{2.5}$$

where $\delta_{m,n}$ is the Kronecker symbol and $\delta_n \neq 0$.

A first step towards finding the measure consists in determining the recurrence coefficients of the orthogonal polynomials. These recurrence coefficients are the elements of a possibly infinite tridiagonal matrix, whose leading principal submatrices provide better and better approximations to the measure. It is well known that the eigenvalues of such a tridiagonal submatrix are the abscissas (*Gauss nodes*) of a *Gauss quadrature formula*, and that the first components of the normalized eigenvectors are proportional to the squares of the corresponding weights (*Gauss weights* or *Christoffel numbers*), the constant of proportionality being $\beta_0 := \mu_0 = \int d\omega(\zeta)$.

We consider a more general setting where the moments μ_k, and hence the functional Φ, may be complex. A polynomial q_n of exact degree n satisfying (2.3)–(2.4) need not exist for every n. In general, this is even true for real moments. We must also emphasize that, in general, the polynomials q_n are *not* complex orthogonal polynomials, even if they exist for all n. For complex orthogonal polynomials one needs a positive measure $d\omega$ and replaces (2.5) by $\int_{-\infty}^{\infty} \overline{q_m(\zeta)} q_n(\zeta) d\omega(\zeta) = \delta_{m,n} \delta_n$.

Therefore, in our setting, q_n is called a *formal orthogonal polynomial (FOP)*. If (2.3) is satisfied and q_n is unique up to scaling, q_n is called *regular*, and n is a *regular index*. Otherwise, we may define a FOP q_n of exact degree n by requiring that either (2.3) holds (in which case q_n is not unique up to scaling and is called *singular*) or that (2.3) holds in relaxed form, for $0 \leq k < n'$ with n' as large as possible (in which case q_n is said to be *deficient*). The existence of singular and deficient FOPs is closely linked to the block structure of the Padé table that belongs to the given moments; see [36].

Clearly, we can normalize a regular q_n to be monic. We set

$$q_n(\zeta) =: \sum_{k=0}^{n} \rho_{k,n} \zeta^k, \quad \text{where } \rho_{n,n} = 1, \tag{2.6}$$

and

$$\dot{\mathbf{q}}_n := [\ \rho_{0,n} \cdots \rho_{n-1,n}\ ]^T \in \mathbb{C}^n, \quad \mathbf{q}_n := \begin{bmatrix} \dot{\mathbf{q}}_n \\ 1 \end{bmatrix} \in \mathbb{C}^{n+1}, \tag{2.7}$$

and let

$$\mathbf{M}_n := \begin{bmatrix} \mu_0 & \mu_1 & \cdots & \mu_{n-1} \\ \mu_1 & & \ddots & \vdots \\ \vdots & \ddots & & \vdots \\ \mu_{n-1} & \cdots & \cdots & \mu_{2n-2} \end{bmatrix} \tag{2.8}$$

be the $n \times n$ leading principal submatrix of the infinite *moment matrix* $\mathbf{M}$, which has Hankel structure. Then condition (2.3) translates into the *Yule-Walker equation*

$$\mathbf{M}_n \dot{\mathbf{q}}_n = -\mathbf{m}_n, \quad \mathbf{m}_n := [\ \mu_n \quad \cdots \quad \mu_{2n-1}\ ]^T \in \mathbb{C}^n. \tag{2.9}$$

Hence, a regular FOP (i.e., a unique solution) exists if and only if $\mathbf{M}_n$ is nonsingular. If it exists, we consider additionally the unique solution $\grave{\mathbf{q}}_n \in \mathbb{C}^n$ of

$$\mathbf{M}_n \grave{\mathbf{q}}_n = \mathbf{e}_n, \quad \mathbf{e}_n := [\ 0 \quad \cdots \quad 0 \quad 1\ ]^T \in \mathbb{R}^n. \tag{2.10}$$

We let

$$\grave{\mathbf{q}}_n =: [\ \grave{\rho}_{0,n} \cdots \grave{\rho}_{n-1,n}\ ]^T, \quad \grave{q}_n(\zeta) =: \sum_{k=0}^{n-1} \grave{\rho}_{k,n} \zeta^k, \tag{2.11}$$

and say that $[\ \grave{q}_n \quad q_n\ ]$ is a normalized nth *regular pair*. (Heinig and Rost [43] call $\grave{\mathbf{q}}_n$ and $\mathbf{q}_n$ 'fundamental solutions'.)

Equation (2.10) means that $\dot{q}_n$, which has degree at most $n - 1$, satisfies

$$\Phi(\zeta^k \dot{q}_n) = 0 \ (0 \leq k < n - 1), \quad \Phi(\zeta^{n-1} \dot{q}_n) = 1. \tag{2.12}$$

Therefore, if $\dot{q}_n$ has exact degree $n - 1$, it is a regular FOP. In general, this need not be the case, however.

There is yet another way to express the orthogonality of the polynomials $\dot{q}_n$ and q_n. In terms of the *symbol*

$$h(\zeta) := \sum_{k=0}^{\infty} \frac{\mu_k}{\zeta^{k+1}}. \tag{2.13}$$

of the Hankel matrix $\mathbf{M}$, the two linear systems (2.10) and (2.9) can be expressed as

$$\underbrace{h(\zeta)\dot{q}_n(\zeta)}_{\in \mathcal{L}^*_{n-2}} = \underbrace{\dot{p}_n(\zeta)}_{\in \mathcal{P}_{n-2}} - \underbrace{g_n(\zeta)}_{\in \mathcal{L}^*_{-n}}, \quad \underbrace{h(\zeta)q_n(\zeta)}_{\in \mathcal{L}^*_{n-1}} = \underbrace{p_n(\zeta)}_{\in \mathcal{P}_{n-1}} - \underbrace{f_n(\zeta)}_{\in \mathcal{L}^*_{-n-1}}, \tag{2.14}$$

where

$$(\dot{p}_n, \dot{q}_n) \in \mathcal{P}_{n-2} \times \mathcal{P}_{n-1}, \quad (p_n, q_n) \in \mathcal{P}_{n-1} \times \mathcal{P}_n, \tag{2.15}$$

$$g_n \in \mathcal{L}^*_{-n}, \quad f_n \in \mathcal{L}^*_{-n-1}, \tag{2.16}$$

$$q_n \text{ monic}, \quad g_n(\zeta) = \zeta^{-n} + O_-(\zeta^{-n-1}). \tag{2.17}$$

This means that (p_n, q_n) and $(\dot{p}_n, \dot{q}_n)$ are (n, n) and $(n - 1, n - 1)$ *Padé forms*, respectively, of the symbol h at $\zeta = \infty$. The rational functions p_n/q_n and $\dot{p}_n/\dot{q}_n$, obtained by canceling common factors, are the corresponding *Padé approximants*[‡]. They lie on the main diagonal of the *Padé table* of h. The Padé form (p_n, q_n) is also called *regular* if the FOP q_n is regular. The Padé form $(\dot{p}_n, \dot{q}_n)$ is the upper-left neighbor of (p_n, q_n). Recall that both are unique up to scaling.

We call the formal power series f_n and g_n in ζ^{-1} the *residuals* of (p_n, q_n) and $(\dot{p}_n, \dot{q}_n)$, respectively. (In the literature, often the shifted series $\zeta^n f_n(\zeta)$ and $\zeta^n g_n(\zeta)$ are called residuals instead.) If we let

$$g_0(\zeta) :\equiv 1, \quad f_0(\zeta) := -h(\zeta), \tag{2.18}$$

the Padé conditions (2.14) can be written as

$$[\, g_0 \quad f_0 \,] \begin{bmatrix} \dot{p}_n & p_n \\ \dot{q}_n & q_n \end{bmatrix} = [\, g_n \quad f_n \,]. \tag{2.19}$$

[‡]Normally, Padé approximants are defined at $z = 0$. Here, substituting $\zeta = 1/z$ and introducing the reflected polynomials $p_n^*(z) := z^n p_n(1/z)$ and $q_n^*(z) := z^n q_n(1/z)$ with respect to $\mathcal{P}_n$, we see from (2.14) that p_n^*/q_n^* is an (n, n) Padé approximant of the formal power series $h^*(z) := h(1/z)$ at $z = 0$. Likewise, $\dot{p}_n^*$ and $\dot{q}_n^*$ (reflected in $\mathcal{P}_{n-1}$) yield an $(n - 1, n - 1)$ Padé approximant of h^*. Since $h^*(0) = 0$, $p_n^*(0) = \dot{p}_n^*(0) = 0$ also.

with g_n and f_n satisfying (2.16). More generally, this defines Padé forms and residuals of a *pair* of formal power series in ζ^{-1},

$$f_0 \in \mathcal{L}_{-1}^*, \quad g_0 \in \mathcal{L}_0^*, \quad \text{with } g_0(\zeta) = 1 + O_-(\zeta^{-1}). \tag{2.20}$$

Postmultiplying (2.19) by $[\ q_n \quad -\grave{q}_n\ ]^T$ we see that

$$g_0(\grave{p}_n q_n - p_n \grave{q}_n) = g_n q_n - f_n \grave{q}_n = 1 + O_-(\zeta^{-1}).$$

Dividing this by g_0 and recalling that the result must be a polynomial, we obtain the following *Frobenius identity* [22]:

$$\Delta_n := \det \begin{bmatrix} \grave{p}_n & p_n \\ \grave{q}_n & q_n \end{bmatrix} = 1 = \grave{p}_n(0)q_n(0) - p_n(0)\grave{q}_n(0). \tag{2.21}$$

Note that the determinant is a polynomial, which is here the constant 1. This 1 reflects the normalization in (2.6) and (2.10). In a different context, (2.21) is often called the *Bézout equation*, since it guarantees that the polynomials $\grave{q}_n$ and q_n are relatively prime. This is just one of a number of equivalent characterizations of regularity summarized in the following lemma.

Lemma 2.1 *The following conditions, in which the polynomials $\grave{p}_n$, $\grave{q}_n$, p_n, and q_n are assumed to satisfy (2.15) and (2.19) with (2.16), but not necessarily the normalization (2.17), are equivalent when $n > 0$:*

(i) *q_n is a regular FOP; i.e., $[\grave{q}_n \quad q_n]$ is a regular pair;*

(ii) *$\mathbf{M}_n$ is nonsingular;*

(iii) *the equations (2.9)–(2.10) have a unique pair of solutions;*

(iv) *the equations (2.9)–(2.10) have a pair of solutions;*

(v) *(p_n, q_n) is a regular Padé form;*

(vi) *every (n, n) Padé form (p_n, q_n) satisfies $\partial q_n = n$, and the residual g_n of every $(n-1, n-1)$ Padé form satisfies $\partial g_n = -n$;*

(vii) *the (n, n) Padé approximant p_n/q_n differs from its upper-left neighbor in the Padé table, the $(n-1, n-1)$ Padé approximant $\grave{p}_n/\grave{q}_n$;*

(viii) *$\grave{p}_n q_n - p_n \grave{q}_n \neq 0 \in \mathcal{P}$;*

(ix) *$\grave{p}_n(0)q_n(0) - p_n(0)\grave{q}_n(0) \neq 0$;*

(x) *the polynomials q_n and $\grave{q}_n$ are relatively prime.*

Proof: Most of these equivalences can be found either explicitly or implicitly in [30]. Those that do not refer to Padé approximation can also be found in [43] where, however, it took some effort to establish the surprising equivalence of (iii) and (iv). Nearly the full list was also given in [37]. Here, we have seen already that (i) $\Leftrightarrow$ (ii) $\Leftrightarrow$ (iii) $\Rightarrow$ (iv) and (i) $\Leftrightarrow$ (v) $\Leftrightarrow$ (vii) $\Leftrightarrow$ (viii) $\Leftrightarrow$ (ix). Moreover, the equivalence of (ix) and (x) follows immediately from the Frobenius identity (2.21). Finally, the equivalence of (iii) and (iv), and the one of (vi) and (vii), can be concluded from the general formula for Padé forms and their residuals [30], [37]. ∎

For $n = 0$, relation (2.19) is still correct if we let

$$\mathring{p}_0(\zeta) :\equiv 1, \quad \mathring{q}_0(\zeta) :\equiv 0, \quad p_0(\zeta) :\equiv 0, \quad q_0(\zeta) :\equiv 1. \qquad (2.22)$$

Then those statements of Lemma 2.1 that make sense for $n = 0$ are true for this index value. Hence, $[\mathring{q}_0 \quad q_0] := [0 \quad 1]$ should be considered as the 0th regular pair.

By multiplying equation (2.19) from the right by the inverse of the 2×2 matrix that appears in it we obtain

$$[g_0 \quad f_0] = [g_n \quad f_n] \begin{bmatrix} q_n & -p_n \\ -\mathring{q}_n & \mathring{p}_n \end{bmatrix}. \qquad (2.23)$$

This shows that a normalized regular pair $[\mathring{q}_n \quad q_n]$ and the corresponding residuals $[g_n \quad f_n]$ allow us to retrieve the data $[g_0 \quad f_0]$; we just have to determine first the numerators $[\mathring{p}_n \quad p_n]$ from the Frobenius identity (2.21), $\mathring{p}_n q_n - p_n \mathring{q}_n = 1$, a task that can be solved with the extended Euclidean algorithm; see §4. Any pair $[g_n \quad f_n]$ with $g_n(\zeta) = \zeta^{-n} + O_-(\zeta^{-n-1})$ will lead to the same first $2n$ moments of $h = -f_0/g_0$, since these are determined by p_n/q_n, in view of the Padé condition (2.14). However, there is a much more surprising result illustrating this dependency, namely an explicit formula for $\mathbf{M}_n^{-1}$ in terms of the regular pair $[\mathring{q}_n \quad q_n]$; see §7.

3 A RECURRENCE FOR FOPS AND PADÉ APPROXIMANTS

In our notation, the Padé condition (2.19), which is the definition of the regular pair $[\mathring{q}_n \quad q_n]$ and the corresponding numerators $[\mathring{p}_n \quad p_n]$ and residuals g_n and f_n, suggests a very general recurrence. Under the unnecessarily restrictive assumption that all q_n are regular this recurrence was given by Gragg *et al.* [29] who used it for a superfast Hankel solver, applicable to strongly regular matrices (i.e., to Hankel matrices whose leading principal submatrices are all nonsingular). This was a modification and improvement of the solver of Brent *et al.* [5] who used Padé forms on a staircase instead of a diagonal and thus needed even more regularity assumptions which, in general, are not fulfilled by positive definite Hankel matrices. In the generality aimed at here the recurrence was established by Cabay and

Meleshko [12], [57]. They also used it for a Hankel solver, though originally only for one of complexity $O(N^2)$, for which they established weak stability.

Theorem 3.1 Let $[\grave{q}_n \quad q_n]$ be a normalized nth regular pair for $[g_0 \quad f_0]$, so that (2.19) and (2.15)–(2.17) hold. Likewise, let $[\grave{q}_k^{(n)} \quad q_k^{(n)}]$ be a normalized kth regular pair for $\zeta^n[g_n \quad f_n]$, so that

$$[g_n \quad f_n] \begin{bmatrix} \grave{p}_k^{(n)} & p_k^{(n)} \\ \grave{q}_k^{(n)} & q_k^{(n)} \end{bmatrix} = [g_k^{(n)} \quad f_k^{(n)}] \tag{3.1}$$

with

$$(\grave{p}_k^{(n)}, \grave{q}_k^{(n)}) \in \mathcal{P}_{k-2} \times \mathcal{P}_{k-1}, \quad (p_k^{(n)}, q_k^{(n)}) \in \mathcal{P}_{k-1} \times \mathcal{P}_k, \tag{3.2}$$

$$g_k^{(n)} \in \mathcal{L}_{-n-k}^*, \quad f_k^{(n)} \in \mathcal{L}_{-n-k-1}^*, \tag{3.3}$$

$$q_k^{(n)} \text{ monic}, \quad g_k^{(n)}(\zeta) = \zeta^{-n-k} + O_-(\zeta^{-n-k-1}). \tag{3.4}$$

Then,

$$\begin{bmatrix} \grave{p}_{n+k} & p_{n+k} \\ \grave{q}_{n+k} & q_{n+k} \end{bmatrix} := \begin{bmatrix} \grave{p}_n & p_n \\ \grave{q}_n & q_n \end{bmatrix} \begin{bmatrix} \grave{p}_k^{(n)} & p_k^{(n)} \\ \grave{q}_k^{(n)} & q_k^{(n)} \end{bmatrix} \tag{3.5}$$

and

$$[g_{n+k} \quad f_{n+k}] := [g_n \quad f_n] \begin{bmatrix} \grave{p}_k^{(n)} & p_k^{(n)} \\ \grave{q}_k^{(n)} & q_k^{(n)} \end{bmatrix} \tag{3.6}$$

yield a normalized $(n+k)$th regular pair $[\grave{q}_{n+k} \quad q_{n+k}]$ for $[g_0 \quad f_0]$, as well as the corresponding numerators $[\grave{p}_{n+k} \quad p_{n+k}]$ and residuals $[g_{n+k} \quad f_{n+k}]$. The latter satisfy

$$[g_{n+k} \quad f_{n+k}] = [g_k^{(n)} \quad f_k^{(n)}]. \tag{3.7}$$

Every normalized regular pair with index greater than n can be constructed in this way.

Proof: It is readily verified that (2.15), (2.17) and (3.2), (3.4) imply that the new polynomials defined by (3.5) satisfy (2.15) and (2.17) with n replaced by $n + k$. Multiplying (2.19) from the right by the 2×2 polynomial matrix from (3.1) and inserting (3.1) and (3.5) shows that also (2.19) is satisfied with n replaced by $n + k$. Finally, comparing (3.1) and (3.6) makes clear that (3.7) holds and that, in view of (3.4), $g_{n+k}(\zeta) = \zeta^{-n-k} + O_-(\zeta^{-n-k-1})$, which means that $(\grave{p}_{n+k}, \grave{q}_{n+k})$ is correctly normalized.

If we give up normalization, then, for every $k > 0$, $(\grave{p}_k^{(n)}, \grave{q}_k^{(n)})$ and $(p_k^{(n)}, q_k^{(n)})$ can be chosen as $(k-1, k-1)$ and (k, k) Padé forms of $[g_n \quad f_n]$, and the recurrence formula (3.5) will still yield an $(n+k-1)$ and an $(n+k)$ Padé form of $[g_0 \quad f_0]$. By assumption q_n is regular, hence the determinant of the first factor on the right side of (3.5) is not the zero polynomial. Consequently, the determinant of the second factor is not the zero polynomial if and only if the determinant on the left side is not the zero polynomial, i.e., $q_k^{(n)}$ is regular if and only if q_{n+k} is regular. If they are regular, normalization yields an $(n+k)$th regular pair $[\, \mathring{q}_{n+k} \quad q_{n+k}\,]$. This shows that every normalized regular pair can be found in this way. ∎

The condition (3.1) for the 'recurrence coefficients', i.e., for the Padé forms $(\mathring{p}_k^{(n)}, \mathring{q}_k^{(n)})$ and $(p_k^{(n)}, q_k^{(n)})$, translates easily into a pair of linear systems for the coefficients of these Padé forms. Let

$$g_n(\zeta) =: \sum_{j=0}^{\infty} \gamma_{j,n} \zeta^{-n-j}, \quad f_n(\zeta) =: \sum_{j=1}^{\infty} \phi_{j,n} \zeta^{-n-j}, \tag{3.8}$$

with $\gamma_{0,n} = 1$, and consider the $2k \times 2k$ *Sylvester matrix*

$$\mathbf{S}_k(g_n, f_n) := \begin{bmatrix} 0 & \cdots & 0 & 1 & 0 & \cdots & \cdots & 0 \\ \vdots & & \iddots & 1 & \gamma_{1,n} & \vdots & & 0 & \phi_{1,n} \\ 0 & & \iddots & \iddots & \vdots & \vdots & \iddots & \iddots & \vdots \\ 1 & & \iddots & & \gamma_{k-1,n} & 0 & \iddots & & \phi_{k-1,n} \\ \gamma_{1,n} & & & \iddots & \gamma_{k,n} & \phi_{1,n} & & \iddots & \phi_{k,n} \\ \vdots & \iddots & \iddots & & \vdots & \vdots & \iddots & \iddots & \vdots \\ \gamma_{k-1,n} & \iddots & & & \vdots & \phi_{k-1,n} & \iddots & & \vdots \\ \gamma_{k,n} & \cdots & & \cdots & \gamma_{2k-1,n} & \phi_{k,n} & \cdots & \cdots & \phi_{2k-1,n} \end{bmatrix}. \tag{3.9}$$

Moreover, let

$$\mathbf{e}_{2k} := [\, 0 \quad \cdots \quad 0 \quad 1\,]^T \in \mathbb{R}^{2k}, \quad \mathbf{f}_{2k,n} := -[\, \phi_{1,n} \cdots \phi_{2k,n}\,]^T, \tag{3.10}$$

and let

$$\mathring{\mathbf{p}}_k^{(n)}, \quad \mathring{\mathbf{q}}_k^{(n)}, \quad \mathbf{p}_k^{(n)} \in \mathbb{C}^k, \quad \mathbf{q}_k^{(n)} = \begin{bmatrix} \mathring{\mathbf{q}}_k^{(n)} \\ 1 \end{bmatrix} \in \mathbb{C}^{k+1} \tag{3.11}$$

be the coefficient vectors of the polynomials $\mathring{p}_k^{(n)}$, $\mathring{q}_k^{(n)}$, $p_k^{(n)}$, and $q_k^{(n)}$, the first one being augmented by a zero component. Then the conditions from (3.1) can be written as one matrix equation

$$\mathbf{S}_k(g_n, f_n) \begin{bmatrix} \mathring{\mathbf{p}}_k^{(n)} & \mathbf{p}_k^{(n)} \\ \mathring{\mathbf{q}}_k^{(n)} & \mathbf{q}_k^{(n)} \end{bmatrix} = [\, \mathbf{e}_{2k} \quad \mathbf{f}_{2k,n}\,]. \tag{3.12}$$

This summarizes two linear systems that are analogous to (2.9) and (2.10), but of double size and with a Sylvester matrix instead of a Hankel matrix.

When $k = 1$, which is the generic case, (3.12) reduces to

$$\begin{bmatrix} 1 & 0 \\ \gamma_{1,n} & \phi_{1,n} \end{bmatrix} \begin{bmatrix} \mathring{\mathbf{p}}_0^{(n)} & \mathbf{p}_0^{(n)} \\ \mathring{\mathbf{q}}_0^{(n)} & \mathbf{q}_0^{(n)} \end{bmatrix} = \begin{bmatrix} 0 & -\phi_{1,n} \\ 1 & -\phi_{2,n} \end{bmatrix}, \tag{3.13}$$

from which it follows that

$$
\begin{bmatrix} \dot{\mathbf{p}}_0^{(n)} & \mathbf{p}_0^{(n)} \\ \dot{\mathbf{q}}_0^{(n)} & \mathbf{q}_0^{(n)} \end{bmatrix} = \begin{bmatrix} 0 & -\phi_{1,n} \\ \dfrac{1}{\phi_{1,n}} & \gamma_{1,n} - \dfrac{\phi_{2,n}}{\phi_{1,n}} \end{bmatrix}. \tag{3.14}
$$

Clearly, this requires that $\phi_{1,n} \neq 0$. In fact, $\mathbf{S}_1(g_n, f_n)$ is nonsingular if and only if this holds.

The two systems (3.12) are actually easily reduced to systems of type (2.9) and (2.10). We just have to form the first $2k$ coefficients of the negative quotient of the residuals, $h_n := -f_n/g_n$. This is also readily verified in terms of matrices. By post-multiplication with the $2k \times 2k$ block reversal matrix

$$
\mathbf{J}_{k,k} := \begin{bmatrix} \mathbf{J}_k & \mathbf{O} \\ \mathbf{O} & \mathbf{J}_k \end{bmatrix}, \quad \mathbf{J}_k := [\delta_{k-i+1,j}]_{i,j=1}^{k}, \tag{3.15}
$$

the $2k \times 2k$ Sylvester matrix $\mathbf{S}_k(f,g)$ of two power series f and g becomes a 1×2 block matrix with Toeplitz (instead of Hankel) blocks and can be written as

$$
\mathbf{S}_k(f,g) = \begin{bmatrix} \mathbf{T}_k(g) & \mathbf{T}_k(f) \\ \mathbf{T}_k(\zeta^{-k}g) & \mathbf{T}_k(\zeta^{-k}f) \end{bmatrix} \mathbf{J}_{k,k}. \tag{3.16}
$$

Here $\mathbf{T}_k(g)$ denotes the $k \times k$ leading principal submatrix of the infinite Toeplitz matrix with symbol g. Note that $\mathbf{T}_k(g)$ and $\mathbf{T}_k(f)$ are lower triangular, while $\mathbf{T}_k(\zeta^{-k}g)$ and $\mathbf{T}_k(\zeta^{-k}f)$ are not. In fact, $\mathbf{T}_k(\zeta^{-k}f)\mathbf{J}_k =: \mathbf{H}_k(f)$ is the $k \times k$ leading principal submatrix of the infinite Hankel matrix with symbol $f(\zeta) = \sum_{k=1}^{\infty} \phi_k \zeta^{-k}$. In view of $\mathbf{T}_k(f)\mathbf{T}_k^{-1}(g) = \mathbf{T}_k^{-1}(g)\mathbf{T}_k(f) = \mathbf{T}_k(f/g)$ and

$$
\mathbf{T}_{2k}^{-1}(g)\mathbf{S}_k(f,g) = \begin{bmatrix} \mathbf{I}_k & \mathbf{T}_k(f/g) \\ \mathbf{O} & \mathbf{T}_k(\zeta^{-k}f/g) \end{bmatrix} \mathbf{J}_{k,k}, \tag{3.17}
$$

the two linear systems (3.12) can be efficiently transformed into reduced systems of type (2.9) and (2.10) for the coefficient vectors of the denominator polynomials $\dot{q}_k^{(n)}$ and $q_k^{(n)}$, respectively. Since $\mathbf{T}_{2k}(g)$ is triangular, there is a simple recursion for the coefficients of the quotient series f/g. Moreover, when k is large, one can use the FFT based fast polynomial division.

Using the matrix $\mathbf{S}_k(f,g)$ we seemingly avoid forming the quotient f/g. However, the structure of $\mathbf{S}_k(f,g)$ suggests to start solving (3.12) with a block Gauss elimination step with the upper-left $k \times k$ block as pivot. If we choose diagonal blocks $\mathbf{J}_k$ in the block upper triangular factor, we still obtain $\mathbf{T}_k(f/g)\mathbf{J}_k$ as the $(1,2)$-block, but the Schur complement is now $\mathbf{T}_k(\zeta^{-k}f) - \mathbf{T}_k(\zeta^{-k}g)\mathbf{T}_k(f/g)$. So the Schur complement does not require exactly the same operations as in (3.17), but similar ones: only the first k coefficients of the quotient are required now, but instead a product and a difference of matrices have to be computed.

241

4 J-FRACTIONS, P-FRACTIONS, AND THE EXTENDED EUCLI-DEAN ALGORITHM FOR POWER SERIES

The recurrences of Theorem 3.1 are generalizations of the classical procedure, known to Chebyshev already [14], for expanding a formal power series h into a continued fraction and of the recurrences for computing and evaluating the numerators and denominators of the convergents of this continued fraction. This classical procedure can be understood as an extension of the Euclidean algorithm. While in the Euclidean algorithm the quotients f_n/g_n are not formed, Magnus [53], [54] did form these quotients in his procedure for expanding h into a P-fraction. However, these are just two minor variations of the same process. In the case where all n are regular indices, the P-fraction reduces to a J-fraction, and the Euclidean algorithm to the Chebyshev algorithm. In this section we want to look at these relationships in detail.

For a moment, assume that for all n, q_n is a regular FOP. Then the recursion (3.5) can be applied for all n with $k = 1$. (Recall that this requires that $\phi_{1,n} \neq 0$ for all n.) Then the triples p_n, q_n, f_n and $\dot{p}_{n+1}, \dot{q}_{n+1}, g_{n+1}$ only differ in normalization:

$$(\dot{p}_{n+1}, \dot{q}_{n+1}) = \frac{1}{\phi_{1,n}}(p_n, q_n), \quad g_{n+1} = \frac{1}{\phi_{1,n}}f_n. \tag{4.1}$$

Hence

$$\begin{bmatrix} p_n & p_{n+1} \\ q_n & q_{n+1} \end{bmatrix} := \begin{bmatrix} p_{n-1} & p_n \\ q_{n-1} & q_n \end{bmatrix} \begin{bmatrix} 0 & -\beta_n \\ 1 & \zeta - \alpha_{n+1} \end{bmatrix} \quad (n \geq 0), \tag{4.2}$$

where

$$\beta_n := \frac{\phi_{1,n}}{\phi_{1,n-1}}, \quad \alpha_{n+1} := \frac{\phi_{2,n}}{\phi_{1,n}} - \frac{\phi_{2,n-1}}{\phi_{1,n-1}}, \tag{4.3}$$

with $\phi_{1,-1} := -1$, $\phi_{k,0} := \phi_k$, so that $\beta_0 = -\phi_1 = \mu_0$, and

$$\begin{bmatrix} p_{-1} & p_0 \\ q_{-1} & q_0 \end{bmatrix} := \begin{bmatrix} -\dot{p}_0 & p_0 \\ -\dot{q}_0 & q_0 \end{bmatrix} = \begin{bmatrix} -1 & 0 \\ 0 & 1 \end{bmatrix}. \tag{4.4}$$

Eliminating the trivial part of (4.2) we find the well known generic diagonal recurrences for Padé forms and FOPs,

$$\begin{aligned} p_{n+1}(\zeta) &= (\zeta - \alpha_{n+1})p_n(\zeta) - \beta_n p_{n-1}(\zeta), \\ q_{n+1}(\zeta) &= (\zeta - \alpha_{n+1})q_n(\zeta) - \beta_n q_{n-1}(\zeta). \end{aligned} \tag{4.5}$$

They reveal that p_n and q_n are the nth numerator and denominator of the *Jacobi fraction (J-fraction)* of $h = -f/g$,

$$h(\zeta) = -\frac{f_0(\zeta)}{g_0(\zeta)} = \cfrac{\beta_0}{\zeta - \alpha_1} - \cfrac{\beta_1}{\zeta - \alpha_2} - \cdots, \tag{4.6}$$

i.e.,

$$\frac{p_n(\zeta)}{q_n(\zeta)} = \frac{\beta_0}{\left|\zeta - \alpha_1\right.} - \cdots - \frac{\beta_{n-1}}{\left|\zeta - \alpha_n\right.}. \tag{4.7}$$

Recall that this J-fraction does not exist if $\phi_{1,n} = 0$ for some n. It may have very large $|\alpha_{n+1}|$ and $|\beta_{n+1}|$ if $|\phi_{1,n}|$ is small.

In exact arithmetic (with no restriction on the size of numbers) it would make sense to treat the case $\phi_{1,n} = 0$ explicitly. Assume that

$$\phi_{0,n} = \cdots = \phi_{k-1,n} = 0, \quad \phi_{k,n} \neq 0. \tag{4.8}$$

Then $\mathbf{S}_1, \ldots, \mathbf{S}_{k-1}$ are singular, but $\mathbf{S}_k$ is nonsingular. In fact, with $f := f_n, g := g_n$ (3.17) reduces then to

$$\mathbf{T}_{2k}^{-1}(g)\mathbf{S}_k(f,g) = \begin{bmatrix} \mathbf{J}_k & \mathbf{O} \\ \mathbf{O} & \mathbf{H}_k(f/g) \end{bmatrix} \tag{4.9}$$

with the nonsingular lower right triangular Hankel matrix $\mathbf{H}_k(f/g)$.

The linear system $\mathbf{H}_k(f/g)\grave{\mathbf{q}}_k^{(n)} = \mathbf{e}_k$ then has the explicit solution $\grave{\mathbf{q}}_k^{(n)} = \mathbf{e}_1/\phi_{k,n}$. Therefore the left system in (3.12) leads to

$$\grave{p}_k^{(n)}(\zeta) \equiv 0, \quad \grave{q}_k^{(n)}(\zeta) \equiv \frac{1}{\phi_{k,n}}. \tag{4.10}$$

Consequently,

$$(\grave{p}_{n+k}, \grave{q}_{n+k}) = \frac{1}{\phi_{k,n}}(p_n, q_n), \quad g_{n+k} = \frac{f_n}{\phi_{k,n}}. \tag{4.11}$$

In this way, we can proceed from any regular Padé form (p_n, q_n) and its upper-left neighbor $(\grave{p}_n, \grave{q}_n)$ to the next one, (p_{n+k}, q_{n+k}) and its upper-left neighbor $(\grave{p}_{n+k}, \grave{q}_{n+k})$, where k is determined by (4.8).

Let $\{n_j\}_{j=0}^J$ $(J \leq \infty)$ be the sequence of all regular indices, starting with $n_0 = 0$, and let

$$k_j := n_{j+1} - n_j \quad (j \geq 1). \tag{4.12}$$

Additionally, we set $n_{-1} := -1$, $k_{-1} := 1$. Then $\phi_{k_j,n_j} \neq 0$ $(\forall j)$, and in analogy to (4.2), recurrence (3.5) becomes

$$\begin{bmatrix} p_{n_j} & p_{n_{j+1}} \\ q_{n_j} & q_{n_{j+1}} \end{bmatrix} := \begin{bmatrix} p_{n_{j-1}} & p_{n_j} \\ q_{n_{j-1}} & q_{n_j} \end{bmatrix} \begin{bmatrix} 0 & -\beta_j \\ 1 & a_{j+1}(\zeta) \end{bmatrix} \quad (j \geq 0), \tag{4.13}$$

with

$$\beta_j := \frac{\phi_{k_j,n_j}}{\phi_{k_{j-1},n_{j-1}}}, \quad a_{j+1}(\zeta) := q_{k_j}^{(n_j)}(\zeta) \tag{4.14}$$

243

and initial conditions (4.4). Equation (4.13) is equivalent to the general three-term recurrences

$$p_{n_{j+1}}(\zeta) = a_{j+1}(\zeta)p_{n_j}(\zeta) - \beta_j\, p_{n_{j-1}}(\zeta)$$
$$q_{n_{j+1}}(\zeta) = a_{j+1}(\zeta)q_{n_j}(\zeta) - \beta_j\, q_{n_{j-1}}(\zeta) \qquad (j \geq 0) \qquad (4.15)$$

with the initial conditions (4.4). These recurrences appear, e.g., in Chebyshev [17], Magnus [53], [54], and Struble [62] (for the FOPs only), Draux [19], Gragg and Lindquist [32], and Gutknecht [36]. They correspond to the Magnus *P-fraction*

$$h(\zeta) = -\frac{f_0(\zeta)}{g_0(\zeta)} = \frac{\beta_0}{\lceil a_1(\zeta)} - \frac{\beta_1}{\lceil a_2(\zeta)} - \cdots, \qquad (4.16)$$

whose successive numerators and denominators are p_{n_j} and q_{n_j} respectively. Often the partial numerators β_j are normalized to be 1, instead of the requirement that the partial denominators a_j be monic.

Of course, the step from the P-fraction to the recurrence formulas is the same as for any other continued fraction. It is often attributed to Euler, but Brezinski [6, p. 80] points out that in Europe it dates back to Wallis (1655), who also invented the term "continued fraction", while in India the recurrence formulas for the numerators and denominators of a continued fraction were known 500 years earlier [6, p. 33]. Hence, the formulas (4.15) are at least as old as the P-fraction. The latter is normally attributed to Magnus [53], [54], who established the possibility to expand any formal power series into such a continued fraction and knew the connection to the diagonal sequence of Padé approximants of that series. However, P-fractions were used as a tool by Chebyshev [17], [14], [13] already[§]. He expanded certain power series in ζ^{-1} into them and was aware of the optimal interpolation property of their convergents. Interestingly, he did not claim to have discovered these continued fractions. In the paper [15] on the discrete polynomial least squares problem, he mentions that the partial denominators of the relevant continued fraction cannot be of "degree two, three, or higher" (hence, the fraction is a J-fraction). Later, in [14] he takes the P-fraction as starting point for the treatment of a generalization of the Padé approximation problem: given are two functions h and H, of which the first is represented by a P-fraction (hence, could be expanded into a series in ζ^{-1}) and the second as a series in ζ^{-1}, which is then transformed into a series in terms of the residuals $f_n = p_n - hq_n$ of the convergents p_n/q_n of the P-fraction. Chebyshev then determines a sequence of polynomials P and Q so that $\partial(P - hQ + H)$ is minimal among all polynomial pairs $(\tilde{P}, \tilde{Q})$ with $\partial\tilde{Q} \leq \partial Q$. (Hence, in the case $H = 0$ the problem simplifies to the Padé problem, where $Q = q_n$ and $P = p_n$ for some regular index n.) The same problem is alluded to in [17]. Finally, in his derivation [13] of what we call the Kronecker algorithm [50], [67], Chebyshev expands a rational function into a P-fraction.

[§]The authors are indebted to Prof. A.A. Goncar for pointing out that P-fractions were known to Chebyshev.

It took a long time before P-fractions became widely known and their importance was appreciated: Gragg alluded to them in [31], Kalman [49] and Gragg and Linquist [32] revealed the connection to the partial realization problem of systems theory[†], Jones and Thron [47] put them into their place within the whole continued fraction theory, Gutknecht [35] generalized them to continued fractions[‡] that correspond to a formal Newton series (which replaces the formal power series when one treats a rational interpolation problem instead of a Padé approximation problem).

Note that (4.4) and (4.13) yield a representation of the P-fraction as an infinite product, whose jth partial product is

$$
\begin{bmatrix} p_{n_j} & p_{n_{j+1}} \\ q_{n_j} & q_{n_{j+1}} \end{bmatrix} = \begin{bmatrix} -1 & 0 \\ 0 & 1 \end{bmatrix} \prod_{i=0}^{j} \begin{bmatrix} 0 & -\beta_i \\ 1 & a_{i+1}(\zeta) \end{bmatrix} \quad (j \geq 0). \tag{4.17}
$$

In view of (4.11), the recurrence (3.6) for the residuals becomes

$$
f_{n_{j+1}}(\zeta) = a_{j+1}(\zeta) f_{n_j}(\zeta) - \beta_j f_{n_{j-1}}(\zeta) \quad (j \geq 0), \tag{4.18}
$$

with initial conditions

$$
f_{n_{-1}}(\zeta) := f_{-1}(\zeta) := -g_0(\zeta), \quad f_{n_0}(\zeta) := f_0(\zeta). \tag{4.19}
$$

In reordered form we can write this as

$$
\beta_j f_{n_{j-1}}(\zeta) = a_{j+1}(\zeta) f_{n_j}(\zeta) - f_{n_{j+1}}(\zeta) \quad (j \geq 0), \tag{4.20}
$$

where

$$
\partial f_{n_{j-1}} = -n_j > \partial f_{n_j} = -n_{j+1} > \partial f_{n_{j+1}} \tag{4.21}
$$

and

$$
k_{j+1} = \partial a_{j+1} = \partial f_{n_{j-1}} - \partial f_{n_j}, \tag{4.22}
$$

[†]In fact, Kalman [49] meant to have discovered P-fractions, whose importance he emphasized highly. He wrote: "Even more surprisingly, our main result, that *every formal power series may be represented by a* [certain new type of] *continued fraction* has been 'targeted' for well over 100 years. Nevertheless, this fact, discovered by a long chain of logical deductions guided by system-theoretic intuition, is quite unknown at present. It will necessitate a total reorganization of a large, but muddled area of mathematical knowledge, concerned with such things as the Cauchy index, the Routh test and Hurwitz determinants, signature of Hankel forms, continued fractions, Padé tables, Lyapunov stability theory for linear systems, quadratic optimization theory, positive real transfer functions, and very many other subjects. At the very least, the 'singular cases' and 'geometric structure' of these problems are at last revealed; it is now possible to study new things beyond the usual blind rearrangements of complicated but known formulas."

[‡]Two types of such *G*eneral continued fractions are introduced in [35]: diagonal G-fractions and staircase G-fractions. The convergents of the former are diagonal entries of the Newton-Padé table, those of the latter are entries on a generalized staircase consisting of the distinct entries on two adjacent diagonals. Of course, there also exists a corresponding variation of the P-fraction with convergents on a generalized staircase of the Padé table.

while $0 \neq \beta_j \in \mathbb{C}$ is just used to make a_{j+1} monic. (If this is not required, we can normalize β_j to 1.)

This reveals that the P-fraction can be constructed by a simple generalization of the *Euclidean algorithm* for polynomials. Recall that the latter starts from two polynomials f_{-1} and f_0, which we may assume to satisfy $\partial f_{-1} > \partial f_0$, and generates a finite sequence of polynomials f_j of decreasing degree according to (4.20) with $0 \neq \beta_j \in \mathbb{C}$ arbitrary and $a_{j+1} \in \mathcal{P}$.

For distinction, the above generalization will be called the *Euclidean algorithm for power series in* ζ^{-1}. Magnus chose to divide through by $f_{n_j}(\zeta)$ in (4.20). Then a_{j+1} is the *polynomial* or *principal part* of the formal Laurent series $\beta_j f_{n_{j-1}}/f_{n_j}$. This fact gave rise to the name P-fraction. The recurrences (4.15) and (4.18) together are called the *extended Euclidean algorithm* for power series in ζ^{-1}.

In the generic case, when the P-fraction becomes a J-fraction and $n_j = j$, $a_j(\zeta) = \zeta - \alpha_j$ ($\forall j$), the recurrence of the Euclidean algorithm simplifies to

$$\beta_n f_{n-1}(\zeta) = (\zeta - \alpha_{n+1}) f_n(\zeta) - f_{n+1}(\zeta) \quad (n \geq 0). \tag{4.23}$$

Recalling that the coefficients β_n and α_{n+1} are given by (4.3) we easily verify that this is nothing but the *Chebyshev algorithm* [16], [25] for computing the recurrence coefficients $\beta_0, \alpha_1, \beta_1, \alpha_2, \ldots$ that belong to given moments $\phi_{k,0} := \phi_k$. In fact, the nontrivial coefficients $\phi_{j,n}$ ($j \geq 1$) of the residual f_n are just the nontrivial coefficients in the $(n+1)$th row of the upper triangular matrix $\Sigma = [\sigma_{i,k}]$ of Gautschi [25] and of the $(n+1)$th column of the lower triangular matrix $S := \Sigma^T$ in [28]. Moreover, the *nongeneric Chebyshev algorithm*, to which the nongeneric modified Chebyshev algorithm of Golub and Gutknecht [28] reduces when applied to the ordinary moments, is mathematically equivalent to the Euclidean algorithm for power series in ζ^{-1}.

Chebyshev [16] used the algorithm for computing the discrete polynomial least squares approximation of a set of given function values. He considered the same problem before in the aforementioned paper [15] where, however, the emphasis is on the J-fraction representation of the solution.

5   TYPES OF LOOK-AHEAD ALGORITHMS

A *look-ahead algorithm* allows us to construct recursively those elements of a sequence whose dependence on the data is well-conditioned. It requires two ingredients: a *general recurrence formula* that connects successive well-conditioned elements of the sequence, and a *look-ahead strategy* based on some sort of condition estimator that enables us to determine which elements are well-conditioned. Here, the elements of the sequence are the regular pairs $[\mathring{q}_n \quad q_n]$ or the corresponding pairs of Padé forms $(\mathring{p}_n, \mathring{q}_n)$, (p_n, q_n) or the residuals $[g_n \quad f_n]$. The general recurrences are those of Theorem 3.1.

The condition of the regular pair can be measured by the condition number of the leading principal submatrix $\mathbf{M}_n$. We postpone the estimate of this condition number and other look-ahead strategies to §7. Here, we want to sketch various algorithms based on Theorem 3.1 for recursively computing a certain regular FOP q_N. These algorithms determine at intermediate stages all or most of the regular pairs $[\dot{q}_n \quad q_n]$ either explicitly or implicitly (in the sense that the regular pairs could be computed easily from quantities found). They may or may not provide further output, like a series of regular diagonal Padé forms or, at little additional work, a block LDU or an inverse block LDU decomposition of the moment matrix.

As for the recurrences along rows of the Padé table [37], [39] and for the corresponding Toeplitz solvers [40], there are three different types of algorithms that can be built on the recurrences of Theorem 3.1. This is true for the generic case already, and also for the nongeneric algorithms that treat exact breakdowns and are strongly linked to the continued fractions discussed in §4. Especially with look-ahead, there are, within each type, many variations with regard to the look-ahead strategy applied, the way the small systems (3.12) are set up and solved, and the output data that are provided. We will use the terminology from the Toeplitz solver literature and speak of Levinson-type, Schur-type, and superfast algorithms, the latter being obtained by combining the Schur type with recursive doubling and fast polynomial multiplication.

A *Levinson-type* algorithm is based on the recurrence (3.5) for the regular Padé forms (p_n, q_n) and their upper left neighbors $(\dot{p}_n, \dot{q}_n)$. It takes as small steps as possible: $k = 1$ in the generic case, and $k > 1$ if $\mathbf{M}_{n+1}$ is singular or nearly singular. The data for the small systems (3.12) are a few initial coefficients of the residuals f_n and g_n, and they are obtained according to the definition (2.19) of the residuals. This amounts to computing, for each coefficient required, an inner product of lengths about n if $g_0(\zeta) \equiv 1$, and two such inner products if g_0 is not a constant. (As in the Toeplitz case [40], there may be a way to reduce the number of these inner products.)

In contrast, a *Schur-type* algorithm is based on the recurrence (3.6) for the residuals g_n and f_n. In its basic form, it also takes small steps whenever possible. For the recurrence we need at most $4k - 1$ SAXPYs of length $\sim n$, but then the coefficients for the small linear system (3.12) of the next step are available too. There are no inner products required, which is an advantage on vector and parallel computers. Primarily, there is also no need to compute the FOPs $\dot{q}_n$ and q_n or even the Padé forms; it suffices to compute the residuals, unless we want to use the FOPs in the condition estimator; see §7. The Euclidean algorithm (4.20) and its generic variant, the Chebyshev algorithm (4.23), are of Schur type.

The *superfast* $O(N \log^2 N)$ *algorithm* is also basically of Schur type, but it has Levinson-type features too. The residuals are no longer updated in small steps based on (3.5) with small k. Instead, for a whole tree of different data, we use (3.5) for updating Padé forms in sometimes long steps, and the Padé conditions (2.19) and (3.1) for computing possibly many coefficients of the residuals. For both these tasks we apply fast convolutions (based on the FFT) whenever they are more efficient than ordinary convolutions. The Padé conditions (which are the definitions of the residuals) yield in view of (3.7) the same results as the residual

247

recurrence (3.6). We start by pretending to take one long step, from 0 to N, which is then divided into two steps of about half the length, from 0 to $n \sim N/2$ and from there to N (so that $k \sim n$). Recall from Theorem 3.1 that (unlike in the Toeplitz solver situation) both subproblems are of the same sort, namely for determining a regular Padé form and its upper left neighbor for a pair of finite power series. Originally, these are f_0 and $g_0(\zeta) \equiv 1$, but at later stages they are replaced. Note, however, that the data for the second subproblem consist of the residuals of the first pair of Padé forms. Therefore, these data can be determined only after the first pair has been computed. The computation of $2k \sim 2n$ coefficients of the residuals g_n and f_n costs two fast convolutions (i.e., four FFTs) of length $\sim 4n$; hence, the complexity of this update is $O(n \log_2 n)$.

This *divide-and-conquer* strategy is now applied recursively to both subproblems. After $\sim \log_2 N$ reductions one ends up with problems of size $k = 1$ (or one realizes that some of these problems are singular or nearly singular and, hence, the reduction cannot be carried as far). In the generic case, we have at the intermediate level ℓ a total of 2^ℓ problems with data vectors of length $2^{1-\ell}N$. After having solved as many problems of half the size we have to compute the new residuals that are the data for solving the second half of these problems, and once these are solved too, we have to update the Padé forms, which will be needed for computing their residuals. Each of these updates is of complexity $O(2^{-\ell}N \log_2(2^{-\ell}N))$, so that the total cost of the residual and Padé form updates at level ℓ is $O(N(\log_2 N - \ell))$. Since ℓ runs from 1 to $\log_2 N$, the total cost for all levels is $O(N \log_2^2 N) = O(N \log^2 N)$.

Due to the data dependency, we cannot go through this algorithm level by level. The first result that is really computed is, in the generic case, for $n = 1$, and this is then updated to $n = 2, 4, 8, 16, \ldots$. Therefore, the process can also be understood as *recursive doubling*. However, its implementation is simplest from the divide-and-conquer point of view, in a programming language that allows recursive procedure calls; see [37].

In each of the three types of algorithms we determine explicitly or implicitly the factors of a product representation

$$
\begin{bmatrix} \grave{p}_{n_J} & p_{n_J} \\ \grave{q}_{n_J} & q_{n_J} \end{bmatrix} = \begin{bmatrix} -1 & 0 \\ 0 & 1 \end{bmatrix} \prod_{j=0}^{J} \begin{bmatrix} \grave{p}_{k_j}^{(n_j)} & p_{k_j}^{(n_j)} \\ \grave{q}_{k_j}^{(n_j)} & q_{k_j}^{(n_j)} \end{bmatrix}
\tag{5.1}
$$

with $N = n_J$. The Levinson-type algorithm evaluates this product from left to right. The Schur-type algorithm evaluates likewise the product

$$
\begin{bmatrix} f_{n_J} & g_{n_J} \end{bmatrix} = \begin{bmatrix} g_0 & f_0 \end{bmatrix} \prod_{j=0}^{J} \begin{bmatrix} \grave{p}_{k_j}^{(n_j)} & p_{k_j}^{(n_j)} \\ \grave{q}_{k_j}^{(n_j)} & q_{k_j}^{(n_j)} \end{bmatrix}
\tag{5.2}
$$

from left to right, and if q_N is sought, the product (5.1) needs to be evaluated additionally, either along with the other one or at the end. A main feature of the superfast algorithm is that it evaluates the first product according to a binary tree, and this evaluation is an $O(N \log_2^2 N)$ process, while the sequential evaluation requires $O(N^2)$ operations.

Superfast algorithms based on recursive doubling were introduced in several papers appearing 1980: Bitmead and Anderson [4], Brent, Gustavson, and Yun [5], and Morf [58]. They used other recurrences, however. As mentioned before, the generic case of the recurrence of Theorem 3.1 appears in Gragg *et al.* [29], also in the context of recursive doubling. Cabay and Labahn [10] and Beckermann and Labahn [2] generalize the idea to diagonal recurrences of various types of vector and matrix Padé problems, where they have to deal with block Hankel systems with possibly singular leading principal submatrices. Gutknecht [37] and Gutknecht and Hochbruck [40], [39] adapt the recurrences to pairs of row sequences in the Padé table, which amounts to dealing with such submatrices of Toeplitz systems.

The Cabay-Meleshko algorithm [12], [57], which was the first that could handle near-singular leading principal Hankel submatrices, is of Levinson type, and hence $O(N^2)$. As we mentioned in the introduction, an alternative way to treat these near-singularities is by the look-ahead Lanczos procedure for polynomials, which was developed in [36] and, later, in [20] and is a generalization of the classical Stieltjes procedure [61]. This look-ahead Stieltjes procedures is also of Levinson type, but in contrast to the Cabay-Meleshko algorithm it yields directly an inverse block LDU decomposition of the moment matrix. However, we will show in §6 how to construct this decomposition with little additional work from the output of the Cabay-Meleshko algorithm.

6 HANKEL MATRIX FACTORIZATION

It is a well known fact that a full set of formal orthogonal polynomials up to degree $N - 1$, say, provides us with a so-called inverse LDU factorization of the $N \times N$ moment matrix $\mathbf{M}_N$, i.e., a UDL factorization of $\mathbf{M}_N^{-1}$. This is readily seen by exploiting the connection between $\mathbf{M}_N$ and the functional Φ of (2.1). In fact, let

$$\mathbf{w} := [\; 1 \quad \omega \quad \cdots \quad \omega^{N-1} \;]^T, \quad \mathbf{z} := [\; 1 \quad \zeta \quad \cdots \quad \zeta^{N-1} \;]^T, \qquad (6.1)$$

$$\mathbf{s} := [\; \sigma_0 \quad \cdots \quad \sigma_{N-1} \;]^T, \quad s(\zeta) := \sum_{j=0}^{N-1} \sigma_j \zeta^j = \mathbf{s}^T \mathbf{z}, \qquad (6.2)$$

$$\mathbf{t} := [\; \tau_0 \quad \cdots \quad \tau_{N-1} \;]^T, \quad t(\zeta) := \sum_{j=0}^{N-1} \tau_j \zeta^j = \mathbf{t}^T \mathbf{z}. \qquad (6.3)$$

Then,

$$\Phi(st) = \mathbf{s}^T \mathbf{M}_N \mathbf{t} \quad (\mathbf{s}, \mathbf{t} \in \mathbb{C}^N). \qquad (6.4)$$

Moreover, if $\{q_n\}_{n=0}^{N-1}$ is a set of normalized regular FOPs, and

$$
\mathbf{R}_N := \begin{bmatrix} \rho_{0,0} & \rho_{0,1} & \cdots & & \rho_{0,N-1} \\ & \rho_{1,1} & \cdots & & \rho_{1,N-1} \\ & & \ddots & & \vdots \\ & & & & \rho_{N-1,N-1} \end{bmatrix} \tag{6.5}
$$

is the unit upper triangular matrix whose columns contain the coefficients of these FOPs, then (2.5) translates into

$$
\mathbf{R}_N^T \mathbf{M}_N \mathbf{R}_N = \mathbf{D}_N := \text{diag}\,[\,\delta_0 \quad \delta_1 \quad \cdots \quad \delta_{N-1}\,], \tag{6.6}
$$

with nonsingular $\mathbf{D}_N$. We call this a *symmetric inverse LDU decomposition of* $\mathbf{M}_N$. If, for some $n \le N - 1$, q_n is not regular, then such a decomposition (with nonsingular diagonal $\mathbf{D}_N$) does not exist. But if q_N is regular, it is always possible to find a *symmetric inverse block LDU decomposition of* $\mathbf{M}_N$ of the form (6.6), with unit upper triangular $\mathbf{R}_N$ and block diagonal $\mathbf{D}_N$.

Let us redefine q_n as the monic polynomial of degree n whose coefficients are given by the $(n + 1)$th column of $\mathbf{R}_N$ appearing in such an inverse block LDU decomposition. We let n_j $(j = 0, \ldots, J - 1)$ denote the index of the polynomial in the first column of the $(j + 1)$th block. Then, in view of (6.4), the relation (6.6) is equivalent to

$$
\Phi(q_m q_{n+k}) = 0 \quad \text{if} \quad m < n_j = n \le n + k < n_{j+1} \tag{6.7}
$$

(where $0 \le j < J$ and $n_J := N$). Here, due to the linearity of Φ, we may replace q_m by ζ^m. Then, in terms of the symbol h, the condition can be expressed as

$$
\underbrace{h(\zeta)q_{n+k}(\zeta)}_{\in \mathcal{L}_{n+k-1}^*} = \underbrace{p_{n+k}(\zeta)}_{\in \mathcal{P}_{n+k-1}} - \underbrace{f_{n+k}(\zeta)}_{\in \mathcal{L}_{-n-1}^*} \quad \text{if} \quad n_j = n \le n + k < n_{j+1}. \tag{6.8}
$$

If $k = 0$ (i.e., $n+k = n = n_j$), the condition is the same as in (2.14); then we know that a monic q_n satisfying (6.7) is uniquely determined, and is an nth regular FOP. Hence, q_{n_j} is regular. Theoretically, in exact arithmetic, we could start a new block of our inverse block LDU decomposition at every regular index, but in practice we may want to choose some of the blocks larger for stability reasons.

However, if $k > 0$, the condition $f_{n+k} \in \mathcal{L}_{-n-1}^*$ is weaker than for a regular or singular FOP of degree $n + k$. This just mirrors that (6.7) is weaker too. We call any such q_{n+k} an *inner* polynomial. In general, it need not be a FOP.

Nevertheless, once a regular FOP q_n and its upper-left neighbor $\grave{q}_n$ are known, any inner polynomial q_{n+k} $(n_j < n + k < n_{j+1})$ can be determined from a formula analogous to the recurrence (3.5). For the formulation of this result we turn again to the slightly more general situation where h is replaced by a pair $[\,g_0 \quad f_0\,] \in \mathcal{L}_0^* \times \mathcal{L}_{-1}^*$.

Theorem 6.1 Let $[\dot{q}_n \quad q_n]$ be a normalized nth regular pair for $[g_0 \quad f_0]$, so that (2.15)–(2.17) and (2.19) hold, and let k be a positive integer. Consider any pair (p_{n+k}, q_{n+k}) constructed according to

$$\begin{bmatrix} p_{n+k} \\ q_{n+k} \end{bmatrix} = \begin{bmatrix} \dot{p}_n & p_n \\ \dot{q}_n & q_n \end{bmatrix} \begin{bmatrix} p_k^{(n)} \\ q_k^{(n)} \end{bmatrix} \tag{6.9}$$

from a pair

$$(p_k^{(n)}, q_k^{(n)}) \in \mathcal{P}_{k-1} \times \mathcal{P}_k, \quad q_k^{(n)} \text{ monic} \tag{6.10}$$

that satisfies

$$f_k^{(n)} := g_n p_k^{(n)} + f_n q_k^{(n)} \in \mathcal{L}_{-n-1}^*. \tag{6.11}$$

Then $(p_{n+k}, q_{n+k}) \in \mathcal{P}_{n+k-1} \times \mathcal{P}_{n+k}$, q_{n+k} is monic of degree $n + k$, and the pair satisfies

$$f_{n+k} := g_0 p_{n+k} + f_0 q_{n+k} = f_k^{(n)} \in \mathcal{L}_{-n-1}^*. \tag{6.12}$$

Condition (6.11) is fulfilled when

$$p_k^{(n)}(\zeta) = -\frac{\zeta^k f_n(\zeta)}{g_n(\zeta)} + O_-(\zeta^{-1}), \quad q_k^{(n)} = \zeta^k, \tag{6.13}$$

i.e., if $q_k^{(n)}$ is chosen as monomial, and $p_k^{(n)}$ is the principal part of the formal Laurent series at ∞ of $-\zeta^k f_n/g_n$.

The computation of $p_k^{(n)}$ in (6.13) requires only the solution of a $k \times k$ triangular Toeplitz system, and as k is growing, these systems are nested, so that the solution of the last one contains the solution for all.

In [37] the quotient of a pair (p_{n+k}, q_{n+k}) satisfying (6.12) is called an *underdetermined* Padé approximant.

Proof of Theorem 6.1: As in the proof of Theorem 3.1, (2.15), (2.17), and (6.10) imply that the pair (p_{n+k}, q_{n+k}) defined by (6.9) is in $\mathcal{P}_{n+k-1} \times \mathcal{P}_{n+k}$, and that q_{n+k} is monic. Multiplying (2.19) by $[\ p_k^{(n)} \quad q_k^{(n)}\ ]^T$ satisfying (6.11) yields (6.12) and $f_{n+k} = f_k^{(n)}$. Finally, by inserting the polynomials from (6.13) into (6.11) and multiplying by $g_n(\zeta) = \zeta^{-n} + O_-(\zeta^{-n-1})$, we see that (6.11) holds. ∎

The next theorem states that the above construction yields not only the inverse block LDU decomposition we were looking for, but also an ordinary block LDU decomposition of the moment matrix.

Theorem 6.2 Given a nonsingular $N \times N$ moment matrix $\mathbf{M}_N := [\mu_{i+j}]_{i,j=0}^{N-1}$, let $0 = n_0 < n_1 < \cdots < n_J = N$ be a subsequence of the indices for which $\mathbf{M}_n$ is nonsingular. Let $\{q_{n_j}\}_{j=0}^{J-1}$ and $\{f_{n_j}\}_{j=0}^{J-1}$ be the corresponding sequences of normalized regular FOPs and residuals, respectively. (Note that q_N, which depends on μ_{2N-1}, is not needed.) Whenever $k_j := n_{j+1} - n_j > 1$, define for $k = 1, \ldots, k_j - 1$ inner

*polynomials q_{n_j+k} and the corresponding residual f_{n_j+k} according to (6.9)–(6.12).
Place the coefficients of the polynomials q_n $(0 \le n < N)$ into the columns of an
$N \times N$ unit upper triangular matrix $\mathbf{R}_N$, and the coefficients of the residuals f_n
$(0 \le n < N)$ (more exactly of their terms in $\zeta^{-1}, \zeta^{-2}, \ldots, \zeta^{-N}$) into an $N \times N$
matrix $\mathbf{F}_N$. Then $\mathbf{F}_N$ is lower block triangular, with blocks of size k_j $(0 \le j < J)$.
Moreover, the following relations hold with a symmetric block diagonal matrix
$\mathbf{D}_N$ with blocks of size k_j:*

$$\mathbf{R}_N^T \mathbf{M}_N \mathbf{R}_N = \mathbf{D}_N, \tag{6.14}$$

$$\mathbf{M}_N \mathbf{R}_N = \mathbf{F}_N = \mathbf{R}_N^{-T} \mathbf{D}_N, \tag{6.15}$$

and

$$\mathbf{M}_N = \mathbf{F}_N \mathbf{D}_N^{-1} \mathbf{F}_N^T. \tag{6.16}$$

Proof: $\mathbf{F}_N$ is lower block triangular since $f_{n+k} \in \mathcal{L}_{-n-1}^*$ if $n = n_j \le n+k < n_{j+1}$.
Equation (6.14) is equivalent to (6.7), and $\mathbf{M}_N \mathbf{R}_N = \mathbf{F}_N$ is a consequence of (6.8).
Since $\mathbf{M}_N$ is symmetric, so is $\mathbf{D}_N$. These relations then imply $\mathbf{F}_N = \mathbf{R}_N^{-T} \mathbf{D}_N$ and
(6.16). ∎

7 LOOK-AHEAD STRATEGIES BASED ON AN INVERSION FORMULA

In §5 we mentioned that a look-ahead algorithm requires two ingredients, a general
recurrence formula and a look-ahead strategy, and that for our problem the latter
can be based on a condition estimate of the leading principal submatrices $\mathbf{M}_n$.
Clearly, we need to avoid any near-singular $\mathbf{M}_n$ since the corresponding normalized
pair $[\mathring{q}_n \quad q_n]$ (and thus also the residuals $[g_n \quad f_n]$) would be ill determined by
the data. However, instead of estimating the condition of $\mathbf{M}_n$, we could control
some other number that is a quantitative indicator for the nonsingularity of $\mathbf{M}_n$.
In fact, also in Gaussian elimination with partial pivoting the condition of the
relevant submatrix (or of its Schur complement) is not estimated directly. Instead,
the relative size of the pivot element is maximized by balancing the matrix and
choosing the largest competing element as pivot. This guarantees that the relevant
submatrix and its Schur complement are not too ill conditioned if the given matrix
is well conditioned.

When solving a Hankel or a Toeplitz system, pivoting would destroy the
structure, but the look-ahead algorithms perform instead some kind of block pivoting. In block Gaussian elimination, one normally monitors the smallest singular
value of the block pivots and the size of the block pivot rows, and, in fact, this
could be used as the basis of a look-ahead strategy [20]. However, Hankel and
Toeplitz systems have the particular feature that directly estimating the condition
of leading principal submatrices is relatively cheap. This is due to the existence of
simple explicit inversion formulas [27], [43], of which the Heinig formula [42], [43,
Theorem 1.1′] is the most appropriate one for our situation. It expresses $\mathbf{M}_n^{-1}$ in
terms of the coefficients of $\mathring{q}_n$ and q_n whenever $\mathbf{M}_n$ is nonsingular.

Theorem 7.1 *The $n \times n$ Hankel matrix $\mathbf{M}_n$ is nonsingular if and only if the corresponding normalized regular pair $[\grave{q}_n \quad q_n]$ exists. If it does, then*

$$\mathbf{M}_n^{-1} = \mathbf{B}(\grave{q}_n, q_n) \quad := \quad \begin{bmatrix} \grave{\rho}_0 & 0 & \cdots & 0 \\ \grave{\rho}_1 & \ddots & \ddots & \vdots \\ \vdots & \ddots & \ddots & 0 \\ \grave{\rho}_{n-1} & \cdots & \grave{\rho}_1 & \grave{\rho}_0 \end{bmatrix} \begin{bmatrix} \rho_1 & \cdots & \rho_{n-1} & 1 \\ \vdots & \ddots & \ddots & 0 \\ \rho_{n-1} & \ddots & \ddots & \vdots \\ 1 & 0 & \cdots & 0 \end{bmatrix}$$
$$- \begin{bmatrix} \rho_0 & 0 & \cdots & 0 \\ \rho_1 & \ddots & \ddots & \vdots \\ \vdots & \ddots & \ddots & 0 \\ \rho_{n-1} & \cdots & \rho_1 & \rho_0 \end{bmatrix} \begin{bmatrix} \grave{\rho}_1 & \cdots & \grave{\rho}_{n-1} & 0 \\ \vdots & \ddots & \ddots & \vdots \\ \grave{\rho}_{n-1} & \ddots & & \vdots \\ 0 & \cdots & \cdots & 0 \end{bmatrix}.$$

$$(7.1)$$

For an elegant proof see Heinig and Rost [43]. The formula can also be derived from the Padé conditions (2.14) and the Frobenius identity (2.21). One might fear that, sometimes, evaluating the formula may be unstable due to catastrophic cancellation. It was shown by Gutknecht and Hochbruck [41] that this cannot happen if $\mathbf{M}_n$ is well conditioned.

Note that $\mathbf{M}_n$ and $\grave{q}_n$ depend only on $\mu_0, \ldots, \mu_{2n-2}$, while q_n depends on $\mu_0, \ldots, \mu_{2n-1}$. Hence, μ_{2n-1} affects q_n, but neither the inverse $\mathbf{M}_n^{-1}$ nor the regularity of q_n.

A matrix that can be represented in the form of the right side of (7.1) in terms of the coefficients of two relatively prime polynomials $\grave{q}_n$ and q_n is a *Bézoutian* or *Bézoutiant*. We choose the latter expression for the noun, the former for the adjective. More exactly, the Bézoutiant $\mathbf{B}$ of any two polynomials $\grave{q}_n, q_n \in \mathcal{P}_n$ is defined as the $n \times n$ matrix with the generating function

$$K_{n-1}(\omega, \zeta) := \frac{q_n(\omega)\grave{q}_n(\zeta) - \grave{q}_n(\omega)q_n(\zeta)}{\omega - \zeta}, \quad (7.2)$$

i.e., the matrix $\mathbf{B}$ for which

$$K_{n-1}(\omega, \zeta) =: \mathbf{w}^T \mathbf{B} \mathbf{z}, \quad (7.3)$$

where

$$\mathbf{w} := [\,1 \quad \omega \quad \cdots \quad \omega^{n-1}\,]^T, \quad \mathbf{z} := [\,1 \quad \zeta \quad \cdots \quad \zeta^{n-1}\,]^T. \quad (7.4)$$

Note that the factor $\omega - \zeta$ cancels in (7.2), and hence, K_{n-1} can be written as the symmetric bilinear form (7.3). By an elementary calculation $\mathbf{B}$ is seen to be equal to the right side of (7.1) if $\partial \grave{q}_n < n$. Otherwise, the diagonal of the last matrix in (7.1) contains the leading coefficient $\grave{\rho}_n$ of $\grave{q}_n$.

There is an extensive literature on Bézoutiants; see, e.g., [24], [44], [45], [46]. Some of the results are particularly easy to understand from the Padé theory point of view. Returning to the previous meaning of $\mathring{q}_n$ and q_n, we obtain the following corollary of Theorem 7.1.

Corollary 7.2 *The inverse of a $n \times n$ Hankel matrix is a Bézoutiant of two relatively prime polynomials $\mathring{q}_n$, q_n with $\partial \mathring{q}_n < \partial q_n = n$, and vice versa.*

Proof: By Theorem 7.1, every inverse of a nonsingular Hankel matrix is a Bézoutiant, and from Lemma 2.1 we know that $\mathring{q}_n$ and q_n are relatively prime. Conversely, given a Bézoutiant of a relatively prime pair $\mathring{q}_n$, q_n with $\partial \mathring{q}_n < \partial q_n = n$, we know from the remarks at the end of §2 how to find $\mu_0, \ldots, \mu_{2n-1}$ so that q_n is a regular FOP for any Φ that has these first $2n$ moments. Applying Theorem 7.1 to the corresponding moment matrix $\mathbf{M}_n$ reproduces the given Bézoutiant. ∎

Heinig's formula can be considered as an explicit form of Corollary 7.2. From the latter it is now an easy step to a theorem of Lander [52].

Corollary 7.3 *Any Bézoutiant of two relatively prime polynomials $\hat{q}_n$, q_n with $\partial \hat{q}_n \leq \partial q_n = n$ is the inverse of an $n \times n$ Hankel matrix.*

Proof: If $\partial \hat{q}_n = n$, define $\mathring{q}_n := \hat{q}_n - \alpha q_n$, where α is such that $\partial \mathring{q}_n < n$. Note that the Bézoutiant does not change. ∎

Note that in Corollary 7.3 there is no claim that the correspondence between Bézoutiants and nonsingular Hankel matrices is one-to-one, while, in Corollary 7.2, we have such a one-to-one relationship.

In a recursive form, a formula very similar to (7.1) has been used in Trench's Hankel solver [63]. However, Trench's formula assumes that $\mathbf{M}_{n-1}$ is nonsingular too. This formula also has a polynomial form, called the *Christoffel-Darboux formula*, whose classical derivation assumes that a full set of regular FOPs exists; see Gragg [31]. For the analogous connection between the Christoffel-Darboux-Szegö formula (based on polynomials orthogonal on the unit circle) and the Gohberg-Semençul formula [27], see Kailath *et al.* [48]. The Christoffel-Darboux formula says that the function K_{n-1} from (7.2) with $\mathring{q}_n(\zeta) := q_{n-1}(\zeta)/\phi_{1,n-1}$ (as in (4.11)) can be defined instead by

$$K_{n-1}(\omega, \zeta) := \sum_{j=0}^{n-1} \frac{q_j(\omega)q_j(\zeta)}{\delta_j} \quad \text{with} \quad \delta_j := \phi_{1,j}. \tag{7.5}$$

This function, on the other hand, is readily seen to be a *reproducing kernel function* for Φ in $\mathcal{P}_{n-1}$, i.e.,

$$s(\zeta) = \Phi(K_{n-1}(\omega, \zeta)s(\zeta)) \quad \text{for } s \in \mathcal{P}_{n-1}, \tag{7.6}$$

where Φ still acts on polynomials in ζ. Using (6.4), the condition (7.6) translates into

$$\mathbf{w}^T \mathbf{s} = \mathbf{w}^T \mathbf{B} \mathbf{M} \mathbf{s} \quad (\forall \mathbf{s}, \mathbf{w} \in \mathbb{C}^n), \tag{7.7}$$

which shows that $\mathbf{M}^{-1} = \mathbf{B}$.

Let us now return to the task of judging the condition of $\mathbf{M}_n$ for $n < N$. Since, for the spectral norm, $||\mathbf{M}_n|| \leq ||\mathbf{M}_N||$, we may, instead of the condition number, estimate the smallest singular value of $\mathbf{M}_n$, which is equal to the norm of $\mathbf{M}_n^{-1}$. By a simple estimate [37, Equation (5.2)] of the Frobenius norm of the right side of (7.1), we obtain

$$||\mathbf{M}_n^{-1}|| \leq ||\mathbf{M}_n^{-1}||_F \leq 2n||\dot{\mathbf{q}}_n|| \, ||\mathbf{q}_n||. \tag{7.8}$$

(The vector norm is the Euclidean one.) Conversely, in view of (2.9) and (2.10), $||\dot{\mathbf{q}}_n||$ and $||\mathbf{q}_n||$ can be bounded in terms of $||\mathbf{M}_n^{-1}||$ and $||\mathbf{M}_n||$. Therefore, $\mathbf{M}_n$ is a well-conditioned submatrix of $\mathbf{M}_N$ if and only if the product $||\dot{\mathbf{q}}_n|| \, ||\mathbf{q}_n||$ is not too large. This is also one of the condition estimator options suggested in [37] for the sawtooth look-ahead algorithm based on regular Padé forms in a row of the Padé table; see [40] for an analogous explicit estimate based on the Gohberg-Semençul formula.

Instead of the product $||\dot{\mathbf{q}}_n|| \, ||\mathbf{q}_n||$ we could control the product

$$(||\dot{\mathbf{p}}_n|| + ||\dot{\mathbf{q}}_n||) \, (||\mathbf{p}_n|| + ||\mathbf{q}_n||) \,, \tag{7.9}$$

in which case the look-ahead strategy would be equivalent to that of Cabay and Meleshko [12], [57], who established the (weak) stability of their algorithm. These authors use a different normalization (namely, they make each of the two factors in (7.9) equal to 1), whence their estimate involves other quantities (which are here normalized to 1). Of course, a bound for the product in (7.9) is also a bound for $||\dot{\mathbf{q}}_n|| \, ||\mathbf{q}_n||$. Conversely, a bound for the latter and for $\mathbf{M}_N$ leads in view of (2.14) to a bound for $||\dot{\mathbf{p}}_n|| \, ||\mathbf{p}_n||$, since each factor is part of a polynomial product, i.e., a convolution.

The disadvantage of the strategy based on (7.9) is that for estimating $||\mathbf{M}_{n+k}^{-1}||$ we need to compute $\dot{\mathbf{q}}_{n+k}$ and $\mathbf{q}_{n+k}$. To avoid this, we can instead estimate the norm of these two vectors in terms of $||\dot{\mathbf{q}}_n||, ||\mathbf{q}_n||, ||\dot{\mathbf{p}}_k^{(n)}||, ||\mathbf{p}_k^{(n)}||, ||\dot{\mathbf{q}}_k^{(n)}||$, and $||\mathbf{q}_k^{(n)}||$ by writing the second row of (3.5) as a multiplication of $[\dot{\mathbf{q}}_n \quad \mathbf{q}_n]$ by a 2×2 block matrix with triangular Toeplitz blocks. The same idea has been applied for an alternative look-ahead strategy in [37] and [40] where, however, the four coefficient vectors that are needed to update $\dot{\mathbf{q}}_n$ and $\mathbf{q}_n$ evolve from a two-point Padé approximation problem. To justify this look-ahead strategy here we point out that, by the aforementioned argument, bounds for $||\dot{\mathbf{q}}_n||, ||\mathbf{q}_n||$, and $||\mathbf{M}_N||$ allow us to derive bounds for the relevant coefficient vectors of the residuals f_n and g_n, and, since the latter has leading coefficient 1, also for those of $h_n = -f_n/g_n$. Hence, a bound for the norm of the moment matrix $\mathbf{H}_{N-n}(h_n)$ of the reduced problem with symbol h_n follows. Using this bound, a given bound for $||\dot{\mathbf{q}}_k^{(n)}||$ and $||\mathbf{q}_k^{(n)}||$ yields again a bound for $||\dot{\mathbf{p}}_k^{(n)}||$ and $||\mathbf{p}_k^{(n)}||$. For theoretical and practical reasons, the given bound should depend on k and, perhaps, on n.

Therefore, on the one hand, prescribing a bound for $||\dot{\mathbf{q}}_k^{(n)}||$ and $||\mathbf{q}_k^{(n)}||$ leads, by applying (7.8) to the leading principal submatrix matrix $\mathbf{H}_k(h_n)$ of the reduced problem, to a bound for the inverse of this submatrix. On the other hand, by the above arguments, it also leads to a bound for the condition number of $\mathbf{M}_{n+k}$, the moment submatrix of the updated problem. This argument does not assume that k is small. Therefore, this second look-ahead strategy is also applicable to recursive doubling and, moreover, to the whole tree of problems that are solved in the superfast algorithm. A subtle detail is that we capitalize upon the fact that, in our normalization of regular pairs, there is no need to renormalize $\dot{\mathbf{q}}_{n+k}$ and $\mathbf{q}_{n+k}$. In contrast, the need of renormalization in the treatment of Cabay and Meleshko did not allow them to extend their stability argument to recursive doubling.

Finally, a third look-ahead strategy consists in prescribing a bound for the smallest singular value (or for the condition number) of $\mathbf{H}_k(h_n)$ (or, in the superfast algorithm, for all small submatrices that play a role at the lowest level). Since such a bound, together with the one for $\|\mathbf{M}_N\|$, leads to a bound for $\|\dot{\mathbf{q}}_k^{(n)}\|$ and $\|\mathbf{q}_k^{(n)}\|$, it seems possible to justify this third strategy by the same arguments as the second one. An analogous strategy was also suggested for the Toeplitz solvers in [37] and [40].

REFERENCES

[1] Antoulas A.C. On recursiveness and related topics in linear systems. *IEEE Trans. Automatic Control*, AC-**31**:1121–1135, 1986.

[2] Beckermann B., Labahn G. A uniform approach to the fast computation of matrix-type Padé approximants. *SIAM J. Matrix Anal. Appl.* To appear.

[3] Berlekamp E.R. *Algebraic Coding Theory*. McGraw-Hill, New York, 1968.

[4] Bitmead R.R., Anderson B.D.O. Asymptotically fast solution of Toeplitz and related systems of linear equations. *Linear Algebra Appl.*, **34**:103–116, 1980.

[5] Brent R.P., Gustavson F.G., Yun C.Y.Y. Fast solution of Toeplitz systems of equations and computation of Padé approximants. *J. Algorithms*, **1**:259–295, 1980.

[6] Brezinski C. *History of Continued Fractions and Padé Approximants*. Springer-Verlag, Berlin, 1991.

[7] Bultheel A. Recursive algorithms for the matrix Padé problem. *Math. Comp.*, **35**:875–892, 1980.

[8] Bultheel A., van Barel M. Padé techniques for model reduction in linear system theory: a survey. *J. Comput. Appl. Math.*, **14**:401–438, 1986.

[9] Bultheel A., van Barel M. Euclid, Padé and Lanczos, another golden braid. Tech. Report TW 188, Dept. Computer Science, Katholieke Universiteit Leuven, Belgium, April 1993.

[10] Cabay S., Labahn G. A superfast algorithm for multi-dimensional Padé systems. *Numerical Algorithms*, **2**:201–224, 1992.

[11] Cabay S., Labahn G., Beckermann B. On the theory and computation of non-perfect Padé-Hermite approximants. *J. Comput. Appl. Math.*, **39**:295–313, 1992.

[12] Cabay S., Meleshko R. A weakly stable algorithm for Padé approximants and the inversion of Hankel matrices. *SIAM J. Matrix Anal. Appl.*, **14**:735–765, 1993.

[13] Chebyshev P. Sur la détermination des fonctions d'après les valeurs qu'elles ont pour certaines valeurs de variables. [*Oeuvres*, vol. **2**:69–82; Russian publication in 1870].

[14] Chebyshev P. Sur le développement des fonctions en séries à l'aide des fractions continues. [*Oeuvres*, vol. **1**:615–636; Russian publication in 1866].

[15] Chebyshev P. Sur les fractions continues. *J. de math. pures et appliquées*, série II, **3**:289–323, 1858. [*Oeuvres*, vol. **1**:201–230; Russian publication in 1855].

[16] Chebyshev P. Sur l'interpolation par la méthode des moindres carrés. *Mém. Acad. Impér. Sciences St. Pétersbourg*, série VII, **1**:1–24, 1859. [*Oeuvres*, vol. **1**:471–498].

[17] Chebyshev P. Sur les fractions continues algébriques. *Jour. de math. pures et appliquées*, série II, **10**:353–358, 1865. [*Oeuvres*, vol. **1**:609–614; Russian publication in 1866].

[18] Dickinson B.W., Morf M., Kailath T. A minimal realization algorithm for matrix sequences. *IEEE Trans. Automat. Contr.*, **16**:31–38, 1974.

[19] Draux A. *Polynômes orthogonaux formels—applications*. Vol. 974 of Lecture Notes in Mathematics, Springer-Verlag, Berlin, 1983.

[20] Freund R., Zha H. A look-ahead algorithm for the solution of general Hankel systems. *Numer. Math.*, **64**:295–321, 1993.

[21] Freund R.W., Gutknecht M.H., Nachtigal N.M. An implementation of the look-ahead Lanczos algorithm for non-Hermitian matrices. *SIAM J. Sci. Comput.*, **14**:137–158, 1993.

[22] Frobenius G. Über Relationen zwischen den Näherungsbrüchen von Potenzreihen. *J. reine ang. Math.*, **90**:1–17, 1881.

[23] Fuhrmann P.A. A matrix Euclidean algorithm and matrix continued fractions. *System Control Letters*, **3**:263–271, 1983.

[24] Fuhrmann P.A., Datta B. On Bezoutians, VanderMonde matrices and the Lienard-Chipart stability criterion. *Linear Algebra Appl.*, **120**:23–37, 1989.

[25] Gautschi W. On generating orthogonal polynomials. *SIAM J. Sci. Stat. Comput.*, **3**:289–317, 1982.

[26] Geddes K.O. Symbolic computation of Padé approximants. *ACM Trans. Math. Software*, 218–233, 1979.

[27] Gohberg I., Semençul A. On the inversion of finite Toeplitz matrices and their continuous analogs (in Russian). *Mat. Issled.*, **2**:201–233, 1972.

[28] Golub G., Gutknecht M. Modified moments for indefinite weight functions. *Numer. Math.*, **57**:607–624, 1990.

[29] Gragg W., Gustavson F., Warner D., Yun D. On fast computation of super-diagonal Padé fractions. *Math. Programming Stud.*, **18**:39–42, 1982.

[30] Gragg W.B. The Padé table and its relation to certain algorithms of numerical analysis. *SIAM Rev.*, **14**:1–62, 1972.

[31] Gragg W.B. Matrix interpretations and applications of the continued fraction algorithm. *Rocky Mountain J. Math.*, **4**:213–225, 1974.

[32] Gragg W.B., Lindquist A. On the partial realization problem. *Linear Algebra Appl.*, **50**:277–319, 1983.

[33] Gutknecht M.H. Block structure and recursiveness in rational interpolation. In *Approximation Theory VII*, pages 93–130. Academic Press, Boston, 1993. E.W. Cheney, C.K. Chui and L.L. Schumaker, eds.

[34] Gutknecht M.H. The multipoint Padé table and general recurrences for rational interpolation. *Acta Appl. Math.*, **33**:165–194, 1993. Also in *Nonlinear Numerical Methods and Rational Approximation II*, pages 109–136. Kluwer, Dordrecht, The Netherlands, 1994. A. Cuyt, ed.

[35] Gutknecht M.H. Continued fractions associated with the Newton-Padé table. *Numer. Math.*, **56**:547–589, 1989.

[36] Gutknecht M.H. A completed theory of the unsymmetric Lanczos process and related algorithms, Part I. *SIAM J. Matrix Anal. Appl.*, **13**:594–639, 1992.

[37] Gutknecht M.H. Stable row recurrences in the Padé table and generically superfast lookahead solvers for non-Hermitian Toeplitz systems. *Linear Algebra Appl.*, **188/189**:351–421, 1993.

[38] Gutknecht M.H. A completed theory of the unsymmetric Lanczos process and related algorithms, Part II. *SIAM J. Matrix Anal. Appl.*, **15**, 1994.

[39] Gutknecht M.H., Hochbruck M. Look-ahead Levinson- and Schur-type recurrences in the Padé table. Tech. report. In preparation.

[40] Gutknecht M.H., Hochbruck M. Look-ahead Levinson and Schur algorithms for non-Hermitian Toeplitz systems. IPS Research Report 93-11, IPS, ETH-Zürich, August 1993.

[41] Gutknecht M.H., Hochbruck M. The stability of inversion formulas for Toeplitz matrices. IPS Research Report 93-13, IPS, ETH-Zürich, October 1993.

[42] Heinig G. *Beiträge zur Spektraltheorie von Operatorbüscheln und zur algebraischen Theorie von Toeplitzmatrizen.* PhD thesis, TH Karl-Marx-Stadt, 1979.

[43] Heinig G., Rost K. *Algebraic Methods for Toeplitz-like Matrices and Operators.* Akademie-Verlag, Berlin, DDR, and Birkhäuser, Basel/Stuttgart, 1984.

[44] Helmke U., Fuhrmann P.A. Bezoutians. *Linear Algebra Appl.*, **122/123/124**:1039–1097, 1989.

[45] Householder A.S. Bigradients and the problem of Routh and Hurwitz. *SIAM Rev.*, **10**:56–66, 1968.

[46] Householder A.S. Bezoutiants, elimination and localization. *SIAM Rev.*, **12**:73–78, 1970.

[47] Jones W.B., Thron W.J. *Continued Fractions, Analytic Theory and Applications.* Addison-Wesley, Reading, Mass., 1980.

[48] Kailath T., Vieira A., Morf M. Inverses of Toeplitz operators, innovations, and orthogonal polynomials. *SIAM Rev.*, **20**:106–119, 1978.

[49] Kalman R.E. On partial realizations, transfer functions, and canonical forms. *Acta Polytech. Scand., Math. Comput. Sci. Ser.*, **31**:9–32, 1979.

[50] Kronecker L. Zur Theorie der Elimination einer Variabel aus zwei algebraischen Gleichungen. *Monatsber. Königl. Preuss. Akad. Wiss.*, 535–600, 1881.

[51] Lanczos C. An iteration method for the solution of the eigenvalue problem of linear differential and integral operators. *J. Res. Nat. Bureau Standards*, **45**:255–281, 1950.

[52] Lander F.I. The Bezoutian and the inversion of Hankel and Toeplitz matrices. *Matem. Issled., Kishinev*, **9**:69–87, 1974.

[53] Magnus A. Certain continued fractions associated with the Padé table. *Math. Z.*, **78**:361–374. 1962.

[54] Magnus A. Expansion of power series into P-fractions. *Math. Z.*, **80**:209–216, 1962.

[55] Massey J.L. Shift-register synthesis and BCH decoding. *IEEE Trans. Inform Theory*, **15**:122–127, 1969.

[56] McEliece R.J., Shearer J.B. A property of Euclid's algorithm and an application to Padé approximation. *SIAM J. Appl. Math.*, **34**:611–615, 1978.

[57] Meleshko R.J. *A stable algorithm for the computation of Padé approximants.* Ph.D. thesis, Department of Computing Science, University of Alberta, Canada, 1990.

[58] Morf M. Doubling algorithms for Toeplitz and related equations. In *Proc. IEEE Internat. Conf. on Acoustics, Speech and Signal Processing.* Denver, CO, April 954–959, 1980.

[59] Parlett B.N. Reduction to tridiagonal form and minimal realizations. *SIAM J. Matrix Anal. Appl.*, **13**:567–593, 1992.

[60] Parlett B.N., Taylor D.R., Liu Z.A. A look-ahead Lanczos algorithm for unsymmetric matrices. *Math. Comp.*, **44**:105–124, 1985.

[61] Stieltjes T.J. Quelques recherches sur la théorie des quadratures dites mécaniques. *Ann. Sci. École Norm. Paris*, Sér. 3, 1:409–426, 1884. [*Oeuvres*, vol. 1:377–396].

[62] Struble G.W. Orthogonal polynomials: variable-signed weight functions. *Numer. Math.*, **5**:88–94, 1963.

[63] Trench W.F. An algorithm for the inversion of finite Hankel matrices. *J. Soc. Indust. Appl. Math.*, **13**:1102–1107, 1965.

[64] van Barel M., Bultheel A. The "look-ahead" philosophy applied to matrix rational interpolation problems. In *Proceedings MTNS-'93.* Regensburg. To appear.

[65] van Barel M., Bultheel A. A canonical matrix continued fraction solution of the minimal (partial) realization problem. *Linear Algebra Appl.*, **122/123/124**:973–1002, 1989.

[66] van Barel M., Bultheel A. A matrix Euclidean algorithm for minimal partial realization. *Linear Algebra Appl.*, **121**:674–682, 1989.

[67] Warner D.D. *Hermite Interpolation with Rational Functions.* Ph.D. thesis, University of California at San Diego, 1974.

International Series of Numerical Mathematics, Vol. 119, ©1994 Birkhäuser

ON THE APPROXIMATION OF UNIVALENT FUNCTIONS BY SUBORDINATE POLYNOMIALS IN THE UNIT DISK

Richard Greiner*   Stephan Ruscheweyh*

Dedicated to Walter Gautschi on the occasion of his 65th anniversary

Abstract. Let f be any conformal map in the unit disk $\mathbb{D}$. We investigate the size of the largest number $\rho = \rho(f, n) \in (0, 1]$, $n \in \mathbb{N}$ such that there exists a univalent polynomial p_n in $\mathbb{D}$ with $f(0) = p_n(0)$ and $f(\rho \mathbb{D}) \subset p_n(\mathbb{D}) \subset f(\mathbb{D})$. Clearly, these numbers are related to the quality of polynomial approximation to conformal maps. In this note we find a sharp uniform bound for $\rho(f, n)$ if f is convex univalent, and discuss some related results and conjectures in the starlike and the general univalent case.

1 INTRODUCTION

Let f, g be two functions holomorphic in the unit disk $\mathbb{D}$. g is called *subordinate to* f (and we write $g \prec f$) when there exists a function ϕ, holomorphic in $\mathbb{D}$, with $|\phi(z)| \leq |z|$ and $f = g \circ \phi$. If f is univalent in $\mathbb{D}$ we have $g \prec f$ iff $g(0) = f(0)$ and $g(\mathbb{D}) \subset f(\mathbb{D})$.

In a recent paper [1] Andrievskij and Ruscheweyh proved:

Theorem 1.1 *There exists a universal constant $c < \infty$ with the following property: for each f univalent in $\mathbb{D}$ there exists a sequence of polynomials p_n, univalent in $\mathbb{D}$ with $p_n(0) = f(0)$, such that for $n \geq 2c$*

$$f_{1-\frac{c}{n}} \prec p_n \prec f. \tag{1.1}$$

Here f_ρ denotes the function $z \mapsto f(\rho z)$, $z \in \mathbb{D}$, for $|\rho| \leq 1$. Although the proof of Theorem 1 was constructive, no explicit bound for c was given. Greiner [7] has shown that one can choose $c \leq 73$, but the general impression is that there is still much room for further improvement. The order of convergence n^{-1}, however, is correct, and we shall see below that π is indeed a lower bound for c.

Theorem 1.1 has an obvious significance in the problem of the speed of approximation to a given univalent function $f : \mathbb{D} \to \Omega$ through polynomials p_n of degree n with $p_n(0) = f(0)$ which map $\mathbb{D}$ one-to-one into Ω. Since this speed will clearly depend on many factors, for instance the geometry of Ω and the smoothness of $\partial\Omega$, it makes sense to discuss this problem for certain subsets $\mathcal{M}$ of the set of all conformal mappings in $\mathbb{D}$, where we can expect better values of the corresponding

*Mathematisches Institut, Universität Würzburg, D-97074 Würzburg, Germany

$c = c(\mathcal{M})$ or, at least, more complete information in comparison to the general case.

In the present article we initiate such investigations, namely we restrict the class of univalent functions considered to the sets $\mathcal{C}, \mathcal{S}_0$ of convex, and starlike (w.r.t. the origin) univalent functions in $\mathbb{D}$.

In the convex case (Theorem 1.2 below) we find a complete solution in the sense that we can determine $c(\mathcal{C})$ and give polynomials p_n, depending on $f \in \mathcal{C}$, for which (1.1) holds with c replaced by $c(\mathcal{C})$. Theorem 1.2 also covers, to a certain extent, a more general question in the present context: namely not only to approximate an univalent f by subordinate univalent polynomials, but by subordinate polynomials with additional special properties, for instance convex univalent ones if f is convex univalent.

We recall a few definitions and results. As in the original proof of Theorem 1.1, the convolution (or Hadamard product) will be the tool for constructing the approximating polynomials. We recall that for $f(z) = \sum_{k=0}^{\infty} a_k z^k$, $g(z) = \sum_{k=0}^{\infty} b_k z^k$ the Hadamard-convolution is defined as $(f * g)(z) := \sum_{k=0}^{\infty} a_k b_k z^k$.

For given f, holomorphic in $\mathbb{D}$ with power series $f(z) = \sum_{k=0}^{\infty} a_k z^k$, the n-th *Cesàro mean of order* $\alpha \geq 0$ is defined by

$$\sigma_n^\alpha(f, z) := \sum_{k=0}^{n} \frac{\binom{n+\alpha-k}{n-k}}{\binom{n+\alpha}{n}} a_k z^k.$$

Using the abbreviation $\sigma_n^\alpha(z)$ for $\sigma_n^\alpha(\frac{1}{1-z}, z)$ one may also write

$$\sigma_n^\alpha(f, z) = (f * \sigma_n^\alpha)(z).$$

Lewis [8] (compare also Salinas and Ruscheweyh [11]) has shown that the Cesàro means $f * \sigma_n^\alpha$ of order $\alpha \geq 1$ are univalent in $\mathbb{D}$, in fact close-to-convex, for f convex. The given lower bound for α is sharp with respect to this property.

Theorem 1.2 *For $f \in \mathcal{C}$, the Cesàro means $f * \sigma_n^\alpha$ of order $\alpha \geq 1$ are univalent in $\mathbb{D}$ and fulfill*

$$f_{\frac{n}{n+\alpha+1}} \prec f * \sigma_n^\alpha \prec f$$

for all $n \in \mathbb{N}$. If $f(\mathbb{D})$ is a half-plane, then, for no $n \in \mathbb{N}$ and no $\alpha \geq 1$, can the number $\frac{n}{n+\alpha+1}$ be replaced by any larger one.

Note that Theorem 1.2 implies, in particular (the case $\alpha = 1$), that

$$c(\mathcal{C}) \leq 2. \tag{1.2}$$

The theory of *maximal polynomial ranges*, which has been introduced by Cordova and Ruscheweyh (compare [2,3,4]), however, admits the conclusion that we have indeed equality in (1.2) (for details see below).

Corollary 1.1 $c(\mathcal{C}) = 2$.

It is due to Egervary [6] (for $\alpha = 3$) and Ruscheweyh [9] (for $\alpha > 3$) that $f * \sigma_n^\alpha$ are convex univalent for these values of α if $f \in \mathcal{C}$. Therefore, we have the following second corollary to Theorem 1.2.

Corollary 1.2 *Let $f \in \mathcal{C}$ and $n \in \mathbb{N}$. Then there exists a convex univalent polynomial p_n of degree $\leq n$ such that*

$$f_{1-\frac{4}{n}} \prec p_n \prec f. \tag{1.3}$$

It is not known, however, whether we can generally replace the number 4 in (1.3) by a smaller one.

The use of Cesàro means of order 1 to obtain Corollary 1.1 was actually suggested by the fact that these means describe the 'maximal polynomial range' for half-plane mappings, the most natural extremal functions in the class $\mathcal{C}$. To be more specific, we briefly describe this concept: given a simply connected domain $\Omega \subset \mathbb{C}$ and a point $w \in \Omega$, the 'maximal range' is

$$\Omega_n := \bigcup p_n(\mathbb{D}),$$

the union of all image-domains $p_n(\mathbb{D})$ where p_n is a polynomial of degree n with $p_n(0) = w$ and $p_n(\mathbb{D}) \subset \Omega$. The general theory shows that in many cases the boundary $\partial \Omega_n \backslash \Omega$ can be described by portions of $p(\partial \mathbb{D})$, where p runs through a very small set of admissable polynomials. In case of a half-plane Ω it can be shown that the essential boundary of Ω_n is described by the first Cesàro mean of the corresponding mapping function f. Fig. 1 demonstrates this (and the contents of Theorem 1.2) for $n = 4$ ($\alpha = 1$). The half-plane is $\operatorname{Re} z > \frac{1}{2}$, the function $f(z) = \frac{1}{1-z}$. The hashed area (dark and light) is Ω_4, and the curved graph which partially coincides with the boundary of Ω_4 is the boundary of $f * \sigma_4^1(\mathbb{D})$. The interior circle (dark-grey hashed) is $f_{2/3}(\mathbb{D})$. Note that the latter set touches the former one in $n = 4$ points: the three cusps and the maximum at $3 = \frac{n+2}{2}$. A similar situation prevails for all n (and this Ω).

This connection of maximal ranges and the present problem can, at least, lead to reasonable guesses for suitable polynomial convolution factors if one has a guess for the extremal domain (univalent function) in a subclass of univalent functions, and if one can solve the maximal range problem for this domain. In the case of starlike functions in $\mathcal{S}_0$ the natural conjecture for the extremal function will be the *Koebe-function* $k(z) := \frac{z}{(1-z)^2}$ which maps $\mathbb{D}$ onto the slit domain $\mathbb{C} \backslash] - \infty, -\frac{1}{4}]$.

Define polynomials λ_n of degree n by

$$\lambda_n(z) := 1 + \frac{\cot \frac{\pi}{2(n+1)}}{2(n+1)} \sum_{k=1}^{n} \frac{n+1-k}{k} \sin \frac{k\pi}{n+1} z^k.$$

According to the results in [4] $k * \lambda_n = z\lambda_n'(z)$ describes the maximal range Ω_n of k as indicated in Fig. 2, which has the same interpretation as Fig. 1. The right hand side of Fig. 2 is an enlargement of the essential portion of the left hand side.

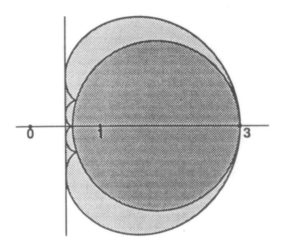

Figure 1: Maximal range and maximal subordination for $n = 4$ and a halfplane.

One can show that

$$k_\epsilon \prec \lambda_n * k \prec k,$$

implies

$$\epsilon \leq e_n := \frac{1 - \sin \frac{\pi}{2(n+1)}}{1 + \sin \frac{\pi}{2(n+1)}}, \tag{1.4}$$

since $k(e_n) = (\lambda_n * k)(1)$. However,

$$k_\epsilon(\mathbb{D}) \not\subset \Omega_n \text{ for } \epsilon > e_n,$$

and therefore e_n can be used to derive a lower bound for $c(\mathcal{S}_0)$, and therefore for c itself. We have

$$e_n \approx 1 - \frac{\pi}{n},$$

and therefore

$$c \geq c(\mathcal{S}_0) \geq \pi. \tag{1.5}$$

Our result concerning $k(z)$ suggests:

Conjecture 1.1 Let f be a univalent mapping from $\mathbb{D}$ onto some domain $\Omega \subset \mathbb{C}$, starlike w.r.t. the origin. Then $f * \lambda_n$ is a univalent polynomial of degree n with

$$f_{e_n} \prec f * \lambda_n \prec f,$$

264

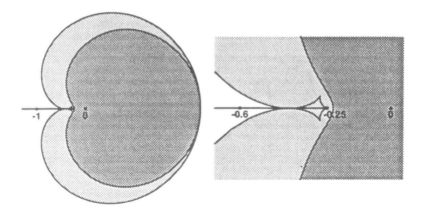

Figure 2: Maximal range and maximal subordination for $n = 4$ and the Koebe domain.

where e_n is as in (1.4). *If f is a rotation of the Koebe function, then e_n cannot be replaced by any greater number.*

Presently we can prove only one part of this conjecture. Since its proof contains an ingredient which can be useful in other situations as well (including the proof of the rest of Conjecture 1.1), we include it here.

Theorem 1.3 *Let f be a starlike function. Then*

$$f * \lambda_n \prec f.$$

The authors performed extensive numerical experiments in the present context. There are examples for univalent functions f for which

$$f_{e_n} \not\prec \lambda_n * f$$

and where the latter functions are apparently not even univalent (the examples being spiral-like functions). This, of course, does not rule out the possibility that we have equality in (1.5). On the other hand, there is some numerical evidence that Cesàro means of order 4, or even 3, of univalent functions could serve as suitable polynomials p_n in Theorem 1.1. If this were so, then we would have $c \leq 2\sqrt{5} = 4.472\ldots$. It would not be wise, however, to make a corresponding conjecture at our present state of knowledge. Perhaps a refined version of Lemma 3.1 below, applicable for any univalent functions instead of close-to-convex ones, and a procedure similiar to that one applied to prove Theorem 1.3 could be successful.

265

2 PROOF OF THEOREM 1.2

We start with the special case $f = h := \frac{1}{1-z}$ and later apply this to obtain the general result, using convolution methods. Note that h acts as an identity for convolution.

Lemma 2.1 *If $\alpha \geq 1$ and $n \in \mathbb{N}$ then $\sigma_n^\alpha \prec h$.*

Proof: We have to show $\operatorname{Re} \sigma_n^\alpha(z) > \frac{1}{2}$ for $z \in \mathbb{D}$. To do so we use the generating function for $\binom{n+\alpha}{n}\sigma_n^\alpha$:

$$\sum_{n=0}^{\infty} \binom{n+\alpha}{n}\sigma_n^\alpha(z)\, x^n = \frac{1}{(1-xz)(1-x)^{\alpha+1}}.$$

From this we obtain

$$\sum_{n=0}^{\infty} \binom{n+\alpha}{n} \left(\sigma_n^\alpha(z) - \frac{1}{2} \right) x^n$$

$$= \frac{1+xz}{2(1-xz)(1-x)^{\alpha+1}}$$

$$= \frac{1}{(1-x)^{\alpha-1}} \left[-\frac{z}{(1-z)^2}\frac{1}{1-x} + \frac{1}{2}\frac{1+z}{1-z}\frac{1}{(1-x)^2} + \left(\frac{z}{1-z}\right)^2 \frac{1}{1-xz} \right]$$

$$= \sum_{n=0}^{\infty} \binom{n+\alpha-2}{n} x^n \cdot \sum_{n=0}^{\infty} \left[-\frac{z}{(1-z)^2} + \frac{n+1}{2}\frac{1+z}{1-z} + \frac{z^{n+2}}{(1-z)^2} \right] x^n.$$

For $z = e^{i\phi}$, $\phi \in \mathbb{R}$, the real part of the expression in square brackets equals

$$\frac{1}{2(1-\cos\phi)} - \frac{\cos(n+1)\phi}{2(1-\cos\phi)} \geq 0.$$

Comparing coefficients and using $\binom{n+\alpha-2}{n} \geq 0$, $\alpha \geq 1$, we deduce as desired $\operatorname{Re}\left(\sigma_n^\alpha(e^{i\phi}) - \frac{1}{2}\right) \geq 0$. ∎

The next lemma is the well-known former Wilf's conjecture ([10, p. 86] or [5, p. 254]).

Lemma 2.2 *Let f, h be convex univalent functions in $\mathbb{D}$, and let $g \prec h$ be holomorphic in $\mathbb{D}$. Then*

$$f * g \prec f * h.$$

The first assertion

$$f * \sigma_n^\alpha \prec f = f * h$$

of Theorem 1.2 follows readily from Lemma 2.1 and Lemma 2.2.

Now we provide a technical lemma which, as a matter of fact, already implies $h_{\frac{n}{n+\alpha+1}} \prec \sigma_n^\alpha$.

Lemma 2.3 *For* $n \in \mathbb{N}$ *and* $\alpha \geq 0$ *define*

$$t_n^\alpha(z) := \frac{\sigma_n^\alpha(z) - h_{\frac{n}{n+\alpha+1}}(z)}{z h'_{\frac{n}{n+\alpha+1}}(z)}.$$

Then, for $\alpha \geq 1$ *and* $z \in \mathbb{D}$,
$$\mathrm{Re}\, t_n^\alpha(z) > 0.$$

Proof: Again we make use of the generating function for $\binom{n+\alpha}{n}\sigma_n^\alpha$, and also of a differential equation satisfied by σ_n^α:

$$z(1-z)\sigma_n^{\alpha\prime}(z) - (n+\alpha+1)\left[\left(1 - \frac{n}{n+\alpha+1}z\right)\sigma_n^\alpha(z) - 1\right] = 0.$$

This implies

$$t_n^\alpha(z) = \frac{1}{(n+\alpha+1)^2}\left[(\alpha+1)(1-z) + n(1-z)^2\right]\sigma_n^{\alpha\prime}(z).$$

Differentiation of the generating function for $\binom{n+\alpha}{n}\sigma_n^\alpha$ with respect to z and afterwards once again with respect to x supplies us with a generating function for $(n+\alpha+1)^2\binom{n+\alpha}{n}t_n^\alpha$, namely

$$\sum_{n=0}^\infty (n+\alpha+1)^2\binom{n+\alpha}{n}t_n^\alpha(z)\, x^n$$

$$= \frac{(\alpha+1)(1-z)x}{(1-xz)^2(1-x)^{\alpha+1}} + \frac{(1-z)^2 x[1 + \alpha x + xz - (\alpha+2)x^2 z]}{(1-xz)^3(1-x)^{\alpha+2}}$$

$$= \frac{x}{(1-x)^{\alpha-1}}\left[\frac{\alpha+1}{(1-x)^3} + \frac{1-\alpha z}{1-z}\frac{1}{(1-x)^2} + \frac{z[(\alpha-1)z-4]}{(1-z)^2}\frac{1}{1-x} + \right.$$
$$\left. \frac{2z^2}{(1-xz)^3} + \frac{z^2(z+3)}{1-z}\frac{1}{(1-xz)^2} - \frac{z^2[(\alpha-1)z-4]}{(1-z)^2}\frac{1}{1-xz}\right]$$

$$= x\sum_{n=0}^\infty \binom{n+\alpha-2}{n}x^n \cdot \sum_{n=0}^\infty \tilde{t}_n^\alpha(z)x^n.$$

Here $\tilde{t}_n^\alpha(z)$ is defined by

$$\tilde{t}_n^\alpha(z) := \left[(\alpha+1)\binom{n+2}{2} + \frac{1-\alpha z}{1-z}\binom{n+1}{1} + \frac{z[(\alpha-1)z-4]}{(1-z)^2}\right] + $$
$$\left[2\binom{n+2}{2} + \frac{z+3}{1-z}\binom{n+1}{1} + \frac{4-(\alpha-1)z}{(1-z)^2}\right]z^{n+2}.$$

As in the proof of Lemma 2.1 it is sufficient to show $\mathrm{Re}\, \tilde{t}_n^\alpha(z) > 0$ for $n \in \mathbb{N}$ and $z \in \mathbb{D}$ to obtain $\mathrm{Re}\, t_n^\alpha(z) > 0$. For $z = e^{i\phi}$, $\phi \in \mathbb{R}$, one verifies after some straightforward computations

$$
\operatorname{Re} \tilde{t}^{\alpha}_{n-1}(e^{i\phi}) \;=\; 2\left[n\cos\frac{(n+1)\phi}{2} + \frac{\sin\frac{n\phi}{2}}{\sin\frac{\phi}{2}} \right]^2 +
$$
$$
\frac{\alpha-1}{2}\left[n(n+2) - \frac{\sin\frac{n\phi}{2}\,\sin\frac{(n+2)\phi}{2}}{(\sin\frac{\phi}{2})^2} \right],
$$

and therefore

$$
\operatorname{Re} \tilde{t}^{\alpha}_{n}(z) > 0, \quad \alpha \ge 1,\; z \in \mathbb{D},
$$

as desired. ∎

To complete the proof of Theorem 1.2 we need another result from convolution theory (compare [10, p. 54] or [5, p. 248]).

Lemma 2.4 *Let f be a convex and g be a starlike function with $f(0) = g(0) = 0$. Let F be any analytic function in $\mathbb{D}$. Then*

$$
\frac{f * gF}{f * g}(\mathbb{D}) \;\subset\; \overline{\mathrm{co}}F(\mathbb{D})
$$

where $\overline{\mathrm{co}}F(\mathbb{D})$ denotes the closure of the convex hull $\mathrm{co}F(\mathbb{D})$.

We recall that $f * \sigma^{\alpha}_n$ is univalent in $\mathbb{D}$ for $\alpha \ge 1$ (compare the paragraph before the statement of Theorem 1.2). Hence to prove the assertion

$$
f_{\frac{n}{n+\alpha+1}} \prec f * \sigma^{\alpha}_n
$$

it suffices to show that

$$
\sigma^{\alpha}_n(f,z) \notin f_{\frac{n}{n+\alpha+1}}(\mathbb{D})
$$

for all z with $|z| = 1$. We use a geometrical consideration which is similiar to the one in [5, p. 253]. For a given $z \in \mathbb{D}$ the number

$$
\operatorname{Re} \frac{\sigma^{\alpha}_n(f,z) - f_{\frac{n}{n+\alpha+1}}(z)}{zf'_{\frac{n}{n+\alpha+1}}(z)} \tag{2.1}
$$

is a nonnegative multiple of the cosine of the angle between the vector from $f_{\frac{n}{n+\alpha+1}}(z)$ to $\sigma^{\alpha}_n(f,z)$ and the outer normal vector on the curve $f_{\frac{n}{n+\alpha+1}}(\partial\mathbb{D})$, at z. Now suppose that (2.1) is positive. Then $\sigma^{\alpha}_n(f,z)$ lies outside of $f_{\frac{n}{n+\alpha+1}}(\mathbb{D})$, since $f_{\frac{n}{n+\alpha+1}}(\mathbb{D})$ is a convex set. So we are left with the proof that (2.1) is positive. For $z \in \mathbb{D}$ we have

$$
\frac{\sigma^{\alpha}_n(f,z) - f_{\frac{n}{n+\alpha+1}}(z)}{zf'_{\frac{n}{n+\alpha+1}}(z)} = \frac{f(z) * \left[\sigma^{\alpha}_n(z) - h_{\frac{n}{n+\alpha+1}}(z)\right]}{z\left[f * h_{\frac{n}{n+\alpha+1}}\right]'(z)} = \frac{f(z) * \left[zh'_{\frac{n}{n+\alpha+1}}(z)\,t^{\alpha}_n(z)\right]}{f(z) * zh'_{\frac{n}{n+\alpha+1}}(z)}.
$$

The function $z \mapsto zh'_{\frac{n}{n+\alpha+1}}(z)$ is starlike. Therefore Lemma 2.4 in combination with Lemma 2.3 show that (2.1) is positive. This completes the proof of Theorem 1.2. ∎

3 PROOF OF THEOREM 1.3

To prove this theorem we use a special characterization of $g \prec f$ in the case that f is univalent in $\mathbb{D}$. Assume $g(0) = f(0) = 0$. Then $g \prec f$ is equivalent to $g(\mathbb{D}) \subset f(\mathbb{D})$ which in turn is equivalent to

$$f(z) \notin g_{|z|}(\mathbb{D})$$

for all $z \in \mathbb{D}\backslash\{0\}$. This leads immediately to the following characterization:
  g is subordinate to a univalent function f with $f(0) = 0$ iff $g(0) = 0$ and

$$g(xz) - f(z) \neq 0$$

holds for all $x \in \mathbb{D}, z \in \mathbb{D}\backslash\{0\}$. Therefore we are done if we show

$$\frac{1}{z}\left(f(z) * [\lambda_n(xz) - h(z)]\right) \neq 0$$

for all $x, z \in \mathbb{D}$.

In order to further reduce the problem, we can apply another result from convolution theory in [10, p. 76]:

Lemma 3.1 *Let f be a starlike function and g be a close-to-convex function. Then for all $z \in \mathbb{D}$*

$$\frac{1}{z}(f * g)(z) \neq 0.$$

In the following lemma we show that $z \mapsto \lambda_n(xz) - h(z)$ is a close-to-convex function for all $x \in \mathbb{D}$. This, via Lemma 3.1, will then prove Theorem 1.3.

Lemma 3.2 *The function*

$$z \mapsto \lambda_n(xz) - h(z)$$

is close-to-convex univalent for all $x \in \mathbb{D}$.

We remark that numerical calculations indicate that $z \mapsto \lambda_n(xz) - h(z)$ is not starlike, hence there is no hope in reversing the application of Lemma 3.1 to get corresponding conclusions as in Theorem 1.3 for close-to-convex functions.

Proof: First we consider

$$\tau_n(z) := z\lambda_n'(z) = \frac{\cot \frac{\pi}{2(n+1)}}{2(n+1)} \sum_{k=1}^{n} (n+1-k) \sin \frac{k\pi}{n+1} \, z^k.$$

Using the formula

$$\sum_{k=1}^{n} \sin kx \, z^k = \frac{z \sin x - z^{n+1} \sin(n+1)x + z^{n+2} \sin nx}{1 - 2z \cos x + z^2}$$

269

for $x = \frac{\pi}{n+1}$ and also the equation which can be obtained from this one by differentiating with respect to z we get after some calculations

$$\tau_n(z) = \left[\cos\frac{\pi}{2(n+1)}\right]^2 \left[l_n(z) - \frac{(1-z^2)(1+z^{n+1})}{(n+1)z}l_n(z)^2\right]$$

where

$$l_n(z) := \frac{z}{1 - 2z\cos\frac{\pi}{n+1} + z^2}.$$

Let g be the starlike mapping $g(z) := \frac{z}{1-z^2}$, which maps $\mathbb{D}$ onto the double-slit domain $\mathbb{C} \setminus \{it : t \in \mathbb{R}, |t| \geq \frac{1}{2}\}$. For $u, t \in \mathbb{R}$ one gets

$$\mathrm{Re}\,\frac{\tau_n(e^{iu}) - k(e^{it})}{g(e^{iu})} = \mathrm{Re}\,\frac{\tau_n(e^{iu})}{g(e^{iu})} = \mathrm{Re}\left[-2i\sin u\,\tau_n(e^{iu})\right]$$

$$= \frac{1}{n+1}\left(2\cos\frac{\pi}{2(n+1)}\frac{\cos\frac{(n+1)u}{2}}{\cos u - \cos\frac{\pi}{n+1}}\sin u\right)^2 \geq 0.$$

This shows

$$\mathrm{Re}\,\frac{z\left(\lambda_n(xz) - h(z)\right)'}{g(xz)} > 0$$

for all $x, z \in \mathbb{D}$. Therefore $z \mapsto \lambda_n(xz) - h(z)$ is in fact close-to-convex. ■

We finally mention another conjecture which, based on the previous method, could complete the proof of Conjecture 1.1. Numerical calculations support the following conjecture:

Conjecture 3.1 *The function*

$$z \mapsto h(e_n xz) - \lambda_n(z)$$

is close-to-convex for all $x \in \mathbb{D}$.

If one could furthermore prove that $f * \lambda_n$ is univalent, for f starlike, then

$$f_{e_n} \prec f * \lambda_n,$$

as in the proof of Theorem 1.3.

REFERENCES

[1] Andrievskij V., Ruscheweyh S. Maximal polynomial subordination to univalent functions in the unit disk. To appear in *Constr. Approx.*.

[2] Cordova A., Ruscheweyh S. On maximal ranges of polynomial spaces in the unit disk. *Constr. Approx.*, 5:309–327, 1989.

[3] Cordova A., Ruscheweyh S. On maximal polynomial ranges in circular domains. *Compl. Var.*, **10**:295–309, 1988.

[4] Cordova A., Ruscheweyh S. On the maximal range problem for slit domains. In *Computational Methods and Function Theory*. Proceedings, Valparaiso, 1989. Lecture Notes in Mathematics. Springer-Verlag, 1990. S. Ruschweyh, E.B. Saff, L.C. Salinas, R.S. Varga, eds.

[5] Duren P.L. *Univalent Functions*. Springer-Verlag, Berlin, 1983.

[6] Egervary E. Abbildungseigenschaften der arithmetischen Mittel der geometrischen Reihe. *Math. Z.*, **42**:221–230, 1937.

[7] Greiner R. *Zur Güte der Approximation schlichter Abbildungen durch maximal subordinierende Polynomfolgen*. Diplomarbeit, Würzburg, 1993.

[8] Lewis J. Applications of a convolution theorem to Jacobi polynomials. *SIAM Math. Anal.*, **10**:1110–1120, 1979.

[9] Ruscheweyh, S. Geometric properties of Cesàro means. *Results in Mathematics*, **22**:739–748, 1992.

[10] Ruscheweyh S. *Convolutions in Geometric Function Theory*. Séminaire de Mathématiques supérieures, NATO Advanced Study Institute, Les Presses de l'Université de Montréal, Montréal, 1982.

[11] Ruscheweyh S., Salinas L.C. Subordination by Cesàro means. *Compl. Var.*, **21**:279–285, 1993.

International Series of Numerical Mathematics, Vol. 119, ©1994 Birkhäuser

SOLUTIONS TO THE ASSOCIATED q-ASKEY-WILSON POLYNOMIAL RECURRENCE RELATION*

Dharma P. Gupta[†]   David R. Masson[†]

In honor of Walter Gautschi on the occasion of his 65th birthday

Abstract. A $_{10}\phi_9$ contiguous relation is used to derive contiguous relations for a very-well-poised $_8\phi_7$. These in turn yield solutions to the associated q-Askey-Wilson polynomial recurrence relation, expressions for the associated continued fraction, the weight function and a q-analogue of a generalized Dougall's theorem. Key words and phrases: *contiguous relations, Askey-Wilson polynomials, continued fraction, Pincherle's theorem, Jacobi matrix, spectrum, Dougall's theorem.*

1 INTRODUCTION

A deeper insight into the properties of classical orthogonal polynomials is obtained by studying their associated cases. For recent work on associated Hermite, Laguerre, ultraspherical, Jacobi, continuous Hahn, continuous dual Hahn, big q-Laguerre and big q-Jacobi polynomials see [2], [4], [8], [9], [11], [12], [14], [15], [23]. The most general case of associated q-Askey-Wilson polynomials [16] and their $q \to 1$ limit, the associated Wilson polynomials, have also been investigated [13], [18].

In this paper we examine the associated q-Askey-Wilson polynomial case using the methods of [18]. All of our parameters are, in general, complex. We essentially follow the notation in Gasper and Rahman [5], except that we omit the designation q for the base in the q-shifted factorials and basic hypergeometric functions. Thus we have, with $|q| < 1$,

$$(a)_0 := 1, \quad (a)_n := \prod_{j=1}^{n}(1 - aq^{j-1}), \quad n > 0, \text{ or } n = \infty,$$

$$(a_1, a_2, \cdots, a_k)_n := \prod_{j=1}^{k}(a_j)_n.$$

The basic hypergeometric series $_{r+1}\phi_r$ is

$$_{r+1}\phi_r \left(\begin{array}{cccc} a_1, & a_2, & \cdots, & a_{r+1} \\ b_1, & b_2, & \cdots, & b_r \end{array} ; z \right) := \sum_{n=0}^{\infty} \frac{(a_1, a_2, \cdots, a_{r+1})_n}{(b_1, b_2, \cdots, b_r, q)_n} z^n.$$

AMS subject classification: 33D45, 40A15, 39A10, 47B39

*Research partially supported by NSERC (Canada)

[†]Department of Mathematics, University of Toronto, Toronto, Ontario, Canada, M5S 1A1

The explicit form of monic q-Askey-Wilson polynomials is given by [1]:

$$P_n(z) = P_n(z; \alpha, \beta, \gamma, \delta) = \frac{(\alpha\beta, \alpha\gamma, \alpha\delta)_n (\alpha\beta\gamma\delta/q)_n}{(2\alpha)^n (\alpha\beta\gamma\delta/q)_{2n}} \tag{1.1}$$

$$\times \; {}_4\phi_3\left(\begin{matrix} q^{-n}, \alpha\beta\gamma\delta q^{n-1}, \alpha u, \alpha/u \\ \alpha\beta, \alpha\gamma, \alpha\delta \end{matrix}; q\right), \qquad z = \frac{u + u^{-1}}{2},$$

which satisfy the three-term recurrence

$$P_{n+1}(z) - (z - a_n)P_n(z) + b_n^2 P_{n-1}(z) = 0, \tag{1.2}$$

$$a_n = a_n(\alpha, \beta, \gamma, \delta) = -A_n - B_n + \frac{\alpha}{2} + \frac{1}{2\alpha},$$

$$b_n^2 = b_n^2(\alpha, \beta, \gamma, \delta) = A_{n-1}B_n,$$

$$A_n = \frac{(1 - \alpha\beta\gamma\delta q^{n-1})(1 - \alpha\beta q^n)(1 - \alpha\gamma q^n)(1 - \alpha\delta q^n)}{2\alpha(1 - \alpha\beta\gamma\delta q^{2n-1})(1 - \alpha\beta\gamma\delta q^{2n})},$$

$$B_n = \frac{\alpha(1 - q^n)(1 - \beta\gamma q^{n-1})(1 - \beta\delta q^{n-1})(1 - \gamma\delta q^{n-1})}{2(1 - \alpha\beta\gamma\delta q^{2n-2})(1 - \alpha\beta\gamma\delta q^{2n-1})}.$$

Associated monic q-Askey-Wilson polynomials are defined [16] by introducing an extra parameter, say ϵ, through a shift of the discrete variable n in the coefficients of (1.2). Here we choose to replace n by $n + (\log \epsilon)/(\log q)$ so that q^n is replaced by ϵq^n. The new polynomial set $\{P_n(z; \epsilon)\}$ is an initial value solution $(X_{-1} = P_{-1} = 0, \; X_0 = P_0 = 1)$ to the difference equation

$$X_{n+1} - (z - a_v)X_n + b_v^2 X_{n-1} = 0, \qquad v := n + (\log \epsilon)/(\log q). \tag{1.3}$$

The associated q-Askey-Wilson polynomials have been studied by Ismail and Rahman [16]. They derive contiguous relations for a very well poised ${}_8\phi_7$ series and then obtain two linearly independent solutions of the three-term recurrence relation for the associated Askey-Wilson polynomials. Explicit representations of two families of associated Askey-Wilson polynomials are given. Corresponding continued fractions are identified and weight functions obtained.

On the other hand Masson [18] studied the associated Wilson polynomials by obtaining ${}_7F_6$ contiguous relations with a procedure (see Wilson [22]) different from that employed by Ismail and Rahman. He discussed explicit forms of solutions to the three-term recurrence satisfied by associated Wilson polynomials and identified minimal solutions for different regions. Using Pincherle's theorem [6],[17], he obtained a number of new continued fraction representations and discussed the spectral properties of the corresponding Jacobi matrix. A weight function was obtained and his method gave a generalization of Dougall's theorem [3] which removed both the terminating and the balancing conditions.

The object of the present article is to study the associated q-Askey-Wilson polynomials using Masson's approach [18]. In spite of the fact that some of the basic results here will overlap with the results obtained by Ismail and Rahman [16], we have to offer a different procedure for obtaining the $_8\phi_7$ contiguous relations, the minimal solution to the associated Askey-Wilson polynomial recurrence, the corresponding continued fraction representations and an extension of Masson's generalization of Dougall's theorem [18] to the basic hypergeometric case. Moreover we again see, as in many previous cases ([7], [8], [9], [18], [19], [21]), that Pincherle's theorem [6],[17] plays a crucial role. The importance of Pincherle's theorem was re-emphasized in 1967 by Gautschi [6].

In §2, following Wilson's method [22], we derive a pair of three-term recurrence relations for very-well-poised balanced and terminating $_{10}\phi_9$ basic hypergeometric functions. From these recurrence relations and using certain known transformations [5] we obtain a number of explicit solutions to (1.3). Minimal solutions for different regions of the complex plane are identified. In §3, the associated continued fractions are obtained. In §4 we derive an expression for the weight function of orthogonality and a Wronskian identity which turns out to be the basic analogue of a generalized Dougall's theorem [18].

2   RECURRENCE RELATION AND SOLUTIONS

In a recent paper [10] we have derived contiguous relations for a very-well-poised balanced $_{10}\phi_9$ basic hypergeometric function

$$\phi := \,_{10}\phi_9 \left(\begin{array}{c} a, q\sqrt{a}, -q\sqrt{a}, b, c, d, e, f, g, h \\ \sqrt{a}, -\sqrt{a}, \dfrac{aq}{b}, \dfrac{aq}{c}, \ldots, \dfrac{aq}{h} \end{array} ; q \right), \qquad (2.1)$$

where one of the numerator parameters, say h, is equal to q^{-n}, $n = 0, 1, \ldots$, $bcdefgh = a^3q^2$ and $|q| < 1$. One such relation is

$$\frac{b(1-c)(1-\frac{a}{c})(1-\frac{aq}{c})(1-\frac{aq}{bd})(1-\frac{aq}{be})(1-\frac{aq}{bf})(1-\frac{aq}{bg})(1-\frac{aq}{bh})}{(1-\frac{cq}{b})} \qquad (2.2)$$

$$\times \left[\phi(b-, c+) - \phi \right]$$

$$- \frac{c(1-b)(1-\frac{a}{b})(1-\frac{aq}{b})(1-\frac{aq}{cd})(1-\frac{aq}{ce})(1-\frac{aq}{cf})(1-\frac{aq}{cg})(1-\frac{aq}{ch})}{(1-\frac{bq}{c})}$$

$$\times \left[\phi(c-, b+) - \phi \right]$$

$$- \frac{aq}{c}(1-\frac{c}{b})(1-\frac{bc}{aq})(1-d)(1-e)(1-f)(1-g)(1-h)\phi = 0.$$

Note that we are using the notation $\phi(b-, c+)$ to mean ϕ in (2.1) with b replaced by b/q and c replaced by cq.
In (2.2), if we write $h = q^{-n}$, $g = a^3q^{n+2}/bcdef$ and let $n \to \infty$, we obtain

$$\frac{bcdef}{a^2q} \frac{(1-c)(1-\frac{a}{c})(1-\frac{aq}{c})(1-\frac{aq}{bd})(1-\frac{aq}{be})(1-\frac{aq}{bf})}{(1-\frac{cq}{b})} \left[\,_8\phi_7(b-, c+) - \,_8\phi_7 \right]$$

275

$$-\frac{bcdef}{a^2q}\frac{(1-b)(1-\frac{a}{b})(1-\frac{aq}{b})(1-\frac{aq}{cd})(1-\frac{aq}{ce})(1-\frac{aq}{cf})}{(1-\frac{bq}{c})}\left[\,{}_8\phi_7(c-,b+)-{}_8\phi_7\right]$$

$$-\frac{aq}{c}(1-\frac{c}{b})(1-\frac{bc}{aq})(1-d)(1-e)(1-f)\,{}_8\phi_7 \;=\; 0\,, \tag{2.3}$$

where

$$
{}_8\phi_7 \;=\; {}_8\phi_7\left(\begin{array}{c} a,q\sqrt{a},-q\sqrt{a},b,c,d,e,f \\[2pt] \sqrt{a},-\sqrt{a},\dfrac{aq}{b},\dfrac{aq}{c},\dfrac{aq}{d},\dfrac{aq}{e},\dfrac{aq}{f} \end{array}\;;\;\dfrac{a^2q^2}{bcdef}\right)
$$

$$= W(a;\,b,c,d,e,f;\frac{a^2q^2}{bcdef})\quad\text{say}\,.$$

For non-terminating ${}_8\phi_7$, the convergence condition is $|a^2q^2/bcdef|<1$. In (2.3) we now put

$$a \;=\; \frac{\alpha\beta\gamma u}{q},\quad b \;=\; q^{-n}/\epsilon,\quad c \;=\; \epsilon s q^{n-1},\quad d \;=\; \alpha u,\quad e \;=\; \beta u,\quad f \;=\; \gamma u,$$
$$s \;=\; \alpha\beta\gamma\delta\,.$$

Renormalizing and simplifying, we obtain the associated Askey-Wilson recurrence

$$X_{n+1}^{(1)} - (z - a_n')X_n^{(1)} + b_n'^2 X_{n-1}^{(1)} \;=\; 0 \tag{2.5}$$

with

$$z \;=\; \frac{u+u^{-1}}{2}\,,$$

$$a_n' \;=\; -A_n' - B_n' + \frac{\alpha}{2} + \frac{1}{2\alpha}$$

$$b_n'^2 \;=\; A_{n-1}'B_n'$$

$$A_n' \;=\; \frac{(1-sq^{n-1}\epsilon)(1-\alpha\beta\epsilon q^n)(1-\alpha\gamma\epsilon q^n)(1-\alpha\delta\epsilon q^n)}{2\alpha(1-sq^{2n-1}\epsilon^2)(1-sq^{2n}\epsilon^2)}\,,$$

$$B_n' \;=\; \frac{\alpha(1-\epsilon q^n)(1-\beta\gamma\epsilon q^{n-1})(1-\beta\delta\epsilon q^{n-1})(1-\gamma\delta\epsilon q^{n-1})}{2(1-s\epsilon^2 q^{2n-2})(1-s\epsilon^2 q^{2n-1})}\,,$$

and a solution of (2.5) is therefore

$$X_n^{(1)}(u) \;=\; \left(\frac{u}{2}\right)^n \frac{\left(su\epsilon q^n/\delta\right)_\infty\left(s\epsilon^2 q^{2n-1}\right)_\infty}{\left(s\epsilon q^{n-1}\right)_\infty\left(\delta\epsilon q^n/u\right)_\infty\left(\alpha\beta\epsilon q^n,\alpha\gamma\epsilon q^n,\beta\gamma\epsilon q^n\right)_\infty} \tag{2.6}$$

$$\times W\left(\frac{\alpha\beta\gamma u}{q}\;;\;\frac{q^{-n}}{\epsilon},\epsilon s q^{n-1},\alpha u,\beta u,\gamma u;\frac{q}{\delta u}\right)\,.$$

The solution $X_n^{(1)}$ is a generalization of the Askey-Wilson solution (1.1) to the $\epsilon=1$ case of (2.5). This can be verified through an application of Watson's formula ([5], III.18, p.242) connecting a terminating ${}_4\phi_3$ with an ${}_8\phi_7$.

A second linearly independent solution of (2.5) may be obtained with the help of the reflection symmetry transformation

$$v \to -v-1, \qquad (\alpha, \beta, \gamma, \delta) \to (q/\alpha, q/\beta, q/\gamma, q/\delta).$$

Using (see also [10])

$$b^2_{-v-1}(q/\alpha, q/\beta, q/\gamma, q/\delta) = b^2_{v+1}(\alpha, \beta, \gamma, \delta)$$
$$a_{-v-1}(q/\alpha, q/\beta, q/\gamma, q/\delta) = a_v(\alpha, \beta, \gamma, \delta)$$

and renormalizing, we arrive at the solution

$$X_n^{(2)}(u) = \left(\frac{u}{2}\right)^n \frac{\left(s\epsilon^2 q^{2n-1}\right)_\infty \left(\epsilon\delta u q^{n+1}\right)_\infty}{\left(\epsilon q^{n+1}\right)_\infty \left(\beta\delta\epsilon q^n, \gamma\delta\epsilon q^n, \alpha\delta\epsilon q^n\right)_\infty \left(\frac{s\epsilon}{\delta u}q^{n-1}\right)_\infty} \qquad (2.7)$$
$$\times W\left(\frac{q^2 u}{\alpha\beta\gamma}; \ \epsilon q^{n+1}, \frac{q^{-n+2}}{\epsilon s}, \frac{qu}{\alpha}, \frac{qu}{\beta}, \frac{qu}{\gamma}; \frac{\delta}{u}\right).$$

In order to obtain a third solution of (2.5) we make the following alterations in (2.2). Interchange c and h, replace $a, b, c, \ldots, g$ by h^2/a, hb/a, $hc/a, \ldots, hg/a$ respectively, put $h = q^{-n}$, n being a positive integer, and reverse the $_{10}\phi_9$ series. This gives

$$\frac{\frac{b}{a}(1-h)(1-\frac{h}{a})(1-\frac{hq}{a})(1-\frac{aq}{bd})(1-\frac{aq}{be})(1-\frac{aq}{bf})(1-\frac{aq}{bg})(1-\frac{aq}{bc})}{(1-\frac{aq}{b})} \qquad (2.8)$$
$$\times \left[c_2\phi_+(b-) - c_1\phi\right]$$
$$-\frac{(1-\frac{bh}{a})(1-\frac{h}{b})(1-\frac{hq}{b})(1-\frac{q}{d})(1-\frac{q}{e})(1-\frac{q}{f})(1-\frac{q}{g})(1-\frac{q}{c})}{(1-\frac{bq}{a})}$$
$$\times \left[c_3\phi_-(b+) - c_1\phi\right]$$
$$-\frac{q}{a}(1-\frac{a}{b})(1-\frac{b}{q})(1-\frac{hd}{a})(1-\frac{he}{a})(1-\frac{hf}{a})(1-\frac{hg}{a})(1-\frac{hc}{a})c_1\phi = 0,$$

where

$$c_1 = \frac{(\sqrt{a})_n(-\sqrt{a})_n\left(\frac{aq}{b}\right)_n\left(\frac{aq}{c}\right)_n \cdots \left(\frac{aq}{g}\right)_n\left(aq^{n+1}\right)_n}{(a)_n(q\sqrt{a})_n(-q\sqrt{a})_n(b)_n(c)_n \cdots (g)_n}(-1)^n q^{n(n-1)/2}$$

$$c_2 = \frac{(q\sqrt{a})_{n-1}(-q\sqrt{a})_{n-1}}{(aq^2)_{n-1}(q^2\sqrt{a})_{n-1}(-q^2\sqrt{a})_{n-1}}$$
$$\times \frac{\left(\frac{aq^3}{b}\right)_{n-1}\left(\frac{aq^2}{c}\right)_{n-1} \cdots \left(\frac{aq^2}{g}\right)_{n-1}\left(aq^{n+2}\right)_{n-1}}{(b)_{n-1}(cq)_{n-1} \cdots (gq)_{n-1}}(-1)^{n-1}q^{(n-1)(n-2)/2}$$

$$c_3 = \frac{\left(\frac{\sqrt{a}}{q}\right)_{n+1}\left(-\frac{\sqrt{a}}{q}\right)_{n+1}\left(\frac{a}{bq}\right)_{n+1}\left(\frac{a}{c}\right)_{n+1} \cdots \left(\frac{a}{g}\right)_{n+1}\left(aq^n\right)_{n+1}}{\left(\frac{a}{q^2}\right)_{n+1}(\sqrt{a})_{n+1}(-\sqrt{a})_{n+1}(b)_{n+1}\left(\frac{c}{q}\right)_{n+1} \cdots \left(\frac{g}{q}\right)_{n+1}}(-1)^{n+1}q^{n(n+1)/2}.$$

Note that $\phi_+(b-)$ and $\phi_-(b+)$ refer to ϕ with (a,c,d,e,f,g,h) replaced by $(aq^2, cq, dq, eq, fq, gq, hq)$ and $(aq^{-2}, c/q, d/q, e/q, f/q, g/q, h/q)$ respectively. Taking the limit $h = q^{-n} \to \infty$, $g = a^3 q^{2+n}/bcdef \to 0$, we obtain

$$\frac{b(1 - \frac{aq}{bd})(1 - \frac{aq}{be})(1 - \frac{aq}{bf})(1 - \frac{aq}{bc})}{a(1 - \frac{aq}{b})} \tag{2.9}$$

$$\times \left[\frac{a^2 q^2}{bcdef} \frac{(1 - aq)(1 - aq^2)(1 - c)(1 - d)(1 - e)(1 - f)}{(1 - \frac{aq}{b})(1 - \frac{aq^2}{b})(1 - \frac{aq}{c})(1 - \frac{aq}{d})(1 - \frac{aq}{e})(1 - \frac{aq}{f})} W_+(b-) - W \right]$$

$$- \frac{(1 - \frac{q}{d})(1 - \frac{q}{e})(1 - \frac{q}{f})(1 - \frac{q}{c})}{(1 - \frac{bq}{a})}$$

$$\times \left[\frac{bcdef}{a^2 q^2} \frac{(1 - \frac{a}{bq})(1 - \frac{a}{b})(1 - \frac{a}{c})(1 - \frac{a}{d})(1 - \frac{a}{e})(1 - \frac{a}{f})}{(1 - \frac{a}{q})(1 - a)(1 - \frac{c}{q})(1 - \frac{d}{q})(1 - \frac{e}{q})(1 - \frac{f}{q})} W_-(b+) - W \right]$$

$$- \frac{q}{a}(1 - \frac{a}{b})(1 - \frac{b}{q})(1 - \frac{a^2 q^2}{bcdef}) W = 0.$$

In (2.9), writing

$$a = \frac{\beta\gamma\delta\epsilon^2}{u} q^{2n}, \quad b = \frac{q}{\alpha u}, \quad c = q^{n+1}\epsilon, \quad d = \beta\delta\epsilon q^n \tag{2.10}$$
$$e = \gamma\delta\epsilon q^n, \quad f = \beta\gamma\epsilon q^n, \quad s = \alpha\beta\gamma\delta$$

and renormalizing, we obtain a third solution to (2.5) given by

$$X_n^{(3)}(u) = \frac{1}{(2u)^n} \frac{(s\epsilon^2 q^{2n})_\infty}{(\epsilon q^{n+1})_\infty}$$

$$\times \frac{(s\epsilon^2 q^{2n-1})_\infty (\frac{\beta\gamma\delta\epsilon}{u} q^n)_\infty (\frac{\epsilon\beta q^{n+1}}{u}, \frac{\epsilon\gamma q^{n+1}}{u}, \frac{\epsilon\delta q^{n+1}}{u})_\infty}{(\epsilon s q^{n-1})_\infty (\alpha\beta\epsilon q^n, \alpha\gamma\epsilon q^n, \alpha\delta\epsilon q^n, \beta\gamma\epsilon q^n, \beta\delta\epsilon q^n, \gamma\delta\epsilon q^n)_\infty (\frac{\beta\gamma\delta\epsilon^2}{u} q^{2n+1})_\infty}$$

$$\times W\left(\frac{\beta\gamma\delta\epsilon^2}{u} q^{2n} ; \frac{q}{\alpha u}, \epsilon q^{n+1}, \beta\delta\epsilon q^n, \gamma\delta\epsilon q^n, \beta\gamma\epsilon q^n; \frac{\alpha}{u} \right). \tag{2.11}$$

A transformation ([5], III.23, p.243) may be applied to $X_n^{(3)}$ to obtain this solution in a form which is symmetric in the parameters $\alpha, \beta, \gamma, \delta$. We get the solution (omitting constant factors)

$$X_n^{(4)}(u) = \frac{1}{(2u)^n} \frac{(\epsilon^2 s q^{2n-1})_\infty (\frac{\alpha\epsilon}{u} q^{n+1}, \frac{\beta\epsilon}{u} q^{n+1}, \frac{\gamma\epsilon}{u} q^{n+1}, \frac{\delta\epsilon}{u} q^{n+1})_\infty}{(\epsilon q^{n+1})_\infty (\frac{q^{n+2}\epsilon}{u^2})_\infty (\alpha\beta\epsilon q^n, \alpha\delta\epsilon q^n, \alpha\gamma\epsilon q^n, \beta\gamma\epsilon q^n, \beta\delta\epsilon q^n, \gamma\delta\epsilon q^n)_\infty}$$

$$\times W\left(\frac{q^{n+1}\epsilon}{u^2}; \epsilon q^{n+1}, \frac{q}{\alpha u}, \frac{q}{\beta u}, \frac{q}{\gamma u}, \frac{q}{\delta u}; s q^{n-1}\epsilon \right). \tag{2.12}$$

The large n behavior of $X_n^{(4)}(u)$ is easily seen to be

$$X_n^{(4)}(u) \overset{n\to\infty}{\sim} \frac{1}{(2u)^n}$$

while the corresponding solution

$$X_n^{(4)}(1/u) \sim \left(\frac{u}{2}\right)^n.$$

It is evident that $X_n^{(4)}(u)$ is a sub-dominant or minimal solution for $|u| > 1$, $|sq^{n-1}\epsilon| < 1$ while $X_n^{(4)}(1/u)$ is sub-dominant (minimal) for $|u| < 1$, $|sq^{n-1}\epsilon| < 1$.

We proceed to get another solution by making the following parameter replacements in (2.9):

$$a = \frac{q^{-2n+1}}{\beta\gamma\delta u\epsilon^2}, \quad b = \frac{\alpha}{u}, \quad c = \frac{q^{-n}}{\epsilon}, \quad d = \frac{1}{\beta\delta\epsilon}q^{-n+1},$$

$$e = \frac{1}{\gamma\delta\epsilon}q^{-n+1}, \quad f = \frac{1}{\beta\gamma\epsilon}q^{-n+1}.$$

(2.14)

Note that this is just the reflection transformation that was used to obtain the solution $X_n^{(2)}$ from $X_n^{(1)}$ but now it is being applied to $X_n^{(3)}$.

After renormalization, omitting constant factors and using the transformation [5, III.23, p.243] we arrive at the following solution

$$X_n^{(5)}(u) = \frac{1}{(2u)^n} \frac{\left(sq^{2n-1}\epsilon^2\right)_\infty \left(u^2\epsilon q^n\right)_\infty}{\left(sq^{n-1}\epsilon\right)_\infty \left(\alpha u\epsilon q^n, \beta u\epsilon q^n, \gamma u\epsilon q^n, \delta u\epsilon q^n\right)_\infty}$$

$$\times W\left(\frac{q^{-n}}{\epsilon u^2}; \frac{q^{-n}}{\epsilon}, \frac{\alpha}{u}, \frac{\beta}{u}, \frac{\gamma}{u}, \frac{\delta}{u}; \frac{q^{2-n}}{s\epsilon}\right).$$

(2.15)

Note that, when $\epsilon = 1$, the solution (2.15) is proportional to the Askey-Wilson polynomial $P_n(z; \alpha, \beta, \gamma, \delta)$ in (1.1) and can be used to give $P_n(z)$ in a form which is explicitly symmetric in its parameters. That is

$$P_n(z) = (2u)^{-n}\frac{(\alpha u, \beta u, \gamma u, \delta u, s/q)_n}{(s/q)_{2n}(u^2)_n}W(q^{-n}/u^2; q^{-n}, \frac{\alpha}{u}, \frac{\beta}{u}, \frac{\gamma}{u}, \frac{\delta}{u}; \frac{q^{2-n}}{\alpha\beta\gamma\delta}).$$

A solution in which the convergence condition is independent of z and also independent of the parameters $\alpha, \beta, \gamma, \delta$ can be obtained from (2.12) by applying the transformation ([5], III.24, p.243). This gives

$$X_n^{(6)}(u) = \frac{1}{(2u)^n} \frac{\left(s\epsilon^2 q^{2n-1}\right)_\infty \left(s\epsilon q^n/\alpha u, s\epsilon q^n/\beta u, s\epsilon q^n/\gamma u, s\epsilon q^n/\delta u\right)_\infty}{\left(\epsilon sq^{n-1}\right)_\infty \left(s\epsilon q^n/u^2\right)_\infty \left(\alpha\beta\epsilon q^n, \alpha\gamma\epsilon q^n, \alpha\delta\epsilon q^n, \beta\gamma\epsilon q^n, \beta\delta\epsilon q^n, \gamma\delta\epsilon q^n\right)_\infty}$$

$$\times W\left(\frac{s\epsilon q^{n-1}}{u^2}; \epsilon sq^{n-1}, \frac{\alpha}{u}, \frac{\beta}{u}, \frac{\gamma}{u}, \frac{\delta}{u}; q^{n+1}\epsilon\right)$$

(2.16).

Since $X_n^{(6)}(u) \overset{n\to\infty}{\sim} \frac{1}{(2u)^n}$, we have that $X_n^{(6)}(u)$ is a sub-dominant(minimal) solution for $|u| > 1$ while $X_n^{(6)}(1/u)$ is a sub-dominant(minimal) solution for $|u| < 1$. Note that $X_n^{(6)}$ is really the same solution as $X_n^{(4)}$, but analytically continued.

Summarizing we have

Theorem 1 *The associated Askey-Wilson equation (2.5) has solutions $X_n^{(i)}(u)$ and $X_n^{(i)}(1/u)$, $i = 1, 2, \ldots, 6$ given by (2.6), (2.7), (2.11), (2.12), (2.15) and (2.16). If $|u| > 1$, then $X_n^{(3)}(u)$, $X_n^{(4)}(u)$ and $X_n^{(6)}(u)$ (which are connected by general $_8\phi_7$ transformations) each represent a minimal solution $X_n^{(s)}(z)$ for the parameter values $\left|\frac{\alpha}{u}\right| < 1$, $|\epsilon s q^{n-1}| < 1$, $|\epsilon q^{n+1}| < 1$ respectively.*

3   CONTINUED FRACTION REPRESENTATIONS

The continued fraction associated with the difference equation (2.5) is

$$CF(z) := z - a_0' + \overset{\infty}{\underset{n=1}{K}} \left(\frac{-b_n'^2}{z - a_n'} \right) . \tag{3.1}$$

If $b_n'^2 \neq 0$, $n \geq 1$, by Pincherle's theorem [6], [17] and Theorem 1 of §2 we have

$$\frac{1}{CF(z)} = \frac{X_0^{(s)}(z)}{b_0'^2 X_{-1}^{(s)}(z)} . \tag{3.2}$$

Using $X_0^{(s)}(z) = X_0^{(4)}(u)$, $X_{-1}^{(s)}(z) = X_{-1}^{(4)}(u)$, $|u| > 1$, $z = \frac{u + u^{-1}}{2}$, we have from (2.12),

$$\frac{1}{CF(z)} = \frac{2}{u} \frac{(1 - \epsilon q/u^2)(1 - s\epsilon^2/q^2)(1 - s\epsilon^2/q)}{(1 - \alpha\epsilon/u)(1 - \beta\epsilon/u)(1 - \gamma\epsilon/u)(1 - \delta\epsilon/u)(1 - s\epsilon/q^2)}$$
$$\times \frac{W\Big(q\epsilon/u^2; \ q\epsilon, q/\alpha u, q/\beta u, q/\gamma u, q/\delta u; s\epsilon/q\Big)}{W\Big(\epsilon/u^2; \ \epsilon, q/\alpha u, q/\beta u, q/\gamma u, q/\delta u; s\epsilon/q^2\Big)} \tag{3.3}$$

for $|u| > 1$ and $|s\epsilon/q^2| < 1$. For $|u| < 1$, $|s\epsilon/q^2| < 1$ we replace u in (3.3) by $1/u$.

Other representations for the continued fraction for different parameter ranges are obtained by taking a different representation for the minimal solution. For example, by taking $X_0^{(s)}(z) = X_0^{(6)}(u)$ and $X_{-1}^{(s)}(z) = X_{-1}^{(6)}(u)$, $|u| > 1$, we obtain the continued fraction representation

$$\frac{1}{CF(z)} = \frac{2}{u} \frac{(1 - s\epsilon^2/q^2)(1 - s\epsilon^2/q)(1 - s\epsilon/u^2 q)}{(1 - \epsilon)(1 - s\epsilon/\alpha u q)(1 - s\epsilon/\beta u q)(1 - s\epsilon/\gamma u q)(1 - s\epsilon/\delta u q)} \tag{3.4}$$
$$\times \frac{W\Big(\epsilon s/u^2 q; \ \epsilon s/q, \alpha/u, \beta/u, \gamma/u, \delta/u; \epsilon q\Big)}{W\Big(s\epsilon/u^2 q^2; \ \epsilon s/q^2, \alpha/u, \beta/u, \gamma/u, \delta/u; \epsilon\Big)}, \quad \text{for} |u| > 1, |\epsilon| < 1.$$

4   WEIGHT FUNCTION

We have seen above that a minimal solution exists for $|u| > 1$ and $|u| < 1$. There is no minimal solution for $z \in [-1, 1]$, ($|u| = 1$), where we have a continuous spectrum for the associated tridiagonal Jacobi matrix J with diagonal $(a_0', a_1', \dots)$ and $(b_1', b_2', \dots)$ above and below the diagonal. For a probability measure $d\omega(x; \epsilon)$ we have, in the case of real orthogonality $(b_{n+1}'^2 > 0, a_n'$ real, $n \geq 0)$ for the associated monic q-Askey-Wilson polynomials $P_n(x; \epsilon)$,

$$\int_{\mathbb{R}} P_n(x; \epsilon) P_m(x; \epsilon) d\omega(x; \epsilon) = \delta_{nm} \prod_{k=1}^{n} b_k'^2 . \tag{4.1}$$

From [21], we have the representation

$$\frac{1}{CF(z)} = \int_{\mathbb{R}} \frac{d\omega(x; \epsilon)}{z - x} = \frac{X_0^{(s)}(z)}{b_0'^2 X_{-1}^{(s)}(z)} \tag{4.2}$$

and for the absolutely continuous part when $x \in [-1, 1]$,

$$\frac{d\omega(x; \epsilon)}{dx} = \frac{1}{2\pi i b_0'^2} \left(\frac{X_0^{(s)}(x - i0)}{X_{-1}^{(s)}(x - i0)} - \frac{X_0^{(s)}(x + i0)}{X_{-1}^{(s)}(x + i0)} \right) \tag{4.3}$$

$$= \frac{1}{2\pi i b_0'^2} \frac{\mathcal{W}\big(X_{-1}^{(s)}(x + i0), X_{-1}^{(s)}(x - i0) \big)}{\left| X_{-1}^{(s)}(x - i0) \right|^2} ,$$

where

$$\mathcal{W}(X_n, Y_n) := X_n Y_{n+1} - X_{n+1} Y_n .$$

From (2.5) we have

$$\mathcal{W}\big(X_{-1}^{(4)}(u), X_{-1}^{(4)}(1/u) \big) = \lim_{n \to \infty} \frac{\mathcal{W}\big(X_n^{(4)}(u), X_n^{(4)}(1/u) \big)}{\displaystyle\prod_{k=0}^{n} b_k'^2} . \tag{4.4}$$

Using (2.5) and (2.12) we then obtain

$$\mathcal{W}\big(X_{-1}^{(4)}(u), X_{-1}^{(4)}(1/u) \big) = 2(u - u^{-1}) \frac{\left(\frac{s\epsilon^2}{q^3} \right)_\infty \left(\frac{s\epsilon^2}{q^2} \right)_\infty}{\left(\frac{\alpha\beta\epsilon}{q}, \frac{\alpha\gamma\epsilon}{q}, \frac{\alpha\delta\epsilon}{q}, \frac{\beta\gamma\epsilon}{q}, \frac{\beta\delta\epsilon}{q}, \frac{\gamma\delta\epsilon}{q}, \frac{s\epsilon}{q^2}, \epsilon \right)_\infty} . \tag{4.5}$$

Thus (4.3) with

$$X_n^{(s)}(x \pm i0) = X_n^{(4)}(e^{\mp i\theta}), \quad x = \cos\theta, \quad u = e^{i\theta}$$

gives for $-1 \leq x \leq 1$, $\left| s\epsilon/q^2 \right| < 1$,

$$\frac{d\omega(x; \epsilon)}{dx} = \frac{2\sqrt{1 - x^2}}{\pi} \frac{(1 - s\epsilon^2/q)(1 - s\epsilon^2/q^2)^2}{(1 - s\epsilon/q^2)} \tag{4.6}$$

$$\times \frac{\left(\epsilon q/u^2, \epsilon q u^2, \alpha\beta\epsilon, \alpha\gamma\epsilon, \alpha\delta\epsilon, \beta\gamma\epsilon, \beta\delta\epsilon, \gamma\delta\epsilon, \epsilon q \right)_\infty}{\left(\alpha\epsilon/u, \alpha\epsilon u, \beta\epsilon/u, \beta\epsilon u, \gamma\epsilon/u, \gamma\epsilon u, \delta\epsilon/u, \delta\epsilon u, s\epsilon/q^2 \right)_\infty}$$

$$\times \frac{1}{\left| W\big(\epsilon/u^2; q/\alpha u, q/\beta u, q/\gamma u, q/\delta u, \epsilon; s\epsilon/q^2 \big) \right|^2} .$$

This checks with the weight function obtained by Ismail and Rahman ([16], (4.31), p.218). When $\epsilon = 1$, this reduces to the Askey-Wilson weight function [1].

We may also deduce conditions for the absence of a discrete spectrum.

From (3.4), no mass points exist if $0 \leq \epsilon < 1$ and $|\alpha|, |\beta|, |\gamma|, |\delta| < |q|^{1/2}$ [16]. These conditions ensure that when $|u| > 1$, we have $|\alpha/u|, |\beta/u|, |\gamma/u|, |\delta/u| < 1$, $|s\epsilon/\alpha uq|, |s\epsilon/\beta uq|, |s\epsilon/\gamma uq|, |s\epsilon/\delta uq| < 1$ and $|s\epsilon/q^2| < 1$. Thus all denominator terms in (3.4) are non-zero including the $W(s\epsilon/u^2q^2; \epsilon s/q^2, \alpha/u, \beta/u, \gamma/u, \delta/u; \epsilon)$ which becomes a sum of positive terms.

Similarly, from (3.3), no mass points exist if $0 \leq \epsilon s/q^2 < 1$ and $|\alpha|, |\beta|, |\gamma|, |\delta| > |q|^{1/2}$, since these conditions imply that when $|u| > 1$ we have $|\epsilon| < |q^2/s| = |q^2/\alpha\beta\gamma\delta| < 1$, $|q/\alpha u| < |q|^{1/2} < 1$ and $|\alpha\epsilon/u| = |s\epsilon/u\beta\gamma\delta| < |s\epsilon|/|q|^{3/2} < |q|^{1/2} < 1$, etc. This makes all denominators in (3.3) non-zero.

If $\epsilon = 1$ (the Askey-Wilson case) the denominator W in (3.3) is equal to one and the discrete spectrum is given explicitly by the zeros of $(\alpha/u, \beta/u, \gamma/u, \delta/u)_\infty$, provided that $|s/q| < 1 < |u|$. This is because

$$(q\alpha/u, q\beta/u, q\gamma/u, q\delta/u)_\infty W(q/u^2; q, q/\alpha u, q/\beta u, q/\gamma u, q/\delta u; s/q)$$

is an entire function of its parameters. Thus for the absence of a discrete spectrum it suffices also to have $\epsilon = 1$, $|\alpha|, |\beta|, |\gamma|, |\delta| < 1$ and $|s/q| < 1$.

Similarly if $s\epsilon/q^2 = 1$, $|\epsilon q| < 1 < |u|$ then, from (3.4), the discrete spectrum is given explicitly by the zeros of $(q/\alpha u, q/\beta u, q/\gamma u, q/\delta u)_\infty$ and there is no discrete spectrum if $s\epsilon/q^2 = 1$, $|\epsilon q| < 1$ and $|\alpha|, |\beta|, |\gamma|, |\delta| > |q|$.

We next derive an identity associated with the continuous spectrum. If we use (4.3) with

$$\frac{d\omega(x; \epsilon)}{dx} = \frac{1}{2\pi i b_0''^2} \left(\frac{X_0^{(6)}(u)}{X_{-1}^{(4)}(u)} - \frac{X_0^{(6)}(1/u)}{X_{-1}^{(4)}(1/u)} \right) , \quad x = \cos\theta, \quad u = e^{i\theta},$$

then the right side simplifies to

$$\frac{1}{2\pi i} \frac{(1 - s\epsilon^2/q^2)(1 - s\epsilon^2/q)(\epsilon q)_\infty}{(s\epsilon/q^2)_\infty} \tag{4.7}$$

$$\times \left[2u \frac{(\epsilon q u^2)_\infty \left(\frac{s\epsilon}{\alpha}u, \frac{s\epsilon}{\beta}u, \frac{s\epsilon}{\gamma}u, \frac{s\epsilon}{\delta}u \right)_\infty W\left(\frac{su^2\epsilon}{q}; \frac{s\epsilon}{q}, \alpha u, \beta u, \gamma u, \delta u; \epsilon q \right)}{(su^2\epsilon)_\infty (\alpha u\epsilon, \beta u\epsilon, \gamma u\epsilon, \delta u\epsilon)_\infty W\left(\epsilon u^2; \epsilon, \frac{qu}{\alpha}, \frac{qu}{\beta}, \frac{qu}{\gamma}, \frac{qu}{\delta}; s\epsilon/q^2 \right)} \right.$$

$$\left. - \frac{2}{u} \frac{(\epsilon q/u^2)_\infty \left(s\epsilon/\alpha u, s\epsilon/\beta u, s\epsilon/\gamma u, s\epsilon/\delta u \right)_\infty W\left(\frac{s}{u^2}\frac{\epsilon}{q}; \frac{s\epsilon}{q}, \frac{\alpha}{u}, \frac{\beta}{u}, \frac{\gamma}{u}, \frac{\delta}{u}; \epsilon q \right)}{\left(\frac{s\epsilon}{u^2} \right)_\infty \left(\frac{\alpha\epsilon}{u}, \frac{\beta\epsilon}{u}, \frac{\gamma\epsilon}{u}, \frac{\delta\epsilon}{u} \right)_\infty W\left(\frac{\epsilon}{u^2}; \epsilon, \frac{q}{\alpha u}, \frac{q}{\beta u}, \frac{q}{\gamma u}, \frac{q}{\delta u}; s\epsilon/q^2 \right)} \right].$$

Equating (4.6) and (4.7), and writing

$$G(\alpha, \beta, \gamma, \delta, \epsilon, u) := \frac{1}{u} \frac{\left(\frac{\epsilon q}{u^2} \right)_\infty \left(\frac{s\epsilon}{\alpha u}, \frac{s\epsilon}{\beta u}, \frac{s\epsilon}{\gamma u}, \frac{s\epsilon}{\delta u} \right)_\infty}{(s\epsilon/u^2)_\infty \left(\frac{\alpha\epsilon}{u}, \frac{\beta\epsilon}{u}, \frac{\gamma\epsilon}{u}, \frac{\delta\epsilon}{u} \right)_\infty} \tag{4.8}$$

$$\times W\left(\frac{s\epsilon}{u^2 q}; \frac{s\epsilon}{q}, \frac{\alpha}{u}, \frac{\beta}{u}, \frac{\gamma}{u}, \frac{\delta}{u}; \epsilon q \right) W\left(\epsilon u^2; \epsilon, \frac{qu}{\alpha}, \frac{qu}{\beta}, \frac{qu}{\gamma}, \frac{qu}{\delta}; s\epsilon/q^2 \right) ,$$

we obtain the identity

$$G(\alpha,\beta,\gamma,\delta,\epsilon,u) - G(\alpha,\beta,\gamma,\delta,\epsilon,1/u) = \left(1/u - u\right)\frac{(1 - s\epsilon^2/q^2)}{(1 - s\epsilon/q^2)} \quad (4.9)$$

$$\times \frac{\left(\alpha\beta\epsilon, \alpha\gamma\epsilon, \alpha\delta\epsilon, \beta\gamma\epsilon, \beta\delta\epsilon, \gamma\delta\epsilon, \epsilon q/u^2, \epsilon q u^2\right)_\infty}{\left(\alpha\epsilon/u, \alpha\epsilon u, \beta\epsilon/u, \beta\epsilon u, \gamma\epsilon/u, \gamma\epsilon u, \delta\epsilon/u, \delta\epsilon u\right)_\infty}$$

which is a q-analogue of Masson's generalization of Dougall's theorem [18]. When $\epsilon = 1$, this gives a q-analogue of Dougall's theorem (see [3]).

We can recover from (4.9) the following identity which we had obtained in our earlier paper ([7], (31), p.723) for $\epsilon = 1$, $|u| = 1$, $s = q^m$; $m = 1, 2, \ldots$, i.e.

$$\left(\frac{q}{\alpha}\right)^{m-3}\left(\frac{1}{u} - u\right)\left[u^{m-2}\Pi_1(u)\Pi_2\left(\frac{1}{u}\right) - u^{2-m}\Pi_1\left(\frac{1}{u}\right)\Pi_2(u)\right] \quad (4.10)$$

$$= (\alpha\beta q^{-1}, \alpha\gamma q^{-1}, \alpha\delta q^{-1})_\infty \left(\frac{q^2}{\alpha\beta}, \frac{q^2}{\alpha\gamma}, \frac{q^2}{\alpha\delta}\right)_\infty \left(u^2, \frac{1}{u^2}\right)_\infty$$

where

$$\Pi_1(u) = \left(\frac{q}{\alpha u}, \frac{q}{\beta u}, \frac{q}{\gamma u}, \frac{q}{\delta u}\right)_\infty, \quad \Pi_2(u) = \left(\frac{\alpha}{u}, \frac{\beta}{u}, \frac{\gamma}{u}, \frac{\delta}{u}\right)_\infty .$$

REFERENCES

[1] Askey R., Wilson J. Some basic hypergeometric orthogonal polynomials that generalize Jacobi polynomials. *Memoirs Amer. Math. Soc.*, **319**:1–55, 1985.

[2] Askey R., Wimp J. Associated Laguerre and Hermite polynomials. *Proc. Roy. Soc. Edinburgh*, Sect. A **96**:15–37, 1984.

[3] Bailey W.N. *Generalized Hypergeometric Series*. Cambridge Univ. Press, London, 1935.

[4] Bustoz J., Ismail M.E.H. The associated ultraspherical polynomials and their q-analogues. *Canad. J. Math.*, **34**:718–736, 1982.

[5] Gasper G., Rahman M. *Basic Hypergeometric Series*. Cambridge Univ. Press, Cambridge, 1990.

[6] Gautschi W. Computational aspects of three-term recurrence relations. *SIAM Rev.*, **9**:24–82, 1967.

[7] Gupta D.P., Masson D.R. Exceptional q-Askey-Wilson polynomials and continued fractions. *Proc. A.M.S.*, **112**:717–727, 1991.

[8] Gupta D.P., Ismail M.E.H., Masson D.R. Associated continuous Hahn polynomials. *Can. J. of Math.*, **43**:1263–1280, 1991.

[9] Gupta D.P., Ismail M.E.H., Masson D.R. Contiguous relations, Basic Hypergeometric functions and orthogonal polynomials II, Associated big q-Jacobi polynomials. *J. of Math. Analysis and Applications*, **171**:477–497, 1992.

[10] Gupta D.P., Masson D.R. Watson's basic analogue of Ramanujan's Entry 40 and its generalization. *SIAM J. Math. Anal.* To appear.

[11] Ismail M.E.H., Letessier J., Valent G. Linear birth and death models and associated Laguerre polynomials. *J. Approx. Theory*, **56**:337–348, 1988.

[12] Ismail M.E.H., Letessier J., Valent G. Quadratic birth and death processes and associated continuous dual Hahn polynomials. *SIAM J. Math. Anal.*, **20**:727–737, 1989.

[13] Ismail M.E.H., Letessier J., Valent G., Wimp J. Two families of associated Wilson polynomials. *Can. J. Math.*, **42**:659–695, 1990.

[14] Ismail M.E.H., Libis C.A. Contiguous relations, basic hypergeometric functions and orthogonal polynomials I. *J. Math. Anal. Appl.*, **141**:349–372, 1989.

[15] Ismail M.E.H., Masson D.R. Two families of orthogonal polynomials related to Jacobi polynomials. *Rocky Mountain J. Math.*, **21**:359–375, 1991.

[16] Ismail M.E.H., Rahman M. Associated Askey-Wilson polynomials. *Trans. Amer. Math. Soc.*, **328**:201–239, 1991.

[17] Jones W.B., Thron W.J. *Continued Fractions: Analytic Theory and Applications*. Addison-Wesley, Reading, Mass., 1980.

[18] Masson D.R. Associated Wilson polynomials. *Constructive Approximation*, **7**:521–534, 1991.

[19] Masson D.R. Wilson polynomials and some continued fractions of Ramanujan. *Rocky Mountain J. of Math.*, **21**:489–499, 1991.

[20] Masson D.R. The rotating harmonic oscillator eigenvalue problems, I. Continued fractions and analytic continuation. *J. Math. Phys.*, **24**:2074–2088, 1983.

[21] Masson D.R. Difference equations, continued fractions, Jacobi Matrices and orthogonal polynomials. In *Non-linear numerical methods and Rational Approximation*, pages 239–257. Dordrecht, Reidel, 1988. A. Cuyt, ed.

[22] Wilson J.A. *Hypergeometric series, recurrence relations and some new orthogonal polynomials*. Ph.D. dissert., University of Wisconsin, Madison, 1978.

[23] Wimp J. Explicit formulas for the associated Jacobi polynomials and some applications. *Canad. J. Math.*, **39**:983–1000, 1987.

International Series of Numerical Mathematics, Vol. 119, ©1994 Birkhäuser

FROM SCHRÖDINGER SPECTRA TO ORTHOGONAL POLYNOMIALS VIA A FUNCTIONAL EQUATION

Arieh Iserles*

Dedicated, with much admiration, to Walter Gautschi on his 65th birthday

Abstract. The main difference between certain spectral problems for linear Schrödinger operators, e.g. the almost Mathieu equation, and three-term recurrence relations for orthogonal polynomials is that in the former the index ranges across $\mathbb{Z}$ and in the latter only across $\mathbb{Z}^+$. We present a technique that, by a mixture of Dirichlet and Taylor expansions, translates the almost Mathieu equation and its generalizations to three term recurrence relations. This opens up the possibility of exploiting the full power of the theory of orthogonal polynomials in the analysis of Schrödinger spectra.

Aforementioned three-term recurrence relations share the property that their coefficients are almost periodic. In the special case when they form a periodic sequence, the support can be explicitly identified by a technique due to Geronimus. The more difficult problem, when the recurrence is almost periodic but fails to be periodic, is still open and we report partial results.

The main promise of the technique of this article is that it can be extended to deal with multivariate extensions of the almost Mathieu equations. However, important theoretical questions need be answered before it can be implemented to the analysis of the spectral problem for Schrödinger operators.

1   THE ALMOST MATHIEU EQUATION AND ORTHOGONAL POLYNOMIALS

The point of departure of our analysis is the *almost Mathieu equation* (also known as the *Harper equation*). We seek $\lambda \in \mathbb{R}$ and $\{a_n\}_{n\in\mathbb{Z}}$ that satisfy

$$a_{n-1} - 2\kappa\cos(\alpha n + \beta)a_n + a_{n+1} = \lambda a_n, \qquad n \in \mathbb{Z}, \qquad (1.1)$$

α, β and $\kappa \neq 0$ being given real constants.

The almost Mathieu equation features in a number of applications [15] and has been already extensively studied [1, 2, 5, 11, 20, 21, 24]. The purpose of our analysis is not to reveal new features of the spectrum of (1.1) *per se*, since the latter is quite comprehensively known. Instead, we intend to demonstrate that the almost Mathieu equation exhibits an intriguing connection with orthogonal polynomials, a connection that lends itself to far-reaching generalizations.

---

*Department of Applied Mathematics and Theoretical Physics, University of Cambridge, England

This is the place to confess that our analysis is incomplete and that many important questions remain unresolved. Inasmuch as we hope to return to this theme in future papers, the present exposition of partial results might be of interest.

Let $\omega = e^{i\alpha}$, $b_1 = \kappa e^{i\beta}$ and $b_2 = \kappa e^{-i\beta} = \bar{b}_1$. We rewrite (1.1) as

$$(b_1\omega^{2n} + b_2)a_n = \omega^n(a_{n-1} - \lambda a_n + a_{n+1}), \qquad n \in \mathbb{Z}, \tag{1.2}$$

and consider the *Dirichlet expansion*

$$y(z) = \sum_{n=-\infty}^{\infty} a_n \exp\left\{b_1^{\frac{1}{2}}\omega^n z\right\}, \qquad z \in \mathbb{C}.$$

The choice of a specific branch of the square root of b_1 is arbitrary. Note that, since $|\omega| = 1$, it is easy to demonstrate that $\{a_n\}_{n\in\mathbb{Z}} \in \ell_1[\mathbb{Z}]$ is sufficient for convergence of the series for all $z \in \mathbb{C}$ [13, 16].

We multiply (1.2) by $\exp\left\{b_1^{\frac{1}{2}}\omega^n z\right\}$ and sum for $n \in \mathbb{Z}$. Since

$$y'(z) = b_1^{\frac{1}{2}} \sum_{n=-\infty}^{\infty} a_n\omega^n e^{b_1^{\frac{1}{2}}\omega^n z}, \qquad y''(z) = b_1 \sum_{n=-\infty}^{\infty} a_n\omega^{2n} e^{b_1^{\frac{1}{2}}\omega^n z},$$

we readily deduce that y obeys the functional differential equation

$$y''(z) + b_2 y(z) = b_1^{-\frac{1}{2}}\left\{\omega^{-1}y'(\omega^{-1}z) - \lambda y'(z) + \omega y'(\omega z)\right\}. \tag{1.3}$$

The solution of (1.3) is determined uniquely by the values of $y(0)$ and $y'(0)$.

Dirichlet expansions have been employed by Derfel and Molchanov [6] to investigate the simplified spectral problem

$$a_{n-1} - 2\kappa q^n a_n + a_{n-1} = \lambda a_n, \qquad n \in \mathbb{Z},$$

where $\kappa \in \mathbb{R} \setminus \{0\}$ and $q > 0$, and they have demonstrated that it obeys another functional differential equation. Both the Derfel–Molchanov equation and (1.3) are a generalization of the *pantograph equation*, that has been extensively analysed in [19] and [16]. However, it is important to emphasize that (1.3) has an important feature that sets it apart from other functional equations of the pantograph type, namely that, unless $\omega \in \mathbb{R}$, its evolution makes sense only for complex z and proceeds along circles of constant $|z|$, emanating from the origin.

Inasmuch as the equation (1.3) can be analyzed directly, our next step entails expanding it in Taylor series. Thus, letting

$$y(z) = \sum_{m=0}^{\infty} \frac{y_m}{m!} z^m,$$

substitution in (1.3) readily yields

$$y_{m+1} = b_1^{-\frac{1}{2}}(2\cos\alpha m - \lambda)y_m - b_2 y_{m-1}, \qquad m = 1, 2, \ldots. \tag{1.4}$$

It is beneficial to treat y_m as a function of the spectral parameter λ and to define

$$\tilde{y}_m(t) := b_2^{-m/2} y_m(-(b_1 b_2)^{1/2} t), \qquad m = 0, 1, \ldots .$$

Brief manipulation affirms that (1.4) is equivalent to

$$\tilde{y}_{m+1}(t) = \left(t + \frac{2}{\kappa} \cos \alpha m \right) \tilde{y}_m(t) - \tilde{y}_{m-1}(t), \qquad m = 0, 1, \ldots . \tag{1.5}$$

To specify the solution of (1.5) in a unique fashion we need to choose $\tilde{y}_0$ and $\tilde{y}_1$, which, of course, corresponds to equipping (1.3) with requisite initial conditions. We note in passing that $t = -\kappa \lambda$, hence real initial values in (1.3) correspond to $\tilde{y}_0, \tilde{y}_1 \in \mathbb{R}$.

Each solution of (1.5) is a linear combination of two linearly independent solutions. Setting $\sigma = 2/\kappa$, we let

$$
\begin{aligned}
r_{-1}(t) &\equiv 0, \\
r_0(t) &\equiv 1, \\
r_{m+1}(t) &= (t + \sigma \cos \alpha m) r_m(t) - r_{m-1}(t), \qquad m = 0, 1, \ldots, \tag{1.6}
\end{aligned}
$$

and

$$
\begin{aligned}
s_{-1}(t) &\equiv 0 \\
s_0(t) &\equiv 1 \\
s_{m+1}(t) &= (t + \sigma \cos \alpha(m+1)) s_m(t) - s_{m-1}(t), \qquad m = 0, 1, \ldots . \tag{1.7}
\end{aligned}
$$

It is trivial to verify that $\{r_m\}_{m \in \mathbb{Z}^+}$ and $\{s_{m+1}\}_{m \in \mathbb{Z}^+}$ are linearly independent, hence they span all solutions of (1.5). Moreover, we observe that each $r_m(t)$ and $s_m(t)$ is an mth degree monic polynomial in t. This is a crucial observation, by virtue of the Favard theorem [4]: given any three-term recurrence relation of the form

$$
\begin{aligned}
p_{-1}(t) &\equiv 0, \\
p_0(t) &\equiv 1, \\
p_{m+1}(t) &= (t + c_m) p_m(t) - d_m p_{m-1}(t), \qquad m = 0, 1, \ldots,
\end{aligned}
$$

the monic polynomial sequence $\{p_m\}_{m \in \mathbb{Z}^+}$ is orthogonal with respect to some Borel measure $d\varphi$, i.e.

$$\int_{\mathbb{R}} p_m(t) p_n(t) \, d\varphi(t) = 0, \qquad m \neq n.$$

We note that, in general, the Favard theorem falls short of producing a unique measure and it is entirely possible that there might exist many Borel measures that produce an identical set of monic orthogonal polynomials. This, however, is ruled out in the present case by the determinacy of the underlying Hamburger moment problem and the latter can be affirmed for both (1.6) and (1.7) by either uniform boundedness of recurrence coefficients or by the Carleman criterion [4, 25].

287

This is the place to state the major open question that arises from the present article. The main goal of the analysis of the almost Mathieu equation is, usually, to determine the essential spectrum of (1.1). Let Ξ be the essential support of $\mathrm{d}\varphi$. What is the connection between Ξ and the essential spectrum of (1.1)? Computer experiments seem to indicate that, after a linear transformation $\lambda = -t/\kappa$, these two sets coincide or, at the very least, possess similar geometry,[†] but a formal proof is not yet available.

The remainder of this paper is devoted to the determination of Ξ. In §2 we demonstrate by using a technique originally due to Geronimus [10] that the essential support can be specified explicitly when α/π is rational. We prove in essence that $\mathrm{d}\varphi$ is a linear combination of Chebyshev measures of the second kind, supported on a set of disjoint intervals.

Various results on irrational α/π are reported in §3. Inasmuch as the general form of $\mathrm{d}\varphi$ is currently a matter for conjecture, we derive a number of results that, besides being of interest on their own merit, are fully consistent with known results about the almost Mathieu operator for irrational α/π. It is known that often (but not always [20]) its spectrum is a Cantor set [2] and our results seem to indicate, at the very least, that Ξ has an 'interesting' geometry.

The equation (1.1) has been extensively studied in the past and, inasmuch as there are few outstanding conjectures, our knowledge of the spectrum of the almost Mathieu operator is quite comprehensive. Indeed, the connection between (1.1) and orthogonal polynomials has been already exploited by converting a Jacobi matrix over $\mathbb{Z}$ into a block-Jacobi matrix over $\mathbb{Z}^+$ with 2×2 blocks [3, 23]. This leads to matrix orthogonal polynomials and the spectrum can be determined by two sets of 'standard' orthogonal polynomials. Another observation linking (1.1) with orthogonal polynomials led in [18] to a very interesting generalization of Chebyshev polynomials.

Although the approach of this paper introduces a new perspective, there is no claim that §2–3 add to the current state of knowledge of Schrödinger spectra. This state of affairs is remedied in §4, where the framework of our discussion undergoes a far-reaching generalization. Firstly, we demonstrate that general periodic potentials with a finite number of Fourier harmonics lend themselves to similar analysis, except that, instead of orthogonal polynomials, the outcome is a generalized eigenvalue problem for a certain matrix pencil. Secondly, we prove that a multivariate extension of the almost Mathieu equation can be 'transformed' by our techniques to a problem in (univariate) orthogonal polynomials.

A possible future application of our analysis, and in particular of §4, might be toward numerical computation of the essential spectrum of (1.1) and of its generalizations. Needless to say, it requires first a clarification of the precise connection between Ξ and the spectrum of (1.1). We do not pursue this in the present paper.

[†]Since both sets sometimes have fractal dimension, naive comparison might well be deceptive.

2 ORTHOGONAL POLYNOMIALS WITH PERIODIC RECURRENCE COEFFICIENTS

The focus of our attention in this section is the three-term recurrence

$$
\begin{aligned}
p_{-1}(t) &\equiv 0, \\
p_0(t) &\equiv 1, \\
p_{m+1}(t) &= (t - \alpha_m)p_m(t) - p_{m-1}(t), \qquad m = 0, 1, \ldots,
\end{aligned}
\tag{2.1}
$$

where the sequence $\{\alpha_m\}_{m \in \mathbb{Z}^+}$ is K-periodic,

$$
\alpha_{m+K} = \alpha_m, \qquad m = 0, 1, \ldots .
\tag{2.2}
$$

Note that both (1.6) and (1.7) assume this form when α/π is rational. Our objective is to determine the Borel measure that renders $\{p_m\}_{m \in \mathbb{Z}^+}$ an orthogonal polynomial system (OPS).

This measure is known and its determination rests upon a technique of Geronimus [10] (see also [8, 12], as well as [9] for a related construction). For completeness, we derive the measure explicitly in the reminder of this section.

Let

$$
q_n(t) := p_{(n+1)K-1}(t), \qquad n = 0, 1, \ldots .
$$

Note that $q_{-1} \equiv 0$ and, moreover, (2.1) and (2.2) imply

$$
p_{nK}(t) = (t - \alpha_0)q_{n-1} - p_{nK-2}(t), \qquad n = 1, 2, \ldots .
\tag{2.3}
$$

We seek polynomials a_ℓ, b_ℓ, $\ell = 0, 1, \ldots, K - 1$, such that

$$
p_{nK+\ell}(t) = a_\ell(t)q_{n-1}(t) - b_\ell(t)p_{nK-2}(t), \qquad \ell = 0, 1, \ldots, K - 1, \quad n = 1, 2, \ldots .
$$

Because of (2.3) and the definition of q_n, we have

$$
\begin{array}{llll}
a_{-1}(t) &\equiv 1, & b_{-1}(t) &\equiv 0, \\
a_0(t) &= t - \alpha_0, & b_0(t) &\equiv 1.
\end{array}
\tag{2.4}
$$

We next substitute in the recurrence relation (2.1) and, by virtue of (2.2), obtain

$$
\begin{aligned}
p_{nK+\ell+1}(t) &= (t - \alpha_{\ell+1})p_{nK+\ell}(t) - p_{nK+\ell-1}(t) \\
&= (t - \alpha_{\ell+1})\{a_\ell(t)q_{n-1}(t) - b_\ell(t)p_{nK-2}(t)\} \\
&\quad - \{a_{\ell-1}(t)q_{n-1}(t) - b_{\ell-1}(t)p_{nK-2}(t)\}.
\end{aligned}
$$

Thus, comparing coefficients, we derive the recurrences

$$
a_{\ell+1}(t) = (t - \alpha_{\ell+1})a_\ell(t) - a_{\ell-1}(t),
\tag{2.5}
$$

$$
b_{\ell+1}(t) = (t - \alpha_{\ell+1})b_\ell(t) - b_{\ell-1}(t), \qquad \ell = 0, 1, \ldots, K - 2,
\tag{2.6}
$$

which, in tandem with (2.4), determine $\{a_\ell, b_\ell\}_{\ell=0}^{K-1}$.

Let $\ell = K - 1$, then

$$p_{nK-2}(t) = \frac{a_{K-1}(t)q_{n-1}(t) - q_n(t)}{b_{K-1}(t)}$$

and, shifting the index,

$$p_{(n+1)K-2}(t) = \frac{a_{K-1}(t)q_n(t) - q_{n+1}(t)}{b_{K-1}(t)}.$$

Substituting both expressions into

$$p_{(n+1)K-2}(t) = a_{K-2}(t)q_{n-1}(t) - b_{K-2}(t)p_{nK-2}(t)$$

yields the recurrence relation

$$q_{n+1}(t) = (a_{K-1}(t) - b_{K-2}(t))q_n(t) - \Delta_{n-2}(t)q_{n-1}(t), \tag{2.7}$$

where

$$\Delta_\ell(t) = \det \begin{bmatrix} a_\ell(t) & a_{\ell+1}(t) \\ b_\ell(t) & b_{\ell+1}(t) \end{bmatrix}, \qquad \ell = 0, 1, \ldots, K-1.$$

We multiply (2.6) by $a_\ell(t)$, (2.5) by $b_\ell(t)$ and subtract from each other. This readily affirms by induction that

$$\Delta_\ell(t) = \Delta_{\ell-1}(t) = \cdots = 1$$

and (2.7) simplifies to

$$q_{n+1}(t) = (a_{K-1}(t) - b_{K-2}(t))q_n(t) - q_{n-1}(t), \qquad n = 0, 1, \ldots. \tag{2.8}$$

Note that (2.8) is consistent with $n = 0$, since $q_{-1} \equiv 0$. To further simplify the recurrence, we set $r(t) := a_{K-1}(t) - b_{K-2}(t)$, hence, letting

$$\tilde{q}_n(x) := \frac{q_n(t)}{q_0(t)}, \qquad n = -1, 0, \ldots,$$

where $x = r(t)$, we obtain the three-term recurrence

$$\begin{aligned}
\tilde{q}_{-1}(x) &\equiv 0, \\
\tilde{q}_0(x) &\equiv 1, \\
\tilde{q}_{n+1}(x) &= x\tilde{q}_n(x) - \tilde{q}_{n-1}(x), \qquad n = 0, 1, \ldots.
\end{aligned}$$

Thus, each $\tilde{q}_n$ is an nth degree monic polynomial and, by virtue of the Favard Theorem, $\{\tilde{q}_n\}_{n \in \mathbb{Z}^+}$ is an OPS. It can be easily identified as a shifted and scaled *Chebyshev polynomial of the second kind*,

$$\tilde{q}_n(x) = 2^n U_n \left(\tfrac{1}{2}x\right), \qquad n = 0, 1, \ldots.$$

We thus deduce that

$$q_n(t) = 2^n q_0(t) U_n \left(\tfrac{1}{2}r(t)\right), \qquad n = 0, 1, \ldots. \tag{2.9}$$

Before identifying the underlying Borel measure, let us 'fill in' the remaining values of p_m. By definition, $p_{nK-1} = q_{n-1}$, $p_{(n+1)K-1} = q_n$, hence the recurrence (2.1) gives

$$
\begin{aligned}
(x - \alpha_1)p_{nK}(t) - p_{nK+1}(t) &= q_{n-1}(t), \\
-p_{nK+\ell-1}(t) + (t - \alpha_{\ell+1})p_{nK+\ell}(t) - p_{nK+\ell+1}(t) &= 0, \qquad \ell = 1, 2, \ldots, K - 3, \\
-p_{(n+1)K-3}(t) + (t - \alpha_{K-1})p_{(n+1)K-2}(t) &= q_n(t).
\end{aligned}
$$

This is a linear system of equations, which we write as

$$A_{K-1}\mathbf{p}_n = \mathbf{q}_n, \tag{2.10}$$

where

$$
A_m = \begin{bmatrix}
t - \alpha_1 & -1 \\
-1 & t - \alpha_2 & -1 \\
& -1 & t - \alpha_3 & -1 \\
& & \ddots & \ddots & \ddots \\
& & & -1 & t - \alpha_{m-1} & -1 \\
& & & & -1 & t - \alpha_m
\end{bmatrix}, \quad m = 1, 2, \ldots, K-1,
$$

$$
\mathbf{p}_n = \begin{bmatrix} p_{nK}(t) \\ p_{nK+1}(t) \\ \vdots \\ p_{(n+1)K-3}(t) \\ p_{(n+1)K-2}(t) \end{bmatrix} \quad \text{and} \quad \mathbf{q}_n = \begin{bmatrix} q_{n-1}(t) \\ 0 \\ \vdots \\ 0 \\ q_n(t) \end{bmatrix}.
$$

We expand the determinant of A_m in its bottom row and rightmost column. This results in a three-term recurrence relation and comparison with (2.4) and (2.6) affirms that $\det A_m = b_m(t)$. Hence, solving (2.10) with Cramer's rule, we deduce that there exist $(K - 2)$-degree polynomials $\tilde{a}_\ell$ and $\tilde{b}_\ell$, $\ell = 0, 1, \ldots, K - 2$, such that

$$p_{nK+\ell}(t) = \frac{\tilde{a}_\ell(t)q_{n-1}(t) + \tilde{b}_\ell(t)q_n(t)}{b_{K-1}(t)}, \qquad \ell = 0, 1, \ldots, K - 2. \tag{2.11}$$

Bearing in mind the definition of a_ℓ and b_ℓ,

$$p_{nK+\ell}(t) = a_\ell(t)q_{n-1}(t) - b_\ell(t)p_{nK-2}(t),$$

we obtain from (2.11) the identity

$$\frac{\tilde{a}_\ell(t)q_{n-1}(t) + \tilde{b}_\ell(t)q_n(t)}{b_{K-1}(t)} = a_\ell(t)q_{n-1}(t) - \frac{b_\ell(t)(\tilde{a}_{K-2}(t)q_{n-2}(t) + \tilde{b}_{K-2}(t)q_{n-1}(t))}{b_{K-1}(t)}.$$

291

We next substitute

$$q_{n-2}(t) = (a_{K-1}(t) - b_{K-2}(t))q_{n-1}(t) - q_n(t)$$

(*pace* (2.8)) and rearrange terms, whereby

$$\{\tilde{b}_\ell(t) - b_\ell(t)\tilde{a}_{K-2}(t)\}q_n(t) = \{-\tilde{a}_\ell + a_\ell(t)b_{K-1}(t) - (a_{K-1}(t)b_\ell(t)$$
$$- b_\ell(t)b_{K-2}(t))\tilde{a}_{K-2}(t) - b_\ell(t)\tilde{b}_{K-2}(t)\}q_{n-1}(t).$$

However, consecutive orthogonal polynomials q_{n-1} and q_n cannot share zeros [4], therefore both sides of the last equality identically vanish and we derive the explicit expressions

$$\tilde{a}_\ell(t) = a_\ell(t)b_{K-1}(t) - (a_{K-1}(t)b_\ell(t) - b_\ell(t)b_{K-2}(t))\tilde{a}_{K-2}(t)$$
$$- b_\ell(t)\tilde{b}_{K-2}(t), \qquad (2.12)$$
$$\tilde{b}_\ell(t) = b_\ell(t)\tilde{a}_{K-2}(t). \qquad (2.13)$$

Letting $\ell = K - 2$ in (2.13) gives $\tilde{b}_{K-2}(t) = b_{K-2}(t)\tilde{a}_{K-2}(t)$ and we substitute this into (2.12). The outcome is

$$\tilde{a}_\ell(t) = a_\ell(t)b_{K-1}(t) - a_{K-1}(t)b_\ell(t)\tilde{a}_{K-2}(t). \qquad (2.14)$$

In particular, $\ell = K - 2$ and the definition of Δ_m result in

$$(1 + a_{K-1}(t)b_{K-2}(t))\tilde{a}_{K-2}(t) = a_{K-2}(t)b_{K-1}(t) = a_{K-1}(t)b_{K-2}(t) + \Delta_{K-2}(t)$$
$$= 1 + a_{K-1}(t)b_{K-2}(t).$$

Since $a_{K-1}b_{K-2} \not\equiv -1$, we conclude that $\tilde{a}_{K-2} \equiv 1$ and substitution in (2.13) and (2.14) yields the explicit formulae

$$\tilde{a}_\ell = \det\begin{bmatrix} a_\ell(t) & a_{K-1}(t) \\ b_\ell(t) & b_{K-1}(t) \end{bmatrix}, \qquad \tilde{b}_\ell(t) = b_\ell(t), \qquad \ell = 0, 1, \ldots, K - 2.$$

Theorem 1 *The OPS $\{p_m\}_{m \in \mathbb{Z}^+}$ has an explicit representation in the form*

$$p_{nK+\ell}(t) = \frac{1}{b_{K-1}(t)}\left\{\det\begin{bmatrix} a_\ell(t) & a_{K-1}(t) \\ b_\ell(t) & b_{K-1}(t) \end{bmatrix}q_{n-1}(t) + b_\ell(t)q_n(t)\right\}, \qquad (2.15)$$

where $n = 0, 1, \ldots, \ell = 0, 1, \ldots, K - 1$ and the OPS $\{q_n\}_{n \in \mathbb{Z}^+}$ satisfies the three-term recurrence (2.8). ∎

Note that letting $\ell = K - 1$ or $\ell = -1$ in (2.15), in tandem with (2.4), results in $p_{(n+1)K-1} = q_n$ and $p_{nK-1} = q_{n-1}$ respectively, as required.

Let $t \in \Xi$. Then, by the discussion preceding the representation (2.9), we know that $\frac{1}{2}r(t) \in [-1, 1]$, and there exists $\theta \in [-\pi, \pi]$ such that $\frac{1}{2}r(t) = \cos\theta$. Since, by the definition of Chebyshev polynomials of the second kind,

$$U_n(\cos\theta) = \frac{\sin(n + 1)\theta}{\sin\theta},$$

(2.9) implies that

$$q_n(t) = 2^n q_0(t) \frac{\sin(n+1)\theta}{\sin\theta}.$$

Substitution into (2.15) results in

$$p_{nK+\ell}(t) = \frac{2^n q_0(t)}{b_{K-1}(t)\sin\theta} \left\{ \det \begin{bmatrix} a_\ell(t) & a_{K-1}(t) \\ b_\ell(t) & b_{K-1}(t) \end{bmatrix} \sin n\theta + b_\ell(t)\sin(n+1)\theta \right\}.$$

We next proceed, following Geronimus [10], to determine the essential support Ξ of the Borel measure $d\varphi$ which corresponds to the OPS $\{p_n\}_{n\in\mathbb{Z}^+}$. We commence by observing that, by virtue of Theorem 1, everything depends on the support of $\tilde{q}_n(x(t)) = 2^n U_n\left(\frac{1}{2}r(t)\right)$, $n = 0, 1, \ldots$. Thus, we seek a Borel measure $d\psi$ such that

$$I_{n,m} = \int_{-\infty}^{\infty} \tilde{q}_n(x(t))\tilde{q}_m(x(t))\,d\psi(t) = 0, \qquad n, m = 0, 1, \ldots, \quad n \neq m.$$

It follows from the defintion of $\tilde{q}_n$ that

$$I_{n,m} = 2^{n+k} \int_{-\infty}^{\infty} U_n\left(\tfrac{1}{2}r(t)\right) U_m\left(\tfrac{1}{2}r(t)\right)\,d\psi(t).$$

Similarly to [10, 8, 12], we seek the inverse function to $x = \frac{1}{2}r(t)$. Let $\xi_1 < \xi_2 < \cdots < \xi_s$ be all the minima and maxima of r in $\mathbb{R}$ (of course, $s \leq K - 2$) and define $\xi_0 = -\infty$, $\xi_{s+1} = \infty$. In each interval $[\xi_j, \xi_{j+1}]$, $j = 0, 1, \ldots, s$, the function $\frac{1}{2}r(t)$ is monotone, hence it possesses a well-defined inverse there. We denote it by $X_j(x)$, hence $\frac{1}{2}r(X_j(x)) = x$. Changing the integration variable, we have

$$\begin{aligned} I_{n,m} &= 2^{n+m} \sum_{j=0}^{s} \int_{\xi_j}^{\xi_{j+1}} U_n\left(\tfrac{1}{2}r(t)\right) U_m\left(\tfrac{1}{2}r(t)\right)\,d\psi(t) \\ &= 2^{n+m} \sum_{j=0}^{s} \int_{\frac{1}{2}r(\xi_j)}^{\frac{1}{2}r(\xi_{j+1})} U_n(x)U_m(x)\,d\psi(X_j(x)). \end{aligned} \tag{2.16}$$

We recall that $\{U_n\}_{n\in\mathbb{Z}^+}$ is an OPS with respect to the Borel measure $(1 - x^2)^{\frac{1}{2}}\,dx$, supported by $x \in [-1, 1]$. Thus, for every $j = 0, 1, \ldots, s$ we distinguish among the following cases:

Case 1: $r(\xi_j) \leq -2$ and $2 \leq r(\xi_{j+1})$.
We stipulate that $d\psi(X_j(x))$ vanishes for all

$$x \in \left[\tfrac{1}{2}r(\xi_j), \tfrac{1}{2}r(\xi_{j+1})\right] \setminus [-1, 1].$$

Since X_j increases monotonically in $[\xi_j, \xi_{j+1}]$, the contribution of this interval to (2.16) is

$$2^{n+m} \int_{-1}^{1} U_n(x)U_m(x)\,d\psi(|X_j(x)|). \tag{2.17}$$

Case 2: $r(\xi_{j+1}) \leq -2$ and $2 \leq r(\xi_j)$.
Likewise, we require that the support of $d\psi(X_j)$ is restricted to $[-1, 1]$. r decreases monotonically within $[\xi_j, \xi_{j+1}]$ and straightforward manipulation affirms that (2.17) represents the contribution of this interval to (2.16).

Case 3: $\min\{r(\xi_j), r(\xi_{j+1})\} > -2$ or $\max\{r(\xi_j), r(\xi_{j+1})\} < 2$.
In that case we cannot fit $[-1, 1]$ into $[\xi_j, \xi_{j+1}]$, hence we stipulate that $d\psi(X_j)$ is not supported in $[\xi_j, \xi_{j+1}]$.

Let $\nu_1 < \nu_2 < \cdots < \nu_r$ be all the indices in $\{0, 1, \ldots, s\}$ such that either Case 1 or Case 2 holds. We require that $r \geq 1$. Then (2.16) reduces to

$$I_{n,m} = \int_{-1}^{1} U_n(x) U_m(x) \sum_{\ell=1}^{r} d\psi(|X_{\nu_\ell}(x)|).$$

Since the Hamburger moment problem for the Chebyshev measure of the second kind is determinate, it follows that necessarily, up to normalization

$$\sum_{\ell=1}^{r} d\psi(|X_{\nu_\ell}(x)|) = (1 - x^2)^{\frac{1}{2}} \, dx, \qquad x \in [-1, 1].$$

Theorem 2 [10] *The orthogonality measure corresponding to the OPS $\{p_m\}_{m \in \mathbb{Z}^+}$ is supported by*

$$\Xi = \mathcal{I}_1 \cup \mathcal{I}_2 \cup \cdots \cup \mathcal{I}_r,$$

where for each $\ell = 1, 2, \ldots, r$ $\mathcal{I}_\ell \subseteq [\xi_{\nu_\ell}, \xi_{\nu_{\ell+1}}]$ *is the unique interval such that* $|r(t)| = 2$ *at its endpoints, as well as possible mass points at the zeros of* $q_0 = p_{K-1}$.

Proof: The proof follows at once from our construction. ∎

Figure 1 displays two examples of the present construction, for different cases of r. In each case Ξ is the union of the 'thick' intervals. Harking back to (1.6) and (1.7), we let $\alpha_m = -\sigma \cos \alpha m$ and $\alpha_m = -\sigma \cos(m+1)\alpha$, $m = 0, 1, \ldots$, respectively, where $\alpha = 2\pi L/K$. Thus, $\{\alpha_m\}_{m \in \mathbb{Z}^+}$ is indeed K-periodic.

We mention in passing that the analysis of this section can be easily extended to recurrences of the form

$$\begin{aligned} p_{-1}(t) &\equiv 0, \\ p_0(t) &\equiv 1, \\ p_{m+1}(t) &= (t - \alpha_m) p_m(t) - \beta_m p_{m-1}(t), \qquad m = 0, 1, \ldots, \end{aligned}$$

where both $\{\alpha_m\}_{m \in \mathbb{Z}^+}$ and $\{\beta_m\}_{m \in \mathbb{Z}^+}$ are K-periodic. This, however, is of little relevance to the theme of this paper.

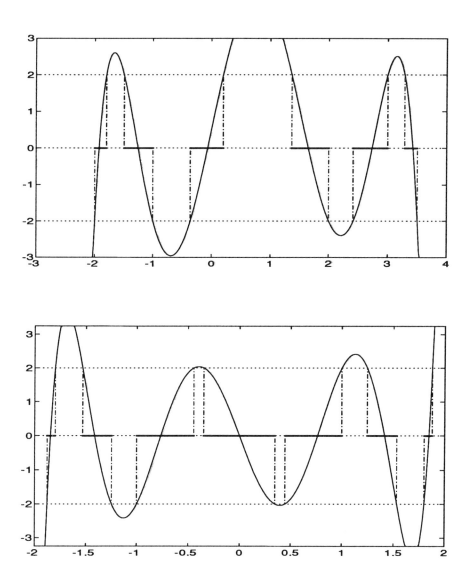

Figure 1: The sets Ξ for two different polynomials r.

3 ORTHOGONAL POLYNOMIALS WITH ALMOST PERIODIC RECURRENCE COEFFICIENTS

The three-term recurrence relations (1.6) and (1.7) assume, α/π being irrational, almost-periodic recurrence coefficients and this state of affairs is even more important in a multivariate generalization of (1.1) in §4. Unfortunately, no general theory exists to cater for orthogonal polynomials with almost periodic recurrence coefficients. The theme of the present section is a preliminary and – in the nature of things – incomplete investigation of the case when an irrational α/π is approximated by rationals. In other words, we commence with the K-periodic recurrence (2.1), except that we will allow the period K to become unbounded.

The motivation for our analysis is an observation which is interesting on its own merit. Denote by σ_K the value of $q_{K-1}(0)$ for $\alpha_\ell = -\cos\frac{2\pi\ell}{K}$, $\ell = 0, 1, \ldots, K-1$, i.e.

$$\sigma_K = \det \begin{bmatrix} \cos\frac{2\pi}{K} & 1 & & & & \\ 1 & \cos\frac{4\pi}{K} & 1 & & & \\ & \ddots & \ddots & \ddots & & \\ & & 1 & \cos\frac{2(K-2)\pi}{K} & 1 \\ & & & 1 & \cos\frac{2(K-1)\pi}{K} \end{bmatrix}, \qquad K = 1, 2, \ldots.$$

Computation indicates that

$$\sigma_{4L} \equiv 0, \qquad \lim_{L\to\infty}\sigma_{4L+2} = 0, \qquad \lim_{L\to\infty}\sigma_{4L+1} = -\lim_{L\to\infty}\sigma_{4L+3} = \frac{2\sqrt{3}}{3}. \qquad (3.1)$$

We prove first that $\sigma_{4L} = 0$ for all $L \geq 1$. Let $K = 4L$ and set $\ell = L$ in

$$\nu_\ell = \cos\frac{2\pi\ell}{K}\nu_{\ell-1} - \nu_{\ell-2}, \qquad \ell = 1, 2, \ldots, 4L-1 \qquad (3.2)$$

(note that (3.2), in tandem with $\nu_{-1} = 0$, $\nu_0 = 1$, yields $\sigma_K = \nu_{K-1}$). This yields $\nu_L + \nu_{L-2} = 0$ and we claim that, in general,

$$\nu_{L+k} + (-1)^k\nu_{L-k-2} = 0, \qquad k = -1, 0, \ldots, L-1. \qquad (3.3)$$

We have already proved (3.3) for $k = 0$ and it is trivially true for $k = -1$. We continue by induction and assume that (3.3) is true for $k = -1, 0, \ldots, s-1$. Letting $\ell = L \pm s$ in (3.2), we have

$$\nu_{L+s} = -\sin\frac{\pi s}{2L}\nu_{L+s-1} - \nu_{L+s-2},$$
$$\nu_{L-s} = +\sin\frac{\pi s}{2L}\nu_{L-s-1} - \nu_{L-s-2}.$$

We multiply the second equation by $(-1)^s$ and add to the first, thus

$$(\nu_{L+s} + (-1)^s\nu_{L-s-2}) = -\sin\frac{\pi s}{2L}(\nu_{L+s-1} + (-1)^{s-1}\nu_{L-s-1})$$
$$- (\nu_{L+s-2} + (-1)^{s-2}\nu_{L-s})$$

and (3.3) follows at once.

Hence, letting $k = L - 1$ in (3.3) and recalling that $\nu_{-1} = 0$, we obtain $\nu_{2L-1} = 0$. Moreover, similarly to (3.3), we can prove that

$$\nu_{3L+k} + (-1)^k \nu_{3L-k-2} = 0, \qquad k = -1, 0, \ldots, L - 1,$$

and $k = L - 1$ gives

$$\sigma_{4L} = \nu_{4L-1} = (-1)^L \nu_{2L-1} = 0.$$

This completes the proof of the first expression in (3.1).

The two other observations in (3.1) are also true and in the sequel we prove them in a generalized setting.

Given $\sigma \in (-1, 1)$, we define

$$
\begin{aligned}
A_{-1}(t) &\equiv 0, & A_0(t) &\equiv 1, \\
A_n(t) &= t\xi_n A_{n-1}(t) - A_{n-2}(t), & n &= 1, 2, \ldots,
\end{aligned}
\qquad (3.4)
$$

where $\{\xi_n\}_{n=1}^{\infty}$ is a given real sequence. To emphasize the dependence on parameters, we write, as and when necessary, $A_n(\cdot) = A_n(\cdot; \xi_1, \xi_2, \ldots, \xi_n)$.

An alternative representation of A_n is

$$
A_n(t) = \det
\begin{bmatrix}
t\xi_1 & 1 & & \\
1 & t\xi_2 & \ddots & \\
& \ddots & \ddots & 1 \\
& & 1 & t\xi_n
\end{bmatrix},
\qquad n = 1, 2, \ldots,
$$

hence A_n is an nth degree polynomial. We observe that

$$
A_n(0) =
\begin{cases}
(-1)^s, & n = 2s; \\
0, & n = 2s + 1.
\end{cases}
$$

Moreover, differentiating with respect to t, we obtain

$$
A_n'(t) = \sum_{k=1}^{n} \det
\begin{bmatrix}
t\xi_1 & 1 & & & & & & & \\
1 & \ddots & \ddots & & & & & & \\
& \ddots & \ddots & \ddots & & & & & \\
& & 1 & t\xi_{k-1} & 1 & & & & \\
& & & 0 & \xi_k & 0 & & & \\
& & & & 1 & t\xi_{k+1} & 1 & & \\
& & & & & \ddots & \ddots & \ddots & \\
& & & & & & \ddots & \ddots & 1 \\
& & & & & & & 1 & t\xi_n
\end{bmatrix},
$$

297

therefore, expanding in the kth row, we derive the identity

$$A'_n(t;\xi_1,\ldots,\xi_n) = \sum_{k=1}^{n} \xi_k A_{k-1}(t;\xi_1,\ldots,\xi_{k-1})A_{n-k}(t;\xi_{k+1},\ldots,x_n). \qquad (3.5)$$

Proposition 3 $A_n^{(r)}(0) = 0$ whenever $n+r$ is odd, hence the polynomial A_n has the same parity as n.

Proof: The proof is by induction on r. The assertion is true for $r = 0$. Moreover, repeatedly differentiating (3.5) with the Leibnitz rule and letting $t = 0$, we obtain

$$A_{2n}^{(2r+1)}(0;\xi_1,\ldots,\xi_{2n}) \quad =$$

$$\sum_{\ell=0}^{2r} \binom{2r}{\ell} \sum_{k=1}^{2n} \xi_k A_{k-1}^{(\ell)}(0;\xi_1,\ldots,\xi_{k-1})A_{2n-k}^{(2r-\ell)}(0;\xi_{k+1},\ldots,\xi_{2n}).$$

But

$$(k-1)+\ell \quad \text{is even} \quad \Longleftrightarrow \quad (2n-k)+(2r-\ell) \quad \text{is odd},$$

therefore for all $\ell = 0,1,\ldots,2r$ and $k = 1,2,\ldots,2n$ the induction hypothesis affirms that at least one of the terms in the product vanishes. A similar argument demonstrates that $A_{2n+1}^{(2r)}(0) = 0$. ∎

We therefore let for all $n = 0,1,\ldots$, $s = 1,2,\ldots$,

$$A_{2n}(t;\xi_s,\ldots,\xi_{2n+s-1}) \quad = \quad \sum_{r=0}^{n} B_{2n,s}^{(2r)}t^{2r}$$

$$A_{2n+1}(t;\xi_s,\ldots,\xi_{2n+s}) \quad = \quad \sum_{r=0}^{n} B_{2n+1,s}^{(2r+1)}t^{2r+1}.$$

Substitution into (3.4) (where we replace ξ_n by ξ_{n+s-1}) results in the recurrences

$$B_{2n,s}^{(2r)} \quad = \quad \xi_{s+2n-1}B_{2n-1,s}^{(2r-1)} - B_{2n-2,s}^{(2r)}, \qquad (3.6)$$

$$B_{2n+1,s}^{(2r+1)} \quad = \quad \xi_{s+2n}B_{2n,s}^{(2r)} - B_{2n-1,s}^{(2r+1)}. \qquad (3.7)$$

Given $q \in \mathbb{C}$, we recall that the q-factorial symbol is defined as

$$(z;q)_n = \prod_{k=0}^{n-1}(1-q^k z), \qquad z \in \mathbb{C}, \quad n \in \mathbb{Z} \cup \{\infty\},$$

whereas the q-binomial reads

$$\begin{bmatrix} n \\ m \end{bmatrix} := \frac{(q;q)_n}{(q;q)_m(q;q)_{n-m}}, \qquad 0 \le m \le n.$$

Lemma 4 Let $\xi_s = q^{\frac{1}{2}s} + q^{-\frac{1}{2}s}$, $s = 1, 2, \ldots$, where $q \in \mathbb{C}$ is given. Then, for every $n = 0, 1, \ldots$, $s = 1, 2, \ldots$ and $r = 0, 1, \ldots, n$,

$$B_{2n,s}^{(2r)} = (-1)^{n+r} q^{-r\left(2n+s-r-\frac{1}{2}\right)} \begin{bmatrix} n+r \\ 2r \end{bmatrix} (-q^{n+s-r}; q)_{2r}, \qquad (3.8)$$

$$B_{2n+1,s}^{(2r+1)} = (-1)^{n+r} q^{-\left(r+\frac{1}{2}\right)(2n+s-r)} \begin{bmatrix} n+r+1 \\ 2r+1 \end{bmatrix} (-q^{n+s-r}; q)_{2r+1}. \quad (3.9)$$

Proof: The proof is by induction on n, using (3.8) and (3.9). Obviously, the assertion of the lemma is true for $n = 0$. Otherwise, for even values,

$$\left(q^{n+\frac{s-1}{2}} + q^{-n-\frac{s-1}{2}}\right) B_{2n-1}^{(2r-1)} - B_{2n-2,s}^{(2r)}$$

$$= (-1)^{n+r} q^{-\left(r-\frac{1}{2}\right)(2n+s-r-1)} \left(q^{n+\frac{s-1}{2}} + q^{-n-\frac{s-1}{2}}\right) \begin{bmatrix} n+r-1 \\ 2r-1 \end{bmatrix} (-q^{n+s-r}; q)_{2r-1}$$

$$- (-1)^{n+r-1} q^{-r\left(2n+s-r-\frac{5}{2}\right)} \begin{bmatrix} n+r-1 \\ 2r \end{bmatrix} (-q^{n+s-r-1}; q)_{2r}$$

$$= (-1)^{n+r} q^{-r\left(2n+s-r-\frac{1}{2}\right)} \frac{(q; q)_{n+r-1}}{(q; q)_{2r}(q; q)_{n-r}} (-q^{n+s-r}; q)_{2r-1}$$

$$\times \left\{ (1 + q^{2n+s-1})(1 - q^{2r}) + q^{2r}(1 - q^{n-r})(1 + q^{n+s-r-1}) \right\}$$

$$= (-1)^{n+r} q^{-r\left(2n+s-r-\frac{1}{2}\right)} \begin{bmatrix} n+r \\ 2r \end{bmatrix} (-q^{n+s-r}; q)_{2r}.$$

This accomplishes a single inductive step for (3.8). We prove (3.9) in an identical manner, by considering odd values of n. ∎

Recall that our interest in the polynomials A_n has been sparked by the observation (3.1). Thus, we require to recover cosine terms, and to this end we choose q of unit modulus.

Proposition 5 Suppose that $q^{n+1} = 1$ and $q^m \neq 1$ for $m = 1, 2, \ldots, n$. Then $B_{2n+1,1}^{(2r+1)} = 0$ for all $0 \leq r \leq \frac{n}{2}$.

Proof: Let $2r \leq n$. We have from (3.9) that

$$B_{2n+1,1}^{(2r+1)} = (-1)^{n+r} q^{\left(r+\frac{1}{2}\right)(r+1)} \frac{(q; q)_{n+r+1}}{(q; q)_{2r+1}(q; q)_{n-r}} (-q^{-r}; q)_{2r+1}.$$

However,

$$\frac{(q; q)_{n+r+1}}{(q; q)_{n-r}} = \prod_{\ell=-r}^{r} (1 - q^\ell) = 0,$$

whereas, because of our restriction on r,

$$(q; q)_{2r+1} = \prod_{\ell=1}^{2r+1} (1 - q^\ell) \neq 0,$$

since q is a root of unity of *primitive* degree $n + 1$. The proposition follows. ∎

Corollary Let $q = \exp \frac{2\pi i m}{n+1}$, where m and n are relatively prime. Then it is true that

$$\lim_{n \to \infty} A_{2n+1}(t) = 0 \qquad (3.10)$$

for every $t \in (-1, 1)$.

Proof: The proof is straightforward, since $A_{2n+1}(t) = \mathcal{O}\left(t^{\frac{1}{2}n}\right)$. ∎

Proposition 6 *Suppose that $\omega = q^{\frac{1}{2}}$ is a root of unity of primitive degree $2n+1$. Then, for all $r = 0, 1, \ldots, n$ it is true that*

$$(-1)^n B_{2n,1}^{(2r)} = \prod_{\ell=1}^{r} \frac{\sin(2\ell-1)\phi}{\sin 2\ell\phi}, \qquad (3.11)$$

where $\phi = \arg \omega$.

Proof: Since $q^{n+\frac{1}{2}} = 1$, it follows from (3.8) that

$$(-1)^n B_{2n,1}^{(2r)} = (-1)^r q^{r\left(r+\frac{1}{2}\right)} \frac{(q;q)_{n+r}}{(q;q)_{2r}(q;q)_{n-r}} (-q^{-r+\frac{1}{2}}; q)_{2r}.$$

But

$$\frac{(q;q)_{n+r}(q^{-r+\frac{1}{2}}; q)_{2r}}{(q;q)_{n-r}} = \prod_{\ell=-r}^{r-1} (1 - q^{2\ell+1}).$$

Moreover, $(q;q)_{2r} \neq 0$ for $r = 0, 1, \ldots, n$, since $2n + 1$ is the primitive degree of the root of unity q, and we deduce that

$$(-1)^n B_{2n,1}^{(2r)} = (-1)^r q^{r\left(r+\frac{1}{2}\right)} \frac{\prod_{\ell=-r}^{r-1}(1 - q^{2\ell+1})}{\prod_{\ell=1}^{2r}(1 - q^\ell)}.$$

But

$$\prod_{\ell=1}^{2r}(1 - q^\ell) = \prod_{\ell=1}^{2r} \omega^\ell(\omega^\ell - \omega^{-\ell}) = q^{r\left(r+\frac{1}{2}\right)} \prod_{\ell=1}^{2r}(\omega^\ell - \omega^{-\ell})$$

and, likewise,

$$\prod_{\ell=-r}^{r-1}(1 - q^{2\ell+1}) = \prod_{\ell=-r}^{r-1} \omega^{2\ell+1}(\omega^{2\ell+1} - \omega^{-2\ell-1}) = (-1)^r \prod_{\ell=0}^{r-1}(\omega^{2\ell+1} - \omega^{-2\ell-1})^2.$$

Consequently,

$$(-1)^n B_{2n,1}^{(2r)} = \frac{\prod_{\ell=0}^{r-1}(\omega^{2\ell+1} - \omega^{-2\ell-1})^2}{\prod_{\ell=1}^{2r}(\omega^\ell - \omega^{-\ell})}.$$

This is precisely the identity (3.11). ∎

Next, we consider progression to a limit as $n \to \infty$; (3.1) is a special case. Thus, suppose that we have a sequence $\Phi = \{\phi_n\}_{n \in \mathcal{I}}$, where $\phi_n = 2\pi m_n/(n+1)$, $m_n \in \mathbb{Z}^+$, $\mathcal{I} \subseteq \mathbb{Z}^+$ is a set of infinite cardinality and

$$\lim_{\substack{n \to \infty \\ n \in \mathcal{I}}} \phi_n = \phi \in [0, 2\pi).$$

Set

$$C(t, \Phi) = \lim_{\substack{n \to \infty \\ n \in \mathcal{I}}} (-1)^{[n/2]} A_n(t; \xi_1^{(n)}, \xi_2^{(n)}, \ldots, \xi_n^{(n)}),$$

where $\xi_\ell^{(n)} = 2\cos\ell\phi_n$, $\ell = 1, 2, \ldots, n$, $n = 1, 2, \ldots$. Thus, $\xi_\ell^{(n)} = q_n^\ell + q_n^{-\ell}$, where $q_n = \exp\frac{4\pi i}{n+1}$. Consequently, according to Proposition 5, if $\mathcal{I}$ consists of only odd indices, $C(t, \Phi) \equiv 0$ necessarily. This proves, therefore, that $\sigma_{2L} \to 0$ in (3.1).

Lemma 7 *Suppose that $\mathcal{I} \subseteq 2\mathbb{Z}^+$ and that $\phi = 0$. Then, provided that $m_n = o(n^{\frac{2}{3}})$, it is true that $C(t, \Phi) = (1 - t^2)^{-\frac{1}{2}}$.*

Proof: Since

$$\frac{\sin(2\ell - 1)\phi_n}{\sin 2\ell\phi_n} = \frac{2\ell - 1}{2\ell} + \mathcal{O}(\phi_n^3),$$

we deduce from (3.11) that

$$(-1)^n B_{2n,1}^{(2r)} = 4^{-r}\binom{2r}{r} + \mathcal{O}(n\phi_n^3).$$

According to the assumption, $\mathcal{O}(n\phi_n^3) = o(1)$ and the lemma follows from

$$\sum_{r=0}^{\infty} \binom{2r}{r}\frac{t^{2r}}{4^r} = \frac{1}{\sqrt{1 - t^2}}. \qquad \blacksquare$$

Letting $t = \frac{1}{2}$ affirms the remaining part of (3.1).

In a manner similar to the last proposition, it is possible to derive an explicit expression for $C(t, \Phi)$, provided that ϕ/π is rational and that $n(\phi - \phi_n)^3 = o(1)$ as $n \to \infty$. The derivation is long and it will be published elsewhere. It suffices to mention here the remarkable sensitivity of $C(t, \Phi)$ to both the choice of $\mathcal{I}$ and to the specific nature of ϕ.

What happens when ϕ/π is irrational? This is, as things stand, an open problem. It is possible to show that, formally,

$$C(t, \Phi) = \sum_{\ell=0}^{\infty} t^{2\ell} \prod_{k=1}^{\ell} \frac{\sin(2k-1)\phi}{2k\phi} = \sum_{\ell=0}^{\infty} t^{2\ell} q^{\frac{1}{4}\ell} \frac{(q^{\frac{1}{2}}; q)_\ell}{(q; q)_\ell},$$

where $q = e^{4i\phi}$. The latter series can be summed up by means of the Gauss–Heine theorem (a.k.a. the q-binomial theorem [7]) for $|q| < 1$ and, after simple manipulation, for $|q| > 1$. Unfortunately, because of a breakdown in Hölder-continuity across $|q| = 1$, it is impossible to deduce its value on the unit circle from the values within and without by means, for example, of the Sokhotsky formula [14].

Clearly, there is much to be done to understand better the behavior of the p_n's when the period K becomes infinite. In this section we have established few results with regard to the values at the origin. They should be regarded as a preliminary foray into an interesting problem in orthogonal polynomial theory *cum* linear algebra and we hope to return to this theme in the future.

4   GENERALIZATIONS

There are two natural ways of generalizing an almost Mathieu equation (1.1), either by specifying a more general periodic potential or by replacing the index by a multi-index. Remarkably, the basic framework of this paper – replacing a doubly-infinite recurrence by a functional equation which, in turn, is replaced by a singly-infinite recurrence – survives both generalizations! In the present section we describe briefly this state of affairs.

First, suppose that the cosine term in (1.1) is replaced by a more general harmonic term and we consider the spectral problem

$$a_{n-1} - 2 \left\{ \sum_{\ell=1}^{m} \kappa_\ell \cos(n\ell\theta + \psi_\ell) \right\} a_n + a_{n+1} = \lambda a_n, \qquad n \in \mathbb{Z}. \qquad (4.1)$$

We assume that $\kappa_1, \kappa_2, \ldots, \kappa_m \in \mathbb{R}$ and, without loss of generality, that $\kappa_m \neq 0$. Letting $q = e^{i\theta}$, we set

$$\kappa_\ell^* = e^{i\psi_\ell} \kappa_\ell, \quad \kappa_{-\ell}^* = e^{-i\psi_\ell} \kappa_\ell, \qquad \ell = 1, 2, \ldots, m,$$

and $\kappa_0^* = 0$. Therefore (4.1) assumes the form

$$q^{mn} a_{n-1} - \left\{ \sum_{\ell=0}^{2m} \kappa_{\ell-m}^* q^{n\ell} \right\} a_n + q^{mn} a_{n+1} = \lambda q^{mn} a_n, \qquad n \in \mathbb{Z}. \qquad (4.2)$$

Let $c \in \mathbb{C} \setminus \{0\}$ and consider the Dirichlet series $y(t) = \sum_{n=-\infty}^{\infty} a_n \exp\{cq^n t\}$. Since, formally,

$$y^{(\ell)}(t) = c^\ell \sum_{n=-\infty}^{\infty} q^{n\ell} a_n e^{cq^n t}, \qquad \ell = 0, 1, \ldots,$$

we obtain from (4.2) the functional differential equation

$$\sum_{\ell=0}^{2m} \kappa_{\ell-m}^* c^{m-\ell} y^{(\ell)}(t) = q^m y^{(m)}(qt) - \lambda y^{(m)}(t) + q^{-m} y^{(m)}(q^{-1}t). \qquad (4.3)$$

The derivation is identical to that of (1.3) and is left to the reader.

In line with §1, we next expand the solution of (4.3) in Taylor series, $y(t) = \sum_{n=0}^{\infty} p_n t^n / n!$. This readily yields

$$\sum_{\ell=0}^{2m} \kappa_{\ell-m}^* c^{m-\ell} p_{n+\ell} = (q^{n+m} + q^{-n-m} - \lambda) p_{n+m}, \qquad n = 0, 1, \ldots,$$

hence, replacing $n + m$ by n,

$$\sum_{\ell=-m}^{m} \kappa_\ell^* c^{-\ell} p_{n+\ell} = (2\cos n\theta - \lambda) p_n, \qquad \ell = m, m+1, \ldots.$$

Finally, we choose $c = \exp\{i\psi_m/m\}$, hence $\kappa_m^* c^{-m} = \kappa_{-m}^* = \kappa_m \in \mathbb{R} \setminus \{0\}$. We thus define $\alpha_\ell = c^{-\ell} \kappa_\ell^* / \kappa_m$, $|\ell| \le m$ and replace λ by $-\lambda \kappa_m$. This results in the recurrence

$$\sum_{\ell=-m}^{m} \alpha_\ell p_{n+\ell} = (\lambda - \beta_n) p_n, \qquad n = m, m+1, \ldots, \qquad (4.4)$$

where

$$\beta_n = -\frac{\cos n\theta}{\kappa_m}, \qquad n = m, m+1, \ldots.$$

Note that, inasmuch as the α_ℓs may be complex, we have $\alpha_{-\ell} = \bar{\alpha}_\ell$, $\ell = 1, 2, \ldots, m$, $\alpha_0 = 0$.

The recurrence (4.4) is spanned by $2m$ linearly independent solutions. However, unless $m = 1$, it is no longer true that, for appropriate choice of $p_0, p_1, \ldots,$ p_{2m-1}, each p_n is a polynomial of degree $n + k$ for some k, independent of n. Indeed, it is easy to verify that the degree of p_n increases roughly as $[n/m]$. Hence, orthogonality is lost. Fortunately, an important feature of orthogonal polynomials, namely that their zeros are eigenvalues of a truncated Jacobi matrix [4], can be generalized to the present framework. It is possible to show that the zeros of p_n are generalized eigenvalues of a specific pencil of 'truncated' matrices and this provides a handle on their location. We expect to address ourselves to this issue in a future publication.

Another generalization of (1.1) allows the index n to be replaced by a multi-index $\mathbf{n} = (n_1, n_2, \ldots, n_d) \in \mathbb{Z}^d$. Thus, let $\mathbf{e}_\ell \in \mathbb{Z}^d$ be the ℓth unit vector, $\ell = 1, 2, \ldots, d$, and consider the spectral problem

$$\sum_{\ell=1}^{d}(a_{\mathbf{n}+\mathbf{e}_\ell} + a_{\mathbf{n}-\mathbf{e}_\ell}) - 2\kappa \cos\left(\sum_{\ell=1}^{d} \alpha_\ell n_\ell + \beta\right) a_{\mathbf{n}} = \lambda a_{\mathbf{n}}, \qquad \mathbf{n} \in \mathbb{Z}^d. \qquad (4.5)$$

In line with §1, we let

$$b_1 = b e^{i\beta}, \qquad b_2 = e^{-i\beta}, \qquad q_\ell = e^{i\alpha_\ell}, \qquad \ell = 1, 2, \ldots, d,$$

whereupon (4.5) becomes

$$\sum_{\ell=1}^{d} \mathbf{q^n}(a_{\mathbf{n}+\mathbf{e}_\ell} + a_{\mathbf{n}-\mathbf{e}_\ell}) - (b_1\mathbf{q^{2n}} + b_2)a_{\mathbf{n}} = \lambda\mathbf{q^n}a_{\mathbf{n}}, \qquad \mathbf{n} \in \mathbb{Z}^d. \qquad (4.6)$$

The last formula employs standard multi-index notation, e.g. $\mathbf{q^n} = q_1^{n_1}q_2^{n_2}\cdots q_d^{n_d}$.
We let formally

$$y(t) = \sum_{\mathbf{n}\in\mathbb{Z}^d} a_{\mathbf{n}}\exp\left\{b_1^{\frac{1}{2}}\mathbf{q^n}t\right\}$$

and note that

$$y'(t) = b_1^{\frac{1}{2}}\sum_{\mathbf{n}\in\mathbb{Z}^d} a_{\mathbf{n}}\mathbf{q^n}\exp\left\{b_1^{\frac{1}{2}}\mathbf{q^n}t\right\},$$

$$y''(t) = b_1^{\frac{1}{2}}\sum_{\mathbf{n}\in\mathbb{Z}^d} a_{\mathbf{n}}\mathbf{q^{2n}}\exp\left\{b_1^{\frac{1}{2}}\mathbf{q^n}t\right\}.$$

Therefore, multiplying (4.6) by $\exp\left\{b_1^{\frac{1}{2}}\mathbf{q^n}t\right\}$ and summing up for $\mathbf{n} \in \mathbb{Z}^d$ yields, after brief manipulation, the complex functional differential equation

$$y''(t) + b_2 y(t) = b_1^{-\frac{1}{2}}\left\{\sum_{\ell=1}^{d}\left(q_\ell^{-1}y'(q_\ell^{-1}t) + q_\ell y'(q_\ell t)\right) - \lambda y'(t)\right\}. \qquad (4.7)$$

Equation (4.7) is of independent interest, being a special case of the equation

$$y''(t) + c_1 y'(t) + c_2 y(t) = \int_0^{2\pi} y(e^{i\theta}t)\,\mathrm{d}\mu(\theta),$$

where $\mathrm{d}\mu$ is a complex-valued Borel measure. This, in turn, is similar to the functional integro-differential equations of the form

$$y''(t) + c_1 y'(t) + c_2 y(t) = \int_0^1 y(qt)\,\mathrm{d}\eta(q),$$

say, where $\mathrm{d}\eta$ is, again, a complex-valued Borel measure. Equations of this kind have been considered by the present author, jointly with Liu [17], with an emphasis on their dynamics and asymptotic behavior. In the present paper, however, we are interested in the spectral problem for (4.7), and to this end we again expand y in Taylor series, $y(t) = \sum_{m=0}^{\infty} y_m t^m/m!$. It is easy to affirm by substitution into (4.7) the three-term recurrence relation

$$y_{m+1} = b_1^{-\frac{1}{2}}\left(2\sum_{\ell=1}^{d}\cos\alpha_\ell m - \lambda\right)y_m - b_2 y_{m-1}, \qquad m = 1, 2, \ldots. \qquad (4.8)$$

Note a most remarkable phenomenon – although (4.5) is d-dimensional, the index in (4.8) lives in $\mathbb{Z}^+$! In other words, the dimensionality of the resultant three-term recurrence is independent of d – it is, instead, expressed as the number of harmonics in the recurrence coefficient. Moreover, inasmuch as (4.8) is more complicated for $d \geq 2$ then its one-dimensional counterpart (1.4), both recurrences display similar qualitative characteristics. In particular, we can use the theory of §2 to cater for the case of $\alpha_1/\pi, \alpha_2/\pi, \ldots, \alpha_d/\pi$ being all rational.

In line with the analysis of §1, we let $\tilde{y}_m(t) = b_2^{\frac{1}{2}m} y_m(-(b_1 b_2)^{\frac{1}{2}} t)$, $m \in \mathbb{Z}^+$, whereupon (4.8) becomes

$$\tilde{y}_{m+1}(t) = \left(t + \frac{2}{\kappa} \sum_{\ell=1}^{d} \cos \alpha_\ell m\right) \tilde{y}_m(t) - \tilde{y}_{m-1}(t), \qquad m = 1, 2, \ldots. \quad (4.9)$$

To recover all solutions of (4.9) we need to consider a linearly independent two-dimensional set of solutions. Letting $r_{-1} = 0$, $r_0 = 1$ and $s_0 = 0$, $s_1 = 1$, we recover, similarly to (1.6–7), two sequences $\{r_m\}_{m \in \mathbb{Z}^+}$ and $\{s_m\}_{m \in \mathbb{Z}^+}$ that span all solutions of (4.9) and such that $\deg r_m = m$, $\deg s_m = m - 1$. In other words, by the Favard theorem both $\{r_m\}_{m \in \mathbb{Z}^+}$ and $\{s_{m+1}\}_{m \in \mathbb{Z}^+}$ are OPS and, in line with our analysis of the one-dimensional almost Mathieu equation (1.1), we are in position to exploit the theory of orthogonal polynomials.

Acknowledgements. I have discussed various aspects of this paper with many colleagues and am delighted to acknowledge their helpful comments. Particular gratitude deserve Grisha Barenblatt (Cambridge), Brad Baxter (Manchester), Martin Buhmann (Zürich), Grisha Derfel (Beer Sheva), Mourad Ismail (Tampa), Herb Keller (Pasadena), Joe Keller (Stanford), Alphonse Magnus (Louvain-la-Neuve) and Walter van Assche (Leuven). Many thanks are also due to the anonymous referee, who has pointed out a serious oversight in an early version of the manuscript.

REFERENCES

[1] Avron J., Simon B. Singular continuous spectrum for a class of almost periodic Jacobi matrices. *Bull. Amer. Math. Soc.*, 6:81–85, 1982.

[2] Bellissard J., Simon B. Cantor spectrum for the almost Mathieu equation. *J. Funct. Anal.*, 48:408–419, 1982.

[3] Berezanskĭ Yu.M. *Expansions in Eigenfunctions of Self-Adjoint Operators.* Amer. Math. Soc., Providence, RI, 1968.

[4] Chihara T.S. *An Introduction to Orthogonal Polynomials.* Gordon & Breach, New York, 1978.

[5] Cycon H.L., Froese R.G., Kirsch W., Simon B. *Schrödinger Operators.* Springer-Verlag, Berlin, 1987.

[6] Derfel G.A., Molchanov S.A. Spectral methods in the theory of differential-functional equations. *Matematicheske Zametki Akad. Nauk USSR*, No. 3 **47**:42–51, 1990.

[7] Gaspar G., Rahman M. *Basic Hypergeometric Series.* Cambridge University Press, Cambridge, 1990.

[8] Geronimo J.S., van Assche W. Orthogonal polynomials with asymptotically periodic recurrence coefficients. *J. Approx. Th.*, **46**:251–283, 1986.

[9] Geronimo J.S., van Assche W. Orthogonal polynomials on several intervals *via* a polynomial mapping. *Transactions Amer. Math. Soc.*, **308**:559–581, 1988.

[10] Geronimus Yu.L. On some finite difference equations and corresponding systems of orthogonal polynomials. *Zap. Mat. Otd. Fiz.-Mat. Fak. i Kharkov Mat. Obsc.*, **25**(4):87–100, 1957.

[11] Gesztesy F., Holden H., Simon B., Zhao Z. Trace formulae and inverse spectral theory for Schrödinger operators. *Bull. Amer. Math. Soc.*, **29**:250–255, 1993.

[12] Grosjean C.C. The measure induced by orthogonal polynomials satisfying a recurrence formula with either constant or periodic coefficients. Part II: Pure or mixed periodic coefficients. *Acad. Koninkl. Acad. Wetensch. Lett. Sch. Kunsten Belgie*, **48**(5):55–94, 1986.

[13] Hardy G.H., Riesz M. *The General Theory of Dirichlet's Series.* Cambridge University Press, Cambridge, 1915.

[14] Henrici P. *Applied and Computational Complex Analysis*, Volume III. Wiley Interscience, New York, 1986.

[15] Hofstadter D.R. Energy levels and wave functions of Bloch electrons in a rational or irrational magnetic field. *Phys. Rev. B*, **14**:2239–2249, 1976.

[16] Iserles A. On the generalized pantograph functional-differential equation. *Europ. J. Appl. Math.*, **4**:1–38, 1993.

[17] Iserles A., Liu Y. On pantograph integro-differential equations. DAMTP Tech. Rep. *1993/NA13*, University of Cambridge 1993. To appear in *J. Integral Eqns & Appls.*

[18] Ismail M.E.H., Mulla F. On the generalized Chebyshev polynomials. *SIAM J. Math. Anal.*, **18**:243–258, 1987.

[19] Kato T., McLeod B. The functional-differential equation $y'(t) = ay(\lambda t) + by(t)$. *Bull. Amer. Math. Soc.*, **77**:891–937, 1971.

[20] Last Y. A relation between a.c. spectrum of ergodic Jacobi matrices and the spectra of periodic approximants. *Comm. Math. Phys.*, **155**:183–192, 1993.

[21] Last Y. Zero measure spectrum for the almost Mathieu operator. Tech. Rep., Technion, Haifa, 1993.

[22] Máté A., Nevai P., van Assche W. The supports of measures associated with orthogonal polynomials and the spectra of the related self-adjoint operators. *Rocky Mntn J. Maths*, **21**:501–527, 1991.

[23] Nikišin E.M. Discrete Sturm–Liouville operators and some problems of function theory. *J. Soviet Math.*, **35**:2679–2744, 1986.

[24] Simon B. Almost periodic Schrödinger operators: A review. *Adv. Appl. Maths*, **3**:463–490, 1982.

[25] Stahl H., Totik V. *General Orthogonal Polynomials*. Cambridge University Press, Cambridge, 1992.

International Series of Numerical Mathematics, Vol. 119, ©1994 Birkhäuser

Inequalities and Monotonicity Properties for Gamma and q-Gamma Functions

Mourad E. H. Ismail[*]   Martin E. Muldoon[†]

To Walter Gautschi on his 65th birthday

Abstract. We prove some new results and unify the proofs of old ones involving complete monotonicity of expressions involving gamma and q–gamma functions, $0 < q < 1$. Each of these results implies the infinite divisibility of a related probability measure. In a few cases, we are able to get simple monotonicity without having complete monotonicity. All of the results lead to inequalities for these functions. Many of these were motivated by the bounds in a 1959 paper by Walter Gautschi. We show that some of the bounds can be extended to complex arguments.

1 INTRODUCTION AND PRELIMINARIES

Among Walter Gautschi's many contributions to mathematics are some interesting inequalities for the gamma function. For example, he shows ([10], [11]) that if $x_k > 0, k = 1, \ldots n$, $x_1 x_2 \cdots x_n = 1$, then the inequality

$$\sum_{k=1}^{n} \frac{1}{\Gamma(x_k)} \leq n \tag{1.1}$$

is true for $n = 1, 2 \ldots, 8$ but not for $n \geq 9$. Here we will be more concerned with an earlier result of Gautschi's [9], the two-sided inequality

$$n^{1-s} < \frac{\Gamma(n+1)}{\Gamma(n+s)} < \exp[(1-s)\,\psi(n+1)], \quad 0 < s < 1,\ n = 1, 2, \ldots \tag{1.2}$$

which still inspires extensions. For example, D. Kershaw [13] proved

$$\exp[(1-s)\psi(x+s^{1/2})] < \frac{\Gamma(x+1)}{\Gamma(x+s)} < \exp[(1-s)\,\psi(x+(s+1)/2)],$$
$$0 < s < 1,\ x > 0, \tag{1.3}$$

and

$$\left[x + \frac{s}{2}\right]^{1-s} < \frac{\Gamma(x+1)}{\Gamma(x+s)} < \left[x - \frac{1}{2} + \left(s + \frac{1}{4}\right)^{\frac{1}{2}}\right]^{1-s}, \quad 0 < s < 1,\ x > 0. \tag{1.4}$$

[*]Department of Mathematics, University of South Florida, Tampa, FL 33620–5700, USA
[†]Department of Mathematics and Statistics, York University, North York, Ont. M3J 1P3, Canada

In all of these inequalities, $\psi(x)$ denotes the logarithmic derivative $\Gamma'(x)/\Gamma(x)$.

Many inequalities for special functions follow from monotonicity properties. Often such inequalities are special cases of the complete monotonicity of related special functions. For example, an inequality of the form $f(x) \geq g(x)$, $x \in [a, \infty)$ with equality if and only if $x = a$, may be a disguised form of the complete monotonicity of $g(\varphi(x))/f(\varphi(x))$ where φ is a nondecreasing function on (a, ∞) and $g(\varphi(a))/f(\varphi(a)) = 1$. Thus, for example, the left–hand inequality in (1.2) and the right–hand one in (1.3) follow from the facts that $x^s \Gamma(x + s)/\Gamma(x + 1)$ and $\exp[(s - 1)\psi(x + (s + 1)/2)]\Gamma(x + s)/\Gamma(x + 1)$ are, respectively, decreasing and increasing functions of x on $(0, \infty)$. Bustoz and Ismail [4] proved that some of the above inequalities for the gamma function follow from the complete monotonicity of certain functions involving the ratio $\Gamma(x + 1)/\Gamma(x + s)$.

Recall that a function f is completely monotonic on an interval I if

$$(-1)^n f^{(n)}(x) \geq 0$$

for $n = 1, 2, \ldots$ on I. We collect some known facts, all either easily proved or contained in [21] or [7], in the following theorem.

Theorem 1.1 *(i) A necessary and sufficient condition that $f(x)$ should be completely monotonic on $(0, \infty)$ is that*

$$f(x) = \int_0^\infty e^{-xt} d\alpha(t),$$

where $\alpha(t)$ is nondecreasing and the integral converges for $0 < x < \infty$.

(ii) $e^{-h(x)}$ is completely monotonic on I if $h'(x)$ is completely monotonic on I.

(iii) A probability distribution supported on a subset of $[0, \infty)$ is infinitely divisible if and only if its Laplace transform (moment generating function) is of the form $e^{-h(x)}$ with $h(0^+) = 0$ and $h'(s)$ is completely monotonic on $(0, \infty)$.

For brevity, we shall use *completely monotonic* to mean *completely monotonic on $(0, \infty)$*.

There is already an extensive and rich literature on inequalities for gamma functions; for references see [16], [17]. One of the objects of the present work is to show that many of these can be extended, using essentially the same methods of proof, to the q–gamma function defined (see, e.g., [8]) by

$$\Gamma_q(x) := (1 - q)^{1-x} \prod_{n=0}^\infty \frac{1 - q^{n+1}}{1 - q^{n+x}}, \quad 0 < q < 1. \tag{1.5}$$

Although the right-hand side of (1.5) is meaningful when $|q| < 1$, our results will require $q^x \in (0, 1)$ for all positive x. This forces $q \in (0, 1)$. As expected

$$\Gamma_q(x) \to \Gamma(x) \quad \text{as } q \to 1^-. \tag{1.6}$$

The most elegant proof of this, due to R. William Gosper, is in Appendix A of Andrews' excellent monograph [3]; see also [8, p. 17]. For a rigorous justification, see [14]. It is worth noting that

$$\Gamma_q(x) \approx (1 - q)^{1-x} \prod_0^\infty (1 - q^{n+1}), \quad \text{as } x \to \infty \text{ if } |q| < 1. \tag{1.7}$$

It seems that most of our results have analogues also for the q–gamma function with $q > 1$, in which case the definition (1.5) must be changed. We do not pursue this question here. For the gamma function, we have the (Mittag–Leffler) sum representation [6]

$$\psi(z) = \frac{\Gamma'(z)}{\Gamma(z)} = -\gamma + \sum_{n=0}^\infty \left(\frac{1}{n+1} - \frac{1}{z+n} \right), \tag{1.8}$$

and the integral representation [6, (1.7.14)]

$$\psi(x) = -\gamma + \int_0^\infty \frac{e^{-t} - e^{-tx}}{1 - e^{-t}} \, dt, \quad \text{Re } x > 0. \tag{1.9}$$

Although (1.9) and (1.8) are equivalent, for Re $x > 0$, it turns out that (1.9) is more useful in proving the kind of inequalities in which we are interested. This situation occurred also in [20].

A corresponding sum representation for the case of the q–gamma function, easily following from (1.5) is

$$\psi_q(x) := \Gamma'_q(x)/\Gamma_q(x) = -\log(1-q) + \log q \sum_{n=0}^\infty q^{n+x}/(1-q^{n+x}), \quad 0 < q < 1. \tag{1.10}$$

Although this representation has been used directly in the proofs of many results for the q–gamma function (in [4], for example) we shall find it more convenient to use the equivalent Stieltjes integral representation

$$\psi_q(x) = -\log(1 - q) - \int_0^\infty \frac{e^{-xt}}{1 - e^{-t}} d\gamma_q(t), \quad 0 < q < 1, \ x > 0, \tag{1.11}$$

where $d\gamma_q(t)$ is a discrete measure with positive masses $-\log q$ at the positive points $-k \log q$, $k = 1, 2, \ldots$. For completeness, and economy of later statements, we include the value $q = 1$ in the definition of $\gamma_q(t)$:

$$\gamma_q(t) = \begin{cases} -\log q \sum_{k=1}^\infty \delta(t + k \log q), & 0 < q < 1. \\ t, \quad q = 1. \end{cases} \tag{1.12}$$

To get the representation (1.11), we expand the denominator of the sum in (1.10) by the binomial theorem and interchange the orders of summation to get

$$\psi_q(x) = -\log(1 - q) + \log q \sum_{k=1}^\infty q^{kx}/(1 - q^k), \tag{1.13}$$

311

which is equivalent to (1.11).

Note that $\psi_q(x)$ can also be expressed as a q–integral [8, p. 19],

$$\psi_q(x) = -\log(1-q) + \frac{\log q}{1-q} \int_0^1 \frac{t^{x-1}}{1-t} d_q(t), \quad 0 < q < 1, \ x > 0, \tag{1.14}$$

just as, from (1.9), $\psi(x)$ can be expressed as an ordinary integral over $[0,1]$:

$$\psi(x) = -\gamma + \int_0^1 \frac{1-t^{x-1}}{1-t} dt, \quad \operatorname{Re} x > 0. \tag{1.15}$$

We will need the following relations which follow easily from the definition of $d\gamma_q(t)$:

$$\int_0^\infty e^{-xt} d\gamma_q(t) = \frac{-q^x \log q}{1-q^x}, \quad 0 < q < 1, \ x > 0, \tag{1.16}$$

and

$$\int_0^\infty \frac{e^{-xt}}{t} d\gamma_q(t) = \sum_{k=1}^\infty \frac{q^{kx}}{k} = -\log(1-q^x), \quad 0 < q < 1, \ x > 0. \tag{1.17}$$

We will use the following lemma in many of our proofs. We remark that it includes results stated in different notations [4, Lemma 3.1] and [12, Lemma 4.1], as well as individual steps proved by *ad hoc* methods in these and other papers.

Lemma 1.2 *Let* $0 < \alpha < 1$. *Then*

$$\alpha e^{(\alpha-1)t} < \frac{\sinh\alpha t}{\sinh t} < \alpha, \quad t > 0. \tag{1.18}$$

The inequalities become equalities when $\alpha = 1$ *and they are reversed when* $\alpha > 1$.

The following lemma will also be useful.

Lemma 1.3 *(i) Let* $f(x)$ *be completely monotonic on* $(0,\infty)$ *and let* $a > 0$. *Then* $f(x) - f(x+a)$ *is completely monotonic on* $(0,\infty)$. *(ii) Let* $f(x) \geq 0$ *and let* $f(x) - f(x+a)$ *be completely monotonic on* $(0,\infty)$ *for each* a *in some right–hand neighborhood of* 0. *Then* $f(x)$ *is completely monotonic on* $(0,\infty)$.

Proof: (i) We have

$$f(x) = \int_0^\infty e^{-xt} d\alpha(t),$$

where $\alpha(t)$ is nondecreasing and the integral converges for $0 < x < \infty$. Hence

$$(-1)^n D_x^n[f(x) - f(x+a)] = \int_0^\infty [e^{-xt} - e^{-(x+a)t}] t^n d\alpha(t) \geq 0.$$

(ii) Under the hypotheses, we find that $-f'(x) = \lim_{a\to 0^+} [f(x) - f(x+a)]/a$ is completely monotonic on $(0,\infty)$. ∎

Remark 1 A feature of the present work is that the similarity between (1.9) and (1.11) makes it possible to unify the proofs of some of our results for the gamma and q–gamma functions.

Remark 2 The integral representation in Theorem 1.1 (i) provides a necessary as well as a sufficient condition for the complete monotonicity of f. This enables us to show that certain monotonic functions are *not* completely monotonic more easily than is done in [2], for example. As in [12], many of our results will assert the complete monotonicity of a function for a certain range of values of a parameter, the complete monotonicity of its derivative for another range, and the complete monotonicity of neither of these for an intermediate range.

Remark 3 In many discussions of completely monotonic functions, the concept of *strict complete monotonicity* is used to indicate the strict inequality $(-1)^n f^{(n)}(x) > 0$. But if, as here, our interval is a half–line, we get such strict inequality in all but trivial cases: a result of J. Dubourdieu [5, p. 98] asserts that for a completely monotonic function on (a, ∞), we have $(-1)^n f^{(n)}(x) > 0$ for $n = 1, 2, \ldots$, unless $f(x)$ is constant.

Remark 4 In §5, we extend some bounds for ratios of gamma functions to complex values of the arguments.

2   GAMMA AND q–GAMMA FUNCTIONS

The following result was proved in [12]:

Theorem 2.1 *Let* $h_\alpha(x) = \log[x^\alpha \Gamma(x)(e/x)^x]$. *Then* $-h_\alpha'(x)$ *is completely monotonic on* $(0, \infty)$ *for* $\alpha \le 1/2$, $h_\alpha'(x)$ *is completely monotonic for* $\alpha \ge 1$, *and neither is completely monotonic for* $1/2 < \alpha < 1$.

The proof in [12] is based on the consequence

$$-h_\alpha'(x) = \int_0^\infty \left[\frac{1}{1 - e^{-t}} - \frac{1}{t} - \alpha \right] e^{-xt} dt,$$

of (1.9) and the fact that the quantity in the square brackets, which has the same sign as $(1 + \alpha t)e^{-t} - 1 + (1 - \alpha t)$, is positive for $\alpha \le 1/2$, negative for $\alpha \ge 1$ and undergoes a change of sign for $1/2 < \alpha < 1$. The next result can be considered a q-analogue of Theorem 2.1.

Theorem 2.2 *Let* $0 < q < 1$ *and let*

$$h_\alpha(x) = \log \left[(1 - q)^x (1 - q^x)^\alpha \Gamma_q(x) \exp \left(\sum_{k=1}^\infty q^{kx} / (k^2 \log q) \right) \right].$$

Then $-h_\alpha'(x)$ *is completely monotonic on* $(0, \infty)$ *for* $\alpha \le 1/2$, $h_\alpha'(x)$ *is completely monotonic on* $(0, \infty)$ *for* $\alpha \ge 1$ *and neither is completely monotonic for* $1/2 < \alpha < 1$.

Proof: It follows from (1.11) that

$$-h'_\alpha(x) = \int_0^\infty \left[\frac{1}{1 - e^{-t}} - \frac{1}{t} - \alpha \right] e^{-xt} d\gamma_q(t),$$

where $d\gamma_q(t)$ is defined by (1.12). As in the proof of [12, Theorem 2.1], the quantity in square brackets is positive for $\alpha \leq 1/2$, negative for $\alpha \geq 1$ and has a change of sign when $1/2 < \alpha < 1$. Thus the results follow from Theorem 1.1 (i). ∎

Remark. To see that Theorem 2.1 is the limiting case of Theorem 2.2 as $q \to 1^-$, we will show that

$$\lim_{q \to 1^-} \frac{(1 - q)^x (1 - q^x)^\alpha \Gamma_q(x) \exp[F(q^x)/\log q]}{(1 - q)^\alpha \exp[F(1)/\log q]} = x^\alpha \Gamma(x)(e/x)^x,$$

where

$$F(x) = \sum_{n=1}^\infty \frac{x^n}{n^2} = -\int_0^x \frac{\log(1 - t)}{t} dt. \tag{2.1}$$

There is no difficulty in seeing that

$$\lim_{q \to 1^-} \frac{(1 - q^x)^\alpha \Gamma_q(x)}{(1 - q)^\alpha} = x^\alpha \Gamma(x),$$

so it remains to show that

$$\lim_{q \to 1^-} \frac{(1 - q)^x \exp[F(q^x)/\log q]}{\exp[F(1)/\log q]} = (e/x)^x.$$

Taking logarithms, this is equivalent to showing that

$$\lim_{q \to 1^-} \left[x \log(1 - q) + \frac{F(q^x) - F(1)}{\log q} \right] = x - x \log x,$$

and this follows from the identity

$$x \log(1 - q) + \frac{F(q^x) - F(1)}{\log q} \equiv -\int_0^x \log \frac{1 - q^t}{1 - q} dt,$$

which is easily checked by differentiation with respect to x.

Applying Lemma 1.3 to the results of the last theorem, we get:

Theorem 2.3 Let $0 < q < 1$, $a > 0$ and let

$$H_\alpha(x) = \log \left[\left(\frac{1 - q^x}{1 - q^{x+a}} \right)^\alpha \frac{\Gamma_q(x)}{\Gamma_q(x + a)} \exp \left(\sum_{k=1}^\infty \frac{q^{kx} - q^{k(x+a)}}{k^2 \log q} \right) \right].$$

Then $-H_\alpha'(x)$ is completely monotonic on $(0, \infty)$ for $\alpha \leq 1/2$, $H_\alpha'(x)$ is completely monotonic on $(0, \infty)$ for $\alpha \geq 1$ and neither is completely monotonic for $1/2 < \alpha < 1$.

In the case $q \to 1^-$, this theorem gives the following corollary, which also follows by applying Lemma 1.3 to the result of Theorem 2.1.

Corollary 2.4 Let $0 < q < 1$, $a > 0$ and let

$$H_\alpha(x) = \log\left(\frac{x^{\alpha-x}\Gamma(x)}{(x+a)^{\alpha-x-a}\Gamma(x+a)}\right).$$

Then $-H_\alpha'(x)$ is completely monotonic on $(0,\infty)$ for $\alpha \le 1/2$, $H_\alpha'(x)$ is completely monotonic on $(0,\infty)$ for $\alpha \ge 1$ and neither is completely monotonic, for $1/2 < \alpha < 1$.

The corollary introduces ratios of gamma functions whose asymptotic behavior [1, 6.1.47]

$$z^{b-a}\frac{\Gamma(z+a)}{\Gamma(z+b)} \sim 1 + \frac{(a-b)(a+b+1)}{2z} + \dots \tag{2.2}$$

suggests that we look at the possible completely monotonic character of such ratios for both gamma and q-gamma functions.

Our first result is suggested by [4, Theorem 3] on $(x+c)^{a-b}\Gamma(x+b)/\Gamma(x+a)$:

Theorem 2.5 Let $a < b \le a+1$ and let

$$g(x) := \left[\frac{1-q^{x+c}}{1-q}\right]^{a-b}\frac{\Gamma_q(x+b)}{\Gamma_q(x+a)}. \tag{2.3}$$

Then $-(\log g(x))'$ is completely monotonic on $(-c,\infty)$ if $0 \le c \le (a+b-1)/2$ and $(\log g(x))'$ is completely monotonic on $(-a,\infty)$ if $c \ge a \ge 0$. Neither is completely monotonic for $(a+b-1)/2 < c < a$.

Remarks. Apart from its last sentence, the limiting case $q \to 1^-$ of this theorem was proved by Bustoz and Ismail [4, Theorem 3]. (Note that in the statement of [4, Theorem 3 (ii)] "$x > \beta$" should read "$x > \alpha$".) Although, by a translation of the variable x, one could assume $a = 0$ and so express the theorem in a simpler form, we prefer to retain the a and b for reasons of symmetry. An interesting consequence of the last assertion of the theorem is that neither h' or $-h'$ is completely monotonic when

$$c = -\frac{1}{2} + \sqrt{ab + \frac{1}{4}},$$

and hence $(a+b-1)/2 < c < a$. Thus although it is true that

$$\frac{d}{dx}\log\left[\frac{\Gamma(x+a)}{\Gamma(x+b)}\left(x - \frac{1}{2} + \sqrt{ab + \frac{1}{4}}\right)^{b-1}\right] \tag{2.4}$$

is positive (a result motivated by the right–hand inequality in (1.4), and proved essentially by the method given for the case $b = 1$ in [4, Theorem 8]), it is not completely monotonic.

Proof of Theorem 2.5: We note that

$$\frac{d}{dx} \log g(x) = (b-a)\frac{q^{x+c} \log q}{1-q^{x+c}} + \psi_q(x+b) - \psi_q(x+a)$$

$$= -\int_0^\infty e^{-xt} \left[\frac{e^{-bt} - e^{-at}}{1-e^{-t}} + (b-a)e^{-ct} \right] d\gamma_q(t),$$

where γ_q is given by (1.12). Now we will use Lemma 1.2, to show that the quantity in square brackets has the appropriate sign. In case $c \le (a+b-1)/2$, we find that the integrand is $e^{-(x+c)t}$ times a function of t which exceeds

$$b - a - \frac{\sinh[(b-a)t/2]}{\sinh(t/2)}$$

and this is positive by Lemma 1.2. When $c \ge a$, the integrand is $e^{-(x+a)t}$ times a function of t which is less than

$$(b-a)e^{(b-a-1)t/2} - \frac{\sinh[(b-a)t/2]}{\sinh(t/2)}$$

and this is negative by Lemma 1.2. On the other hand, if $(a+b-1)/2 < c < a$, the quantity in square brackets is positive for t close to 0 and negative for large t. ∎

The ranges of a, b and c in Theorem 2.5 fail to cover some interesting cases. For example, when $b = c = 0$, it gives $-\log g(x)$ completely monotonic for $a \le 0$, but gives no information for $a > 0$. Hence it is of interest to record the following result:

Theorem 2.6 Let $0 < q < 1$ and let

$$h(x) = \log \left[\left(\frac{1-q^x}{1-q} \right)^a \frac{\Gamma_q(x)}{\Gamma_q(x+a)} \right]. \tag{2.5}$$

Then $h'(x)$ is completely monotonic on $(0, \infty)$ for $a \ge 1$.

Proof: From (1.11), we have

$$h'(x) = \int_0^\infty e^{-xt} \left[\frac{e^{-at}-1}{1-e^{-t}} + a \right] d\gamma_q(t)$$
$$= \int_0^\infty e^{-xt+t(1-a)/2} \left[ae^{(a-1)t/2} - \frac{\sinh(at/2)}{\sinh(t/2)} \right] d\gamma_q(t) \tag{2.6}$$

and we see from Lemma 2.2 that the quantity in square brackets is positive for $a > 1$. ∎

3 PSI AND q–PSI FUNCTIONS

The psi function has particularly simple monotonicity properties. For example, it follows from (1.8) that

$$\psi'(x) = \sum_{n=0}^{\infty}(n+x)^{-2}, \tag{3.1}$$

so $\psi'(x)$ is completely monotonic on $(0,\infty)$. G. Ronning [20] showed that, for $0 < \alpha < 1$, $\psi'(x) - \alpha\psi'(\alpha x) < 0$, $0 < x < \infty$. More generally, we show:

Theorem 3.1 Let $0 < \alpha < 1$, $0 < q < 1$. Then the function $\psi_q(x) - \psi_{q^{1/\alpha}}(\alpha x)$ is completely monotonic, i.e.,

$$(-1)^n[\psi_q^{(n)}(x) - \alpha^n\psi_{q^{1/\alpha}}^{(n)}(\alpha x)] > 0, \ 0 < x < \infty, \ n = 0, 1, \ldots . \tag{3.2}$$

Proof: Using (1.9) and (1.11) we see that

$$\psi_q(x) - \psi_{q^{1/\alpha}}(\alpha x) = \frac{1}{\alpha}\int_0^{\infty} e^{-xt}\left[-\frac{\alpha}{1-e^{-t}} + \frac{1}{1-e^{-t/\alpha}}\right]d\gamma_q(t). \tag{3.3}$$

The quantity in square brackets is seen to be equal to

$$\frac{\alpha e^{(1-1/\alpha)t/2}}{1-e^{-t/\alpha}}\left[\frac{1}{\alpha}e^{(t/2)(1/\alpha-1)} - \frac{\sinh(t/(2\alpha))}{\sinh(t/2)}\right],$$

and the term in square brackets is seen to be positive on using the left–hand inequality in Lemma 1.2 (with α replaced by $1/\alpha$). The result follows. ∎

Gautschi's and Kershaw's inequalities (1.2) and (1.3) suggest that we consider ratios of the form

$$\frac{\Gamma_q(x+a)}{\Gamma_q(x+b)}\exp\left[(b-a)\psi_q(x+c)\right]. \tag{3.4}$$

We have the following result.

Theorem 3.2 Let $0 < a < b$, $0 < q \leq 1$ and let $h(x)$ denote the logarithm of the function in (3.4). Then, if $c \geq (a+b)/2$, $-h'(x)$ is completely monotonic on $(-a,\infty)$ and if $c \leq a$, $h'(x)$ is completely monotonic on $(-c,\infty)$. Neither h' or $-h'$ is completely monotonic for $a < c < (a+b)/2$.

Bustoz and Ismail have this result in the case $q = 1$, $b = 1$, $c = (a+b)/2$.

Proof: We have, from (1.9) and (1.11),

$$h'(x) = -\int_0^{\infty}\frac{e^{-(x+(a+b)/2)t}}{1-e^{-t}}\left[2\sinh\frac{b-a}{2}t - (b-a)te^{((a+b)/2-c)t}\right]d\gamma_q(t). \tag{3.5}$$

Using the inequality $\sinh\theta > \theta$, $\theta > 0$, we see that the quantity in square brackets is positive when $c \geq (a+b)/2$. When $c \leq a$, the quantity in square brackets is negative on account of the inequality $e^{-2x} > 1 - 2x$, $x > 0$. On the other hand, when $a < c < (a + b)/2$ this quantity is negative for small t and positive for large t. This completes the proof. An interesting consequence of the last assertion of the theorem is that neither h' or $-h'$ is completely monotonic when c is the geometric mean of a and b. Thus although it is true that

$$\frac{d}{dx} \log \left[\frac{\Gamma(x + a)}{\Gamma(x + b)} \exp \left[(b - a)\psi(x + \sqrt{ab}) \right] \right] \tag{3.6}$$

is positive (the result is suggested by the left–hand inequality in (1.3) and the proof is essentially that given for the case $b = 1$ in [4, Theorem 8]), neither it nor its negative is completely monotonic. ∎

A q–analogue of inequalities in (1.3) and (1.4) runs as follows:

Theorem 3.3 *For $0 < q \leq 1$, we have*

$$\left(\frac{1 - q^{x+s/2}}{1 - q} \right)^{1-s} < \frac{\Gamma_q(x + 1)}{\Gamma_q(x + s)} < \exp[(1-s)\,\psi_q(x+(s+1)/2)], \quad 0 < s < 1, \tag{3.7}$$

where the left–hand inequality holds for $x > -s/2$ and the right–hand one holds for $x > -s$.

Proof: Theorem 2.5 shows that

$$\left[\frac{1 - q^{x+s/2}}{1 - q} \right]^{s-1} \frac{\Gamma_q(x + 1)}{\Gamma_q(x + s)}$$

decreases on $(-s/2, \infty)$ to its limiting value 1. This gives the left–hand part of (3.7). Theorem 3.3 shows that

$$\exp(1 - s)\psi_q \left(x + \frac{s+1}{2} \right) \frac{\Gamma_q(x + s)}{\Gamma_q(x + 1)}$$

decreases on $(-s, \infty)$ to its limiting value 1. This gives the right–hand part of (3.7).

A result of H. Alzer [2, Theorem 1] suggests dealing with products of q–gamma functions and exponentials of *derivatives* of q–psi functions. In this connection, we prove:

Theorem 3.4 *Let $0 < q < 1$, $0 < s < 1$ and*

$$g_\alpha(x) = (1 - q)^x (1 - q^x)^{1/2} \Gamma_q(x) \exp \left[F(q^x)/\log q - \frac{1}{12} \psi'_q(x + \alpha) \right],$$

where F is given by (2.1). Then $(\log g_\alpha)'$ is completely monotonic on $(0, \infty)$ for $\alpha \geq 1/2$, $-(\log g_\alpha)'$ is completely monotonic on $(0, \infty)$ for $\alpha \leq 0$, and neither is completely monotonic on $(0, \infty)$ for $0 < \alpha < 1/2$.

318

Proof: We have

$$\frac{d}{dx}g_\alpha(x) = -\int_0^\infty e^{-xt}p_\alpha(t)d\gamma_q t$$

where

$$p_\alpha(t) = \frac{12 - t^2e^{-\alpha t}}{12(1 - e^{-t})} - \frac{1}{2} - \frac{1}{t}.$$

Now, for $\alpha \geq 1/2$, we have $p_\alpha(t) > 0$ for $t > 0$ [2, p. 339]. Also when $\alpha \leq 0$, we have $p_\alpha(t) < 0$ for $t > 0$ and when $0 < \alpha < 1/2$, we have $p_\alpha(t) > 0$ for t close to 0 and $p_\alpha(t) < 0$ for large t. Thus we get the required complete monotonicity properties of g_α.

If we combine this theorem with Lemma 1.3 we get the following extension of [2, Theorem 1] to which it reduces when $q \to 1^-$.

Corollary 3.5 *Let $0 < q < 1$, $0 < s < 1$ and*

$$
\begin{aligned}
f_\alpha(x) &= g_\alpha(x + s)/g_\alpha(x + 1) \\
&= \left[\frac{(1 - q)^{s-1}(1 - q^{x+s})^{1/2}\Gamma_q(x + s)}{(1 - q^{x+1})^{1/2}\Gamma_q(x + 1)}\right. \\
&\quad \times \left.\exp\left\{\frac{F(q^{x+s}) - F(q^{x+1})}{\log q} + \frac{1}{12}[\psi'_q(x + 1 + \alpha) - \psi'_q(x + s + \alpha)]\right\}\right].
\end{aligned}
$$

Then $(\log f)'$ is completely monotonic on $(0, \infty)$ for $\alpha \geq 1/2$, $-(\log f)'$ is completely monotonic on $(0, \infty)$ for $\alpha \leq 0$, and neither is completely monotonic on $(0, \infty)$ for $0 < \alpha < 1/2$.

The limiting case $q \to 1^-$ of this last Corollary runs as follows:

Corollary 3.6 *Let $0 < s < 1$ and*

$$
\begin{aligned}
f_\alpha(x) &= \frac{(x + 1)^{x+1/2}\Gamma_q(x + s)}{(x + s)^{x+s-1/2}\Gamma_q(x + 1)} \\
&\quad \times \exp\left\{s - 1 + \frac{1}{12}[\psi'(x + 1 + \alpha) - \psi'(x + s + \alpha)]\right\}.
\end{aligned}
$$

Then $(\log f)'$ is completely monotonic on $(0, \infty)$ for $\alpha \geq 1/2$, $-(\log f)'$ is completely monotonic on $(0, \infty)$ for $\alpha \leq 0$, and neither is completely monotonic on $(0, \infty)$ for $0 < \alpha < 1/2$.

This recovers, in a slightly extended form, the main result [2, Theorem 1] of H. Alzer.

4 FURTHER PRODUCTS AND QUOTIENTS

Results of Bustoz and Ismail [4, Theorem 6] concerning the ratio

$$\frac{\Gamma(x+a)\Gamma(x+b)}{\Gamma(x)\Gamma(x+a+b)},$$

suggest the consideration of ratios

$$\frac{\Gamma_q(x+a_1)\Gamma_q(x+a_2)\ldots\Gamma_q(x+a_n)}{\Gamma_q(x+b_1)\Gamma_q(x+b_2)\ldots\Gamma_q(x+b_n)}, \tag{4.1}$$

where

$$a_1 + a_2 + \ldots + a_n = b_1 + b_2 + \ldots + b_n = s. \tag{4.2}$$

If we treat the a_i as given, each choice of the b_i may be thought of as a partition of the sum of the a_i. An extreme case would consist of taking the b_i equal to each other; another extreme case would be to take all except one of the b_i to be 0. Both of these lead to completely monotonic functions:

Theorem 4.1 *Let* $a_1, \ldots, a_n$ *be positive numbers, let* $n\bar{a} = a_1 + \ldots a_n$, *and* $0 < q \le 1$. *Then both*

$$-\frac{d}{dx}\log\frac{\Gamma_q(x+a_1)\Gamma_q(x+a_2)\ldots\Gamma_q(x+a_n)}{\Gamma_q(x+\bar{a})^n} \tag{4.3}$$

and

$$\frac{d}{dx}\log\frac{\Gamma_q(x+a_1)\Gamma_q(x+a_2)\ldots\Gamma_q(x+a_n)}{\Gamma_q(x)^{n-1}\Gamma_q(x+a_1+a_2+\ldots+a_n)} \tag{4.4}$$

are completely monotonic on $(0, \infty)$.

Proof: Using (1.9) and (1.11), we find that these expressions may be written

$$\int_0^\infty \frac{e^{-xt}}{1-e^{-t}}\left[ne^{-\bar{a}t} - e^{-a_1 t} - \ldots - e^{-a_n t}\right]d\gamma_q(t) \tag{4.5}$$

and

$$\int_0^\infty \frac{e^{-xt}}{1-e^{-t}}\left[n-1+e^{-(a_1+\ldots a_n)t} - e^{-a_1 t} - \ldots - e^{-a_n t}\right]d\gamma_q(t). \tag{4.6}$$

In the first case the quantity in square brackets is positive by Jensen's theorem while in the second case its positivity follows from

$$n-1+z_1 z_2 \ldots z_n - (z_1 + z_2 + \ldots + z_n) \ge 0, \quad 0 \le z_i < 1. \tag{4.7}$$

This is clear since it may be established by induction on n that

$$n-1+z_1 z_2 \ldots z_n - (z_1 + z_2 + \ldots + z_n) = \sum_{j=2}^n (1-z_j)(1-z_1 z_2 \ldots z_{j-1}). \tag{4.8}$$

■

The positivity of the quantity (4.3) leads to the inequality

$$\Gamma_q(1 + a_1)\Gamma_q(1 + a_2)\ldots\Gamma_q(1 + a_n) \geq \Gamma_q(1 + \bar{a})^n, \ a_i > 0,$$

which is known already in the case $q = 1$ [17, p. 285].

5 BOUNDS IN THE COMPLEX PLANE

A generalization of the Phragmén–Lindelöf theorem is used in [19, pp. 68–70] to show that

$$\left|\frac{\Gamma(s + c)}{\Gamma(s)}\right| \leq |s|^c, \ 0 \leq c \leq 1, \ \mathrm{Re}(s) \geq (1 - c)/2. \tag{5.1}$$

Here we show that the same method can be used to get bounds for more complicated functions involving gamma functions.

The Phragmén–Lindelöf theorem runs as follows [19, p. 59]:

Theorem 5.1 *Let $f(z)$ be analytic in the strip $S(\alpha, \beta) = \{z | z = x + iy, \ \alpha < x < \beta\}$. Let us assume $|f(z)| \leq 1$ on the boundaries $x = \alpha$ and $x = \beta$ and moreover,*

$$|f(z)| < Ce^{e^{k|y|}}$$

for some $C > 0$ and $0 < k < \pi/(\beta - \alpha)$. Then $|f(z)| \leq 1$ throughout the strip $S(\alpha, \beta)$.

By Theorem 4.1, the function

$$f(x) := \frac{\Gamma(x + a)\Gamma(x + b)}{\Gamma(x)\Gamma(x + a + b)}, \quad a, b \geq 0. \tag{5.2}$$

is increasing on the interval $(0, \infty)$ to its limit 1. Hence we have

$$\left|\frac{\Gamma(x + a)\Gamma(x + b)}{\Gamma(x)\Gamma(x + a + b)}\right| \leq 1, \ x > 0. \tag{5.3}$$

We have, in fact:

Theorem 5.2 *We have, for $0 \leq a \leq 1, b \geq 0$,*

$$\left|\frac{\Gamma(s + a)\Gamma(s + b)}{\Gamma(s)\Gamma(s + a + b)}\right| \leq 1, \ \mathrm{Re}\,s > \frac{1 - a - b}{2}. \tag{5.4}$$

Proof: The method follows the proof of [19, Theorem A, p. 68]. Since the assertion is trivial for $a = 0$, we may as well choose $a > 0$. We choose a complex number $s = \sigma + i\tau$ satisfying the hypotheses of the theorem and let

$$f(z) = \frac{\Gamma(a + 2\sigma - z)\Gamma(b + z)}{\Gamma(z)\Gamma(a + b + 2\sigma - z)}.$$

321

Clearly

$$|f(s)| = \left| \frac{\Gamma(s+a)\Gamma(s+b)}{\Gamma(s)\Gamma(s+a+b)} \right|$$

so we have to show that $|f(s)| \leq 1$. Now with $\alpha = (a-1)/2 + \sigma$, $\beta = a/2 + \sigma$, and using $\overline{\Gamma(z)} = \Gamma(\bar{z})$, we get

$$|f(\alpha + it)| = \frac{|\alpha + it|}{|b + \alpha + it|} \leq 1$$

where we have used Re $s \geq (1 - a - b)/2$. Also

$$|f(\beta + it)| = 1,$$

and (2.2) shows that the growth condition of Theorem 5.1 is satisfied. Using this theorem, we find that $|f(z)| \leq 1$, for $\alpha \leq$ Re $z \leq \beta$, so, in particular, $|f(s)| \leq 1$.

Acknowledgments. The authors' work was supported partially by NSF grant DMS 9203659 and by NSERC Canada grant A5199. We are grateful to Ruiming Zhang for pointing out the relevance of reference [19]. We thank a referee for corrections and suggestions.

REFERENCES

[1] Abramowitz M., Stegun I.A., eds. *Handbook of Mathematical Functions, with Formulas, Graphs and Mathematical Tables.* National Bureau of Standards, Applied Mathematics Series **55**, Washington, 1964.

[2] Alzer H. Some gamma function inequalities. *Math. Comp.*, **60**:337–346, 1993.

[3] Andrews G.E. *q-Series: Their Development and Application in Analysis, Number Theory, Combinatorics, Physics, and Computer Algebra.* Regional Conference Series in Mathematics Number **66**, American Mathematical Society, Providence, 1985.

[4] Bustoz J., Ismail M.E.H. On gamma function inequalities. *Math. Comp.*, **47**:659–667, 1986.

[5] Dubourdieu J. Sur un théorème de M. S. Bernstein relatif à la transformation de Laplace–Stieltjes. *Compositio Math.*, **7**:96–111, 1939–40.

[6] Erdélyi A., Magnus W., Oberhettinger F., Tricomi F.G. *Higher Transcendental Functions*, vol. 1. McGraw-Hill, New York, 1953.

[7] Feller W. *An Introduction to Probability Theory and its Applications*, vol. 2. Wiley, New York, 1966.

[8] Gasper G., Rahman M. *Basic Hypergeometric Series.* Cambridge University Press, Cambridge, 1990.

[9] Gautschi W. Some elementary inequalities relating to the gamma and incomplete gamma function. *J. Math. Phys.*, **38**:77–81, 1959.

[10] Gautschi W. A harmonic mean inequality for the gamma function. *SIAM J. Math. Anal.*, **5**:278–281, 1974.

[11] Gautschi W. Some mean value inequalities for the gamma function. *SIAM J. Math. Anal.*, **5**:282–292, 1974.

[12] Ismail M.E.H., Lorch L., Muldoon M.E. Completely monotonic functions associated with the gamma function and its q-analogues. *J. Math. Anal. Appl.*, **116**:1–9, 1986.

[13] Kershaw D. Some extensions of W. Gautschi's inequalities for the gamma function. *Math. Comp.*, **41**:607–611, 1983.

[14] Koornwinder T.H. Jacobi functions as limit cases of q–ultraspherical polynomials. *J. Math. Anal. Appl.*, **148**:44–54, 1990.

[15] Laforgia A. Further inequalities for the gamma function. *Math. Comp.*, **42**:597–600, 1984.

[16] Marshal A.W., Olkin I. *Inequalities: Theory of Majorization and Applications.* Academic Press, New York, 1979.

[17] Mitronović D.S. *Analytic Inequalities.* Springer-Verlag, Berlin, 1971.

[18] Muldoon M.E. Some monotonicity properties and characterizations of the gamma function. *Aequationes Math.*, **18**:54–63, 1978.

[19] Rademacher H. *Topics in Analytic Number Theory.* Springer–Verlag, Berlin–New York, 1973.

[20] Ronning G. On the curvature of the trigamma function. *J. Comp. Appl. Math.*, **15**:397–399, 1986.

[21] Widder D.V. *The Laplace Transform.* Princeton University Press, Princeton, 1941.

International Series of Numerical Mathematics, Vol. 119, ©1994 Birkhäuser

BOUNDS FOR REMAINDER TERMS IN SZEGÖ QUADRATURE ON THE UNIT CIRCLE

William B. Jones[*]   Haakon Waadeland[†]

Dedicated to Walter Gautschi on the occasion of his 65th birthday

Abstract. This paper deals with Szegö quadrature for integration around the unit circle in the complex plane. Nodes for the quadrature formulas are the zeros $\zeta_j^{(n)}(w_n)$, $j = 1, 2, \ldots, n$, of para-orthogonal Szegö polynomials $B_n(z, w_n)$ in z of degree n. The parameter w_n is a complex constant satisfying $|w_n| = 1$. Results are described for convergence of the quadrature formulas as $n \to \infty$ and upper bounds for the remainder term that results when the value of the integral is replaced by an n-point quadrature approximation. The upper bounds for remainder terms apply to integrals that represent Carathéodory functions and real parts of such integrals. The latter are Poisson integrals used to represent harmonic functions determined by boundary values on the unit circle.

1 INTRODUCTION

A comprehensive, historical survey of Gauss–Christoffel quadrature, containing an extensive bibliography was given in [4] by W. Gautschi. Gauss–Christoffel quadrature formulas have also been used in the development of moment theory (e.g., [1] and [6]). An expository survey of Szegö quadrature formulas for integration around the unit circle in the complex plane $\mathbb{C}$ is contained in [7]. There quadrature is used to solve the trigonometric moment problem. The present paper gives sharp, computable upper bounds for the remainder terms in Szegö quadrature related to Carathéodory functions and Poisson integrals. Further recent results on Szegö quadrature on the unit circle can be found in [2], [3] and [10].

Let $\Phi(-\pi, \pi)$ denote the *family of distribution functions* $\psi(\theta)$ (i.e., functions that are bounded and non-decreasing with infinitely many points of increase on the real interval $(-\pi, \pi)$. For integers p and q, with $p \leq q$, $\bigwedge\limits_{p,q}$ denotes the class of *Laurent polynomials*

---

[*]Department of Mathematics, University of Colorado, Boulder, Colorado 80309–0395, USA. Research supported in part by the U.S. National Science Foundation under Grant No. DMS–9302584

[†]Department of Mathematics & Statistics, University of Trondheim (AVH), N-7055 Dragvoll, Norway. Research supported by the Norwegian Research Council

$$\bigwedge_{p,q} := \left[g(\zeta) = \sum_{j=p}^{q} c_j \zeta^j : c_j \in \mathbb{C}, \text{ for } p \le j \le q \right]. \tag{1.1}$$

An *n-point Szegö quadrature formula with respect to* (ψ, w_n), $\psi \in \Phi(-\pi, \pi)$, $|w_n| = 1$, $w_n \in \mathbb{C}$, has the form

$$\int_{-\pi}^{\pi} g(e^{i\theta}) d\psi(\theta) = \sum_{j=1}^{n} \lambda_j^{(n)}(w_n) g(\zeta_j^{(n)}(w_n)) + e_n(g, \psi, w_n), \tag{1.2}$$

where the remainder term $e_n(g, \psi, w_n)$ satisfies

$$e_n(g, \psi, w_n) = 0 \text{ for all } g(\zeta) \in \bigwedge_{-(n-1),n-1}. \tag{1.3}$$

The *nodes* $\zeta_j^{(n)}(w_n)$ are the zeros of the nth degree para-orthogonal Szegö polynomial

$$B_n(\zeta, w_n) = \rho_n(\psi, \zeta) + w_n \rho_n^*(\psi; \zeta), \tag{1.4}$$

where $\{\rho_n(\psi; \zeta)\}_{n=0}^{\infty}$ is the sequence of monic Szegö polynomials orthogonal on the unit circle with respect to ψ, and $\rho_n^*(\psi; z) := z^n \overline{\rho_n(\psi; 1/\bar{z})}$. The n zeros $\zeta_j^{(n)}(w_n)$, $j = 1, \ldots, n$, are simple and satisfy $|\zeta_j^{(n)}(w_n)| = 1$. The weights $\lambda_j^{(n)}(w_n)$ satisfy

$$\sum_{j=1}^{n} \lambda_j^{(n)}(w_n) = \mu_0 := \int_{-\pi}^{\pi} d\psi(\theta) > 0 \text{ and } \lambda_j^{(n)}(w_n) > 0, \ j = 1, 2, \ldots, n. \tag{1.5}$$

The convergence result,

$$\lim_{n \to \infty} e_n(g, \psi, w_n) = 0, \text{ if } g(\zeta) \text{ is continuous on } T := [\zeta \in \mathbb{C} : |\zeta| = 1], \tag{1.6}$$

was given in the recent paper [2].

The focus of the present paper is on computable upper bounds of the remainder term $|e_n(K(\zeta, z), \psi, w_n)|$, when Szegö quadrature is applied to the function

$$K(\zeta, z) := \frac{\zeta + z}{\zeta - z}, \ \zeta \in T, \ z \in D := [u \in \mathbb{C} : |z| < 1]. \tag{1.7}$$

The associated integral

$$f(\psi, z) := \int_{-\pi}^{\pi} \frac{e^{i\theta} + z}{e^{i\theta} - z} d\psi(\theta), \ z \in D \tag{1.8}$$

belongs to the family $\mathcal{C}$ of *normalized Carathéodory functions*

$$\mathcal{C} := [h(z) : h(z) \text{ is analytic and } \operatorname{Re} f(z) > 0 \text{ for } |z| < 1, f(0) > 0]. \tag{1.9}$$

Since $\psi(\theta)$ is assumed to have infinitely many points of increase, functions $f(\psi, z)$ of the form (1.8) satisfy

$$f(\psi, z) \in \mathcal{C}_a := \mathcal{C} - \mathcal{C}_b, \tag{1.10a}$$

where

$$\mathcal{C}_b := \left[h(z) : h(z) = \sum_{j=1}^{n} \lambda_j K(e^{i\theta_j}, z), \text{ for } \lambda_j > 0 \text{ and} \right.$$
$$\left. -\pi \le \theta_j < \theta_{j+1} < \pi, \ j = 1, \dots, n, \ n = 1, 2, 3, \dots \right]. \tag{1.10b}$$

There is a one-to-one correspondence between the Carathéodory functions $f \in \mathcal{C}_a$ and the distribution functions $\psi \in \Phi(-\pi, \pi)$. It is readily seen that

$$\operatorname{Re}\left(\frac{e^{i\theta} + z}{e^{i\theta} - z}\right) = \frac{1 - r^2}{1 - 2r\cos(\varphi - \theta) + r^2}, \text{ where } z = re^{i\varphi}, \ r := |z| < 1. \tag{1.11}$$

Hence if

$$f(\psi, z) = u(\psi, z) + iv(\psi, z), \ u(\psi, z) := \operatorname{Re} f(\psi, z), \ v(\psi, z) := \operatorname{Im} f(\psi, z), \tag{1.12}$$

then

$$u(\psi, re^{i\varphi}) = \int_{-\pi}^{\pi} \frac{(1 - r^2)d\psi(\theta)}{1 - 2r\cos(\varphi - \theta) + r^2}, \text{ for } r := |z| < 1, \ -\pi < \varphi = \arg z \le \pi. \tag{1.13}$$

In particular, if $U(e^{i\theta}) > 0$ and $U(e^{i\theta})$ is continuous on $[-\pi, \pi]$, except for at most a finite number of jump discontinuities, then with

$$\psi(\theta) := \frac{1}{2\pi} \int_{-\pi}^{\theta} U(e^{it})dt \in \Phi(-\pi, \pi), \tag{1.14a}$$

one has

$$u(\psi, re^{i\varphi}) = \frac{1 - r^2}{2\pi} \int_{-\pi}^{\pi} \frac{U(e^{i\theta})d\theta}{1 - 2r\cos(\theta - \varphi) + r^2}, \ 0 \le r < 1, \ -\pi < \varphi \le \pi, \tag{1.14b}$$

$u(\psi, z)$ is harmonic for $|z| = |re^{i\varphi}| < 1$ and

$$\lim_{\substack{r \to 1^- \\ \varphi \to \theta}} u(\psi, re^{i\varphi}) = U(e^{i\theta}) \tag{1.14c}$$

327

for all θ where $U(e^{i\theta})$ is continuous. A proof of (1.14c) (with weaker conditions assumed for $U(e^{i\theta})$) can be found in Theorem 17.5.1 of [5].

It follows from (1.2), (1.8) and (1.12) that

$$
\begin{aligned}
u(\psi, z) &= \int_{-\pi}^{\pi} \frac{(1 - r^2) d\psi(\theta)}{1 - 2r\cos(\varphi - \theta) + r^2} \\
&= \sum_{j=1}^{n} \lambda_j^{(n)}(w_n) \operatorname{Re} K(\zeta_j^{(n)}(w_n), z) + \operatorname{Re} e_n(K(\zeta, z), \psi, w_n).
\end{aligned}
\tag{1.15}
$$

Since $|\operatorname{Re}(e_n(K(\zeta, z), \psi, w_n))| \leq |e_n(K(\zeta, z), \psi, w_n)|$, upper bounds for the remainder term $|e_n(K(\zeta, z), \psi, w_n)|$ are also bounds for the remainder term $|\operatorname{Re} e_n(K(\zeta, z), \psi, w_n)|$. Such upper bounds are given in §3, making use of Cara-théodory linear fractional approximant sequences (**CLFASs**). Some useful background on CLFASs is described in §2. Section 4 contains an elementary example that illustrates the main results of the paper. In the remainder of this introduction we summarize notation and background on Szegö polynomials and para-orthogonal Szegö polynomials that are subsequently used.

Let $\psi(\theta) \in \Phi(-\pi, \pi)$ be a given distribution function. Its moment sequence $\{\mu_n\}_{n=-\infty}^{\infty}$ is *Hermitian positive definite*; that is, if

$$
\mu_n := \int_{-\pi}^{\pi} e^{-in\theta} d\psi(\theta), \quad n = 0, \pm 1, \pm 2, \ldots,
\tag{1.16}
$$

then

$$
\mu_n = \bar{\mu}_{-n} \text{ and } \Delta_n :=
\begin{vmatrix}
\mu_0 & \mu_{-1} & \cdots & \mu_{-n} \\
\mu_1 & \mu_0 & \cdots & \mu_{-n+1} \\
\vdots & \vdots & \ddots & \vdots \\
\mu_n & \mu_{n-1} & \cdots & \mu_0
\end{vmatrix} > 0, \quad n = 0, 1, 2, \ldots, \quad (\Delta_{-1} := 1).
\tag{1.17}
$$

An inner product

$$
(f, g)_{\psi} := \int_{-\pi}^{\pi} f(e^{i\theta}) \overline{g(e^{i\theta})} d\psi(\theta)
\tag{1.18}
$$

is defined for $f(z), g(z)$ continuous on $T := [z \in \mathbb{C} : |z| = 1]$. The *monic Szegö polynomial sequence* $\{\rho_n(\psi; z)\}_{n=0}^{\infty}$ satisfies the orthogonality conditions

$$
(\rho_n(\psi; z), z^m)_{\psi} =
\begin{cases}
0, & 0 \leq m \leq n-1 \\
\Delta_n / \Delta_{n-1}, & m = n,
\end{cases} \quad n = 1, 2, 3, \ldots.
\tag{1.19}
$$

The *reciprocal sequence* $\{\rho_n^*(\psi; z)\}_{n=0}^{\infty}$, defined by

$$
\rho_n^*(\psi; z) := z^n \overline{\rho_n(\psi; 1/\bar{z})}, \quad n = 0, 1, 2, \ldots,
\tag{1.20}
$$

328

satisfies orthogonality conditions

$$(\rho_n^*(\psi; z), z^m)_\psi = \begin{cases} \Delta_n/\Delta_{n-1}, & m = 0 \\ 0, & 1 \leq m \leq n, \end{cases} \quad n = 1, 2, 3, \ldots . \quad (1.21)$$

These polynomials satisfy recurrence relations

$$\rho_0(\psi; z) = \rho_0^*(\psi; z) = 1 \quad (1.22a)$$

$$\rho_n(\psi, z) = z\rho_{n-1}(\psi; z) + \delta_n \rho_{n-1}^*(\psi; z), \quad n = 1, 2, 3, \ldots \quad (1.22b)$$

$$\rho_n^*(\psi; z) = \bar{\delta}_n z\rho_{n-1}(\psi; z) + \rho_{n-1}^*(\psi; z), \quad n = 1, 2, 3, \ldots . \quad (1.22c)$$

where

$$\delta_n := \rho_n(\psi; 0), \quad n = 0, 1, 2, \ldots, \quad (1.23)$$

are called *reflection coefficients*. Applying orthogonality to (1.22b) yields

$$\delta_n = -\frac{(z\rho_{n-1}(\psi; z), 1)_\psi}{(\rho_{n-1}^*(\psi; z), 1)_\psi} = -\frac{\sum_{j=0}^{n-1} q_j^{(n-1)} \mu_{-j-1}}{\sum_{j=0}^{n-1} q_j^{(n-1)} \mu_{j+1-n}}, \quad n = 1, 2, 3, \ldots, \quad (1.24)$$

where $\sum_{j=0}^{n-1} q_j^{(n-1)} z^j := \rho_{n-1}(\psi; z)$. Starting with $\delta_0 = 1$ and (1.22a) we can obtain $\rho_1(\psi; z)$ and $\rho_1^*(\psi; z)$ from (1.22b,c) with $n = 1$. This gives $\delta_1 = \rho_1(\psi, 0)$, $q_0^{(1)}$, and $q_1^{(1)} = 1$. By alternately applying (1.22) and (1.24), we can compute $\{\delta_n\}$, $\{q_n(\psi; z)\}$ and $\{q_n^*(\psi; z)\}$. This procedure is called the *Levinson algorithm*.

A sequence $\{X_n(z)\}$ is called a *para-orthogonal polynomial sequence with respect to* $\psi(\theta)$ if for each $n \geq 0$,

$$X_n(z) \text{ is a polynomial in } z \text{ of degree } n \text{ and} \quad (1.25a)$$

$$(X_n(z), 1)_\psi \neq 0, \ (X_n(z), z^m)_\psi = 0 \text{ for } 1 \leq m \leq n-1, \ (X_n(z), z^n) \neq 0. \quad (1.25b)$$

For $0 \neq \kappa \in \mathbb{C}$, a polynomial $X(z)$ is called *κ-invariant* if

$$X^*(z) := (z^{\deg X}) \overline{X(1/\bar{z})} = \kappa X(z), \quad z \in \mathbb{C}. \quad (1.26)$$

A sequence of polynomials $\{X_n(z)\}$ is said to be $\{\kappa_n\}$-*invariant* if $X_n(z)$ is κ_n-invariant for each $n \geq 0$. It is shown in Theorem 6.1 of [7] that if $\{c_n\}, \{w_n\}$ and $\{\kappa_n\}$ are sequences of complex numbers satisfying

$$c_n \neq 0, \ |w_n| = 1 \text{ and } \kappa_n = \frac{\bar{c}_n}{c_n} \bar{w}_n, \quad n = 0, 1, 2, \ldots, \quad (1.27)$$

and if $\{B_n(z, w_n)\}$ is defined by (1.4), then $\{c_n B_n(z, w_n)\}$ *is a* $\{\kappa_n\}$-*invariant, para-orthogonal sequence of polynomials with respect to* $\psi(\theta)$. Moreover, by Theorem 6.2 in [7], the n zeros $\zeta_1^{(n)}(w_n), \ldots, \zeta_n^{(n)}(w_n)$ of $B_n(z, w_n)$ are simple and satisfy

$$|\zeta_j^{(n)}(w_n)| = 1, \quad j = 1, \ldots, n. \quad (1.28)$$

329

The n-point Szegő quadrature formula (1.2), (1.3) is proven in Theorem 7.1 of [7], where the *weights* $\lambda_j^{(n)}(w_n)$ are defined by

$$\lambda_j^{(n)}(w_n) := \int_{-\pi}^{\pi} L_{n,j}(e^{i\theta}, w_n) d\psi(\theta), \ j = 1, 2, \ldots, n, \tag{1.29a}$$

where

$$L_{n,j}(z, w_n) := \frac{B_n(z, w_n)}{(z - \zeta_j^{(n)}(w_n)) B_n'(\zeta_n^{(n)}(w_n), w_n)}, \ j = 1, 2, \ldots, n. \tag{1.29b}$$

Associated Szegő polynomial sequences $\{\pi_n(\psi; z)\}$ and $\{\omega_n(\psi, z)\}$ are defined by $\omega_0(\psi; z) := \mu_0 =: -\pi_0(\psi; z)$ and

$$\pi_n(\psi; z) := \int_{-\pi}^{\pi} \frac{z + e^{i\theta}}{z - e^{i\theta}} (\rho_n(\psi; e^{i\theta}) - \rho_n(\psi; z)) d\psi(\theta), \ n = 1, 2, 3, \ldots \tag{1.30a}$$

$$\omega_n(\psi; z) := \int_{-\pi}^{\pi} \frac{z + e^{i\theta}}{z - e^{i\theta}} \left(\frac{z^n}{e^{in\theta}} \rho_n^*(\psi; e^{i\theta}) - \rho_n^*(\psi; z) \right) d\psi(\theta), \tag{1.30b}$$

$$n = 1, 2, 3, \ldots .$$

They satisfy $\pi_0(\psi, z) := -\delta_0$, $\omega_0(\psi, z) := \delta_0$,

$$\pi_n^*(\psi; z) := z^n \overline{\pi_n(\psi; 1/\bar{z})} = -\omega_n(\psi; z), \ n = 0, 1, 2, \ldots, \tag{1.31a}$$

$$\pi_n(\psi; z) = z\pi_{n-1}(\psi; z) + \delta_n \omega_{n-1}(\psi; z), \ n = 1, 2, 3, \ldots, \tag{1.31b}$$

$$\omega_n(\psi; z) = \bar{\delta}_n z \pi_{n-1}(\psi; z) + \omega_{n-1}(\psi; z), \ n = 1, 2, 3, \ldots, \tag{1.31c}$$

(see, e.g., Theorem 4.1 in [7]). The rational approximants

$$R_n(z, w_n) := \frac{\pi_n(\psi; z) + w_n \omega_n(\psi; z)}{\rho_n(\psi; z) + w_n \rho_n^*(\psi; z)} = \frac{A_n(z, w_n)}{B_n(z, w_n)}, \tag{1.32a}$$

where

$$A_n(z, w_n) := \pi_n(\psi; z) + w_n \omega_n(\psi; z), \ B_n(z, w_n) \text{ defined by (1.4)}, \tag{1.32b}$$

can be expressed as

$$R_n(z, w_n) = \sum_{j=1}^{n} \lambda_j^{(n)}(w_n) \frac{\zeta_j^{(n)}(w_n) + z}{\zeta_j^{(n)}(w_n) - z}, \ n = 1, 2, 3, \ldots \tag{1.33}$$

((7.22) in [7]) and they satisfy

$$\lim_{n \to \infty} R_n(z, w_n) = f(\psi, z) := \int_{-\pi}^{\pi} \frac{e^{i\theta} + z}{e^{i\theta} - z} d\psi(\theta), \ |z| < 1. \tag{1.34}$$

Therefore $R_n(z, w_n)$ is the n-point Szegő quadrature approximation of the integral in (1.34). We then have, in the terminology of (1.2) and (1.7),

$$f(\psi, z) = R_n(z, w_n) + e_n(K(\zeta, z), \psi, w_n), \tag{1.35}$$

and so the remainder term $e_n(K(\zeta, z), \psi, w_n)$ tends to zero as $n \to \infty$.

2 CARATHÉODORY LINEAR FRACTIONAL APPROXIMANT SEQUENCES (CLFASs)

Let D_0, D and T be subsets of $\mathbb{C}$ defined by

$$D_0 := [u \in \mathbb{C} : 0 < |u| < 1], \ D := [u \in \mathbb{C} : |u| < 1], \ T := [u \in \mathbb{C} : |u| = 1]. \quad (2.1)$$

Let $\{\delta_j\}_{j=0}^{\infty}$, $\{w_j\}_{j=0}^{\infty}$ and z satisfy the conditions

$$\delta_0 > 0, \ \delta_j \in D \text{ for } j \geq 1; \ w_j \in T \text{ for } j \geq 0; \ z \in D_0. \quad (2.2)$$

A *Carathéodory linear fractional approximant sequence* (**CLFAS**) $F = F(\{\delta_j\}, \{w_j\}, z)$ is an ordered pair

$$F = \langle\langle\{\delta_j\}, \{w_j\}, z\rangle, \{f_n\}\rangle, \quad (2.3a)$$

where the *nth approximant* f_n is defined by

$$f_n = v_n(F) := T_n(F, w_n), \quad n = 0, 1, 2, \ldots, \quad (2.3b)$$

$$T_0(F, w) := t_0^F(w), \ T_n(F, w) := T_{n-1}(F, t_n^F(w)), \ n = 1, 2, 3, \ldots, \quad (2.3c)$$

$$t_0^F(w) = \delta_0 \frac{1-w}{1+w}, \ t_j^F(w) := z \frac{\bar{\delta}_j + w}{1 + \delta_j w}, \ j = 1, 2, 3, \ldots . \quad (2.3d)$$

The CLFAS F is said to *converge to a value* $v(F) \in \mathbb{C}$, if $\{v_n(F)\}$ converges to $v(F)$. The numbers δ_j are called the *elements of F* and the w_j are called *converging factors of F*. To indicate explicitly the association with F of its elements and converging factors, we may write $\delta_j(F)$ and $w_j(F)$, respectively. To indicate explicit association with z, we may write $F(z), v_n(F(z))$ and $v(F(z)) = \lim_{n \to \infty} v_n(F(z))$. It is known (Theorem 2.1) that every CLFAS $F(z)$ converges to a normalized Carathéodory function $v(F(z)) \in \mathcal{C}_c$ (see (1.10)).

For given $z \in D_0$ and $\{w_j\}$ satisfying (2.2), we define a family $\mathcal{F} = \mathcal{F}(\{w_j\}, z)$ of CLFASs by

$$\mathcal{F} = \mathcal{F}(\{w_j\}, z) := [\text{CLFASs } F : \delta_0(F) > 0, \ |\delta_j(F)| < 1 \text{ for } j \geq 1, \quad (2.4)$$
$$w_j(F) = w_j \text{ for } j \geq 0].$$

For simplicity we may write $\mathcal{F}$ instead of $\mathcal{F}(\{w_j\}, z)$, when the association with $(\{w_j\}, z)$ is clearly understood. For each $F \in \mathcal{F}$ and each integer $n \geq 0$, we define the subfamily $\mathcal{F}_n(F)$ of $\mathcal{F}$ by

$$\mathcal{F}_n(F) := [G \in \mathcal{F} : \delta_j(G) = \delta_j(F) \text{ for } j = 0, 1, \ldots, n]. \quad (2.5)$$

If $G \in \mathcal{F}_n(F)$, then G is said to be *n-equivalent to F in $\mathcal{F}$*. $\mathcal{F}_n(F)$ is called the *n-equivalent subfamily of F in $\mathcal{F}$*.

Our main interest is to determine sharp upper bounds for the truncation error $|v(F(z)) - v_n(F(z))|$ that results from replacing the true value $v(F(z))$ by its nth approximant $v_n(F(z))$. To give precise meaning to the word "sharp" used above, we assume that a CLFAS $F(z)$ under consideration belongs to a family $\mathcal{F} = \mathcal{F}(\{w_j\}, z)$ of CLFASs. Then for each integer $n \geq 0$, the nth *limit region for* $F(z)$ with respect to $\mathcal{F}$, denoted by $L_n(F(z), \mathcal{F})$, is defined by

$$L_n(F(z), \mathcal{F}) := c[v(G(z)) : G(z) \in \mathcal{F}_n(F(z))], \qquad (2.6)$$

where $c(S)$ denotes the *closure* of the set $S \subseteq \mathbb{C}$.

We call

$$\beta_n(F(z), \mathcal{F}) := \sup[|v(G(z)) - v_n(F(z))| : G(z) \in L_n(F(z), \mathcal{F})] \qquad (2.7)$$

the *best upper bound of the truncation error* $|v(F(z)) - v_n(F(z))|$ *for* $v_n(F(z))$ *with respect to* $\mathcal{F}$; more simply, $\beta_n(F(z), \mathcal{F})$ is called the *best truncation error bound for* $v_n(F(z))$ *with respect to* $\mathcal{F}$. Clearly $|v(F(z)) - v_n(F(z))| \leq \beta_n(F(z), \mathcal{F})$, since $v(F(z)) \in L_n(F(z), \mathcal{F})$. The term "best" is used for $\beta_n(F(z), \mathcal{F})$, since we assume that our knowledge of $F(z)$ is limited to the following: (a) $F(z) \in \mathcal{F}$, and (b) the only known elements of $F(z)$ are $\delta_j(F(z))$, $j = 0, 1, \ldots, n$. Therefore in (2.7) we are comparing $v_n(F(z))$ with all of the possible values $v(G(z))$ that could be $v(F(z))$.

Before giving an explicit expression for $\beta_n(F(z), \mathcal{F})$ in §3, we summarize a few properties of CLFASs that are subsequently used. Associated with a CLFAS $F = F(z)$ of the form (2.3) is a continued fraction,

$$\delta_0(F) - \frac{2\delta_0(F)}{1} + \frac{1}{\overline{\delta_1(F)}z} + \frac{(1 - |\delta_1(F)|^2)z}{\delta_1(F)} + \frac{1}{\overline{\delta_2(F)}z} + \frac{(1 - |\delta_2(F)|^2)z}{\delta_2(F)} + \cdots, \tag{2.8}$$

called a *positive Perron–Carathéodory continued fraction* (PPC-fraction). We denote the PPC-fraction (2.8) by PPC $(\{\delta_j(F)\}, z)$. Its *generating sequence of linear fractional transformations* (l.f.t.) $\{s_n^F(w)\}$ is defined by

$$s_0^F(w) := \delta_0(F) + w, \quad s_{2j}^F(w) := \frac{1}{\overline{\delta_j(F)}z + w}, \quad j = 1, 2, 3, \ldots, \tag{2.9a}$$

$$s_1^F(w) := \frac{-2\delta_0(F)}{1 + w}, \quad s_{2j+1}^F(w) := \frac{(1 - |\delta_j(F)|^2)z}{\delta_j(F) + w}, \quad j = 1, 2, 3, \ldots. \tag{2.9b}$$

The nth *approximant* $S_n(F(z), 0)$ of the PPC $(\{\delta_j(F)\}, z)$ is defined by

$$S_0(F(z), w) := s_0^F(w), \quad S_n(F(z), w) := S_{n-1}(F(z), s_n^F(w)), \quad n = 1, 2, 3, \ldots. \tag{2.10}$$

332

The nth numerator $P_n(F, z)$ and denominator $Q_n(F, z)$ of a PPC $(\{\delta_j(F)\}, z)$ are defined by *difference equations*

$$P_0(F, z) := \delta_0(F), Q_0(F, z) := 1, P_1(F, z) := -\delta_0(F), Q_1(F, z) := 1, \quad (2.11a)$$

$$P_{2n}(F, z) := \overline{\delta_n(F)} z P_{2n-1}(F, z) + P_{2n-2}(F, z), \quad n = 1, 2, 3, \ldots, \quad (2.11b)$$

$$Q_{2n}(F, z) := \overline{\delta_n(F)} z Q_{2n-1}(F, z) + Q_{2n-2}(F, z), \quad n = 1, 2, 3, \ldots, \quad (2.11c)$$

$$P_{2n+1}(F, z) := \delta_n(F) P_{2n}(F, z) + (1 - |\delta_n(F)|^2) z P_{2n-1}(F, z), \quad (2.11d)$$
$$n = 1, 2, 3, \ldots,$$

$$Q_{2n+1}(F, z) := \delta_n(F) Q_{2m}(F, z) + (1 - |\delta_n(F)|^2) z Q_{2n-1}(F, z), \quad (2.11e)$$
$$n = 1, 2, 3, \ldots .$$

Using (1.22), (1.31) and (2.11), one can verify that

$$P_{2n}(F, z) = \omega_n(\psi, z), \quad Q_{2n}(F, z) = \rho_n^*(\psi, z), \quad n = 0, 1, 2, \ldots, \quad (2.12a)$$

$$P_{2n+1}(F, z) = \pi_n(\psi, z), \quad Q_{2n+1}(F, z) = \rho_n(\psi, z), \quad n = 0, 1, 2, \ldots . \quad (2.12b)$$

Moreover, by continued fraction properties (Theorem 2.1 in [9]) we can write

$$S_n(F, w) = \frac{P_n(F, z) + w P_{n-1}(F, z)}{Q_n(F, z) + w Q_{n-1}(F, z)}, \quad n = 0, 1, 2, \ldots, \quad (2.13)$$

and

$$P_{2n}(F, z) Q_{2n-1}(F, z) - P_{2n-1}(F, z) Q_{2n}(F, z) = 2\delta_0(F) \prod_{j=1}^{n-1} (1 - |\delta_j(F)|^2) z^{n-1},$$
$$(2.14a)$$

$$P_{2n+1}(F, z) Q_{2n}(F, z) - P_{2n}(F, z) Q_{2n+1}(F, z) = -2\delta_0(F) \prod_{j=1}^{n} (1 - |\delta_j(F)|^2) z^n.$$
$$(2.14b)$$

It follows from (2.3), (2.9) and (2.12) that

$$t_0^F(w) = s_0^F \circ s_1^F(w^{-1}), t_j^F(w) = [s_{2j}^F \circ s_{2j+1}^F(w^{-1})]^{-1}, \quad j = 1, 2, 3, \ldots, \quad (2.15)$$

and hence, for all $F \in \mathcal{F}$ and $n = 0, 1, 2, \ldots,$

$$T_n(F, w) = S_{2n+1}(F, w^{-1}) = \frac{P_{2n+1}(F, z) w + P_{2n}(F, z)}{Q_{2n+1}(F, z) w + Q_{2n}(F, z)}$$
$$= \frac{\pi_n(\psi, z) + w^{-1} \omega_n(\psi, z)}{\rho_n(\psi, z) + w^{-1} \rho_n^*(\psi, z)}. \quad (2.16)$$

From (1.32) and (2.16) we obtain, for $n = 1, 2, 3, \ldots,$

$$v_n(F(z)) := T_n(F(z), w_n) = R_n(z, w_n^{-1}) = \sum_{j=1}^{n} \lambda_j^{(n)}(w_n^{-1}) \frac{\zeta_j^{(n)}(w_j^{-1}) + z}{\zeta_j^{(n)}(w_j^{-1}) + z}. \quad (2.17)$$

The connection between CLFASs and Szegő quadrature applied to $K(\zeta, z)$ in (1.7) can be seen from (2.17).

The following result summarizes convergence properties of CLFASs.

Theorem 2.1 Let $\psi(\theta) \in (-\pi, \pi)$ be a given distribution function and let $\{\mu_n\}_{n=-\infty}^{\infty}$ denote its Hermitian positive definite bisequence of moments (1.16). Let $\{\delta_j\}_{j=0}^{\infty}$ denote the corresponding reflection coefficients (1.24). Then:

(A)

$$\delta_0 > 0, \ \delta_j \in \mathbb{C} \text{ and } |\delta_j| < 1 \text{ for } j = 1, 2, 3, \dots . \tag{2.18}$$

(B) Let $\{w_j\}$ be a given sequence of complex numbers satisfying (2.2). For $z \in D_0$ (see (2.1)), let F denote the CLFAS

$$F = F(z) = F(\{\delta_j\}, \{w_j\}, z) \quad (\text{see (2.3)}). \tag{2.19}$$

Then, for $0 < |z| < 1$, F converges to the normalized Carathéodory function of type $\mathcal{C}_a$ (see (1.10))

$$f(\psi, z) = v(F(z)) = \lim_{n \to \infty} v_n(F(z)) = \int_{-\pi}^{\pi} \frac{e^{i\theta} + z}{e^{i\theta} - z} \, d\psi(\theta). \tag{2.20}$$

A proof of Theorem 2.1 can be obtained by combining (2.3) and (2.17), with Theorem 3.2, Theorem 7.3(A) and Theorem 10.2 in [7].

In (2.1) of [8] the sequence $\{Q_n(F, z)\}_{n=0}^{\infty}$ is shown to satisfy the Christoffel–Darboux formulas, for $n = 0, 1, 2, \dots$,

$$\sum_{j=0}^{n} K_j^2 Q_{2j+1}(F, x)\overline{Q_{2j+1}(F, y)} =$$

$$\frac{K_n^2[Q_{2n}(F, x)\overline{Q_{2n}(F, y)} - x\bar{y}Q_{2n+1}(F, x)\overline{Q_{2n+1}(F, y)}]}{1 - x\bar{y}}, \tag{2.21a}$$

where

$$K_n := \sqrt{\Delta_{n-1}/\Delta_n}, \quad n = 0, 1, 2, \dots \text{ (see (1.17))}. \tag{2.21b}$$

It is also shown [8, (2.5)] that

$$K_n^2 = \frac{K_{n-1}^2}{1 - |\delta_n(F)|^2} = \frac{K_0^2}{\prod_{j=1}^{n}(1 - |\delta_j(F)|^2)}, \quad n = 1, 2, 3, \dots . \tag{2.22}$$

Setting $x = y = z$ in (2.21a) and using (2.22) yields the inequalities

$$|Q_{2n}(F, z)|^2 - |zQ_{2n+1}(F, z)|^2 \geq (1 - |z|^2) \prod_{j=1}^{n}(1 - |\delta_j(F)|^2), \quad n = 1, 2, 3, \dots . \tag{2.23}$$

3   UPPER BOUNDS FOR THE REMAINDER TERM OF SZEGÖ QUADRATURE

In this section we consider a given distribution function $\psi(\theta) \in \Phi(-\pi, \pi)$. Associated with ψ are a Carathéodory function

$$f(\psi, z) := \int_{-\pi}^{\pi} \frac{e^{i\theta} + z}{e^{i\theta} - z} \, d\psi(\theta), \quad |z| < 1, \tag{3.1}$$

a positive definite Hermitian bisequence of moments $\{\mu_n\}_{-\infty}^{\infty}$, and a sequence of reflection coefficients $\{\delta_n\}_0^{\infty}$. In addition we consider a given sequence of converging factors $\{w_j\}_0^{\infty}$ satisfying

$$w_j \in \mathbb{C} \text{ and } |w_j| = 1, \quad j = 0, 1, 2, \ldots . \tag{3.2}$$

Associated with $(\{\delta_j\}, \{w_j\}, z)$ are Szegö quadrature formulas

$$\int_{-\pi}^{\pi} \frac{e^{i\theta} + z}{e^{i\theta} - z} \, d\psi(\theta) = \sum_{j=1}^{n} \lambda_j^{(n)}(w_n^{-1}) K(\zeta_j^{(n)}(w_n^{-1}), z) + e_n(K, \psi, w_n^{-1}), \tag{3.3}$$

$$n = 1, 2, 3, \ldots,$$

where $K(\zeta, z)$ denotes the kernel function (1.7), and $\{\lambda_j^{(n)}(w_n^{-1})\}$ and $\{\zeta_j^{(n)}(w_n^{-1})\}$ denote the weights and nodes of the n-point Szegö quadrature with respect to $\psi(\theta)$ and w_n^{-1}. The remainder term for the n-point formula (3.3) is denoted by $e_n(K, \psi, w_n^{-1})$. Also associated with $(\{\delta_j\}, \{w_j\}, z)$ is the CLFAS

$$F = F(z) = F(\{\delta_j\}, \{w_j\}, z) \quad \text{(see (2.3))}. \tag{3.4}$$

It follows from (2.17) and Theorem 2.1 that

$$e_n(K, \psi, w_n^{-1}) = v(F(z)) - v_n(F(z)). \tag{3.5}$$

In this section we determine sharp, computable upper bounds for $|v(F(z)) - v_n(F(z))| = |e_n(K, \psi, w_n^{-1})|$.

Theorem 3.1   Let $F = F(z) = F(\{\delta_j(F)\}, \{w_j\}, z)$ be a given CLFAS and let $\mathcal{F} = \mathcal{F}(\{w_j\}, z)$ denote the family of CLFASs (2.4), where $0 < |z| < 1$. Then for each integer $n \geq 1$, the best truncation error bound $\beta_n(F, \mathcal{F})$ for $v_n(F(z))$ with respect to $\mathcal{F}$ is given by

$$\beta_n(F, z) =$$

$$\frac{2\delta_0(F) \prod_{j=1}^{n}(1 - |\delta_j(F)|^2) \left[\left| 1 + w_n \frac{Q_{2n+1}(F,z)}{Q_{2n}(F,z)} |z|^2 \right| + |z| \left| 1 + w_n \frac{Q_{2n+1}(F,z)}{Q_{2n}(F,z)} \right| \right] |z|^n}{|Q_{2n}(F,z)|^2 \left[1 - |z|^2 \left| \frac{Q_{2n+1}(F,z)}{Q_{2n}(F,z)} \right|^2 \right] \left| 1 + w_n \frac{Q_{2n+1}(F,z)}{Q_{2n}(F,z)} \right|}. \tag{3.6}$$

335

Before proving Theorem 3.1 we note by (2.12) that $Q_{2n+1}(F,z)/Q_{2n}(F,z)$ is a Blaschke product

$$\frac{Q_{2n+1}(F,z)}{Q_{2n}(F,z)} = \prod_{j=1}^{n} \left(\frac{z-z_j}{1-\bar{z}_j z} \right), \quad \text{where } |z_j| < 1 \text{ for } j = 1, 2, \ldots, n, \qquad (3.7)$$

since by (2.12) $Q_{2n+1}(F,z)$ is the monic Szegö polynomial $\rho_n(\psi; z)$ of degree n, whose zeros z_j all lie in the open disk $|z| < 1$, and since $Q_{2n}(F,z)$ is the reciprocal polynomial (1.20). Therefore

$$\left| \frac{Q_{2n+1}(F,z)}{Q_{2n}(F,z)} \right| < 1, \quad \text{for } |z| < 1. \qquad (3.8)$$

Thus the middle factor in the denominator of (3.6) is positive.

Proof of Theorem 3.1: For each $G \in \mathcal{F}_n(F)$, it follows from (2.5) and (2.16) that

$$v(G) - v_n(F) = T_n(F, v(G^{(n)})) - T_n(F, w_n), \qquad (3.9a)$$

where

$$v(G^{(n)}) := \lim_{m \to \infty} t_{n+1}^G \circ t_{n+2}^G \circ \cdots \circ t_{n+m}^G (w_{n+m}) \qquad (3.9b)$$

Let

$$V(z) := [u \in \mathbb{C} : |u| \le |z|] \text{ and } V_0(z) := [u \in \mathbb{C} : |u| < |z|]. \qquad (3.10)$$

From (2.2) and (2.3d) it is readily shown that, for $G \in \mathcal{F}$,

$$|t_j^G(w_j)| = |z|, \quad j = 1, 2, 3, \ldots, \qquad (3.11)$$

and

$$t_j^G(V_0(z)) \subset [u \in \mathbb{C} : |u - \Gamma_j(G)| \le \rho_j(G)] \subset V_0(z), \ j = 1, 2, 3, \ldots, \qquad (3.12a)$$

where

$$\Gamma_j(G) := \frac{z\overline{\delta_j(G)}(1 - |z|^2)}{1 - |z|^2|\delta_j(G)|^2} \text{ and } \rho_j(G) := \frac{|z|^2(1 - |\delta_j(G)|^2)}{1 - |z|^2|\delta_j(G)|^2}. \qquad (3.12b)$$

From (3.9b), (3.10), (3.11) and (3.12), one can conclude that

$$|v(G^{(n)})| \le |z|, \text{ for } G \in \mathcal{F}_n(F) \text{ and } n = 0, 1, 2, \ldots . \qquad (3.13)$$

Therefore (2.6), (2.7), (3.9) and (3.13) imply that

$$\beta_n(F, \mathcal{F}) := \sup[|v(G) - v_n(F)| : G \in \mathcal{F}_n(F)]$$
$$\le \sup_{|\zeta| \le |z|} |T_n(F, \zeta) - T_n(F, w_n)|. \qquad (3.14)$$

336

From (2.16), (3.9) and (2.14b) we obtain

$$T_n(F, \zeta) - T_n(F, w_n) = \frac{P_{2n+1}(F, z)\zeta + P_{2n}(F, z)}{Q_{2n+1}(F, z)\zeta + Q_{2n}(F, z)} - \frac{P_{2n+1}(F, z)w_n + P_{2n}(F, z)]}{Q_{2n+1}(F, z)w_n + Q_{2n}(F, z)}$$
$$= \frac{(\zeta - w_n)[P_{2n+1}(F, z)Q_{2n}(F, z) - P_{2n}(F, z)Q_{2n+1}(F, z)]}{[Q_{2n+1}(F, z)\zeta + Q_{2n}(F, z)][Q_{2n+1}(F, z)w_n + Q_{2n}(F, z)]},$$

and hence

$$T_n(F, \zeta) - T_n(w_n) = \frac{-2\delta_0(F)\prod_{j=1}^n(1 - |\delta_j(F)|^2)z^n(\zeta - w_n)}{[Q_{2n}(F, z)]^2[1 + w_n b_n(F)][1 + \zeta b_n(F)]}, \tag{3.15a}$$

where

$$b_n(F) := Q_{2n+1}(F, z)/Q_{2n}(F, z). \tag{3.15b}$$

In view of (3.14) and (3.15a), we seek

$$M_n(F(z)) := \max_{|\zeta| \le |z|} \left| \frac{\zeta - w_n}{1 + b_n(F)\zeta} \right|. \tag{3.16}$$

It is readily shown that, if

$$h(\zeta) := \frac{\zeta - w_n}{1 + b_n(F)\zeta}, \tag{3.17}$$

then

$$h(V(z)) = [u \in \mathbb{C} : |u - c_n(F)| \le r_n(F)], \tag{3.18a}$$

where

$$c_n(F) := -\frac{w_n + \overline{b_n(F)}|z|^2}{1 - |b_n(F)|^2|z|^2}, \quad r_n(F) := \frac{|z||1 + b_n(F)w_n|}{1 - |b_n(F)|^2|z|^2}. \tag{3.18b}$$

It follows that

$$M_n(F(z)) = |c_n(F)| + r_n(F) = \frac{|1 + w_n b_n(F)|z|^2| + |z||1 + w_n b_n(F)|}{1 - |b_n(F)|^2|z|^2}. \tag{3.19}$$

Combining (3.14), (3.15), (3.16) and (3.19), we see that the right side of (3.6) is an upper bound of $\beta_n(F, \mathcal{F})$. We now show that this upper bound is sharp. For that purpose we consider the number

$$W_n(F(z)) := c_n(F) + e^{i\gamma}r_n(F), \text{ where } \gamma := \arg(c_n(F)). \tag{3.20}$$

which satisfies

$$M_n(F(z)) = |W_n(F(z))| \text{ and } W_n(F(z)) = h(\zeta_0) = \frac{\zeta_0 - w_n}{1 + b_n(F)\zeta_0}, \tag{3.21}$$

where

$$\zeta_0 := \frac{w_n + W_n(F(z))}{1 - b_n(F)W_n((F(z))}, \quad \zeta_0 =: ze^{i\alpha}, \ |\zeta_0| = |z|, \ 0 \le \alpha < 2\pi. \tag{3.22}$$

We define

$$\zeta_\varepsilon := (1 - \varepsilon)\zeta_0 = (1 - \varepsilon)ze^{i\alpha}, \ \text{for } 0 < \varepsilon < 1 \tag{3.23}$$

and CLFASs $G_\varepsilon(z) \in \mathcal{F}_n(F(z))$, for $0 \le \varepsilon < 1$, by

$$\delta_j(G_\varepsilon(z)) := \begin{cases} \delta_j(F), & 0 \le j \le n, \\ (1 - \varepsilon)e^{-i\alpha}, & j = n + 1 \\ 0, & j \ge n + 2. \end{cases} \tag{3.24}$$

It follows that

$$\begin{aligned} v(G_\varepsilon^{(n)}(z)) &:= \lim_{m \to \infty} t_{n+1}^{G_\varepsilon} \circ t_{n+2}^{G_\varepsilon} \circ \cdots \circ t_{n+m}^{G_\varepsilon}(w_{n+m}) \\ &= \lim_{m \to \infty} z \frac{\overline{\delta_{n+1}(G_\varepsilon(z))} + w_{n+m}z^{m-1}}{1 + \delta_{n+1}(G_\varepsilon(z))w_{n+m}z^{m-1}} = (1 - \varepsilon)ze^{i\alpha} = \zeta_\varepsilon. \end{aligned} \tag{3.25}$$

Therefore

$$M_n(F(z)) = \max_{|\zeta| \le |z|} \left| \frac{\zeta - w_n}{1 + b_n(F)\zeta} \right| = \left| \frac{\zeta_0 + w_n}{1 + b_n(F)\zeta_0} \right| = \lim_{\varepsilon \to 0^-} \left| \frac{\zeta_\varepsilon + w_n}{1 + b_n(F)\zeta_\varepsilon} \right|, \tag{3.26}$$

and

$$\lim_{\varepsilon \to 0^+} v(G_\varepsilon(z)) = \lim_{\varepsilon \to 0^+} T_n(F, v(G_\varepsilon^{(n)}(z))) = T_n(F, \zeta_0). \tag{3.27}$$

We conclude from (3.16), (3.21), (3.22) and (3.27) that the upper bound of $\beta_n(F, \mathcal{F})$ on the right side of (3.6) is attained by the following limit

$$\lim_{\varepsilon \to 0^+} |v(G_\varepsilon(z)) - v_n(F)|, \ \text{where } G_\varepsilon(z) \in \mathcal{F}_n(F(z)). \tag{3.28}$$

This completes the proof of the theorem. ∎

We conclude this section by deriving several upper bounds of $\beta_n(F, \mathcal{F})$ in (3.6), which may not be as sharp as (3.6) but have simpler expressions. To further simplify expressions we utilize the notation $b_n(F)$ given in (3.15b).

Theorem 3.2 *Let the hypotheses of Theorem 3.1 be satisfied and let $b_n(F)$ be defined by (3.15b). Then:*
(A)

$$\beta_n(F, \mathcal{F}) \le \hat{\beta}_n(F, \mathcal{F}) := \frac{2\delta_0(F) \prod_{j=1}^{n}(1 - |\delta_j(F)|^2)(1 + |z|)|z|^n}{|Q_{2n}(F, z)|^2(1 - |zb_n(F)|)|1 + w_n b_n(F)|}, \tag{3.29}$$

and

$$\beta_n(F, \mathcal{F}) = \hat{\beta}_n(F, \mathcal{F}), \ \text{if } w_n b_n(F) > 0. \tag{3.30}$$

(B)

$$\beta_n(F, \mathcal{F}) \leq \tilde{\beta}_n(F, \mathcal{F}) := \frac{2\delta_0(F)[|1 + w_n b_n(F)|z|^2| + |z||1 + w_n b_n(F)|]|z|^n}{(1 - |z|^2)|1 + w_n b_n(F)|}, \tag{3.31}$$

and

$$\beta_n(F, \mathcal{F}) \leq \tilde{\beta}_n(F, \mathcal{F}) \leq \beta_n^*(F, \mathcal{F}) := \frac{8\delta_0(F)|z|^n}{(1 - |z|^2)|1 + w_n b_n(F)|}. \tag{3.32}$$

Proof: (A) By (3.16)

$$M_n(F(z)) := \max_{|\zeta| \leq |z|} \left| \frac{\zeta - w_n}{1 + b_n(F)\zeta} \right| \leq \frac{1 + |z|}{1 - |z b_n(F)|}.$$

Using this with (3.6), (3.14) and (3.15) yields (3.29). If, in addition $w_n b_n(F) > 0$, then by (3.16)

$$M_n(F(z)) = \frac{(1 + w_n b_n(F)|z|^2) + |z|(1 + w_n b_n(F))}{1 - (w_n b_n(F))^2 |z|^2} = \frac{1 + |z|}{1 - |z b_n(F)|}.$$

Using this with (3.6), (3.14) and (3.15) yields (3.30).
(B) We obtain (3.31) from (3.6) and the inequality (2.23). Then (3.32) follows from (3.31), $|w_n| = 1$, $|z| < 1$ and $|b_n(F)| < 1$. ∎

4 APPLICATIONS

To illustrate the Szegö quadrature formulas and truncation error bounds, we describe an elementary application.

Let $\psi(\theta)$ denote the absolutely continuous distribution function with derivative

$$\psi'(\theta) := \frac{1}{2\pi} \operatorname{Re}(1 + e^{i\theta}) = \frac{1}{2\pi}(1 + \cos\theta), \ -\pi \leq \theta \leq \pi. \tag{4.1}$$

The associated Carathéodory function $f(\psi, z)$ is given by

$$f(\psi, z) := \int_{-\pi}^{\pi} \frac{e^{i\theta} + z}{e^{i\theta} - z} \frac{1}{2\pi}(1 + \cos\theta)d\theta = 1 + z, \ |z| < 1. \tag{4.2}$$

Since

$$f(\psi, z) = \mu_0 + 2\sum_{k=1}^{\infty} \mu_k z^k, \ \text{for } |z| < 1, \tag{4.3a}$$

and

$$f(\psi, z) = -\mu_0 - 2 \sum_{k=1}^{\infty} \mu_{-k} z^{-k}, \text{ for } |z| > 1, \tag{4.3b}$$

the moment bisequence $\{\mu_k\}_{k=-\infty}^{\infty}$ is given by

$$\mu_0 = 1, \ \mu_1 = \mu_{-1} = \frac{1}{2}, \ \mu_k = 0 \text{ for } |k| \geq 2. \tag{4.4}$$

From this we obtain reflection coefficients δ_n (see (1.23) and (1.24)) given by

$$\delta_n = \frac{(-1)^n}{n+1} \tag{4.5}$$

and Szegö polynomials, reciprocal polynomials and associated polynomials, for small values of n, given by

$$\rho_0^*(\psi; z) = Q_0(F, z) = 1, \quad \rho_0(\psi; z) = Q_1(F, z) = 1, \tag{4.6a}$$

$$\rho_1^*(\psi; z) = Q_2(F, z) = 1 - \frac{z}{2}, \ \rho_1(\psi; z) = Q_3(F, z) = z - \frac{1}{2}, \tag{4.6b}$$

$$\rho_2^*(\psi; z) = Q_4(F, z) = \frac{1}{3} z^2 - \frac{2}{3} z + 1,$$

$$\rho_2(\psi; z) = Q_5(F, z) = z^2 - \frac{2}{3} z + \frac{1}{3}, \tag{4.6c}$$

$$\omega_0(\psi; z) = P_0(F, z) = 1, \quad \pi_0(\psi; z) = P_1(F, z) = -1, \tag{4.7a}$$

$$\omega_1(\psi; z) = P_2(F, z) = 1 + \frac{1}{2} z, \ \pi_1(\psi; z) = P_3(F, z) = -\frac{1}{2} - z, \tag{4.7b}$$

$$\omega_2(\psi; z) = P_4(F, z) = 1 + \frac{1}{3} z - \frac{1}{3} z^2,$$

$$\pi_2(\psi; z) = P_5(F, z) = \frac{1}{3} - \frac{1}{3} z - z^2. \tag{4.7c}$$

Thus by (1.4) and (1.32), with $w := 1$, we obtain

$$A_0(z, 1) = 0, \ A_1(z, 1) = \frac{1}{2} (1 - z), \ A_2(z, 1) = \frac{4}{3} (1 - z^2), \tag{4.8a}$$

$$B_0(z, 1) = 2, \ B_1(z, 1) = \frac{1}{2} (z + 1), \ B_2(z, 1) = \frac{4}{3} (z^2 - z + 1), \tag{4.8b}$$

$$R_1(z, 1) := \frac{A_1(z, 1)}{B_1(z, 1)} = \frac{1 - z}{1 + z} \text{ and } R_2(z) := \frac{A_2(z, 1)}{B_2(z, 1)} = \frac{1 - z^2}{z^2 - z + 1}. \tag{4.8c}$$

From (4.8b) we obtain

$$B_2(z, 1) = \frac{4}{3} (z - e^{i\pi/3})(z - e^{-i\pi/3}), \tag{4.9}$$

so the nodes for the 2-point Szegö quadrature are

$$\zeta_1^{(2)}(1) = e^{i\pi/3} \text{ and } \zeta_2^{(2)}(1) = e^{-i\pi/3}. \tag{4.10}$$

From (1.29) we have

$$L_{2,1}(z,1) = \frac{z - e^{-i\pi/3}}{i\sqrt{3}} \text{ and } L_{2,2}(z,1) = \frac{z - e^{i\pi/3}}{-i\sqrt{3}}, \tag{4.11a}$$

and hence for the weights $\lambda_j^{(2)}(1)$ we have

$$\lambda_1^{(2)}(1) = \lambda_2^{(2)}(1) = \frac{1}{2}. \tag{4.11b}$$

With the notation of (1.32), (1.33), (2.17), (4.6) and (4.7), we write

$$\begin{aligned}
R_2(z,1) = T_2(z,1) &= \frac{1}{2}\left(\frac{e^{i\pi/3} + z}{e^{i\pi/3} - z}\right) + \frac{1}{2}\left(\frac{e^{-i\pi/3} + z}{e^{-i\pi/3} - z}\right) \quad \binom{\text{Szegö}}{\text{quadrature}} \\
&= \frac{P_5(F,z) + P_4(F,z)}{Q_5(F,z) + Q_4(F,z)} = \frac{1 - z^2}{1 - z + z^2}.
\end{aligned} \tag{4.12}$$

Thus the actual truncation error for the 2-point Szegö quadrature is given by

$$\begin{aligned}
|f(\psi,z) - R_2(z,1)| = |e_2(K,\psi,1)| &= \left|(1+z) - \left(\frac{1-z^2}{1-z+z^2}\right)\right| \\
&= \frac{|1+z| \cdot |z|^2}{|1 - z + z^2|}.
\end{aligned} \tag{4.13}$$

Applying Theorems 3.1 and 3.2, we obtain the following upper bounds of (4.13):

$$\beta_2(F,\mathcal{F}) = \frac{12|z|^2[|1 + b_2(F)||z|^2| + |z| \cdot |1 + b_2(F)|]}{|z^2 - 2z + 3|^2 \cdot |1 + b_2(F)|[1 - |z|^2|b_2(F)|^2]}, \tag{4.14a}$$

where

$$b_2(F) = \frac{3z^2 - 2z + 1}{z^2 - 2z + 3}, \tag{4.14b}$$

$$\hat{\beta}_2(F,z) = \frac{12|z|^2(1 + |z|)}{|z^2 - 2z + 3| \cdot |1 + b_2(F)|(1 - |z| \cdot |b_2(F)|)}, \tag{4.15}$$

$$\tilde{\beta}_2(F,z) = \frac{2|z|^2[|1 + |z|^2 b_2(F)| + |z| \cdot |1 + b_2(F)|]}{(1 - |z|^2)|1 + b_2(F)|}, \tag{4.16}$$

and

$$\beta_2^*(F,z) = \frac{8|z|^2}{(1 - |z|^2)|1 + b_2(F)|}. \tag{4.17}$$

REFERENCES

[1] Akhiezer N.I. *The Classical Moment Problem and Some Related Questions in Analysis*. Hafner, New York, 1964. (Translated by N. Kemmer.)

[2] Bultheel A., González–Vera P., Hendriksen E., Njåstad O. Orthogonality and quadrature on the unit circle. *IMACS Annals on Comp. and Appl. Math.*, 9:205–210, 1991.

[3] Bultheel A., González–Vera P., Hendriksen E., Njåstad O. Quadrature formulas on the unit circle and two-point Padé approximation. (Submitted).

[4] Gautschi Walter. A survey of Gauss–Christoffel quadrature formulae. In *E.B. Christoffel; The Influence of his Work in Mathematics and the Physical Sciences*, pages 72–147. Birkhäuser Verlag, Basel, 1981. P.L. Butzer and F. Fehér, eds.

[5] Hille Einar. *Analytic Function Theory*, Vol. II. Ginn and Company, Boston, 1962.

[6] Jones William B., Njåstad Olav, Thron W.J. Orthogonal Laurent polynomials and the strong Hamburger moment problem. *J. Math. Anal. and Appl.*, 98(2):528–554, 1984.

[7] Jones William B., Njåstad Olav, Thron W.J. Moment theory, orthogonal polynomials, quadrature, and continued fractions associated with the unit circle. *Bull. London Math. Soc.*, 21:113–152, 1989.

[8] Jones William B., Thron W.J. A constructive proof of convergence of the even approximants of positive PC-fractions. *Rocky Mountain J. of Math.*, 19(1):199–210, 1989.

[9] Jones William B., Thron W.J. *Continued Fractions: Analytic Theory and Applications*. Encyclopedia of Mathematics and its Applications, **11**. Addison–Wesley Publishing Company, Reading, Mass. (1980). Distributed by Cambridge University Press, New York.

[10] Waadeland Haakon. A Szegö quadrature formula for the Poisson formula. *Comp. and Appl. Math.*, I, pages 479–486. Elsevier Science Publishers B.V. (North–Holland), IMACS 1992. C. Brezinski and U. Kulish, eds.

International Series of Numerical Mathematics, Vol. 119, ©1994 Birkhäuser

APPROXIMATION OF INTEGRALS BY ARITHMETIC MEANS AND RELATED MATTERS

Jacob Korevaar*

Dedicated to Walter Gautschi on his sixty-fifth birthday

Abstract. This is a survey of recent work by a small group investigating Chebyshev-type quadratures for large numbers of nodes and related potential theory. Besides the author, the participants were A.B.J. Kuijlaars, J.L.H. Meyers and M.A. Monterie. Much of the work concerns nice "surfaces" in $\mathbf{R}^d$, $d \geq 1$, equipped with normalized "area" measure. Fundamental results of S.N. Bernstein for the interval $[-1, 1]$ are extended and applied. It is shown that minimum-norm formulas exhibit massive coalescence of nodes. Other results involve domains of product type including the sphere. On the sphere, good N-tuples of nodes correspond to configurations of N point charges $1/N$ for which the electrostatic field is very small on the compact subsets of the unit ball ("Faraday cage effect"). By "Several complex variables" it becomes plausible that this field can be made as small as $\exp(-cN^{1/2})$ in the case of S^2. This observation supports our conjecture that there exist N-tuples of distinct nodes on S^2 which give Chebyshev-type formulas that are polynomially exact to degree $p \sim cN^{1/2}$ (so that there are "spherical p-designs" consisting of $N = \mathcal{O}(p^2)$ points).

1 CHEBYSHEV-TYPE QUADRATURE

Let E be a compact set in $\mathbf{R}^d$ and let σ be a positive measure on E of total mass $\sigma(E) = 1$. Important special cases are

- the interval $I = [-1, 1]$ with $d\sigma = dx/2$,

- the unit circle $C = C(0, 1)$ in $\mathbf{R}^2 \simeq \mathbf{C}$ with $d\sigma = ds/2\pi$,

- the unit sphere $S = S(0, 1)$ in $\mathbf{R}^3$ with $d\sigma = d\lambda/4\pi$.

A *Chebyshev-type quadrature formula* for E and σ (of order N) is a numerical integration formula which gives the same weight to each of the (not necessarily distinct) nodes $\zeta_1, \ldots, \zeta_N$:

$$\int_E f(x) \, d\sigma(x) \approx \frac{1}{N} \sum_{j=1}^N f(\zeta_j), \quad \zeta_j \in E. \tag{1}$$

*Faculty of Mathematics and Computer Science, University of Amsterdam, Plantage Muidergracht 24, 1018 TV Amsterdam, The Netherlands; korevaar@fwi.uva.nl

In other words, integrals are approximated by arithmetic means of function values. The study of such quadrature formulas was initiated by P.L. Chebyshev in 1874. There are extensive surveys by W. Gautschi [8] and K.-J. Förster [6].

Question 1.1 How should one choose N-tuples

$$Z_N = (\zeta_1, \ldots, \zeta_N) \tag{2}$$

of Chebyshev nodes in order to get a very small quadrature remainder

$$R(f, Z_N) = \int_E f(x) \, d\sigma(x) - \frac{1}{N} \sum_{j=1}^{N} f(\zeta_j) \tag{3}$$

for a large class of functions f?

We would call an N-tuple Z_N "good" if the corresponding quadrature remainder vanishes or is very small for all polynomials $f(x) = f(x_1, \ldots, x_d)$ up to relatively high degree $p = p_N$. If formula (1) is *exact* [that is, $R(f, Z_N) = 0$] for all polynomials of degree $\leq p$, we speak of a formula *of degree p*.

Example 1.2 (*the circle*) Chebyshev-type quadrature is more natural for some pairs (E, σ) than for others. For example, relative to normalized Lebesgue measure, Chebyshev-type quadrature appears to be natural for circle and sphere, but not for $[-1, 1]$. In fact, using as nodes the vertices $\zeta_j = e^{i\phi_j}$, $j = 1, \ldots, N$, of a regular N-gon inscribed in C, the quadrature formula

$$\int_C f(x_1, x_2) \, d\sigma = \frac{1}{2\pi} \int_0^{2\pi} f(\cos\phi, \sin\phi) \, d\phi \approx \frac{1}{N} \sum_{j=1}^{N} f(\cos\phi_j, \sin\phi_j)$$

turns out to be polynomially exact to degree $p = N - 1$. (Think of Nth roots of unity and trigonometric polynomials of order p.)

For the sphere, one would also like to have formulas of high degree p. Too bad that there are no regular polytopes with more than 20 vertices!

Gauss-type formulas. The best known quadrature formula for $I = [-1, 1]$ is the classical m-point Gauss formula for the measure $d\sigma(x) = dx/2$:

$$\int_{-1}^{1} f(x) \, dx/2 \approx \sum_{k=1}^{m} \lambda_k f(\alpha_k). \tag{4}$$

Here the nodes $\alpha_1 > \cdots > \alpha_m$, $\alpha_k = \alpha_k(m)$, are the zeros of the Legendre polynomial $P_m(x)$. The coefficients or Cotes-Christoffel numbers $\lambda_k = \lambda_k(m)$ are given by the formula

$$\lambda_k = \frac{1}{(1 - \alpha_k^2) P_m'(\alpha_k)^2} \approx \frac{\pi}{2m} \sqrt{1 - \alpha_k^2}.$$

For $m \geq 3$ formula (4) is not a Chebyshev-type formula. However, the Gauss-type formula for the measure $d\sigma(x) = dx/\pi\sqrt{1-x^2}$ on I does have equal coefficients for every m:

$$\int_{-1}^{1} f(x)\frac{dx}{\pi\sqrt{1-x^2}} = \int_{0}^{\pi} f(\cos\phi)\,d\phi/\pi \approx \frac{1}{m}\sum_{k=1}^{m} f(\xi_k) \qquad (5)$$

(Mehler 1864). Here the nodes ξ_k are the zeros of the Chebyshev polynomial $T_m(x) = \cos(m\arccos x)$. The Gauss formulas for (I,σ) are related to the corresponding orthogonal polynomials and the m-point formulas are exact to degree $2m-1$.

Minimal number of nodes in Chebyshev-type formulas of degree p. In most of our work, E is a simple "surface" (such as the interval I or the sphere S) equipped with normalized "area" measure σ. By fundamental work of Seymour and Zaslavsky [24] for arcwise connected sets, there exist Chebyshev-type quadrature formulas for (E,σ) of every degree p.

Question 1.3 What is the minimal number $N = N_E(p)$ of nodes for which one can have Chebyshev-type quadrature (1) of polynomial degree p? We are mostly interested in the order of $N_E(p)$ as $p \to \infty$.

2   THE INTERVAL [-1,1]

In the thirties, S.N. Bernstein treated Question 1.3 for the interval $I = [-1,1]$ with $d\sigma = dx/2$. He found that

$$N_I(p) \asymp (\text{"is exactly of order"}) \ p^2 \ \text{ as } \ p \to \infty; \qquad (6)$$

his precise results may be formulated as follows.

Theorem A [2,4] *Suppose that the real nodes $x_1 \geq \cdots \geq x_N$ are such that*

$$\int_{-1}^{1} f(x)\,dx/2 = \frac{1}{N}\sum_{j=1}^{N} f(x_j)$$

for all polynomials $f(x)$ of degree $\leq p$, where $p = 2m - 1$ or $p = 2m$. Then $x_1 \geq \alpha_1(m)$ and

$$N \geq \frac{1}{\lambda_1(m)},$$

where $\alpha_1(m)$ is the largest zero of the Legendre polynomial $P_m(x)$ and $\lambda_1(m)$ is the corresponding Christoffel number in the Gauss formula (4).
    One has $\lambda_1(m) < 4/m^2$, so that $N_I(p) > m^2/4 \geq p^2/16$.

Theorem B [3] *Let m be a positive integer and let M be an even integer,*

$$M \geq N_0(m) = 2[2\sqrt{2}(m+1)(m+4) + 1] \sim 4\sqrt{2}m^2,$$

345

where $[s]$ denotes the integral part of s. Then there exist points $t_k \in (-1, 1)$, $t_1 > \cdots > t_{2m-1}$, $t_{2m-k} = -t_k$, and positive integers $M_k = M_{2m-k}$ with $\sum_{k=1}^{2m-1} M_k = M$ such that

$$\int_{-1}^{1} f(x)\,dx/2 = \frac{1}{M} \sum_{k=1}^{2m-1} M_k f(t_k) \tag{7}$$

for all polynomials $f(x)$ of degree $p \leq 2m - 1$.

Taking $M = N_0(m)$, one may conclude that $N_I(p) = \mathcal{O}(p^2)$.

Comments on Theorem A. The author [12] has obtained the monotonicity result

$$m(m + 1)\lambda_k(m) \nearrow J_0'(j_k)^{-2} \quad \text{for} \quad (k \leq)\ m \nearrow \infty,$$

where j_k is the kth positive zero of the Bessel function $J_0(t)$. Hence, in particular,

$$\lambda_1(m) < J_0'(j_1)^{-2}/m(m+1) \approx 3.71/m^2.$$

A similar inequality was obtained by different means in [7].

More important, Gautschi and others, see [8], have observed that Theorem A immediately extends to *arbitrary probability measures* σ on $[-1, 1]$.

Comments on Theorem B. The proof of Theorem B is difficult. Bernstein first obtained a fairly general condition ensuring that symmetric nodes t_k generate a positive quadrature formula of degree $2m-1$. Having this condition, he could perturb the m-point Gauss formula (4) to obtain a $(2m-1)$-point symmetric quadrature formula of degree $2m-1$ with positive rational coefficients having common denominator M. The resulting formula (7) is a Chebyshev-type formula of order M in which the node t_k occurs with multiplicity M_k. However, it has recently been shown by Kuijlaars [19] that the multiple nodes in (7) can be split into simple nodes without losing the polynomial exactness to degree $2m-1$. Nevertheless, the occurrence of multiple nodes seems to be natural, cf. Theorem 2.2 on minimum norm formulas below.

Numerical results of Kuijlaars indicate that the minimal $M \sim 4\sqrt{2}m^2$ in Theorem B can be reduced to about $.5m^2$ [20]. One would conjecture that there is a constant c such that $N_I(p) \sim cp^2$ as $p \to \infty$.

For the ultraspherical measures $d\sigma = c_\alpha(1 - x^2)^\alpha dx$ with $\alpha > 0$, Kuijlaars has shown that $N_{I,\sigma}(p)$ is of order $p^{2+2\alpha}$ [19].

Supplement to Theorem A. By Theorem A, an Nth-order Chebyshev-type formula for $(I, dx/2)$ cannot be exact for all polynomials of degree $p \geq 4\sqrt{N}$. In this context we have proved:

Theorem 2.1 [16] *For every N and every N-tuple $X_N = (x_1, \ldots, x_N)$ of real nodes x_j, there exists a polynomial g of degree q (close to but) $< 4\sqrt{N}$ with* $\sup_{[-1,1]} |g| = 1$ *for which*

$$R(g, X_N) = \int_{-1}^{1} g(x)\,dx/2 - \frac{1}{N} \sum_{j=1}^{N} g(x_j) > \frac{1}{100N}.$$

Minimum-norm quadrature. For good Chebyshev-type quadrature on $[-1, 1]$, various authors have proposed to use N-tuples X_N of nodes which minimize a quadratic expression involving a certain number of nonzero remainders $\Delta_k = R(x^k, X_N)$. In such N-tuples one will in general find some coincident nodes, cf. [9,1,8]. Using Theorem 2.1 and complex analysis, we have found massive coalescence of nodes in minimum norm quadrature for $(I, dx/2)$:

Theorem 2.2 [16] *For fixed* $r \in (0, 1)$ *and* $4\sqrt{N} \le p = p_N \le \infty$, *let the* N-*tuple* X_N *of nodes* $x_j \in [-1, 1]$ *be chosen so as to minimize the function*

$$G_r(x_1, \ldots, x_N) = \sum_{k=1}^{p} \Delta_k^2 r^{2k}, \quad \Delta_k = R(x^k, X_N).$$

Then the number of distinct nodes does not exceed $C(r)\sqrt{N}$ *with a constant* $C(r)$ *independent of* N *and* p. *Similarly, if the nodes minimize* G_r *with* $r = 1$, *and if* $0 < \lambda < 1$, *then* $\le c(\lambda)\sqrt{N}$ *of the approximately* λN *nodes on* $[-\lambda, \lambda]$ *are distinct.*

3   DOMAINS OF PRODUCT TYPE

For the interval $I = [-1, 1]$ with $d\sigma = dx/2$, Bernstein's Theorem B provides a Chebyshev-type quadrature formula of degree $p = 2m - 1$ and order $N = N_0(m) = \mathcal{O}(p^2)$. For the unit circle $C = C(0, 1)$ with $d\sigma = ds/2$, Example 1.2 gives a Chebyshev-type formula of degree p and order $N = p + 1$. Combining these results, one obtains Chebyshev-type quadrature formulas for pairs (E, σ) which can be represented as *Cartesian products* of intervals and/or circles by suitable parametrization. In a number of cases the resulting upper bound for $N_E(p)$ will be of the right order as $p \to \infty$, cf. [17].

Example 3.1 For the *square* $Q = I^2$, the *cylindrical surface* $CS = I \times C$ and the closed *unit disc* $D = "I \times C"$ one finds

$$N_Q(p) \asymp p^4, \quad N_{CS}(p) \asymp p^3, \quad N_D(p) \asymp p^3. \tag{8}$$

For the disc one may use the parametrization

$$x = \sqrt{(t+1)/2}\cos\phi, \ y = \sqrt{(t+1)/2}\sin\phi, \ -1 \le t \le 1, \ 0 \le \phi < 2\pi,$$

so that the area element takes the form $d\sigma = dxdy/\pi = rdrd\phi/\pi = dtd\phi/4\pi$. The upper bounds in (8) follow immediately from Theorem B. That the resulting order estimates are sharp may be derived from suitable adaptations of Theorem A, see [17].

Example 3.2 (*the sphere*) The unit sphere S in $\mathbf{R}^3$ has the parametrization

$$x = \sin\theta\cos\phi, \ y = \sin\theta\sin\phi, \ z = \cos\theta, \quad 0 \le \theta \le \pi, \ 0 \le \phi < 2\pi$$

347

and the normalized area element

$$d\sigma = \sin\theta \, d\theta d\phi/4\pi = |dz| \, d\phi/4\pi.$$

Restricted to S, every polynomial $f(x, y, z)$ of degree $p \le 2m - 1$ is equal to a linear combination $F(\cos\theta, \phi) = F(z, \phi)$ of spherical harmonics of order $n \le p$. If S is considered as $I \times C$, the product method gives the quadrature formula

$$\int_S f(x, y, z) \, d\sigma = \int_{-1}^{1} (dz/2) \int_0^{2\pi} F(z, \phi) \, d\phi/2\pi \approx \frac{1}{2mN_0(m)} \sum_{k=1}^{2m-1} \sum_{l=1}^{2mM_k} F(t_k, \phi_{kl}), \tag{9}$$

where t_k and M_k are as in Theorem B with $M = N_0(m)$, while $\phi_{kl} = (l-1)2\pi/2mM_k$. Using the basis elements

$$F(z, \phi) = (1 - z^2)^{|s|/2} P_n^{(|s|)}(z) e^{is\phi}, \quad -n \le s \le n,$$

of the linear space of the spherical harmonics of order n, $0 \le n \le 2m-1$, one readily sees that formula (9) is exact for all polynomials $f(x, y, z)$ of degree $p \le 2m - 1$. It follows that $N_S(p) = \mathcal{O}(p^3)$.

On the other hand, if one substitutes $f(x, y, z) = g(z)$, one finds that any Chebyshev-type quadrature formula of order N and degree p for S, with nodes (x_j, y_j, z_j), implies one for the interval $-1 \le z \le 1$ with nodes z_j. Thus, by Theorem A, $N \ge p^2/16$, and hence for all $p \ge 1$,

$$p^2/16 \le N_S(p) \le cp^3 \qquad (c > 0). \tag{10}$$

It is an *important open problem* to determine the exact order of $N_S(p)$. We have found a good deal of support for the following:

Conjecture 3.3 (cf. Section 4 and [13]): $\qquad N_S(p) \asymp p^2$.

Spherical designs. In combinatorics, J.J. Seidel et al. have introduced so-called spherical p-designs. These may be described as configurations of N distinct points $\zeta_j \in S$ for which the Chebyshev-type formula (1) with $E = S$ has polynomial degree (at least) p, see e.g. [5,23].

By Example 3.2, the special points $\tilde{\zeta}_j$, $j = 1, \dots, 2mN_0(m)$, given by $z = t_k$, $\phi = \phi_{kl}$, form a spherical $(2m - 1)$-design of size $N = 2mN_0(m) \sim 8\sqrt{2}m^3$. It follows that there exist spherical p-designs consisting of $\mathcal{O}(p^3)$ points. This upper bound improves an earlier one by Rabau and Bajnok [22].

Remark 3.4 The preceding results have extensions to higher dimensions [17]. For the case of the *torus* $T = $ "$C \times C$" in $\mathbf{R}^3$, Kuijlaars has shown that $N_T(p)$ has the order p^2 which one would predict. However, there is no parametrization which makes T equal to $C \times C$ with the correct area element, and therefore the proof was not easy; it required a new Chebyshev-type modification of Mehler's formula (5), see [21].

4 THE SPHERE AND AN ELECTRON PROBLEM

The Faraday cage phenomenon of electrostatics. Let E be a bounded hollow conductor in $\mathbf{R}^3$ with smooth outer boundary, e.g. a sphere. Consider a positive charge distribution on E, of total charge 1 (say), in the most stable equilibrium. That is, the potential energy should be minimal. Then there is no (measurable) electrostatic field inside E. The electrostatic potential is constant on E and throughout its interior. Assuming *continuous* charge distribution, classical potential theory (culminating in the work of Frostman) provides a very satisfactory explanation.

Question 4.1 Can one explain the Faraday cage phenomenon on the basis of a model in which the charge 1 on E is made up of N equal *point charges* $1/N$, where N is large?

For now, we restrict ourselves to the case of the unit sphere S. If the charges are placed at the points of the N-tuple $Z_N \subset S$ (2), the electrostatic field $\mathcal{E}(x) = \mathcal{E}(x, Z_N)$ is equal to *minus the gradient* of the following potential $U(x)$, which we normalize so that $U(0) = 0$:

$$U(x) = U(x, Z_N) = \frac{1}{N} \sum_{j=1}^{N} \frac{1}{|\zeta_j - x|} - 1. \tag{11}$$

Ignoring the desideratum of minimal potential energy

$$V(Z_N) = \frac{1}{N^2} \sum_{j,k=1,\, j\neq k}^{N} \frac{1}{|\zeta_j - \zeta_k|}, \tag{12}$$

we ask the following:

Question 4.2 How small can one make $\sup |\mathcal{E}(x)|$ or $\sup |U(x)|$ on balls $B(0,r)$ with $r < 1$ by judicious choice of N-tuples Z_N on S?

One may use a Poisson integral representation for U to show that $\mathcal{E}$ and U have essentially the same order of smallness on the balls $B(0,r)$. Observe that $U(x, Z_N)$ can be interpreted as a Chebyshev-type quadrature remainder:

$$\int_S \frac{1}{|\zeta - x|} \, d\sigma(\zeta) - \frac{1}{N} \sum_{j=1}^{N} \frac{1}{|\zeta_j - x|} = -U(x, Z_N) \qquad (|x| < 1). \tag{13}$$

Setting $x = r\zeta$, $\zeta \in S$ and expanding in a series of spherical harmonics, one may use formula (13) to deduce the following:

Equivalence principle [18,13]. For the unit sphere S, good N-tuples of nodes for Chebyshev-type quadrature correspond to configurations of N point charges $1/N$ on S for which the electrostatic potential and field are very small on the balls $B(0,r)$. We give two quantitative versions, always for $E = S$:

Theorem 4.3 (i) *For a given N-tuple Z_N, formula (1) is polynomially exact to degree p if and only if $\mathcal{E}(x, Z_N) = \mathcal{O}(|x|^p)$ as $x \to 0$.*

(ii) *For a given constant $\alpha > 0$ and a given family of N-tuples Z_N with $N \to \infty$, one has an estimate*

$$\sup_{|x|=r} |U(x, Z_N)| = \mathcal{O}\{\exp(-c_1 N^\alpha)\} \tag{14}$$

for $r \in (0, 1)$ and a constant $c_1 = c_1(r) > 0$ if and only if there are constants $c_2, c_3 > 0$ such that for polynomials f,

$$\sup_{\deg f \le c_2 N^\alpha, \sup_S |f| \le 1} |R(f, Z_N)| = \mathcal{O}\{\exp(-c_3 N^\alpha)\}. \tag{15}$$

Counting linearly independent spherical harmonics of order $\le \sqrt{N}$, one can show that for given Z_N, there is always a nonnegative polynomial f of degree $\le 2\sqrt{N}$ with $\sup_S |f| = 1$ which vanishes on Z_N. Estimating $R(f, Z_N) = \int_S f$ from below and using a precise form of the Equivalence Principle, one thus obtains the

Corollary 4.4 [18] *For any N-tuple Z_N on S,*

$$\sup_{|x|=r} |U(x, Z_N)| > \frac{1}{4(\sqrt{N} + 1)^3} r^{2\sqrt{N}}, \qquad 0 < r < 1. \tag{16}$$

There is a similar inequality for the logarithmic potential in the electron problem for the unit circle, and there the order r^N of the lower bound is attained, even for the case of minimal potential energy. It is plausible that the order of the lower bound in (16) is also sharp, which means that Conjecture 3.3 should be true.

How would we find good N-tuples of Chebyshev-type nodes on S for large N? Minimization of the potential energy (12) does not give a good handle on the problem, but the situation improves if we introduce a parameter r from $(0, 1)$ and minimize the related function

$$G_r(\zeta_1, \dots, \zeta_N) = \frac{1}{N^2} \sum_{j,k=1, j \neq k}^{N} \frac{1}{|\zeta_j - r\zeta_k|}, \qquad \zeta_j \in S. \tag{17}$$

For minimizing Z_N the potential $U(r\zeta)$, $\zeta \in S$ is stationary at each point ζ_k. The minimizing points $\zeta_k \in S$ are well distributed, and it is very likely that they are well separated. One may then consider partial derivatives of the complexified potential and use results from "Several complex variables" on the smallness of analytic functions with many well-separated real zero points to deduce that (14) is true with $\alpha = 1/2$, see [13].

Corollary 4.5 [13] Suppose that the minimizing N-tuples Z_N for the functions $G_r(\zeta_1, \dots, \zeta_N)$ (with fixed r) satisfy the following plausible separation condition:

$$|\zeta_j - \zeta_k| \ge 1/\sqrt{N}, \qquad j, k = 1, \dots, N, \quad j \neq k.$$

Then the minimizing N-tuples Z_N are good N-tuples of nodes for Chebyshev-type quadrature on S in the sense that (15) will hold with the optimal exponent $\alpha = 1/2$.

Remark 4.6 (*planar electrostatic fields*) Let Γ be a Jordan curve of class $C^{3+\epsilon}$ in the z-plane different from a circle and let K be an arbitrary closed domain in its interior. Then for charges $1/N$ at the points of a Fekete N-tuple on Γ (such a configuration corresponds to minimal potential energy),

$$\sup_{z \in K} |\mathcal{E}(z, Z_N)| \sim \frac{c}{N} \quad \text{as } N \to \infty \tag{18}$$

with a constant $c = c(\Gamma, K) > 0$, cf. [14,10,15]. Monterie is working on the case of curves with corners, as well as on an analog of (18) for smooth surfaces in $\mathbf{R}^3$ with N replaced by $\sqrt{N}$, cf. [11]. Could it be that the sphere in $\mathbf{R}^3$ is as exceptional as the circle in $\mathbf{R}^2$?

REFERENCES

[1] Anderson L.A., Gautschi W. Optimal weighted Chebyshev-type quadrature formulas. *Calcolo*, **12**:211–248, 1975.

[2] Bernstein S.N. Sur les formules de quadrature de Cotes et Tchebycheff. *C.R. Acad. Sci. URSS* (Dokl. Akad. Nauk SSSR), N.S., **14**:323–327, 1937.

[3] Bernstein S.N. On quadrature formulas with positive coefficients (Russian). *Izv. Akad. Nauk SSSR*, Ser. Mat. 1, No. **4**:479–503, 1937. See also the announcements in *C.R. Acad. Sci. Paris*, **204**:1294–1296 and 1526–1529, 1937.

[4] Bernstein S.N. Sur un système d'équations indéterminées. *J. Math. Pures Appl.*, (9) **17**:179–186, 1938.

[5] Delsarte P., Goethals J.M., Seidel J.J. Spherical codes and designs. *Geom. Dedicata*, **6**:363–388, 1977.

[6] Förster K.-J. Variance in quadrature – a survey. In *Numerical Integration IV*, pages 91–110. Birkhäuser, Basel, 1993. H. Brass and G. Hämmerlin, eds.

[7] Gatteschi L., Vinardi G. Sul grado di precisione di formule di quadratura del tipo di Tchebycheff. *Calcolo*, **15**:59–85, 1978.

[8] Gautschi W. Advances in Chebyshev quadrature. In *Numerical Analysis*, (Proc. 6th Dundee Conf.). Lect. Notes in Math., **506**:100–121, Springer, Berlin, 1976. G.A. Watson, ed.

[9] Gautschi W., Yanagiwara H. On Chebyshev-type quadratures. *Math. Comp.*, **28**:125–134, 1974.

[10] Korevaar J. Equilibrium distributions of electrons on roundish plane conductors. *Nederl. Akad. Wetensch. Proc. Ser. A 77 = Indag. Math.*, **36**:423–456, 1974.

[11] Korevaar J. Problems of equilibrium points on the sphere and electrostatic fields. Univ. of Amsterdam, Math. Dept. Report 1976-03.

[12] Korevaar J. Behavior of Cotes numbers and other constants, with an application to Chebyshev-type quadrature. *Indag. Math. N.S.*, **3**:391–402, 1992.

[13] Korevaar J. Chebyshev-type quadratures: use of complex analysis and potential theory. In *Proceedings Complex Potential Theory*, Montréal, 1993. Kluwer, Dordrecht. To appear in 1994.

[14] Korevaar J., Geveci T. Fields due to electrons on an analytic curve. *SIAM J. Math. Anal.*, **2**:445–453, 1971.

[15] Korevaar J., Kortram R.A. Equilibrium distributions of electrons on smooth plane conductors. *Nederl. Akad. Wetensch. Proc. Ser. A*, **86**:203–219, 1983.

[16] Korevaar J., Meyers J.L.H. Massive coalescence of nodes in optimal Chebyshev-type quadrature on [−1, 1]. *Indag. Math. N.S.*, **4**:327–338, 1993.

[17] Korevaar J., Meyers J.L.H. Chebyshev-type quadrature on multidimensional domains. Univ. of Amsterdam, Math. Dept. Report 1993-01. To appear in *J. Approx. Theory*, 1994.

[18] Korevaar J., Meyers J.L.H. Spherical Faraday cage for the case of equal point charges and Chebyshev-type quadrature on the sphere. *ITSF*, **1**:105–117, 1993.

[19] Kuijlaars A.B.J. The minimal number of nodes in Chebyshev type quadrature formulas. *Indag. Math. N.S.*, **4**:339–362, 1993.

[20] Kuijlaars A.B.J. Chebyshev type quadrature for Jacobi weight functions. To appear in *J. Comput. Appl. Math.*

[21] Kuijlaars A.B.J. Chebyshev-type quadrature and partial sums of the exponential series. Univ. of Amsterdam, Math. Dept. Report 1993-07. To appear in *Math. Comp.*

[22] Rabau P., Bajnok B. Bounds on the number of nodes in Chebyshev-type quadrature formulas. *J. Approx. Theory*, **67**:199–215, 1991.

[23] Seidel J.J. Isometric embeddings and geometric designs. To appear in *Trends in Discrete Mathematics*.

[24] Seymour P.D., Zaslavsky T. Averaging sets: a generalization of mean values and spherical designs. *Adv. Math.*, **52**:213–240, 1984.

International Series of Numerical Mathematics, Vol. 119, ©1994 Birkhäuser

Using Quasi-Interpolants in a Result of Favard

Thomas Kunkle*

With best wishes to Professor Gautschi on the happy occasion of his sixty-fifth birthday

Abstract. A proof of Favard can be restructured using quasi-interpolants of the type discussed in these proceedings [6] and his result strengthened.

1 INTRODUCTION

In a paper in this volume [6], Lei and Cheney discuss the quasi-interpolant; that is, a linear operator of the form

$$L : f \mapsto \sum_i f(y_i) g_i$$

for some collection $\{g_i\}$ of d-variate functions and a set $\{y_i\}$ of points in $\mathbb{R}^d$. Of particular interest are those L satisfying $L^2 = L$ and $Lp = p$ for each p in the space Π_n of all d-variate polynomials of total degree n or less.

One such L is obtained as follows. Fix n a natural number, and suppose that, for every i in the index set, X_i is a finite set of points in $\mathbb{R}^d$ with the property that, for any function f, there is a unique polynomial $P_i f$ in Π_n agreeing with f on X_i. Let $\{\theta_i\}$ be a partition of unity on $\mathbb{R}^d$. Then the quasi-interpolant

$$L : f \mapsto \sum_i \theta_i P_i f \qquad (1.1)$$

is a projector and its restriction to Π_n is the identity. One can think of Lf as a moving average of polynomials, since the weights $\{\theta_i\}$ are not constant.

Just such an operator occurs implicitly in the original proof of a univariate extension theorem by Favard [4,2]. The purpose of this note is to demonstrate that, when this proof is modified to make explicit use of quasi-interpolants, a stronger result is immediately obtained. Furthermore, quasi-interpolants allow one to prove similar multivariate extension theorems [5].

Favard's result is the following, except that originally, the nth derivative of the extension Lf was only piecewise continuous.

---

*Dept. of Mathematics, College of Charleston, Charleston, SC 29424, USA; kunkle@math.cofc.edu

Theorem 1.2 *Let* $\mathbf{m} := \{m_i : i \in \mathbb{Z}\}$ *be a strictly increasing, bi-infinite sequence of real numbers diverging to* $\pm\infty$, *and let* f *be a real-valued function defined on* $\mathbf{m}$. *Then there exists an infinitely differentiable extension* Lf *of* f *to* $\mathbb{R}$ *with the property that, on each interval* $[m_i, m_{i+1}]$,

$$|D^n Lf| \leq C \max\Big\{ \big|[m_s, \ldots, m_{s+n}]f\big| : i - n < s \leq i \Big\} \tag{1.3}$$

where C *is a constant independent of both* f *and* $\mathbf{m}$

Here $[m_s, \ldots, m_{s+n}]f$ denotes the divided difference of f at the real numbers $\{m_s, \ldots, m_{s+n}\}$. By Rolle's Theorem, any smooth function agreeing with f on $\mathbf{m}$ would have an nth derivative at least as large in the uniform norm as the numbers

$$n! \big|[m_s, \ldots, m_{s+n}]f\big|.$$

The extension promised by Theorem 1.2 has an nth derivative that is no more than a constant times this necessary size. In several papers [1,2,3], de Boor has estimated this constant as well as

$$\sup_{f,\mathbf{m}} \frac{\inf\{\|D^n F\|_\infty : F \in L^{(n)}_\infty, F = f \text{ on } \mathbf{m}\}}{\max\{n! |[m_i, \ldots, m_{i+n}]f| : i \in \mathbb{Z}\}},$$

and investigated the optimization problem:

Minimize the norm of $F \in L^{(n)}_p$ *such that* $F = f$ *on* $\mathbf{m}$.

2   PROOF OF THEOREM 1.2

Define $P_i f$ to be the polynomial of degree less than n agreeing with f at the real numbers $\{m_i, \ldots, m_{i+n-1}\}$. We will take for L a quasi-interpolant of the form (1.1), with functions $\{\theta_i\}$ which are to be constructed.

Choose a function $\theta \in C^\infty[0, 1]$ that satisfies

$$\theta(0) = 1, \quad \theta(1) = 0,$$
$$\forall k \in \mathbb{N}, \quad D^k \theta(0) = D^k \theta(1) = 0,$$

and

$$\theta + \theta(1 - \cdot) \equiv 1. \tag{2.1}$$

(Here, $\theta(1 - \cdot)$ represents the function given by the rule

$$\theta(1 - \cdot) : [0, 1] \to \mathbb{R} : x \mapsto \theta(1 - x).$$

The dot "$\cdot$" is used similarly below, enabling us to make a clear distinction between a function such as θ and its output $\theta(x)$ at some point x in its domain.)

For every i, set I_i to be the closed interval $[m_i, m_{i+1}]$, and define $|I_i|$ to be its length. Choose $k = k_i$ so that I_k is the largest of the intervals

$$I_{i-1}, \cdots, I_{i+n-2}.$$

If this leaves more than one possibility for k, take k minimal. Choose $l = l_i$ so that I_l is the largest of

$$I_i, \cdots, I_{i+n-1},$$

with the same rule in case of a tie. Note also that when k and l are viewed as functions of i, we have $k_i \leq l_i = k_{i+1}$. Abbreviating k_i and l_i to k and l, construct the weight function θ_i as follows. If $k = l$, define θ_i to be identically zero. If, instead, $k < l$, set

$$\theta_i := \begin{cases} \theta\left(\dfrac{m_{k+1} - \cdot}{m_{k+1} - m_k}\right) & \text{on } [m_k, m_{k+1}]; \\ 1 & \text{on } [m_{k+1}, m_l]; \\ \theta\left(\dfrac{\cdot - m_l}{m_{l+1} - m_l}\right) & \text{on } [m_l, m_{l+1}]; \\ 0 & \text{elsewhere.} \end{cases}$$

Consider the resulting sequence $\{\theta_i\}_{-\infty}^{\infty}$. After deleting those terms that are identically zero, for every adjacent pair in the list

$$\ldots, \theta_j, \theta_h, \ldots$$

θ_j decreases to zero and θ_h increases to one on the same interval. From (2.1) it follows that $\{\theta_i\}$ is a partition of unity. By θ_i's construction, $\theta_i(m_j) = 0$ unless $i \leq j < i + n$. Thus

$$Lf := \sum_i \theta_i P_i f$$

agrees with f on $\mathbf{m}$. Since each θ_i is infinitely differentiable, so is Lf. It remains only to prove (1.3).

Restricting our attention to the interval I_i, if only one of the functions θ_j is nonzero, then that weight function is identically one and $Lf = P_j f$. Thus $D^n Lf = 0$ and (1.3) is trivially satisfied. If, instead, there are two weight functions, θ_j and θ_h $(j < h)$, with support on this interval, then θ_j decreases to zero and θ_h increases to one on I_i. This means that $l_j = i = k_h$, i.e., that I_i is the largest of the intervals

$$I_j, \ldots, I_{j+n-1}$$

and also the largest of

$$I_{h-1}, \ldots, I_{h+n-2}.$$

Consequently,

$$h - 1 \leq i \leq j + n - 1, \tag{2.2}$$

and

$$(m_t - m_s) \leq (t - s)|I_i| \quad \text{when } j \leq s \leq t \leq h + n - 1. \tag{2.3}$$

Set ψ_s to be the polynomial

$$\psi_s := (\cdot - m_s) \cdots (\cdot - m_{s+n-2}).$$

On I_i, Lf can be written

$$\begin{aligned}
Lf &= \theta_j P_j f + \theta_h P_h f \\
&= P_j f + \theta_h (P_h - P_j) f \\
&= P_j f + \theta_h \sum_{s=j}^{h-1} (P_{s+1} - P_s) f \\
&= P_j f + \sum_{s=j}^{h-1} (m_{s+n} - m_s) \theta_h \psi_{s+1} [m_s, \cdots, m_{s+n}] f.
\end{aligned} \tag{2.4}$$

We now arrive at a bound for $\|D^n(\theta_h \psi_{s+1})\|$, where $\|\cdot\|$ is the max-norm on I_i. Trivially, for some constant c_1 and for $0 \leq l \leq n$,

$$\|D^l \theta_h\| = c_1 |I_i|^{-l}.$$

Consider $D^{n-l} \psi_{s+1}$. When $l = 0$, this is zero; otherwise it is the sum of products of $l - 1$ terms of the form $(\cdot - m_t)$ with $j < t < h + n - 1$. On the interval of interest, each of these terms is bounded by the larger of $|m_i - m_t|$ and $|m_{i+1} - m_t|$. By (2.2) and (2.3), this is bounded by $2n|I_i|$. Therefore, with constants c_2 and c_3 depending only on n and θ,

$$\|D^{n-l} \psi_{s+1}\| \leq c_2 |I_i|^{l-1}$$

and

$$\|D^n(\theta_h \psi_{s+1})\| \leq c_3 |I_i|^{-1}. \tag{2.5}$$

Differentiating (2.4), applying (2.3) to the term $(m_{s+n} - m_s)$, and combining the result with (2.5) gives (1.3), completing the proof of the theorem. ∎

In his proof, Favard uses interpolating polynomials of degree n when, as we've seen, $n - 1$ suffices. However, the greatest difference between the proof of this theorem and Favard's proof is the use of an operator of the form (1.1). Favard writes his extension on each interval I_i as the sum of polynomials and piecewise polynomial "smoothing functions," both of degree n.

Acknowledgement. The author extends his thanks to the Gautschis, the organizers, and Prof. Jörg Peters for their hospitality.

REFERENCES

[1] de Boor C. A smooth and local interpolant with "small" kth derivative. In *Numerical Solutions of Boundary Value Problems for Ordinary Differential Equations*, pages 177–197. Academic Press, New York, 1974. A. K. Aziz, ed.

[2] de Boor C. How small can one make the derivatives of an interpolating function? *J. Approx. Theory*, **13**:105–116, 1975.

[3] de Boor C. On "best" interpolation. *J. Approx. Theory*, **16**:28–42, 1976.

[4] Favard J. Sur l'interpolation. *J. Math. Pures Appl.*, **19**:281–300, 1940.

[5] Kunkle T. Lagrange interpolation on a lattice: bounding derivatives by divided differences. *J. Approx. Theory*, **71**:94–103, 1992.

[6] Lei J., Cheney E.W. Quasi-interpolation on irregular points I. In these proceedings.

International Series of Numerical Mathematics, Vol. 119, ©1994 Birkhäuser

Null Rules and Orthogonal Expansions

Dirk P. Laurie*

To Walter Gautschi, on achieving $n = 65$

Abstract. Any quadrature formula with positive coefficients induces a discrete inner product with respect to which orthogonal polynomials may be defined. The discrete linear functionals that evaluate the coefficients in the expansion of a function in terms of these polynomials are *null rules,* i.e. they map polynomials of sufficiently low degree to zero. Moreover, they are orthogonal to one another and to the quadrature rule itself with respect to the dual inner product.

It is suggested that this way of obtaining null rules is preferable to the current method of using null rules that are orthogonal to one another in the Euclidean sense, but not to the quadrature rule. One advantage of the orthogonal expansion approach is that polynomial approximants can explicitly be constructed, leading to non-linear null rules that behave less capriciously than linear ones.

1 INTRODUCTION

The estimation of the error incurred by a quadrature formula using only the function values already computed is a process best described by the proverb 'You cannot eat your cake and have it'. If one can obtain an error estimate correct to even one significant digit, that error estimate could have been used to obtain an approximate integral with one significant digit more, for which no error estimate is available. In practice one has to make do with error estimates that in the case of smooth functions overestimate the true error by several orders of magnitude.

The most obvious way of constructing an error estimate is to apply to a function f two different quadrature formulas A and B, where A is more accurate than B, and to use the difference $Ef = Af - Bf$ as an error estimate. One may expect that Ef should be a good estimate of the error in Bf, but what one really would like to have is a good estimate of the error in Af.

Usually A is *interpolatory,* which means that it integrates polynomials up to degree $n - 1$ exactly (where n is the number of points used). If B reuses $n - 1$ of the points in A and is also interpolatory, the difference Ef is zero whenever f is a polynomial of degree up to $n - 2$. The functional E is called a *null rule* of degree $n - 2$ on the points of the quadrature formula A.

Various devices have been tried to extract an error estimate for Af from the single number Ef.

---

*Potchefstroom University for Christian Higher Education, P. O. Box 1174, Vanderbijlpark, South Africa; wskdpl@puknet.puk.ac.za

• In QUADPACK ([13], Section 3.4.1), A is a $2m + 1$ point Kronrod rule of degree $3m + 1$ or $3m + 2$, and B is an m point Gaussian rule of degree $2m - 1$. The heuristic employed there is based on the assumption that as the length h of the integration interval tends to 0, the error of a rule of degree d is roughly proportional to h^d. Therefore the relative error of the Kronrod rule should be roughly proportional to that of the Gaussian rule raised to the power $\frac{3}{2}$. We call this relationship "asymptotic behavior".

An interesting feature of the QUADPACK estimate is that a test is made to diagnose whether the expected asymptotic behavior is in fact present.

This test requires the computation of $Rf := A|f - \bar{f}|$ where $\bar{f}$ is the estimate for the mean value of f obtained from the Kronrod rule A. We show in § 4 that Rf is a discretization of an error bound for Af. It is zero if and only if f is constant, and can therefore be thought of as a *non-linear* null rule of degree zero. In § 4 we introduce non-linear null rules of higher degree.

Since E is a null rule of high degree, it is expected that in the presence of asymptotic behavior, $|Ef|$ will be much smaller than Rf. The heuristic used in QUADPACK is that if $r > 0.005$, where $r = |Ef|/Rf$, asymptotic behavior is not present; the error estimate is then Rf. Otherwise the error estimate is $(200r)^{3/2}Rf$. This expression can be written as $200^{3/2}r^{1/2}|Ef|$. So Ef has been decreased by the factor $r^{1/2}$ but at the same time increased by the factor $200^{3/2}$. One may think of the number $200^{3/2}$ as a pessimism factor to guard against fortuitous underestimates.

• The usual application for error estimates occurs during adaptive quadrature, where the typical step is the subdivision of the interval with the largest error estimate. Laurie [12] uses the information on the parent interval as well as its two descendants to obtain the error formula $rE_i f$, $i = 1, 2$ for the two subintervals, where A_i and E_i are obtained by scaling A and E respectively to the two subintervals, and r is given by

$$r = \left| \frac{Af - (A_1 f + A_2 f)}{Ef - (E_1 f + E_2 f)} \right|.$$

He shows that $r|E_1 f + E_2 f|$ is a bound for the error in $A_1 f + A_2 f$ when Af and Bf both converge monotonically when compounded and Af converges no slower than Bf. However, Espelid and Sørevik [5] give numerical experiments that show that for some of the families of integrands used to test automatic quadrature routines, these assumptions fail often enough to render the unqualified use of the estimate $rE_i f$ unreliable.

The QUADPACK idea was greatly elaborated by Berntsen and Espelid [1]. Their approach is based on the fact that there exist not merely two null rules on n points, but an $n - 1$ dimensional linear space of null rules. It is possible to find a basis $\{N_0, \ldots, N_{n-2}\}$ for this space in which the null rule N_d is of degree d. Berntsen and Espelid choose a basis in which the coefficients comprising the N_d

are mutually orthogonal in $\mathbb{R}^n$ and of the same norm as the coefficients of the quadrature rule A. If the function f produces random numbers of mean zero, then the expected value of $N_d f$ is zero, and its variance does not depend on d. Any steady decrease in $N_d f$ with increasing d can only be explained as evidence of regularity in the integrand f.

Three possible types of behavior of the sequence $\{N_d f, \ d = 0, \ldots, n - 2\}$ are distinguished heuristically:

Non-asymptotic: the numbers $\{N_d f\}$ do not decrease monotonically. In this case, no attempt is made to extrapolate the error estimate;

Weakly asymptotic: the numbers $\{N_d f\}$ do decrease monotonically, but not rapidly so. That is, the ratios N_{d+1}/N_d are not all less than some predetermined cutoff point. In this case, the error estimate is extrapolated for one further stage past the null rule of highest degree;

Strongly asymptotic: the numbers $\{N_d f\}$ decrease monotonically and rapidly. In this case, the error estimate is extrapolated up to the degree of the quadrature rule.

There are some other heuristics (e.g. allowance for "phase effects") in Berntsen and Espelid's application of null rules that are too technical to be described here, although we later briefly refer to them when presenting examples. As in the case of QUADPACK, a pessimism factor is slapped onto the extrapolated estimate, and this factor, together with the criterion for distinguishing between weak and strong asymptotic behavior, is tuned on a battery of test integrals to get the best trade-off between reliability and efficiency.

The state of the art in the application of null rules can be summarized as follows:

- The null rules are related to the quadrature rule only inasmuch they are defined on the same set of points;

- The information provided by null rules is used to decide how polynomial-like the integrand is behaving;

- The final decision on how to extract an error estimate from the numbers produced by the null rules is made heuristically and with the aid of a substantial amount of numerical experimentation.

A totally different approach to the problem of delivering an approximate integral with error estimate comes from the work of Clenshaw and Curtis [2]. Their idea is to expand the integrand in Chebyshev polynomials and to integrate the resulting series term by term. Scrutiny of the partial sums of the series then provides one with an error estimate (once again, some heuristics are involved). The coefficients in the expansion are approximated by Chebyshev-Lobatto rules [3], which use the endpoints together with optimally placed points inside the interval. The points and weights of these rules are expressible in closed form in terms of cosines.

The Clenshaw-Curtis algorithm was one of the first automatic quadratures, and shared an important common feature with Romberg's algorithm (at the time, its only serious competitor): when going from an $n + 1$ point to a $2n + 1$ point formula, all previous points are reused, with weights half of what they were before. A later refinement by Filippi [6] uses Chebyshev polynomials of the second kind: for these polynomials, the weights at the end points are zero.

The purpose of the present paper is to point out a connection between linear null rules and orthogonal expansions, and to exploit that connection to obtain non-linear null rules that behave less erratically than linear ones. The main points are the following.

- Any quadrature formula Q with positive weights is a Gaussian formula with respect to a suitably defined inner product. We say that the formula *induces* the inner product.

- The formula for the d-th coefficient in an orthogonal expansion with respect to the inner product induced by a quadrature formula Q is a null rule of degree $d - 1$ on the points of Q.

- These null rules are orthogonal and of equal norm in a suitably chosen inner product space (usually not the same space as that induced by Q).

- The Clenshaw-Curtis and Filippi methods are special cases of a general class of methods that can be defined for any quadrature formula with positive weights.

- For any interpolatory quadrature formula, it is possible to construct a sequence of non-linear null rules with monotonic behavior, each of which can be seen as an approximation to a bound on the error.

2   NULL RULES VIA ORTHOGONAL EXPANSIONS

Let Q be any n point quadrature formula given by

$$Qf = \sum_{j=1}^{n} w_j f(x_j), \qquad (2.1)$$

with positive weights $\{w_j\}$. In this section we are not concerned with the original integral If that is supposed to be evaluated by the use of Q. It does not matter whether Q is a good or a bad formula for that integral: all that we need is the positivity of the weights. In that case, Q can be viewed as the n point Gaussian formula for the integral

$$I_w f = \int_a^b f(x) w(x) \, dx \qquad (2.2)$$

362

where (a, b) is any open interval containing all the points x_j, and

$$w(x) = \sum_{j=1}^{n} w_j \delta(x - x_j), \tag{2.3}$$

where δ denotes the Dirac delta.

The first n monic orthogonal polynomials with respect to this weight can be computed by the Stieltjes algorithm (advocated by Gautschi [7, 8])

$$
\begin{align}
p_{-1}(x) &= 0; \tag{2.4} \\
p_0(x) &= 1; \tag{2.5} \\
p_{k+1}(x) &= (x - a_k)p_k(x) - b_k p_{k-1}(x), \ k = 0, \ldots, n-1, \tag{2.6}
\end{align}
$$

where the coefficients a_k, b_k are given by

$$a_k = \frac{Q(x.p_k^2)}{Q(p_k^2)}, \ k = 0, \ldots, n-1; \tag{2.7}$$

$$b_k = \frac{Q(p_k^2)}{Q(p_{k-1}^2)}, \ k = 1, \ldots, n-1. \tag{2.8}$$

Here the notation $p.q$ is used for the pointwise product of functions, i.e. $(p.q)(x) = p(x)q(x)$, with the obvious meaning $p^2 = p.p$, and x is used for the identity function $x(t) = t$. The quantity b_0 may be any finite number: we follow Gautschi [7] in putting $b_0 = Qp_0$.

The formulas (2.7) and (2.8), though often perfectly satisfactory, can in some cases deliver inaccurate results. It is also possible to compute the coefficients a_k, b_k from the points x_j and weights w_j by using orthogonal matrix transformations: such algorithms have been proposed by several authors [4, 10, 14]. These algorithms theoretically have better numerical properties than the use of (2.7) and (2.8).

The orthogonal polynomials p_k can be used to obtain an orthogonal sequence of null rules. We first define what we mean by that:

Definition 1 *A discrete functional N defined on the points $x_1, \ldots, x_n$ is a **null rule** of degree k if $Nf = 0$ when f is a polynomial of degree $\leq k$, but $Nf \neq 0$ for at least one polynomial f of degree $k + 1$.*

A *linear* null rule N can be expressed in terms of a coefficient vector $\nu_1, \ldots, \nu_n$ such that

$$Nf = \sum_{j=1}^{n} \nu_j f(x_j).$$

We use the notation **N** to denote the coefficient vector of the discrete linear functional N. Thus, Nf can also be seen as the $\mathbb{R}^n$ inner product of **N** and the vector **f** obtained by evaluating f at the points $x_1, \ldots, x_n$.

It is also useful to work with inner products and norms with respect to a positive definite matrix.

Definition 2 *The* **W**-*inner product of two vectors* **a** *and* **b** *in* $\mathbb{R}^n$ *is given by*

$$(\mathbf{a}, \mathbf{b})_{\mathbf{W}} = \mathbf{a}^T \mathbf{W} \mathbf{b},$$

where **W** *is any positive definite matrix.*

It is well known that the **W**-inner product satisfies all the axioms for an inner product, and we may therefore use the **W**-norm of a vector

$$\|\mathbf{a}\|_{\mathbf{W}} = (\mathbf{a}, \mathbf{a})_{\mathbf{W}}^{1/2}.$$

When we say that two linear null rules are **W**-*orthogonal*, we mean that their coefficient vectors are **W**-orthogonal, and when we speak of the **W**-norm of a linear null rule, we mean the **W**-norm of its coefficient vector. In the sequel, **W** will always denote the matrix $\mathbf{W} = \text{diag}\,[w_j]$.

The inverse of **W** is also positive definite. The inner product $(\cdot, \cdot)_{\mathbf{W}^{-1}}$ is called the dual inner product to $(\cdot, \cdot)_{\mathbf{W}}$.

Theorem 1 *The discrete linear functionals* N_k *defined by*

$$N_k f = \frac{Q(p_k.f)}{\sqrt{Q(p_k.p_k)}} = Q(p_k^2)^{-1/2} \sum_{j=1}^{n} w_j p_k(x_j) f(x_j), \ k = 0, \ldots, n-1 \qquad (2.9)$$

are null rules of degree $k-1$, $k = 1, \ldots, n-1$, *and are mutually* $\mathbf{W}^{-1}$-*orthonormal.*

Proof: Since $N_k f$ is a constant multiple of $Q(p_k.f)$, it follows from the orthogonality of p_k that $N_k f = 0$ when f is a polynomial of degree less then k. On the other hand, $N_k p_k$ is nonzero because $Q(p_k^2) > 0$ for $k < n$. Therefore N_k is a null rule of degree $k - 1$.

The $\mathbf{W}^{-1}$-inner product of N_k and N_l is

$$
\begin{aligned}
(\mathbf{N}_k, \mathbf{N}_l)_{\mathbf{W}^{-1}} &= Q(p_k^2)^{-1/2} Q(p_l^2)^{-1/2} \sum_{j=1}^{n} w_j^{-1}(w_j p_k(x_j))(w_j p_l(x_j)) \\
&= Q(p_k^2)^{-1/2} Q(p_l^2)^{-1/2} \sum_{j=1}^{n} w_j p_k(x_j) p_l(x_j) \\
&= Q(p_k^2)^{-1/2} Q(p_l^2)^{-1/2} Q(p_k.p_l) \\
&= \delta_l^k,
\end{aligned}
$$

where δ_l^k is the Kronecker delta. ∎

Remark 1 Note that $N_k f$ is simply the k-th coefficient in the expansion of f in terms of the polynomials orthonormal with respect to the weight w in (2.3). Since w is a discrete weight function,

$$f(x_j) = \sum_{k=0}^{n-1} N_k f. p_k(x_j), \ j = 1, \ldots, n.$$

This property is exploited in the following section.

Remark 2   Since Q is a multiple of N_0, it is therefore also $\mathbf{W}^{-1}$-orthogonal to the null rules N_k, $k = 1, \ldots, n-1$. An analogous property does not hold for the Berntsen-Espelid orthogonalization of null rules: i.e. the quadrature rule Q is not in general orthogonal in $\mathbb{R}^n$ to a sequence of null rules that are mutually orthogonal in $\mathbb{R}^n$.

3   INTERPOLATORY QUADRATURE VIA NULL RULES

Suppose that we know the expansion of f in terms of the polynomials p_k orthogonal with respect to the discrete weight w in (2.3), i.e. the numbers $N_k f$ obtained by evaluating the null rules (2.9), and we wish to obtain an interpolatory formula for a (possibly different) integral

$$If = \int_a^b f(x) \, d\sigma(x). \tag{3.1}$$

We can do so provided that the *modified moments*

$$\mu_k = \int_a^b p_k(x) \, d\sigma(x) \tag{3.2}$$

of the polynomials p_k defined in (2.4–2.6) exist. The desired approximation $Q_I f$ can then be written as the finite series

$$Q_I f = \sum_{k=0}^{n-1} \mu_k Q(p_k^2)^{-1/2} N_k f. \tag{3.3}$$

The formula Q_I is an interpolatory formula because, whenever f is a polynomial of degree not greater than $n - 1$, the expansion $f(x) = \sum_{k=0}^{n-1} Q(p_k^2)^{-1/2}(N_k f) p_k(x)$ is an identity. Therefore, for such f, the quadrature formula Q_I gives the exact result.

This approach to interpolatory quadrature is useful mainly when Q is not an interpolatory quadrature formula for I, because the the partial sums in (3.3) provide useful additional information. The Clenshaw-Curtis [2] and Filippi [6] quadrature methods are particular examples for which the recurrence coefficients and modified moments are known analytically.

Remark 3   The approach to interpolatory quadrature given here can be used as a basis for the $\mathcal{O}(n^2)$ computation of weights for arbitrary nodes. First, we generate the null rules induced by the given nodes and any positive set of weights; then the modified moments are obtained by running the Sack-Donovan-Wheeler algorithm [15, 16, 8] for computing recurrence coefficients from modified moments in the reverse direction; and finally the coefficients are computed from $Q_I = \sum_{k=0}^{n-1} \mu_k Q(p_k^2)^{-1/2} N_k$. It remains to be seen whether this procedure has any advantage over known $\mathcal{O}(n^2)$ methods, i.e. fast Vandermonde solvers [9] or methods based on orthogonal matrix operations [11].

4 NON-LINEAR NULL RULES

If Q is the Gaussian formula for I, all the modified moments μ_k, $k = 1, \ldots, n-1$ are zero, and so all the information about f is concentrated in the single number $N_0 f$. How can one use the null rules to provide error estimates even when all the modified moments are zero? One possibility is to apply the same principle as the degree zero null rule in QUADPACK ([13], equation 3.4.3), but to use the expansion provided by the null rules N_k. The idea is very simple: let f_d be any function such that $Qf = I f_d$. Then $I f - Q f = I(f - f_d)$, leading to the error bound

$$|If - Qf| \leq I(|f - f_d|). \tag{4.1}$$

If Q is an interpolatory rule of degree at least d, such a function is

$$f_d(x) = \sum_{k=0}^{d} Q(p_k^2)^{-1/2}(N_k f)p_k(x). \tag{4.2}$$

Here $Qf = I f_d$ because both are equal to $Q f_d$: in the case of Qf, because $Q p_k = 0$, $k = 1, \ldots, n$; and in the case of $I f_d$, because Q has degree at least d. The error bound (4.1) can be approximated by the *non-linear* null rule of degree d

$$E_d f = Q(|f - f_d|). \tag{4.3}$$

For $d = 0$, this is precisely the non-linear null rule R used in QUADPACK. This is not the only possibility, though: from Schwarz's inequality

$$|I(f - f_d)| \leq \sqrt{I((f - f_d)^2)\, I(1)}$$

we obtain another non-linear null rule of degree d :

$$H_d f = \left(b_0 Q((f - f_d)^2)\right)^{1/2}. \tag{4.4}$$

Theorem 2 *The non-linear null rule H_d defined in (4.4) is of degree d and the sequence $H_0, \ldots, H_{n-2}$ is non-increasing.*

Proof: When f is a polynomial of degree d, then $f_d = f$ and therefore $H_d f = 0$. When $f = p_{d+1}$, then $f_d = 0$, but for $d \leq n-2$, f does not vanish at all the nodes x_j and therefore $H_d f > 0$. This shows that H_d is a null rule of degree d.

Letting $\mathbf{f}$ denote the vector obtained by evaluating f at the points $x_1, \ldots, x_n$, we have the expansion of $\mathbf{f}$ in terms of the set of $\mathbf{W}$-orthonormal vectors

$$\mathbf{f} = \sum_{k=0}^{n-1} Q(p_k^2)^{-1/2}(N_k f)\mathbf{p}_k \tag{4.5}$$

of which the vectors $\mathbf{f}_d$ are simply the partial sums. Therefore

$$Q(f - f_d)^2 = \|\mathbf{f} - \mathbf{f}_d\|_{\mathbf{W}}^2 = \sum_{k=d+1}^{n-1} (N_k f)^2$$

which is non-increasing in d. From (4.4) it follows that $H_d f$ is also non-increasing in d. ∎

5 NUMERICAL EXAMPLES

We give numerical examples for linear and non-linear null rules for four functions: an entire function, a function with poles in the complex plane close to the integration interval, a function with an interior jump discontinuity in the first derivative, and a function with infinite value at one of the end points. In each case, Q is the 11 point Gauss-Legendre rule rule over $(-1, 1)$.

The purpose of these examples is not to present a new strategy for estimating errors, but to point out that the non-linear null rules defined above have significant advantages over linear null rules when used as building blocks for such a strategy. The four columns in each table are the following.

1. The linear null rules B_d of Berntsen and Espelid, which are orthogonal and of the same norm as Q in $\mathbb{R}^n$. Since the sign of $B_d f$ depends on the algorithm used to generate the orthogonal basis, we show $|B_d|$.

2. The linear null rules $N_{d+1} f$, which are $\mathbf{W}^{-1}$-orthonormal, are multiplied by $b_0 = \sqrt{2}$, which gives them the same $\mathbf{W}^{-1}$ norm as Q. For these null rules, we show the signs, which are not arbitrary.

3. The non-linear null rules $E_d f$ of (4.3).

4. The non-linear null rules $H_d f$ of (4.4).

Example 1
Integrand: $f(x) = \exp(x)$; actual error: $|If - Qf| = 9 \times 10^{-016}$ (i.e. at machine roundoff level).

d	$|B_d f|$	$\sqrt{b_0} N_{d+1} f$	$E_d f$	$H_d f$
0	1.68654950604058	1.27437176633797	1.12151712756238	1.31503970796580
1	0.40430392252036	0.32003888454899	0.27629185279866	0.32450891310937
2	0.06576759015832	0.05325945290008	0.04564592449842	0.05367631753447
3	0.00807541607854	0.00664341876591	0.00576914693052	0.00667665641212
4	0.00079488268854	0.00066307538339	0.00051584851143	0.00066537804758
5	0.00006516218042	0.00005516734142	0.00004854445140	0.00005530806580
6	0.00000456317921	0.00000393523061	0.00000330281609	0.00000394291549
7	0.00000027734464	0.00000024567499	0.00000021179472	0.00000024605404
8	0.00000001468788	0.00000001363557	0.00000001223083	0.00000001365251
9	0.00000000065427	0.00000000067994	0.00000000067103	0.00000000067994

All four null rule sequences behave in a similar way on this easiest of examples.

Example 2

Integrand: $f(x) = 1/(1 + 100x^2)$; actual error: $|If - Qf| = -0.07014347757568$.

d	$|B_d f|$	$\sqrt{b_0}N_{d+1}f$	$E_d f$	$H_d f$
0	0.00000000000000	0.00000000000000	0.44640472921741	0.65576128702730
1	0.32660672374274	−0.35251620923670	0.44640472921741	0.65576128702730
2	0.00000000000000	0.00000000000000	0.43214153836495	0.55295134305750
3	0.28490505261225	0.31125988721890	0.43214153836495	0.55295134305750
4	0.00000000000000	0.00000000000000	0.35996795935691	0.45702567805055
5	0.25493909552301	−0.28031635981901	0.35996795935691	0.45702567805055
6	0.00000000000000	0.00000000000000	0.31202087635484	0.36096427637009
7	0.23650170551631	0.26014620757453	0.31202087635484	0.36096427637009
8	0.00000000000000	0.00000000000000	0.24696004853825	0.25023820551621
9	0.24079138868836	−0.25023820551621	0.24696004853825	0.25023820551621

What Berntsen and Espelid call a "phase effect" can clearly be seen here: the linear null rules show zero coefficients because the integrand is an even function. The non-linear null rules give better-behaved sequences, although the effect is still there.

Example 3

Integrand: $f(x) = |x - 0.3|$; actual error: $|If - Qf| = 0.00459097043866$.

d	$|B_d f|$	$\sqrt{b_0}N_{d+1}f$	$E_d f$	$H_d f$
0	0.57606043156926	−0.50565834185326	0.59458410981475	0.71778866797969
1	0.56001786065261	0.46774938655800	0.43252390178779	0.50944107921747
2	0.17339131739874	0.16833542607286	0.16661287324567	0.20184331687941
3	0.05383931384166	−0.04201669955262	0.08322372242611	0.11137283734264
4	0.08295339832401	−0.08190292636191	0.08606377571517	0.10314313286135
5	0.01038913889482	−0.01551091070246	0.05219838057032	0.06269303398154
6	0.04120783225819	0.04037242553454	0.04884484173953	0.06074395574039
7	0.02519225861125	0.02861087122808	0.03970654965948	0.04538607072053
8	0.01780909151593	−0.01653316031528	0.03073904941826	0.03523227870886
9	0.02993765993357	−0.03111218528178	0.03070461111831	0.03111218528178

The linear null rules again show similar behavior. The absence of monotonicity would persuade the Berntsen-Espelid algorithm to diagnose this sequence as non-asymptotic, despite the fact that the coefficients show a decreasing tendency. The non-linear null rules E_d also suffer a loss of monotonicity. The null rules H_d, as they must, show monotonic decrease.

Example 4

Integrand: $f(x) = \log(1+x)$; exact error: $|If - Qf| = 0.00956147643807$.

d	$|B_d f|$	$\sqrt{b_0} N_{d+1} f$	$E_d f$	$H_d f$
0	2.57760497279556	1.71538949188603	1.44673030988730	1.94371983293663
1	1.11409609013247	−0.72358327767017	0.63052633466498	0.91404905780718
2	0.63963322656647	0.41471788281131	0.35745476906187	0.55849164752419
3	0.40841747382460	−0.26949548666125	0.24936611281177	0.37406148963878
4	0.27286597321144	0.18629759325899	0.18249192141209	0.25941121930248
5	0.18441819246581	−0.13234677200397	0.12893045266099	0.18051977023558
6	0.12261217054579	0.09403493217040	0.09061344819887	0.12276693115835
7	0.07749040416594	−0.06471013331858	0.06197927241796	0.07892497017894
8	0.04382449594087	0.04068469278620	0.03920665723835	0.04518572300670
9	0.01891755832833	−0.01965973897412	0.01940219352710	0.01965973897412

This example is another one where all four null rules behave in a similar way. It clearly shows the pitfalls of extrapolation on the order of the null rule, and the reason why pessimism factors are applied: the coefficients seem to decrease regularly, yet the last null rule is in each case only about twice as large as the true error.

6 CONCLUSIONS

We have shown that the null rules based on the expansion of the integrand in terms of the polynomials orthogonal with respect to the inner product induced by a quadrature rule Q are viable alternatives to the null rules orthogonal in $\mathbb{R}^n$ that have been used so far. From a purely practical point of view, they seem to produce similar results, but they have theoretical advantages. The first two points in the state of the art given in the introduction can now be modified to read as follows.

- These null rules are not only mutually $\mathbf{W}^{-1}$-orthogonal, but are also $\mathbf{W}^{-1}$-orthogonal to the quadrature rule Q.

- The information provided by these null rules can be used to construct actual approximation polynomials which lead to non-linear null rules that behave monotonically.

The practical application of the non-linear null rules requires more investigation. It is likely that the last of the three points will remain unchanged: "The final decision on how to extract an error estimate from the numbers produced by the null rules is made heuristically and with the aid of a substantial amount of numerical experimentation."

REFERENCES

[1] Berntsen J., Espelid T.O. Error estimation in automatic quadrature routines. *ACM Trans. Math. Software*, **17**:233–252, 1991.

[2] Clenshaw C.W., Curtis A.R. A method for numerical integration on an automatic computer. *Numer. Math.*, **2**:197–205, 1960.

[3] Davis P., Rabinowitz P. *Methods of Numerical Integration.* Academic Press, New York, second edition, 1984.

[4] de Boor C., Golub G.H. The numerically stable reconstruction of a Jacobi matrix from spectral data. *Lin. Alg. Appl.*, **21**:245–260, 1978.

[5] Espelid T.O., Sørevik T. A discussion of a new error estimate for adaptive quadrature. *BIT*, **29**:283–294, 1989.

[6] Filippi S. Angenäherte Tschebyscheff-Approximation einer Stammfunktion — eine Modifikation eines Verfahren von Clenshaw und Curtis. *Numer. Math.*, **6**:320–328, 1964.

[7] Gautschi W. On generating orthogonal polynomials. *SIAM J. Scient. Statist. Comput.*, **3**:289–317, 1982.

[8] Gautschi W. Questions of numerical condition related to polynomials. In *Studies in Numerical Analysis*, pages 140–177. Math. Assoc. Amer., 1984. G.H. Golub, ed.

[9] Golub G.H., Van Loan C. *Matrix Computations.* Johns Hopkins University Press, Baltimore, second edition, 1989.

[10] Gragg W.B., Harrod W.J. The numerically stable reconstruction of Jacobi matrices from spectral data. *Numer. Math.*, **44**:317–335, 1984.

[11] Kautsky J., Elhay S. Calculation of the weights of interpolatory quadratures. *Math. Comput.*, **40**:407–422, 1982.

[12] Laurie D.P. Sharper error estimates in adaptive quadrature. *BIT*, **23**:258–261, 1983.

[13] Piessens R., de Doncker-Kapenga E., Überhuber C.W., Kahaner D.K. *QUADPACK: A Subroutine Package for Automatic Integration.* Springer, Berlin, 1983.

[14] Reichel L. Construction of polynomials that are orthogonal with respect to a discrete bilinear form. *Adv. Computat. Math*, **1**:241–258, 1993.

[15] Sack R.A., Donovan A.F. An algorithm for Gaussian quadrature given modified moments. *Numer. Math.*, **18**:465–478, 1972.

[16] Wheeler J.C. Modified moments and Gaussian quadratures. *Rocky Mountain J. Math.*, **4**:287–296, 1974.

International Series of Numerical Mathematics, Vol. 119, ©1994 Birkhäuser

SOME DIFFERENTIAL EQUATIONS SATISFIED BY HYPERGEOMETRIC FUNCTIONS

Jean Letessier*   Galliano Valent*   Jet Wimp†

With regard and affection, this article is dedicated to Walter Gautschi on the occasion of his sixty-fifth birthday

Abstract. We derive some differential equations satisfied by hypergeometric functions, some of whose numerator parameters exceed some denominator parameters by unity. The result is a consequence of a general theory we establish concerning differential equations satisfied by combinations of derivatives of solutions of known differential equations.

Keywords: *differential equations, hypergeometric functions.*

1 INTRODUCTION

In this paper, we use the notation

$$
{}_pF_q \left(\begin{array}{c} a_1, a_2, \ldots, a_p \\ b_1, b_2, \ldots, b_q \end{array} ; x \right) \equiv {}_pF_q \left(\begin{array}{c} \{a_p\} \\ \{b_q\} \end{array} ; x \right) \tag{1.1}
$$

for the standard hypergeometric function with p numerator parameters and q denominator parameters. We always assume that the parameters are such that the function makes sense, i.e., that $b_j \neq 0, -1, -2, \ldots$; $j = 1, 2, \ldots q$. For the properties of this function, see the reference [3, chap.4]. The series converges for all x when $p \leq q$, and for $|x| < 1$ when $p = q + 1$. The series diverges for all non-zero x when $p > q + 1$ unless it terminates, as it does when one of the a_j is zero or a negative integer. In this paper, we are interested only in the convergent cases.

It is shown in the cited reference that the function (1.1) satisfies a linear homogeneous differential equation of order $\rho = \max\{p, q + 1\}$,

$$
\{\delta(\delta + b_1 - 1) \ldots (\delta + b_q - 1) - x(\delta + a_1) \ldots (\delta + a_p)\}y = 0 \tag{1.2}
$$

where δ denotes the operator $x\, d/dx$.

*Laboratoire de Physique Théorique et Hautes Energies, Unité associée au CNRS UA 280, Université PARIS 7, Tour 24, 5è ét., Place Jussieu, F-75251 CEDEX 05

†Department of Mathematics and Computer Sciences, Drexel University, Philadelphia, PA 19104, USA

Investigations by several authors have shown that in special cases, the order of the differential equation may reduce. Hendriksen and van Rossum [4] showed that

$$
{}_2F_2 \left(\begin{array}{c} s, r+1 \\ 2, r \end{array} ; x \right)
\tag{1.3}
$$

satisfies a second, rather than a third order differential equation. In [6, 7], Letessier encountered the two functions

$$
{}_3F_2 \left(\begin{array}{c} a, b, e+1 \\ c, e \end{array} ; x \right), \qquad {}_2F_2 \left(\begin{array}{c} a, e+1 \\ c, e \end{array} ; x \right).
\tag{1.4}
$$

He was able to show that both these functions satisfied differential equations of order *two*, one less than expected.

Even more surprising were some cases studied in [13] and [5]. Both of those authors treated polynomials of the form

$$
{}_4F_3 \left(\begin{array}{c} -n, n+\lambda, e_1+1, e_2+1 \\ b, e_1, e_2 \end{array} ; x \right).
\tag{1.5}
$$

Those authors found that these polynomials also satisfied differential equations of order two, *two* less than expected.

The purpose of this paper is to generalize these various cases and to apply our general theory to some interesting special hypergeometric functions.

Our convention for sums and products in this paper is that empty products are interpreted as unity, empty sums as zero.

2   GENERAL THEORY

Theorem 1 *Let f satisfy a homogeneous linear differential equation of exact order s,*

$$
\mathcal{L}\{f\} = 0
$$
$$
\mathcal{L} \equiv \mathcal{L}(D) = \sum_{m=0}^{s} \mu_m D^m, \quad \mu_s \neq 0.
\tag{2.1}
$$

Let

$$
g = Xf' + cf, \quad X = ax + b,
\tag{2.2}
$$

a, b, c, general parameters. Then g satisfies the homogeneous linear differential equation of order s,

$$
\mathcal{K}\{g\} = 0
$$
$$
\mathcal{K} = \sum_{m=0}^{s} \rho_m D^m,
\tag{2.3}
$$

where

$$\rho_s = \mu_s X^{c/a-1} \mathcal{L}\{X^{-c/a}\}. \tag{2.4}$$

Furthermore, the equation reduces or becomes trivial ($\rho_s \equiv 0$) for at most a finite number of values of c.

Proof: Define

$$\begin{aligned} R &= X^{c/a} \mathcal{L}\{X^{-c/a}\}, \\ S &= X^{c/a} D\, \mathcal{L}\{X^{-c/a}\} = R' - \frac{c}{X} R, \end{aligned} \tag{2.5}$$

and

$$l_k = \frac{(-1)^k}{X^k} \prod_{j=0}^{k-1} c_j, \qquad c_j = c + ja. \tag{2.6}$$

An easy computation shows that for any operator

$$\mathcal{M}(D) = \sum_{m=0}^{M} \tau_m D^m, \tag{2.7}$$

we have

$$\sum_{m=0}^{M} \tau_m l_m = X^{c/a} \mathcal{M}\{X^{-c/a}\}. \tag{2.8}$$

Define the operator $\mathcal{L}^*$ by

$$\mathcal{L}^*(D) = D\, \mathcal{L}(D) = \sum_{m=0}^{s+1} \sigma_m D^m. \tag{2.9}$$

We find

$$\begin{aligned} \sigma_m &= \mu_{m-1} + \mu'_m, \qquad m = 0,1,2,\ldots,s+1, \\ \mu_{-1} &= \mu_{s+1} = 0. \end{aligned} \tag{2.10}$$

Both $\mathcal{L}^*(D), \mathcal{L}(D)$ annihilate f, i.e.,

$$\sum_{m=0}^{s} \mu_m f^{(m)} = \sum_{m=0}^{s+1} \sigma_m f^{(m)} = 0. \tag{2.11}$$

Now differentiate (2.2) repeatedly to get

$$\begin{aligned} g_r &= X f_{r+1} + c_r f_r, \qquad r = 0,1,2,\ldots, \\ g_r &= g^{(r)}, \qquad f_r = f^{(r)}. \end{aligned} \tag{2.12}$$

We consider (2.12) as a difference equation in f_r. Solving in the usual manner gives

$$f_r = l_r \left\{ f - \sum_{m=0}^{r-1} \frac{g_m}{c_m l_m} \right\}, \qquad r = 0,1,2,\ldots. \tag{2.13}$$

373

Multiply (2.13) by μ_r and sum from $r = 0$ to s, then multiply (2.13) by σ_r and sum from $r = 0$ to $s + 1$. Using $(2.5), (2.7), (2.8), (2.11)$ and interchanging the order of summation, we get the two equations

$$R f = \sum_{m=0}^{s-1} \frac{g_m}{c_m l_m} \sum_{j=m}^{s-1} \mu_{j+1} l_{j+1} , \tag{2.14}$$

$$S f = \sum_{m=0}^{s} \frac{g_m}{c_m l_m} \sum_{j=m}^{s} \sigma_{j+1} l_{j+1} . \tag{2.15}$$

Eliminating f from these two equations gives the differential equation (2.3) with the values of the coefficients,

$$\rho_m = \frac{1}{c_m l_m} \left\{ S \sum_{j=m}^{s-1} \mu_{j+1} l_{j+1} - R \sum_{j=m}^{s} \sigma_{j+1} l_{j+1} \right\} . \tag{2.16}$$

Setting $m = s$ and using (2.10) gives the value of the leading coefficient (2.4) . Written out, this coefficient is

$$\rho_s = \frac{\mu_s}{X} \sum_{m=0}^{s} \frac{\mu_m (-1)^m \prod_{j=0}^{m-1} c_j}{X^m} . \tag{2.17}$$

This is a polynomial of degree $s - 1$ in c. For general a and b, it cannot vanish identically since the leading coefficient is non-zero. The theorem is proved. ∎

Note that we can also obtain a simple formula for ρ_0, namely,

$$\rho_0 = \frac{1}{c} \left(R \mu_0' - S \mu_0 \right) . \tag{2.18}$$

We can give a more general result, but it requires a different sort of proof, and does not yield an explicit formula for the coefficients in the differential equation.

Theorem 2 *Let*

$$g = c f + \sum_{j=1}^{l} \xi_j f^{(j)}, \tag{2.19}$$

where the ξ_j are differentiable functions of x, and f satisfies the differential equation (2.1). Then g satisfies a similar differential equation of order s whose leading coefficient vanishes for at most a finite number of values of c.

Proof: As before, differentiate equation (2.19) repeatedly for $1, 2, \ldots, s$ times. Replace each $f^{(j)}$ in this system where $j \geq s$ by its value in terms of $f, f', \ldots, f^{(s-1)}$. This gives $s + 1$ equations in the s unknown $f, f', \ldots, f^{(s-1)}$. The augmented determinant of this system, whose left hand column is $[g, g', \ldots, g^{(s)}]^T$, must vanish.

Expanding this determinant by minors of the first column and equating the result to zero gives the differential equation.

It is easy to see that the coefficient of $g^{(s)}$ in this system is a polynomial in c and of the form $(-1)^s\{c^s + O(c^{s-1})\}, c \to \infty$. Thus it can vanish identically in x for at most a finite number of values of c. ∎

One can be even more general. Let

$$g = \sum_{j=0}^{m} \xi_j \prod_{m=1}^{r_j} f_m^{(k_m)}, \tag{2.20}$$

where the ξ_j are differentiable functions of x, and f_i satisfies a homogeneous differential equation of order s_i. Then g will satisfy a differential equation of order at most $s_1 \cdot s_2 \cdot s_3 \ldots s_\tau, \tau = \max r_j$. However in this general setting, the order of the equation for g may reduce if the the differential equation for the f_i are reducible, or not distinct, or if the ξ_j are themselves solutions of the equations for the f_i. One just has to construct the equation and see what happens. For example, it is well known that if f, g satisfy differential equations of order 2, their product satisfies a differential equation of order 4, a fact that probably goes back to Orr[10]. However, if the differential equations are the same, their product satisfies an equation of order only 3.

3 DIFFERENTIAL EQUATIONS FOR HYPERGEOMETRIC FUNCTIONS

Define

$$g_r = {}_{p+r}F_{q+r}\left(\begin{array}{c} \{a_p\}, \{e_r + 1\} \\ \{b_q\}, \{e_r\} \end{array} ; x \right). \tag{3.1}$$

Theorem 3 *The function g_r satisfies a linear homogeneous differential equation of exact degree $\rho = \max\{p, q + 1\}$.*

Proof: The proof is by induction on r. We have

$$g_r = \frac{x}{e_r} g'_{r-1} + g_{r-1}, \qquad r = 1, 2, \ldots . \tag{3.2}$$

At the rth stage, $a = 1/e_r, b = 0, c = 1, s = \rho = \max\{p, q + 1\}$. The leading coefficient cannot vanish if $e_j \neq -1, -2, \ldots, -(s-1), j = 1, 2, \ldots, r$. This condition is already assured by our convention on hypergeometric functions. The proof is completed by observing the triviality of the case $r = 0$. ∎

Remark. The conclusion of Theorem 3 also prevails when

$$g_r = {}_{p+r}F_{q+r}\left(\begin{array}{c} \{a_p\}, \{e_r + m_r\} \\ \{b_q\}, \{e_r\} \end{array} ; x \right), \tag{3.3}$$

the m_j being positive integers. The proof is a corollary of Theorem 2. ∎

In certain prominent cases, it is desirable to have a closed form expression for the coefficients of the differential equation, or at least, a more manageable way of computing them than offered by Theorem 1. We will treat the situation

$$g = {}_{p+1}F_p\left(\begin{array}{c}\{a_p\}, e+1 \\ \{b_{p-1}+1\}, e\end{array}; x\right). \tag{3.4}$$

(The results are more æsthetically pleasing if we work with the parameter set $\{b_{p-1}+1\}$ rather than $\{b_{p-1}\}$.) We define the symmetric functions $S_l^{(p)}$ by

$$\begin{aligned}
\prod_{i=1}^{p}(n+a_i) &= \sum_{i=0}^{p}S_l^{(p)}(\{a_p\})n^{p-l}, \\
S_0^{(p)}(\{a_p\}) &= 1, \\
S_l^{(p)}(\{a_p\}) &= \prod_{i=1}^{p}a_i.
\end{aligned} \tag{3.5}$$

For convenience, define

$$_{p+1}F_p\left(\begin{array}{c}\{a_p\}, e+1 \\ \{b_{p-1}+1\}, e\end{array}; x\right) = \sum_{n=0}^{\infty}C_n x^n. \tag{3.6}$$

Lemma 1 We have the three-term identity

$$-e(n+1)\prod_{i=1}^{p-1}(b_i-e)\prod_{i=1}^{p-1}(b_i+n+1)C_{n+1} + \prod_{i=1}^{p}(a_i-e)\prod_{i=1}^{p}(a_i+n-1)C_{n-1}$$

$$+\left\{\prod_{i=1}^{p}(a_i-e)\left[-n\prod_{i=1}^{p-1}(b_i+n)+\sum_{k=1}^{p-1}n^k\sum_{l=0}^{p-1-k}S_l^{(p-1)}(\{b_{p-1}\})(-e)^{p-1-k-l}\right]\right.$$

$$+e\prod_{i=1}^{p-1}(b_i-e)\left[\prod_{i=1}^{p}(a_i+n)+\sum_{k=1}^{p}n^k\sum_{l=0}^{p-k}S_l^{(p)}(\{a_p\})(-e)^{p-1-k-l}+\frac{1}{e}\prod_{i=1}^{p}a_i\right]\right\}C_n$$

$$= 0, \tag{3.7}$$

for $n = 0, 1, 2, \ldots$, with $C_{-1} = 0, C_0 = 1$.

Proof: We can write

$$\begin{aligned}
\frac{C_{n+1}}{C_n} &= \frac{(n+e+1)\prod_{i=1}^{p}(a_i+n)}{(n+1)(n+e)\prod_{i=1}^{p-1}(b_i+n+1)}, \\
\frac{C_{n-1}}{C_n} &= \frac{n(n+e-1)\prod_{i=1}^{p-1}(b_i+n)}{(n+e)\prod_{i=1}^{p}(a_i+n-1)}.
\end{aligned} \tag{3.8}$$

376

Inserting these expressions into (3.7) and multiplying by a factor of $(n + e)$ gives

$$
-e(n + e + 1) \prod_{i=1}^{p-1}(b_i - e) \prod_{i=1}^{p}(a_i + n) + n(n + e - 1) \prod_{i=1}^{p}(a_i - e) \prod_{i=1}^{p-1}(b_i + n)
$$

$$
+ (n+e) \left\{ \prod_{i=1}^{p}(a_i - e) \left[-n \prod_{i=1}^{p-1}(b_i + n) + \sum_{k=1}^{p-1} n^k \sum_{l=0}^{p-1-k} S_l^{(p-1)}(\{b_{p-1}\})(-e)^{p-1-k-l} \right] \right.
$$

$$
\left. + \prod_{i=1}^{p-1}(b_i - e) \left[-e \prod_{i=1}^{p}(a_i + n) + e \sum_{k=1}^{p} n^k \sum_{l=0}^{p-k} S_l^{(p)}(\{a_p\})(-e)^{p-1-k-l} + \prod_{i=1}^{p} a_i \right] \right\}.
$$

$$
\tag{3.9}
$$

After a little rewriting, this becomes

$$
\prod_{i=1}^{p}(a_i - e) \left\{ -n \prod_{i=1}^{p-1}(b_i + n) + (n + e) \sum_{k=1}^{p-1} n^k \sum_{l=0}^{p-1-k} S_l^{(p-1)}(\{b_{p-1}\})(-e)^{p-1-k-l} \right\}
$$

$$
+ \prod_{i=1}^{p-1}(b_i - e) \left\{ -e \prod_{i=1}^{p}(a_i + n) + (n + e) \prod_{i=1}^{p} a_i \right.
$$

$$
\left. + e(n + e) \sum_{k=1}^{p} n^k \sum_{l=0}^{p-k} S_l^{(p)}(\{a_p\})(-e)^{p-1-k-l} \right\}. \tag{3.10}
$$

To simplify the quantity inside the first set of brackets in (3.10) we expand the difference

$$
n \left[\prod_{i=1}^{p-1}(b_i + n) - \prod_{i=1}^{p-1}(b_i - e) \right], \tag{3.11}
$$

using (3.5) to get

$$
n \sum_{l=0}^{p-1} S_l^{(p-1)}(\{b_{p-1}\}) \left[n^{p-l-1} - (-e)^{p-l-1} \right]. \tag{3.12}
$$

Factoring out an $(n + e)$ and reversing the order of summation gives

$$
(n + e) \sum_{k=1}^{p-1} n^k \sum_{l=0}^{p-k-1} S_l^{(p-1)}(\{b_{p-1}\})(-e)^{p-1-k-l}. \tag{3.13}
$$

Thus we have

$$
(n + e) \sum_{k=1}^{p-1} n^k \sum_{l=0}^{p-k-1} S_l^{(p-1)}(\{b_{p-1}\})(-e)^{p-1-k-l} - n \prod_{i=1}^{p-1}(b_i + n) = n \prod_{i=1}^{p-1}(b_i - e).
$$

$$
\tag{3.14}
$$

Inserting this and the corresponding expression involving the parameters $\{a_p\}$ into (3.10) shows that the expression is 0, and the lemma follows. ∎

Theorem 4 *The hypergeometric function* $g = {}_{p+1}F_p\left(\begin{array}{c}\{a_p\}, e+1 \\ \{b_{p-1}+1\}, e\end{array}; x\right)$ *satisfies the pth order differential equation*

$$
\left[\prod_{i=1}^{p}(a_i - e)\left\{x\prod_{i=1}^{p}(\delta + a_i) - \delta\prod_{i=1}^{p-1}(\delta + b_i)\right.\right.
$$

$$
\left.+ \sum_{k=1}^{p-1}\delta^k\sum_{l=0}^{p-1-k}S_l^{(p-1)}(\{b_{p-1}\})(-e)^{p-1-k-l}\right\}
$$

$$
+e\prod_{i=1}^{p-1}(b_i - e)\left\{\prod_{i=1}^{p}(\delta + a_i) - \frac{1}{x}\delta\prod_{i=1}^{p-1}(\delta + b_i)\right.
$$

$$
\left.\left.+ \sum_{k=1}^{p}\delta^k\sum_{l=0}^{p-k}S_l^{(p)}(\{a_p\})(-e)^{p-1-k-l} + \frac{1}{e}\prod_{i=1}^{p}a_i\right\}\right]g = 0. \tag{3.15}
$$

Proof: Multiply the identity (3.7) by x^n and sum from $n = 0$ to ∞. The first term gives

$$
-e\prod_{i=1}^{p-1}(b_i - e)\sum_{n=0}^{\infty}(n+1)\prod_{i=1}^{p-1}(b_i + n + 1)C_{n+1}x^n
$$

$$
= -\frac{e}{x}\prod_{i=1}^{p-1}(b_i - e)\sum_{n=0}^{\infty}n\prod_{i=1}^{p-1}(b_i + n)C_n x^n \tag{3.16}
$$

$$
= -\frac{e}{x}\prod_{i=1}^{p-1}(b_i - e)\delta\prod_{i=1}^{p-1}(\delta + b_i)g.
$$

The second term gives

$$
\prod_{i=1}^{p}(a_i - e)\sum_{n=0}^{\infty}\prod_{i=1}^{p}(a_i + n - 1)C_{n-1}x^n
$$

$$
= x\prod_{i=1}^{p}(a_i - e)\sum_{n=0}^{\infty}\prod_{i=1}^{p}(a_i + n)C_n x^n \tag{3.17}
$$

$$
= \prod_{i=1}^{p}(a_i - e)x\prod_{i=1}^{p}(\delta + a_i)g.
$$

The term involving C_n is equal to

$$\sum_{n=0}^{\infty}\{\ldots\}C_n x^n =$$

$$\prod_{i=1}^{p}(a_i - e)\left[-\delta\prod_{i=1}^{p-1}(\delta + b_i) + \sum_{k=1}^{p-1}\delta^k\sum_{l=0}^{p-1-k}S_l^{(p-1)}(\{b_{p-1}\})(-e)^{p-1-k-l}\right]g$$

$$+ e\prod_{i=1}^{p-1}(b_i - e)\left[\prod_{i=1}^{p}(\delta + a_i) + \sum_{k=1}^{p}\delta^k\sum_{l=0}^{p-k}S_l^{(p)}(\{a_p\})(-e)^{p-1-k-l} + \frac{1}{e}\prod_{i=1}^{p}a_i\right]g.$$

$$(3.18)$$

The last three equations prove the theorem. ∎

It is well known that the differential equation for $_pF_q\left(\begin{array}{c}\{a_p\}\\\{b_q + 1\}\end{array}; x\right)$ has singular points at 0 and ∞ when $p \neq q + 1$, and an additional singular point at 1 when $p = q + 1$. In the former case, the point 0 is a regular singular point and ∞ is an irregular singular point. In the latter case, all the singular points are regular. A strange feature of the equation for g is that it possesses not only the expected singular points $0, 1, \infty$, but an *additional singular point*. We do not yet understand the significance of this. The coefficient of $y^{(p)}$ in the above differential equation is easily seen to be

$$x^{p-1}(x - 1)\left[\prod_{i=1}^{p}(a_i - e)x + e\prod_{i=1}^{p-1}(b_i - e)\right].$$

$$(3.19)$$

The rogue singular point, of course, is the zero of the second linear factor. When e is allowed to approach ∞, g approaches $_pF_{p-1}\left(\begin{array}{c}\{a_p\}\\\{b_{p-1} + 1\}\end{array}; x\right)$ and the singular point disappears, or rather, it coincides with the singular point at 0. The method of Frobenius will yield simple expansions about this singular point analogous to the known expansions about $0, 1$, and ∞. (Meijer, in a series of papers [8] gave the expansions about ∞. Nørlund [9] and Bühring [1], [2] have worked on the vastly more difficult problem of deriving expansions about 1.)

We consider the important special case $p = 2$. The differential equation satisfied by

$$y = {}_3F_2\left(\begin{array}{c}a_1, a_2, e + 1\\b + 1, e\end{array}; x\right),$$

$$(3.20)$$

is

$$x(x - 1)\Big\{(a_1 - e)(a_2 - e)x + e(b - e)\Big\}y'' + \Big\{(a_1 - e)(a_2 - e)(a_1 + a_2 + 1)x^2$$

$$+\Big[e(a_1 + a_2 + 1)(2b - e) - (b + 1)(a_1 a_2 + e^2) + a_1 a_2\Big]x + (b + 1)e(e - b)\Big\}y'$$

$$+ a_1 a_2\Big\{(a_1 - e)(a_2 - e)x + (e + 1)(b - e)\Big\}y = 0.$$

$$(3.21)$$

This equation may be written as a Heun-type equation,

$$y'' + \left[\frac{\gamma}{x} - \frac{\delta}{1-x} - \frac{\varepsilon k^2}{1-k^2 x}x\right]y' + \frac{\alpha\beta k^2 x + s}{x(1-x)(1-k^2 x)}y = 0$$

(3.22)

$$\alpha + \beta = \gamma + \delta + \varepsilon + 1,$$

where the coefficients are given by

$$\alpha = a_1, \beta = a_2, \gamma = b+1, \delta = a_1 + a_2 - b + 1, \varepsilon = -1$$

$$k^2 = \frac{(a_1-e)(a_2-e)}{e(e-b)}, s = -a_1 a_2 \frac{e+1}{e}$$

(3.23)

Heun's equation is of growing importance in the theory of orthogonal polynomials, (see [11], [12]). When one seeks Frobenius-type expansions around the singular point $x = 0$, one finds that the coefficients in the expansions satisfy three-term recurrence relations, and these coefficients turn out to be orthogonal polynomials (in some special parameter). A moment problem which is indeterminate appears in certain fourth order differential equations which generalize the Heun equation, and the orthogonal polynomials which occur in these problems do not have a unique measure. The polynomials are related to quadratic birth and death processes. See the cited references for more information.

The differential equation for the function $g = {}_{p+1}F_q \left(\begin{matrix} \{a_p\}, e+1 \\ \{b_{q-1}+1\}, e \end{matrix}; x\right), q > p$, is easily obtained from the equation (3.15) by confluence. For instance, to obtain the differential equation for ${}_pF_q \left(\begin{matrix} \{a_{p-1}\}, e+1 \\ \{b_{q-1}+1\}, e \end{matrix}; x\right)$, one replaces x by x/a_p and lets $a_p \to \infty$. We leave the details to the reader.

Note added in proof. A referee has pointed out that ${}_{p+1}F_{q+1}$ with $a_1 = 1 + b_1$ can be expressed as a sum $AF_1 + BxF_2$, where F_1 and F_2 are ${}_pF_q$'s, and thus is a linear combination of solutions of different differential equations, each of order $max\{p, q+1\}$. However, it is surprising that the linear combination satisfies an equation of the same order, since the hypergeometric functions involved are different.

REFERENCES

[1] Bühring W. The behavior at unit argument of the hypergeometric function ${}_3F_2$. *SIAM J. Math. Anal.*, **18**:1227-1234, 1987.

[2] Bühring W. Generalized hypergeometric functions at unit argument. *Proc. Amer. Math. Soc.*, **114**:145-153, 1992.

[3] Erdélyi A. (ed). *Higher Transcendental Functions*. 3 v., McGraw-Hill, New York, Toronto, London, 1953.

[4] Hendriksen E., van Rossum H. A Padé-type approach to non-classical orthogonal polynomials. *J. Math. Anal. Appl.*, **106**:237-248, 1985.

[5] Koornwinder Tom H. Orthogonal polynomials with weight functions $(1 - x)^\alpha (1+x)^\beta + M\delta(x+1) + N\delta(x-1)$. *Can. Math. Bull.*, **27** (2):205-214, 1984.

[6] Letessier Jean. Some results on co-recursive associated Laguerre and Jacobi polynomials. *J. Comp. Appl. Math.*, **49**:127–136, 1993.

[7] Letessier Jean. Co-recursive associated Jacobi polynomials. To appear in *J.Comp. Appl. Math.*

[8] Meijer C.S. On the G-function. *Proc. Kon. Akad. v. Wetensch.*, **49**, 1946; 7 papers on the subject appear in this volume.

[9] Nørlund N. Hypergeometric functions. *Acta Math.*, **94**:289-349, 1955.

[10] Orr W. McF. On the product $J_m(x)J_n(x)$. *Proc. Camb. Philos. Soc.*, **10**:93-100, 1900.

[11] Valent G. An integral transform involving Heun functions and a related eigenvalue problem. *SIAM J. Math. Anal.*, **17**:688-703, 1986.

[12] Valent G. Associated Stieljes-Carlitz polynomials and a generalization of Heun's differential equation. To appear.

[13] Wimp Jet, Kiesel H. Non-linear recurrence relations and some derived orthogonal polynomials. To appear.

International Series of Numerical Mathematics, Vol. 119, ©1994 Birkhäuser

ON MEAN CONVERGENCE OF LAGRANGE-KRONROD INTERPOLATION

Shikang Li*

Dedicated to my mentor Walter Gautschi on his 65th birthday

Abstract. It is known that Lagrange interpolation to any continuous function on $[-1, 1]$ at the zeros of an orthogonal polynomial converges in the mean. In this paper we study mean convergence of Lagrange-Kronrod interpolation where the nodes are the n zeros of an orthogonal polynomial of degree n and the $n + 1$ zeros of the associated Stieltjes polynomial of degree $n + 1$. A sufficient condition for mean convergence, due to P. Erdös and P. Turán, is shown to hold for the first-, the third- and fourth-kind Chebyshev weight functions.

1 INTRODUCTION

Let $\pi_n(\, \cdot \, ; w)$, $n \geq 1$, denote the nth-degree orthogonal polynomial on $[-1, 1]$ with respect to a positive weight function w. For a given positive integer n, one can construct the Stieltjes polynomial $\hat{\pi}_{n+1}(\cdot) = \pi_{n+1}(\, \cdot \, ; \, \pi_n w)$ of degree $n + 1$ orthogonal to all lower-degree polynomials with respect to the weight function $\hat{w}_n = \pi_n w$. This kind of polynomial was first considered by Stieltjes for the weight function $w \equiv 1$. Many years later, Szegö [9] generalized the problem to construct orthogonal polynomials relative to the Gegenbauer weight function $w(x) = (1 - x^2)^{\mu - 1/2}$, $\mu > \frac{1}{2}$, and showed that the zeros $\hat{\tau}_j = \hat{\tau}_j^{(n)}$ of $\hat{\pi}_{n+1}$ are simple, inside $(-1, 1)$, and interlace with those of π_n whenever $0 \leq \mu \leq 2$. The zeros of π_n are denoted by $\tau_i = \tau_i^{(n)}$. In 1964 Kronrod [6], [7] introduced an extended quadrature formula of the form

$$\int_{-1}^{1} f(x) \, dx \approx \sum_{i=1}^{n} \lambda_i^{(n)} f(\tau_i^{(n)}) + \sum_{j=1}^{n+1} \mu_j^{(n)} f(\hat{\tau}_j^{(n)}). \tag{1.1}$$

This formula was again generalized to integrals with an arbitrary positive weight function. One can view (1.1) as an approximation of $\int_{-1}^{1} f(x) \, dx$ by the integral of the polynomial $(\hat{L}_{2n+1} f)(\cdot)$ of degree less than or equal to $2n$ interpolating f on the union of the nodes $\{\tau_i\}$ and $\{\hat{\tau}_j\}$. Clearly this can be generalized to any positive weight function. There are many papers in the literature on Gauss-Kronrod extension formulas (see [4]). Less attention has been given to the extended interpolation

*Department of Mathematics, Southeastern Louisiana University, Hammond, Louisiana 70402, USA; fmat1627@selu.edu

itself. The Gauss-Kronrod extension formula was introduced to estimate the error of the corresponding Gauss quadrature formula. The same purpose can be served by the extended interpolation for Lagrange interpolation. Recently, A. Bellen [1] and W. Gautschi [5] have studied mean convergence of Lagrange interpolation on an extended set of nodes that includes, in addition to the n zeros of the orthogonal polynomial π_n of degree n relative to some positive weight function w, other $n+1$ nodes, which are the zeros of the orthogonal polynomial $\hat{\pi}_{n+1}$ of degree $n+1$ corresponding to the weight function $\hat{w}_n = \pi_n^2 w$ (not $\hat{w}_n = \pi_n w$, as in Gauss-Kronrod quadrature). Bellen developed a sufficient criterion for mean convergence of this kind of extended Lagrange interpolation. Gautschi showed that this condition fails for Chebyshev weight functions of the first, third and fourth kind. In this paper we will investigate the convergence of the proposed Lagrange-Kronrod interpolation of a continuous function f, that is, to see if $\lim_{n\to\infty} \|f - \hat{L}_{2n+1}(f)\|_w = 0$ for all $f \in C[-1, 1]$. A sufficient condition for convergence in the mean can be observed from the proof of Theorem Ia in Erdös and Turán's article [3] in terms of the function

$$L(n; w) := \sum_{i=1}^{n}\sum_{k=1}^{n}\left|\int_{-1}^{1} l_i(t)l_k(t)w(t)\,dt\right| + \sum_{i=1}^{n}\sum_{k=1}^{n+1}\left|\int_{-1}^{1} l_i(t)\hat{l}_k(t)w(t)\,dt\right|$$
$$+ \sum_{i=1}^{n+1}\sum_{k=1}^{n+1}\left|\int_{-1}^{1} \hat{l}_i(t)\hat{l}_k(t)w(t)\,dt\right|, \qquad (1.2)$$

where l_i and $\hat{l}_k$ are the elementary Lagrange interpolation polynomials for the point set $\{\tau_i\} \cup \{\hat{\tau}_j\}$,

$$l_i(t) = \frac{\pi_n(t)\hat{\pi}_{n+1}(t)}{(t - \tau_i)\pi_n'(\tau_i)\hat{\pi}_{n+1}(\tau_i)}, \quad i = 1, 2, \ldots, n, \qquad (1.3)$$

$$\hat{l}_k(t) = \frac{\hat{\pi}_{n+1}(t)\pi_n(t)}{(t - \hat{\tau}_k)\hat{\pi}_{n+1}'(\hat{\tau}_k)\pi_n(\hat{\tau}_k)}, \quad k = 1, 2, \ldots, n + 1. \qquad (1.4)$$

We have mean convergence,

$$\lim_{n\to\infty} \|f - \hat{L}_{2n+1}(f)\|_w = 0, \quad \text{for all } f \in C[-1, 1],$$

of the extended interpolation if $\{L(n; w)\}$ is uniformly bounded.

In the case of the second-kind Chebyshev weight function $w(t) = (1 - t^2)^{1/2}$, $-1 < t < 1$, the interpolation polynomial $\hat{L}_{2n+1}(f)$ is the same as the one considered by Bellen and Gautschi for this weight function. They have shown that mean convergence of the extended Lagrange interpolation holds for this case. Here, mean convergence is shown to hold for the extended Lagrange-Kronrod interpolation polynomial for the first-kind and third-kind Chebyshev weight functions in Section 2 and 3, respectively. Mean convergence for the extended Lagrange interpolation polynomial corresponding to the fourth-kind Chebyshev weight function is related to that of the extended Lagrange interpolant associated with the third-kind Chebyshev weight function.

2 FIRST-KIND CHEBYSHEV WEIGHT FUNCTION

In this section we consider the extended Lagrange-Kronrod interpolation of a continuous function f on the interval $[-1, 1]$ using the nodes $\{\tau_i\}$ and $\{\hat{\tau}_j\}$, where τ_i are the zeros of the nth-degree orthogonal polynomial π_n relative to the weight function $w_1(t) = (1 - t^2)^{-1/2}$, $-1 < t < 1$, and $\hat{\tau}_j$ are the zeros of the orthogonal polynomial $\hat{\pi}_{n+1}$ of degree $n + 1$ relative to the weight function $w = \pi_n w_1$. In this case there are explicit formulas for π_n and $\hat{\pi}_{n+1}$ (see [8]). Their zeros also can be determined exactly. We state the respective results in the following two lemmas.

Lemma 2.1 *The nth-degree orthogonal polynomial π_n relative to the first-kind Chebyshev weight function is given by*

$$\pi_n(t) = T_n(t), \quad T_n(\cos\theta) = \cos n\theta. \tag{2.1}$$

The Stieltjes polynomial $\hat{\pi}_{n+1}$ of degree $n + 1$ orthogonal to all polynomials of lower degree with respect to the weight function $w = \pi_n w_1$ can be expressed as

$$\hat{\pi}_{n+1}(t) = T_{n+1}(t) - T_{n-1}(t) = 2(t^2 - 1)U_{n-1}(t) \quad for \ n \geq 2, \tag{2.2}$$

where $U_{n-1}(\cos\theta) = \sin n\theta / \sin\theta$ and $U_{n-1}(t)$ is the second-kind Chebyshev polynomial of degree $n - 1$.

Lemma 2.2 *The zeros τ_i of π_n are given by*

$$\tau_i = \cos\theta_i, \quad \theta_i = \frac{(2i - 1)\pi}{2n}, \quad i = 1, 2, \ldots, n, \tag{2.3}$$

the zeros $\hat{\tau}_k$ of $\hat{\pi}_{n+1}$ by

$$\hat{\tau}_k = \cos\hat{\theta}_k, \quad \hat{\theta}_k = \frac{(k - 1)\pi}{n}, \quad k = 1, 2, \ldots, n + 1. \tag{2.4}$$

Moreover, the nodes $\{\tau_i\}$ and $\{\hat{\tau}_k\}$ interlace.

In the proof of the following theorem we also use the n-point Gauss-Chebyshev quadrature formula (see [2])

$$\int_{-1}^{1} f(t)(1 - t^2)^{-1/2}\, dt = \frac{\pi}{n}\sum_{i=1}^{n} f(\tau_i) + \frac{f^{(2n)}(\xi)}{(2n)!}\int_{-1}^{1} \omega_n^2(t)(1 - t^2)^{-1/2}\, dt, \tag{2.5}$$

where $\omega_n^2(t) = \prod_{i=1}^{n}(t - \tau_i)$ is the nodal polynomial (the monic Chebyshev polynomial of the first kind) and ξ is a number that lies in the smallest interval which contains the nodes τ_i. It can be shown without difficulty that

$$\int_{-1}^{1} \omega_n^2(t)(1 - t^2)^{-1/2}\, dt = \frac{\pi}{2^{2n-1}}, \quad n \geq 1. \tag{2.6}$$

Theorem 2.3 *The sequence $\{L(n;\ w)\}$ for the interpolation of a continuous function on $[-1, 1]$ using the nodes τ_i and $\hat{\tau}_k$, when w is the first-kind Chebyshev weight, is uniformly bounded. In fact,*

$$L(n;\ w) = \frac{(11n^2 - 4n + 1)\pi}{8n^2}, \quad n \geq 2. \tag{2.7}$$

As a consequence of Theorem 2.3 and the sufficient condition for mean convergence of the interpolant mentioned in Section 1, mean convergence holds for the interpolant corresponding to the first-kind Chebyshev weight function.

Proof of Theorem 2.3: We will proceed to evaluate the three integrals in (1.2) for the first-kind Chebyshev weight function. In this case the fundamental extended Lagrange polynomials can be expressed by

$$l_i(t) = \frac{T_n(t)(T_{n+1}(t) - T_{n-1}(t))}{(t - \tau_i)T_n'(\tau_i)(T_{n+1}(\tau_i) - T_{n-1}(\tau_i))}, \quad i = 1, 2, \ldots, n, \tag{2.8}$$

$$\hat{l}_k(t) = \frac{T_n(t)(T_{n+1}(t) - T_{n-1}(t))}{(t - \hat{\tau}_k)(T_{n+1}'(\hat{\tau}_k) - T_{n-1}'(\hat{\tau}_k))T_n(\hat{\tau}_k)}, \quad k = 1, 2, \ldots, n+1. \tag{2.9}$$

The first integral in (1.2) differs from the integral that we now evaluate by a constant factor. We have, for $n \geq 2$,

$$\int_{-1}^{1} \frac{T_n^2(t)}{(t - \tau_i)(t - \tau_k)}(T_{n+1}(t) - T_{n-1}(t))^2 w_1(t)\, dt$$

$$= \frac{1}{2} \int_{-1}^{1} \frac{T_n^2(t)}{(t - \tau_i)(t - \tau_k)}(T_{2n+2}(t) - 2T_{2n}(t) + T_{2n-2}(t) - 2T_2(t) + 2)w_1(t)\, dt$$

$$= \frac{1}{2} \int_{-1}^{1} \frac{T_n^2(t)}{(t - \tau_i)(t - \tau_k)}(T_{2n-2}(t) - 2T_2(t) + 2)w_1(t)\, dt$$

$$= \int_{-1}^{1} T_{2n-2}^2(t)w_1(t)\, dt + \int_{-1}^{1} \frac{T_n^2(t)}{(t - \tau_i)(t - \tau_k)}(1 - T_2(t))w_1(t)\, dt$$

$$= \frac{\pi}{2} + \int_{-1}^{1} \frac{T_n^2(t)}{(t - \tau_i)(t - \tau_k)}(1 - T_2(t))w_1(t)\, dt.$$

In the above calculation we have repeatedly used the well-known identity

$$T_n(t)T_m(t) = \frac{1}{2}(T_{n+m}(t) + T_{|n-m|}(t)) \tag{2.10}$$

and the orthogonality of $\{T_n\}$. To compute the last integral in the above sequence of equalities, we apply the Gauss-Chebyshev quadrature formula (2.5) with remainder term.

Since

$$\frac{1}{(2n)!} \left[\frac{T_n^2(t)}{(t - \tau_i)(t - \tau_k)} (1 - T_2(t)) \right]^{(2n)} = -2^{2n-1},$$

we get, if $i \neq k$,

$$\int_{-1}^{1} \frac{T_n^2(t)}{(t - \tau_i)(t - \tau_k)} (1 - T_2(t)) w_1(t) \, dt = -\pi, \tag{2.11}$$

and if $i = k$,

$$\int_{-1}^{1} \frac{T_n^2(t)}{(t - \tau_i)(t - \tau_k)} (1 - T_2(t)) w_1(t) \, dt$$

$$= \frac{\pi}{n} (T_n'(\tau_i))^2 (1 - T_2(\tau_i)) - 2^{2n-1} \int_{-1}^{1} \left[\frac{1}{2^{n-1}} T_n(t) \right]^2 w_1(t) \, dt \tag{2.12}$$

$$= \frac{n\pi}{\sin^2 \theta_i} (2 - 2\cos^2 \theta_i) - \pi$$

$$= (2n - 1)\pi.$$

The next-to-last equality is obtained by using

$$T_n'(\tau_i) = \frac{(-1)^{i-1} n}{\sin \theta_i}. \tag{2.13}$$

The formulas (2.8), (2.11), (2.12) and (2.13), together with

$$T_{n+1}(\tau_i) - T_{n-1}(\tau_i) = 2(-1)^i \sin \theta_i, \tag{2.14}$$

yield

$$\int_{-1}^{1} l_i(t) l_k(t) w_1(t) \, dt = \begin{cases} -\frac{\pi}{8n^2} & \text{if } i \neq k, \\ \frac{(4n-1)\pi}{8n^2} & \text{if } i = k. \end{cases} \tag{2.15}$$

This enables us to evaluate the first double sum in the formula (1.2):

$$\sum_{i=1}^{n} \sum_{k=1}^{n} \left| \int_{-1}^{1} l_i(t) l_k(t) w_1(t) \, dt \right| = \sum_{i=1}^{n} \sum_{\substack{k=1 \\ k \neq i}}^{n} \frac{\pi}{8n^2} + \sum_{i=1}^{n} \frac{(4n-1)\pi}{8n^2}$$

$$= \frac{(5n-2)\pi}{8n}. \tag{2.16}$$

Next, we compute the integral in the second double sum in the formula (1.2). For these computations the formulas

$$T_n(\hat{\tau}_k) = (-1)^{k-1}, \quad T_{n+1}'(\hat{\tau}_k) - T_{n-1}'(\hat{\tau}_k) = \begin{cases} 4n & \text{if } k = 1, \\ 2(-1)^{k-1} n & \text{if } 1 < k < n+1, \\ 4(-1)^n n & \text{if } k = n+1, \end{cases} \tag{2.17}$$

are needed. The orthogonality of $\{T_n\}$ and formula (2.10) are used to derive

$$\int_{-1}^{1} \frac{T_n^2(t)(T_{n+1}(t) - T_{n-1}(t))^2}{(t - \tau_i)(t - \hat{\tau}_k)} w_1(t)\, dt = -\frac{\pi}{2}. \tag{2.18}$$

We apply formulas (2.8), (2.9), (2.13), (2.14), (2.17) and (2.18) to get

$$\int_{-1}^{1} l_i(t)\hat{l}_k(t)w_1(t)\, dt = \begin{cases} \frac{\pi}{8n^2} & \text{if } k \neq 1,\ k \neq n+1, \\ \frac{\pi}{16n^2} & \text{otherwise.} \end{cases} \tag{2.19}$$

As a consequence of (2.19),

$$\sum_{i=1}^{n} \sum_{k=1}^{n+1} \left| \int_{-1}^{1} l_i(t)\hat{l}_k(t)w_1(t)\, dt \right| = \frac{\pi}{8}. \tag{2.20}$$

It remains to find the last double sum in formula (1.2) to complete our proof. The main task is to evaluate the integral in the double sum for the different nodes. First we ignore the constant multiple in the integral and compute as follows (the orthogonality of $\{T_n\}$ is used whenever appropriate):

$$\int_{-1}^{1} \frac{T_n^2(t)(T_{n+1}(t) - T_{n-1}(t))^2}{(t - \hat{\tau}_i)(t - \hat{\tau}_k)} w_1(t)\, dt$$

$$= \frac{1}{2} \int_{-1}^{1} (T_{2n}(t) + 1) \frac{(T_{n+1}(t) - T_{n-1}(t))^2}{(t - \hat{\tau}_i)(t - \hat{\tau}_k)} w_1(t)\, dt \tag{2.21}$$

$$= \int_{-1}^{1} T_{2n}^2(t)w_1(t)\, dt + 2 \int_{-1}^{1} \frac{(t^2 - 1)^2 U_{n-1}^2(t)}{(t - \hat{\tau}_i)(t - \hat{\tau}_k)} w_1(t)\, dt.$$

The first integral in the last formula is $\frac{\pi}{2}$; to evaluate the second, we need to consider six different cases. For convenience, let us denote the integral by I.

Case 1: $\hat{\tau}_i = \hat{\tau}_k = 1$.
In this case the integral I can be expressed as

$$2 \int_{0}^{\pi} \frac{\sin^2 n\theta}{\sin^2 \theta}\, d\theta - \frac{\pi}{2}$$

by using the trigonometric substitution $t = \cos\theta$ and the formulas

$$U_{n-1}(\cos\theta) = \frac{\sin n\theta}{\sin\theta}, \qquad \int_{0}^{\pi} \sin^2 n\theta\, d\theta = \frac{\pi}{2}.$$

The recurrence formula

$$\int_{0}^{\pi} \frac{\sin^2(k+1)\theta}{\sin^2\theta}\, d\theta = \int_{0}^{\pi} \frac{\sin^2 k\theta}{\sin^2\theta}\, d\theta + \pi \tag{2.22}$$

produces

$$\int_0^\pi \frac{\sin^2 n\theta}{\sin^2 \theta}\, d\theta = n\pi.$$

Consequently, we have

$$\int_{-1}^1 \frac{(t^2-1)^2 U_{n-1}^2(t)}{(t-\hat{\tau}_i)(t-\hat{\tau}_k)} w_1(t)\, dt = \frac{(4n-1)\pi}{2}. \tag{2.23}$$

Case 2: $\hat{\tau}_i = \hat{\tau}_k = -1$.
In this case the integral I has the same value as in Case 1.

Case 3: $\hat{\tau}_i = 1$, $k \neq i$ or $\hat{\tau}_k = 1$, $i \neq k$.
If $\hat{\tau}_i = 1$, $k \neq i$, then

$$\begin{aligned}
\int_{-1}^1 \frac{(t^2-1)^2 U_{n-1}^2(t)}{(t-\hat{\tau}_i)(t-\hat{\tau}_k)} w_1(t)\, dt &= -\int_{-1}^1 \frac{(1+t) U_{n-1}^2(t)}{t-\hat{\tau}_k}(1-t^2)^{1/2}\, dt \\
&= -\int_{-1}^1 U_{n-1}^2(t)(1-t^2)^{1/2}\, dt \tag{2.24} \\
&= -\int_0^\pi \sin^2 n\theta\, d\theta = -\frac{\pi}{2}.
\end{aligned}$$

The same result is obtained if $\hat{\tau}_k = 1$ and $i \neq k$.

Case 4: $\hat{\tau}_i = -1$, $k \neq i$ or $\hat{\tau}_k = -1$, $i \neq k$.
Similarly to Case 3, one shows that

$$\int_{-1}^1 \frac{(t^2-1)^2 U_{n-1}^2(t)}{(t-\hat{\tau}_i)(t-\hat{\tau}_k)} w_1(t)\, dt = -\frac{\pi}{2}.$$

Case 5: $\hat{\tau}_i = \hat{\tau}_k$, but $\hat{\tau}_i \neq -1$ and $\hat{\tau}_k \neq 1$.
First observe that

$$\int_{-1}^1 \frac{(t^2-1)^2 U_{n-1}^2(t)}{(t-\hat{\tau}_i)(t-\hat{\tau}_k)} w_1(t)\, dt = \int_{-1}^1 \frac{(1-t^2) U_{n-1}^2(t)}{(t-\hat{\tau}_i)(t-\hat{\tau}_k)}(1-t^2)^{1/2}\, dt,$$

then apply the $(n-1)$-point Gauss-Chebyshev formula of the second kind with remainder term (see [2]) and note that

$$\frac{1}{(2n-2)!}\left[\frac{(1-t^2) U_{n-1}^2(t)}{(t-\hat{\tau}_i)^2}\right]^{(2n-2)} = -2^{2n-2}.$$

We obtain, for $2 \leq i \leq n$,

$$\begin{aligned}
&\int_{-1}^1 \frac{(1-t^2) U_{n-1}^2(t)}{(t-\hat{\tau}_i)^2}(1-t^2)^{1/2}\, dt \\
&= \frac{\pi}{n}(1-\hat{\tau}_i^2)^2 (U_{n-1}'(\hat{\tau}_i))^2 - 2^{2n-2}\int_{-1}^1 \left[\frac{1}{2^{n-1}} U_{n-1}(t)\right]^2 (1-t^2)^{1/2}\, dt,
\end{aligned}$$

which together with

$$U'_{n-1}(\hat{\tau}_i) = \frac{(-1)^i n}{1 - \hat{\tau}_i^2}$$

yields

$$\int_{-1}^{1} \frac{(1-t^2)U_{n-1}^2(t)}{(t-\hat{\tau}_i)^2}(1-t^2)^{1/2}\,dt = \frac{(2n-1)\pi}{2}. \qquad (2.25)$$

Case 6: $\hat{\tau}_i \neq \hat{\tau}_k$, $\hat{\tau}_i \neq -1, 1$ and $\hat{\tau}_k \neq -1, 1$.

In this case the integral I can again be computed by using the $(n-1)$-point Gauss-Chebyshev formula of the second kind with remainder term. The result is given by the following formula:

$$\int_{-1}^{1} \frac{(t^2-1)^2 U_{n-1}^2(t)}{(t-\hat{\tau}_i)(t-\hat{\tau}_k)}w_1(t)\,dt = \int_{-1}^{1} \frac{(1-t^2)U_{n-1}^2(t)}{(t-\hat{\tau}_i)(t-\hat{\tau}_k)}(1-t^2)^{1/2}\,dt = -\frac{\pi}{2}. \quad (2.26)$$

By combining all the results obtained above, the last double sum in (1.2) can be readily computed:

$$\sum_{i=1}^{n+1}\sum_{k=1}^{n+1}\left|\int_{-1}^{1}\hat{l}_i(t)\hat{l}_k(t)w_1(t)\,dt\right| = \frac{(5n^2-2n+1)\pi}{8n^2}. \qquad (2.27)$$

Finally, we can use (2.16), (2.20) and (2.27) to complete the proof. ∎

3 THIRD-KIND CHEBYSHEV WEIGHT FUNCTION

In this section we will repeat the same study on the mean convergence of the extended Lagrange-Kronrod interpolant of a continuous function f on the interval $[-1, 1]$ related to the third-kind Chebyshev weight function,

$$w(t) = w_3(t), \quad w_3(t) = (1-t)^{-1/2}(1+t)^{1/2}, \quad -1 < t < 1.$$

The corresponding orthogonal polynomial of degree n is

$$\pi_n(t) = V_n(t), \quad V_n(\cos\theta) = \frac{\cos(n+\frac{1}{2})\theta}{\cos\frac{1}{2}\theta}. \qquad (3.1)$$

It is well known (see [8]) that

$$\hat{\pi}_{n+1}(t) = T_{n+1}(t) - T_n(t). \qquad (3.2)$$

The nodes τ_i and $\hat{\tau}_k$ of the extended Lagrange-Kronrod interpolation are the zeros of the orthogonal polynomial π_n of degree n and the zeros of the Stieltjes polynomial $\hat{\pi}_{n+1}$ of degree $n+1$, respectively. They can be explicitly determined by

$$\tau_i = \cos\theta_i, \quad \theta_i = \frac{(2i-1)\pi}{2n+1}, \quad i = 1, 2, \ldots, n; \qquad (3.3)$$

$$\hat{\tau}_k = \cos\hat{\theta}_k, \quad \hat{\theta}_k = \frac{2(k-1)\pi}{2n+1}, \quad k = 1, 2, \ldots, n+1. \qquad (3.4)$$

Our main result of this section is stated in the following theorem.

Theorem 3.1 *The sequence $\{L(n; w)\}$ for the interpolation of a continuous function on $[-1, 1]$ using the nodes τ_i and $\hat{\tau}_k$ when w is the third-kind Chebyshev weight function is uniformly bounded. Indeed,*

$$L(n; w_3) = \frac{\pi}{2n+1} + \frac{\pi}{2n+1}\sum_{i=1}^{n}(1 + \tau_i) + \frac{\pi}{2n+1}\sum_{k=2}^{n+1}(1 + \hat{\tau}_k). \qquad (3.5)$$

Again, it follows from the above theorem that mean convergence holds for the extended Lagrange-Kronrod interpolation related to the third-kind Chebyshev weight function.

Proof of Theorem 3.1: We will begin the proof with the computation of the integral in the first double sum in (1.2) when $w(t) = w_3(t)$. Using

$$(\hat{\pi}_{n+1}(t))^2 = \frac{1}{2}(T_{2n+2}(t) + T_{2n}(t)) - T_{2n+1}(t) - T_1(t) + 1 \qquad (3.6)$$

and the orthogonality of $\{T_n\}$, the integral (disregarding a constant multiple) can be transformed to

$$\int_{-1}^{1} \frac{\pi_n^2(t)\hat{\pi}_{n+1}^2(t)}{(t - \tau_i)(t - \tau_k)} w_3(t)\, dt = \int_{-1}^{1} \frac{(1 + t)\pi_n^2(t)\hat{\pi}_{n+1}^2(t)}{(t - \tau_i)(t - \tau_k)} w_1(t)\, dt$$

$$= \int_{-1}^{1} \frac{(1 + t)(1 - T_1(t))\pi_n^2(t)}{(t - \tau_i)(t - \tau_k)} w_1(t)\, dt \qquad (3.7)$$

$$= \int_{-1}^{1} \frac{(1 - t)\pi_n^2(t)}{(t - \tau_i)(t - \tau_k)} w_3(t)\, dt.$$

The integrand in the last integral is a polynomial of degree $2n - 1$; the exact value can be found if one applies the n-point Gauss-Chebyshev quadrature of the third kind. Indeed,

$$\int_{-1}^{1} \frac{(1 - t)\pi_n^2(t)}{(t - \tau_i)(t - \tau_k)} w_3(t)\, dt = \begin{cases} 0 & \text{if } i \neq k, \\ \frac{2\pi(1-\tau_i^2)}{2n+1}(V_n'(\tau_i))^2 & \text{if } i = k. \end{cases} \qquad (3.8)$$

A direct computation gives

$$\hat{\pi}_{n+1}(\tau_i) = 2(-1)^i\sqrt{\frac{1 - \tau_i}{2}}.$$

The definition of $l_i(t)$, the above formula and (3.8) yield

$$\int_{-1}^{1} l_i^2(t)w_3(t)\, dt = \frac{\pi}{2n+1}(1 + \tau_i). \qquad (3.9)$$

391

It follows that

$$\sum_{i=1}^{n}\sum_{k=1}^{n}\left|\int_{-1}^{1}l_i(t)l_k(t)w_3(t)\,dt\right| = \frac{\pi}{2n+1}\sum_{i=1}^{n}(1+\tau_i). \qquad (3.10)$$

Next we compute

$$\int_{-1}^{1}l_i(t)\hat{l}_k(t)w_3(t)\,dt.$$

One can check easily, using trigonometric identities, that

$$(1+t)\pi_n(t) = T_{n+1}(t) + T_n(t). \qquad (3.11)$$

Up to a constant multiple, the integral in question is

$$\int_{-1}^{1}\frac{\pi_n^2(t)\hat{\pi}_{n+1}^2(t)}{(t-\tau_i)(t-\hat{\tau}_k)}w_3(t)\,dt$$

$$= \int_{-1}^{1}\frac{\pi_n(t)\hat{\pi}_{n+1}(t)}{(t-\tau_i)(t-\hat{\tau}_k)}(1+t)\pi_n(t)\hat{\pi}_{n+1}(t)w_1(t)\,dt$$

$$= \int_{-1}^{1}\frac{\pi_n(t)\hat{\pi}_{n+1}(t)}{(t-\tau_i)(t-\hat{\tau}_k)}(T_{n+1}(t)+T_n(t))(T_{n+1}(t)-T_n(t))w_1(t)\,dt$$

$$= \frac{1}{2}\int_{-1}^{1}\frac{\pi_n(t)\hat{\pi}_{n+1}(t)}{(t-\tau_i)(t-\hat{\tau}_k)}(T_{2n+2}(t)-T_{2n}(t))w_1(t)\,dt$$

$$= 0.$$

The last equality follows from the orthogonality of $\{T_n\}$. The above identities yield directly

$$\sum_{i=1}^{n}\sum_{k=1}^{n+1}\left|\int_{-1}^{1}l_i(t)\hat{l}_k(t)w_3(t)\,dt\right| = 0. \qquad (3.12)$$

The computation of

$$\int_{-1}^{1}\hat{l}_i(t)\hat{l}_k(t)w_3(t)\,dt$$

is more complicated. By noting that

$$(1+t)\pi_n^2(t) = 1 + T_{2n+1}(t)$$

and utilizing the orthogonality of $\{T_n\}$, the following equalities can be derived without difficulty:

$$\int_{-1}^{1}\frac{\hat{\pi}_{n+1}^2(t)}{(t-\hat{\tau}_i)(t-\hat{\tau}_k)}\pi_n^2(t)w_3(t)\,dt = \int_{-1}^{1}\frac{\hat{\pi}_{n+1}^2(t)}{(t-\hat{\tau}_i)(t-\hat{\tau}_k)}(1+T_{2n+1}(t))w_1(t)\,dt$$

$$= \int_{-1}^{1}\frac{\hat{\pi}_{n+1}^2(t)}{(t-\hat{\tau}_i)(t-\hat{\tau}_k)}w_1(t)\,dt$$

$$= \int_{-1}^{1}\frac{(1-t)^2W_n^2(t)}{(t-\hat{\tau}_i)(t-\hat{\tau}_k)}w_1(t)\,dt, \qquad (3.13)$$

392

where $W_n(t)$ is the n-th degree orthogonal polynomial associated with the fourth-kind Chebyshev weight function $w_4(t) = (1-t)^{1/2}(1+t)^{-1/2}$ and

$$W_n(t) = \frac{\sin(n+\frac{1}{2})\theta}{\sin\frac{1}{2}\theta}. \tag{3.14}$$

To compute the last integral in (3.13), we need to consider the four following cases.

Case 1: $\hat{\tau}_i = \hat{\tau}_k = 1$.
In this case, the integral can be reduced to

$$\int_{-1}^{1} W_n^2(t)w_1(t)\, dt = \int_0^\pi \frac{\sin^2(n+\frac{1}{2})\theta}{\sin^2\frac{1}{2}\theta}\, d\theta. \tag{3.15}$$

The trigonometric integral can be shown to satisfy the recurrence formula

$$\int_0^\pi \frac{\sin^2(k+1+\frac{1}{2})\theta}{\sin^2\frac{1}{2}\theta}\, d\theta = \int_0^\pi \frac{\sin^2(k+\frac{1}{2})\theta}{\sin^2\frac{1}{2}\theta}\, d\theta + 2\pi.$$

Consequently,

$$\int_0^\pi \frac{\sin^2(n+\frac{1}{2})\theta}{\sin^2\frac{1}{2}\theta}\, d\theta = (2n+1)\pi. \tag{3.16}$$

Moreover, direct computation produces

$$\pi_n(\hat{\tau}_k) = \frac{(-1)^{k-1}}{\cos\frac{1}{2}\hat{\theta}_k}, \quad \hat{\pi}'_{n+1}(\hat{\tau}_k) = \begin{cases} 2n+1 & \text{if } k = 1, \\ \frac{(-1)^{k-1}(2n+1)}{2\cos\frac{1}{2}\hat{\theta}_k} & \text{otherwise.} \end{cases} \tag{3.17}$$

Hence,

$$\int_{-1}^{1} \hat{l}_i(t)\hat{l}_k(t)w_3(t)\, dt = \frac{\pi}{2n+1}. \tag{3.18}$$

Case 2: $\hat{\tau}_i = 1$ and $\hat{\tau}_k \neq 1$, or $\hat{\tau}_k = 1$ and $\hat{\tau}_i \neq 1$.
If $\hat{\tau}_i = 1$ and $\hat{\tau}_k \neq 1$,

$$\int_{-1}^{1} \frac{(1-t)^2 W_n^2(t)}{(t-\hat{\tau}_i)(t-\hat{\tau}_k)}w_1(t)\, dt = -\int_{-1}^{1} \frac{W_n^2(t)}{t-\hat{\tau}_k}w_4(t)\, dt = 0. \tag{3.19}$$

The last equality follows from the orthogonality of $\{W_n\}$. This leads to

$$\int_{-1}^{1} \hat{l}_i(t)\hat{l}_k(t)w_3(t)\, dt = 0. \tag{3.20}$$

393

Case 3: $\hat{\tau}_i \neq \hat{\tau}_k$, and $\hat{\tau}_k \neq 1$, $\hat{\tau}_i \neq 1$.
In this case,

$$\int_{-1}^{1} \frac{(1-t)^2 W_n^2(t)}{(t - \hat{\tau}_i)(t - \hat{\tau}_k)} w_1(t)\, dt = \int_{-1}^{1} \frac{(1-t) W_n^2(t)}{(t - \hat{\tau}_i)(t - \hat{\tau}_k)} w_4(t)\, dt.$$

The last integral turns out to be zero by using the orthogonality of $\{W_n\}$.

Case 4: $\hat{\tau}_i = \hat{\tau}_k$, and $\hat{\tau}_i \neq 1$.
Again applying the n-point Gauss-Chebyshev quadrature formula of the fourth-kind, we get

$$\int_{-1}^{1} \frac{(1-t) W_n^2(t)}{(t - \hat{\tau}_i)^2} w_4(t)\, dt = (1 - \hat{\tau}_i)(W_n'(\hat{\tau}_i))^2 \gamma_i, \tag{3.21}$$

where

$$\gamma_i = \frac{(2n+1)\pi}{1 - \hat{\tau}_i^2} (W_n'(\hat{\tau}_i))^{-2}.$$

Hence,

$$\int_{-1}^{1} \frac{(1-t) W_n^2(t)}{(t - \hat{\tau}_i)^2} w_4(t)\, dt = \frac{(2n+1)\pi}{1 + \hat{\tau}_i}. \tag{3.22}$$

Now we use (3.17) and the above formula to obtain

$$\int_{-1}^{1} \hat{l}_i^2(t) w_3(t)\, dt = \frac{\pi(1 + \hat{\tau}_i)}{2n+1}. \tag{3.23}$$

The above discussion and results enable us to find the value for the last double sum in (1.2):

$$\sum_{i=1}^{n+1} \sum_{k=1}^{n+1} \left| \int_{-1}^{1} \hat{l}_i(t) \hat{l}_k(t) w_3(t)\, dt \right| = \frac{\pi}{2n+1} + \sum_{k=2}^{n+1} \frac{\pi(1 + \hat{\tau}_k)}{2n+1}. \tag{3.24}$$

Finally the formula (3.5) follows readily from (3.10), (3.12) and (3.24). ∎

For the fourth-kind Chebyshev weight function, we observe that

$$\pi_n^{(\frac{1}{2}, -\frac{1}{2})}(t) = (-1)^n \pi_n^{(-\frac{1}{2}, \frac{1}{2})}(t), \quad \hat{\pi}_{n+1}^{(\frac{1}{2}, -\frac{1}{2})}(t) = (-1)^{n+1} \pi_n^{(-\frac{1}{2}, \frac{1}{2})}(t)$$

and

$$\tau_i^{(\frac{1}{2}, -\frac{1}{2})} = -\tau_{n+1-i}^{(\frac{1}{2}, -\frac{1}{2})}, \ i = 1, 2, \ldots, n, \quad \hat{\tau}_k^{(\frac{1}{2}, -\frac{1}{2})} = -\hat{\tau}_{n+2-k}^{(\frac{1}{2}, -\frac{1}{2})}, \ k = 1, 2, \ldots, n+1,$$

where the first superscript and the second superscript of each variable correspond to the exponents of $1 - t$ and $1 + t$ in the weight function, respectively. The same notations are used below. The following equalities can then be shown:

$$\int_{-1}^{1} l_i^{(\frac{1}{2}, -\frac{1}{2})}(t) l_k^{(\frac{1}{2}, -\frac{1}{2})}(t) w_4(t) \, dt = \int_{-1}^{1} l_{n+1-i}^{(-\frac{1}{2}, \frac{1}{2})}(t) l_{n+1-k}^{(-\frac{1}{2}, \frac{1}{2})}(t) w_3(t) \, dt,$$

$$\int_{-1}^{1} l_i^{(\frac{1}{2}, -\frac{1}{2})}(t) \hat{l}_k^{(\frac{1}{2}, -\frac{1}{2})}(t) w_4(t) \, dt = \int_{-1}^{1} l_{n+1-i}^{(-\frac{1}{2}, \frac{1}{2})}(t) \hat{l}_{n+2-k}^{(-\frac{1}{2}, \frac{1}{2})}(t) w_3(t) \, dt,$$

and

$$\int_{-1}^{1} \hat{l}_i^{(\frac{1}{2}, -\frac{1}{2})}(t) \hat{l}_k^{(\frac{1}{2}, -\frac{1}{2})}(t) w_4(t) \, dt = \int_{-1}^{1} \hat{l}_{n+2-i}^{(-\frac{1}{2}, \frac{1}{2})}(t) \hat{l}_{n+2-k}^{(-\frac{1}{2}, \frac{1}{2})}(t) w_3(t) \, dt.$$

Therefore, $L(n; w_4) = L(n; w_3)$. We conclude that mean convergence also holds for the extended Lagrange-Kronrod interpolation associated with the fourth-kind Chebyshev weight function.

The question whether mean convergence of the extended Lagrange-Kronrod interpolation related to the Jacobi-type weight function

$$w(t) = (1 - t)^{\alpha}(1 + t)^{\beta}, \quad \alpha > -1, \ \beta > -1,$$

holds is worth studying. The same question can be asked for the Bernstein-Szegö type weight functions.

Acknowledgment. The author thanks the referee for his valuable suggestions and thoughtful comments. The author is also indebted to Professor S. Notaris for his helpful discussion.

REFERENCES

[1] Bellen A. Alcuni problemi aperti sulla convergenza in media dell'interpolazione Lagrangiana estesa. *Rend. Ist. Mat Univ. Trieste*, **20**:1–9, 1988.

[2] Davis P.J., Rabinowitz P. *Methods of numerical integration*. Academic Press, Orlando, Fl., 1984.

[3] Erdös P., Turán P. On interpolation I. *Ann. of Math.*, **38**:142–155, 1937.

[4] Gautschi W. Gauss-Kronrod quadrature–A survey. In *Numerical Methods and Approximation Theory III*, pages 39–66. Faculty of Electronic Engineering, Univ. Niš, Niš, 1988. G. V. Milovanović, ed.

[5] Gautschi W. On mean convergence of extended Lagrange interpolation. *J. Comput. Appl. Math.*, **43**:19–35, 1992.

[6] Kronrod A.S. *Nodes and weights for quadrature formulae*. Sixteen-place tables (Russian). Izdat. "Nauka", Moscow, 1964.

[7] Kronrod A.S. Integration with control of accuracy (Russian). *Dokl. Akad. Nauk SSSR*, **154**:283–286, 1964.

[8] Monegato G. Stieltjes polynomials and related quadrature rules. *SIAM Review*, **24**:137–158, 1982.

[9] Szegö G. Über gewisse orthogonale Polynome, die zu einer oszillierenden Belegungsfunktion gehören. *Math. Ann.*, **110**:501–513, 1934.

International Series of Numerical Mathematics, Vol. 119, ©1994 Birkhäuser

FINITE-PART INTEGRALS AND THE
EULER-MACLAURIN EXPANSION*

J. N. Lyness†

*This article is dedicated to Walter Gautschi
on the occasion of his 65th birthday*

Abstract. The context of this note is the discretization error made by the m-panel trapezoidal rule when the integrand has an algebraic singularity at one end, say $x = 0$, of the finite integration interval. In the absence of a singularity, this error is described by the classical Euler-Maclaurin summation formula, which is an asymptotic expansion in inverse integer powers of m. When an integrable singularity (x^α with $\alpha > -1$) is present, Navot's generalization is valid. This introduces negative fractional powers of m into the expansion. Ninham has generalized this result to noninteger α satisfying $\alpha < -1$. In this note, we extend these results to all α by providing the nontrivial extension to negative integer α. This expansion differs from the previous expansions by the introduction of a term $\log m$.

1 INTRODUCTION

The classical Euler-Maclaurin summation formula (see (2.14) below with $\beta = 1$) is an asymptotic expansion in integer powers of $1/m$ of the discretization error of the m-panel trapezoidal rule. This formula is valid when $f(x)$, together with its early derivatives, is continuous over the integration interval.

A major generalization to integrable functions of the form $f(x) = x^\alpha g(x)$, where $g(x)$ is regular and $\alpha > -1$, was established by Navot in 1961. Later, Ninham (1966) generalized the result to the cases in which $\alpha < -1$ but is not an integer. He established that the expansion was formally the same but that divergent integrals should be interpreted as Hadamard finite-part integrals. In this paper we complete the theory by covering the negative integer case. We also provide much simpler proofs for the other cases than those provided by Navot and by Ninham.

The ultimate result in this paper is Theorem 3.2. This establishes an asymptotic expansion for the m-panel offset trapezoidal rule $R^{(m)}(\beta)f$, which is valid for all integrand functions of the form $x^\alpha g(x)$, where $g(x)$ is regular and α may take any value, whether or not the integral being approximated is convergent. This expansion may include terms in m^{-j}, $m^{-(j+\alpha)}$, and $\log m$ for both positive and

*This work was supported in part by the Office of Scientific Computing, U.S. Department of Energy, under Contract W-31-109-Eng-38

†Mathematics and Computer Science Division, Argonne National Laboratory, Argonne, IL 60439

negative integer j. Integral representations are given for all terms in the expansion and for the remainder term.

2 CLASSICAL BACKGROUND MATERIAL

In this section, we recall some elementary results about Hadamard's finite-part integral and about the Euler-Maclaurin asymptotic expansion. In each case, we treat the integration interval $[0,1]$. Background material about these is available in many references, including, for example, Davis and Rabinowitz (1984), pp. 11, 188, and 136, and Diligenti and Monegato (1994). For our purposes we treat only finite-part integrals of the form

$$If = fp \int_0^1 x^\alpha g(x)dx, \tag{2.1}$$

where $g(x)$ is regular; specifically, $g(x) \in C^{q-1}[0,1]$ with $\alpha \in R$, $q \in Z^+$, and $q > -\alpha$.

We recall that any convergent conventional integral coincides with the corresponding finite-part integral and that finite-part integrals may be combined and manipulated in much the same way as conventional integrals. An exception is the classical change of variable rule, which needs minor modification; details may be found in Monegato (1994).

We now provide a two-stage elementary definition of the finite-part integrals (2.1) above.

Definition 2.1 *For all real α, the Hadamard finite-part integral If_α of the function $f_\alpha(x) = x^\alpha$ over the unit interval $[0,1]$ is defined by*

$$
\begin{aligned}
If_\alpha &= \int_0^1 f_\alpha(x)dx & \alpha > -1 \\
&= -\int_1^\infty f_\alpha(x)dx & \alpha < -1 \\
&= 0 & \alpha = -1.
\end{aligned}
\tag{2.2}
$$

Clearly, If_α coincides with $f_\alpha(1)$ when this is defined. For values of α for which the first integral in (2.2) exists, the second does not, and vice versa.

Unless α is a nonnegative integer, $f_\alpha^{(s)}(x)$ is a constant multiple of $f_{\alpha-s}(x)$. A simple result, needed in a later proof, is the following lemma.

Lemma 2.2

$$
\begin{aligned}
\int_1^m f_\alpha^{(s)}(x)dx &= (m^{\alpha-s+1} - 1)If_\alpha^{(s)} & \alpha - s \neq -1 \\
\int_1^m f_\alpha^{(s)}(x)dx &= \frac{\log m}{\log 2} \int_1^2 f_\alpha^{(s)}(x)dx & \alpha - s = -1.
\end{aligned}
\tag{2.3}
$$

It is trivial to verify this. The underlying reason that this result and the subsequent theory is simple is that the integrand is a homogeneous function of degree $\alpha - s$ about the origin.

Extending Definition 2.1 to functions of the form $x^\alpha g(x)$, where $g(x) \in C^\ell[0,1]$, requires only the expression of $g(x)$ as a Taylor expansion about the origin and term-by-term integration of a finite series. Specifically, we may set

$$g(x) = \sum_{j=0}^{\ell-1} g^{(j)}(0)x^j/j! + G_\ell(x) \tag{2.4}$$

and

$$f(x) = x^\alpha g(x) = \sum_{j=0}^{\ell-1} \kappa_{\alpha+j} f_{\alpha+j}(x) + x^\alpha G_\ell(x), \tag{2.5}$$

where

$$\kappa_{\alpha+j} = g^{(j)}(0)/j!. \tag{2.6}$$

When $\ell + \alpha + 1 > 0$, the remainder term

$$F_{\alpha;\ell}(x) = x^\alpha G_\ell(x) = f(x) - \sum_{j=0}^{\ell-1} \kappa_{\alpha+j} f_{\alpha+j}(x) \tag{2.7}$$

is integrable over [0,1]. Term-by-term integration of (2.5) above leads to the following definition.

Definition 2.3 *Let $f(x) = x^\alpha g(x)$ and $g(x) \in C^{q-1}[0,1]$ with $q + \alpha > 0$; then*

$$If = \sum_{j=0}^{q-1} \kappa_{\alpha+j} If_{\alpha+j} + \int_0^1 (f(x) - \sum_{j=0}^{q-1} \kappa_{\alpha+j} f_{\alpha+j}(x))dx, \tag{2.8}$$

where $\kappa_{\alpha+j} = g^{(j)}(0)/j!$ and $If_{\alpha+j}$ is defined in Definition 2.1 above.

We now turn to the Euler-Maclaurin expansion.

Definition 2.4 *The offset trapezoidal rule for the interval [0,1] is denoted by*

$$R^{(m)}(\beta)f = \frac{1}{m} \sum_{j=0}^{m-1} f\left(\frac{j+\beta}{m}\right). \tag{2.9}$$

The Euler-Maclaurin expansion involves coefficients and kernel functions

$$c_s(\beta) = B_s(\beta)/s! \qquad\qquad s \geq 0, \tag{2.10}$$
$$h_s(\beta, t) = (B_s(\beta) - \bar{B}_s(\beta - t))/s! \qquad s \geq 1, \tag{2.11}$$

where $B_s(x)$ is the Bernoulli polynomial and $\bar{B}_s(x)$ the corresponding Bernoulli function, which coincides with $B_s(x)$ in the interval $(0,1)$ and has unit period. Note that

$$h_p(\beta, t) = h_p(\beta, t+1) \tag{2.12}$$

$$c_p(\beta) = \int_0^1 h_p(\beta, t)dt. \tag{2.13}$$

Theorem 2.5 (**The Euler-Maclaurin asymptotic expansion for regular** $f(x)$.) *Let $f(x)$ and its first p derivatives be integrable in $[0,1]$, and let $p \geq 1$. Then*

$$R^{(m)}(\beta)f = \sum_{s=0}^{p-1} \frac{c_s(\beta)}{m^s} \int_0^1 f^{(s)}(x)dx + \frac{1}{m^p} \int_0^1 h_p(\beta, mt)f^{(p)}(t)dt. \tag{2.14}$$

Note that $c_0(\beta) = 1$, and so the first term in the summation is simply If.

3 THE EULER-MACLAURIN EXPANSION FOR x^α

In this paper, we treat functions that have an algebraic singularity at an end point of our integration interval, $[0,1]$. We restrict our attention to functions of the form

$$f(x) = x^\alpha g(x), \tag{3.1}$$

where $g(x)$ is $C^p[0,1]$. First we treat the monomial

$$f_\alpha(x) = x^\alpha; \tag{3.2}$$

and then, as the theory is linear, we concatenate individual terms of this type, using a Taylor expansion for $g(x)$, to obtain the corresponding expansion for $f(x)$ in (3.1).

The basic theorem in this paper is the following.

Theorem 3.1 (**The Euler-Maclaurin expansion for x^α.**) *Let α be real, and let p be a nonnegative integer satisfying $p + \alpha + 1 \geq 0$; then*

$$R^{(m)}(\beta)f_\alpha = \frac{A_{\alpha+1}(\beta; f_\alpha)}{m^{\alpha+1}} + \delta_{\alpha+1,0}\log m + \sum_{s=0}^{p-1} \frac{c_s(\beta)If_\alpha^{(s)}}{m^s} + E_p^{(m)}(\beta)f_\alpha, \tag{3.3}$$

where

$$A_{\alpha+1}(\beta; f_\alpha) = f_\alpha(\beta) - \sum_{s=0}^{p-1} c_s(\beta)If_\alpha^{(s)} + \int_1^\infty h_p(\beta, t)f_\alpha^p(t)dt \tag{3.4}$$

and

$$E_p^{(m)}(\beta)f_\alpha = -\frac{1}{m^{\alpha+1}} \int_m^\infty h_p(\beta, t)f_\alpha^{(p)}(t)dt = O(m^{-p}). \tag{3.5}$$

The integrals in (3.3) are defined in (2.2) as either conventional or Hadamard finite-part integrals. Simpler expressions for (3.4) are established in Section 4.

The proof given below, valid for all α, involves only elementary algebra and calculus. It exploits the fact that the result may be considered to be a matter of scaling. Comments on previous proofs of less general versions of this theorem are made in Section 5.

Proof: As a preliminary, we apply (2.14) translated to the interval $[j, j+1]$ with $m = 1$ to the function $f_\alpha(x)$. This gives

$$
f_\alpha(j + \beta) = (j + \beta)^\alpha \;=\; \sum_{s=0}^{p-1} c_s(\beta) \int_j^{j+1} f_\alpha^{(s)}(x)dx
$$
$$
+ \int_j^{j+1} h_p(\beta, t) f_\alpha^{(p)}(t)dt \quad j = 1, 2, \ldots, m - 1. \,(3.6)
$$

Note that we have used the periodicity property (2.12) of $h_p(\beta, t)$.

We now substitute (3.6) into a scaled version of Definition 2.4 as follows.

$$
R^{(m)}(\beta)f_\alpha \;=\; \frac{1}{m} \sum_{j=0}^{m-1} f_\alpha\left(\frac{j + \beta}{m}\right)
$$
$$
= \frac{1}{m^{1+\alpha}} \sum_{j=0}^{m-1} f_\alpha(j + \beta)
$$
$$
= \frac{1}{m^{1+\alpha}} f_\alpha(\beta) + \sum_{s=1}^{p-1} \frac{c_s(\beta)}{m^{1+\alpha}} \int_1^m f_\alpha^{(s)}(x)dx
$$
$$
+ \frac{1}{m^{1+\alpha}} \int_1^m h_p(\beta, t) f_\alpha^{(p)}(t)dt \tag{3.7}
$$

Here we have isolated the term with $j = 0$. We have collected together integrals over $[j, j+1]$ to form an integral over $[1, m]$. In the final term we have used once more the periodicity of the kernel function $h_p(\beta, t)$. Up to this point, no singularity is involved , and any nonnegative integer p is permitted. All manipulation has been conventional. Our final manipulative steps are to apply Lemma 2.2 to reexpress the integrals whose upper limit is m, and to split the final integral into two parts. We find, when α is not an integer,

$$
R^{(m)}(\beta)f_\alpha \;=\; \frac{1}{m^{1+\alpha}} f_\alpha(\beta) + \sum_{s=1}^{p-1} \frac{c_s(\beta)}{m^{1+\alpha}} (m^{\alpha-s+1} - 1) I\!f_\alpha^{(s)}
$$
$$
+ \frac{1}{m^{1+\alpha}} \int_1^\infty h_p(\beta, t) f_\alpha^{(p)} dt
$$
$$
- \frac{1}{m^{\alpha+1}} \int_m^\infty h_p(\beta, t) f_\alpha^{(p)}(t)dt \quad p > -1 - \alpha. \tag{3.8}
$$

Note that the condition on p is needed only to validate splitting the final integral, extracting a part of order $m^{-(1+\alpha)}$, and leaving another part, the remainder term of lower order.

When α is an integer, (3.8) may require adjustment. Specifically, in the sum over index s in (3.7), any term in which $\alpha - s = -1$ should have been replaced by the second member of (2.3) instead of the first. Since s is nonnegative, this adjustment is needed, if at all, only when $\alpha \leq -1$. In these cases, one should remove the current term with $s = \alpha + 1$ and introduce the second member of (2.3), to obtain a single additional term, namely,

$$\frac{c_{1+\alpha}(\beta)}{m^{\alpha+1}} \frac{\log m}{\log 2} \int_1^2 f_\alpha^{\alpha+1}(x) dx.$$

Moreover, when α is nonnegative, the integrand is zero. Thus, this additional term is nonzero only when $\alpha = -1$, when it reduces to $\log m$.

Collecting together terms of specific orders in m, we recover the expressions given in the statement of the theorem. To establish the theorem, we need to show that the term (3.5) is of the order stated and that (3.4) is not dependent on p. Since $f_\alpha(x) = x^\alpha$, we see

$$E_p^{(m)}(\beta)f_\alpha = \frac{1}{m^{\alpha+1}} \int_m^\infty h_p(\beta,t) \frac{\alpha!}{(\alpha - p)!} t^{\alpha-p} dt \qquad p > -\alpha - 1.$$

An elementary calculation shows that

$$|E_p^{(m)}(\beta)f_\alpha| \leq \frac{\alpha}{(\alpha - p + 1)!} \max_t |h_p(\beta,t)| m^{-p},$$

establishing the correct order.

This dependence on p in the expression (3.4) is illusory. Since $h_p(\beta,t)$ is periodic in t and

$$\int_j^{j+1} h_p(\beta,t) dt = c_p(t),$$

integration by parts yields the same expression for $A_{\alpha+1}(\beta; f_\alpha)$ with p replaced by $p + 1$. (We cannot replace p in this expression by any integer less than $-\alpha - 1$. Besides not being justified, the resulting (incorrect) expression contains a divergent integral.) ∎

This theorem asserts that the expansion, with this value of $A_{\alpha+1}$, holds for all finite p exceeding $-\alpha - 1$.

For smaller values of p, we find directly from Theorem 2.5 that

$$R^{(m)}(\beta)f_\alpha = \sum_{s=0}^{p-1} \frac{c_s(\beta)}{m^s} If_\alpha^{(s)} + \tilde{E}_p^{(m)}(\beta)f_\alpha$$

402

with

$$\tilde{E}_p^{(m)}(\beta])f_\alpha = \frac{1}{m^p}\int_0^1 h_p(\beta, mt)f_\alpha^{(p)}(t)dt \qquad p < -\alpha - 1.$$

It follows that, when p and $\bar{p}$ are integers satisfying $p < -\alpha - 1 < \bar{p}$, the forms of remainder terms are connected by

$$\tilde{E}_p^{(m)}(\beta)f_\alpha = \sum_{s=p}^{\bar{p}} \frac{c_s(\beta)}{m^s}If_\alpha^{(s)} + \frac{A_{\alpha+1}(\beta, f_\alpha)}{m^{\alpha+1}} + E_{\bar{p}}^{(m)}(\beta)f_\alpha.$$

When α is a nonnegative integer, the expression for $A_{\alpha+1}(\beta; f_\alpha)$ reduces to zero, and $f_\alpha^{(s)} = 0$ for $s \geq \alpha + 1$. The result in the statement of the theorem follows directly from Theorem 2.5.

It is a simple step from the Euler-Maclaurin expansion for $f_\alpha(x)$ in this theorem to the corresponding expansion for $f_\alpha(x)g(x)$ when $g(x)$ is regular. We simply follow our definition of the Hadamard finite-part integral by expanding $g(x)$ in a Taylor series. This gives expansion (2.5). We may apply already available versions of the Euler-Maclaurin expansion to each term in this expansion. We shall apply Theorem 2.5 to the final term. To this end we require that $g(x)$ have sufficient continuity that the final term in (2.5) differentiated p times is integrable. We treat

$$g(x) \in C^{p+q-1}[0,1] \text{ with } p+q+\alpha > 0; \quad p \geq 1; \quad q \geq 0. \qquad (3.9)$$

It is convenient to differentiate (2.5) term by term to obtain

$$f^{(s)}(x) = \sum_{j=0}^{p+q-1} \kappa_{\alpha+j}f_{\alpha+j}^{(s)}(x) + F_{\alpha;p+q}^{(s)}(x) \qquad s = 0, 1, \ldots, p \qquad (3.10)$$

and integrate term by term to obtain

$$If^{(s)} = \sum_{j=0}^{p+q-1} \kappa_{\alpha+j}If_{\alpha+j}^{(s)} + \int_0^1 F_{\alpha;p+q}^{(s)}(x)dx \qquad s = 0, 1, \ldots, p. \qquad (3.11)$$

These formulas are helpful in establishing the following theorem.

Theorem 3.2 Let α be real, p a positive integer, and q a nonnegative integer satisfying $p + q + \alpha > 0$. Let $f(x) = x^\alpha g(x)$, where

$$g(x) \in C^{p+q-1}[0,1].$$

Then

$$R^{(m)}(\beta)f = \sum_{j=0}^{p+q-1} \frac{A_{\alpha+1+j}(\beta; f)}{m^{\alpha+1+j}} + \kappa_{-1}\ln m + \sum_{s=0}^{p-1} \frac{c_s(\beta)}{m^s}If^{(s)}$$

$$+ O(m^{-p}), \qquad (3.12)$$

where κ_{-1} is taken to be zero unless α is a negative integer and

$$A_{\alpha+1+j}(\beta; f) = \kappa_{\alpha+j} A_{\alpha+1+j}(\beta; f_{\alpha+j}). \tag{3.13}$$

Proof: When $g(x) \in C^{p+q-1}[0,1]$ with $p + q + \alpha > 0$, it follows that the p-th derivative of $F_{\alpha;p+q}(x)$ is integrable in $[0,1]$. Hence, we may apply Theorem 2.5 to this function. Applying Theorem 3.1 to the other terms in (2.5), we find

$$
\begin{aligned}
R^{(m)}(\beta)f &= \sum_{j=0}^{p+q-1} \kappa_{\alpha+j} R^{(m)}(\beta) f_{\alpha+j} + R^{(m)}(\beta) F_{\alpha;p+q} \\
&= \sum_{j=0}^{p+q-1} \frac{\kappa_{\alpha+j} A_{\alpha+j+1}}{m^{\alpha+j+1}} + \sum_{j=0}^{p+q-1} \kappa_{\alpha+j} \delta_{\alpha+j+1,0} \ln m \\
&\quad + \sum_{s=0}^{p-1} \frac{c_s(\beta)}{m^s} \sum_{j=0}^{p+q-1} \kappa_{\alpha+j} I f_{\alpha+j}^{(s)} + \sum_{j=0}^{p+q-1} \kappa_{\alpha+j} E_p^{(m)}(\beta) f_{\alpha+j} \\
&\quad + \sum_{s=0}^{p-1} \frac{c_s(\beta)}{m^s} \int_0^1 F_{\alpha;p+q}^{(s)}(x)dx + \frac{1}{m^p} \int_0^1 h_p(\beta; mt) F_{\alpha;p+q}^{(p)}(t)dt. \tag{3.14}
\end{aligned}
$$

In view of (3.5) and (2.14), the fourth and the sixth term here are $O(m^{-p})$ and belong in the remainder term in (3.12). Also belonging in this remainder term are those elements in the summation in the first expression for which $\alpha + j + 1 \geq p$. The third and fifth expressions here can be reduced, by using (3.11), to the sum over index s in (3.12). Finally, the second term above exists only when α is an integer and $j = -\alpha - 1$. This gives the second term in (3.12). Thus, all terms in (3.14) are accounted for; together they give rise to the right-hand side of (3.12). This establishes the theorem. ∎

4 NUMERICAL VALUES OF THE EXPANSION COEFFICIENTS

It was shown by Navot and by Ninham that, when $\alpha \neq$ integer,

$$A_{\alpha+1}(\beta; f_\alpha) = \zeta(-\alpha, \beta), \tag{4.1}$$

where ζ is the Hurwitz zeta function, defined by

$$\zeta(-\alpha, \beta) = \beta^\alpha + (\beta+1)^\alpha + \dots \qquad \beta > 0; \quad \alpha < -1, \tag{4.2}$$

and by analytic continuation for all values of α other than -1, where this analytic function has its only singularity, a pole of order 1.

A simple proof of (4.1) valid when $\alpha > -1$ may be obtained by applying a version of the Euler-Maclaurin expansion (2.14) with $m = 1$, $f(x) = x^\alpha$ to an interval $[k, k+1]$. This procedure gives

$$f_\alpha(k+\beta) = (k+\beta)^\alpha = \sum_{s=0}^{p-1} c_s(\beta) \int_k^{k+1} f_\alpha^{(s)}(x)dx + \int_k^{k+1} h_p(\beta, t) f_\alpha^{(p)}(t)dt. \quad (4.3)$$

Set $\alpha < 1$, sum (4.3) over all positive integer k, and add $f_\alpha(\beta)$ to both sides. The result is $\zeta(-\alpha, \beta)$ on the left and Definition (3.4) for $A_{\alpha+1}(\beta; f_\alpha)$ on the right.

For α a positive integer, one may use (3.4) and (2.14) with $p > \alpha$ and $m = 1$ to show directly

$$A_{\alpha+1}(\beta; f_\alpha) = 0 \qquad \alpha = 0, 1, 2, \ldots \quad . \quad (4.4)$$

Analytic continuation of the Hurwitz zeta function establishes (4.1) above for all α except $\alpha = -1$.

The reader will have noticed that, because of the singularity, no information about $A_0(\beta; f_{-1})$ has been forthcoming. This turns out to be the most intractable of the coefficients, and, as will appear below, it is the only one for which a numerical value is required in the extrapolation application. In Theorem 4.1 below, this value is expressed in terms of the Euler-Mascheroni constant

$$\gamma = \lim_{m \to \infty} \left(\sum_{j=1}^m \frac{1}{j} - \ln m \right) \approx 0.5772 \quad (4.5)$$

and a finite harmonic function defined by

$$H(\beta) = \sum_{n \geq 1} \left(\frac{1}{n} - \frac{1}{n+\beta} \right). \quad (4.6)$$

This function appears in Knuth (1973) where it is not given a name; when $\beta = p/q$ and p and q are positive integers satisfying $0 < p < q$, Knuth (p. 94) shows that

$$H(\beta) = \frac{1}{\beta} - \frac{1}{2}\pi \cot \beta\pi - \ln 2q + 2\sum_{1 \leq n < q/2} \cos 2\pi n\beta \ln \sin \frac{n}{q}\pi. \quad (4.7)$$

Explicit forms of these expressions are listed by Knuth (p. 616) for all p, q satisfying $0 < p < q \leq 6$. For example,

$$H(1) = 1; \qquad H(\tfrac{1}{2}) = 2 - 2\ln 2.$$

Theorem 4.1

$$A_0(\beta; f_{-1}) = \gamma + \frac{1}{\beta} - H(\beta), \quad (4.8)$$

where γ is the Euler-Mascheroni constant (4.5) above, and $H(\beta)$ is Knuth's finite harmonic function (4.6) above.

Proof: We exploit the expansion in which A_0 occurs, namely, (3.3) with $\alpha = -1$. This gives

$$R^{(m)}(\beta)f_{-1} = \frac{1}{m}\sum_{j=0}^{m-1}\frac{m}{j+\beta} = A_0(\beta; f_{-1}) + \log m$$
$$+ \sum_{s=0}^{p-1}\frac{c_s(\beta)If_{-1}^{(s)}}{m^s} + E_p^{(m)}(\beta)f_{-1}.$$

Since $If_{-1} = 0$, the $s = 0$ term in the sum is zero. Taking the limit as m becomes infinite, we find

$$A_0(\beta; f_{-1}) = \lim_{m\to\infty}\left(\sum_{j=0}^{m-1}\frac{1}{j+\beta} - \log m\right).$$

Elementary manipulation of this, involving (4.5) and (4.6) above, leads directly to the result in the theorem. ∎

Examples include

$$A_0(1; f_{-1}) = \gamma; \qquad A_0(\tfrac{1}{2}; f_{-1}) = \gamma + 2\ln 2.$$

5 CONCLUDING REMARKS

It is beyond the scope of this paper to discuss in any detail the ways in which this expansion may be used for extrapolation or to examine its relationship with known quadrature formulas. We content ourselves by making a few points of a general nature.

When a finite-part integral is being approximated using extrapolation and so $\alpha \leq -1$, one is extracting the coefficient of a higher-order term in the expansion instead of the principal term. This is akin to numerical differentiation, and one should expect a corresponding increase in the amplification of noise level error in the calculation.

When, in addition, α is a negative integer, say $\alpha = -N$, the expansion takes the following form.

$$R^{(m)}(\beta)f = \kappa_{-N}A_{1-N}m^{N-1} + \ldots + \kappa_{-2}A_{-1}m + \kappa_{-1}\ln m$$
$$+\kappa_{-1}A_0 + If + \frac{c_1(\beta)}{m}If^{(1)} + \ldots$$

We may extrapolate in the usual way. But when doing so, we shall be obliged to extract estimates of two coefficients, namely, κ_{-1} and $\kappa_{-1}A_0 + If$. Then, to obtain If, we have to separate it from $\kappa_{-1}A_0$. This step requires both the extrapolated value of κ_{-1} and the numerical value of constant $A_0(\beta; f_{-1})$. For some values of β, the latter is provided by Theorem 4.1. This is an unusual situation in the practice of extrapolation. Usually only simple data, such as the value of α and data relating to the structure of the expansion, are required.

The significant result of this paper is Theorem 3.2. Various proofs for the conventional case $\alpha > -1$, have appeared. The original proof by Navot (1961) involves several pages of detailed algebraic manipulation. A subsequent proof by Lyness and Ninham (1967) is shorter but involves generalized functions in an essential way. A somewhat pedestrian proof using the generalized zeta function has been given by Lyness (1971). None of these proofs is particularly illuminating. The reader finds, at the end, that the result has been established. For the finite-part integral case with noninteger α, Ninham's proof involves a Fourier decomposition of the integrand, as well as the use of generalized functions. That paper comprises an elegant but lengthy application of a little known theory contained in a twelve-page paper devoted to this single result.

The proof of Section 3 of this paper is more general and more straightforward than any of the proofs mentioned above.

Acknowledgment. It is a pleasure to acknowledge the significant help and encouragement of Dr. G. Monegato, without which the underlying research of this paper would not have been undertaken.

REFERENCES

[1] Davis P.J., Rabinowitz P. *Methods of Numerical Integration*, 2nd edition. Academic Press, London, 1984.

[2] Diligenti M., Monegato G. Finite-part integrals: their occurrence and computation. To appear in *Rend. Circolo Mat. di Palermo*, 1994.

[3] Knuth D.E. *The Art of Computer Programming*, Volume 1. Addison-Wesley, London, 1973.

[4] Lyness J.N. The calculation of Fourier coefficients by the Mobius inversion of the Poisson summation formula — Part III: Functions having algebraic singularities. *Math. Comp.*, 25:483–494, 1971.

[5] Lyness J.N., Ninham B. W. Numerical quadrature and asymptotic expansions. *Math. Comp.*, 21:162–178, 1967.

[6] Monegato G. Numerical evaluation of hypersingular integrals. *JCAM*, 50, 1994.

[7] Navot I. An extension of the Euler-Maclaurin summation formula to functions with a branch singularity. *J. Math. and Phys.*, 40:271–276, 1961.

[8] Ninham B.W. Generalised functions and divergent integrals. *Num. Math.*, 8:444–457, 1966.

International Series of Numerical Mathematics, Vol. 119, ©1994 Birkhäuser

APPROXIMATION OF FUNCTIONS BY EXTENDED LAGRANGE INTERPOLATION

G. Mastroianni*

Dedicated to Professor Walter Gautschi on his 65th birthday

Abstract. We consider some extended interpolation processes based on the zeros of the product of two orthogonal polynomials and on some additional points near ± 1. We derive necessary conditions for the convergence in the L_p weighted spaces. These conditions involve the weight of the orthogonal polynomials and the distribution of the interpolation points.

1 INTRODUCTION AND MAIN RESULTS

We consider $X = \{x_{m,1}, \ldots, x_{m,m}; \; m = 1, 2, \ldots\} \subseteq [-1, 1]$ a matrix of distinct knots. Any further matrix we can obtain by adding a finite number of new distinct nodes in each row of X will be called an *extended matrix*. That is, if $y_{n,i}$, $i = 1, \ldots, n$ are $n = n(m)$ knots with $y_{n,i} \neq y_{n,j} \neq x_{m,i}$, then

$$Z = \{y_{n,1}, \ldots, y_{n,n}, x_{m,1}, \ldots, x_{m,m}\}_{m=1,2,\ldots}$$

will be an extended matrix associated with X. In the same way, we denote by $\mathcal{L}_{m+n}(f, x)$, the extended Lagrange polynomial interpolating the function $f \in C^0([-1, 1])$ in $x_{m,1}, \ldots, x_{m,m}, y_{n,1}, \ldots, y_{n,n}$. Such extended matrices and/or the corresponding extended interpolation processes can offer wide applications. For example, the extended matrices are often applied in the search of the numerical solutions of singular integral equations solvable with the "quadrature" method [17, 13]. Other fields of applications are the usual numerical quadrature [12, 21], the numerical evaluation of interpolation errors [11] and the developing of new approximation procedures. Another important application consists in the correction of not well–based interpolation.

In this section, we briefly illustrate this latter aspect. Let us consider the Jacobi weight $v^{\alpha,\beta}(x) = (1-x)^\alpha(1+x)^\beta$, $|x| < 1$, $\alpha, \beta > -1$, and the corresponding system of orthogonal polynomials $\{p_m(v^{\alpha,\beta})\}_{m \in \mathbb{N}}$. Suppose that the elements of X are the zeros of Jacobi polynomials and $\mathcal{L}_m(v^{\alpha,\beta}, f)$ the Lagrange polynomial based on the zeros of $p_m(v^{\alpha,\beta})$.

*Dipartimento di Matematica, Universitá della Basilicata, Via Nazario Sauro, 85100 Potenza, Italy

Letting

$$u(x) = \prod_{k=0}^{r} | \tau_k - x |^{\gamma_k}, \quad -1 = \tau_0 < \tau_1 < \cdots < \tau_{r-1} < \tau_r = 1, \ \gamma_k > -1, \quad (1.1)$$

be a generalized Jacobi weight ($u \in GJ$), we can define $L_{u,p}$, $1 \le p < \infty$ ($L_{u,p} \equiv L_p$ if $u = 1$), as the set of the functions f such that

$$\|fu\|_p^p := \int_{-1}^{1} | f(x)u(x) |^p \, dx < \infty.$$

In the following, $\|f\| = \|f\|_\infty$ is the uniform norm of f and $\|\mathcal{L}_m(v^{\alpha,\beta})\|_{u,p} = \sup_{\|f\|=1} \|\mathcal{L}_m(v^{\alpha,\beta}, f)u\|_p$ represents the norm of the operator $\mathcal{L}_m(v^{\alpha,\beta}) : C^0 \mapsto L_{u,p}$.

The well known result of P. Nevai [22] which states that

$$\sup_m \|\mathcal{L}_m(v^{\alpha,\beta}, f)\|_{u,p} < \infty$$

if and only if

$$\frac{1}{\sqrt{v^{\alpha+\frac{1}{2},\beta+\frac{1}{2}}}} \in L_{u,p}, \ 1 \le p < \infty \quad (1.2)$$

can now applied. In fact, assuming (1.2) we obtain:

$$\forall f \in C^0 ([-1,1]) \quad \lim_m \|f - \mathcal{L}_m(v^{\alpha,\beta}, f)u\|_p = 0.$$

Conversely, when the elements of the matrix X are the zeros of Jacobi polynomials with α, β that do not satisfy (1.2), then for some $g \in C^0 ([-1,1])$,

$$\lim_m \|g - \mathcal{L}_m(v^{\alpha,\beta}, g)u\|_p = \infty.$$

Denote by $\mathcal{L}_{m,r,s}(v^{\alpha,\beta}, f)$ the Lagrange polynomial interpolating f on

$$\bar{x}_{-s+1}, \ldots, \bar{x}_{-1}, \bar{x}_0, x_1, \ldots, x_m, \bar{x}_{m+1}, \ldots, \bar{x}_{m+r},$$

where

$$p_m(v^{\alpha,\beta}, x_i) = 0, \ i = 1, \ldots, m,$$

$$\bar{x}_{-s+j} = x_1 - (s - j + 1)\frac{1 + x_1}{s}, \ j = 1, \ldots, s,$$

$$\bar{x}_{m+i} = x_m + i\frac{1 - x_m}{r}; \ i = 1, \ldots, r,$$

$$(1.3)$$

and r, s are fixed integers. Using [16, Theorem 3.1, p. 388] we obtain

$$\forall u \in L_p, \quad \sup_m \|\mathcal{L}_{m,r,s}(v^{\alpha,\beta})\|_{u,p} < \infty$$

410

if and only if

$$\frac{\sqrt{v^{\alpha+\frac{1}{2},\beta+\frac{1}{2}}}}{v^{r,s}} \in L_1 \text{ and } \frac{v^{r,s}}{\sqrt{v^{\alpha+\frac{1}{2},\beta+\frac{1}{2}}}} \in L_{u,p}. \tag{1.4}$$

Therefore, if (1.2) doesn't hold, we can choose r and s so that (1.4) holds. Thus every divergent interpolation process based on the zeros of Jacobi polynomials can be complemented by means of an extended interpolation procedure. Similar results still hold in several different contexts [15, 16].

In this paper a more general approach is proposed. Let σ and σ^* be two arbitrary weight functions and denote by $\{p_m(\sigma)\}$, $\{p_n(\sigma^*)\}$ the associated sequences of orthogonal polynomials. Let us suppose that $p_m(\sigma)p_n(\sigma^*)$ have $m+n$ distinct zeros in $(-1,1)$, $n = n(m)$, $t_1 < \cdots < t_{m+n}$; then we consider the Lagrange polynomial $\mathcal{L}_M(f) = \mathcal{L}_M(\sigma,\sigma^*,f)$, $M = m+n+r+s$, interpolating a given function f at the knots

$$-1 \leq \bar{t}_{1-s} < \cdots < \bar{t}_0 < t_1 < \cdots < t_{m+n} < \bar{t}_{m+n+1} < \cdots < \bar{t}_{m+n+r} \leq 1, \tag{1.5}$$

where r and s are fixed integer numbers and the $\bar{t}_j$ are equidistant, defined as in (1.3). In the following we present a few examples of choices of the nodes $t_1, \ldots, t_{m+n}$, with $n = m+1$. For instance, given an arbitrary weight function σ, we can consider the roots of the polynomial $p_m(\sigma)p_{m+1}(\sigma), m \in \mathbb{N}$.

Setting $\sigma_1(x) = (1-x)\sigma(x)$, $\sigma_2(x) = (1+x)\sigma(x)$, $\sigma_3(x) = (1-x^2)\sigma(x)$, ($\sigma$ arbitrary), we can take the zeros of $p_{m+1}(\sigma)p_m(\sigma_3)$ or $p_m(\sigma_1)p_m(\sigma_2)$ (see for instance [9]). In the case in which σ represents a generalized Jacobi weight (as in (1.1)), several approaches were already proposed to determine the sufficient conditions to allow the convergence for the extended Hermite and Lagrange interpolation [2, 3, 4, 5, 6, 8, 10]. On the other hand, the necessary conditions for the convergence are indeed poorly clarified in the previous papers.

A first condition, including only the weight function σ, σ^* and u, can be derived by a recent result [18] and can be summarized as follows.

Theorem 1.1 Let $\mathcal{L}_M(f) = \mathcal{L}_M(\sigma,\sigma^*;t)$, $M = m+n+r+s$, be the Lagrange polynomial interpolating the function $f \in C^0([-1,1])$ at points (1.5). Furthermore, let u,σ be two arbitrary weights and $u \in L_p$, $1 \leq p < \infty$. If σ^* is a generalized Jacobi weight, then a positive constant $C \neq C(M)$ exists such that, for $1 \leq p < \infty$

$$\sup_M \|\mathcal{L}_M f\|_{u,p} \geq C \left[\int_{-1}^1 \left(\frac{(1-x)^r(1+x)^s u(x)}{\sqrt{\sigma(x)\sqrt{1-x^2}}\sqrt{\sigma^*(x)\sqrt{1-x^2}}} \right)^p dx \right]^{\frac{1}{p}}. \tag{1.6}$$

A direct consequence of Theorem 1.1 is the fact that if for some $p_0 \geq 1$ the right–side of (1.6) is infinity, then a continuous function g exists such that

$$\lim_m \|(g - \mathcal{L}_M g)u\|_p = \infty, \quad p \geq p_0.$$

411

It appears interesting to stress that an ordinary interpolation process based on the zeros of $p_m(\sigma)$ converge in $L_{\sqrt{\sigma},2}$ for each σ and for every continuous function; this favorable situation in general is not valid for the extended interpolation. Sometimes condition (1.6) is also sufficient. In fact, let us consider the interpolation process $\tilde{\mathcal{L}}_M f$, $M = 2m + r + s + 1$ based on the roots t_k, $k = 1, \ldots, 2m + 1$ of $p_{m+1}(\sigma)p_m(\sigma_3)$, where $\sigma_3(x) = (1 - x^2)\sigma(x)$, $\sigma(x) = v^{\alpha,\beta}(x) \prod_{k=1}^{r} |c_k - x|^{\Gamma_k}$, $\alpha, \beta, \Gamma_k > -1$, $-1 < c_1 < \cdots < c_r < 1$, and on the $r + s$ additional nodes $\bar{t}_j$ (as in (1.3)).

From Theorem 1.1 and from [8, Theorem 3.1] we have:

Corollary 1.1 *Let $\tilde{\mathcal{L}}_M f$ be the previous Lagrange polynomial and let u be defined by (1.1). Let us assume $u \in L_p$, $1 \leq p < \infty$, and $\sigma_3/v^{r,s} \in L_1$. Then, for every continuous function f,*

$$\lim_m \left\| \left(f - \tilde{\mathcal{L}}_M f \right) u \right\|_p = 0$$

if and only if

$$\frac{v^{r,s}}{\sigma_3} \in L_{u,p}, \quad p \in [1, \infty).$$

In particular, $\forall p \in [1, \infty)$

$$\lim_m \left\| \left(f - \tilde{\mathcal{L}}_{2m+3} f \right) \sigma \right\|_p = 0.$$

The same result holds when the $2m$ roots of $p_m(\sigma_1)p_m(\sigma_2)$, $\sigma_1(x) = \sigma(x)(1 - x)$, $\sigma_2(x) = \sigma(x)(1 + x)$, take the place of the zeros of $p_{m+1}(\sigma)p_m(\sigma_3)$. Generally, however, condition (1.6) is not sufficient for the convergence, and some other assumptions are required. Indeed, in $[-1, 1]$ let $\sigma(x) = v^{\alpha,\beta}(x) \prod_{k=1}^{r} |c_k - x|^{\Gamma_k}$, $\alpha, \beta, \Gamma_k > -1$, $-1 < c_1 < \ldots < c_r < 1$, and let us consider the interpolating process $L_M^* f = L_M^*(\sigma, \sigma; f)$, $M = 2m + 1$, based on the zeros of $p_m(\sigma)p_{m+1}(\sigma)$. The following theorem holds.

Theorem 1.2 *Let $f \in C^0([-1, 1])$, let $L_M^* f = L_M^*(\sigma, \sigma; f)$ be the previous Lagrange polynomial and let $u \in L_p$ be defined by (1.1). Then, for $1 \leq p < \infty$*

$$\lim_m \left\| [f - L_M^*(\sigma, \sigma; f)] u \right\|_p = 0$$

if and only if

$$\frac{1}{v^{1/2,1/2}} \quad and \quad \frac{1}{\sigma v^{1/2,1/2}} \in L_{u,p}, \quad 1 \leq p < \infty.$$

The Theorem 1.2 completes the Theorem 3.4 in [8] and the Theorem 2.3 in [18]: in both theorems the condition $v^{-1/2,-1/2} \in L_{u,p}$ is only sufficient. In the particular case in which the interpolation points are the zeros of $T_m T_{m+1}$, then the following simple formula is true:

$$L_{2m+1}(fx) = T_m(x)T_{m+1}(x) \left[\frac{1}{m+1} \sum_{k=1}^{m+1} \frac{f(x_{m+1,k})}{x - x_{m+1,k}} - \frac{1}{m} \sum_{k=1}^{m} \frac{f(x_{m,k})}{x - x_{m,k}} \right],$$

(1.7)

where

$$x_{n,k} = -\cos\frac{2k-1}{2n}\pi, \quad k = 1, ..., n.$$

However, using the Theorem 1.2 with $u(x) = (1 - x^2)^{-\frac{1}{2p}}$, $\sigma(x) = (1 - x^2)^{-\frac{1}{2}}$, condition $1/v^{\frac{1}{2},\frac{1}{2}} \in L_{u,p}$ is not satisfied for $p \geq 1$. Thus, there exists a continuous function g such that

$$\lim_m \int_{-1}^{1} |g(x) - L_{2m+1}(g, x)|^p \frac{dx}{\sqrt{1 - x^2}} = \infty, \quad \forall p \geq 1.$$

Obviously, (1.7) can again be convergent in $L_{u,p}$, if we assume a greater regularity of the function f.

In [11], among other articles, W. Gautschi considers the polynomial sequence $\{\pi_m\}$, defined as

$$\int_{-1}^{1} \pi_{m+1}(x)T_m^2(x) \frac{x^k}{\sqrt{1 - x^2}} dx = 0, \quad k = 0, 1, \dots, m, \quad m > 0,$$

where T_m is the m-th Chebyshev polynomial, and he shows that $\pi_{m+1}T_m$ has $2m + 1$ distinct zeros $\{z_i\}_{i=1}^{2m+1}$ belonging to $(-1,1)$. He considers the interpolating $\mathcal{L}_{2m+1}f$ process based on $z_1, \dots, z_{2m+1}$ and studies the convergence in $L_{\sqrt{\sigma},2}$, $\sigma(x) = (1 - x^2)^{-\frac{1}{2}}$. We remark that the sequence $\{\pi_m\}$ is identical to the polynomial sequence $\{p_m(w)\}$ orthogonal to the Bernstein–Szegö weight

$$w(x) = \frac{1}{(1 - \frac{9}{8}x^2)\sqrt{1 - x^2}},$$

(see [1, p.205, (13.9)]), and therefore (1.6) is fulfilled. But, unfortunately, it is possible to show (see §2) that the interpolating process studied by W. Gautschi is equivalent to (1.7) and the considerations given above are true.

It is worth noticing that in the special case of Corollary 1.1, the interpolation points have a "good" distribution. More precisely, if we denote by $t_k = \cos\theta_{2m+1,k}$, $k = 1, \dots, 2m + 1$, the zeros of $p_m(\sigma)p_m(\sigma_3)$ with $\sigma \in GJ$ and $\sigma_3 = (1 - x^2)\sigma(x)$ we have [9]

$$|\theta_{2m+1,k} - \theta_{2m+1,k+1}| \sim m^{-1}, \quad \text{i.e.} \min_{j,k} |t_j - t_k| \sim m^{-2}.$$

For the zeros $z_i, i = 1, \dots, 2m + 1$, used by W. Gautschi or in Theorem 1.2 we have

$$\min_{j,k} |z_j - z_k| \sim m^{-3}.$$

413

On the other hand, condition (1.6) appears independent of the distribution of interpolation points and is related with the $\underline{\lim}$ of each $p_m(\sigma)$ and $p_m(\sigma^*)$ separately (see the proof of Theorem 1.1 in §2).

The following theorem shows that the "good" distribution of nodes is, in some sense, a necessary condition for the $L_{u,p}$ convergence of general interpolating processes.

Theorem 1.3 Let $f \in C^0([-1,1])$, $X = \{x_{m,i}, \ i = 1, \ldots, m\}_{m=1,2,\ldots}$ be an arbitrary matrix of knots and $\{\mathcal{L}_m f\}$ be the corresponding Lagrange polynomial sequence. Assume that for some $i \in \{1.\ldots,m\}$ and for each $m = 1,2,\ldots$

$$| x_{m,i+1} - x_{m,i} | \sim \frac{\alpha_m^\tau}{m^{2+\lambda}}, \quad \lambda, \tau \geq 0, \quad \alpha_m = o(1), \quad m \to \infty.$$

Then, for every $u \in GJ$ with $u \in L_p$, $u^{-1} \in L_q$, $q^{-1} + p^{-1} = 1$, $1 < p < \infty$

$$\sup_{\|f\|=1} \|(\mathcal{L}_m f)u\|_p \geq c \frac{m^\lambda}{\alpha_m^\tau} \left(\int_{x_{m,i}}^{x_{m,i+1}} u^p(x)dx \right)^{\frac{1}{p}},$$

where $c > 0$ is a constant independent of m.

We think that the previous theorem can be improved. However, this theorem already shows that when the interpolation points are very close (λ big) then we obtain a bad interpolation process.

In conclusion, we can obtain a good extended interpolation process when the condition (1.6) is satisfied and the interpolation knots have a "good" distribution in the meaning previously indicated.

2 PROOFS

First we state the following theorem, which is an easy consequence of Theorem 2.2 in [18].

Theorem 2.1 Assume $f \in C^0([-1,1])$. Let $M = m+n+r+s$ and $L_M f$ be the Lagrange polynomial that interpolates the function f on the zeros of $p_m(\sigma)p_n(\sigma^*)$ and on $r+s$ additional knots $\bar{t}_j$ (see (1.5)). Then for each of the weights σ, σ^*, u and w, we obtain

$$\|L_M\|_{u,p} \geq C \, \|p_m(\sigma)B_s A_r p_m(\sigma^*)w\|_1^{-1} \|A_r B_s p_m(\sigma)p_m(\sigma^*)u\|_p \qquad (2.1)$$

where $C > 0$ is independent of m, $A_r(x) = \prod_{i=1}^{r}(x - \bar{t}_{m+i})$ and $B_s(x) = \prod_{j=0}^{s-1}(x - \bar{t}_j)$.

Proof of Theorem 1.1: Starting from (2.1), with $w = \sqrt{\sigma \, \sigma^*}$, we obtain

$$\|p_m(\sigma)p_m(\sigma^*)A_rB_s\sqrt{\sigma \, \sigma^*}\|_1 \leq 2^{r+s}\|p_m(\sigma)p_n(\sigma^*)\sqrt{\sigma \, \sigma^*}\|_1$$
$$\leq 2^{r+s}\|p_m^2(\sigma)\sqrt{\sigma}\|_2\|p_n^2(\sigma^*)\sqrt{\sigma^*}\|_2 = 2^{r+s};$$
$$\|L_M\|_{u,p} \geq C \, \|A_rB_sp_m(\sigma)p_n(\sigma^*)u\|_p.$$

By hypothesis, σ^* is a generalized Jacobi weight and, only for simplicity, we assume $\sigma^*(x) = v^{\gamma,\delta}(x)|x|^\rho \in L_p$. Denoting by $x_i = x_{n,i}(\sigma^*) = \cos\theta_i$, $x_i < x_{i+1}$, the zeros of $p_n(\sigma^*)$, it follows that $|\theta_i - \theta_{i+1}| \sim m^{-1}$ and (see [23, p.171], for every $x \in [-1,1]$ we have

$$|p_n(\sigma*,x)| \sim \frac{n|\theta - \theta_d|}{\sqrt{\sigma_n^*(x)}}, \qquad (2.2)$$

where $\cos\theta = x$, $\cos\theta_d = x_d$ is the closest knot to x and

$$\sigma_m^*(x) = (\sqrt{1-x} + m^{-1})^{2\gamma+1}(|x| + m^{-1})^\rho(\sqrt{1+x} + m^{-1})^{2\delta+1}.$$

Now, following [18, p.439], for a fixed "small" $\delta > 0$ let

$$\delta_{n,k} = \frac{\delta}{3}\frac{\sqrt{1 - x_{n,k}^2}}{n} \sim \delta(\theta - \theta_{n,k}),$$

and define the sets

$$I_{1n} = [-1, x_{n,1}] \cup [x_{n,n}, 1] \cup [-\frac{1}{n}, +\frac{1}{n}]$$
$$I_{2n} = \cup_{k=1}^n [x_{n,k} - \delta_{n,k}, x_{n,k} + \delta_{n,k}], \qquad I_n = I_{1n} \cup I_{2n}.$$

Notice that

$$meas I_n \leq \frac{c}{n^2} + \frac{2}{n} + \frac{2}{3}\delta < \delta, \qquad n > n_0.$$

For all $x \in I_n' = I \backslash I_n$ the following lower bounds hold

$$(A_rB_s)(x) \geq v^{r,s}(x),$$
$$|p_n(\sigma^*,x)| \geq \frac{c\delta}{\sqrt{\sigma^*(x)}\sqrt{1-x^2}}, \qquad \text{(from} \quad (2.2)\text{)}.$$

Therefore,

$$\|L_M\|_{u,p} \geq \|p_n(\sigma^*)p_n(\sigma)A_rB_s\|_{L_p(I_n')} \geq C \, \delta \left\|\frac{v^{r,s}p_m(\sigma)u}{\sqrt{\sigma^*}v^{\frac{1}{2},\frac{1}{2}}}\right\|_{L_p(I_n')}.$$

Since $meas I_n' = 2 - \delta$, and because of the absolute continuity of the integral we can choose δ such that

$$\left\|\frac{v^{r,s}p_m(\sigma)u}{\sqrt{\sigma^*}v^{\frac{1}{2},\frac{1}{2}}}\right\|_{L_p(I_n')} \geq \frac{1}{2}\left\|\frac{v^{r,s}p_m(\sigma)u}{\sqrt{\sigma^*}v^{\frac{1}{2},\frac{1}{2}}}\right\|_p,$$

i.e.

$$||L_M||_{u,p} \geq \frac{C\delta}{2} \left\| \frac{v^{r,s}p_m(\sigma)u}{\sqrt{\sigma^* v^{\frac{1}{2},\frac{1}{2}}}} \right\|_p.$$

Finally, for every measurable function g and for every weight function σ we have the result [19] that

$$\underline{\lim}_n ||p_n(\sigma)g||_p \geq C \left\| \frac{g}{\sqrt{\sigma v^{\frac{1}{2},\frac{1}{2}}}} \right\|_p$$

and the theorem follows. ∎

Corollary 1.1 follows immediately by Theorem 3.1 in [7], where the sufficiency of the condition is proved, and by Theorem 1.1.

Proof of Theorem 1.2: In Theorem 3.4 of [8] it is proved that $1/v^{\frac{1}{2},\frac{1}{2}}$, $1/\sigma v^{\frac{1}{2},\frac{1}{2}}$ $\in L_{u,p}$ is a sufficient condition for obtaining convergence. By Theorem 1.1 it follows that $1/\sigma v^{\frac{1}{2},\frac{1}{2}} \in L_{u,p}$ is a necessary condition. Therefore it remains to prove that $1/v^{\frac{1}{2},\frac{1}{2}} \in L_{u,p}$ is also a necessary condition. Now we assume $\sup_m ||L_M||_{u,p} < \infty$, and $\frac{u}{\sigma v^{\frac{1}{2},\frac{1}{2}}} \in L_p$, and we show that also $u/v^{\frac{1}{2},\frac{1}{2}} \in L_p$. If the exponents α, β of the weight σ are positive then $u/\sigma v^{\frac{1}{2},\frac{1}{2}} \in L_p$ implies $u/v^{\frac{1}{2},\frac{1}{2}} \in L_p$. So suppose $\alpha, \beta \leq 0$. In that case, using the previous notations, we have [9]

$$L_{2m+1}(f,x) = p_{m+1}p_m(x)\frac{\gamma_m(\sigma)}{\gamma_{m+1}(\sigma)} \left[H_{m+1}(\sigma, f, x) - H_m(\sigma, f, x) \right],$$

where

$$H_n(\sigma, f, x) = \sum_{k=1}^{n} \frac{\lambda_{n,k}(\sigma)f(x_{m,k})}{x - x_{n,k}}, \quad x_{n,k} = x_{n,k}(\sigma),$$

$$\lambda_{n,k}(\sigma) = \left[\sum_{i=1}^{n} p_i^2(\sigma, x_{n,k}) \right]^{-1}.$$

Now, let $g = g_m$ be the continuous function defined as follows : $g(x_{m+1,k}) = 0$, $k = 1, ..., m+1$, $g(x_{m,k}) = 1$, $k = 1, ..., m$ and g is a linear function on the intervals $[x_{m+1,k}, x_{m,k}]$ and $[x_{m,k}, x_{m+1,k+1}]$. Obviously $||g||_\infty = 1$ and

$$||(L_{2m+1}g)u||_p \leq ||L_{2m+1}||_{u,p} < \infty.$$

Moreover,

$$|L_{2m+1}(g,x)| = \left| p_{m+1}(\sigma, x)\frac{\gamma_m(\sigma)}{\gamma_{m+1}(\sigma)}p_m(\sigma, x) \sum_{k=1}^{m} \frac{\lambda_{m,k}(\sigma)}{x - x_k} \right|$$

$$= \frac{\gamma_m(\sigma)}{\gamma_{m+1}(\sigma)} \left| p_{m+1}(\sigma, x) \sum_{k=1}^{m} \lambda_{m,k}(\sigma)\frac{p_m(\sigma, x) - p_m(\sigma, x_k)}{x - x_k} \right|$$

$$= \frac{\gamma_m(\sigma)}{\gamma_{m+1}(\sigma)}\left|p_{m+1}(\sigma,x)\int_{-1}^{1}\frac{p_m(\sigma,x)-p_m(\sigma,t)}{x-t}\sigma(t)dt\right|$$

$$= \frac{\gamma_m(\sigma)}{\gamma_{m+1}(\sigma)}|p_{m+1}(\sigma,x)q_{m-1}(x)|,$$

where $\{q_m\} = \{q_m(\overline{\sigma})\}$ is the sequence of orthonormal polynomials with respect to the weight (see [8])

$$\overline{\sigma}(x) = \frac{\sigma(x)}{Q_0(x)^2 + \pi^2\sigma^2(x)}, \qquad Q_0(x) = \int_{-1}^{1}\frac{\sigma(t)}{t-x}dt.$$

Since $\frac{\gamma_m(\sigma)}{\gamma_{m+1}(\sigma)} \sim 1$, we have

$$\|(L_{2m+1}g)u\|_p \sim \|p_{m+1}(\sigma)q_{m-1}(\overline{\sigma})u\|_p.$$

By proceeding exactly as in the proof of the Theorem 1.1 (i.e. introducing δ, I_m, I'_M, and using (2.2)) we obtain the following estimate

$$\overline{\lim}_m\|(L_{2m+1}g)u\|_p \geq \overline{\lim}_m\delta\left\|\frac{q_{m-1}(\overline{\sigma})u}{\sqrt{\sigma}v^{\frac{1}{2},\frac{1}{2}}}\right\|_p.$$

Moreover by [19]

$$\underline{\lim}_m\left\|\frac{q_{m-1}(\overline{\sigma})u}{\sqrt{\sigma}v^{\frac{1}{2},\frac{1}{2}}}\right\|_p \geq C\left\|\frac{u}{\sqrt{\sigma}v^{\frac{1}{2},\frac{1}{2}}\sqrt{\overline{\sigma}}v^{\frac{1}{2},\frac{1}{2}}}\right\|_p, \qquad 1 \leq p < \infty.$$

But

$$\frac{u}{\sqrt{\sigma}v^{\frac{1}{2},\frac{1}{2}}\sqrt{\overline{\sigma}}v^{\frac{1}{2},\frac{1}{2}}} = \frac{u}{v^{\frac{1}{2},\frac{1}{2}}}\left[\sqrt{\frac{Q_0^2 + \pi^2\sigma^2}{\sigma^2}}\right] \geq \frac{\pi}{v^{\frac{1}{2},\frac{1}{2}}}$$

and the theorem holds. ∎

Remark. The Lagrange polynomial considered by W. Gautschi, is based on the zeros of $T_m(x)p_{m+1}(w,x)$, $w(x) = (1-(8/9)x^2)^{-1}(1-x^2)^{-1/2}$ and can be written as

$$L_{2m+1}(f,x) = \frac{p_{m+1}(w,x)T_m(x)}{2}$$

$$\times \left[\sum_{k=1}^{m+1}\lambda_{m,k}(w)\left(1-\frac{T_{m-2}(x_k)}{T_m(x_k)}\right)\frac{f(x_k)}{x-x_k} - \frac{\pi}{3m}\sum_{k=1}^{m}\frac{f(t_k)}{x-t_k}\right],$$

$$p_{m+1}(w,x_k) = 0, \qquad k = 1,...,m+1,$$

$$T_m(t_k) = 0, \qquad k = 1,...,m.$$

417

We can then apply to it the machinery used in the proof of the Theorem 1.2 and derive that $u/v^{\frac{1}{2},\frac{1}{2}} \in L_p$ is a necessary condition. Moreover, recalling that $|T_m(x)| \sim \frac{m|x-t_d|}{\sqrt{1-x^2}}$ [23, p.171], where t_d is the closest zeros of T_m to x, and that $|T(x_k)| = \sqrt{\frac{1-x_k^2}{1+(8/9)x_k^2}}$ [11], it also follows that $|x_k - t_d| \sim \frac{1-x_k^2}{m}$.

Proof of Theorem 1.3: Setting $I_m = [x_{m,i}, x_{m,i+1}]$ and $|I_m| = x_{m,i+1} - x_{m,i}$, it follows by hypothesis that $\frac{m^{2+\lambda}}{\alpha_m^\tau} \sim |I_m|^{-1}$. We define $f = f_m$ to be a continuous function in $[-1,1]$, linear in $[x_{m,i-1}, x_{m,i}]$ and in $[x_{m,i}, x_{m,i+1}]$ and such that $f(x_{m,i\pm1}) = 0$ and $f(x) = 0$, $x \notin (x_{m,i-1}, x_{m,i+1})$. Notice that $\|f\| = 1$. Because $u \in L_p$ and $u^{-1} \in L_q$, we have

$$
\begin{aligned}
\frac{m^{2+\lambda}}{\alpha_m^\tau} &\leq C\,|I_m|^{-1} = C\,|I_m|^{-1}[f(x_{m,i} - f(x_{m+1,i})] \\
&= C\,|I_m|^{-1}[L_m(f, x_{m,i}) - L_m(f, x_{m+1,i})] \\
&= C\,|I_m|^{-1} \int_{I_m} (L_m f)'(x)dx \\
&\leq C\,|I_m|^{-1} \left(\int_{I_m} |(L_m f)'(x)u(x)|^p dx \right)^{1/p} \left(\int_{I_m} u^{-q}(x)dx \right)^{1/q}.
\end{aligned}
$$

Since [20, p.313]

$$
\left(\int_{I_m} u^p(x)dx \right)^{1/p} \left(\int_{I_m} u^{-q}(x)dx \right)^{1/q} \leq C\,|I_m|,
$$

we have

$$
\begin{aligned}
\frac{m^{2+\lambda}}{\alpha_m^\tau} \left(\int_{I_m} u^p(x)dx \right)^{1/p} &\leq C \left(\int_{I_m} |(L_m f)'(x)u(x)|^p dx \right)^{1/p} \\
&\leq C \left(\int_{-1}^{1} |(L_m f)'(x)u(x)|^p dx \right)^{1/p}
\end{aligned}
$$

Finally, by the Markov inequality [14]

$$
\frac{m^\lambda}{\alpha_m^\tau} \left(\int_{I_m} u^p(x)dx \right)^{1/p} \leq C\,\|(L_m f)u\|_p \leq \|L_m\|_{u,p}
$$

and the theorem follows. ∎

REFERENCES

[1] Chihara T.S. *An Introduction to Orthogonal Polynomials.* Gordon and Breach, New York, London, Paris 1978.

[2] Criscuolo G., Della Vecchia B., Mastroianni G. Hermite interpolation and mean convergence of its derivatives. *Calcolo,* **28**:111–127, 1991.

[3] Criscuolo G., Della Vecchia B., Mastroianni G. Extended Hermite interpolation on Jacobi zeros and mean convergence of its derivatives. *Facta Univ., Ser. Math. Inform.,* **7**:111–127, 1992.

[4] Criscuolo G., Della Vecchia B., Mastroianni G. Extended Hermite interpolation with additional nodes and mean convergence of its derivatives. *Rendiconti di Matematica,* Serie VII, **12**:709–728, 1992.

[5] Criscuolo G., Della Vecchia B., Mastroianni G. Hermite–Fejér and Hermite interpolation. In *Approximation Theory, Spline Functions and Applications,* pages 317–331. Kluwer Acad. Publ. S.P. Singh, ed.

[6] Della Vecchia B., Mastroianni G., Vertesi P. Weighted L^p approximation by Hermite interpolation of higher order plus endpoints. To appear in *Studia Sci. Hungar.,* 1994.

[7] Criscuolo G., Mastroianni G., Nevai P. Associated Jacobi functions and polynomials. *J. Math. Anal. and Appl.,* **158**:15–34, 1991.

[8] Criscuolo G., Mastroianni G., Nevai P. Mean convergence of the derivatives of extended Lagrange interpolation with additional nodes. *Math. Nachr.,* **163**:73–92, 1993.

[9] Criscuolo G., Mastroianni G., Occorsio D. Convergence of extended Lagrange interpolation. *Math. Comp.,* **55**:197–212, 1990.

[10] Della Vecchia B., Mastroianni G. On Hermite interpolation. To appear in *Calcolo.*

[11] Gautschi W. On mean convergence of extended Lagrange interpolation. *J. Comp. and Appl. Math.,* **43**:19–35, 1992.

[12] Gautschi W., Notaris S.E. Gauss–Kronrod quadrature formulae for weight functions of Bernstein–Szegö type. *J. Comp. and Appl. Math.,* **25**:199–22, 1989.

[13] Junghanns P., Silbermann B. Numerical analysis of the quadrature method for solving linear and nonlinear singular integral equations. Preprint, *Technical University Chemnitz,* 1988.

[14] Lubinsky D.S., Nevai P. Markov–Bernstein inequalities revisited. *Approx. Theory Appl.,* **3**:98–119, 1987.

[15] Mastroianni G. Uniform convergence of derivatives of Lagrange interpolation. *J Comp. and Appl. Math.,* **43**(2):37–51, 1992.

[16] Mastroianni G., Nevai P. Mean convergence of derivatives of Lagrange interpolation. *J. Comp. and Appl. Math.*, **34**:385–396, 1991.

[17] Mastroianni G., Prössdorf S. A quadrature method for Cauchy integral equations with weakly singular perturbation kernel. *J. Integral Equations Appl.*, **4**:205–228, 1992.

[18] Mastroianni G., Vértesi P. Mean convergence of Lagrange interpolation on arbitrary system of nodes. *Acta Sci. Math.*, (Szeged), **57**:429–441, 1993.

[19] Maté A., Nevai P., Totik V. Necessary condition for weighted mean convergence of Fourier series in orthogonal polynomials. *J.Approx. Theory*, **46**:314–322, 1986.

[20] Mikhlin S.G., Prössdorf S. *Singular Integral Operators*. Springer–Verlag, Berlin Heidelberg, New York, Tokyo, 1986.

[21] Monegato G. Stieltjes polynomials and related quadrature rules. *SIAM Rev.*, **24**:137–158, 1982.

[22] Nevai P. Mean convergence of Lagrange interpolation III. *Trans. Amer. Soc.*, **282**:669–698, 1984.

[23] Nevai P. *Orthogonal Polynomials*. Mem. Amer. Math. Soc. No. 213, 1979.

International Series of Numerical Mathematics, Vol. 119, ©1994 Birkhäuser

ERROR ESTIMATES FOR GAUSS-LAGUERRE AND GAUSS-HERMITE QUADRATURE FORMULAS*

G. Mastroianni[†]   G. Monegato[‡]

Dedicated to Walter Gautschi on the occasion of his 65th birthday

Abstract. New error estimates are derived for Gauss-Laguerre and Gauss-Hermite $m-$point quadrature formulas; they are of the type $O(m^{-r/2})\|x^{r/2}f^{(r)}x^\alpha e^{-x}\|_{L_1}$ and $O(m^{-r/2})\|f^{(r)}e^{-qx^2}\|_{L_1}$, $0 < q < 1$, respectively, for functions f in suitable function classes.

1 INTRODUCTION

The numerical evaluation or discretization of integrals over infinite intervals is in general a more difficult task than the corresponding finite-interval case. Among the quadrature formulas for infinite intervals, the Gauss-Laguerre and Gauss-Hermite rules

$$\int_0^\infty x^\alpha e^{-x} f(x) dx = \sum_{k=1}^m \lambda_{m,k}^L f(x_{m,k}^L) + R_m^L(f),$$

$$\int_{-\infty}^\infty e^{-x^2} f(x) dx = \sum_{k=1}^m \lambda_{m,k}^H f(x_{m,k}^H) + R_m^H(f)$$

are certainly the most important and widely used in applications.

Uspensky [19] seems to have been the first to have proved the convergence of these formulas, by assuming

$$|f(x)| \le c \frac{e^x}{x^{\alpha+1+\rho}}, \quad x \to \infty,$$

and

$$|f(x)| \le c \frac{e^{x^2}}{|x|^{1+\rho}}, \quad x \to \infty,$$

---

*Work performed under the auspices of the Ministero dell'Università e della Ricerca Scientifica e Tecnologica of Italy

[†]Dipartimento di Matematica, Università della Basilicata, Italia; mastroianni@pzvx85.cineca.it

[‡]Dipartimento di Matematica, Politecnico di Torino, Italia; monegato@itopoli.bitnet

respectively, for some $\rho > 0$. When f is an entire function, and its Taylor coefficients satisfy certain conditions, Lubinsky [12] has proved geometric convergence for both types of Gaussian rules.

When $f \in C^r[0, \infty)$, and $f \in C^r(-\infty, \infty)$ in the Hermite case, and additional growth conditions are satisfied on $f^{(r)}$, known results of approximation theory can be used to derive fairly easily (see for instance [17],[15]) error bounds of the type $O(m^{-r/2})$. In the Laguerre case, however, there are applications (for example certain Wiener-Hopf integral equations, (see [8],[15]) where the function f has an extra singular behavior at the origin. For instance, in a recent paper (see [15]) we have considered the case of a function $f \in C^p[0, \infty) \cap C^r(0, \infty)$, $r > p \geq 0$, such that $x^i f^{(p+i)}(x) \in C[0, \infty)$, $i = 1, ..., r - p$, and have derived the bounds

$$|R_m^L(f)| = \begin{cases} O(m^{-r/2})E_{m-r-1}(\Phi^{(r)}; e^{-x/2}) & \text{if } r \leq 2p+1, \\ O(m^{-p-1}\log m)E_{m-r-1}(\Phi^{(r)}; e^{-x/2}) & \text{if } r = 2p+2, \\ O(m^{-p-1})E_{m-r-1}(\Phi^{(r)}; e^{-x/2}) & \text{if } r \geq 2p+3, \end{cases}$$

where $\Phi(x) = x^{r-p}f(x)$ and $E_n(g; w) = \inf_{p_n} \|w(f - p_n)\|_\infty$, the infimum being taken over all polynomials of degree $\leq n$.

In this paper we present some new estimates. In the Laguerre case they are valid for functions f not necessarily in $C^r(0, \infty)$: it is sufficient that $f^{(r-1)}$ is locally absolutely continuous ($f^{(r-1)} \in AC_{\text{loc}}$) and that $x^{r/2+\alpha} f^{(r)}(x)e^{-x} \in L_1$. For example, one of the error estimates we obtain in §2 is of the form

$$|R_m^L(f)| \leq cm^{-r/2}\|x^{r/2}f^{(r)}x^\alpha e^{-x}\|_{L_1}.$$

In §3 we derive an analogous error estimate for the Gauss-Hermite formula; it holds whenever $f^{(r)} \in AC_{\text{loc}}$ and $f^{(r)}(x)e^{-qx^2} \in L_1$ for some $0 < q < 1$.

The estimates we present in this paper are of the same type as the corresponding ones recently obtained in [14] for the Gauss-Jacobi formulas and certain associated product rules.

2   ERROR BOUNDS FOR THE GAUSS-LAGUERRE FORMULA

Let $w_\alpha(x) := x^\alpha e^{-x}$, $\alpha > -1$, $x > 0$. Denote by $\{p_m(w_\alpha; x)\}$ the associated (generalized) Laguerre polynomials (see [4],[18]), and let $\{x_k := x_{m,k}(w_\alpha)\}_{k=1}^m$ be the m zeros of $p_m(w_\alpha; x)$ ordered as follows:

$$0 < x_1 < x_2 < \cdots < x_{m-1} < x_m < \infty.$$

Consider the Gauss-Laguerre formula

$$\int_0^\infty f(x)w_\alpha(x)dx = \sum_{k=1}^m \lambda_{m,k}(w_\alpha)f(x_k) + R_m(f; w_\alpha). \tag{2.1}$$

In the following, $L_1 := L_1(0, \infty)$, and we denote by AC_{loc} the space of functions which are absolutely continuous on every closed subset of $(0, \infty)$, and $\phi(x) := \sqrt{x}$. The symbols c, c_0, c_1, c_2 will denote positive constants taking different values on different occurrences.

For the remainder term $R_m(f; w_\alpha)$ in (2.1) we derive the following estimates.

Theorem 1 Let $f^{(r-1)} \in AC_{\text{loc}}$ and $\phi^r f^{(r)} w_\alpha \in L_1$ for some $r \geq 1$. Then we have

$$|R_m(f; w_\alpha)| \leq cm^{-r/2} \|\phi^r f^{(r)} w_\alpha\|_{L_1}. \tag{2.2}$$

If we assume $f^{(r-1)} \in AC_{\text{loc}}$ and $\phi^{2r} f^{(r)} w_\alpha \in L_1$, then we have the bound

$$|R_m(f; w_\alpha)| \leq c_1 \{ \|\phi^{2r} f^{(r)} w_\alpha\|_{L_1(0, x_1)} + m^{-r/2} \|\phi^r f^{(r)} w_\alpha\|_{L_1(x_1, cx_m)} \tag{2.3}$$

$$+ m^{-r/2} \|\phi^{2r} f^{(r)} w_\alpha\|_{L_1(cx_m, \infty)} \}$$

with constants c_1 and c independent of m and f.

Proof: First we assume $f \in AC_{\text{loc}}$, that is, $r = 1$, and define

$$\Gamma_t(x) = (x - t)^0_+ = \begin{cases} 1 & \text{if } x > t, \\ 0 & \text{if } x \leq t. \end{cases}$$

From Peano's theorem (see for instance [2, p.286]) we have

$$|R_m(f; w_\alpha)| \leq \int_0^\infty |R_m(\Gamma_t; w_\alpha)| |f'(t)| dt. \tag{2.4}$$

We also know (see [4, p.105]) that there exist two polynomials $Q_t^+(x)$, $Q_t^-(x)$ of degree $2m - 2$ such that

$$Q_t^-(x) \leq \Gamma_t(x) \leq Q_t^+(x)$$

and

$$\int_0^\infty [Q_t^+(x) - Q_t^-(x)] w_\alpha(x) dx = \lambda_m(w_\alpha; t).$$

Therefore, setting

$$A := \int_0^\infty [\Gamma_t(x) - Q_t^-(x)] w_\alpha(x) dx,$$

$$B := \sum_{k=1}^m \lambda_{m,k}(w_\alpha) [\Gamma_t(x_k) - Q_t^-(x_k)],$$

we can write

$$|R_m(\Gamma_t; w_\alpha)| = |R_m(\Gamma_t - Q_t^-; w_\alpha)| = |A - B|.$$

Since

$$0 < A \le \int_0^\infty [Q_t^+(x) - Q_t^-(x)]w_\alpha(x)dx = \lambda_m(w_\alpha; t)$$

and

$$0 < B \le \sum_{k=1}^m \lambda_{m,k}(w_\alpha)[Q_t^+(x_k) - Q_t^-(x_k)] = \lambda_m(w_\alpha; t),$$

we also have

$$|R_m(\Gamma_t; w_\alpha)| \le \lambda_m(w_\alpha; t), \ t \ge 0. \tag{2.5}$$

For $0 \le t < x_1$ we obtain

$$|R_m(\Gamma_t; w_\alpha)| = |\int_0^\infty (x - t)_+^0 w_\alpha(x)dx - \sum_{k=1}^m \lambda_{m,k}(w_\alpha)| \tag{2.6}$$

$$= |\int_t^\infty w_\alpha(x)dx - \int_0^\infty w_\alpha(x)dx| = \int_0^t x^\alpha e^{-x}dx$$

$$\le \frac{t^{\alpha+1}}{\alpha+1} \le \frac{e^{x_1}}{\alpha+1}tw_\alpha(t) \le \frac{e^{\alpha+1}}{\alpha+1}tw_\alpha(t).$$

The latter bound follows by a result in [18, Eq.(6.31.12)].

When $x_m \le t < \infty$

$$|R_m(\Gamma_t; w_\alpha)| = \int_t^\infty x^\alpha e^{-x}dx \le c(\frac{t}{x_m})^\mu w_\alpha(t), \ \forall \mu \in R_+. \tag{2.7}$$

For $x_1 \le t < x_m$ we proceed as follows. First we notice that we have (see [6],[9, p.231])

$$\lambda_m(w_\alpha; t) \le c(\alpha)\sqrt{\frac{t}{m}}w_\alpha(t) \tag{2.8}$$

for $c_1(\alpha)m^{-1} \le t \le c_2(\alpha)m$, where $c_1(\alpha)$ and $c_2(\alpha)$ are two (unknown) constants depending only upon α. Thus, from (2.5) we can state the bound

$$|R_m(\Gamma_t; w_\alpha)| \le c(\alpha)\sqrt{\frac{t}{m}}w_\alpha(t), \qquad c_1(\alpha)m^{-1} \le t \le c_2(\alpha)m. \tag{2.9}$$

We recall (see [18, Eq.(6.31.13)]) that*

$$x_1 \sim m^{-1} \qquad \text{and} \qquad x_m \sim m;$$

---

*The relation $a_m \sim b_m$ here means that $|a_m/b_m|$ remains between positive bounds as $m \to \infty$, which may depend on α but not on m.

however we do not know the values of the constants $c_1(\alpha)$ and $c_2(\alpha)$. Therefore, we still need to consider the cases

$$x_1 \leq t < c_1(\alpha)m^{-1} \qquad \text{and} \qquad c_2(\alpha)m < t < x_m,$$

which we cannot exclude a priori.

For $x_1 \leq t < c_1(\alpha)m^{-1}$, let $x_1 \leq x_d \leq t < x_{d+1}$ for some $d \geq 1$. Notice that since $x_d \sim \frac{(d+1)^2}{m}$ (see [18, Eq. (6.31.13)]), d is certainly bounded by a constant independent of m. Then we examine the representation

$$R_m(\Gamma_t; w_\alpha) = \int_t^\infty w_\alpha(x)dx - \sum_{k=d+1}^m \lambda_{m,k}(w_\alpha),$$

i.e.,

$$R_m(\Gamma_t; w_\alpha) = -\int_0^t w_\alpha(x)dx + \sum_{k=1}^d \lambda_{m,k}(w_\alpha).$$

We have

$$|R_m(\Gamma_t; w_\alpha)| \leq \int_0^t w_\alpha(x)dx + \sum_{k=1}^d \lambda_{m,k}(w_\alpha).$$

Furthermore, by the Markov-Stieltjes inequality (see [4, p.32],[13, p.222]) we also have

$$\sum_{k=1}^d \lambda_{m,k}(w_\alpha) \leq \int_0^{x_{d+1}} w_\alpha(x)dx.$$

Therefore, since

$$\int_0^t w_\alpha(x)dx \leq ctw_\alpha(t)$$

for the t-values currently considered, and

$$x_{d+1}w_\alpha(x_{d+1}) \leq c_0tw_\alpha(t),$$

we obtain

$$|R_m(\Gamma_t; w_\alpha)| \leq c_1tw_\alpha(t) \leq c_2\sqrt{\frac{t}{m}}w_\alpha(t). \qquad (2.10)$$

For $c_2(\alpha)m < t < x_m$ we proceed similarly. Having set $x_d \leq t < x_{d+1} \leq x_m$, we recall that (see [13, p.222])

$$\sum_{k=1}^d \lambda_{m,k}(w_\alpha) \geq \int_0^{x_d} w_\alpha(x)dx,$$

425

which implies

$$\sum_{k=d+1}^{m} \lambda_{m,k}(w_\alpha) \leq \int_{x_d}^{\infty} w_\alpha(x)dx.$$

Thus we obtain

$$|R_m(\Gamma_t; w_\alpha)| \leq \int_t^{\infty} w_\alpha(x)dx + \int_{x_d}^{\infty} w_\alpha(x)dx \leq c(w_\alpha(t) + w_\alpha(x_d));$$

hence

$$|R_m(\Gamma_t; w_\alpha)| \leq c_1 w_\alpha(t) \leq c_2 \sqrt{\frac{t}{m}} w_\alpha(t) \tag{2.11}$$

since $w_\alpha(x_d) \leq c_0 w_\alpha(t)$.

Finally, from (2.6), (2.7) with x_m replaced by m, (2.9), (2.10) and (2.11) there follows

$$|R_m(\Gamma_t; w_\alpha)| \leq c \begin{cases} tw_\alpha(t) & \text{if } 0 \leq t < x_1, \\ \sqrt{\frac{t}{m}} w_\alpha(t) & \text{if } x_1 \leq t < x_m, \\ (\frac{t}{m})^\mu w_\alpha(t) & \text{if } x_m \leq t < \infty, \forall \mu \in R_+. \end{cases} \tag{2.12}$$

By inserting (2.12) into (2.4), we obtain

$$|R_m(f; w_\alpha)| \leq c[\int_0^{x_1} t|f'(t)|w_\alpha(t)dt + \frac{1}{\sqrt{m}} \int_{x_1}^{x_m} \sqrt{t}|f'(t)|w_\alpha(t)dt$$

$$+\frac{1}{m^\mu} \int_{x_m}^{\infty} t^\mu|f'(t)|w_\alpha(t)dt],$$

that is,

$$|R_m(f; w_\alpha)| \leq c[\|\phi^2 f' w_\alpha\|_{L_1(0,x_1)} + \frac{1}{\sqrt{m}} \|\phi f' w_\alpha\|_{L_1(x_1,x_m)} \tag{2.13}$$

$$+\frac{1}{m^\mu} \|\phi^{2\mu} f' w_\alpha\|_{L_1(x_m,\infty)}].$$

If in (2.13) we set $\mu = \frac{1}{2}$, then (2.3) with $r = 1$ (and a slightly sharper third term) follows. Moreover, since $x_1 \sim m^{-1}$ and $\phi f' w_\alpha \in L_1$, we have

$$|R_m(f; w_\alpha)| \leq \frac{c}{\sqrt{m}} \|\phi f' w_\alpha\|_{L_1}, \tag{2.14}$$

i.e., (2.2) with $r = 1$.

To obtain (2.2) for any $r > 1$, it is sufficient to insert into (2.14) a result recently proved in [9]. Indeed, given any polynomial $P \in \Pi_{2m-2}$, and having defined

$$Q(x) = \int_0^x P(t)dt \in \Pi_{2m-1},$$

we have

$$|R_m(f; w_\alpha)| = |R_m(f - Q; w_\alpha)| \leq \frac{c}{\sqrt{m}} \|\phi(f - Q)'w_\alpha\|_{L_1};$$

hence

$$|R_m(f; w_\alpha)| \leq \frac{c_1}{\sqrt{2m-1}} E_{2m-2}(w_{\alpha+1/2}; f'),$$

where

$$E_n(w; g) = \inf_{p \in \Pi_n} \|(g - p)w\|_{L_1}.$$

Moreover, from [9, Theorem 1[†]] we know that

$$E_n(w_\alpha; g) \leq \frac{c}{n^{k/2}} \|\phi^k g^{(k)} w_\alpha\|_{L_1},$$

with c depending only on k, whenever $g \in AC_{\text{loc}}$ and $\phi^k g^{(k)} w_\alpha \in L_1$. Therefore, we have (2.2).

Next we assume $f' \in AC_{\text{loc}}$ and apply Peano's theorem once more, this time taking $\Gamma_{t,1} = (x - t)_+$. We have

$$|R_m(f; w_\alpha)| \leq \int_0^\infty |R_m(\Gamma_{t,1}; w_\alpha)| |f''(t)| dt. \qquad (2.15)$$

To estimate $|R_m(\Gamma_{t,1}; w_\alpha)|$ we use (2.14). Thus, setting

$$q_t^\pm(x) = \int_0^x Q_t^\pm(y) dy \in \Pi_{2m-1},$$

with $Q_t^\pm$ as defined immediately after (2.4), and noting that

$$\phi \Gamma_{t,1}' w_\alpha \in L_1,$$

we obtain

$$|R_m(\Gamma_{t,1}; w_\alpha)| = |R_m(\Gamma_{t,1} - q_t^-; w_\alpha)| \leq \frac{c}{\sqrt{m}} \|\phi(\Gamma_{t,1} - q_t^-)'w_\alpha\|_{L_1} \qquad (2.16)$$

$$= \frac{c}{\sqrt{m}} \|\phi(\Gamma_t - Q_t^-)w_\alpha\|_{L_1} \leq \frac{c}{\sqrt{m}} \|\phi(Q_t^+ - Q_t^-)w_\alpha\|_{L_1} = \frac{c}{\sqrt{m}} \lambda_m(w_{\alpha+1/2}; t).$$

By proceeding as in the previous case, when $t \in [0, x_1) \cup [x_m, \infty)$, we derive the bounds

$$|R_m(\Gamma_{t,1}; w_\alpha)| \leq c \begin{cases} t^2 w_\alpha(t) & \text{if } 0 \leq t < x_1, \\ m^{-\mu} w_{\alpha+\mu+1}(t) & \text{if } x_m \leq t < \infty, \forall \mu \in R_+. \end{cases}$$

[†]Since our estimates are in L_1 the theorem holds for any $\alpha > -1$ (this is remarked in the first line of the proof given in [9]).

For $c_1(\alpha)m^{-1} \leq t \leq c_2(\alpha)m$ it is sufficient to use in (2.16) the inequality (see (2.8))

$$\lambda_m(w_{\alpha+1/2};t) \leq c\frac{t}{\sqrt{m}}w_\alpha(t).$$

The cases $x_1 \leq t < c_1(\alpha)m^{-1}$ and $c_2(\alpha)m < t < x_m$ can be examined as we did in the corresponding situation for $R_m(\Gamma_t;w_\alpha)$. Actually, for the sake of generality we consider $R_m(\Gamma_{t,j};w_\alpha)$, with $\Gamma_{t,j}(x) = \frac{1}{j!}(x-t)^j_+$ and $1 \leq j \leq r$.

When $x_1 \leq x_d \leq t < c_1(\alpha)m^{-1}$, $t < x_{d+1}$, we have

$$|R_m(\Gamma_{t,j};w_\alpha)| = |\int_t^\infty \frac{(x-t)^j}{j!}w_\alpha(x)dx - \sum_{k=d+1}^m \frac{(x_k-t)^j}{j!}\lambda_{m,k}(w_\alpha)| \qquad (2.17)$$

$$= \frac{1}{j!}|\int_0^t (t-x)^j w_\alpha(x)dx - \sum_{k=1}^d (t-x_k)^j\lambda_{m,k}(w_\alpha)|$$

$$\leq \frac{t^j}{j!}[\int_0^t w_\alpha(x)dx + \sum_{k=1}^d \lambda_{m,k}(w_\alpha)] \leq ct^{j+1}w_\alpha(t).$$

When $c_2(\alpha)m < t < x_{d+1} \leq x_m$, $x_d \leq t$, then

$$|R_m(\Gamma_{t,j};w_\alpha)| = \frac{1}{j!}|\int_t^\infty (x-t)^j w_\alpha(x)dx - \sum_{k=d+1}^m (x_k-t)^j\lambda_{m,k}(w_\alpha)|$$

$$< \frac{w_\alpha(t)}{j!}\int_0^\infty u^j(1+\frac{u}{t})^\alpha e^{-u}du + \frac{1}{j!}\sum_{k=d+1}^m x_k^j\lambda_{m,k}(w_\alpha).$$

Recalling the Markov-Stieltjes inequality (see [13, p.222])

$$\sum_{k=1}^d x_k^j\lambda_{m,k}(w_\alpha) \geq \int_0^{x_d} x^j w_\alpha(x)dx,$$

we obtain

$$|R_m(\Gamma_{t,j};w_\alpha)| \leq c_1[w_\alpha(t) + t^j w_\alpha(t)]. \qquad (2.18)$$

Thus we have

$$|R_m(\Gamma_{t,1};w_\alpha)| \leq c\begin{cases} t^2 w_\alpha(t) & \text{if } 0 \leq t < x_1, \\ \frac{t}{m}w_\alpha(t) & \text{if } x_1 \leq t \leq c_2(\alpha)x_m, \\ \frac{t^{\mu+1}}{m^\mu}w_\alpha(t) & \text{if } c_2(\alpha)x_m < t < \infty, \forall\mu \in R_+. \end{cases} \qquad (2.19)$$

Taking $\mu = 1$ in (2.19), we obtain from (2.15)

$$|R_m(f;w_\alpha)| \leq c[\|\phi^4 f''w_\alpha\|_{L_1(0,x_1)} + \frac{1}{m}\|\phi^2 f''w_\alpha\|_{L_1(x_1,c_2x_m)}]$$

$$+\frac{1}{m}\|\phi^4 f'' w_\alpha\|_{L_1(c_2 x_m, \infty)}],$$

i.e., (2.3) with $r = 2$.

When $f'' \in AC_{\text{loc}}$, we take $Q_t^\pm$ as defined previously, but now of degree $2m - 4$; then we consider the polynomials $q_t^\pm \in \Pi_{2m-2}$ such that

$$\frac{d^2}{dx^2} q_t^\pm(x) = Q_t^\pm(x),$$

and proceed as before. By iterating the process, we obtain (2.3). ∎

Remark 1 If we would have examined the remainder term in (2.1) following a standard procedure, i.e., assuming $f \in C^r[0, \infty)$ and considering the equality

$$R_m(f; w_\alpha) = R_m(f - P_m; w_\alpha),$$

where P_m is an ad hoc polynomial of degree m (see for instance [15, Remark 1]), we would have obtained the bound

$$R_m(f; w_\alpha) = O(m^{-r/2}).$$

Our estimate (2.2) is of the form

$$|R_m(f; w_\alpha)| \le cm^{-r/2}\|t^{r/2} f^{(r)} w_\alpha\|_{L_1}$$

and holds under the much weaker conditions $f^{(r-1)} \in AC_{\text{loc}}$ and $t^{r/2} f^{(r)} w_\alpha \in L_1$. This means, for instance, that when $f(t) \sim c + t^\beta$ as $t \to 0$, with $\frac{r}{2} - 1 < \beta < r$, $r \ge 1$, β not an integer, and is sufficiently smooth elsewhere, and we take for example $\alpha = 0$, we still have

$$|R_m(f; w_0)| = O(m^{-r/2}).$$

Notice that in this case $f(t)$ does not even belong to $C^{\lceil \beta \rceil}[0, \infty)$.

To appreciate the usefulness of (2.3), consider for example the function $f(x) = x^{\frac{1}{2}} \sin \frac{1}{x}$ and the Gauss-Laguerre formula with $\alpha = 0$. The derivative $f^{(1)}$ has at the origin a singularity of the type $x^{-\frac{3}{2}} \cos \frac{1}{x}$; hence (2.2) cannot be applied. We can however use the bound (2.3). After noticing that in $(0, x_1)$ we have

$$\|x f^{(1)}(x) w_0(x)\|_{L_1(0, x_1)} \le cm^{-\frac{1}{2} + \epsilon} \|x^{\frac{1}{2} + \epsilon} f^{(1)}(x) w_0(x)\|_{L_1(0, x_1)}$$

with $\epsilon > 0$ as small as we like, we would obtain for our function f the bound $|R_m(f; w_0)| = O(m^{-\frac{1}{2} + \epsilon})$.

3 ERROR BOUNDS FOR THE GAUSS-HERMITE FORMULA

Let $w(x) = e^{-x^2}$, $x \in R$, and consider the associated Gauss-Hermite quadrature formula

$$\int_{-\infty}^{\infty} f(x)w(x)dx = \sum_{k=1}^{m} \lambda_{m,k}(w)f(x_k) + R_m(f;w), \qquad (3.1)$$

where the nodes $\{x_k\}_{k=1}^{m}$, with $x_k = x_{m+1-k}$, are assumed to be ordered as follows:

$$-\infty < x_1 < x_2 < \cdots < x_{m-1} < x_m < \infty.$$

In the following, $L_1 \equiv L_1(-\infty, \infty)$ and $w_q \equiv e^{-qx^2}$ with $0 < q < 1$.

Theorem 2 Let $f^{(r-1)} \in AC_{\mathrm{loc}}$ and $f^{(r)}e^{\epsilon x^2}w \in L_1$ for some $r \geq 1$ and some $0 < \epsilon < 1$. Then we have

$$|R_m(f;w)| \leq cm^{-r/2}\|f^{(r)}e^{\epsilon x^2}w\|_{L_1}, \qquad (3.2)$$

where c is a constant independent of m and f.

Proof: This proof is also based on the classical Peano's representation of the remainder term. For simplicity we describe only the first two cases $r = 1, 2$, since the subsequent ones are very similar.

For $r = 1$ we have

$$R_m(f;w) = \int_{-\infty}^{\infty} R_m(\Gamma_t; w)f'(t)dt.$$

For the Hermite weight too[†] there exist two polynomials $Q_t^+(x)$, $Q_t^-(x)$ of degree $2m - 2$ such that

$$Q_t^-(x) \leq \Gamma_t(x) \leq Q_t^+(x)$$

and

$$\int_{-\infty}^{\infty} [Q_t^+(x) - Q_t^-(x)]w(x)dx = \lambda_m(w;t).$$

Therefore, following the proof of Theorem 1, we can in this case also derive the bound

$$|R_m(\Gamma_t; w)| \leq \lambda_m(w;t), \quad t \in R.$$

Furthermore, it is known (see [7],[16, p.92]) that there exists a constant $\delta > 0$ such that

$$\lambda_m(w;t) \leq c\frac{w(t)}{\sqrt{m}}$$

[†]Actually, this is true for any nonnegative weight function with finite moments (see [4, p.105]).

whenever $-\delta x_m \leq t \leq \delta x_m$. Thus, in this interval we have

$$|R_m(\Gamma_t; w)| \leq c \frac{w(t)}{\sqrt{m}}. \tag{3.3}$$

However, it is not known if $\delta < 1$ or not. In any case, when $x_m \leq t < \infty$ a direct calculation gives[‡]

$$R_m(\Gamma_t; w) = \int_t^\infty e^{-x^2} dx \leq \frac{w(t)}{t} \leq \frac{w(t)}{x_m} \leq c \frac{w(t)}{\sqrt{m}},$$

which is similar to (3.3).

If $\delta x_m < t < x_m$, let $x_d \leq t < x_{d+1}$. We have

$$R_m(\Gamma_t; w) = \int_t^\infty e^{-x^2} dx - \sum_{k=d+1}^m \lambda_{m,k}(w).$$

Since (see [7],[13])

$$\int_{-\infty}^{x_d} e^{-x^2} dx \leq \sum_{k=1}^d \lambda_{m,k}(w) \leq \int_{-\infty}^{x_{d+1}} e^{-x^2} dx,$$

we also have

$$\sum_{k=d+1}^m \lambda_{m,k}(w) \leq \int_{x_d}^\infty e^{-x^2} dx,$$

and

$$|R_m(\Gamma_t; w)| \leq \int_t^\infty e^{-x^2} dx + \int_{x_d}^\infty e^{-x^2} dx.$$

It remains only to examine the last term. We have

$$\int_{x_d}^\infty e^{-x^2} dx \leq \frac{e^{-x_d^2}}{x_d} \leq c \frac{e^{-x_d^2}}{\sqrt{m}} = ce^{p^2 t^2 - x_d^2} \frac{e^{-p^2 t^2}}{\sqrt{m}}.$$

Recalling that $x_{d+1} - x_d = o(1)$ as $m \to \infty$, we have for all m sufficiently large

$$\delta x_m < x_d \leq t < x_{d+1} < x_m;$$

furthermore, $(1 - \delta)x_m \sim \sqrt{m}$ and $(1 - p)t \sim \sqrt{m}$ for $0 < p < 1$. Therefore, for all m sufficiently large,

$$pt < x_d \quad \text{for all} \quad 0 < p < 1$$

[‡] We recall that $x_m \sim \sqrt{m}$ (see [18, §6.32]).

and

$$e^{p^2 t^2 - x_d^2} \le c_1.$$

This last inequality implies that the bound

$$|R_m(\Gamma_t; w)| \le c_0 \frac{e^{-qt^2}}{\sqrt{m}}$$

holds for any $0 < q < 1$ and all integers m.

The cases $-\infty < t < x_1$ and $x_1 \le t < \delta x_1$ are analogous. Thus, for the ϵ given in Theorem 2, we have

$$|R_m(f; w)| \le \frac{c}{\sqrt{m}} \int_{-\infty}^{\infty} e^{-(1-\epsilon)t^2} |f'(t)| dt, \tag{3.4}$$

which is (3.2) for $r = 1$.

Next we consider the representation

$$R_m(f; w) = \int_{-\infty}^{\infty} R_m(\Gamma_{t,1}; w) f''(t) dt$$

and examine the kernel $R_m(\Gamma_{t,1}; w)$. Following (2.16) we derive first the bounds

$$R_m(\Gamma_{t,1}; w) = R_m(\Gamma_{t,1} - q_t^-; w) \le \frac{c}{\sqrt{m}} \|(\Gamma_t - Q_t^-) w_q\|_{L_1} \le \frac{c_1}{\sqrt{m}} \lambda_m(w_q; t), \tag{3.5}$$

where q_t^- is defined as in the Laguerre case.

For $-\delta_q x_m \le t \le \delta_q x_m$, $\delta_q > 0$, we have

$$\lambda_m(w_q; t) \le c \frac{w_q(t)}{\sqrt{m}}. \tag{3.6}$$

When $x_m \le t < \infty$,

$$|R_m(\Gamma_{t,1}; w_q)| = \int_t^{\infty} (x - t) e^{-qx^2} dx \tag{3.7}$$

$$< \int_t^{\infty} x e^{-qx^2} dx \le c w_q(t) \le c \frac{w_{q_1}(t)}{m^\mu}, \ \forall \mu \in R_+,$$

with $q_1 < q$ as close as we like to q.

Finally, since we do not know if $\delta_q \ge 1$ or not, we need to examine also the case $\delta_q x_m < t < x_m$, that is,

$$R_m(\Gamma_{t,1}; w_q) = \int_t^{\infty} (x - t) e^{-qx^2} dx - \sum_{k=d+1}^{m} (x_k - t) \lambda_{m,k}(w_q),$$

where, for all m sufficiently large, we have

$$\delta_q x_m < x_d \leq t < x_{d+1} < x_m.$$

Notice that

$$|R_m(\Gamma_{t,1}; w_q)| < \int_t^\infty x e^{-qx^2} dx + \sum_{k=d+1}^m x_k \lambda_{m,k}(w_q)$$

and

$$\sum_{k=d+1}^m x_k \lambda_{m,k}(w_q) \leq \int_{x_d}^\infty x e^{-qx^2} dx,$$

so that

$$|R_m(\Gamma_{t,1}; w_q)| \leq c[w_q(t) + w_q(x_d)] \qquad (3.8)$$

$$\leq c_1 \frac{w_{q_1}(t)}{m^\mu}, \ \forall \mu \in R_+.$$

The cases $-\infty < t \leq -x_m$ and $-x_m < t < -\delta_q x_m$ can be treated similarly. Thus, by taking $\mu = 1$ in (3.7) and (3.8), and recalling (3.5) and (3.6), we obtain

$$|R_m(\Gamma_{t,1}; w)| \leq c \frac{w_{q_1}(t)}{m}, \ t \in R,$$

with $q_1 < q$ as close as we like to q. Since q can be chosen as close as we like to 1, we have (3.2) for $r = 2$. ∎

Acknowledgement. The authors are grateful to the referee for his very thorough report.

REFERENCES

[1] Criscuolo G., Della Vecchia B., Lubinsky D.S., Mastroianni G. Functions of the second kind for Freud weights and series expansions of Hilbert transform. To appear.

[2] Davis P.J., Rabinowitz P. *Methods of Numerical Integration*, 2nd edition. Academic Press, New York, 1984.

[3] De Vore R.A., Scott L.R. Error bounds for Gaussian quadrature and weighted-L_1 polynomial approximation. *SIAM J. Numer. Anal.*, **21**:400–412, 1984.

[4] Freud G. *Orthogonal Polynomials*. Akadémiai Kiadó, Budapest, 1971.

[5] Freud G. A certain class of orthogonal polynomials (in Russian). *Mat. Zametki*, **9**:511–520, 1971.

[6] Freud G. On the theory of onesided weighted L_1-approximation by polynomials. In *Lineare Operatoren und Approximation II*. Internat. Ser. Numer. Math., **25**:285–303, Birkhäuser-Verlag, Basel, 1974. P.L.Butzer et al., eds.

[7] Freud G. Extension of the Dirichlet-Jordan criterion to a general class of orthogonal polynomial expansions. *Acta Math. Acad. Sci. Hungar.*, **25**:109–122, 1974.

[8] Graham I.G., Mendes W.R. Nystrom-product integration for Wiener-Hopf equations with applications to radiative transfer. *IMA J. Numer. Anal.*, **9**:261–284, 1989.

[9] Joó I., Ky N.X. Answer to a problem of Paul Turán. *Ann. Univ. Sci. Budapest. Sect. Math.*, **31**:229–241, 1991.

[10] Knopfmacher A., Lubinsky D.S. Mean convergence of Lagrange interpolation for Freud's weights with application to product integration rules. *J. Comp. Appl. Math.*, **17**:79–103, 1987.

[11] Levin A.L., Lubinsky D.S. Christoffel functions, orthogonal polynomials and Nevai's conjecture for Freud weights. *Constr. Approx.*, **8**:461–533, 1992.

[12] Lubinsky D.S. Geometric convergence of Lagrangian interpolation and numerical integration rules over unbounded contours and intervals. *J. Approx. Theory*, **39**:338–360, 1983.

[13] Lubinsky D.S., Rabinowitz P. Rates of convergence of Gaussian quadrature for singular integrands. *Math. Comp.*, **43**:219–242, 1984.

[14] Mastroianni G., Vértesi P. Error estimates of product quadrature formulae. In *Numerical Integration IV*. Internat. Ser. Numer. Math., **112**:241–252, Birkhäuser Verlag, Basel, 1993. H. Brass, G. Hämmerlin, eds.

[15] Mastroianni G., Monegato G. Convergence of product integration rules over $(0, \infty)$ for functions with weak singularities at the origin. To appear in *Math. Comp.*.

[16] Nevai P., Freud G. Orthogonal polynomials and Christoffel functions. A case study. *J. Approx. Theory*, **48**:3–167, 1986.

[17] Smith W.E., Sloan I.H., Opie A.H. Product integration over infinite intervals I. Rules based on the zeros of Hermite polynomials. *Math. Comp.*, **40**:519–535, 1983.

[18] Szegö G., *Orthogonal Polynomials*. Amer. Math. Soc. Colloq. Publ., 23, Amer. Math. Soc., Providence, R.I., 1975.

[19] Uspensky J.V. On the convergence of quadrature formulas related to an infinite interval. *Trans. Amer. Math. Soc.*, **30**:542–559, 1928.

International Series of Numerical Mathematics, Vol. 119, ©1994 Birkhäuser

Using the Matrix Refinement Equation for the Construction of Wavelets II: Smooth Wavelets on [0,1]

Charles A. Micchelli[*]   Yuesheng Xu[†]

Dedicated to Walter Gautschi on the occasion of his 65th birthday

Abstract. This paper continues the work in [4] on constructing orthogonal bases on the interval [0, 1] by using the matrix refinement equation and the two basic operations of translation and scale. We call the elements of these bases wavelets. Here we amplify on the applicability of our method and construct smooth wavelets with and without boundary conditions. That is, we describe a procedure to recursively generate orthonormal bases with any prescribed number of continuous derivatives. As a caveat to the reader we reiterate our remark above that this paper is a *continuation* of our work in [4] and therefore some familiarity with [4] is assumed.

1 INTRODUCTION

This paper continues our recent work [4] on constructing wavelets on the interval [0, 1] by using the matrix refinement equation (cf., [2]) and the two basic operations of translation and scale. A familiarity with the results of [4] would be helpful. At least it should be realized that the multivariate wavelets constructed in [4] on invariant sets (see [4] for a definition of this notion) were **discontinuous**. Nonetheless, we showed in the case that our invariant set was [0, 1] that our technique could be modified to yield, at least, **continuous** wavelets. For the potential use of these wavelets we will extend this method here to achieve any prescribed degree of smoothness. Moreover, we will show that the same approach can be used to construct wavelets satisfying homogeneous boundary conditions. This is done in the last section of this paper while in the next section we deal with the problem of constructing smooth wavelets on [0, 1] from a refinable curve.

Several other techniques for wavelet construction on a finite interval have been proposed; see the references in [4]. It is *not* our intention here to compare these methods but rather merely describe in precise mathematical terms how the technique used in [4] can be modified to yield smooth wavelets. Perhaps on another occasion we will explore applications of the wavelets described here to certain problems of numerical computation.

[*]IBM T. J. Watson Research Center, P. O. Box 218, Yorktown Height, New York 10598
[†]Department of Mathematics, North Dakota State University, Fargo, ND 58105

2 SMOOTH WAVELETS ON [0,1]

There are two essential ingredients in our wavelet construction: a finite dimensional **refinable space** and two **refinement operators** generated by the basic operation of shift and scale. In general terms, we start with our refinable space and then apply our refinement operators to this space. A new space is formed of twice the dimension. Discontinuities of functions and derivatives are introduced at the point $t = \frac{1}{2}$. We now make a change of basis to isolate precisely the undesirable nonsmooth functions in our basis. We then throw them away and retain only the smooth basis functions. This process of "thinning" or "weeding" the nonsmooth components is then followed by another refinement and so on. In other words by alternating the refining procedure with a thinning operation we generate our wavelet basis. The remainder of this paper contains the details of this construction. The first item on our agenda is a description of the space.

We begin with n functions $f_1, \ldots, f_n \in C^\ell[0, 1]$, where $n \geq 1$ and $0 \leq \ell \leq n - 1$. The corresponding curve

$$\mathbf{f}(t) := (f_1(t), \ldots, f_n(t))^T, \quad 0 \leq t \leq 1 \tag{2.1}$$

is assumed to be **refinable,** that is, there are two $n \times n$ matrices

$$A_\epsilon, \quad \epsilon \in \{0, 1\}$$

such that $\mathbf{f}$ satisfies the **matrix refinement equation**

$$\mathbf{f}\left(\frac{t+\epsilon}{2}\right) = A_\epsilon^T \mathbf{f}(t), \quad t \in [0, 1], \quad \epsilon \in \{0, 1\}. \tag{2.2}$$

The components of the refinable curve $\mathbf{f}$ span the finite dimensional linear space

$$F_0 = \{\mathbf{c}^T \mathbf{f} : \mathbf{c} \in \mathbb{R}^n\}. \tag{2.3}$$

Conditions on the matrices $A_\epsilon, \epsilon \in \{0, 1\}$ are given in [3] to guarantee that the matrix refinement equation (2.2) has a nontrivial **refinable curve f** whose components are in $C^\ell[0, 1]$. As a **consequence** of this hypothesis on $\mathbf{f}$, two important facts follow which we record here for later use:

(a) Let $\mathbf{v} := (1, 1, \ldots, 1)^T \in \mathbb{R}^n$. Then there is some **nonzero constant** u such that

$$\mathbf{v}^T \mathbf{f} = u. \tag{2.4}$$

(b) Suppose

$$\mathbf{f}^{(\ell)} = (f_1^{(\ell)}, \ldots, f_n^{(\ell)})^T$$

is a nontrivial curve. Then

$$\Pi_\ell := \{\text{all polynomials of degree} \leq \ell\} \subseteq F_0. \tag{2.5}$$

To construct smooth wavelets based on the refinable curve **f** we make the following additional assumptions:

(i) $f_1, \ldots, f_n$ are orthonormal. That is, for $i, j = 1, \ldots, n$

$$(f_i, f_j) := \int_0^1 f_i(t) f_j(t) dt = \delta_{ij}.$$

(ii) The derivatives of $f_1, f_2, \ldots, f_n$ of order ℓ at zero and one exist and there exist signs $\epsilon_1, \ldots, \epsilon_n \in \{0, 1\}$ such that for $i = 1, 2, \ldots, n$ and $j = 0, 1, \ldots, \ell$

$$f_i^{(j)}(0) = (-1)^{\epsilon_i + j} f_i^{(j)}(1).$$

Examples of such refinable curves are easily constructed. For this purpose, we observe that the curve **f** is refinable provided that the space F_0 is invariant under each of the maps G_ϵ given by

$$(G_\epsilon g)(t) := g\left(\frac{t + \epsilon}{2}\right), \quad t \in [0, 1], \quad \epsilon \in \{0, 1\}. \tag{2.6}$$

That is,

$$G_\epsilon(F_0) \subseteq F_0, \quad \epsilon \in \{0, 1\}.$$

Clearly, the choice $F_0 = \Pi_{n-1}$ satisfies this inclusion. For instance, whenever for each $j = 1, 2, \ldots, n$, f_j is a polynomial of exact degree $j - 1$, F_0 is indeed Π_{n-1} and hence **f** is refinable. Moreover, if we choose $f_1, \ldots, f_n$ to be the orthonormal polynomials associated with a given weight function $w(t)$, $t \in [0, 1]$, which satisfies the equation

$$w(t) = w(1 - t), \quad t \in [0, 1]$$

then it follows for $i = 1, \ldots, n$, $t \in [0, 1]$ that

$$f_i(1 - t) = (-1)^{i-1} f_i(t), \quad t \in [0, 1].$$

In particular, condition (ii) is also met with the choice $\epsilon_i = i - 1$, $i = 1, 2, \ldots n$.

Now that we have the space we turn to a description of the refinement operators. We follow [4] and introduce for each $\epsilon \in \{0, 1\}$ the linear operator

$$(T_\epsilon g)(t) = \begin{cases} g(2t), & 0 \leq t \leq \frac{1}{2} \\ (-1)^\epsilon g(2t - 1), & \frac{1}{2} < t \leq 1. \end{cases} \tag{2.7}$$

We showed in [4] that the adjoint of T_ϵ, acting on $L^2[0, 1]$, is given by

$$T_\epsilon^* = \frac{1}{2}(G_0 + (-1)^\epsilon G_1), \quad \epsilon \in \{0, 1\} \tag{2.8}$$

and also

$$T_{\epsilon'}^* T_\epsilon = \delta_{\epsilon'\epsilon} I, \quad \epsilon', \epsilon \in \{0, 1\}. \tag{2.9}$$

These formulas are special cases of equations (4.3) and (4.6) of [4], respectively.

We also remark that whenever $\mathbf{f}$ is refinable and satisfies condition (2.2) then the curve

$$\hat{\mathbf{f}} := (T_0 f_1, \ldots, T_0 f_n, T_1 f_1, \ldots, T_1 f_n)^T$$

in $\mathbb{R}^{2n}$ is also refinable. In fact, $\hat{\mathbf{f}}$ satisfies the refinement equation

$$\hat{\mathbf{f}}\left(\frac{t+\epsilon}{2}\right) = \hat{A}_\epsilon^T \hat{\mathbf{f}}(t), \quad t \in [0,1], \quad \epsilon \in \{0,1\},$$

where

$$\hat{A}_\epsilon := \frac{1}{2}\left(\begin{array}{cc} A_0^T + A_1^T & A_0^T - A_1^T \\ (-1)^\epsilon(A_0^T + A_1^T) & (-1)^\epsilon(A_0^T - A_1^T) \end{array}\right).$$

The difficulty we encounter in the use of the refinement operators T_ϵ, $\epsilon \in \{0,1\}$ is that they are **destroyers of smoothness**, as one can see they squeeze a function defined on $[0,1]$ to first $[0, \frac{1}{2}]$ and then $[\frac{1}{2}, 1]$. In the process, the point $t = \frac{1}{2}$ is likely to become a discontinuity of the resulting function and/or its derivatives. We want to **isolate** the number of "**nonsmooth components**" but at the same time retain sufficient properties of the "**smooth components**" so that we can repeat the process. As our thoughts on this dilemma clarified we decided to address the situation through the following notion. We say that a finite dimensional linear subspace F_0 has an $(\ell + 1)$st degree **convenient basis** $\{f_1, \ldots, f_n\}$ provided that (i), (ii) hold and also that

(iii) the vectors

$$\mathbf{f}^{(j)}(1) := (f_1^{(j)}(1), \ldots, f_n^{(j)}(1))^T, \quad j = 0, 1, \ldots, \ell,$$

are linearly independent. Equivalently, according to property (ii) the vectors

$$\mathbf{f}^{(j)}(0) := (f_1^{(j)}(0), \ldots, f_n^{(j)}(0))^T, \quad j = 0, 1, \ldots, \ell,$$

are linearly independent.

Note that if $\{f_1, \ldots, f_n\}$ is an $(\ell + 1)$st degree convenient basis then any permutation of these functions is likewise a convenient basis. Moreover, in this definition, we neither require that

$$\mathbf{f} = (f_1, \ldots, f_n)^T$$

is refinable nor has components in $C^\ell[0,1]$. However, if $\mathbf{f}$ indeed satisfies *both* of these conditions and in addition, $\mathbf{f}^{(\ell)}$ is nontrivial then (iii) automatically holds. To see this, we suppose that there is a vector

$$\mathbf{c} = (c_0, \ldots, c_\ell)^T \in \mathbb{R}^{\ell+1}$$

such that

$$\sum_{j=0}^{\ell} c_j \mathbf{f}^{(j)}(1) = \mathbf{0}.$$

Hence, for all $g \in F_0$ we have

$$\sum_{j=0}^{\ell} c_j g^{(j)}(1) = 0, \tag{2.10}$$

and, in particular, by property (b), (2.10) holds for all $g \in \Pi_\ell$. From this conclusion it follows that $\mathbf{c} = \mathbf{0}$, thereby establishing that (iii) holds.

We have now assembled three components essential to the construction we wish to describe in this paper: **the refinable space, the refinement operators** and finally **convenient bases**. Remember that it is the alternation of refinement and weeding that will alternately produce our desired smooth wavelets. Unfortunately, weeding is a tedious task. To find our "weeds" (nonsmooth components) we need to make a change of basis. For this purpose we use some simple facts about vectors and matrices. We state them next; however, the reader can skip them and go immediately to Proposition 2.3 which precisely describes how to weed. It is only in the proof of Proposition 2.3 that the next two lemmas are used.

Lemma 2.1 *Let N and q_ϵ, $\epsilon \in \{0, 1\}$ be positive integers with $q_0 + q_1 \leq N$. Let M_ϵ, Q_ϵ and R_ϵ be $N \times q_\epsilon$, $N \times q_\epsilon$ and $q_\epsilon \times q_\epsilon$ matrices, respectively, satisfying*

$$M_\epsilon = Q_\epsilon R_\epsilon, \quad \epsilon \in \{0, 1\}.$$

Suppose that (M_0, M_1) is of rank $q_0 + q_1$ and R_ϵ is nonsingular. Then, (Q_0, Q_1) is of rank $q_0 + q_1$.

Proof: Let $(Q_\epsilon)_j$ denote the jth column of the matrix Q_ϵ. We prove that

$$(Q_0)_j, \quad j = 1, 2, \ldots, q_0, \quad (Q_1)_j, \quad j = 1, 2, \ldots, q_1$$

are linearly independent. To this end, we assume that there are constants $a_{\epsilon,j}$ such that

$$\sum_{\epsilon \in \{0,1\}} \sum_{j=1}^{q_\epsilon} a_{\epsilon,j} (Q_\epsilon)_j = 0.$$

By the hypothesis of this lemma, we have

$$(Q_\epsilon)_j = M_\epsilon (R_\epsilon^{-1})_j, \quad j = 1, 2, \ldots, q_\epsilon, \quad \epsilon \in \{0, 1\}.$$

Hence, letting

$$\mathbf{x}_\epsilon = \sum_{j=1}^{q_\epsilon} a_{\epsilon,j} (R_\epsilon^{-1})_j, \quad \epsilon \in \{0, 1\},$$

439

a vector in $\mathbb{R}^{q_\epsilon}$, we conclude from the above equation for $a_{\epsilon,j}$ that

$$(M_0, M_1) \begin{pmatrix} \mathbf{x}_0 \\ \mathbf{x}_1 \end{pmatrix} = 0.$$

The hypothesis on the matrix $M := (M_0, M_1)$ implies that $\mathbf{x}_0 = \mathbf{x}_1 = 0$. Since R_ϵ is nonsingular, we conclude that $a_{\epsilon,j} = 0$, for $j = 1, 2, \ldots, q_\epsilon$, $\epsilon \in \{0, 1\}$. ∎

Lemma 2.2 *Let two sets of vectors*

$$\{\mathbf{w}_1, \ldots, \mathbf{w}_n, \mathbf{x}_1, \ldots, \mathbf{x}_m\}$$

and

$$\{\mathbf{y}_1, \ldots, \mathbf{y}_m, \mathbf{z}_1, \ldots, \mathbf{z}_n\}$$

be orthonormal and span the same space, and the set of vectors

$$\{\mathbf{w}_1, \ldots, \mathbf{w}_n, \mathbf{y}_1, \ldots, \mathbf{y}_m\}$$

be linearly independent. Then,

$$\{\mathbf{x}_1, \ldots, \mathbf{x}_m, \mathbf{z}_1, \ldots, \mathbf{z}_n\}$$

is linearly independent.

Proof: Assume to the contrary that the set of vectors $\{\mathbf{x}_1, \ldots, \mathbf{x}_m, \mathbf{z}_1, \ldots, \mathbf{z}_n\}$ is linearly dependent. Then there are constants $a_1, \ldots, a_m, b_1, \ldots, b_n$, not all zero, such that

$$\sum_{i=1}^{m} a_i \mathbf{x}_i + \sum_{i=1}^{n} b_i \mathbf{z}_i = 0.$$

First we note that at least there is one b_i not zero, since $\{\mathbf{x}_1, \ldots, \mathbf{x}_m\}$ is orthonormal. Represent $\mathbf{z}_i$ in terms of the orthonormal basis $\{\mathbf{w}_1, \ldots, \mathbf{w}_n, \mathbf{x}_1, \ldots, \mathbf{x}_m\}$, i.e.,

$$\mathbf{z}_i = \sum_{j=1}^{n} \left(\mathbf{z}_i^T \mathbf{w}_j \right) \mathbf{w}_j + \sum_{j=1}^{m} \left(\mathbf{z}_i^T \mathbf{x}_j \right) \mathbf{x}_j, \quad i = 1, 2, \ldots, n.$$

Substituting this equation for $\mathbf{z}_i$ in the above equation and simplifying yields the formula

$$\sum_{i=1}^{m} \left(a_i + \sum_{k=1}^{n} b_k \left(\mathbf{z}_k^T \mathbf{x}_i \right) \right) \mathbf{x}_i + \sum_{j=1}^{n} \left(\sum_{k=1}^{n} b_k \left(\mathbf{z}_k^T \mathbf{w}_j \right) \right) \mathbf{w}_j = 0.$$

This equation implies that

$$\sum_{k=1}^{n} b_k \left(\mathbf{z}_k^T \mathbf{w}_j \right) = 0, \quad j = 1, 2, \ldots, n.$$

440

Since the last equation has a nontrivial solution $\mathbf{b} := (b_1, \ldots, b_n)^T$, we conclude that the coefficient matrix $(\mathbf{z}_k^T \mathbf{w}_j)_{k,j=1,2,\ldots,n}$ is singular. Hence, the homogeneous linear system

$$\sum_{j=1}^{n} d_j (\mathbf{z}_k^T \mathbf{w}_j) = 0, \quad k = 1, 2, \ldots, n$$

has a nontrivial solution $\mathbf{d} := (d_1, \ldots, d_n)^T$. That is, there exists a nontrivial linear combination

$$\mathbf{w}_0 := \sum_{j=1}^{n} d_j \mathbf{w}_j \in Z^{\perp},$$

where $Z := \mathrm{span}\{z_1, \ldots, z_n\}$. However, by hypothesis

$$Z \perp \mathrm{span}\{y_1, \ldots, y_m\},$$

and thus,

$$\sum_{j=1}^{n} d_j \mathbf{w}_j = \sum_{j=1}^{m} e_j \mathbf{y}_j,$$

for some constants $e_1, \ldots, e_m$. This contradicts the hypothesis that the set of vectors $\{\mathbf{w}_1, \ldots, \mathbf{w}_n, \mathbf{y}_1, \ldots, \mathbf{y}_m\}$ is linearly independent. ∎

We now have all that we need to proceed to the next step that gives mathematical substance to "weeding". When you read this result, keep in mind that its purpose is to isolate the nonsmooth functions created by using the refinement operators. Later, as we shall explain this result will be used repeatedly.

We find the following notation helpful. For any function g defined on $[0,1]$ with left and right limits at $\frac{1}{2}$ we set

$$\left[g\left(\frac{1}{2} \right) \right] := g\left(\frac{1}{2}^{+} \right) - g\left(\frac{1}{2}^{-} \right).$$

Proposition 2.3 *Let $N \geq 2$ and $0 \leq \ell \leq N - 1$. Suppose G is a subspace of $C^{\ell}[0,1]$ which has an $(\ell + 1)$st degree convenient basis $\{g_0, \ldots, g_{N-1}\}$. Then the linear space*

$$H := T_0 G \oplus T_1 G$$

has a basis

$$\{h_0, \ldots, h_{2N-1}\} \subseteq C^{\ell}\left([0,1] \setminus \left\{ \frac{1}{2} \right\} \right)$$

such that the functions $h_{\ell+1}, \ldots, h_{2N-1}$ of $C^{\ell}[0,1]$ form an $(\ell + 1)$st-degree convenient basis and the $(\ell + 1) \times (\ell + 1)$ matrix

$$\left(\left[h_i^{(j)}\left(\frac{1}{2} \right) \right] \right)_{i,j=0,1,\ldots,\ell} \tag{2.11}$$

is nonsingular.

In other words, the application of the operators T_0 and T_1 to a subspace G of $C^\ell[0,1]$ with an $(\ell+1)$st degree convenient basis produces exactly $\ell+1$ functions which are **not** in $C^\ell[0,1]$. The linear span of the remainder of the functions compose a "thinned" subpace of $C^\ell[0,1]$ which is again an $(\ell+1)$st convenient basis. In fact, if $k_0,\ldots,k_{2N-1}$ is any other basis of H which has these properties then it is easy to see that

$$\text{span}\{k_{\ell+1},\ldots,k_{2N-1}\} = \text{span}\{h_{\ell+1},\ldots,h_{2N-1}\}.$$

For this reason, when G satisfies the hypothesis of Proposition 2.3, we set

$$T_0 G \oplus_{\ell+1} T_1 G := \text{span}\{h_{\ell+1},\ldots,h_{2N-1}\}. \tag{2.12}$$

Hence, we conclude under this condition on G that

$$\dim(T_0 G \oplus_{\ell+1} T_1 G) = 2\dim G - \ell - 1, \tag{2.13}$$

In particular, when $\ell = N - 1$, it follows that

$$\dim(T_0 G \oplus_{\ell+1} T_1 G) = \dim G$$

and thus in this case

$$T_0 G \oplus_{\ell+1} T_1 G = G.$$

Let us now prove this Proposition.

Proof: This result was proved in [4] when $\ell = 0$. Therefore, we assume that $\ell \geq 1$. However, a very slight modification of this proof also holds when $\ell = 0$.

The set of functions

$$\mathbb{H} = \{T_0 g_1,\ldots,T_0 g_n, T_1 g_1,\ldots,T_1 g_n\}$$

span the linear space H. In fact, according to our hypothesis (i) and property (2.9) of the linear operators T_ϵ, $\epsilon \in \{0,1\}$ these functions form an orthonormal basis of H. We wish to order these functions in a specific fashion dictated by the signs $\epsilon_1,\ldots,\epsilon_N$ appearing in (ii). To this end, we choose $\epsilon'' \in \{0,1\}$ and let

$$\mathbb{G}_{\epsilon''} := \{g_i : \epsilon_i = \epsilon'', \ 0 \leq i \leq N-1\}.$$

Thus, $\mathbb{G}_0$ and $\mathbb{G}_1$ partition the set

$$\mathbb{G} := \{g_0, g_1,\ldots,g_{N-1}\}$$

into disjoint subsets.

Next, let us see the effect of the operator $T_{\epsilon'}$ on elements of $\mathbb{G}_{\epsilon''}$. To this end, fix $\epsilon \in \{0,1\}$ and collect the functions $T_{\epsilon'} g$, $g \in \mathbb{G}_{\epsilon''}$ where $\epsilon' + \epsilon'' = \epsilon$ mod 2 into a set which, we denote by $\mathbb{U}_\epsilon$. That is, we set

$$\mathbb{U}_0 := T_0 \mathbb{G}_0 \cup T_1 \mathbb{G}_1$$

and
$$\mathbb{U}_1 := T_0\mathbb{G}_1 \cup T_1\mathbb{G}_0.$$

Since
$$T_0\mathbb{G}_\epsilon \cap T_1\mathbb{G}_{\epsilon'} = \phi \quad \text{for} \ \epsilon, \epsilon' \in \{0,1\},$$

we see that $\#\mathbb{U}_\epsilon = N$ and also by (2.9) and our hypothesis on $\mathbb{G}$ we conclude that $\mathbb{U}_0 \perp \mathbb{U}_1$. Moreover, for a typical element $u = T_{\epsilon'}g$, $g \in \mathbb{G}_{\epsilon''}$, in $\mathbb{U}_\epsilon$ we have for $j = 0, 1, \ldots, \ell$
$$u^{(j)}(0) = 2^j g^{(j)}(0) = (-1)^{\epsilon+j} u^{(j)}(1)$$

and also
$$\left[u^{(j)}\left(\frac{1}{2}\right) \right] = 2^j g^{(j)}(1)[(-1)^{\epsilon+j} - 1].$$

With these formulas in hand, it is an easy matter to order the elements of
$$\mathbb{U}_\epsilon = \{u_{0,\epsilon}, \ldots, u_{N-1,\epsilon}\}$$

so that for $i = 0, 1, \ldots, N-1$, $j = 0, 1, \ldots, \ell$
$$u_{i,\epsilon}^{(j)}(0) = (-1)^{\epsilon+j} u_{i,\epsilon}^{(j)}(1) = 2^j g_i^{(j)}(0) \tag{2.14}$$

and
$$\left[u_{i,\epsilon}^{(j)}\left(\frac{1}{2}\right) \right] = 2^j [(-1)^{\epsilon+j} - 1]g_i^{(j)}(1). \tag{2.15}$$

The next step in the proof consists in making a change of basis in $\operatorname{span}\mathbb{U}_\epsilon$. For this purpose, choose for each $\epsilon \in \{0,1\}$ an $N \times N$ orthogonal matrix
$$Y_\epsilon = ((Y_\epsilon)_{ij})_{i,j=0,1,\ldots,N-1}$$

(that is, its transpose is its inverse) and consider the functions
$$h_{i,\epsilon}(t) = \sum_{r=0}^{N-1} (Y_\epsilon)_{ir} u_{r,\epsilon}(t), \quad i = 0, 1, \ldots, N-1, \ \epsilon \in \{0,1\}, \tag{2.16}$$

where we use the notation
$$\mathbb{H} := \{h_{0,\epsilon}, \ldots, h_{N-1,\epsilon}, \ \epsilon \in \{0,1\}\}.$$

We (jointly) order the functions in the set $\mathbb{H} := \mathbb{H}_0 \cup \mathbb{H}_1$ in the following way. For each $i = 0, 1, \ldots, 2N - 1$ we set $i = m_i + \hat{\epsilon}_i N$ for some $m_i \in \{0, 1, \ldots, N - 1\}$ and $\epsilon_i \in \{0,1\}$. That is,
$$m_i = \begin{cases} i, & 0 \le i \le N - 1, \\ \\ i - N, & N \le i \le 2N - 1 \end{cases}$$

443

and

$$\hat{\epsilon}_i = \begin{cases} 0, & 0 \le i \le N-1, \\ 1, & N \le i \le 2N-1. \end{cases}$$

Then the functions

$$h_i(x) := h_{m_i, \hat{\epsilon}_i}(x), \quad i = 0, 1, \ldots, 2N-1$$

are orthonormal on $[0,1]$ and for $j = 0, 1, \ldots, \ell$, $\epsilon \in \{0,1\}$ and $t \in [0,1]$ they satisfy the equation

$$\mathbf{h}_\epsilon^{(j)}(t) = Y_\epsilon \mathbf{u}_\epsilon^{(j)}(t),$$

where

$$\mathbf{h}_\epsilon(t) = (h_{0,\epsilon}(t), \cdots, h_{N-1,\epsilon}(t))^T, \quad \epsilon \in \{0,1\}, \quad t \in [0,1]$$

and

$$\mathbf{u}_\epsilon(t) = (u_{0,\epsilon}(t), \cdots, u_{N-1,\epsilon}(t))^T, \quad \epsilon \in \{0,1\}, \quad t \in [0,1].$$

Thus, equation (2.14) implies that

$$\mathbf{h}_\epsilon^{(j)}(1) = Y_\epsilon \mathbf{u}_\epsilon^{(j)}(1) = (-1)^{\epsilon+j} \mathbf{h}_\epsilon^{(j)}(0).$$

In other words the set of functions $\mathbb{H}$ satisfies both properties (i) and (ii) where $n = 2N$, $\epsilon_i = \epsilon$, $i = 1, 2, \ldots, 2N$. To see that property (iii) holds as well we observe by equation (2.14) that for $j = 0, 1, \ldots, \ell$,

$$\mathbf{h}^{(j)}(0) = 2^j L \mathbf{g}^{(j)}(0),$$

where L is the $2N \times N$ matrix

$$L = \begin{pmatrix} Y_0 \\ Y_1 \end{pmatrix},$$

$$\mathbf{h}(t) = (h_0(t), \ldots, h_{2N-1}(t))^T$$

and

$$\mathbf{g}(t) = (g_0(t), \ldots, g_{N-1}(t))^T.$$

Since rank $L = N$ we conclude that the set $\mathbb{H}$ satisfies property (iii) as well.

So far we have not specified a choice for the matrix Y_ϵ. This is done next to achieve the additional condition that

$$\{h_{\ell+1}, \ldots, h_{2N-1}\} \subseteq C^\ell[0,1]$$

and that the matrix (2.11) is nonsingular. Specifically, we choose Y_ϵ in the following way.

Let p_0, p_1 be the largest odd, even number $\le \ell$, respectively. For each $\epsilon \in \{0,1\}$, we let

$$p_\epsilon' := \frac{1}{2}(p_\epsilon + \epsilon + 1)$$

and consider the $N \times p'_\epsilon$ matrix

$$M_\epsilon = ((M_\epsilon)_{ij})$$

defined by

$$(M_\epsilon)_{it} = \left[u_{i,\epsilon}^{(1-\epsilon+2t)} \left(\frac{1}{2} \right) \right], \quad i = 0, 1, \ldots, N-1, \quad t = 0, 1, \ldots, p'_\epsilon - 1.$$

According to formula (2.15) we have for $i = 0, 1, \ldots, N-1$, $t = 0, 1, \ldots, p'_\epsilon - 1$,

$$(M_\epsilon)_{it} = -2^{2-\epsilon+2t} g_i^{(1-\epsilon+2t)}(1)$$

and therefore M_ϵ has rank p'_ϵ.

Thus, by the QR factorization, (cf. [1], Theorem 2.6.1, p. 112) there is an $N \times p'_\epsilon$ matrix Q_ϵ whose columns are orthonormal and there is a $p'_\epsilon \times p'_\epsilon$ nonsingular upper triangular matrix R_ϵ such that

$$M_\epsilon := Q_\epsilon R_\epsilon, \quad \epsilon \in \{0, 1\}. \tag{2.17}$$

Next, we show how to choose an $N \times (N - p'_\epsilon)$ matrix $\hat{Q}_\epsilon$ such that $(Q_\epsilon, \hat{Q}_\epsilon)$ is an $N \times N$ orthogonal matrix and since $p'_0 + p'_1 = \ell + 1$, $(\hat{Q}_0, \hat{Q}_1)$ is an $N \times (2N - \ell - 1)$ matrix of rank N. By hypothesis (iii), the matrix (M_0, M_1) is of rank $\ell + 1$. Hence, Lemma 2.1 guarantees that the set of vectors consisting of the columns of Q_0 and Q_1 is linearly independent. Therefore, the space Q spanned by the columns of Q_0 and Q_1 is of $\ell + 1$ dimension and there exist $N - \ell - 1$ orthonormal vectors $\mathbf{z}_1, \ldots, \mathbf{z}_{N-\ell-1}$ that spans $Q^\perp$ in $\mathbb{R}^N$. Using the **Gram-Schmidt** process, we get two orthonormal bases for Q:

$$\{(Q_0)_0, \ldots, (Q_0)_{p'_0 - 1}, (\hat{Q}_0)_0, \ldots, (\hat{Q}_0)_{p'_1 - 1}\}$$

and

$$\{(Q_1)_0, \ldots, (Q_1)_{p'_1 - 1}, (\hat{Q}_1)_0, \ldots, (\hat{Q}_1)_{p'_0 - 1}\}.$$

By Lemma 2.2, we conclude that

$$(\hat{Q}_0)_0, \ldots, (\hat{Q}_0)_{p'_1 - 1}, (\hat{Q}_1)_0, \ldots, (\hat{Q}_1)_{p'_0 - 1}, \mathbf{z}_1, \ldots, \mathbf{z}_{N-\ell-1}$$

are linearly independent. For $\epsilon \in \{0, 1\}$ we introduce the $N \times (N - p'_\epsilon)$ matrices

$$\hat{Q}_\epsilon := ((\hat{Q}_\epsilon)_0, \ldots, (\hat{Q}_\epsilon)_{p'_{1-\epsilon} - 1}, \mathbf{z}_1, \ldots, \mathbf{z}_{N-\ell-1}).$$

Then, matrices $(\hat{Q}_0, \hat{Q}_1)$ and $(Q_\epsilon, \hat{Q}_\epsilon)$ have the desired properties. Let P_ϵ be the $N \times p'_\epsilon$ matrix

$$P_\epsilon := \left(\begin{array}{c} R_\epsilon \\ 0 \end{array} \right),$$

where 0 represents the $(N - p'_\epsilon) \times p'_\epsilon$ zero matrix, and let $Y_\epsilon^T := (Q_\epsilon, \hat{Q}_\epsilon)$, an $N \times N$ matrix for each $\epsilon \in \{0, 1\}$. Then Y_ϵ is an orthogonal matrix such that

$$M_\epsilon = Y_\epsilon^T P_\epsilon, \quad P_\epsilon = Y_\epsilon M_\epsilon.$$

We now consider the functions $h_{i,\epsilon}$ given by formula (2.16). Observe that by equations (2.15), (2.16) and (2.17) that for $i = 0, 1, \ldots, N - 1$

$$\left[h_{i,\epsilon}^{(j)} \left(\tfrac{1}{2} \right) \right] = \begin{cases} 0, & \epsilon + j = \text{even} \\ (P_\epsilon)_{it}, & \epsilon + j = 2t + 1, t = 0, 1, \ldots, p'_\epsilon - 1. \end{cases} \tag{2.18}$$

Consequently, for $i \in \{p'_\epsilon, \ldots, N - 1\}$ and $\epsilon \in \{0, 1\}$ we have $h_{i,\epsilon} \in C^\ell[0, 1]$. There are $\ell + 1$ remaining functions

$$\{h_{0,0}, \ldots, h_{p'_0 - 1, 0}, h_{0,1}, \ldots, h_{p'_1 - 1, 1}\}.$$

For these functions, we observe that the $(\ell + 1) \times (\ell + 1)$ matrix

$$J := \begin{pmatrix} \left[h_{0,0} \left(\tfrac{1}{2} \right) \right] & \left[h_{0,0}^{(1)} \left(\tfrac{1}{2} \right) \right] & \cdots & \left[h_{0,0}^{(\ell)} \left(\tfrac{1}{2} \right) \right] \\ \vdots & \vdots & & \vdots \\ \left[h_{p'_0 - 1, 0} \left(\tfrac{1}{2} \right) \right] & \left[h_{p'_0 - 1, 0}^{(1)} \left(\tfrac{1}{2} \right) \right] & \cdots & \left[h_{p'_0 - 1, 0}^{(\ell)} \left(\tfrac{1}{2} \right) \right] \\ \left[h_{0,1} \left(\tfrac{1}{2} \right) \right] & \left[h_{0,1}^{(1)} \left(\tfrac{1}{2} \right) \right] & \cdots & \left[h_{0,1}^{(\ell)} \left(\tfrac{1}{2} \right) \right] \\ \vdots & \vdots & & \vdots \\ \left[h_{p'_1 - 1, 1} \left(\tfrac{1}{2} \right) \right] & \left[h_{p'_1 - 1, 1}^{(1)} \left(\tfrac{1}{2} \right) \right] & \cdots & \left[h_{p'_1 - 1, 1}^{(\ell)} \left(\tfrac{1}{2} \right) \right] \end{pmatrix}$$

is invertible. In fact, using (2.18), the definition of P_ϵ and by reordering the columns of J (an odd number of exchanges is needed) we can write it in the block matrix form

$$\begin{pmatrix} R_0 & 0 \\ 0 & R_1 \end{pmatrix},$$

Therefore

$$\det J = - \det R_0 \det R_1 \neq 0,$$

because R_ϵ is nonsingular.

There remains to prove that for $j = 0, 1, \ldots, \ell$ the vectors

$$\mathbf{x}^j := \left(h_{p'_0, 0}^{(j)}(1), \ldots, h_{N-1, 0}^{(j)}(1), h_{p'_1, 1}^{(j)}(1), \ldots, h_{N-1, 1}^{(j)}(1) \right)^T$$

are linearly independent. To this end, suppose there are constants $c_0, \ldots, c_\ell$ such that

$$c_0 \mathbf{x}^0 + \cdots + c_\ell \mathbf{x}^\ell = 0$$

From the hypothesis of this proposition, equations (2.14) and (2.16) we get for $j = 0, 1, \ldots, \ell$, $i = 0, 1, \ldots, N-1$ and $\epsilon \in \{0, 1\}$ that

$$h_{i,\epsilon}^{(j)}(1) = \sum_{r=0}^{N-1} (Y\epsilon)_{ir}(-1)^{\epsilon_r + \epsilon} 2^j g_r^{(j)}(1).$$

Set $\mathbf{y} = (y_0, \ldots, y_{N-1})$ where

$$y_r = \sum_{j=0}^{\ell} c_j (-1)^{\epsilon_r} 2^j g_r^{(j)}(1), \quad r = 0, 1, \ldots, N-1,$$

and let Y_ϵ' be the $(N - p_\epsilon') \times N$ submatrix of Y_ϵ consisting of its last $N - p_\epsilon'$ rows. Then, equation (2.18) is equivalent to the statement that

$$Y_0' \mathbf{y} = Y_1' \mathbf{y} = 0.$$

There are $2N - \ell - 1$ homogeneous equations here for the vector $\mathbf{y} \in \mathbb{R}^N$. Since, by the construction of Y_ϵ,

$$Y := \begin{pmatrix} Y_0' \\ Y_1' \end{pmatrix} = (\hat{Q}_0, \hat{Q}_1)^T,$$

the $(2N - \ell - 1) \times (2N)$ matrix Y has a zero null space. Then, we conclude that $\mathbf{y} = \mathbf{0}$. Therefore,

$$\mathbf{c} := (c_0, c_1, \ldots, c_\ell) = 0,$$

because our hypothesis insures that for $j = 0, 1, \ldots, \ell$ the vectors $\mathbf{g}^{(j)}(1)$ are linearly independent. ∎

We remark that whenever

$$U = \mathrm{span}\{u_0, \ldots, u_{N-1}\} \subseteq C^\ell \left([0,1] \setminus \left\{\frac{1}{2}\right\}\right)$$

is an N-dimensional subspace such that the matrix

$$J := \left(\left[u_i^{(j)}\left(\frac{1}{2}\right)\right]\right), \quad i = 0, 1, \ldots, N-1, \quad j = 0, 1, \ldots, \ell$$

is of rank ℓ. Then there is always a basis $\{h_0, \ldots, h_{N-1}\}$ of U such that

$$h_{\ell+1}, \ldots, h_{N-1} \in C^\ell[0,1],$$

and the matrix

$$\left(\left[h_i^{(j)}\left(\frac{1}{2}\right)\right]\right)_{i,j=0,1,\ldots,\ell}$$

447

is nonsingular. Moreover, any other bases $\{k_0, \ldots, k_{N-1}\}$ of U with these two properties satisfies

$$\text{span}\{k_{\ell+1}, \ldots, k_{N-1}\} = \text{span}\{h_{\ell+1}, \ldots, h_{N-1}\}.$$

For this reason, we define

$$U^{\ell+1} := \text{span}\{h_{\ell+1}, \ldots, h_{N-1}\}$$

as the "component" of U in $C^\ell[0, 1]$. Therefore,

$$\dim U^{\ell+1} = \dim U - \ell - 1.$$

and also, when G is a subspace of $C^\ell[0, 1]$ which satisfies the hypotheses of Proposition 2.3 we use the notation $T_0 G \oplus_{\ell+1} T_1 G$ and $(T_0 G \oplus T_1 G)^{\ell+1}$ interchangeably.

With this last bit of notation and Proposition 2.3 all our preparations have been made. We can now turn to our goal of generating orthonormal bases on $[0, 1]$ by shifts and scales. For this purpose, we employ the method of multiresolution. That is, we introduce a nested scale of spaces that exhaust $L^2[0, 1]$. Then we study the scale of orthogonal complements and demonstrate how the previous facts we assembled lead us to a recursive procedure for the generation of the orthogonal complements. So let us now introduce our scale of spaces:

$$F_{k+1} = T_0 F_k \oplus_\ell T_1 F_k, \quad k = 0, 1, \ldots$$

where F_0 has an ℓ-th degree convenient basis $\{f_1, \ldots, f_n\}$, $0 \le \ell \le n$, which determines a refinable curve $\mathbf{f}$ (note that now we have decided to speak of ℓ-th degree convenient basis rather than $(\ell + 1)$st degree convenient basis as we did previously). According to (2.13) we have for $k = 0, 1, \ldots,$

$$\dim F_{k+1} = 2\dim F_k - \ell$$

and therefore

$$\dim F_k = 2^k \dim F_0 - (2^k - 1)\ell.$$

Also, we note in passing that when $F_0 = \Pi_{n-1}$ then F_k is the space of spline functions of degree $n - 1$ with knots of multiplicity $n - \ell$ at $j/2^k$, $j = 1, \ldots, 2^k - 1$. As a result, when $\mathbf{f} \in C^{\ell-1}[0, 1]$ for $1 \le \ell \le n - 1$ then F_k contains spline functions of **degree** ℓ with simple knots at $j/2^k$, $j = 1, \ldots, 2^k - 1$. This means that

$$\overline{\bigcup_{k=0}^{\infty} F_k} = L^2[0, 1]. \tag{2.19}$$

This is the second part of the next result.

Theorem 2.4 *Let $\mathbf{f} = (f_1, \ldots, f_n)^T$ be a refinable curve such that $\mathbf{f} \in C^{\ell-1}[0,1]$ for some ℓ, $1 \le \ell \le n - 1$, $\mathbf{f}^{(\ell-1)}$ is nontrivial and $\{f_1, \ldots, f_n\}$ is an ℓ-th degree convenient basis for F_0. Then*

$$F_k \subseteq F_{k+1}, \quad k = 0, 1, \ldots \tag{2.20}$$

and equation (2.19) holds.

Proof: The proof is similar to the proof of Theorem 3.2 in [4]. Let us review some of the essential details here.

So far, we have proved (2.19). For the proof of (2.20) we use Proposition 2.3 to conclude that there are functions

$$h_0, \ldots, h_{\ell-1} \in C^{\ell-1}\left([0,1] \setminus \left\{\frac{1}{2}\right\}\right)$$

such that the matrix

$$J := \left(\left[h_i^{(j)}\left(\frac{1}{2}\right)\right]\right)_{i,j=0,1,\ldots,\ell-1} \tag{2.21}$$

is nonsingular and

$$T_0 F_0 \oplus T_1 F_0 = F_1 \oplus \mathrm{span}\{h_0, \ldots, h_{\ell-1}\}.$$

By the refinement equation (2.2) it follows that

$$F_0 \subseteq T_0 F_0 \oplus T_1 F_0$$

(see [4, Lemma 2.1]) and hence

$$F_0 \subseteq F_1 \oplus \mathrm{span}\{h_0, \ldots, h_{\ell-1}\}.$$

Keeping in mind that $F_0, F_1 \subseteq C^{\ell-1}[0,1]$ and the matrix J in (2.21) is nonsingular, it follows that $F_0 \subseteq F_1$.

The general case of (2.20) is proved by induction on k. Suppose (2.20) holds for $k = m$, i.e., $F_m \subseteq F_{m+1}$. We consider the case when $k = m + 1$. Then, as above, by Proposition 2.3 there are functions

$$h_{0,m+1}, \ldots, h_{\ell-1,m+1} \in C^{\ell-1}\left([0,1] \setminus \left\{\frac{1}{2}\right\}\right)$$

such that the matrix

$$J_{m+1} := \left(\left[h_{i,m+1}^{(j)}\left(\frac{1}{2}\right)\right]\right)_{i,j=0,1,\ldots,\ell-1} \tag{2.22}$$

is nonsingular and

$$F_{m+2} \oplus \mathrm{span}\{h_{0,m+1}, \ldots, h_{\ell-1,m+1}\} = T_0 F_{m+1} \oplus T_1 F_{m+1}.$$

Therefore, the induction hypothesis implies that

$$F_{m+2} \oplus \text{span}\{h_{0,m+1}, \ldots, h_{\ell-1,m+1}\} \quad \supseteq \quad T_0 F_m \oplus T_1 F_m$$
$$\supseteq \quad T_0 F_m \oplus_\ell T_1 F_m = F_{m+1}.$$

Since F_{m+1}, F_{m+2} are subspaces of $C^{\ell-1}[0,1]$ and the matrix (2.22) is nonsingular, we conclude from the above inclusion that

$$F_{m+1} \subseteq F_{m+2}.$$

This advances the induction hypothesis and proves (2.20). ∎

Next, we let W_k be the orthogonal complement of F_k in F_{k+1} (as subspaces of $L^2[0,1]$). Clearly,

$$\dim W_k = \dim F_{k+1} - \dim F_k$$
$$= 2^k \dim F_0 - 2^k \ell.$$

The next theorem gives some information about how to generate W_k recursively in k. For this purpose, we choose functions

$$h_{i,k} \in T_0 F_k \oplus T_1 F_k, \quad i = 0, 1, \ldots, \ell - 1,$$

such that

$$T_0 F_k \oplus T_1 F_k = F_{k+1} \oplus \text{span}\{h_{0,k}, \ldots, h_{\ell-1,k}\}$$

and the matrix

$$J_k := \left(\left[h_{j,k}^{(j)} \left(\frac{1}{2} \right) \right] \right)_{i,j=0,1,\ldots,\ell-1} \tag{2.23}$$

is nonsingular.

Theorem 2.5 *Let* $\mathbf{f} = (f_1, \ldots, f_n)^T$ *be a refinable curve such that* $\mathbf{f} \in C^{\ell-1}[0,1]$ *for some* ℓ, $1 \leq \ell \leq n - 1$, $\mathbf{f}^{(\ell-1)}$ *is nontrivial and* $\{f_1, \ldots, f_n\}$ *is an* ℓ-*th degree convenient basis for* F_0. *Then, for* $k = 1, 2, \ldots,$ *we have*

$$W_k \oplus \text{span}\{h_{0,k}, \ldots, h_{\ell-1,k}\} = T_0 W_{k-1} \oplus T_1 W_{k-1} \oplus \text{span}\{h_{0,k-1}, \ldots, h_{\ell-1,k-1}\}. \tag{2.24}$$

In particular,

$$W_k = (T_0 W_{k-1} \oplus T_1 W_{k-1} \oplus \text{span}\{h_{0,k-1}, \ldots, h_{\ell-1,k-1}\})^\ell. \tag{2.25}$$

Proof: The proof is similar to the proof of Theorem 3.3 of [4] and is by induction on k. Suppose that we have already proved that

$$F_k = F_{k-1} \oplus W_{k-1}$$

and

$$F_{k-1} \perp W_{k-1}.$$

Let V_k be the right hand side of equation (2.25) and set

$$H_k := \text{span}\{h_{0,k}, \ldots, h_{\ell-1,k}\}.$$

That is, V_k is the subspace of $C^{\ell-1}[0,1]$ such that

$$V_k \oplus H_k = T_0 W_{k-1} \oplus T_1 W_{k-1} \oplus H_{k-1}.$$

Thus,

$$
\begin{aligned}
F_k \oplus V_k \oplus H_k &= F_k \oplus T_0 W_{k-1} \oplus T_1 W_{k-1} \oplus H_{k-1} \\
&= T_0 F_{k-1} \oplus T_1 F_{k-1} \oplus T_0 W_{k-1} \oplus T_1 W_{k-1} \\
&= T_0 F_k \oplus T_1 F_k \\
&= F_{k+1} \oplus H_k.
\end{aligned}
$$

Since F_k, V_k, and F_{k+1} are subspaces of $C^{\ell-1}[0,1]$ and, the matrix J_k is nonsingular (see (2.23)), we conclude that

$$F_k \oplus V_k = F_{k+1}.$$

Finally, we observe that

$$
\begin{aligned}
V_k &\subseteq T_0 W_{k-1} \oplus T_1 W_{k-1} \oplus H_{k-1} \\
&\subseteq (T_0 F_k \oplus T_1 F_k) \perp F_k.
\end{aligned}
$$

That is, $V_k = W_k$. ∎

3 SPLINE WAVELETS ON [0,1] SATISFYING HOMOGENEOUS BOUNDARY CONDITIONS

In this section, we continue with the point of view of the previous section and use the linear maps T_ϵ, $\epsilon \in \{0,1\}$ to generate smooth wavelets recursively. In this section, the difference is that we restrict ourselves to certain spline spaces which satisfy homogenous boundary conditions. As a consequence, the resulting smooth wavelets satisfy the same boundary conditions.

The flow of mathematical presentation in this section is similar to our earlier one. However, there are differences. First, we abandon refinable curves and find contentment with the use of spline functions. This allows us to bypass at least initially, the refinement operators and describe directly our scale of spline spaces satisfying homogeneous boundary conditions. Then, as in section two we study the scale of orthogonal complements. It is here that the refinement operators reappear. In fact, we bring them back into play by giving an alternative description of our spline spaces in terms of the refinement operators. From this vantage point we can then easily generate our orthogonal complements recursively.

To define the spline spaces that are of interest to us we begin with two integers $\ell \geq 1$, $n \geq 2$ which we use to partition the interval $[0, 1]$ into subintervals

$$I_m = \left[\frac{m}{N+1}, \frac{m+1}{N+1}\right], \quad m = 0, 1, \ldots, N := n + \ell - 2. \tag{3.1}$$

On this partition resides the spline space

$$S := \{s \in C^{n-2}[0, 1] : \ s|_{I_m} \in \Pi_{n-1}, \ m = 0, 1, \ldots, N\}. \tag{3.2}$$

Also, we denote by S_0 the subspace of S defined by

$$S_0 := \{s \in S : \ s^{(j)}(0) = s^{(j)}(1) = 0, \ j = 0, 1, \ldots, n - 2\}.$$

It is well-known that $\dim S_0 = \ell$. In fact, a basis of S_0 can be constructed from the cardinal B-spline. To this end, let M be the n-fold convolution of the characteristic function of the interval $[0, 1)$. Then M is a spline function of degree $n - 1$ with knots at integers and it vanishes off the interval $[0, n)$. Moreover, the functions

$$B_i(x) = M((N+1)x - i), \quad i = 0, 1, \ldots, \ell - 1, \tag{3.3}$$

span S_0.

For each integer $k \geq 1$, we use the points

$$C_k := \left\{2^{-k}\left(j + \frac{i}{N+1}\right) : \ j = 0, 1, \ldots, 2^k - 1, \ i = 0, 1, \ldots, N\right\}$$

to partition the interval $[0, 1]$. The corresponding spline space of degree $n - 1$ with knots at these points which have their derivatives up to order $n - 2$ zero at zero and one will be denoted by S_k. Thus,

$$S_k = \{s \in C^{n-2}[0, 1] : \quad s|_{I_{m,k}} \in \Pi_{n-1}, m = 0, 1, \ldots, 2^k(N+1) - 1,$$
$$s^{(j)}(0) = s^{(j)}(1) = 0, j = 0, 1, \ldots, n - 2\}$$

where

$$I_{m,k} := \left[\frac{m}{2^k(N+1)}, \frac{m+1}{2^k(N+1)}\right], \quad m = 0, 1, \ldots, 2^k(N+1) - 1.$$

As a consequence, we have

$$\dim S_k = (N+1)2^k - n + 1 \tag{3.4}$$

and

$$S_k \subset S_{k+1}, \quad k = 1, 2, \ldots. \tag{3.5}$$

Inclusion (3.5) follows from the fact that the knot sets C_k, $k = 1, 2, \ldots$ are also nested, viz.

$$C_k \subset C_{k+1}, \quad k = 1, 2, \ldots.$$

In fact, if we set

$$\phi_\epsilon(t) = \frac{t + \epsilon}{2}, \quad t \in [0, 1], \quad \epsilon \in \{0, 1\} \tag{3.6}$$

then

$$C_{k+1} = \phi_0(C_k) \cup \phi_1(C_k).$$

We consider S_k as a subspace of $L^2[0, 1]$ and let W_k be the orthogonal complement of S_k as a subspace of S_{k+1}; symbolically we write

$$S_{k+1} = S_k \oplus^\perp W_k, \quad k = 0, 1, \ldots. \tag{3.7}$$

Clearly, it follows from (3.4) and (3.7) that

$$\dim W_k = \dim S_{k+1} - \dim S_k = 2^k(N + 1). \tag{3.8}$$

We now have two scales of spaces S_k and W_k for which we want to provide recursive formulas in k for their generation. To this end, we again make use of the refinement operators

$$(T_\epsilon g)(t) = \begin{cases} g(2t), & 0 \le t \le \frac{1}{2}, \\ (-1)^\epsilon g(2t - 1), & \frac{1}{2} < t \le 1. \end{cases}$$

Also, we set

$$H_0^{n-1}[0, 1] := \{g \in H^{n-1}[0, 1] : \quad g^{(i)}(0) = g^{(i)}(1) = 0, \quad i = 0, 1, \ldots, n - 2\},$$

where $H^{n-1}[0, 1]$ is the usual Hilbert space of functions on $[0, 1]$ with $n - 2$ continuous derivatives and the $(n - 1)$st derivative is in $L^2[0, 1]$. Clearly, $H_0^{n-1}[0, 1]$ contains the spline spaces S_k, $k = 0, 1, \ldots$.

Our first task is to develop a recursive formula for S_k. This will then naturally lead us to a similar formula for W_k. Therefore we must be concerned about how the refinement operators act on $H_0^{n-1}[0, 1]$. This is described in the next lemma.

Lemma 3.1 *If* $g \in H_0^{n-1}[0, 1]$ *then* $T_\epsilon g \in H_0^{n-1}[0, 1]$, $\epsilon \in \{0, 1\}$ *and*

$$(T_\epsilon g)^{(j)}\left(\frac{1}{2}\right) = 0, \quad j = 0, 1, \ldots, n - 2.$$

Proof: Since $g \in H_0^{n-1}[0, 1]$ we have

$$g^{(j)}(0) = g^{(j)}(1) = 0, \quad j = 0, 1, \ldots, n - 2.$$

Therefore, from the definition of T_ϵ we get for $j = 0, 1, \ldots, n - 2$

$$(T_\epsilon g)^{(j)}\left(\frac{1^-}{2}\right) = 2^j g^{(j)}(1) = 0,$$

$$(T_\epsilon g)^{(j)} \left(\frac{1}{2}^+ \right) = (-1)^\epsilon 2^j g^{(j)}(0) = 0,$$

$$(T_\epsilon g)^{(j)}(0) = 2^j g^{(j)}(0) = 0,$$

and

$$(T_\epsilon g)^{(j)}(1) = (-1)^\epsilon 2^j g^{(j)}(1) = 0.$$

■

Next, we show how to generate a basis in S_k recursively in k. Specifically, we will show that by applying the refinement operators to an orthonormal basis of S_k and then adding a (universal) set of $n-1$ orthonormal functions in S_0 (which are independent of k) there results a basis for S_{k+1}. For this purpose, we observe that when $\ell \geq n-1$ there are at least $n-1$ B-splines $B_i(x)$ (see (3.3)), which contain $\frac{1}{2}$ in the **interior** of its support. In fact, when $\frac{1}{2} \in C_0$ then there are exactly $n-1$ such B-splines while for $\frac{1}{2} \notin C_0$ there are n in number (and so we may discard either the left most or right most B-spline which is nonzero at $\frac{1}{2}$). Label these $n-1$ B-splines

$$B_j(x), B_{j+1}(x), \ldots, B_{j+n-2}(x).$$

When $\frac{1}{2} \notin C_0$, j can be chosen to be one of two consecutive integers.

These functions are the universal set of $n-1$ functions that we need. To demonstrate that this is indeed the case we make a preliminary observation about these functions.

Lemma 3.2 *Suppose that* $\ell \geq n-1$. *Let* $B_j, B_{j+1}, \ldots, B_{j+n-2} \in S_0$ *have* $\frac{1}{2}$ *in the interior of their support. Then the* $(n-1) \times (n-1)$ *matrix*

$$B := \left(B_{j+r}^{(s)} \left(\frac{1}{2} \right) \right)_{r,s=0,1,\ldots,n-2}$$

is nonsingular.

Proof: Suppose to the contrary that for some constants $c_0, c_1, \ldots, c_{n-2}$ the spline function

$$s(t) = \sum_{r=0}^{n-2} c_r B_{j+r}(t)$$

satisfies the equations

$$s^{(i)} \left(\frac{1}{2} \right) = 0, \quad i = 0, 1, \ldots, n-2.$$

We only consider the case $\frac{1}{2} \notin C_0$ and j has its least value (the other two cases are argued similarly). Let $x \in (\frac{1}{2}, 1]$ be the right most point in the support of s. Then we also have

$$s^{(i)}(x) = 0, \quad i = 0, 1, \ldots, n-2$$

and s has at most $n - 2$ knots in the interval $(\frac{1}{2}, x)$. Since a B-spline has the least number of $n - 1$ knots in the interior of its support we conclude that $s(t) = 0$ for $t \in (\frac{1}{2}, 1)$. From this conclusion it easily follows that $c_0 = \cdots = c_{n-2} = 0$. $\blacksquare$

Next, we choose orthonormal functions $f_0, \ldots, f_{n-2}$ which form a basis for the subspace

$$\text{span}\{B_{j+r} : \quad r = 0, 1, \ldots, n - 2\}.$$

Clearly, these functions retain the property that the $(n - 1) \times (n - 1)$ matrix

$$F := \left(f_s^{(r)} \left(\frac{1}{2} \right) \right)_{r,s=0,1,\ldots,n-2} \tag{3.9}$$

is nonsingular.

Our first result of this section proves the claim made earlier. Namely, starting with an orthonormal basis for S_k, then applying the refinement operators and finally adding a universal set of $n - 1$ orthonormal functions from S_0 gives a basis for S_{k+1}.

Theorem 3.3 Let $\ell \geq n - 1$, $k \geq 0$ and set $d_k := \dim S_k$. If

$$\mathbb{F}_k = \{f_{k,1}, \ldots, f_{k,d_k}\}$$

form an orthonormal basis of S_k then the functions

$$\{f_0, \ldots, f_{n-2}\} \cup T_0 \mathbb{F}_k \cup T_1 \mathbb{F}_k \tag{3.10}$$

form a basis of S_{k+1}.

Proof: According to the fact that

$$T_\epsilon^* T_{\epsilon'} = \delta_{\epsilon \epsilon'} I,$$

see (2.9), it is clear that the functions in the set

$$T_0 \mathbb{F}_k \cup T_1 \mathbb{F}_k$$

are orthonormal. As a consequence of Lemma 3.2 and formula (3.7) we see that

$$T_\epsilon S_k \subseteq S_{k+1}. \tag{3.12}$$

This inclusion implies that all the functions in (3.10) are in S_{k+1}. Also, in total there are

$$n - 1 + 2\dim S_k$$

functions in (3.10) which, by formula (3.4), is $\dim S_{k+1}$. Therefore it suffices to show that the two sets $\{f_0, \ldots, f_{n-2}\}$ and $T_0 \mathbb{F}_k \cup T_1 \mathbb{F}_k$ are linearly independent.

This is obvious since by Lemma 3.1 all functions f in the second set satisfy the condition

$$f^{(i)}(0) = 0, \quad i = 0, 1, \ldots, n - 2, \tag{3.13}$$

while by Lemma 3.2 the only function in

$$\text{span}\{f_0, \ldots, f_{n-2}\}$$

which satisfies (3.13) is identically zero. ∎

The next result gives a recursive formula for the generating the subspaces W_k, $k = 0, 1, \ldots$. It demonstrates that starting with an orthonormal basis of W_0 in $L^2[0, 1]$ it is simple to compute an orthonormal basis for W_k recursively, merely apply the refinement operators to an orthonormal basis of W_{k-1}. This is indeed a simple recursive procedure.

Theorem 3.4 *Assume $\ell \geq n - 1$ and W_k is the orthogonal complement of S_k in S_{k+1}. Then*

$$W_k = T_0 W_{k-1} \oplus^\perp T_1 W_{k-1}, \quad k = 1, 2, \ldots, \tag{3.14}$$

$$\overline{\bigcup_{k=0}^{\infty} S_k} = H_0^{n-1}[0, 1] \tag{3.15}$$

and

$$S_0 \oplus^\perp W = H_0^{n-1}[0, 1], \tag{3.16}$$

where

$$W := \bigoplus_{k=0}^{\infty}{}^\perp W_k.$$

Proof: Equation (3.15) follows from standard approximation properties of spline spaces, cf. [5]. From this equation and the definition of W_k, equation (3.16) follows directly. Thus, it remains to prove (3.14) which we do next. To this end, we observe by the definition of S_k that

$$G_\epsilon S_{k+1} \subseteq S_k, \quad \epsilon \in \{0, 1\}, \quad k = 1, 2, \ldots,$$

where G_ϵ is the linear operator given by (2.6). Thus, formula (2.8) implies that

$$T_\epsilon^* S_{k+1} \subseteq S_k, \quad \epsilon \in \{0, 1\}, \quad k = 1, 2, \ldots . \tag{3.17}$$

Now, apply T_ϵ^* to both sides of inclusion (3.12) and use formula (2.9) (when $\epsilon' = \epsilon$) to get

$$S_k \subseteq T_\epsilon^* S_{k+1}. \tag{3.18}$$

That is, combining (3.17) and (3.18) we get

$$S_k = T_\epsilon^* S_{k+1}, \quad \epsilon \in \{0, 1\}.$$

Let W_k be the orthogonal complement of S_k in S_{k+1} and set

$$V := T_0 W_k \oplus T_1 W_k \tag{3.19}$$

From formula (2.9) we see that

$$T_0 W_k \perp T_1 W_k. \tag{3.20}$$

Also, using inclusion (3.12) we have

$$V \subseteq T_0 S_{k+1} \oplus T_1 S_{k+1} \subseteq S_{k+2}. \tag{3.21}$$

According to (3.18) we have

$$W_k \perp S_k \subseteq T_\epsilon^* S_{k+1}, \quad \epsilon \in \{0,1\},$$

which insures that

$$T_\epsilon W_k \perp S_{k+1}, \quad \epsilon \in \{0,1\}.$$

This guarantees that

$$V \perp S_{k+1}, \tag{3.22}$$

and so by inclusion (3.21) we get

$$V \subseteq W_{k+1}. \tag{3.23}$$

However, by definition (3.19) and formula (3.8) we have

$$\dim V = 2 \dim W_k = 2^{k+1}(N+1) = \dim W_{k+1}.$$

Combining this equation with (3.23) proves $V = W_{k+1}$. ∎

REFERENCES

[1] Horn R.A., Johnson C.R. *Matrix Analysis*. Cambridge University Press, Cambridge, 1985.

[2] Micchelli C.A. *Mathematical Aspects of Geometric Modeling*. CBMS Series, SIAM, Philadelphia. To appear in 1994.

[3] Micchelli C.A., Prautzsch H. Uniform refinement of curves. *Linear Algebra Appl.*, **114/115**:841-870, 1989.

[4] Micchelli C.A., Xu Y. Using the matrix refinement equation for the construction of wavelets on invariant sets. *Appl. Comp. Harmonic Anal.* To appear.

[5] Schumaker L.L. *Spline Functions: Basic Theory*. Wiley-Interscience, New York, 1981.

457

International Series of Numerical Mathematics, Vol. 119, ©1994 Birkhäuser

SUMMATION OF SERIES AND GAUSSIAN QUADRATURES[*]

Gradimir V. Milovanović[†]

Dedicated to Walter Gautschi on the occasion of his 65th birthday

Abstract. In 1985, Gautschi and the author constructed Gaussian quadrature formulae on $(0, +\infty)$ involving Einstein and Fermi functions as weights and applied them to the summation of slowly convergent series which can be represented in terms of the derivative of a Laplace transform, or in terms of the Laplace transform itself. A problem that may arise in this procedure is the determination of the respective inverse Laplace transform. For the class of slowly convergent series $\sum_{k=1}^{+\infty} (\pm 1)^k a_k$ with $a_k = k^{\nu-1} R(k)$, where $0 < \nu \le 1$ and $R(s)$ is a rational function, Gautschi recently solved this problem. In the present paper, using complex integration and constructing Gauss-Christoffel quadratures on $(0, +\infty)$ with respect to the weight functions $w_1(t) = 1/\cosh^2 t$ and $w_2(t) = \sinh t/\cosh^2 t$, we reduce the series $\sum_{k=m}^{+\infty} f(k)$ and $\sum_{k=m}^{+\infty} (-1)^k f(k)$ to weighted integrals of f involving weights w_1 and w_2, respectively. We illustrate this method with a few numerical examples.

1 INTRODUCTION

We consider the summation of series of the type

$$T_m = \sum_{k=m}^{+\infty} a_k \tag{1.1}$$

and

$$S_m = \sum_{k=m}^{+\infty} (-1)^k a_k, \tag{1.2}$$

where $m \in \mathbb{Z}$.

Methods of summation can be found, for example, in the books of Henrici [11], Lindelöf [12], and Mitrinović and Kečkić [13].

---

[*]1991 *Mathematics Subject Classification.* Primary 40A25; Secondary 30E20, 65D32, 33C45

[†]Faculty of Electronic Engineering, Department of Mathematics, University of Niš, P. O. Box 73, 18000 Niš, Yugoslavia

In a joint paper with Gautschi [9] we considered the construction of Gaussian quadrature formulae on $(0, +\infty)$ with respect to the weight functions

$$w(t) = \varepsilon(t) = \frac{t}{e^t - 1} \qquad \text{(Einstein's function)}$$

and

$$w(t) = \varphi(t) = \frac{1}{e^t + 1} \qquad \text{(Fermi's function)}$$

and showed that these formulae can be applied to sum slowly convergent series of the form T_1 and S_1 whose general term is expressible in terms of the derivative of a Laplace transform, or in terms of the Laplace transform itself. Namely, if $a_k = F'(k)$, where

$$F(s) = \int_0^{+\infty} e^{-st} f(t)\, dt, \qquad \text{Re}\, s \geq 1,$$

we have

$$T_1 = \sum_{k=1}^{+\infty} F'(k) = -\int_0^{+\infty} f(t)\varepsilon(t)\, dt$$

and

$$S_1 = \sum_{k=1}^{+\infty} (-1)^k F'(k) = \int_0^{+\infty} f(t) t\varphi(t)\, dt.$$

Also, for $a_k = F(k)$, we have

$$S_1 = \sum_{k=1}^{+\infty} (-1)^k F(k) = -\int_0^{+\infty} f(t)\varphi(t)\, dt.$$

If the series T_1 and S_1 are slowly convergent and the respective function f on the right of the equalities above is smooth, then low-order Gaussian quadrature[‡]

$$\int_0^{+\infty} g(t)w(t)\, dt = \sum_{\nu=1}^{n} \lambda_\nu g(\tau_\nu) + R_n(g),$$

with $w(t) = \varepsilon(t)$ and $w(t) = \varphi(t)$, applied to the integrals on the right, provides a possible summation procedure. Numerical examples show fast convergence of this procedure (see [9, §4]). In the sequel we refer to this procedure as the "Laplace transform method." A problem which arises with this procedure is the determination of the original function f for a given series. For some other applications see Gautschi [6] and [7].

[‡]The functions $t \mapsto \varepsilon(t)$ and $t \mapsto \varphi(t)$ arise in solid state physics. Integrals with respect to the measure $d\lambda(t) = \varepsilon(t)^r\, dt$, $r = 1$ and $r = 2$, are widely used in phonon statistics and lattice specific heats and occur also in the study of radiative recombination processes. Similarly, integrals with $\varphi(t)$ are encountered in the dynamics of electrons in metals.

In [6], Gautschi treated the case when $a_k = k^{\nu-1}R(k)$, where $0 < \nu \leq 1$ and $R(\,\cdot\,)$ is a rational function $R(s) = P(s)/Q(s)$, with P, Q real polynomials of degrees $\deg P \leq \deg Q$. By interpreting the terms in T_1 and S_1 again as Laplace transforms at integer values, Gautschi expressed the sum of the series as a weighted integral over $\mathbb{R}_+$ of certain special functions related to the incomplete gamma function. The weighting involves the product of a fractional power and either Einstein's function $\varepsilon(t)$ (for T_1) or Fermi's function $\varphi(t)$ (for S_1). The case $\nu = 1$ of purely rational series complements some traditional techniques of summations via quadratures (cf. [11, §7.2II]).

In particular, Gautschi [6] analyzed examples with $a_k = k^{-1/2}/(k+a)^m$, where $\operatorname{Re} a \geq 0$ and $m \geq 1$. The series T_1 with $a = m = 1$ appeared in a study of spirals given by Davis [1] (see also Gautschi [8]).

In this paper we give an alternative summation/integration procedure for the series (1.1) and (1.2) when $a_k = f(k)$, where $z \mapsto f(z)$ is an analytic function in the region

$$\{z \in \mathbb{C} \mid \operatorname{Re} z \geq \alpha, \ m-1 < \alpha < m\}. \tag{1.3}$$

Our method requires the indefinite integral F of f chosen so as to satisfy certain decay properties ((C1) – (C3) below). Using contour integration over a rectangle in the complex plane, we are able to reduce T_m and S_m to a problem of quadrature on $(0, +\infty)$ with respect to the weight functions

$$w_1(t) = \frac{1}{\cosh^2 t} \quad \text{and} \quad w_2(t) = \frac{\sinh t}{\cosh^2 t}, \tag{1.4}$$

respectively.

The contour integration is discussed in §2. The generation of the recursion coefficients in the three-term recurrence relation for the (monic) orthogonal polynomials $\pi_k(\,\cdot\,) = \pi_k(\,\cdot\,; w_p)$, $k = 0, 1, \dots$, with respect to the weight function $w_p(t)$, $p = 1, 2$, is discussed in §3. Numerical examples are presented in §4.

2 PRELIMINARIES

Assume that f and g are analytic functions in a certain domain D of the complex plane with singularities $a_1, a_2, \dots$ and $b_1, b_2, \dots$, respectively, in a region $G = \operatorname{int} \Gamma \, (\subset D)$, where Γ is a closed contour. Then by Cauchy's residue theorem, we have

$$\frac{1}{2\pi i} \oint_\Gamma f(z)g(z)\,dz = \sum_\nu \operatorname*{Res}_{z=a_\nu} \Big(f(z)g(z)\Big) + \sum_\nu \operatorname*{Res}_{z=b_\nu} \Big(f(z)g(z)\Big). \tag{2.1}$$

Let

$$G = \Big\{z \in \mathbb{C} \mid \alpha \leq \operatorname{Re} z \leq \beta, \ |\operatorname{Im} z| \leq \frac{\delta}{\pi}\Big\},$$

461

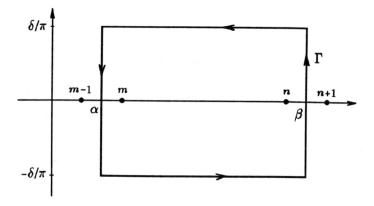

Figure 2.1: The contour of integration

where $m - 1 < \alpha < m$, $n < \beta < n+1$ ($m, n \in \mathbb{Z}, m \leq n$), $\Gamma = \partial G$ (see Figure 2.1), and $g(z) = \pi / \tan \pi z$. Then from (2.1) it immediately follows that (cf. [13, p. 212])

$$T_{m,n} = \sum_{\nu=m}^{n} f(\nu) = \frac{1}{2\pi i} \oint_{\Gamma} f(z) \frac{\pi}{\tan \pi z}\, dz - \sum_{\nu} \operatorname*{Res}_{z=a_\nu} \left(f(z) \frac{\pi}{\tan \pi z} \right).$$

Similarly, for $g(z) = \pi / \sin \pi z$ we have

$$S_{m,n} = \sum_{\nu=m}^{n} (-1)^{\nu} f(\nu) = \frac{1}{2\pi i} \oint_{\Gamma} f(z) \frac{\pi}{\sin \pi z}\, dz - \sum_{\nu} \operatorname*{Res}_{z=a_\nu} \left(f(z) \frac{\pi}{\sin \pi z} \right).$$

For a holomorphic function $z \mapsto f(z)$ in G, the last formulae become

$$T_{m,n} = \frac{1}{2\pi i} \oint_{\Gamma} f(z) \frac{\pi}{\tan \pi z}\, dz \quad \text{and} \quad S_{m,n} = \frac{1}{2\pi i} \oint_{\Gamma} f(z) \frac{\pi}{\sin \pi z}\, dz. \tag{2.2}$$

After integration by parts, formulae (2.2) reduce to

$$T_{m,n} = \frac{1}{2\pi i} \oint_{\Gamma} \left(\frac{\pi}{\sin \pi z} \right)^2 F(z)\, dz \tag{2.3}$$

and

$$S_{m,n} = \frac{1}{2\pi i} \oint_{\Gamma} \left(\frac{\pi}{\sin \pi z} \right)^2 \cos \pi z\, F(z)\, dz, \tag{2.4}$$

where $z \mapsto F(z)$ is an integral of $z \mapsto f(z)$.

Assume now the following conditions for the function $z \mapsto F(z)$ (cf. [12, p. 57]):

(C1) F is a holomorphic function in the region (1.3);

(C2) $\displaystyle \lim_{|t| \to +\infty} e^{-c|t|} F(x + it/\pi) = 0$, uniformly for $x \geq \alpha$;

(C3) $\quad \lim\limits_{x \to +\infty} \int_{-\infty}^{+\infty} e^{-c|t|} \left| F\left(x + it/\pi\right)\right| dt = 0,$

where $c = 2$ or $c = 1$, when we consider $T_{m,n}$ or $S_{n,m}$, respectively.

Set $\alpha = m - 1/2$ and $\beta = n + 1/2$.

On the lines $z = x \pm i(\delta/\pi)$, we have that

$$\left| \frac{\pi}{\sin \pi z} \right| = \left| \frac{2i\pi}{e^{i\pi z} - e^{-i\pi z}} \right| = \frac{2\pi e^{-\delta}}{\left|1 - e^{-2\delta} e^{\pm i 2\pi x}\right|} \leq \frac{2\pi e^{-\delta}}{1 - e^{-2\delta}}$$

and also

$$|\cos \pi z| = \frac{1}{2} e^{\delta} \left|1 + e^{-2\delta} e^{\pm i 2\pi x}\right| \leq \frac{1}{2} e^{\delta} (1 + e^{-2\delta}).$$

Therefore, under condition (C2), the integrals on the lines $z = x \pm i(\delta/\pi)$, $\alpha \leq x \leq \beta$,

$$\frac{1}{2\pi i} \int_{\alpha \pm i(\delta/\pi)}^{\beta \pm i(\delta/\pi)} \left(\frac{\pi}{\sin \pi z} \right)^2 F(z)\, dz, \quad \frac{1}{2\pi i} \int_{\alpha \pm i(\delta/\pi)}^{\beta \pm i(\delta/\pi)} \left(\frac{\pi}{\sin \pi z} \right)^2 \cos \pi z\, F(z)\, dz,$$

tend to zero when $\delta \to +\infty$.

For $z = \beta + iy$ we have

$$\sin \pi z = (-1)^n \cosh \pi y, \qquad \cos \pi z = i(-1)^{n+1} \sinh \pi y,$$

$$\left| \frac{\pi}{\sin \pi z} \right|^2 = \frac{\pi^2}{\cosh^2 \pi y} = \frac{4\pi^2}{(e^{\pi y} + e^{-\pi y})^2} \leq 4\pi^2 e^{-2\pi|y|},$$

and

$$\left| \frac{\pi}{\sin \pi z} \right|^2 |\cos \pi z| \leq 2\pi^2 e^{-\pi|y|},$$

so that

$$\left| \frac{1}{2\pi i} \int_{\beta - i(\delta/\pi)}^{\beta + i(\delta/\pi)} \left(\frac{\pi}{\sin \pi z} \right)^2 F(z)\, dz \right| \leq 2 \int_{-\delta}^{\delta} e^{-2|t|} \left| F\left(\beta + it/\pi\right)\right| dt$$

and

$$\left| \frac{1}{2\pi i} \int_{\beta - i(\delta/\pi)}^{\beta + i(\delta/\pi)} \left(\frac{\pi}{\sin \pi z} \right)^2 \cos \pi z\, F(z)\, dz \right| \leq \int_{-\delta}^{\delta} e^{-|t|} \left| F\left(\beta + it/\pi\right)\right| dt.$$

When $\delta \to +\infty$ and $n \to +\infty$ (i.e., $\beta \to +\infty$), because of (C3), the previous integrals tend to zero.

Thus, when $\delta \to +\infty$ and $n \to +\infty$, the integrals in (2.3) and (2.4) over Γ reduce to integrals along the line $z = \alpha + iy$ $(-\infty < y < +\infty)$, so that

$$T_m = T_{m,\infty} = -\frac{1}{2\pi i} \int_{\alpha - i\infty}^{\alpha + i\infty} \left(\frac{\pi}{\sin \pi z} \right)^2 F(z)\, dz \qquad (2.5)$$

and

$$S_m = S_{m,\infty} = -\frac{1}{2\pi i} \int_{\alpha-i\infty}^{\alpha+i\infty} \left(\frac{\pi}{\sin \pi z}\right)^2 \cos \pi z\, F(z)\, dz. \qquad (2.6)$$

Equality (2.5) can be reduced to

$$T_m = -\frac{1}{2} \int_{-\infty}^{+\infty} \frac{1}{\cosh^2 t} F(\alpha + it/\pi)\, dt,$$

i.e.,

$$T_m = \int_0^{+\infty} \Phi(\alpha, t/\pi)\, w_1(t)\, dt, \qquad (2.7)$$

where w_1 is defined in (1.4) and

$$\Phi(x, y) = -\frac{1}{2}\left[F(x + iy) + F(x - iy)\right]. \qquad (2.8)$$

Similarly, (2.6) reduces to

$$S_m = \int_0^{+\infty} \Psi(\alpha, t/\pi)\, w_2(t)\, dt, \qquad (2.9)$$

where w_2 is also defined in (1.4) and

$$\Psi(x, y) = \frac{(-1)^m}{2i}\left[F(x + iy) - F(x - iy)\right]. \qquad (2.10)$$

Formulae (2.7) and (2.9) suggest that Gaussian quadrature be applied to the integrals on the right, using the weight functions w_1 and w_2, respectively. The required orthogonal polynomials can be computed using the discretized Stieltjes procedure (see Gautschi [3, §2.2], [4–5]).

Instead of reducing integration to the positive half-line, one might keep integration over the full real line and note that

$$T_m = \int_{-\infty}^{+\infty} \Phi(\alpha, t/(2\pi)) \frac{e^{-t}}{(1 + e^{-t})^2}\, dt \qquad (2.11)$$

and

$$S_m = \int_{-\infty}^{+\infty} \Psi(\alpha, t/(2\pi)) \sinh(t/2) \frac{e^{-t}}{(1 + e^{-t})^2}\, dt. \qquad (2.12)$$

Here, the weight function is $t \mapsto w(t) = e^{-t}/(1 + e^{-t})^2$, the logistic weight, for which the recursion coefficients for the respective orthogonal polynomials are explicitly known[†]:

$$\alpha_k = 0, \quad \beta_0 = 1, \quad \beta_k = \frac{k^4 \pi^2}{4k^2 - 1}, \quad k = 1, 2, \dots .$$

Thus, no procedure is required to generate the recursion coefficients. Some comments on the convergence of the corresponding Gaussian quadrature will be given in §4.

[†]The referee pointed out this fact.

3 GENERATION OF THE RECURSION COEFFICIENTS

Let $\pi_k(\cdot) = \pi_k(\cdot\,; w_p)$, $k = 0, 1, \ldots$, be the (monic) polynomials orthogonal with respect to the weight function $t \mapsto w_p(t)$, $p = 1, 2$, on $(0, +\infty)$, where

$$w_1(t) = \frac{1}{\cosh^2 t} \quad \text{and} \quad w_2(t) = \frac{\sinh t}{\cosh^2 t}. \tag{3.1}$$

They satisfy the three-term recurrence relation

$$\pi_{k+1}(t) = (t - \alpha_k)\pi_k(t) - \beta_k \pi_{k-1}(t), \quad k = 0, 1, 2, \ldots,$$

$$\pi_0(t) = 1, \quad \pi_{-1}(t) = 0,$$

where

$$\alpha_k = \alpha_k(w_p), \quad \beta_k = \beta_k(w_p) \quad \left(\beta_0(w_p) = \int_0^{+\infty} w_p(t)\,dt \right).$$

Knowing the first n of these coefficients, α_k, β_k, $k = 0, 1, \ldots, n-1$, one can easily obtain the k-point Gaussian quadrature formula

$$\int_0^{+\infty} g(t)w_p(t)\,dt = \sum_{\nu=1}^{k} \lambda_\nu g(\tau_\nu) + R_{k,p}(g), \quad R_{k,p}(\mathcal{P}_{2k-1}) \equiv 0, \tag{3.2}$$

for any k with $1 \leq k \leq n$. The nodes $\tau_\nu = \tau_{\nu,p}^{(k)}$, indeed, are the eigenvalues of the symmetric tridiagonal Jacobi matrix

$$J_k(w_p) = \begin{bmatrix} \alpha_0 & \sqrt{\beta_1} & & & & 0 \\ \sqrt{\beta_1} & \alpha_1 & \sqrt{\beta_2} & & & \\ & \sqrt{\beta_2} & \alpha_2 & \ddots & & \\ & & \ddots & \ddots & \sqrt{\beta_{k-1}} \\ 0 & & & \sqrt{\beta_{k-1}} & \alpha_{k-1} \end{bmatrix},$$

while the weights $\lambda_\nu = \lambda_{\nu,p}^{(k)}$ are given by $\lambda_\nu = \beta_0 v_{\nu,1}^2$ in terms of the first components $v_{\nu,1}$ of the corresponding normalized eigenvectors (cf. Gautschi [2, §5.1] and Golub and Welsch [10]).

In order to construct the recursion coefficients, we use the discretized Stieltjes procedure as in [9], with the discretization based on the Gauss-Laguerre quadrature rule,

$$\int_0^{+\infty} p(t)w_1(t)\,dt = \int_0^{+\infty} p(t/2)\,\frac{2}{(1 + e^{-t})^2}\,e^{-t}\,dt$$

$$\cong \sum_{k=1}^{N} \lambda_k^L p(\tau_k^L / 2)\,\frac{2}{(1 + e^{-\tau_k^L})^2}$$

and

$$\int_0^{+\infty} p(t) w_2(t)\, dt = \int_0^{+\infty} p(t) \frac{2(1 - e^{-2t})}{(1 + e^{-2t})^2} e^{-t}\, dt$$

$$\cong \sum_{k=1}^{N} \lambda_k^L p(\tau_k^L) \frac{2(1 - e^{-2\tau_k^L})}{(1 + e^{-2\tau_k^L})^2},$$

where $p \in \mathcal{P}$. Here, τ_k^L and λ_k^L are the parameters of the N-point Gauss-Laguerre quadrature formula.

The first 40 recursion coefficients ($n = 40$) can be obtained accurately to 30 decimal digits with $N = 520$ for w_1 and $N = 720$ for w_2. We used the MICROVAX 3400 in Q-arithmetic (machine precision $\approx 1.93 \times 10^{-34}$). The same results were obtained also by a discretization procedure based on the composite Fejér quadrature rule, decomposing the interval of integration into four subintervals, $[0, +\infty] = [0, 10] \cup [10, 100] \cup [100, 500] \cup [500, +\infty]$ and using $N = 280$ points on each subinterval in both cases. We tabulate the results to 29 decimals in Table 1 and 2 of the Appendix.

Using (2.7)–(2.10) and (3.2) we can state the following results:

Theorem 3.1 *Let F be an integral of f such that the conditions* (C1), (C2), (C3) *are satisfied with $c = 2$. If $\lambda_\nu = \lambda_{\nu,1}^{(n)}$ and $\tau_\nu = \tau_{\nu,1}^{(n)}$ are the parameters of the n-point Gaussian quadrature* (3.2) *with weight function w_1, then*

$$T_m = \sum_{k=m}^{+\infty} f(k) = \sum_{\nu=1}^{n} \lambda_{\nu,1}^{(n)} \Phi\left(m - \tfrac{1}{2}, \tfrac{1}{\pi}\tau_{\nu,1}^{(n)}\right) + R_{n,1}(\Phi),$$

where Φ is defined by (2.8).

Theorem 3.2 *Let F be an integral of f such that the conditions* (C1), (C2), (C3) *are satisfied with $c = 1$. If $\lambda_\nu = \lambda_{\nu,2}^{(n)}$ and $\tau_\nu = \tau_{\nu,2}^{(n)}$ are the parameters of the n-point Gaussian quadrature* (3.2) *with weight function w_2, then*

$$S_m = \sum_{k=m}^{+\infty} (-1)^k f(k) = \sum_{\nu=1}^{n} \lambda_{\nu,2}^{(n)} \Psi\left(m - \tfrac{1}{2}, \tfrac{1}{\pi}\tau_{\nu,2}^{(n)}\right) + R_{n,2}(\Psi),$$

where Ψ is defined by (2.10).

Remark 3.1 If $\lambda_\nu = \lambda_\nu^{(n)}$ and $\tau_\nu = \tau_\nu^{(n)}$ are the parameters of the n-point Gaussian quadrature with respect to the logistic weight w, then according to (2.11) and (2.12),

$$T_m = \sum_{k=m}^{+\infty} f(k) \approx \sum_{\nu=1}^{n} \lambda_\nu \Phi\left(m - \tfrac{1}{2}, \tfrac{1}{2\pi}\tau_\nu\right) \tag{3.3}$$

and

$$S_m = \sum_{k=m}^{+\infty} (-1)^k f(k) = \sum_{\nu=1}^{n} \lambda_\nu \Psi\left(m - \tfrac{1}{2}, \tfrac{1}{2\pi}\tau_\nu\right) \sinh(\tau_\nu/2), \tag{3.4}$$

where Φ and Ψ are defined by (2.8) and (2.10), respectively. In Example 4.1 below, we will also consider these formulae.

4 NUMERICAL EXAMPLES

In this section we illustrate our method, using a few examples from [9] and [6]. All computations were done in Q-arithmetic on the MICROVAX 3400 computer.

Example 4.1 Consider

$$T_1 = \sum_{k=1}^{+\infty} \frac{1}{(k+1)^2} = \frac{\pi^2}{6} - 1 \quad \text{and} \quad S_1 = \sum_{k=1}^{+\infty} \frac{(-1)^k}{(k+1)^2} = \frac{\pi^2}{12} - 1.$$

Here, $f(z) = (z+1)^{-2}$, and $F(z) = -(z+1)^{-1}$, the integration constant being zero on account of the condition (C3). Thus,

$$\Phi(x, y) = \operatorname{Re} \frac{1}{z+1} = \frac{x+1}{(x+1)^2 + y^2}$$

and

$$\Psi(x, y) = \operatorname{Im} \frac{1}{z+1} = \frac{-y}{(x+1)^2 + y^2}.$$

Table 4.1 shows the n-point approximations $T_1(n)$ and $S_1(n)$ to T_1 and S_1, respectively, together with the relative errors $r_n(T_1)$ and $r_n(S_1)$, for $n = 5(5)40$. In each entry the first digit in error is underlined. (Numbers in parentheses indicate decimal exponents.)

Table 4.1

Gaussian approximation of the sums T_1 and S_1 and relative errors

n	$T_1(n)$	$r_n(T_1)$	$S_1(n)$	$r_n(S_1)$
5	.644934149	1.3(−7)	−.1775520	1.1(−4)
10	.644934066776	1.1(−10)	−.17753303	3.5(−7)
15	.644934066848158	1.1(−13)	−.17753296569	5.0(−9)
20	.64493406684822733	1.4(−15)	−.1775329665917	8.9(−11)
25	.6449340668482264405	6.2(−18)	−.177532966575286	3.4(−12)
30	.644934066848226436307	2.6(−19)	−.177532966575929	2.4(−13)
35	.64493406684822643647604	5.6(−21)	−.1775329665758832	2.0(−14)
40	.64493406684822643647233	1.3(−22)	−.1775329665758870	1.5(−15)

The corresponding relative errors in the "Laplace transform method" [9] applied to T_1 are given in Table 4.4. The recursion coefficients α_k, β_k for orthogonal polynomials $\pi_k(\,\cdot\,;\varepsilon)$ were also calculated with 30 correct decimal digits.

467

As can be seen, for smaller values of n (≤ 15) we obtained better results than in the "Laplace transform method". Furthermore, these results can be significantly improved if we apply this method to sum the series T_m, $m > 1$. That is, we use

$$T_1 = \sum_{k=1}^{m-1} \frac{1}{(k+1)^2} + T_m, \qquad T_m = \sum_{k=m}^{+\infty} \frac{1}{(k+1)^2}. \tag{4.1}$$

Then, for $m = 2(1)5$ we obtain results whose relative errors are presented in Table 4.2. Also, in Table 4.3 we present the corresponding results for the sum S_1 expressed in a similar way.

Table 4.2

Relative errors in Gaussian approximation of the sum T_1
expressed in the form (4.1) for $m = 2(1)5$

n	$m = 2$	$m = 3$	$m = 4$	$m = 5$
5	5.4(−9)	1.9(−10)	8.6(−12)	3.7(−13)
10	1.1(−13)	1.7(−16)	7.9(−18)	2.0(−19)
15	3.8(−17)	3.7(−20)	1.1(−22)	3.8(−25)
20	4.0(−20)	1.2(−24)	1.9(−27)	2.3(−29)
25	1.1(−22)	2.0(−27)	2.6(−30)	2.5(−33)
30	1.4(−25)	1.1(−31)	2.2(−33)	
35	3.2(−27)	2.4(−32)		
40	3.6(−30)			

The rapidly increasing speed of convergence of the summation process as m increases is due to the poles $\pm i\left(m + \frac{1}{2}\right)\pi$ of $\Phi\left(m - \frac{1}{2}, \frac{t}{\pi}\right)$ moving away from the real line.

Table 4.3

Relative errors in Gaussian approximation of the sum
$$S_1 = \sum_{k=1}^{m-1} (-1)^k (k+1)^{-1} + S_m \text{ for } m = 2(1)5$$

n	$m = 2$	$m = 3$	$m = 4$	$m = 5$
5	1.9(−6)	2.2(−7)	1.5(−8)	4.5(−10)
10	1.9(−9)	2.3(−12)	1.0(−12)	1.1(−14)
15	8.1(−13)	1.9(−15)	3.2(−16)	9.2(−18)
20	6.6(−14)	1.1(−16)	6.2(−19)	7.6(−21)
25	6.2(−16)	6.8(−19)	2.5(−21)	1.3(−23)
30	2.4(−18)	9.7(−21)	1.2(−24)	1.3(−26)
35	1.1(−19)	4.3(−23)	1.3(−25)	2.1(−28)
40	5.6(−21)	1.3(−24)	1.1(−27)	9.8(−31)

Table 4.4

Relative errors in Gaussian approximation of the sum T_1
using the "Laplace transform method" for $m = 1(1)3$

n	$m = 1$	$m = 2$	$m = 3$
5	3.0(−4)	8.4(−3)	3.0(−2)
10	1.1(−8)	1.8(−5)	3.8(−4)
15	3.2(−13)	2.8(−8)	3.7(−6)
20	8.0(−18)	3.9(−11)	3.1(−8)
25	1.8(−22)	5.1(−14)	2.5(−10)
30	3.9(−27)	6.3(−17)	1.9(−12)
35	8.7(−32)	7.6(−20)	1.4(−14)
40	4.6(−33)	8.8(−23)	1.0(−16)

It is interesting to note that a similar approach with the "Laplace transform method" does not lead to acceleration of convergence. For example, in the case of (4.1), we have that

$$T_m = \sum_{k=1}^{+\infty} \frac{1}{(k+m)^2} = \int_0^{+\infty} \varepsilon(t)e^{-mt}\, dt.$$

Then, applying Gaussian quadrature to the integral on the right, using $w(t) = \varepsilon(t)$ as a weight function on $(0, +\infty)$, we can obtain approximations for the sum T_1 for different values of n and m. The corresponding relative errors for $n = 5(5)40$ and $m = 1(1)3$ are presented in Table 4.4. As we can see, the convergence of the process (as m increases) slows down considerably. The reason for this is the behavior of the function $t \mapsto e^{-mt}$, which tends to a discontinuous function when $m \to +\infty$. On the other hand, the function is entire, which explains why the results in Table 4.4, when $m = 1$, are ultimately much better.

It is interesting to mention that Gaussian quadrature over the full real line with respect to the logistic function (cf. (3.3) and (3.4)) converges considerably more slowly than shown in Tables 4.1 – 4.3 for one-sided integration, even though the poles of the integrand have a distance twice as large from the real line. The reason, probably, is that these poles are now centered over the integral of integration, whereas in (2.7) and (2.9) they are located over the left endpoint of the interval. Numerical results for $n = 5(5)40$ and $m = 1(1)5$ are given in Table 4.5.

Example 4.2 The application of the "Laplace transform method" to the series

$$\sum_{k=1}^{+\infty}(k-1)k^{-3}\exp(-1/k) = .34291894384460978096183767790 2 \qquad (4.2)$$

leads to an integration of the Bessel function $t \mapsto J_0(2\sqrt{t})$. Here, however, we work

469

with the exponential function $F(z) = -e^{-1/z}/z$, i.e.,

$$\Phi(x,y) = \frac{1}{r^2}e^{-x/r^2}\left(x\cos\frac{y}{r^2} + y\sin\frac{y}{r^2}\right), \qquad r^2 = x^2 + y^2.$$

As to accuracy, a similar situation prevails as in the previous example. Table 4.6 shows the relative errors in Gaussian approximations for $n = 2(4)18$ and $m = 1(1)3$.

Table 4.5
Relative errors in Gaussian approximation of the sum T_1 and S_1
with respect to the logistic weight for $m = 1(1)5$

n		$m = 1$	$m = 2$	$m = 3$	$m = 4$	$m = 5$
5	T_1	4.7(−5)	5.2(−7)	1.9(−8)	1.5(−9)	1.8(−10)
	S_1	1.1(−3)	1.1(−3)	8.2(−4)	6.3(−4)	4.8(−4)
10	T_1	1.1(−6)	1.2(−9)	6.2(−12)	8.0(−14)	2.0(−15)
	S_1	4.1(−6)	1.3(−7)	1.3(−7)	1.1(−7)	1.0(−7)
15	T_1	1.1(−7)	2.8(−11)	3.4(−14)	1.2(−16)	8.9(−19)
	S_1	4.0(−7)	1.2(−10)	1.7(−11)	1.6(−11)	1.5(−11)
20	T_1	2.1(−8)	1.8(−12)	7.5(−16)	9.4(−19)	2.7(−21)
	S_1	7.5(−8)	6.5(−12)	5.1(−15)	2.2(−15)	2.1(−15)
25	T_1	5.5(−9)	2.1(−13)	3.7(−17)	2.0(−20)	2.6(−23)
	S_1	2.0(−8)	7.5(−13)	1.4(−16)	3.8(−19)	2.9(−19)
30	T_1	1.9(−9)	3.5(−14)	3.1(−18)	8.6(−22)	5.6(−25)
	S_1	7.0(−9)	1.3(−13)	1.1(−17)	3.1(−21)	4.3(−23)
35	T_1	7.7(−10)	7.7(−15)	3.8(−19)	5.8(−23)	2.1(−26)
	S_1	2.8(−9)	2.8(−14)	1.4(−18)	2.1(−22)	7.1(−26)
40	T_1	3.5(−10)	2.1(−15)	6.1(−20)	5.5(−24)	1.2(−27)
	S_1	1.3(−9)	7.5(−15)	2.2(−19)	2.0(−23)	4.4(−27)

Table 4.6
Relative errors in Gaussian approximation of the sum (4.2)

n	$m = 1$	$m = 2$	$m = 3$
2	2.9(−3)	1.2(−5)	2.1(−8)
6	1.3(−4)	3.7(−8)	1.2(−10)
10	1.8(−5)	3.7(−11)	9.9(−14)
14	1.2(−6)	1.2(−12)	1.2(−16)
18	1.3(−7)	8.5(−15)	6.6(−19)

Example 4.3 Consider now

$$T_1(a) = \sum_{k=1}^{+\infty} \frac{1}{\sqrt{k}(k+a)}. \tag{4.3}$$

Table 4.7
Relative errors in the "method of Laplace transform"
for the series (4.3) with a = 8.

$n=5$	$n=10$	$n=15$	$n=20$	$n=25$	$n=30$	$n=35$	$n=40$
1.4(−1)	2.3(−2)	1.5(−3)	1.9(−4)	2.5(−5)	2.1(−6)	2.5(−7)	2.6(−8)

This series with $a = 1$ appeared in a study of spirals (see Davis [1]) and defines the "Theodorus constant." Gautschi [8, p. 69] mentioned that the first 1 000 000 terms of the series $T_1(1)$ give the result 1.8580..., i.e., $T_1(1) \approx 1.86$ (only 3-digit accuracy). Using the "method of Laplace transform," Gautschi (see [6, Example 5.1] and [8]) calculated (4.3) for $a = .5, 1., 2., 4., 8., 16.,$ and 32. As a increases, the convergence of the Gauss quadrature formula slows down considerably. For example, when $a = 8$, we have results with relative errors presented in Table 4.7.

In order to achieve better accuracy, when a is large, Gautschi [6] used "stratified" summation by letting $k = \lambda + \kappa a_0$ and summing over all $\kappa \geq 0$ for $\lambda = 1, 2, \ldots, a_0$, where $a_0 = \lfloor a \rfloor$ denotes the largest integer $\leq a$ $(a = a_0 + a_1, a_0 \geq 1, 0 \leq a_1 < 1)$.

Here, we can apply directly our method to (4.3) with

$$F(z) = \frac{2}{\sqrt{a}} \left(\arctan \sqrt{\frac{z}{a}} - \frac{\pi}{2} \right),$$

where the integration constant is taken so that $F(\infty) = 0$. For computing the arctan function in the complex plane $(z^2 \neq -1)$ we use the formula

$$\arctan z = \frac{1}{2} \arg(u + iv) + \frac{i}{4} \log \frac{x^2 + (y+1)^2}{x^2 + (y-1)^2},$$

where $z = x + iy$, $u = 1 - x^2 - y^2$, $v = 2x$.

As before, we can represent (4.3) in the form

$$T_1(a) = \sum_{k=1}^{m-1} \frac{1}{\sqrt{k}(k+a)} + T_m(a), \qquad T_m(a) = \sum_{k=m}^{+\infty} \frac{1}{\sqrt{k}(k+a)}, \qquad (4.4)$$

and then use Gaussian quadrature to calculate $T_m(a)$. Relative errors in approximations for $T_1(a)$, when $m = 4$ and $a = p_s$, $s = 0(1)7$, where $p_0 = .5$ and $p_{s+1} = 2p_s$, are displayed in Table 4.8.

As we can see from Table 4.8, the method presented is very efficient. Moreover, its convergence is slightly faster if the parameter a is larger. The exact sums $T_1(a)$ (to 30 significant digits), as determined by Gaussian quadrature, are, respectively,

$$T_1(.5) = 2.13441664298623726110148952804,$$
$$T_1(1) = 1.86002507922119030718069591572,$$
$$T_1(2) = 1.53968051235330201287501841998,$$
$$T_1(4) = 1.21827401466989084582915976291,$$
$$T_1(8) = 0.93137293400310387168575138 9665,$$
$$T_1(16) = 0.69493171464104559016304607 1669,$$
$$T_1(32) = 0.50992651702721134803613196 7602,$$
$$T_1(64) = 0.36993169824967113220994236 4907.$$

Table 4.8

Relative errors in Gaussian approximation
of the sum (4.4) for $m = 4$

n	$a = .5$	$a = 1.$	$a = 2.$	$a = 4.$
5	1.4(−11)	8.4(−12)	4.5(−12)	2.6(−12)
10	6.8(−18)	4.4(−18)	2.2(−18)	1.2(−18)
15	5.4(−22)	2.7(−22)	1.6(−22)	1.0(−22)
20	1.2(−25)	5.9(−26)	3.3(−26)	2.0(−26)
25	1.0(−28)	5.2(−29)	3.0(−29)	1.9(−29)
30	1.1(−31)	5.7(−32)	3.3(−32)	2.0(−32)
n	$a = 8.$	$a = 16.$	$a = 32.$	$a = 64.$
5	1.7(−12)	1.1(−12)	7.6(−13)	5.2(−13)
10	7.7(−19)	5.1(−19)	3.4(−19)	2.4(−19)
15	6.7(−23)	4.5(−23)	3.0(−23)	2.1(−23)
20	1.3(−26)	8.7(−27)	5.9(−27)	4.1(−27)
25	1.2(−29)	8.1(−30)	5.5(−30)	3.8(−30)
30	1.3(−32)	9.0(−33)	6.2(−33)	3.6(−33)

Remark 4.1 The hyperbolic sine can be included as a factor in the integrand of S_m, so that only the first weight function w_1 is used. Some further investigations including this approach will be given elsewhere.

Acknowledgment. I would like to thank the referee for his valuable suggestions improving the presentation of the paper and for Remark 4.1.

REFERENCES

[1] Davis P.J. *Spirals: from Theodorus to Chaos.* A & K Peters, Wellesley, MA, 1993.

[2] Gautschi W. A survey of Gauss-Christoffel quadrature formulae. In *E.B. Christoffel – The Influence of his Work in Mathematics and the Physical*

Sciences, pages 72–147. Birkhäuser Verlag, Basel, 1981. P.L. Butzer and F. Fehér, eds.

[3] Gautschi W. On generating orthogonal polynomials. *SIAM J. Sci. Stat. Comput.*, **3**:289–317, 1982.

[4] Gautschi W. Computational aspects of orthogonal polynomials. In *Orthogonal Polynomials – Theory and Practice*, pages 181–216. NATO ASI Series, Series C: Mathematical and Physical Sciences, Vol. **294**, Kluwer, Dordrecht, 1990. P. Nevai, ed.

[5] Gautschi W. Computational problems and applications of orthogonal polynomials. In *Orthogonal Polynomials and Their Applications*, pages 61–71. IMACS Annals on Computing and Applied Mathematics, Vol. **9**, Baltzer, Basel, 1991. C. Brezinski, L. Gori and A. Ronveaux, eds.

[6] Gautschi W. A class of slowly convergent series and their summation by Gaussian quadrature. *Math. Comp.*, **57**:309–324, 1991.

[7] Gautschi W. On certain slowly convergent series occurring in plate contact problems. *Math. Comp.*, **57**:325–338, 1991.

[8] Gautschi W. *The Spiral of Theodorus, special functions, and numerical analysis.* Supplement A in [1].

[9] Gautschi W., Milovanović G.V. Gaussian quadrature involving Einstein and Fermi functions with an application to summation of series. *Math. Comp.*, **44**:177–190, 1985.

[10] Golub G.H., Welsch J.H. Calculation of Gauss quadrature rules. *Math. Comp.*, **23**:221–230, 1969.

[11] Henrici P. *Applied and Computational Complex Analysis*, Vol. 1. Wiley, New York, 1984.

[12] Lindelöf E. *Le calcul des résidus.* Gauthier–Villars, Paris, 1905.

[13] Mitrinović D.S., Kečkić J.D. *The Cauchy Method of Residues – Theory and Applications.* Reidel, Dordrecht, 1984.

Appendix Recursion coefficients α_k, β_k for the (monic) polynomials $\pi_k(\,\cdot\,;w_1)$ and $\pi_k(\,\cdot\,;w_2)$ orthogonal on $[0,+\infty]$ with respect to the weight functions $w_1(t) = 1/\cosh^2 t$ and $w_2(t) = \sinh t/\cosh^2 t$.

Table A.1
Recursion coefficients for the polynomials $\{\pi_k(\,\cdot\,;w_1)\}$

k	alpha(k)	beta(k)
0	6.9314718055994530941723212146D−01	1.0000000000000000000000000000D+00
1	1.5939617276162479667832968293D+00	3.4201401950591179356910505700D−01
2	2.5704790635153506023229623757D+00	1.0982490302711024757750615776D+00
3	3.5593114896246001274923935698D+00	2.3594667503510884842534739092D+00
4	4.5521493508450628251971332577D+00	4.1206409259734225312996126403D+00
5	5.5470312425421284751577289507D+00	6.3814456471249773626945182034D+00
6	6.5431525681679808686029097276D+00	9.1416968963426452858867320270D+00
7	7.5400892325620730871114672774D+00	1.2401381399063813475148435922D+01
8	8.5375932538428070942983060280D+00	1.6160548266582688987382097742D+01
9	9.5355097763213333094562689617D+00	2.0419257322046748675142843925D+01
10	1.0533736923204318781615154727D+01	2.5177563621944731971785459237D+01
11	1.1532204728042289304076302587D+01	3.0435514377967418893654074886D+01
12	1.2530863397285539709205379979D+01	3.6193149336866216766280298026D+01
13	1.3529676424709201161977948095D+01	4.2450501914096420694349014313D+01
14	1.4528616366238649759936923691D+01	4.9207600334152388529884708177D+01
15	1.5527662144549085001093840486D+01	5.6464468600539404421339117817D+01
16	1.6526797270963617088725743755D+01	6.4221127279508584288069879207D+01
17	1.7526008637918384115438928070D+01	7.2477594123026429127768796077D+01
18	1.8525285677873595322626240259D+01	8.1233884562576556315784371707D+01
19	1.9524619764282356143811287725D+01	9.0490012101807598077869428931D+01
20	2.0524003776469000808420287625D+01	1.0024598863057796230602500654D+02
21	2.1523431777972483795565020695D+01	1.1050182467790393695842356506D+02
22	2.2522898774997739104475221939D+01	1.2125752961722091282304977559D+02
23	2.3522400532434740169895471727D+01	1.3251311183419835936824676638D+02
24	2.4521933431914333844086960942D+01	1.4426857886494552420053970101D+02
25	2.5521494361008953916301117690D+01	1.5652393751063399744666521992D+02
26	2.6521080625816019321035035488D+01	1.6927919393319937224868498934D+02
27	2.7520689881310473079490224088D+01	1.8253435373575359768739746871D+02
28	2.8520320075351817558103962351D+01	1.9628942203055681831758711401D+02
29	2.9519969403292101885517710969D+01	2.1054440349679927174793684049D+02
30	3.0519636270892725508068645610D+01	2.2529930242998360956857484232D+02
31	3.1519319263811160102194446841D+01	2.4055412278434155283089849728D+02
32	3.2519017122325364860097489319D+01	2.5630886820944080908989180435D+02
33	3.3518728720265847977149048624D+01	2.7256354208191992660956976966D+02
34	3.4518453047352129107974392492D+01	2.8931817453311628056421932118D+02
35	3.5518189194302192384745772760D+01	3.0657268747321520934747766482D+02
36	3.6517936340214859483708742306D+01	3.2432716461243855083544879391D+02
37	3.7517693741826232595406177242D+01	3.4258158147970247439504427768D+02
38	3.8517460724319973131421574908D+01	3.6133594043910298476082499375D+02
39	3.9517236673432689751095699863D+01	3.8059024370452926742931339305D+02

Table A.2
Recursion coefficients for the polynomials $\{\pi_k(\,\cdot\,;w_2)\}$

k	alpha(k)	beta(k)
0	1.5707963267948966192313216916D+00	1.00000000000000000000000000000D+00
1	3.3371537318859924271186691475D+00	1.1964612764365364055097913098D+00
2	5.2620526190569131940940182735D+00	4.3671477023239855599943524385D+00
3	7.2219253825195963241917294956D+00	9.4920030157217803141931237018D+00
4	9.1955961335463704468573492230D+00	1.6599821474981044395118009643D+01
5	1.1176787153959861421766477155D+01	2.5695925696948527548388764699D+01
6	1.3162545671480695161080532645D+01	3.6782876995537958236321971782D+01
7	1.5151280183457372636272981837D+01	4.9862797030082115498885468165D+01
8	1.7142074910355614719206187525D+01	6.4937217691157557639041385008D+01
9	1.9134367428880868802960977055D+01	8.2007184973877795823095758742D+01
10	2.1127790670570973020159149723D+01	1.0107343286446229617992854696D+02
11	2.3122092802305852725111608578D+01	1.2213650371804518096526015893D+02
12	2.5117094183073749853776676037D+01	1.4519681646446703916664836504D+02
13	2.7112662673152088266615190672D+01	1.7025470480128094752412337012D+02
14	2.9108698612086965724633219820D+01	1.9731044019102761578748694484D+02
15	3.1105125237247274403041559211D+01	2.2636424699020615288701984186D+02
16	3.3101882342887042357068459953D+01	2.5741631307631730571242796003D+02
17	3.5098921957678223189144970437D+01	2.9046679760473857817906878653D+02
18	3.7096205324045538864657425358D+01	3.2551583679630579995868145078D+02
19	3.9093700741028742066090000630D+01	3.6256354832163887272429999003D+02
20	4.1091381993825721133124054539D+01	4.0161003466701202740268160150D+02
21	4.3089227190472453205414608922D+01	4.426553857522931078729024438D+02
22	4.5087217886538866344915524963D+01	4.8569968099341757939040819938D+02
23	4.7085338417168533273674616542D+01	5.3074299094721482202656092833D+02
24	4.9083575380786676535285072721D+01	5.7778537863805553910813201974D+02
25	5.1081917235375518997646639470D+01	6.2682690063889639352450194411D+02
26	5.3080353979410943839777149039D+01	6.7786760796037296068125701887D+02
27	5.5078876897247722033269544524D+01	7.3090754678815650675485255235D+02
28	5.7077478354113316331839369353D+01	7.8594675909912964021774835024D+02
29	5.9076151629678729094404765489D+01	8.4298528317988766942549950784D+02
30	6.1074890781911762948790067278D+01	9.0202315406585390658662356570D+02
31	6.3073690534909888422271725032D+01	9.6306040391537984246753922954D+02
32	6.5072546185876646089606555624D+01	1.0260970623302243321695515516D+03
33	6.70714535274955073745363412D+01	1.0911331566315193667444829680D+03
34	6.9070408782780794385036513587D+01	1.1581687120985568583801970567D+03
35	7.1069408550097444925803391953D+01	1.2272037521763441123907032577D+03
36	7.3068449756530880218659125979D+01	1.2982382986567826423869436105D+03
37	7.5067529618145579567604035523D+01	1.3712723718374572499127807181D+03
38	7.7066645605962108797192785006D+01	1.4463059906613287130107636415D+03
39	7.9065795416704862652340841220D+01	1.5233391728400654792796299133D+03

International Series of Numerical Mathematics, Vol. 119, ©1994 Birkhäuser

ON A SINGULAR PERTURBATION PROBLEM

K.-C. Ng[†]   R. Wong[‡]

Dedicated to Professor Walter Gautschi on the occasion of his 65th birthday

Abstract. A uniformly valid asymptotic approximation is constructed for the solution to the initial value problem

$$\ddot{v} + \varepsilon t \dot{v} + v = 0, \qquad v(0) = 0, \qquad \dot{v}(0) = 1,$$

as $\varepsilon \to 0$. From this, it is deduced that if $\varepsilon t \to 0$ then

$$v(t, \varepsilon) \sim e^{-\varepsilon t^2/4} \sin t,$$

and if $\varepsilon t \to \infty$ then

$$v(t, \varepsilon) \sim \left(\sin \frac{\pi}{2\varepsilon} \right) e^{-1/2\varepsilon} \frac{\sqrt{2}}{(\varepsilon t)^{1/\varepsilon}}.$$

1 INTRODUCTION

In [5], Simmonds and Mann give the following open problem in an exercise, and indicate that they do not know the complete answer: construct a uniformly valid first approximation to the linear oscillator with damping proportional to time

$$\ddot{v} + \varepsilon t \dot{v} + v = 0, \qquad v(0) = 0, \qquad \dot{v}(0) = 1. \tag{1}$$

If one naively proceeds with the perturbation series

$$v(t, \varepsilon) = v_0(t) + \varepsilon v_1(t) + \varepsilon^2 v_2(t) + \cdots,$$

as suggested in [5], one readily finds that $v_0(t) = \sin t$ and

$$v_1(t) = \frac{1}{4} \sin t - \frac{1}{4} t \cos t - \frac{1}{4} t^2 \sin t.$$

But, by the usual method of multiplying the differential equation in (1) by $\dot{v}$ and integrating from 0 to t, one can easily show that the solution v is bounded for all positive t. Since $v_1(t)$ contains secular terms $t \cos t$ and $t^2 \sin t$, the error due to

---

[†]Department of Mathematics, Tamkang University, Taipei, Taiwan, ROC
[‡]Department of Applied Mathematics, University of Manitoba, Winnipeg, Manitoba, Canada R3T 2N2

truncation after two terms cannot be uniformly small in t. If one tries the multiple scale method, and assumes a perturbation series of the form

$$v(t, \varepsilon) = v_0(t, \tau), + \varepsilon v_1(t, \tau) + \varepsilon^2 v_2(t, \tau) + \cdots$$

with $\tau = \varepsilon t$, one obtains

$$\frac{\partial^2 v_0}{\partial t^2} + v_0 = 0,$$

$$\frac{\partial^2 v_1}{\partial t^2} + v_1 = -\frac{2 \partial^2 v_0}{\partial \tau \partial t} - t \frac{\partial v_0}{\partial t}.$$

The solution to the first equation is $v_0 = A(\tau) \sin t + B(\tau) \cos t$. Substituting this expression in the right-hand side of the second equation gives:

$$\frac{\partial^2 v_1}{\partial t^2} + v_1 = -[2A'(\tau) + tA(\tau)] \cos t + [2B'(\tau) + tB(\tau)] \sin t.$$

To eliminate secular terms, the usual procedure is now to set the quantities inside the square brackets to zero. Since these quantities involve the variable t, the solutions $A(\tau)$ and $B(\tau)$ to the resulting equations depend on t, contradicting the fact that $A(\tau)$ and $B(\tau)$ are functions of τ only. This indicates that the choice of the scale $\tau = \varepsilon t$ is inappropriate. A similar consequence results, if one chooses the scale $\tau = \sqrt{\varepsilon} t$.

In this note we will show that the exact solution to equation (1) can be obtained in terms of parabolic cylinder functions, and that a uniformly valid first approximation to this solution can be given in terms of Airy functions. As consequences, we will prove that as $\varepsilon \to 0$,

$$v(t, \varepsilon) \sim e^{-\varepsilon t^2/4} \sin t \qquad \text{if } \varepsilon t \to 0, \qquad (2)$$

and

$$v(t, \varepsilon) \sim \left(\sin \frac{\pi}{2\varepsilon} \right) e^{-1/2\varepsilon} \frac{\sqrt{2}}{(\varepsilon t)^{1/\varepsilon}} \qquad \text{if } \varepsilon t \to \infty. \qquad (3)$$

Furthermore, we will show that there is a turning point at $t = 2/\varepsilon$, and that in a neighborhood of the turning point we have

$$v(t, \varepsilon) \sim \sqrt{\pi} e^{-\varepsilon t^2/4} \varepsilon^{-\frac{1}{6}} \left\{ \sin \frac{\pi}{2\varepsilon} \cdot \text{Bi} \left[\frac{1}{2} \varepsilon^{1/3} \left(t - \frac{2}{\varepsilon} \right) \right] \right.$$

$$\left. - \cos \frac{\pi}{2\varepsilon} \cdot \text{Ai} \left[\frac{1}{2} \varepsilon^{1/3} \left(t - \frac{2}{\varepsilon} \right) \right] \right\}, \qquad (4)$$

as $\varepsilon \to 0$. More precisely, equation (4) holds when $|t - \frac{2}{\varepsilon}| = o(\varepsilon^{-\frac{2}{3}})$.

2 EXACT SOLUTION

Let v be the solution to (1), and put

$$v(t, \varepsilon) = e^{-\varepsilon t^2/4} W(t, \varepsilon). \tag{5}$$

It is easily verified that W satisfies the equation

$$\ddot{W} + \left[\left(1 - \frac{\varepsilon}{2}\right) - \frac{\varepsilon^2}{4} t^2 \right] W = 0. \tag{6}$$

Set $x = \sqrt{\varepsilon} t$ and $u(x, \varepsilon) = W(x/\sqrt{\varepsilon}, \varepsilon)$. Then u satisfies Weber's equation

$$\frac{d^2 u}{dx^2} - \left(\frac{1}{4} x^2 + a \right) u = 0 \tag{7}$$

and the initial conditions $u(0, \varepsilon) = 0$ and $(du/dx)(0, \varepsilon) = 1/\sqrt{\varepsilon}$. In (7), the parameter a is given by

$$a = a_\varepsilon = \frac{1}{2} - \frac{1}{\varepsilon}. \tag{8}$$

It is well known that two linearly independent solutions to (7) are $U(a, x)$ and $V(a, x)$, characterized by the asymptotic behavior

$$U(a, x) \sim x^{-a-\frac{1}{2}} e^{-x^2/4}, \qquad \text{as } x \to \infty.$$

$$V(a, x) \sim \sqrt{\frac{2}{\pi}} x^{a-\frac{1}{2}} e^{x^2/4}, \qquad \text{as } x \to \infty.$$

In terms of Whittaker's notation for parabolic cylinder functions $D_n(x)$, we have $U(a, x) = D_{-a-\frac{1}{2}}(x)$ and

$$V(a, x) = \frac{1}{\pi} \Gamma\left(\frac{1}{2} + a\right) \{\sin \pi a \cdot D_{-a-\frac{1}{2}}(x) + D_{-a-\frac{1}{2}}(-x)\}.$$

For the definition and properties of parabolic cylinder functions, we refer to [1, Chap. 19] or the Introduction to [2]. Since in our case the parameter a in equation (8) is negative, we will take the suggestion of Olver [3, p.134], and use in place of $V(a, x)$ its multiple $\bar{U}(a, x)$, given by

$$\bar{U}(a, x) = \Gamma\left(\frac{1}{2} - a\right) V(a, x) = \tan \pi a \cdot U(a, x) + \sec \pi a \cdot U(a, -x). \tag{9}$$

Since $u(x, \varepsilon)$ is a solution of (7), it is expressible as a linear combination of $U(a_\varepsilon, x)$ and $\bar{U}(a_\varepsilon, x)$. In view of the initial conditions, it is found that

$$u(x, \varepsilon) = -\frac{1}{2\sqrt{\varepsilon}} \frac{(1 + \sin \pi a_\varepsilon) U(a_\varepsilon, x) - \cos \pi a_\varepsilon \cdot \bar{U}(a_\varepsilon, x)}{U(a_\varepsilon - 1, 0)}, \tag{10}$$

479

where use has been made of the result $(dU/dx)(a,0) = -U(a-1,0)$. Since $u(x,\varepsilon) = W(t,\varepsilon)$, the exact solution to equation (1) is given by

$$v(t,\varepsilon) = e^{-\varepsilon t^2/4} u(x,\varepsilon), \qquad x = \sqrt{\varepsilon} t, \qquad (11)$$

on account of (5).

Before leaving this section, we recall two asymptotic formulas for $U(a,x)$ and $\bar{U}(a,x)$. For $x \geq 0$ we write

$$x = 2\sqrt{|a|}\xi, \qquad \eta = (4|a|)^{2/3}\tau, \qquad (12)$$

and

$$\tau = \begin{cases} -\left[\frac{3}{2}\left(\frac{1}{4}\cos^{-1}\xi - \frac{1}{4}\xi\sqrt{1-\xi^2}\right)\right]^{2/3}, & 0 \leq \xi \leq 1, \\ \left[\frac{3}{2}\left(\frac{1}{4}\xi\sqrt{\xi^2-1} - \frac{1}{4}\cosh^{-1}\xi\right)\right]^{2/3}, & \xi \geq 1. \end{cases} \qquad (13)$$

Then for $x \geq 0$ and as $a \to -\infty$,

$$U(a,x) \sim 2^{-\frac{1}{4}-\frac{1}{2}a}\Gamma\left(\frac{1}{4} - \frac{1}{2}a\right)\left(\frac{\eta}{\xi^2-1}\right)^{\frac{1}{4}} \mathrm{Ai}\,(\eta), \qquad (14)$$

$$\bar{U}(a,x) \sim 2^{-\frac{1}{4}-\frac{1}{2}a}\Gamma\left(\frac{1}{4} - \frac{1}{2}a\right)\left(\frac{\eta}{\xi^2-1}\right)^{\frac{1}{4}} \mathrm{Bi}\,(\eta); \qquad (15)$$

see [1, p.689] or [3, Eqs. (8.11) and (11.22)].

3 ASYMPTOTIC BEHAVIOR

To obtain a uniformly valid asymptotic approximation to the solution $v(t,\varepsilon)$, we first note that when $x = 0$ and a is replaced by $a_\varepsilon - 1$, we have, from (8) and (12), $\xi = 0$ and

$$\eta = -\left[\frac{3\pi}{4}\left(\frac{1}{\varepsilon} + \frac{1}{2}\right)\right]^{2/3}.$$

Hence, $\eta \to -\infty$ and

$$\mathrm{Ai}\,(\eta) \sim \frac{1}{\pi^{1/2}(-\eta)^{1/4}} \cos\left(\frac{\pi}{2\varepsilon}\right), \qquad \text{as } \varepsilon \to 0^+; \qquad (16)$$

see [4, p.103]. From (14), it follows that

$$U(a_\varepsilon - 1,0) \sim 2^{\frac{1}{4}-\frac{1}{2}a_\varepsilon}\Gamma\left(\frac{1}{4} - \frac{1}{2}a_\varepsilon + \frac{1}{2}\right)\frac{1}{\sqrt{\pi}} \cos\left(\frac{\pi}{2\varepsilon}\right). \qquad (17)$$

Since $1 + \sin \pi a_\varepsilon = 2\cos^2\left(\frac{\pi}{2\varepsilon}\right)$ and $\cos \pi a_\varepsilon = 2\sin\left(\frac{\pi}{2\varepsilon}\right)\cos\left(\frac{\pi}{2\varepsilon}\right)$, inserting (14), (15) and (17) in (10), we obtain from (11)

$$v(t,\varepsilon) \sim \sqrt{\pi}e^{-\varepsilon t^2/4}\left(\frac{\eta}{\xi^2-1}\right)^{\frac{1}{4}}\left[\sin\frac{\pi}{2\varepsilon} \cdot \mathrm{Bi}\,(\eta) - \cos\frac{\pi}{2\varepsilon} \cdot \mathrm{Ai}\,(\eta)\right], \qquad (18)$$

where η and ξ are given in (12) and (13). In (18), we have also made use of the result

$$\frac{\Gamma(x+a)}{\Gamma(x+b)} \sim x^{a-b}, \qquad \text{as } x \to \infty; \tag{19}$$

see [4, p.119]. The asymptotic formulas (2), (3) and (4) will now be deduced from (18).

First, we prove (2). If $\varepsilon t \to 0$, then $\sqrt{\varepsilon}x \to 0$ and $\xi \to 0$. Consequently, it follows from (8) and (12) that

$$\eta \sim -\left(\frac{3}{2}\right)^{2/3}\left(\frac{\pi - 4\xi}{2\varepsilon} - \frac{\pi}{4}\right)^{2/3}, \qquad \text{as } \varepsilon \to 0.$$

In particular, we have $\eta \to -\infty$,

$$\mathrm{Ai}\,(\eta) \sim \frac{1}{\pi^{1/2}(-\eta)^{1/4}} \sin\left(\frac{\pi - 4\xi}{2\varepsilon}\right) \tag{20}$$

and

$$\mathrm{Bi}\,(\eta) \sim \frac{1}{\pi^{1/2}(-\eta)^{1/4}} \cos\left(\frac{\pi - 4\xi}{2\varepsilon}\right); \tag{21}$$

see [4, pp.393-394]. Since $\xi \to 0$, inserting (20) and (21) in (18) gives

$$v(t,\varepsilon) \sim e^{-\varepsilon t^2/4}\left[\sin\frac{\pi}{2\varepsilon} \cdot \cos\left(\frac{\pi}{2\varepsilon} - \frac{4\xi}{2\varepsilon}\right) - \cos\frac{\pi}{2\varepsilon} \cdot \sin\left(\frac{\pi}{2\varepsilon} - \frac{4\xi}{2\varepsilon}\right)\right].$$

The desired result in (2) now follows from the addition formula for the sine function and the fact that $2\xi/\varepsilon \sim t$.

Next, we prove (3). If $\varepsilon t \to \infty$, then $\sqrt{\varepsilon}x \to \infty$ and $\xi \to \infty$. Since

$$\cosh^{-1}\xi \sim \ln 2\xi - \frac{1}{4}\xi^{-2} - \frac{3}{32}\xi^{-4} - \cdots$$

as $\xi \to \infty$ (see [1, p.88], we have

$$\eta \sim \left[\frac{3}{2}|a|(\xi^2 - \ln 2\xi - \frac{1}{2})\right]^{2/3}. \tag{22}$$

From the asymptotic formulas [4, pp.393-394]

$$\mathrm{Ai}\,(\eta) \sim \frac{1}{2\pi^{1/2}\eta^{1/4}} \exp\left(-\frac{2}{3}\eta^{3/2}\right), \qquad \eta \to \infty,$$

and

$$\mathrm{Bi}\,(\eta) \sim \frac{1}{\pi^{1/2}\eta^{1/4}} \exp\left(\frac{2}{3}\eta^{3/2}\right), \qquad \eta \to \infty,$$

it follows from (18) that

$$v(t,\varepsilon) \sim e^{-\varepsilon t^2/4}\xi^{-1/2}\Big[\sin\frac{\pi}{2\varepsilon}\exp\Big(\frac{2}{3}\eta^{3/2}\Big)$$
$$-\frac{1}{2}\cos\frac{\pi}{2\varepsilon}\exp\Big(-\frac{2}{3}\eta^{3/2}\Big)\Big]. \tag{23}$$

Since $x = \sqrt{\varepsilon}t$, equation (22) gives

$$\frac{2}{3}\eta^{3/2} \sim \frac{1}{4}\varepsilon t^2 + a\log\sqrt{\varepsilon}t - \frac{a}{2}\log(-a) + \frac{a}{2}.$$

Disregarding the exponentially small term in (23), the required result (3) is now obtained by noting that $\xi = \sqrt{\varepsilon}t/2\sqrt{|a|}$.

Finally, we consider the case when t is in a neighborhood of the turning point $t = 2/\varepsilon$. Let

$$f(\xi) \equiv \xi\sqrt{\xi^2 - 1} - \cosh^{-1}\xi, \qquad \xi \geq 1.$$

Simple calculation gives $f'(\xi) = 2\sqrt{\xi^2 - 1}$. From (13), we have

$$\tau = \Big[\frac{3}{8}f(\xi)\Big]^{2/3}$$

and

$$\frac{d\tau}{d\xi} = \frac{1}{3^{1/3}}\frac{\sqrt{\xi^2 - 1}}{[f(\xi)]^{1/3}}.$$

By l'Hospital's rule, it can be shown that

$$\Big(\frac{d\tau}{d\xi}\Big)^3\Big|_{\xi=1^+} = \frac{1}{2}.$$

In the same manner, it can be shown that

$$\Big(\frac{d\tau}{d\xi}\Big)^3\Big|_{\xi=1^-} = \frac{1}{2}.$$

Thus, expanding τ into a Taylor series near $\xi = 1$, we have from (12)

$$\eta = (4|a|)^{2/3}\Big\{\tau(1) + \frac{1}{2^{1/3}}(\xi - 1) + \cdots\Big\}.$$

Since $\tau(1) = 0$, if $|a|^{2/3}(\xi - 1)^2$ is small, or equivalently if $t = \frac{2}{\varepsilon}[1 + o(\varepsilon^{1/3})]$, then it follows from (18) that

$$v(t,\varepsilon) \sim \sqrt{\pi}e^{-\varepsilon t^2/2}|a|^{1/6}\Big\{\sin\frac{\pi}{2\varepsilon}\cdot\mathrm{Bi}\,[2|a|^{2/3}(\xi - 1)]$$
$$-\cos\frac{\pi}{2\varepsilon}\cdot\mathrm{Ai}\,[2|a|^{2/3}(\xi - 1)]\Big\}. \tag{24}$$

The required asymptotic approximation (4) is obtained by observing that

$$|a|^{2/3}(\xi - 1) = \frac{1}{2}\varepsilon^{1/3}\Big(t - \frac{2}{\varepsilon}\Big) + O(\varepsilon^{1/3}), \qquad \text{as } \varepsilon \to 0.$$

4 REMARKS

1. The asymptotic formula (2) can be obtained by using the WKB analysis. Indeed, if we make the change of variable $\tau = \varepsilon t/\sqrt{4-2\varepsilon}$ in equation (6) and put $W(t) = y(\tau)$, then it is easily verified that y satisfies

$$\frac{d^2y}{d\tau^2} + \lambda_\varepsilon^2(1-\tau^2)y = 0$$

where $\lambda_\varepsilon = \frac{2}{\varepsilon} - 1$. The WKB approximation [5, p.79, (6.19)] gives

$$
\begin{aligned}
y(\tau) \sim (1-\tau^2)^{-1/4}[&A\cos(\lambda_\varepsilon \int_0^\tau \sqrt{1-s^2}ds) \\
&+ B\sin(\lambda_\varepsilon \int_0^\tau \sqrt{1-s^2}ds)],
\end{aligned}
\tag{25}
$$

as $\lambda_\varepsilon \to +\infty$, for $|\tau| < 1$. The initial conditions $y(0) = 0$ and $(dy/d\tau)(0) = 2\sqrt{\frac{1}{\varepsilon} - \frac{1}{2}}/\sqrt{\varepsilon}$ imply $A = 0$ and $B = (1 - \frac{\varepsilon}{2})^{-1/2}$. Since $B \sim 1$ and $\lambda_\varepsilon \tau \sim t$ as $\varepsilon \to 0$, and since

$$\sin(\lambda_\varepsilon \int_0^\tau \sqrt{1-s^2}ds) \sim \sin \lambda_\varepsilon \tau \qquad \text{as } \varepsilon t \to 0,$$

formula (2) follows immediately from (25).

2. The above method can also be applied to the initial value problem

$$\ddot{v} - \varepsilon t \dot{v} + v = 0, \qquad v(0) = 0, \qquad \dot{v}(0) = 1. \tag{26}$$

The exact solution is

$$v(t,\varepsilon) = -e^{\varepsilon t^2/4}\frac{(1+\sin \pi a)U(a,\sqrt{\varepsilon}t) - \cos \pi a \cdot \bar{U}(a,\sqrt{\varepsilon}t)}{2\sqrt{\varepsilon}U(a-1,0)}, \tag{27}$$

where

$$a = a_\varepsilon = -\frac{1}{\varepsilon} - \frac{1}{2}. \tag{28}$$

The uniformly valid first approximation is given by

$$v(t,\varepsilon) \sim \sqrt{\pi}e^{\varepsilon t^2/4}\left(\frac{\eta}{\xi^2-1}\right)^{1/4}[\sin \frac{\pi}{2\varepsilon} \cdot \text{Ai}\,(\eta) + \cos \frac{\pi}{2\varepsilon} \cdot \text{Bi}\,(\eta)], \tag{29}$$

where ξ and η are defined as in (12) with a being given by (28).

REFERENCES

[1] Abramowitz A., Stegun I.A., eds. *Handbook of Mathematical Functions.* NBS Appl. Math. Series 55, Washington, D.C., 1964.

[2] Miller J.C.P. *Tables of Weber parabolic cylinder functions.* H. M. Stationery Office, London, 1955.

[3] Olver F.W.J. Uniform asymptotic expansions for Weber parabolic cylinder functions of large orders. *J. Res. Nat. Bur. Standards* Sect. B, **63**:131–169, 1959.

[4] Olver F.W.J. *Asymptotics and Special Functions.* Academic Press, New York, 1974.

[5] Simmonds J.G., Mann J.E. Jr. *A First Look at Perturbation Theory.* Robert E. Krieger Publishing Company, Malabar, Florida, 1986.

International Series of Numerical Mathematics, Vol. 119, ©1994 Birkhäuser

An Overview of Results on the Existence or Nonexistence and the Error Term of Gauss-Kronrod Quadrature Formulae

Sotirios E. Notaris*

Dedicated to Walter Gautschi on the occasion of his 65th birthday

Abstract.   Kronrod in 1964, trying to estimate economically the error of the n-point Gauss-Legendre quadrature formula, developed a new formula by adding to the n Gauss nodes $n + 1$ new ones, which are determined, together with all weights, such that the new formula has maximum degree of exactness. It turns out that the new nodes are the zeros of a polynomial orthogonal with respect to a variable-sign weight function. This polynomial was considered for the first time by Stieltjes in 1894. Important for the new formula, now appropriately called the Gauss-Kronrod quadrature formula, are properties such as the interlacing of the Gauss nodes with the new nodes, the inclusion of all nodes in the interior of the interval of integration, and the positivity of all quadrature weights. We review, for classical and nonclassical weight functions, the existence or nonexistence and the error term of Gauss-Kronrod formulae having one or more of the aforementioned properties.

1   INTRODUCTION

The Gauss quadrature formula for the Legendre weight function $w(t) = 1$ on $[-1, 1]$ has the form

$$\int_{-1}^{1} f(t)dt = \sum_{\nu=1}^{n} \gamma_\nu f(\tau_\nu) + R_n^G(f), \tag{1.1}$$

where $\tau_\nu = \tau_\nu^{(n)}$ are the zeros of the nth-degree (monic) Legendre polynomial π_n, and $\gamma_\nu = \gamma_\nu^{(n)}$ are the Christoffel numbers. It is known that (1.1) has degree of exactness $d_n^G = 2n - 1$, i.e., $R_n^G(f) = 0$ for all $f \in \mathbb{P}_{2n-1}$.

Let $Q_n^G(f) = \sum_{\nu=1}^{n} \gamma_\nu f(\tau_\nu)$. A practical way to estimate the error of (1.1) is the following. Consider two formulae, one with n points and another with m, where $m > n$. Then

$$|R_n^G(f)| \simeq |Q_n^G(f) - Q_m^G(f)|. \tag{1.2}$$

The disadvantage of this method lies in the number of function evaluations required in order to obtain a reasonable estimate for $R_n^G(f)$. For example, with $m = n + 1$

---

*Department of Mathematics, University of Missouri-Columbia, Columbia, MO 65211, USA; *Current address*: 1 Xenokratous St., 10675 Athens, Greece

additional evaluations of the function f (at the new nodes $\tau_\nu^{(n+1)}$), the degree of exactness goes from $2n - 1$ to $2n + 1$, only a slight improvement.

Motivated by this, Kronrod in 1964 (cf. [15, 16]) started with the n Gauss nodes τ_ν and added $n + 1$ new ones, obtaining what is now called the Gauss-Kronrod quadrature formula,

$$\int_{-1}^1 f(t)dt = \sum_{\nu=1}^n \sigma_\nu f(\tau_\nu) + \sum_{\mu=1}^{n+1} \sigma_\mu^* f(\tau_\mu^*) + R_n^K(f). \tag{1.3}$$

The new nodes $\tau_\mu^* = \tau_\mu^{*(n)}$ and all weights $\sigma_\nu = \sigma_\nu^{(n)}$, $\sigma_\mu^* = \sigma_\mu^{*(n)}$ are chosen such that (1.3) has maximum degree of exactness. Since there are $3n + 2$ degrees of freedom, one expects to get $d_n^K = 3n + 1$ (at least). The advantage of using $Q_n^K(f) = \sum_{\nu=1}^n \sigma_\nu f(\tau_\nu) + \sum_{\mu=1}^{n+1} \sigma_\mu^* f(\tau_\mu^*)$ instead of $Q_m^G(f) = Q_{n+1}^G(f)$ in (1.2) is clear. With the same number of additional evaluations of the function (i.e., $n + 1$, at the new nodes τ_μ^*) the degree of exactness goes much higher than $2n + 1$.

Let $\pi_{n+1}^*(t) = \prod_{\mu=1}^{n+1}(t - \tau_\mu^*)$. It can be shown (see [8, Corollary]) that (1.3) has degree of exactness $3n + 1$ if and only if π_{n+1}^* satisfies the orthogonality condition

$$\int_{-1}^1 \pi_{n+1}^*(t)t^k \pi_n(t)dt = 0, \quad k = 0, 1, \ldots, n, \tag{1.4}$$

that is, π_{n+1}^* is orthogonal to all polynomials of lower degree relative to the variable-sign "weight function" $w^*(t) = \pi_n(t)$. This kind of orthogonality was first considered by Stieltjes in 1894, through his work on continued fractions. It goes without saying that the theory of orthogonal polynomials relative to a (nonnegative) weight function cannot be applied here. In particular, there are no general results about the reality of the τ_μ^*, not to mention their inclusion in $(-1,1)$, both of which are important for the practical application of (1.3). Nevertheless, Stieltjes conjectured that π_{n+1}^* has $n + 1$ real and simple zeros, all contained in $(-1,1)$, and interlacing with the zeros of π_n (cf. [2, v. 2, pp. 439-441]).

The connection between the Gauss-Kronrod formula (1.3) and the polynomial π_{n+1}^*, now appropriately called Stieltjes polynomial, was first noticed by Mysovskih in [24], and independently by Barrucand in [3].

In recent years, Gauss-Kronrod formulae have attracted considerable interest from both the computational and the mathematical point of view, the former in view of their potential use in packages for automatic integration, and the latter because of the mathematical problems they pose. It is therefore both useful and interesting to examine questions which relate to the computational feasibility of these formulae, such as the interlacing of the nodes τ_ν, τ_μ^*, the inclusion of all nodes in the interior of the interval of integration, and the positivity of all quadrature weights. The present paper summarizes, for classical and nonclassical weight functions, results on the existence and nonexistence of Gauss-Kronrod formulae having one or more of the interlacing, inclusion and positivity properties. Moreover, the error terms of these formulae are reviewed. Previous surveys on the subject can be found in [8, 22].

2 EXISTENCE AND NONEXISTENCE RESULTS

Consider the Gauss-Kronrod formula for the (nonnegative) weight function w on $[a, b]$,

$$\int_a^b f(t)w(t)dt = \sum_{\nu=1}^{n} \sigma_\nu f(\tau_\nu) + \sum_{\mu=1}^{n+1} \sigma_\mu^* f(\tau_\mu^*) + R_n^K(f), \qquad (2.1)$$

where $\tau_\nu = \tau_\nu^{(n)}$ are the zeros of the nth-degree (monic) orthogonal polynomial $\pi_n(\cdot) = \pi_n(\cdot; w)$ relative to w, and, as in (1.3), $\tau_\mu^* = \tau_\mu^{*(n)}$, $\sigma_\nu = \sigma_\nu^{(n)}$, $\sigma_\mu^* = \sigma_\mu^{*(n)}$ are chosen such that (2.1) has maximum degree of exactness $d_n^K = 3n + 1$ (at least). In this case, the orthogonality condition (1.4) takes the form

$$\int_a^b \pi_{n+1}^*(t)t^k \pi_n(t)w(t)dt = 0, \quad k = 0, 1, \ldots, n, \qquad (2.2)$$

where $\pi_{n+1}^*(\cdot) = \pi_{n+1}^*(\cdot; w)$ is defined the same way as before. We say that the nodes τ_ν, τ_μ^* in (2.1) interlace if they are all real and, when ordered decreasingly, satisfy

$$\tau_{n+1}^* < \tau_n < \tau_n^* < \cdots < \tau_2^* < \tau_1 < \tau_1^*. \qquad (2.3)$$

The weights in (2.1) are given by

$$\sigma_\nu = \gamma_\nu + \frac{\|\pi_n\|^2}{\pi_n'(\tau_\nu)\pi_{n+1}^*(\tau_\nu)}, \quad \nu = 1, 2, \ldots, n,$$

$$\sigma_\mu^* = \frac{\|\pi_n\|^2}{\pi_n(\tau_\mu^*)\pi_{n+1}^{*\prime}(\tau_\mu^*)}, \quad \mu = 1, 2, \ldots, n+1, \qquad (2.4)$$

where $\gamma_\nu = \gamma_\nu^{(n)}$ are the Christoffel numbers in the Gauss formula for the weight function w, and $\| \cdot \|$ is the weighted L_2-norm. Moreover, the interlacing property (2.3) is equivalent to $\sigma_\mu^* > 0$, $\mu = 1, 2, \ldots, n+1$ (see [18, Thms. 1 and 2]). For each $n \geq 1$, we consider the following properties for the formula (2.1).

(a) The nodes τ_ν, τ_μ^* interlace.
(b) All nodes τ_ν, τ_μ^* are contained in (a, b).
(c) All weights σ_ν, σ_μ^* are positive.
(d) All nodes τ_ν, τ_μ^*, without necessarily satisfying (a) and/or (b), are real.

In the following we review, for various weight functions, what has been proved, disproved, or conjectured with regard to the corresponding Gauss-Kronrod formulae having one or more of these properties.

2.1 Classical Weight Functions

The first existence result relates to the Legendre weight function $w(t) = 1$ on $[-1, 1]$. Szegö in 1935 (cf. [35]), following Stieltjes's conjectures (see the introduction), expanded the Stieltjes polynomial in a Chebyshev series, and, using some results from the theory of reciprocal power series, proved that all the expansion coefficients are negative, except for the first one which is positive, and also that the sum of these coefficients is zero. This allowed him to conclude properties (a) and (b) for all $n \geq 1$. Much later, Monegato, relying on Szegö's work, added property (c) for each $n \geq 1$ (see [19]).

Szegö extended his analysis to the case of the Gegenbauer weight function $w_\lambda(t) = (1 - t^2)^{\lambda - 1/2}$ on $[-1, 1]$, $\lambda > -1/2$, whose special case, with $\lambda = 1/2$, is the Legendre weight. He showed that properties (a) and (b) hold for all $n \geq 1$, when $0 < \lambda \leq 2$ (cf. [35, §3]). Monegato noted in [19] that, as in the Legendre case, property (c) is true for each $n \geq 1$, when $0 \leq \lambda \leq 1$. Szegö was unable to determine what happens for $\lambda > 2$, while for $\lambda \leq 0$, already for $n = 2$, two of the τ_μ^* are outside of $(-1, 1)$.

Gautschi and Notaris in [9] tried to close these gaps. Given n, one can compute the precise interval $(\lambda_n^p, \Lambda_n^p)$ of λ such that property (p) is valid, where $p =$ a,b,c,d. This can be done by varying λ and monitoring the motion of the nodes and weights in (2.1). Property (a) breaks down when a node τ_ν collides with a node τ_μ^*. This can be detected from the vanishing of the resultant of $\pi_n^{(\lambda)}(\cdot) = \pi_n(\cdot; w_\lambda)$ and $\pi_{n+1}^{*(\lambda)}(\cdot) = \pi_{n+1}^*(\cdot; w_\lambda)$. Similarly, property (d) ceases to hold when two nodes τ_μ^* collide and split to a pair of conjugate complex nodes, or equivalently, when the discriminant of $\pi_{n+1}^{*(\lambda)}$ vanishes. For properties (b) and (c) things are simpler. Assuming the interlacing property, the former amounts to finding when $\tau_1^* = 1$ and $\tau_{n+1}^* = -1$, while for the latter it suffices to find a $\sigma_\nu = 0$ (the interlacing property is equivalent to the positivity of the σ_μ^*; cf. [18, Thm. 1]). Gautschi and Notaris carried out this project analytically for $n = 1(1)4$, and numerically for $n = 5(1)20(4)40$. The corresponding values of λ_n^p, Λ_n^p, $p =$ a,b,c,d, are given in [9, Tables 2.1 and A.1].

Nonexistence results have been obtained by Monegato, and also by Notaris. Monegato proved (cf. [21, Thm. 1]) that the Gauss-Kronrod formula for the weight function w_λ, having properties (c) and (d), and degree of exactness $[2rn+l]$, where $r > 1$ and l is an integer, does not exist for all $n \geq 1$ when λ is sufficiently large.[†]

On the other hand, Notaris started from the limit formula that connects the Gegenbauer and Hermite orthogonal polynomials, and showed (cf. [26, §2]) that the corresponding Stieltjes polynomials $\pi_{n+1}^{*(\lambda)}$ and π_{n+1}^{*H} satisfy an analogous formula, namely, $\lim_{\lambda \to \infty} \lambda^{(n+1)/2} \pi_{n+1}^{*(\lambda)}(\lambda^{-1/2} t) = \pi_{n+1}^{*H}(t)$, $n \geq 1$. Then, based on this formula and a nonexistence result of Kahaner and Monegato for the Hermite weight function (cf. [14, Corollary]), he proved that the Gauss-Kronrod formula for the weight function w_λ and $n \neq 1, 2, 4$, having properties (c) and (d), does not exist if $\lambda > \lambda_n$, where λ_n is a constant sufficiently large.

Regarding the Jacobi weight function $w^{(\alpha, \beta)}(t) = (1 - t)^\alpha (1 + t)^\beta$ on $[-1, 1]$, $\alpha > -1$, $\beta > -1$, the only results that have been proved concern (besides the Chebyshev weights, which are discussed below) the special cases $w^{(\alpha, 1/2)}$ and

---

[†]The proof has an error, but can be repaired.

$w^{(-1/2,\beta)}$. For the first it can be shown that $\pi_{n+1}^{*(\alpha,1/2)}(2t^2-1) = 2^{n+1}\pi_{2n+2}^{*(\alpha,\alpha)}(t)$, $n \geq 1$ (see [22, Eq. (32)]). Then the nodes and weights in the Gauss-Kronrod formula for $w^{(\alpha,1/2)}$ can be expressed in terms of the corresponding ones for $w^{(\alpha,\alpha)}$, i.e., the Gegenbauer weight (see [9, Thm. 5.1]). Therefore, all results of the previous paragraphs, concerning $w^{(\alpha,\alpha)}$, can be applied to arrive at conclusions for $w^{(\alpha,1/2)}$. On the other hand, for the weight function $w^{(-1/2,\beta)}$, Rabinowitz has shown (cf. [32, pp. 74-75][‡]) that property (b) is false for $-1/2 < \beta < 1/2$ when n is even, and for $1/2 < \beta \leq 3/2$ when n is odd. Moreover, since interchanging α and β in $w^{(\alpha,\beta)}$ causes only a change in the sign of the nodes of the respective Gauss-Kronrod formula, the weights corresponding to nodes symmetric with respect to the origin being the same, one can derive for $w^{(1/2,\beta)}$ and $w^{(\alpha,-1/2)}$ results analogous to those obtained for $w^{(\alpha,1/2)}$ and $w^{(-1/2,\beta)}$.

Gautschi and Notaris in [9] extended their investigations to the Jacobi weight function, and using the same methods as for the Gegenbauer weight, they delineated, for a given n, areas in the (α, β)-plane where each of the properties (a), (b) and (c) (and also (d) when $n = 1$) holds. Their findings, explicitly for $n = 1$ and in the form of graphs for $n = 2(1)10$, are given in [9, p. 239 and Figure 4.1].

Also, Notaris obtained a nonexistence result. First, starting from the limit formula that connects the Jacobi and Laguerre orthogonal polynomials, he showed (cf. [26, §3]) that an analogous formula holds for the corresponding Stieltjes polynomials $\pi_{n+1}^{*(\alpha,\beta)}$ and $\pi_{n+1}^{*(\alpha)}$, namely,

$$\lim_{\beta \to \infty} (\beta/2)^{n+1}\pi_{n+1}^{*(\alpha,\beta)}(1 - 2\beta^{-1}t) = (-1)^{n+1}\pi_{n+1}^{*(\alpha)}(t), \ n \geq 1.$$

This formula, together with a nonexistence result of Kahaner and Monegato for the Laguerre weight function (cf. [14, Theorem]), led him to conclude that the Gauss-Kronrod formula for the weight function $w^{(\alpha,\beta)}$, α fixed, $-1 < \alpha \leq 1$, and $n \geq 23$ ($n > 1$ when $\alpha = 0$), having properties (c) and (d), does not exist if $\beta > \beta_{\alpha,n}$, where $\beta_{\alpha,n}$ is a constant sufficiently large.

When $|\alpha| = |\beta| = 1/2$ in $w^{(\alpha,\beta)}$, we obtain the Chebyshev weight functions of the four kinds, for which the Gauss-Kronrod formula has a special form. Specifically, for $\alpha = \beta = -1/2$, i.e., the Chebyshev weight of the first kind, the Gauss-Kronrod formula is the three-point Gauss formula when $n = 1$, and the $(2n + 1)$-point Gauss-Lobatto formula when $n \geq 2$, for the same weight. For $\alpha = \beta = 1/2$, i.e., the Chebyshev weight of the second kind, we get the $(2n + 1)$-point Gauss formula for that weight. Finally, for $\alpha = \mp 1/2$, $\beta = \pm 1/2$, i.e., the Chebyshev weight of the third or fourth kind, the Gauss-Kronrod formula is the $(2n + 1)$-point Gauss-Radau formula for the same weight, with the preassigned node at 1 and -1, respectively. Furthermore, the degree of exactness for each of these formulae is higher than the expected $3n + 1$, in particular, 5 for $n = 1$ and $4n - 1$ for $n \geq 2$ when $\alpha = \beta = -1/2$, $4n + 1$ when $\alpha = \beta = 1/2$, and $4n$ when

---

[‡]The superscript $\mu + 1/2$ in Eq. (68) and in the discussion following Eq. (69) should read $\mu - 1/2$.

$\alpha = \mp 1/2$, $\beta = \pm 1/2$. All this has been noted first by Mysovskih in [24], and later on by Monegato in [18, §4], and [22, pp. 152-153]. In addition, Monegato in [18, §4] showed that in the cases $\alpha = \beta = \pm 1/2$, Kronrod's idea can be iterated to produce a sequence of quadrature formulae, in which the nodes and weights are explicitly known.

Very little yet has been proved for the Hermite weight function $w_H(t) = e^{-t^2}$ on $(-\infty, \infty)$ and the Laguerre weight function $w^{(\alpha)}(t) = t^\alpha e^{-t}$ on $[0, \infty)$, $\alpha > -1$. Both have been examined numerically for n up to 20. For the former, it has been found (cf. [18, §3], [34]) that the Gauss-Kronrod formula has property (d) only for $n = 1, 2, 4$. On the other hand, for the Laguerre weight, with $\alpha = 0$, numerical calculations indicate that the Gauss-Kronrod formula has real nodes, but one is negative, when $n = 1$, and has some complex nodes for $2 \le n \le 20$ (see [18, §3]).

A nonexistence result has been obtained by Kahaner and Monegato in [14]. They proved that the Gauss-Kronrod formula for the weight function $w^{(\alpha)}$, $-1 < \alpha \le 1$, having properties (c) and (d), does not exist if $n \ge 23$ ($n > 1$ when $\alpha = 0$). As a consequence of this, they concluded that the corresponding formula for the weight function w_H, having properties (c) and (d), does not exist if $n \ne 1, 2, 4$.[§] The nonexistence result for the Laguerre weight remains true for n sufficiently large, even if the degree of the Gauss-Kronrod formula is lowered to $[2rn + l]$, where $r > 1$ and l is an integer (see [21, Thm. 2]).

2.2 Nonclassical Weight Functions

A class of weight functions for which the Gauss-Kronrod formulae have been extensively studied by several authors are the so-called Bernstein-Szegö weight functions defined by $w^{(\pm 1/2)}(t) = (1 - t^2)^{\pm 1/2}/\rho(t)$, $w^{(\pm 1/2, \mp 1/2)}(t) = (1 - t)^{\pm 1/2}(1 + t)^{\mp 1/2}/\rho(t)$ on $[-1, 1]$, where $\rho(t)$ is a polynomial of arbitrary degree that remains positive on $[-1, 1]$. The associated orthogonal and Stieltjes polynomials are linear combinations of Chebyshev polynomials of the four kinds. As a result of this, the corresponding Gauss-Kronrod formulae have properties (a), (b) and (c) for almost all n, exceptions occurring only for n below a specific value. In addition, the degree of exactness of these formulae is of the order $O(4n)$ instead of the usual $O(3n)$. To be precise, for the weight function $w_0^{(1/2)}(t) = (1 + \gamma)^2(1 - t^2)^{1/2}/[(1 + \gamma)^2 - 4\gamma t^2]$ on $[-1, 1]$, $-1 < \gamma \le 1$, which appeared for the first time in work of Geronimus (cf. [12]) and belongs to the above class of weight functions, property (b) was shown by Monegato in [22, p. 146], and properties (a) and (c) by Gautschi and Rivlin in [11], for all $n \ge 1$. (Property (b) for $n = 1$ is not shown in [22, p. 146], but can easily be verified.) Similarly, for the weight function $w^{(1/2)}$, with ρ a quadratic polynomial, the orthogonal and Stieltjes polynomials have been studied by Monegato and Palamara Orsi (cf. [23]). All four weights $w^{(\pm 1/2)}$, $w^{(\pm 1/2, \mp 1/2)}$, with ρ a

---

[§]This nonexistence result is stated in [14, Corollary] as follows: "Extended Gauss-Hermite rules with positive weights (and real nodes) only exist for $n = 1, 2, 4$". This is not quite accurate, since for $n = 4$ two of the σ_ν's in (2.1), relative to w_H, are negative.

quadratic polynomial, have been considered by Gautschi and Notaris in [10]. Properties (a), (b) and (c) have been established for almost all n, and the exceptions, which occur for $n \leq 3$, have been carefully identified. Also, the precise degree of exactness of the quadrature formulae in question has been determined. The general case of the weights $w^{(\pm 1/2)}$, $w^{(\pm 1/2, \mp 1/2)}$, with ρ a polynomial of arbitrary degree l, has been treated by Notaris in [25]. It was shown that properties (a), (b) and (c) hold for all $n \geq l + 2$ if $w = w^{(-1/2)}$, for all $n \geq l$ if $w = w^{(1/2)}$, and for all $n \geq l + 1$ if $w = w^{(\pm 1/2, \mp 1/2)}$. The corresponding degrees of exactness are respectively $4n - l - 1, 4n - l + 1$, and $4n - l$. For the weight $w^{(1/2)}$, with ρ a polynomial of arbitrary degree, Peherstorfer in [28] constructed a sequence of quadrature formulae by iterating Kronrod's idea. More precisely, he considered an n-point interpolatory quadrature formula whose nodes are the zeros of $\pi_n(\cdot; w^{(1/2)}(t)/q(t))$, where $q(t)$ is a polynomial of arbitrary degree m that remains positive on $[-1, 1]$. If N^* is a natural number satisfying $2^{N^* - 1}(n + 1) \geq 2^{N^*}(l + m) + 1 - l$, then Peherstorfer showed that the quadrature formula in question admits N^* Kronrod extensions, all having properties (b) and (c). In addition, the nodes of the Nth Kronrod extension interlace with those of the $(N - 1)$st extension, and the Nth Kronrod extension has degree of exactness $2^N[2(n + 1) - (l + m)] + l - 3,$[†] $N = 1, 2, \ldots, N^*$.

Peherstorfer in [29] extended his work to weight functions of the form $w(t) = (1 - t^2)^{1/2}|D(e^{i\theta})|^2$, $t = \cos\theta$, $\theta \in [0, \pi]$, where $D(z)$ is analytic, $D(z) \neq 0$ for $|z| \leq 1$, and D takes on real values for real z. He begins with an analysis of the asymptotic behavior of the associated functions of the second kind, which subsequently is used, in view of the connection between functions of the second kind and Stieltjes polynomials, to show that the corresponding Gauss-Kronrod formula has properties (a), (b) and (c) for all $n \geq n_0$, where n_0 is a constant not explicitly specified. The weight function in consideration includes as a special case the Bernstein-Szegö weight $w^{(1/2)}$, with ρ a polynomial of arbitrary degree, for which the same results were obtained in [25, 28], with the addition that n_0 was specified there.

Going even further, Peherstorfer in [30] considered polynomials orthogonal on the unit circle, and studied the asymptotic behavior of the associated functions of the second kind. This allowed him to arrive at conclusions for the asymptotic behavior of functions of the second kind associated with polynomials orthogonal on the interval $[-1, 1]$. Subsequently, using the connection between functions of the second kind and Stieltjes polynomials, he showed that if a weight function w on $[-1, 1]$ satisfies $\sqrt{1 - t^2}w(t) > 0$ for $-1 \leq t \leq 1$, and $\sqrt{1 - t^2}w(t) \in C^2[-1, 1]$, then the Stieltjes polynomial $\pi_{n+1}^*(\cdot; w^*)$, where $w^*(t) = (1 - t^2)w(t)$, is asymptotically equal to $\pi_{n+1}(\cdot; w)$. In addition, he proved that the Gauss-Kronrod formula for the weight function w^* has properties (a), (b) and (c) for all $n \geq n_0$, where n_0 is a constant not explicitly specified.[‡]

For the weight function $\gamma w^{(\alpha)}(t) = |t|^\gamma (1 - t^2)^\alpha$ on $[-1, 1]$, $\alpha > -1$, $\gamma > -1$,

---

[†] In [28, Theorem(c)], the 4 in the subscript of the displayed inclusion relation should be replaced by 3.

[‡] In [30, Theorem 4.2(a)], the subscript μ extends up to $n + 1$.

Gautschi and Notaris have shown (cf. [9, Thm. 5.2]) that the corresponding Gauss-Kronrod formula with n odd can be obtained from the one for the Jacobi weight $w^{(\alpha,(1+\gamma)/2)}$. Therefore, whatever is known for the latter can be applied to arrive at conclusions for the former.

Numerical computations of Caliò, Gautschi and Marchetti in [4, Examples 5.2 and 5.3] indicate that the Gauss-Kronrod formulae for the weight functions $w(t) = t^\alpha \ln(1/t)$ on $[0,1]$, $\alpha = 0, \pm 1/2$, have properties (a), (b) and (c) for all $n \geq 1$, with the exception of the weight with $\alpha = -1/2$ and n odd, for which property (b) (but not (d)) seems to be false. However, none of this has been proved yet.

3 ERROR TERM

The Gauss-Kronrod formula (2.1) can be obtained by applying the idea of Markov (cf. [17]). To be precise, we interpolate the function f at the simple nodes τ_ν and the double nodes τ_μ^*, and then integrate the resulting Hermite interpolation polynomial $p_{3n+1}(\cdot; f; \tau_\nu, \tau_\mu^*, \tau_\mu^*)$ of degree at most $3n + 1$. Evidently, the quadrature formula derived that way contains the values $f'(\tau_\mu^*)$. To arrive at formula (2.1), we require that the weights corresponding to these values be all 0, and this requirement leads to the orthogonality condition (2.2).

Integrating the interpolation error yields an expression for the error term of formula (2.1),

$$R_n^K(f) = \frac{1}{(3n+2)!} \int_a^b [\pi_{n+1}^*(t)]^2 f^{(3n+2)}(\xi(t)) \pi_n(t) w(t) dt, \qquad (3.1)$$

assuming that $f \in C^{3n+2}[a,b]$. This formula was applied by Monegato in the case of the Gegenbauer weight function $w_\lambda(t) = (1 - t^2)^{\lambda-1/2}$ on $[-1,1]$, $0 < \lambda < 1$ (see [20, §3]), in conjunction with $|\pi_{n+1}^{*(\lambda)}(t)| < 2^{-(n-1)}$, $-1 \leq t \leq 1$ (which follows directly from the work of Szegö in [35]) and bounds for $|\pi_n^{(\lambda)}(t)|$ (cf. [36, Eqs. (4.7.9) and (7.33.1)]), in order to derive a bound for $|R_n^K(f)|$ in terms of $\|f^{(3n+2+k)}\|_\infty$, where $k = 0$ for n even and $k = 1$ for n odd. This bound was improved and extended to the case $1 < \lambda < 2$ by Rabinowitz in [31, §4]. In the same paper, Rabinowitz proved that the Gauss-Kronrod formula for w_λ, $0 < \lambda \leq 2$, $\lambda \neq 1$, has precise degree of exactness $3n+1$ for n even and $3n+2$ for n odd. When $\lambda = 0$ or $\lambda = 1$, i.e., for the Chebyshev weight of the first or second kind, as already stated in the previous section, the degree of exactness is $4n - 1$ (5 for $n = 1$) and $4n + 1$, respectively. Also, Rabinowitz, based on a result of Akrivis and Förster in [1] and work of Szegö in [35], showed that the Gauss-Kronrod formula for the weight w_λ, $0 < \lambda < 1$, and $n \geq 2$, is nondefinite (see [33]).

Ehrich in [5] considers the Gauss-Kronrod formula for the Legendre weight function $w(t) = 1$ on $[-1,1]$ and is concerned with bounds for the error term of the form $|R_n^K(f)| \leq c_s(R_n^K)\|f^{(s)}\|_\infty$, where $c_s(R_n^K) = \sup_{\substack{f \in C^s[-1,1] \\ \|f^{(s)}\|_\infty \leq 1}} |R_n^K(f)|$. In particular, using results of Szegö in [35] and Rabinowitz in [31], he computes an upper bound for $c_{3n+2+k}(R_n^K)$, where $k = 0$ for n even and $k = 1$ for n odd, and examines its quality. Subsequently, he shows that there are better quadrature formulae using the same number of nodes, by comparing $c_{3n+2+k}(R_n^K)$ with $c_{3n+2+k}(R_{2n+1}^G)$ of the corresponding Gauss formula with $2n + 1$ points. Going further, he obtains in [6] upper bounds for c_s, $s < 3n + 2 + k$, useful for classes of functions of lower continuity. In the same paper, he extends the results in [5] to the case of the Gegenbauer weight function w_λ, $0 < \lambda < 1$. The quality of the bounds obtained in both cases is examined as well.

Also, Ehrich uses the results in [5] in order to derive bounds for the error term appropriate for analytic functions (cf. [7]). Specifically, if f is analytic in a simply connected region containing a closed contour C in its interior which surrounds $[-1, 1]$, he obtains bounds for R_n^K of the form $|R_n^K(f)| \leq l(C)c(n, \delta) \max_{z \in C} |f(z)|$, where $l(C)$ is the length of C and $\delta > 0$ is a lower bound for the distance between any point on C and any point in $[-1, 1]$. In the same paper, he studies the behavior of R_n^K when applied to Chebyshev polynomials of the first and second kind.

A different approach to the error term of the Gauss-Kronrod formula relative to the Legendre weight function is taken by Notaris in [27]. Let f be a holomorphic function in $C_r = \{z \in \mathbb{C} : |z| < r\}$, $r > 1$. Then $f(z) = \sum_{k=0}^\infty a_k z^k$, $z \in C_r$. Define $|f|_r = \sup\{|a_k|r^k : k \in \mathbf{N}_0$ and $R_n^K(t^k) \neq 0\}$, and set $X_r = \{f : f$ holomorphic in C_r and $|f|_r < \infty\}$. It is easy to show that $|\cdot|_r$ is a seminorm on X_r, and that the error term R_n^K of the Gauss-Kronrod formula for the Legendre weight function is a continuous linear functional on $(X_r, |\cdot|_r)$. This leads to bounds for R_n^K of the form $|R_n^K(f)| \leq \|R_n^K\|_r |f|_r$, the right-hand side of which can be optimized as a function of r, that is, if $f \in X_R$, then $|R_n^K(f)| \leq \inf_{1 < r \leq R}(\|R_n^K\|_r |f|_r)$. Another bound for R_n^K can be obtained if $|f|_r$ is estimated by $\max_{|z|=r} |f(z)|$ (cf. [13, Eq. (4.2)]). Notaris, following the analysis of Hämmerlin in [13] for the error term of the Gauss-Legendre formula, gives an upper bound for $\|R_n^K\|_r$, $2 \leq n \leq 30$. (When $n = 1$, the Gauss-Kronrod formula for the Legendre weight function is the 3-point Gauss formula for the same weight function.) Notaris's method can also be extended to $n > 30$, after the computation of appropriate constants.

REFERENCES

[1] Akrivis G., Förster K.-J. On the definiteness of quadrature formulae of Clenshaw-Curtis type. *Computing*, **33**:363–366, 1984.

[2] Baillaud B., Bourget H. *Correspondance d'Hermite et de Stieltjes I, II.* Gauthier-Villars, Paris, 1905.

[3] Barrucand P. Intégration numérique, abscisse de Kronrod-Patterson et polynômes de Szegö. *C. R. Acad. Sci. Paris*, **270**:336–338, 1970.

[4] Caliò F., Gautschi W., Marchetti E. On computing Gauss-Kronrod quadrature formulae. *Math. Comp.*, **47**:639–650, 1986.

[5] Ehrich S. Error bounds for Gauss-Kronrod quadrature formulae. *Math. Comp.*, **62**:295–304, 1994.

[6] Ehrich S. Einige neue Ergebnisse zu den Fehlerkonstanten der Gauß-Kronrod-Quadraturformel. *Z. Angew. Math. Mech.*, **73**:882–886, 1993.

[7] Ehrich S. Gauss-Kronrod quadrature error estimates for analytic functions. *Z. Angew. Math. Mech.* To appear.

[8] Gautschi W. Gauss-Kronrod quadrature—a survey. In *Numerical Methods and Approximation Theory III*, pages 39–66. Faculty of Electronic Engineering, Univ. Niš, Niš, 1988. G.V. Milovanović, ed.

[9] Gautschi W., Notaris S.E. An algebraic study of Gauss-Kronrod quadrature formulae for Jacobi weight functions. *Math. Comp.*, **51**:231–248, 1988.

[10] Gautschi W., Notaris S.E. Gauss-Kronrod quadrature formulae for weight functions of Bernstein-Szegö type. *J. Comput. Appl. Math.*, **25**:199–224, 1989; erratum in *J. Comput. Appl. Math.*, **27**:429, 1989.

[11] Gautschi W., Rivlin T.J. A family of Gauss-Kronrod quadrature formulae. *Math. Comp.*, **51**:749–754, 1988.

[12] Geronimus J. On a set of polynomials. *Ann. of Math.*, **31**:681–686, 1930.

[13] Hämmerlin G. Fehlerabschätzung bei numerischer Integration nach Gauss. In *Methoden und Verfahren der mathematischen Physik*, pages 153–163, v. 6. Bibliographisches Institut, Mannheim, Wien, Zürich, 1972. B. Brosowski and E. Martensen, eds.

[14] Kahaner D.K., Monegato G. Nonexistence of extended Gauss-Laguerre and Gauss-Hermite quadrature rules with positive weights. *Z. Angew. Math. Phys.*, **29**:983–986, 1978.

[15] Kronrod A.S. Integration with control of accuracy (Russian). *Dokl. Akad. Nauk SSSR*, **154**:283–286, 1964.

[16] Kronrod A.S. *Nodes and weights for quadrature formulae. Sixteen-place tables* (Russian). Izdat "Nauka", Moscow, 1964. English translation: Consultants Bureau, New York, 1965.

[17] Markov A. Sur la méthode de Gauss pour le calcul approché des intégrales. *Math. Ann.*, **25**:427–432, 1885.

[18] Monegato G. A note on extended Gaussian quadrature rules. *Math. Comp.*, **30**:812–817, 1976.

[19] Monegato G. Positivity of the weights of extended Gauss-Legendre quadrature rules. *Math. Comp.*, **32**:243–245, 1978.

[20] Monegato G. Some remarks on the construction of extended Gaussian quadrature rules. *Math. Comp.*, **32**:247–252, 1978.

[21] Monegato G. An overview of results and questions related to Kronrod schemes. In *Numerische Integration*, Internat. Ser. Numer. Math., v. 45, pages 231–240. Birkhäuser, Basel, 1979. G. Hämmerlin, ed.

[22] Monegato G. Stieltjes polynomials and related quadrature rules. *SIAM Rev.*, **24**:137–158, 1982.

[23] Monegato G., Palamara Orsi A. On a set of polynomials of Geronimus. *Boll. Un. Mat. Ital.*, **4-B**:491–501, 1985.

[24] Mysovskih I.P. A special case of quadrature formulae containing preassigned nodes (Russian). *Vesci Akad. Navuk BSSR Ser. Fiz.-Tehn. Navuk*, **4**:125–127, 1964.

[25] Notaris S.E. Gauss-Kronrod quadrature formulae for weight functions of Bernstein-Szegö type, II. *J. Comput. Appl. Math.*, **29**:161–169, 1990.

[26] Notaris S.E. Some new formulae for the Stieltjes polynomials relative to classical weight functions. *SIAM J. Numer. Anal.*, **28**:1196–1206, 1991.

[27] Notaris S.E. Error bounds for Gauss-Kronrod quadrature formulae of analytic functions. *Numer. Math.*, **64**:371–380, 1993.

[28] Peherstorfer F. Weight functions admitting repeated positive Kronrod quadrature. *BIT*, **30**:145–151, 1990.

[29] Peherstorfer F. On Stieltjes polynomials and Gauss-Kronrod quadrature. *Math. Comp.*, **55**:649–664, 1990.

[30] Peherstorfer F. On the asymptotic behaviour of functions of the second kind and Stieltjes polynomials and on the Gauss-Kronrod quadrature formulas. *J. Approx. Theory*, **70**:156–190, 1992.

[31] Rabinowitz P. The exact degree of precision of generalized Gauss-Kronrod integration rules. *Math. Comp.*, **35**:1275–1283, 1980; erratum in *Math. Comp.*, **46**:226, footnote, 1986.

[32] Rabinowitz P. Gauss-Kronrod integration rules for Cauchy principal value integrals. *Math. Comp.*, **41**:63–78, 1983; erratum in *Math. Comp.*, **45**:277, 1985.

[33] Rabinowitz P. On the definiteness of Gauss-Kronrod integration rules. *Math. Comp.*, **46**:225–227, 1986.

[34] Ramskiĭ J.S. The improvement of a certain quadrature formula of Gauss type (Russian). *Vyčisl. Prikl. Mat. (Kiev) Vyp.*, **22**:143–146, 1974.

[35] Szegö G. Über gewisse orthogonale Polynome, die zu einer oszillierenden Belegungsfunktion gehören. *Math. Ann.*, **110**:501–513, 1935. Collected Papers,
v. 2, pages 545–557. R. Askey, ed.

[36] Szegö G. *Orthogonal Polynomials.* Colloquium Publications, v. 23, 4th ed.,
American Mathematical Society, Providence, R.I., 1975.

International Series of Numerical Mathematics, Vol. 119, ©1994 Birkhäuser

THE GENERALIZED EXPONENTIAL INTEGRAL

F.W.J. Olver[*]

Dedicated, in friendship, to Walter Gautschi on the occasion of his 65th birthday, and in recognition of his numerous important contributions to the theory of special functions

Abstract. This paper concerns the role of the generalized exponential integral in recently-developed theories of exponentially-improved asymptotic expansions and the Stokes phenomenon. The first part describes the asymptotic behavior of the integral when both the argument and order are large in absolute value. The second part shows how to increase the accuracy of asymptotic expansions of solutions of linear differential equations of the second order by re-expanding the remainder terms in series of generalized exponential integrals.

1 INTRODUCTION

For real or complex values of the variables p and z, the generalized exponential integral ("g.e.i.") is defined by

$$E_p(z) = z^{p-1}\Gamma(1-p, z) = z^{p-1}\int_z^\infty \frac{e^{-t}}{t^p}dt, \quad z \neq 0, \tag{1.1}$$

with the proviso that the integration path does not pass through the origin. Unless p is a nonpositive integer, $E_p(z)$ is a multiple-valued function of z with branch-points at $z = 0$ and ∞. As a function of p, each branch of $E_p(z)$ is entire. The ordinary exponential integral corresponds to the case $p = 1$.

Recently, the g.e.i. has assumed an important role in a branch of asymptotic analysis called "exponential asymptotics". To explain this subject, we turn to the asymptotic expansion of $E_p(z)$ itself. For fixed p and large $|z|$ we have the well-known expansion

$$E_p(z) \sim \frac{e^{-z}}{z}\sum_{s=0}^\infty (-)^s \frac{p(p+1)\cdots(p+s-1)}{z^s}, \quad |\mathrm{ph}\, z| \leq \frac{3}{2}\pi - \delta, \tag{1.2}$$

where δ is an arbitrary small positive constant. The meaning of this expansion is that assigned by Poincaré: for each fixed nonnegative integer n

$$E_p(z) = \frac{e^{-z}}{z}\left\{\sum_{s=0}^{n-1}(-)^s \frac{p(p+1)\cdots(p+s-1)}{z^s} + R_n(z)\right\}, \tag{1.3}$$

[*]Institute for Physical Science and Technology and Department of Mathematics, University of Maryland, College Park, MD 20742

where

$$R_n(z) = O(z^{-n}) \quad \text{uniformly} \quad \text{as} \quad z \to \infty \quad \text{in the sector} \quad |\text{ph } z| \leq \frac{3}{2}\pi - \delta. \quad (1.4)$$

This result is easily established, for example, by repeated integration of (1.1) by parts. The restriction of z to a sector in the final expansion is typical. Indeed, if a Poincaré-type asymptotic expansion of an analytic function is valid for all values of ph z, then it is necessarily a Laurent expansion and therefore converges for all sufficiently large $|z|$ [13, pp.18-19].

Another important standard feature of the expansion (1.2) is that for any given values of p and z the accuracy obtainable by summing this series is limited. Excluding the case in which p is a nonpositive integer, the best that we can do is to terminate the sum just prior to the numerically smallest term. After that, the incorporation of additional terms only makes matters worse. On the other hand, if we take fixed number of terms in the expansion and keep increasing the value of $|z|$, then because

$$R_n(z) \sim (-)^n p(p+1) \cdots (p+n-1)z^{-n}$$

the relative error decreases in an algebraic manner.

The object of exponential asymptotics is to increase the accuracy obtainable from asymptotic approximations or asymptotic expansions of Poincaré type. When successful, it generally leads to approximations in which the relative error is exponentially, rather than algebraically, small as the absolute value of the asymptotic variable z tends to infinity. The earliest work on the subject appears to be that of T.J. Stieltjes in 1886 [17]. The formal work on "converging factors" by J.R. Airey [1] and J.C.P. Miller [11] is also in this vein, as is R.B. Dingle's formal theory of "terminants" [8]. Outlines and comparisons of these investigations will be found in [16].

2   RE-EXPANSION IN SERIES OF ELEMENTARY FUNCTIONS

The general way in which greater accuracy is extracted from an asymptotic expansion is by appropriate re-expansion of the remainder term. In the case of the expansion (1.3) an explicit formula for $R_n(z)$ is available from the partial integration procedure, and is given by

$$R_n(z) = (-)^n \frac{p(p+1) \cdots (p+n-1)}{z^n} z e^z E_{n+p}(z). \quad (2.1)$$

The next step is to truncate the expansion either at or close to its optimal stage, that is, just before the numerically smallest term. Since the ratio of the $(n+1)^{th}$ term to the nth term is $-(p+n-1)/z$ this means that we choose n so that $|p+n-1| = |z|$, approximately. More precisely, if $\hat{n}$ is the positive real number

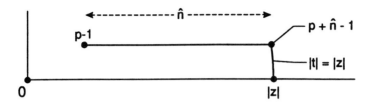

Figure 2.1: t-plane.

that satisfies $|p + \hat{n} - 1| = |z|$, then we set $n = \text{int}[\hat{n}]$. Thus n is now a function of $|z|$, albeit discontinuous, that tends to infinity with $|z|$.

From Figure 2.1, it is clear that $|p + \hat{n} - 1 - |z||$ is bounded as $|z| \to \infty$ (in fact it tends to $|\text{Im}\, p|$). And since

$$|p + n - |z|| \leq |p + \hat{n} - 1 - |z|| + 1,$$

it follows that $|p + n - |z||$ is bounded, also. On replacing $p + n$ by p in (2.1) we see that the desired re-expansion of $R_n(z)$ can be found if we are able to expand $E_p(z)$ when $|z|$ and $|p|$ are both large and $|p - |z||$ is bounded.

Accordingly, we set

$$z = \rho e^{i\theta}, \quad p = \rho + \alpha, \tag{2.2}$$

where $\rho (> 0)$ is large, and $|\alpha|$ is bounded. We also employ another well-known integral representation for the g.e.i., given by[†]

$$E_p(z) = \frac{z^{p-1} e^{-z}}{\Gamma(p)} \int_0^\infty \frac{e^{-zt} t^{p-1}}{1+t} dt, \quad \text{Re}\, p > 0, \quad |\text{ph}\, z| < \frac{1}{2}\pi, \tag{2.3}$$

and for later convenience we renormalize the definition of $E_p(z)$ in the following way:

$$E_p(z) = \frac{2\pi}{\Gamma(p)} z^{p-1} F_p(z). \tag{2.4}$$

With these notational changes we arrive that

$$F_p(z) = \frac{e^{-z}}{2\pi} \int_0^\infty \frac{e^{-zt} t^{p-1}}{1+t} dt = \frac{e^{-z}}{2\pi} \int_0^\infty \frac{\exp(-\rho e^{i\theta} t) t^{\rho+\alpha-1}}{1+t} dt. \tag{2.5}$$

For large ρ the last integral has a saddle-point at $t = e^{-i\theta}$. To apply Laplace's method, we rotate the integration path through an angle $-\theta$. We can do this directly when $-\frac{1}{2}\pi < \theta < \frac{1}{2}\pi$, and by analytic continuation beyond $\theta = \pm\frac{1}{2}\pi$,

---

[†]Formula (2.3) can be obtained from (1.1) with the aid of the convolution formula for the Laplace transform.

but because the integrand has a singularity at $t = -1$, of necessity we restrict $-\pi < \theta < \pi$. Setting $t = \tau e^{-i\theta}$ on the new path, we arrive at

$$F_p(z) = \frac{e^{-i(\rho+\alpha)\theta}e^{-z}}{2\pi} \int_0^\infty \frac{e^{-\rho\tau}\tau^{\rho+\alpha-1}}{1+\tau e^{-i\theta}}\,d\tau.$$

The saddle-point is now located at $\tau = 1$, and expanding the integral about this point in the usual way, we arrive at

$$F_p(z) \sim \frac{e^{-i(\rho+\alpha)\theta}}{1+e^{-i\theta}} \frac{e^{-\rho-z}}{(2\pi\rho)^{1/2}} \sum_{s=0}^\infty \frac{a_{2s}(\theta,\alpha)}{\rho^s} \tag{2.6}$$

as $\rho \to \infty$, uniformly with respect to bounded $|\alpha|$ and $\theta \in [-\pi+\delta, \pi-\delta]$, δ again denoting an arbitrary small positive constant. For this range of θ the coefficients $a_{2s}(\theta,\alpha)$ are continuous functions of θ and α. For example, we have

$$a_0(\theta,\alpha) = 1, \quad a_2(\theta,\alpha) = \frac{1}{12} + \frac{1}{2}\alpha(\alpha-1) - \frac{\alpha}{1+e^{i\theta}} + \frac{1}{(1+e^{i\theta})^2}.$$

To return to the original remainder term, we have from (2.1) and (2.4)

$$R_n(z) = (-)^n \frac{2\pi}{\Gamma(p)} z^p e^z F_{n+p}(z). \tag{2.7}$$

To apply (2.6) we replace p by $n + p$. This simply has the effect of changing the second of (2.2) into

$$p + n = \rho + \alpha, \tag{2.8}$$

except that p is now fixed. In this way we arrive at

$$R_n(z) \sim (-)^n \frac{(2\pi)^{1/2}}{\Gamma(p)} \frac{e^{i(p-\rho-\alpha)\theta}}{1+e^{-i\theta}} \rho^{p-(1/2)} e^{-\rho} \sum_{s=0}^\infty \frac{a_{2s}(\theta,\alpha)}{\rho^s} \tag{2.9}$$

as $z \to \infty$, uniformly with respect to ph $z \in [-\pi+\delta, \pi-\delta]$ and bounded $|\alpha|$ in (2.8). This is the desired expansion. It includes the optimal value of n, as defined earlier in this section.

The most interesting feature of the expansion (2.9) is the factor $e^{-\rho}$, that is, $e^{-|z|}$. This is indeed exponentially small as $|z| \to \infty$; furthermore, this applies uniformly in the sector $|\text{ph } z| \le \pi - \delta$. It is worth reflecting at this stage where the exponential smallness of $R_n(z)$ originates. It is essentially due to the increase of n with $|z|$; thus for each increase of a unit in the value of $|z|$ we take one more term in the expansion (1.3). Now the numerically smallest term of the sum in (1.2) is $(-)^n \Gamma(p+n)/\{\Gamma(p)z^n\}$, with $n = \text{int}[\hat{n}]$ and $\hat{n}$ determined by $|p+\hat{n}-1| = |z|$; compare the opening paragraph of this section. For large n and fixed p this term is approximately

$$(-)^n \sqrt{2\pi} n^{n+p-(1/2)} e^{-n} z^{-n} / \Gamma(p),$$

and since (2.8) applies, with $|\alpha|$ bounded, the last expression is asymptotic to

$$(-)^n \sqrt{2\pi} e^{-|z|} |z|^{|z|+\alpha-(1/2)} z^{-|z|+p-\alpha} / \Gamma(p),$$

that is,

$$(-)^n \sqrt{2\pi} e^{i(p-\alpha-|z|)\theta} |z|^{p-(1/2)} e^{-|z|} / \Gamma(p). \qquad (2.10)$$

This estimate of the numerically smallest term cannot be used for the remainder term, since in the present circumstances neighboring terms are comparable with the smallest term. However, there is obviously a strong link between (2.9) and (2.10). Indeed, the leading term of (2.9) differs from (2.10) only by the factor $1/(1 + e^{-i\theta})$.

The previous paragraph points to the source of the exponential smallness of $R_n(z)$. But our actual analysis has taken us beyond this. We have not merely found the order of magnitude of $R_n(z)$ when (2.8) applies with $|\alpha|$ bounded, we have re-expanded $R_n(z)$ in another asymptotic series. If we truncate the sum in (2.9) after m terms, then for fixed m the absolute error in this approximation to $R_n(z)$ is uniformly $O(e^{-\rho} \rho^{p-(1/2)-m})$ as $\rho \to \infty$. We therefore say that the combination of (1.3) and (2.9) yields an *exponential improvement* over the original Poincaré form (1.2).

3   RE-EXPANSION IN TERMS OF THE ERROR FUNCTION

The analysis of §2 applies when $|\theta| \le \pi - \delta$ $(< \pi)$. Everything breaks down as $\theta \to \pm\pi$. Thus in (2.6) and (2.9) the factor $1/(1 + e^{-i\theta})$ and the coefficients $a_{2s}(\theta, \alpha)$, other than $a_0(\theta, \alpha)$, all become infinite. However, in applications that we shall be making and also because of interest in the g.e.i. itself, we seek approximations for $F_p(z)$, or equivalently $E_p(z)$ or $\Gamma(1 - p, z)$, that are uniform with respect to θ in intervals containing $\theta = \pi$ or $-\pi$.

From the standpoint of Laplace's method for contour integrals, or the method of steepest descents, the basic problem is the coalescence of the saddle-point at $t = e^{-i\theta}$ in the integral in (2.5) with the simple pole at $t = -1$. Thanks to the researches of B.L. van der Waerden [19], N. Bleistein [7] and D.S. Jones [10], this is a well-understood problem in asymptotics, and it is possible to construct uniform asymptotic expansions with the aid of the error function.[‡] We shall not enter into the details of the analysis: they can be found in [14] and [15] (together with the proofs of all results quoted in §2). The final result is expressible as follows.

As in §2, set

[‡]Asymptotic expansions for incomplete Gamma functions in terms of the error function have been provided by N.M. Temme [18]. However, they do not apply to the combination of variables p and z that we are considering here.

$$z = \rho e^{i\theta}, \quad p = \rho + \alpha, \tag{3.1}$$

Then as $\rho \to \infty$ we have

$$F_p(z) \sim ie^{-p\pi i} \left[\frac{1}{2}\text{erfc}\left\{ c(\theta)\sqrt{\frac{1}{2}\rho} \right\} - i\frac{e^{-\rho\{c(\theta)\}^2/2}}{(2\pi\rho)^{1/2}} \sum_{s=0}^{\infty} \frac{h_{2s}(\theta,\alpha)}{\rho^s} \right], \tag{3.2}$$

uniformly with respect to $\theta \in [-\pi+\delta, 3\pi-\delta]$[§] and bounded values of $|\alpha|$. In this expansion erfc denotes the complementary error function and $c(\theta)$ is defined by

$$c(\theta) = \sqrt{2\{1 + e^{i\theta} + i(\theta - \pi)\}}, \tag{3.3}$$

the branch of the square root being chosen so that

$$c(\theta) \sim \pi - \theta \quad \text{as} \quad \theta \to \pi,$$

and determined by continuity elsewhere; see Figs.3.1 and 3.2. The coefficients $h_{2s}(\theta,\alpha)$ are continuous functions of θ and α throughout the region of validity of (3.2). Thus unlike the $a_{2s}(\theta,\alpha)$ they are continuous at $\theta = \pi$. However, the two sets of coefficients are related as follows:

$$h_{2s}(\theta,\alpha) = \frac{e^{i\alpha(\pi-\theta)}}{1 + e^{-i\theta}} a_{2s}(\theta,\alpha) - (-)^s i \frac{1 \cdot 3 \cdot 5 \cdots (2s-1)}{\{c(\theta)\}^{2s+1}}.$$

(At $\theta = \pi$ the singularities in the two terms on the right-hand side cancel each other.)

As in §2 the corresponding expansion for the original remainder term $R_n(z)$ is found on replacing n by $n + p$ in (3.2) and substituting into (2.7). We observe that owing to the factor

$$e^{-\rho\{c(\theta)\}^2/2} \equiv e^{-\rho - z - i\rho(\theta - \pi)}$$

exponential improvement of (1.2) is achieved throughout the interval $-\pi + \delta \leq \theta \leq \frac{3}{2}\pi - \delta$.

Next, the extent of the θ-interval of validity of (3.2) is noteworthy. In particular, it includes that of the expansion (2.6). In fact, (2.6) can be derived as a special case of (3.2), as follows. When $-\pi + \delta \leq \theta \leq \pi - \delta$ we see from Fig.3.2 that $c(\theta)$ lies in the right half-plane and is bounded away from the origin. Accordingly, the argument $c(\theta)\sqrt{\frac{1}{2}\rho}$ of the complementary error function also lies in the right half-plane and tends to infinity with ρ. We may therefore substitute for erfc by means of its own asymptotic expansion, given by

---

[§]A similar expansion applies when $\theta \in [-3\pi + \delta, \pi - \delta]$.

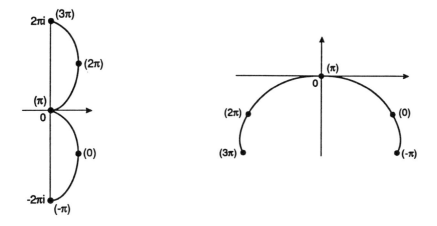

Figure 3.1: $\frac{1}{2}\{c(\theta)\}^2$-plane. (Values of θ in parentheses.)

Figure 3.2: $c(\theta)$-plane. (Values of θ in parentheses.)

$$\operatorname{erfc}(\zeta) \sim \frac{e^{-\zeta^2}}{\sqrt{\pi}\zeta} \sum_{s=0}^{\infty} (-)^s \frac{1 \cdot 3 \cdots (2s-1)}{(2\zeta^2)^s}, \quad |\mathrm{ph}\,\zeta| \le \frac{3}{4}\pi - \delta, \qquad (3.4)$$

with $\zeta = c(\theta)\sqrt{\frac{1}{2}\rho}$. Rearrangement of the result yields (2.6).

4   STOKES' PHENOMENON

Let us now consider the expansion (3.2) when $\theta \in [\pi + \delta, 3\pi - \delta]$. In this case $c(\theta)$ and ζ both lie in the left half-plane; compare again Fig.3.2. The expansion (3.4) is inapplicable, but because

$$\operatorname{erfc}(\zeta) = 2 - \operatorname{erfc}(-\zeta)$$

we have instead

$$\operatorname{erfc}(\zeta) \sim 2 + \frac{e^{-\zeta^2}}{\sqrt{\pi}\zeta} \sum_{s=0}^{\infty} (-)^s \frac{1 \cdot 3 \cdots (2s-1)}{(2\zeta^2)^s}, \quad |\mathrm{ph}\,(-\zeta)| \le \frac{3}{4}\pi - \delta. \qquad (4.1)$$

On substituting into (3.2) we find that

$$F_p(z) = ie^{-ip\pi} + \text{something exponentially small ("ses")}, \quad \pi + \delta \le \theta \le 3\pi - \delta,$$

and hence from (2.7)

$$R_n(z) = 2\pi i \frac{e^{-ip\pi}}{\Gamma(p)} z^p e^z \{1 + (\text{ses})\},$$

valid when (2.8) holds with fixed p and bounded $|\alpha|$. Now consider (1.3). When $\theta \equiv \text{ph } z = \pi + \delta$, $|e^z|$ is exponentially small. As θ increases, $|e^z|$ increases, but remains exponentially small until θ passes through $\frac{3}{2}\pi$, after which it becomes the *dominant* contribution to the right-hand side of (1.3). (This is not inconsistent with (1.2), however, because (1.2) requires ph $z \leq \frac{3}{2}\pi - \delta$.) In fact (1.3) has transformed into another well-known asymptotic expansion, given by*

$$E_p(z) \sim 2\pi i \frac{e^{-ip\pi}}{\Gamma(p)} z^{p-1} + \frac{e^{-z}}{z} \sum_{s=0}^{\infty} (-)^s \frac{p(p+1)\cdots(p+s-1)}{z^s},$$

$$\frac{1}{2}\pi + \delta \leq \text{ph } z \leq \frac{7}{2}\pi - \delta. \qquad (4.2)$$

There has long been an element of mystery concerning overlapping sectors of validity for differing asymptotic expansions of the same analytic function – in this case the expansions (1.2) and (4.2). The mystery does not center on any inconsistency. In the common interval of validity $\frac{1}{2}\pi + \delta \leq \text{ph } z \leq \frac{3}{2}\pi - \delta$ the contribution of the term in z^{p-1} is exponentially small compared with that of the infinite series and is therefore absorbable in any of the remainder terms associated with this series. The mystery is how the change from one form of expansion to the other ought to take place, because the change introduces a discontinuity in form whereas there is no discontinuity in $E_p(z)$ itself. This is of course an example of the celebrated Stokes phenomenon. A remarkable feature of the expansion (3.2) is that it furnishes a *smooth* transition from one asymptotic expansion to the other. When $\rho \equiv |z|$ is large, the change takes place rapidly as θ passes through the critical value π, because it stems almost entirely from the behavior of the term $\text{erfc}\{c(\theta)\sqrt{\frac{1}{2}\rho}\}$; compare Figs.4.1 and 4.2. Nevertheless the change is smooth. To put the matter in a more colorful way, the contribution $2\pi i e^{-ip\pi} z^{p-1}/\Gamma(p)$ is born in a natural manner as θ approaches and then passes beyond the value π.

It was a quest by M.V. Berry for a more satisfactory explanation of the Stokes phenomenon that has resurrected the subject of exponential asymptotics. In Berry's original paper in 1989 [2] he supplied a general analysis of the Stokes phenomenon, and proceeded to apply this to various functions, including the g.e.i. Besides providing a smooth interpretation of the phenomenon, Berry demonstrated the improvement in the accuracy of his new approximations by means of numerical examples. In the case of the g.e.i. the results we have derived in this section go beyond those of Berry. In the first place our results have been established rigorously. Secondly, Berry's approximation applies only in the immediate vicinity of the Stokes line (ph $z = \pi$). Thirdly, our re-expansion (3.2) is considerably more

---

*The expansion (4.2) is easily obtained by combining (1.2) with the connection formula $E_p(z) = \{2\pi i e^{-ip\pi} z^{p-1}/\Gamma(p)\} + E_p(ze^{-2\pi i})$.

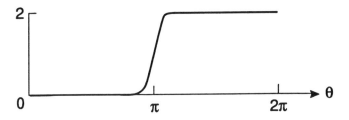

Figure 4.1: Graph of $|\mathrm{erfc}\{4c(\theta)\}|$.

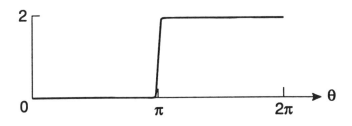

Figure 4.2: Graph of $|\mathrm{erfc}\{16c(\theta)\}|$.

accurate than Berry's approximation, even near the Stokes line. However, these observations in no way detract from the immense value of Berry's pioneering work, both in [2] and in numerous papers published subsequently by him and his students on other aspects of exponential asymptotics, for example, [3] - [6] and [9].

5 APPLICATIONS TO OTHER FUNCTIONS

The analyses for the g.e.i. given in §§2 and 3 above were greatly facilitated by the availability of the simple explicit formulas (2.1) and (2.3) for the remainder term in the original asymptotic expansion (1.3). Asymptotic expansions of other functions invariably have more complicated remainder terms and it is often much more difficult to re-expand them. It is again possible to seek re-expansions in terms of the error function, and indeed this was Berry's approach. I have found it more elegant mathematically to follow an earlier lead of Dingle and employ series of g.e.i.'s as the basic form of approximation, leaving the connection to the error function as an implicit consequence of the known asymptotic behavior of the g.e.i. described in §3 above. As an illustration, I will describe the re-expansion of asymptotic solutions of ordinary linear differential equations of the second order in the neighborhood of an irregular singularity of rank one. Several important special functions satisfy such equations, including Bessel functions and, more generally, confluent hypergeometric functions.

Without loss of generality we may suppose that the singularity is located at $z = \infty$. Then the equations we are considering have the general form

$$\frac{d^2w}{dz^2} + f(z)\frac{dw}{dz} + g(z)w = 0, \tag{5.1}$$

where the coefficients $f(z)$ and $g(z)$ are analytic at infinity, that is, they can be expanded in power series

$$f(z) = \sum_{s=0}^{\infty} \frac{f_s}{z^s}, \quad g(z) = \sum_{s=0}^{\infty} \frac{g_s}{z^s},$$

that converge for all sufficiently large values of $|z|$. And again, without loss of generality, we may assume that the roots λ_1 and λ_2 of the characteristic equation

$$\lambda^2 + f_0\lambda + g_0 = 0$$

are distinct. For notational convenience we may then normalize equation (5.1) so that $\lambda_2 - \lambda_1 = 1$. (This simply means that we replace the original independent variable z by $z/(\lambda_2 - \lambda_1)$.)

Asymptotic solutions of (5.1) are well known in these circumstances, and are given by [13, Chap. 7]

$$w_1(z) = e^{\lambda_1 z} z^{\mu_1} \sum_{s=0}^{n-1} \frac{a_{s,1}}{z^s} + R_n^{(1)}(z), \tag{5.2}$$

$$w_2(z) = e^{\lambda_2 z} z^{\mu_2} \sum_{s=0}^{n-1} \frac{a_{s,2}}{z^s} + R_n^{(2)}(z), \tag{5.3}$$

where for fixed n and large $|z|$

$$R_n^{(1)}(z) = e^{\lambda_1 z} z^{\mu_1} O(z^{-n}), \quad |\mathrm{ph}\, z| \le \frac{3}{2}\pi - \delta, \tag{5.4}$$

$$R_n^{(2)}(z) = e^{\lambda_2 z} z^{\mu_2} O(z^{-n}), \quad -\frac{1}{2}\pi + \delta \le \mathrm{ph}\, z \le \frac{5}{2}\pi - \delta. \tag{5.5}$$

Here the indices μ_1 and μ_2 are given by

$$\mu_1 = f_1\lambda_1 + g_1, \quad \mu_2 = -(f_1\lambda_2 + g_1),$$

and the coefficients $a_{s,1}$ and $a_{s,2}$ are determined recursively by

$$-sa_{s,1} = (s - \mu_1)(s - 1 - \mu_1)a_{s-1,1} + \sum_{j=1}^{s} \{\lambda_1 f_{j+1} + g_{j+1} - (s - j - \mu_1)f_j\} a_{s-j,1},$$

$$sa_{s,2} = (s - \mu_2)(s - 1 - \mu_2)a_{s-1,2} + \sum_{j=1}^{s} \{\lambda_2 f_{j+1} + g_{j+1} - (s - j - \mu_2)f_j\} a_{s-j,2},$$

beginning with $a_{0,1} = a_{0,2} = 1$.

Theorem. *Let m be an arbitrary fixed nonnegative integer. Then*

$$R_n^{(1)}(z) = (-)^{n-1} i e^{(\mu_2 - \mu_1)\pi i} e^{\lambda_2 z} z^{\mu_2} \left\{ C_1 \sum_{s=0}^{m-1} (-)^s a_{s,2} \frac{F_{n+\mu_2-\mu_1-s}(z)}{z^s} + R_{m,n}^{(1)}(z) \right\},$$

$$\tag{5.6}$$

$$R_n^{(2)}(z) = (-)^n i e^{(\mu_2 - \mu_1)\pi i} e^{\lambda_1 z} z^{\mu_1} \left\{ C_2 \sum_{s=0}^{m-1} (-)^s a_{s,1} \frac{F_{n+\mu_1-\mu_2-s}(ze^{-\pi i})}{z^s} + R_{m,n}^{(2)}(z) \right\},$$

$$\tag{5.7}$$

where C_1 and C_2 are the coefficients in the connection formulas

$$w_1(z) = e^{2\pi\mu_1 i} w_1(ze^{-2\pi i}) + C_1 w_2(z), \tag{5.8}$$

$$w_2(z) = e^{-2\pi\mu_2 i} w_2(ze^{2\pi i}) + C_2 w_1(z), \tag{5.9}$$

and as $z \to \infty$ with $|n - |z||$ bounded

$$R_{m,n}^{(1)}(z) = O(e^{-|z|-z} z^{-m}), \quad |\text{ph } z| \le \pi, \tag{5.10}$$

$$R_{m,n}^{(1)}(z) = O(z^{-m}), \quad \pi \le |\text{ph } z| \le \frac{5}{2}\pi - \delta, \tag{5.11}$$

$$R_{m,n}^{(2)}(z) = O(e^{-|z|+z} z^{-m}), \quad 0 \le \text{ph } z \le 2\pi, \tag{5.12}$$

$$R_{m,n}^{(2)}(z) = O(z^{-m}), \quad -\frac{3}{2}\pi + \delta \le \text{ph } z \le 0$$

$$\text{and } 2\pi \le \text{ph } z \le \frac{7}{2}\pi - \delta, \tag{5.13}$$

uniformly with respect to ph z in each case.

This result is proved by A.B. Olde Daalhuis and the present writer in [12]. Several observations may be made, as follows.

(i) The condition that $|n - |z||$ be bounded as $z \to \infty$ includes the situation in which n is chosen optimally in (5.2) or (5.3). This is a consequence of the asymptotic expansions

$$a_{s,1} \sim (-)^s \frac{e^{(\mu_2 - \mu_1)\pi i}}{2\pi i} C_1 \sum_{j=0}^{\infty} (-)^j a_{j,2} \Gamma(s + \mu_2 - \mu_1 - j), \qquad (5.14)$$

$$a_{s,2} \sim -\frac{1}{2\pi i} C_2 \sum_{j=0}^{\infty} a_{j,1} \Gamma(s + \mu_1 - \mu_2 - j), \qquad (5.15)$$

also proved in [12].

(ii) The sector of validity in the z-plane for the expansion of each solution has a total angle $5\pi - 2\delta$, compared with $3\pi - 2\delta$ in the original Poincaré expansions (5.2) to (5.5). Furthermore, uniform exponential improvement is achieved in the central parts of these sectors, given by $|ph\, z| \leq \pi$ for the first solution, and $0 \leq ph\, z \leq 2\pi$ for the second solution.

(iii) The Stokes phenomenon for the solution $w_1(z)$ in the vicinities of the Stokes lines $ph\, z = \pm\pi$ is explained smoothly by the behavior of the F-functions in the expansion (5.6) in these regions (§§3,4). Similarly for the solution $w_2(z)$ in the vicinities of $ph\, z = 0$ and 2π.

(iv) The expansion given by (5.2), (5.6), (5.10) and (5.11) includes three expansion of Poincaré type as particular cases; compare the discussion in §4 leading up to the expansion (4.2). Similarly for the expansion of the solution $w_2(z)$.

(v) The coefficients $a_{s,1}$ in the original Poincaré expansion for the first solution appear as coefficients in the re-expansion of the remainder term for the second solution, and *vice versa*. This phenomenon is sometimes called *resurgence* [5]. The same phenomenon also occurs in the expansions (5.14) and (5.15).

6 SUMMARY

We have re-expanded the remainder term in the well-known asymptotic expansion for the generalized exponential integral, both in a series of elementary functions, and in series involving the complementary error function. Advantages of these re-expansions include increased accuracy, increased regions of validity, and a smooth interpretation of the Stokes phenomenon.

We have also considered the well-known asymptotic solutions of second-order linear differential equations in the neighborhood of an irregular singularity of rank one, and indicated how to re-expand the remainder terms in series of generalized exponential integrals. These re-expansions enjoy similar advantages to those of the re-expansions of the generalized exponential integral itself.

Acknowledgment. This research was supported by the National Science Foundation under Grant DMS 92-08690. The writer is pleased to acknowledge the assistance of M.A. McClain of the National Institute of Science and Technology in the computations for the diagrams.

REFERENCES

[1] Airey J.R. The "converging factor" in asymptotic series and the calculation of Bessel, Laguerre and other functions. *Philos. Mag.* [7], **24**:521–552, 1937.

[2] Berry M.V. Uniform asymptotic smoothing of Stokes's discontinuities. *Proc. Roy. Soc. London* Ser. A, **422**:7–21, 1989.

[3] Berry M.V. Waves near Stokes lines. *Proc. Roy. Soc. London* Ser. A, **427**:265–280, 1990.

[4] Berry M.V. Infinitely many Stokes smoothings in the gamma function. *Proc. Roy. Soc. London* Ser. A, **434**:465–472, 1991.

[5] Berry M.V., Howls C.J. Hyperasymptotics. *Proc. Roy. Soc. London* Ser. A, **430**:653–668, 1990.

[6] Berry M.V., Howls C.J. Hyperasymptotics for integrals with saddles. *Proc. Roy. Soc. London* Ser. A, **434**:657–675, 1991.

[7] Bleistein N. Uniform asymptotic expansions of integrals with stationary point near algebraic singularity. *Comm. Pure Appl. Math.*, **19**:353–370, 1966.

[8] Dingle R.B. *Asymptotic Expansions: Their Derivation and Interpretation.* Academic Press, London, 1973.

[9] Howls C.J. Hyperasymptotics for integrals with finite endpoints. *Proc. Roy. Soc. London* Ser. A, **439**:373–396, 1992.

[10] Jones D.S. Asymptotic behavior of integrals. *SIAM Rev.*, **14**:286–317, 1972.

[11] Miller J.C.P. A method for the determination of converging factors, applied to the asymptotic expansions for the parabolic cylinder functions. *Proc. Cambridge Philos. Soc.*, **48**:243–254, 1952.

[12] Olde Daalhuis A.B., Olver F.W.J. Exponentially-improved asymptotic solutions of ordinary differential equations. II Irregular singularities of rank one. *Proc. Roy. Soc. London* Ser. A, **445**:39–56, 1994.

[13] Olver F.W.J. *Asymptotics and Special Functions.* Academic Press, New York, 1974.

[14] Olver F.W.J. On Stokes' phenomenon and converging factors. In *Asymptotic and Computational Analysis*, Lecture Notes in Pure and Applied Mathematics, **124**:329–355. Marcel Dekker, New York, 1990. R. Wong, ed.

[15] Olver F.W.J. Uniform, exponentially improved, asymptotic expansions for the generalized exponential integral. *SIAM J. Math. Anal.*, **22**:1460–1474, 1991.

[16] Olver F.W.J. Converging factors. In *Wave Asymptotics*, pages 54–68. Cambridge University Press, 1992. P.A. Martin and G.R. Wickham, eds.

[17] Stieltjes T.J. Recherches sur quelques séries semi-convergentes. *Ann. Sci. École Norm. Sup.* [3], **3**:201–258, 1886. Reprinted in *Complete Works*. Noordhoff, Groningen, **2**:2–58, 1918.

[18] Temme N.M. The asymptotic expansion of the incomplete Gamma functions. *SIAM J. Math. Anal.*, **10**:757–766, 1979.

[19] van der Waerden B.L. On the method of saddle points. *Appl. Sci. Res. Ser. B*, **2**:33–45, 1951.

International Series of Numerical Mathematics, Vol. 119, ©1994 Birkhäuser

ON THE COMPUTATION OF THE ZEROS
OF THE BESSEL POLYNOMIALS*

Lionello Pasquini[†]

Dedicated to Walter Gautschi

Abstract. A general method for simultaneously finding the zeros of polynomial solutions to second order linear homogeneous ordinary differential equations and the alternative method of finding the eigenvalues of the Jacobi matrices associated with those polynomials, are compared with each other. Special attention is given to the case of the classical Bessel polynomials. Well-known results on the behavior in the complex plane of the zeros of the Bessel polynomials are taken into account, and an unexpected, and rather interesting, ill-conditioning of associated Jacobi matrices is documented.

0 BASIC NOTATION

The basic notation that will be used in this article is as follows:

$\mathbb{C}$ the set of all complex numbers
$\mathbb{C}^n$ the n-dimensional complex linear space
$\mathbb{N}$ the set of non-negative integers
$\mathbb{N}^+$ the set of positive integers
$\mathbb{P}$ the linear space of all polynomials
$\mathbb{P}^m$ the linear space of the polynomials of degree less than or equal to m.

1 INTRODUCTION

In this article, the study of a method presented in [34] for simultaneously finding the zeros of polynomial solutions to a differential equation of the type

$$Lu = 0, \ L := \sum_{i=0}^{2} a_{2-i} \, d^{2-i}/dz^{2-i}, \ a_j \in \mathbb{P} \ (j = 0, 1, 2), \ a_2 \neq 0, \qquad (1.1)$$

is improved on. It is shown that the method works under hypotheses weaker than those assumed in [34] and a study of the nonsingularity of the roots of the nonlinear equations which the method is based upon is made.

---

*The preparation of this paper has been performed within the activities of the National Research Project: "Analisi Numerica e Matematica Computazionale"
[†]Dipartimento di Matematica Pura ed Applicata, Università Dell'Aquila, 67100 L'Aquila, Italy

Then the method is applied to the computation of the zeros of the well-known Bessel polynomials. The numerical results one obtains are compared with those one gets by the classical alternative method based upon the computation of the eigenvalues of the associated Jacobi matrices by the QR method.

The outline of the paper is as follows. Section 2 is dedicated to the presentation of the method in [34] and to its improvement. The method is referred to as the general method to point out one of its merits. As a matter of fact, it applies with no modification or adaptation to an extremely large class of interesting cases in Approximation Theory. In §3, the particular case we deal with in this paper, that of the Bessel polynomials, is briefly introduced along with some information and references which will be used in the following sections. In §4, the second method, the one based upon the QR method, is presented with special attention to its application to the particular case of the Bessel polynomials. This second method is briefly referred to as the alternative method. In §5, numerical results are discussed. The two methods are compared with each other and an interesting ill-conditioning of numerical problems related to the alternative method is carefully documented. Finally, §6 is devoted to proofs of previously stated theorems and to some additional results that we include for the sake of completeness.

2   THE GENERAL METHOD

The general method essentially consists in solving the nonlinear equation defined by

$$\mathbf{F}_n(\mathbf{z}) = 0, \ \mathbf{z} = (z_1, z_2, \dots, z_n) \in A \subset \mathbb{C}^n, \ n \in \mathbb{N}^+, \tag{2.1a}$$

$$A := \{\mathbf{z} \in \mathbb{C}^n : i \neq j \Rightarrow z_i \neq z_j; \ a_2(z_i) \neq 0, \ 1 \leq i \leq n\}, \tag{2.1b}$$

$$\mathbf{F}_n : A \subset \mathbb{C}^n \to \mathbb{C}^n, \ \mathbf{F}_n(\mathbf{z}) := (F_{n1}(\mathbf{z}), F_{n2}(\mathbf{z}), \dots, F_{nn}(\mathbf{z})), \tag{2.1c}$$

$$F_{ni}(\mathbf{z}) := R(z_i) - \sum_{j=1,n; j \neq i} 1/(z_i - z_j), \ i = 1, 2, \dots, n, \tag{2.1d}$$

$$R := -a_1/(2a_2). \tag{2.1e}$$

The following theorem points out the link between equation (2.1) and the problem we are interested in. The assertion in the theorem was already stated in [34] and in [35]. The proof was omitted in [34] and set out only in a particular case in [35]. It is outlined in §6 for the sake of completeness.

Theorem 2.1 *Suppose that for some* $n \in \mathbb{N}^+$, *one has*

$$Lp_n = 0, \tag{2.2a}$$

$$p_n(z) := z^n + c_1 z^{n-1} + \cdots + c_n := (z - \zeta_{n1})(z - \zeta_{n2}) \cdots (z - \zeta_{nn}). \tag{2.2b}$$

Also, suppose that the zeros of the polynomial p_n *satisfy the conditions*

$$\zeta_{ni} \notin Z_2 := \{z \in \mathbb{C} : a_2(z) = 0\}, \ i = 1, 2, \dots, n, \tag{2.3a}$$

512

$$\zeta_{ni} \neq \zeta_{nj} \quad \text{if} \quad i \neq j. \tag{2.3b}$$

Then $\vec{\zeta}_n := (\zeta_{n1}, \zeta_{n2}, \ldots, \zeta_{nn})$ is a root of equation (2.1). Conversely, suppose that for some $n \in \mathbb{N}^+$, $\vec{\zeta}_n = (\zeta_{n1}, \zeta_{n2}, \ldots, \zeta_{nn}) \in A \subset \mathbb{C}^n$ is a root of equation (2.1). Then there exists a suitable $a_0 \in \mathbb{P}$ such that (2.2) is satisfied.

Theorem 2.1 states that in practice, finding roots of equation (2.1) is equivalent to finding the zeros of the nth degree polynomial solutions to a differential equation like the one in (1.1). Note that (2.1) does not depend on a_0. Thus, it can be used even if the a_0 making (1.1) admit nth degree polynomial solutions is unknown. Moreover, the second part of the theorem points out that we can use (2.1), even if we do not know if such an a_0 exists, to find out if given a_2 and a_1 allow (1.1) to admit nth degree polynomial solutions.

These observations suggest the study of an effective method for solving (2.1).

Remark 2.1 Of course, from now on, any necessary condition for the existence of polynomial solutions of degree at least one to the differential equation in (1.1) has to be considered as included in the set of minimal hypotheses. Also, it appears quite obvious that the condition

$$\ker(L) \cap \mathbb{P}^n = \text{span}\{p_n\} \tag{2.4}$$

will have to be added to (2.2) and (2.3) and will play a crucial role in the sequel. In fact, this condition is necessary to prevent the stiff case of nonisolated roots of (2.1) lying on curves in $\mathbb{C}^n$.

A quite careful analysis of the nonsingularity of the roots of (2.1) is summarized in the following two theorems. Before stating them, let us introduce some more notation and make remarks and appropriate assumptions.

From now on, the condition

$$a_j \neq 0, \quad j = 0, 1, 2, \tag{2.5}$$

is assumed in (1.1) in place of $a_2 \neq 0$. The stronger condition in (2.5) simplifies the exposition. Moreover, it must be noted that the differential equations excluded by (2.5) either can be transformed into others satisfying (2.5), or are of no interest in this paper.

We now set

$$d_j := \text{degree of } a_j, \quad j = 0, 1, 2, \quad k := \max_{j=1,2}(d_j - j), \tag{2.6}$$

$$a_j(z) := \sum_{h=0}^{k+j} \alpha_{j,h} z^{k+j-h}, \quad j = 0, 1, 2, \tag{2.7}$$

and introduce the polynomials

$$\gamma_h \in \mathbb{P}^2, \gamma_h(z) := z(\alpha_{2,h}(z-1) + \alpha_{1,h}) \ (h = 0, 1, \ldots, k+2, \ \alpha_{1,k+2} := 0). \tag{2.8}$$

Remark 2.2 The condition

$$k := \max_{j=1,2}(d_j - j) = \max_{j=0,1}(d_j - j) = \max_{j=0,2}(d_j - j) = \max_{j=0,1,2}(d_j - j) \geq 0 \qquad (2.9)$$

is easily seen to be a necessary condition for the differential equation defined by (1.1) and (2.6) to have polynomial solutions of degree at least one. Note that (2.9) implies $d_j \leq k + j$, $j = 0, 1, 2$, and this justifies the notation used in (2.7). Also, note that (2.9) implies $d_j = k + j$ for at least two values of j, that is to say: $|\alpha_{j,0}| + |\alpha_{i,0}| > 0$, for $j \neq i$.

Remark 2.3 The practical implication of (2.4) can be summed up by saying that the general method is essentially a method for simultaneously approaching the zeros of the *lower* degree polynomial solutions to a differential equation like the one in (1.1). In fact, (2.4) allows the existence of a second polynomial solution p_m, $p_m(z) = z^m + b_1 z^{m-1} + \cdots + b_m$, provided that its degree m is greater than n.

Now we are in a position to state the two theorems announced above.

Theorem 2.2 *Let conditions* (2.2) *and* (2.3) *be satisfied for some* $n \in \mathbb{N}^+$. *Let* $\vec{\zeta}_n \in A \subset \mathbb{C}^n$ *be a root of nonlinear equation* $\mathbf{F}_n(\mathbf{z}) = 0$ *and let* $\mathbf{J}_{\mathbf{F}_n}(\vec{\zeta}_n)$ *be the Jacobian matrix of the function* $\mathbf{F}_n$ *at the point* $\vec{\zeta}_n$. *In the case* $k = 0$, *the inequality*

$$\det \mathbf{J}_{\mathbf{F}_n}(\vec{\zeta}_n) \neq 0 \qquad (2.10)$$

is equivalent to condition (2.4). *In the case* $k > 0$, *inequality* (2.10) *is equivalent to the condition*

$$Lw \in \mathbf{Q}_{nk}, \quad w \in \mathbb{P}^{n-1} \Rightarrow w = 0, \qquad (2.11)$$

where $\mathbf{Q}_{nk} := \{P \in \mathbb{P}^{n+k-1} : P = q p_n, \ q \in \mathbb{P}^{k-1}\}$.

Theorem 2.3 *Let conditions* (2.2) *and* (2.3) *be satisfied. Then* $n - d - k \leq \mathrm{rank}(\mathbf{J}_{\mathbf{F}_n}(\vec{\zeta}_n)) \leq n - d$, *where* $d := \dim(\ker(L) \cap \mathbb{P}^{n-1})$.

These two theorems suggest using Newton's method to solve equation (2.1).

The choice of Newton's method is quite obvious in the case $k = 0$. For this case, in fact, (2.10) is a consequence of hypotheses (2.2), (2.3) and (2.4), which actually have to be considered as completely reasonable in this context (see Remark 2.1).

In the case $k > 0$, (2.4) has to be replaced by the stronger condition (2.11) to guarantee (2.10). However, (2.11) can be regarded as a generically satisfied condition on the parameters $\alpha_{j,h}$ in (2.7). To see this, we can write the equation $Lw = q p_n$, $w \in \mathbb{P}^{n-1}$, $q \in \mathbb{P}^{k-1}$, consider the linear system of $n + k$ equations in the $n + k$ coefficients of w and q defined by that equation, and then observe that (2.11) is equivalent to the nonsingularity of the coefficient matrix of this linear system. To conclude, note that the coefficients c_i in p_n, and consequently the entries in the coefficient matrix, can be regarded as rational functions of the parameters $\alpha_{j,h}$ in (2.7).

On the other hand, it must be said that in no case among the many we tried, except for that of a counterexample we constructed on purpose, did a root turn out to be singular. Anyway, should singular roots occur, Theorem 2.3 gives information about the nature of the singularity, yielding an upper bound for the rank deficiency. Newton's method at singular points has been extensively studied—above all in the case of rank deficiency equal to 1 or small with respect to the matrix size—by many authors (see e.g. [5,6,7,18,19,20,37,38,39]) and techniques can be applied to retain quadratic convergence.

Remark 2.4 It is worth stressing the importance of inequality (2.10). Indeed, since $\mathbf{F}_n$ is an analytic function on the A of (2.1), (2.10) implies that Newton's method will define a sequence $\{\mathbf{z}^{(j)}\}_{j=0,1,\ldots}$ convergent to a root if the starting point $\mathbf{z}^{(0)}$ is good enough. In addition, it implies that the convergence will be of quadratic type. Moreover, it must be pointed out that the quadratic convergence of the sequence $\{\mathbf{z}^{(j)}\}_{j=0,1,\ldots}$ ensures favorable stability properties for its computed version $\{\mathbf{z}^{(j)^*}\}_{j=0,1,\ldots}$. As a matter of fact, the error can be practically reduced, for j large enough, to that due to the roundoff errors generated within a single iteration in the neighborhood of the approached root (see e.g. [24, §4.11]). From the stability properties of $\{\mathbf{z}^{(j)^*}\}_{j=0,1,\ldots}$, it follows that the general method possesses a remarkable capability for yielding appreciably accurate results (see for instance, §6 and [35]).

Further motivations for using Newton's method result from the two observations below.

As a consequence of the symmetry and the simple definition of the function $\mathbf{F}_n$, the Jacobian matrix is symmetric and simply defined as well. The definitions of the entries $(\mathbf{J}_{\mathbf{F}_n}(\mathbf{z}))_{ij}$ in the Jacobian matrix $\mathbf{J}_{\mathbf{F}_n}(\mathbf{z})$ at the point $\mathbf{z}$ can be derived by (2.1d) and are as follows:

$$(\mathbf{J}_{\mathbf{F}_n}(\mathbf{z}))_{ii} := R'(z_i) + \sum_{j=1,n; j \neq i} 1/(z_i - z_j)^2,$$

$$(\mathbf{J}_{\mathbf{F}_n}(\mathbf{z}))_{ij} := (\mathbf{J}_{\mathbf{F}_n}(\mathbf{z}))_{ji} := -1/(z_i - z_j)^2, \ i \neq j. \tag{2.12}$$

It will be seen in Remark 2.15 that when systems of polynomials with real zeros are treated, $\mathbf{J}_{\mathbf{F}_n}(\mathbf{z})$ has further properties of remarkable interest from a computational point of view as well.

It can be seen by experiment that constructing suitable starting points $\mathbf{z}^{(0)}$ is not difficult at all. On the other hand, in all the particular cases we selected to test the method (see Remark 2.13), we easily constructed procedures able to generate starting points good enough to keep very small the number of iterations required to get the maximum attainable precision, even in the case of large values of n (see §6 and [35] for some examples). To do so, results in the literature were taken into account.

Theorems stating sufficient conditions for hypotheses (2.2), (2.3), and (2.4) to be satisfied and some final remarks conclude this section.

Although information about hypotheses (2.2), (2.3), (2.4) and (2.9) can be found in the literature when polynomials belonging to systems already investigated have to be treated, it is worth reporting below some theorems from [34], as they are of interest at least in principle. In fact, the general method applies to any polynomial solution to a differential equation like the one in (1.1), provided only that the above recalled hypotheses are satisfied. In addition, it should be said that, due to the weaker conditions assumed in this paper, Theorems 2.9 and 2.10 are only analogous (not exact) to the results in [34]. The proofs are not reported in §6 because they can be easily obtained from proofs in [34], even in the cases of the two improved theorems that we mentioned. We refer the reader to [34] for further details.

Theorem 2.4 *Let v be any nontrivial solution to (1.1). If multiple zeros of v exist, they belong to the set Z_2 in (2.3a).*

Theorem 2.5 *Let v be any nontrivial solution to (1.1). Let $Z_j := \{z \in \mathbb{C} : a_j(z) = 0\}$, $Z_v := \{z \in \mathbb{C} : v(z) = 0\}$. Let $\zeta \in Z_2 - Z_1$. Then ζ cannot belong to Z_v if one of the following two conditions is satisfied: 1) $a_2'(\zeta) = 0$; 2) $a_2'(\zeta) \neq 0$ and $z_\zeta = -a_1(\zeta)/a_2'(\zeta)$ is not a positive integer.*

Theorem 2.6 *Let v be any nontrivial solution to (1.1). Let $Z_j := \{z \in \mathbb{C} : a_j(z) = 0\}$, $Z_v := \{z \in \mathbb{C} : v(z) = 0\}$. Let $\zeta \in Z_2 - Z_1$. Then ζ cannot belong to Z_v unless $v'(\zeta) = 0$.*

Theorem 2.7 *In order that the polynomial $P_n(z) := b_0 z^n + b_1 z^{n-1} + \cdots + b_n$ be a solution to (1.1), it is necessary and sufficient that*

$$\sum_{j=0}^{i} (\alpha_{0,i-j} + \gamma_{i-j}(n-j)) b_j = 0, \quad i = 0, 1, \ldots, n+k, \tag{2.13}$$

where it is understood that $b_j = 0$ if $j > n$, $\alpha_{j,h} = 0$ if $h > k+j$, and $\gamma_h = 0$ if $h > k+2$.

Theorem 2.8 *In order that nontrivial solutions to (1.1) exist in $\mathbb{P}$, it is necessary that $\alpha_{0,0} = -\gamma_0(m)$ for at least one non-negative integer m. The condition*

$$\alpha_{0,0} = -\gamma_0(n) \tag{2.14}$$

is necessary for (1.1) to have a polynomial solution of degree n.

Theorem 2.9 *Let $k = 0$. Suppose that condition (2.14) is satisfied and assume that the algebraic equation in the unknown z*

$$\alpha_{0,0} = -\gamma_0(z) \tag{2.15}$$

has no root less than n in $\mathbb{N}$. Then conditions (2.2) and (2.4) hold for suitable c_i's.

In the following statement, the matrices $\mathbf{M}_m$ ($m = 1, 2, \ldots, k$) are defined as follows. Consider conditions (2.13) as a linear homogeneous system of $n + k + 1$ equations in the $n + 1$ unknowns b_j's and denote by $\mathbf{C}_n$ the coefficient matrix of this set of linear homogeneous equations. $\mathbf{M}_m$ ($m = 1, 2, \ldots, k$) denotes the $(n+1) \times (n+1)$ submatrix of $\mathbf{C}_n$ formed by the rows of $\mathbf{C}_n$ with indices $i = 1, 2, \ldots, n$ and $i = n + m$.

Theorem 2.10 *Let $k > 0$. Suppose that condition (2.14) is satisfied and assume that the algebraic equation (2.15) has no root less than n in $\mathbb{N}$. Then no nontrivial solution exists in $\mathbb{P}^n$ if the conditions*

$$\det \mathbf{M}_m = 0, \quad m = 1, 2, \ldots, k, \tag{2.16}$$

are not satisfied. If conditions (2.16) are satisfied, then conditions (2.2) and (2.4) hold for suitable c_i's.

Remark 2.5 Note that condition (2.3a) implies (2.3b) by virtue of Theorem 2.4. Also, note that Theorems 2.5 and 2.6 point out the importance of the condition $Z_2 \cap Z_1 = \emptyset$. In fact, if this condition is satisfied, condition (2.3a) is generically satisfied (Theorem 2.5) and it is certainly satisfied if the ζ_{nh}'s are known to be simple (Theorem 2.6).

Remark 2.6 Theorems 2.2, 2.3, 2.9 and 2.10 show that the case $k = 0$ is the most favorable one and give information that is getting worse as k increases. It is worth observing that the case $k = 0$ includes, besides the ones defining the classical systems of Jacobi, Laguerre and Hermite, other families of differential equations (1.1). Among these, the generalizations of the just mentioned Jacobi, Laguerre and Hermite families [44, §6.72 and §6.73] and the family defining the Bessel system and its generalizations (see §3 and, e.g., [23,3]) deserve to be mentioned.

Remark 2.7 Let us go back to consider (2.1) as a tool to find out if, for given a_2 and a_1, a compatible a_0 exists, i.e. an a_0 making (1.1) have nth degree polynomial solutions. Conditions (2.13) can be regarded as a set of k algebraic equations $f_m(\vec{\alpha}) = 0$, $m = 1, 2, \ldots, k$, $\vec{\alpha} = (\alpha_{0,1}, \alpha_{0,2}, \ldots, \alpha_{0,k})$, in the k unknowns $\alpha_{0,m}$, $m = 1, 2, \ldots, k$. Thus, Theorems 2.9 and 2.10 say that compatible a_0's exist in the cases $k = 0$ and $k = 1$. What about the case $k > 1$? Can one say that compatible a_0's exist for a generic choice of the couple (a_2, a_1)? Note that f_m is a $(n+1)$-degree polynomial and $f_m(\vec{\alpha}) = \alpha_{0,1}^n \alpha_{0,m} +$ terms of degree less than $n+1$.

Remark 2.8 Besides the integer k in (2.9), the algebraic equation in (2.15) also plays an important role in Theorems 2.9 and 2.10. It is assumed that (2.15) has no root less than n in $\mathbb{N}$ to guarantee condition (2.4) and this points out the importance of the subset $\mathbb{N}^*$ of $\mathbb{N}^+$ defined by

$$\mathbb{N}^* := \{n \in \mathbb{N}^+ : \gamma_0(m) \neq \gamma_0(n), m \in \mathbb{N}, m < n\}. \tag{2.17}$$

Remark 2.9 Note that the algebraic equation in (2.15) never vanishes by virtue of what has been said in Remark 2.2, and is of degree two at most.

Remark 2.10 Note that the set $\mathbb{N}^+ \setminus \mathbb{N}^*$, when it is not the empty set, consists of a finite number of consecutive positive integers at most.

Remark 2.11 Note that the integer k in (2.9), the algebraic equation in (2.15) and the set in (2.17) can be defined without knowing a_0.

Remark 2.12 The last two remarks suggest the following questions. Assume $k > 0$ (since only in this case are the questions still unanswered, at least as far as we know). Are there couples (a_2, a_1) for which compatible a_0's, depending on n, exist for each $n \in \mathbb{N}^*$? If such couples exist, what is their particular meaning, if any, and what are the properties of the sequences of polynomials they define by means of (1.1)? In particular, is condition (2.3) satisfied for each $n \in \mathbb{N}^*$, so that the general method can be applied to investigate the behavior of their zeros? Note that when these questions have a positive answer (this is the case if $k = 0$), using the general method is simpler. In fact, we do not have to modify the definition of the rational function R in (2.1e) when we want to change the polynomial of the sequence. We only have to assign the proper value to n.

Remark 2.13 The general method has been intensively tested. Besides the one of the Bessel polynomials, which is discussed in this paper, other cases have been taken from the literature to represent different aspects. Among them, we have chosen the system in [12, 13, 14] to represent the case $k = 1$ and to try another case in which the zeros are in the complex plane, a system in [27, §5] to represent the case of a system of polynomials which are not known to satisfy a three-term recurrence but a five-term one (k is equal to 5 in this case), the Littlejohn-Shore system in [25,41] to represent the case of a system orthogonal with respect to a weight function defined by means of the Dirac distribution (k is equal to 3 in this case), the Jacobi, Laguerre and Hermite systems ($k = 0$).

Remark 2.14 In the last cases quoted in Remark 2.13, the zeros ζ_{nh}, $h = 1, 2, \ldots, n$, are known to lie on the real line and the general method can be applied in $\mathbb{R}^n$, instead of in $\mathbb{C}^n$, to save computational cost. In particular, when symmetry about the origin occurs, a cheaper version based upon a reduced equation in $\mathbb{R}^\nu$ ($\nu = [n/2]$, $[x]$ = the greatest integer $\leq x$) can be used (see [34, 35] for details).

Remark 2.15 In all the cases with real zeros quoted in Remark 2.13, the Jacobian matrices turn out to be Stieltjes matrices.

Remark 2.16 Finally, we sum up the merits of the general method. In our opinion, they are the following ones.

i) It is general and of wide use. As a matter of fact, it can be applied with no modification to any (lower-degree) polynomial solution of a differential equation like the one in (1.1), (2.2), (2.3) and (2.4).

ii) In particular, it can be viewed as an unifying procedure for simultaneously finding the zeros of the polynomials which verify a classical, semi-classical or vectorial [28,47] condition of orthogonality.

iii) It is considerably sharp.

iv) It is easy to use.

Remark 2.17 An implementation of the general method, carried out using the FORTRAN 77 Standard language and related to the general equation in (1.1), is available on request.

3 THE BESSEL POLYNOMIALS

The Bessel polynomials were introduced by Bochner in [1] and then thoroughly studied in [23]. Subsequently, many authors (see e.g. [2, 3, 4, 8, 21, 22, 26, 29, 33, 42, 43, 46]) investigated asymptotic properties and the location of their zeros, because of the great interest in the Bessel polynomials for their many applications (see e.g. [23], where it is shown that they occur naturally in the theory of travelling spherical waves). We cannot attempt a survey of so many remarkable contributions to this subject and we just recall the information we shall need in the following sections. This concerns the differential equation (1.1) and the recurrence relation—which will be used to apply the general method and the alternative method respectively— and effective results about regions of inclusion of the zeros ζ_{ni} $(i = 1, 2, \ldots, n)$ [3, 42], which will be used in §6 to document the ill-conditioning of associated Jacobi matrices.

3.1 Differential equation and recurrence relation

The differential equation (1.1) and the recurrence relation can be written [23] as follows

$$z^2 u'' + 2(z+1)u' - n(n+1)u = 0, \tag{3.1}$$

$$p_0(z) = 1, \quad p_1(z) = (z+1)p_0(z),$$

$$p_n(z) = (2n-1)zp_{n-1}(z) + p_{n-2}(z), \quad n = 2, 3 \ldots . \tag{3.2a}$$

We rewrite the three-term recurrence in the form

$$zp_{n-1}(z) = \sigma_n p_n(z) + \delta_n p_{n-1}(z) - \sigma_n p_{n-2}(z),$$

$$\sigma_1 := 1, \ \delta_1 := -1, \ p_{-1}(z) := 0; \quad \sigma_n := 1/(2n-1), \ \delta_n := 0, \ n = 2, 3 \ldots , \tag{3.2b}$$

as it is more convenient for our purpose.

3.2. Inclusion regions

The information about the inclusion regions we shall use in §6 is contained in the following theorems.

Theorem 3.1 [42] *All zeros of the nth degree Bessel polynomial lie in the cardioidal region*

$$C(n) := \{z = \rho e^{i\theta} \in \mathbb{C} : 0 < \rho < (1 - \cos\theta)/(n+1)\} \cup \{-2/(n+1)\}.$$

Theorem 3.2 [3] *For $n \geq 2$, all zeros of the nth degree Bessel polynomial belong to the sector*

$$S(n) := \{z = \rho e^{i\theta} \in \mathbb{C} : |\theta| > \cos^{-1}(-1/n), -\pi < \theta \leq \pi\}.$$

Theorem 3.3 [3] *For $n \geq 1$, all zeros of the nth degree Bessel polynomial belong to the infinite region*

$$D(n) := \{z \in \mathbb{C} : |z| > 1/(n + 2/3)\}.$$

4   THE ALTERNATIVE METHOD

A general outline of the alternative method can be given by saying that it consists in transforming the original problem of finding the zeros of a polynomial to that of finding the eigenvalues of a suitable matrix associated with the polynomial itself. An example of a matrix associated with an assigned polynomial is the so-called companion matrix, which can be derived from the usual representation of the polynomial in the power basis or, more generally, from any its Taylor expansion. Also, banded matrices can be associated in many ways to systems of polynomials which solve difference equations, i.e. which solve recurrence relations. Once the associated matrix has been defined, the QR method is recommended to find its eigenvalues, and it will be convenient to choose for each case the particular version of this powerful method that better fits the particular matrix.

For example, if one is interested in a system of polynomials satisfying the three-term recurrence relation

$$zp_{n-1}(z) = \sigma_n p_n(z) + \delta_n p_{n-1}(z) + \tau_n p_{n-2}(z), \tag{4.1}$$

the matrix associated with the nth degree polynomial p_n of the system can be chosen as the Jacobi matrix

$$\mathbf{J}_n = \begin{vmatrix} \delta_1 & \sigma_1 & & & & \\ \tau_1 & \delta_2 & \sigma_2 & & & \\ & \tau_2 & \delta_3 & \sigma_3 & & \\ & & \cdot & \cdot & \cdot & \\ & & & \cdot & \cdot & \cdot \\ & & & \tau_{n-2} & \delta_{n-1} & \sigma_{n-1} \\ & & & & \tau_{n-1} & \delta_n \end{vmatrix}. \tag{4.2}$$

In fact, one has

$$z\mathbf{p}_n(z) = \mathbf{J}_n \mathbf{p}_n(z) + cp_n(z)\mathbf{e}_n,$$

where $\mathbf{p}_n(z) = [p_0(z), p_1(z), \ldots, p_{n-1}(z)]^T$, $\mathbf{e}_n = [0, 0, \ldots, 1]^T$, and c is a suitable constant. Recommended algorithms for finding the eigenvalues of a matrix like that in (4.2) are those in [9, 10, 31], in case the matrix is symmetric, and those in [30, 32], when the matrix is unsymmetric.

The Bessel system leads to a particular case (see (3.2b) and (4.1)). In this event, one has in (4.2)

$$\delta_1 := -1; \ \delta_h := 0, \ h \geq 2; \quad \sigma_h = -\tau_{h-1} := 1/(2h-1), \quad (4.3)$$

and the routine HQR in [30] is the most appropriate tool to use.

Remark 4.1 The matrix in (4.2) is not the only tridiagonal matrix one can use. Indeed, one can transform (4.2) into other tridiagonal matrices by similarity transformations defined by nonsingular diagonal matrices. Note that any transformed matrix $\mathbf{J}'_n = \{\tau'_h, \delta'_h, \sigma'_h\}$ obtained in this way, is such that $\delta'_h = \delta_h$ ($h = 1, 2, \ldots, n$), and $\tau'_h \sigma'_h = \tau_h \sigma_h$ ($h = 1, 2, \ldots, n-1$).

5 THE COMPARISON

In this section, the performances of the two methods described in §2 and §4 are discussed and compared with each other in the case of the Bessel polynomials.

The reason why we chose to discuss the case of Bessel polynomials in this paper, is that it showed an interesting ill-conditioning of Jacobi matrices which, in our opinion, deserves to be closely investigated. Comparisons related to other cases, including the ones quoted in Remark 2.13, will be discussed in [36].

All tests were carried out by a VAX 6410 machine, using FORTRAN in double precision. The resulting machine precision will be assumed, and referred to, as the basic machine precision. In fact, some tests, all related to the alternative method and to large values of n, were repeated using different machines and machine precisions to better document the above mentioned ill-conditioning.

Concerning the application of the general method, it was very easy to provide Newton's method with suitable starting points $\mathbf{z}^{(0)}$. As a matter of fact, the following procedure that we adjusted experimentally, proved itself to be very effective. Let

$$c = n/(n([n/2]+1)) \text{ if } n \text{ is even}, \ c = (n+1)/(n([n/2]+1)) \text{ if } n \text{ is odd}, \rho = c2^{-1/2}.$$

The procedure assumes $z_h^{(0)} = x_h^{(0)} + iy_h^{(0)}$ ($h = 1, 2, \ldots, n$), where

$$x_{j+s}^{(0)} = x_{j+s+1}^{(0)} = -\rho\cos(j\pi/(2([n/2]+1))), \ j = 2m-1, \ m = 1, 2, \ldots, [n/2],$$

$$y_{j+s}^{(0)} = -\rho\sin(j\pi/(2([n/2]+1))), \ y_{j+s+1}^{(0)} = -y_{j+s}^{(0)}, \ j = 2m-1, \ m = 1, 2, \ldots, [n/2],$$

with $s = 0$ if n is even, and $s = 1$ if n is odd. The real coordinate $z_1^{(0)} = -\rho$ is added if n is odd.

Such defined $\mathbf{z}^{(0)}$'s proved themselves good enough to keep appreciably low, and practically independent of n, the number of iterations of Newton's method required to get the maximum accuracy attainable by the basic machine precision. For $5 \leq n \leq 205$, that number increased from 6 to 8.

It is not necessary to apply the theorems in §2 (Theorems 2.5, 2.4 and 2.9) to assert that conditions (2.2), (2.3) and (2.4) are satisfied for every $n > 0$. Indeed, such an assertion follows from the literature. We note only that one has $k = 0$ in the case of the Bessel polynomials (see (1.1), (2.6) and (3.1)), so that quadratic convergence is ensured by virtue of Theorem 2.2.

There is nothing else to add concerning the application of the general method, except to say that no problem arose in the range of the degrees that we tried ($n \leq 205$). Consequently, it is conceivable that the method continues to work for appreciably larger values of n. For even larger values of n, an improvement of the procedure which generates the starting point $z^{(0)}$ might be necessary, but ultimately, the general method should work until the roundoff error generated within a single iteration of Newton's method is reasonably smaller than the distance between the zeros ζ_{nh}'s.

Coming now to the alternative method, we first tried the matrix defined in (4.2) and (4.3), which can be considered as the most natural one as it is derived directly from recurrence relation (3.2), and we used the HQR routine in [30] and the basic machine precision. The results agree with those obtained from the general method only for small values of n, but then a phenomenon of bifurcation of the computed eigenvalues from the curve apparently described by the zeros determined by the general method appears, and becomes more and more evident as n increases (see Figures 1, 2 and 3). In addition, for $n = 40$, the output is clearly incompatible with known theoretical results (two real zeros).

The failure of the alternative method in the case $n = 40$, and both the surprising regularity and the shape of the bifurcated curves in the figures quoted above, lead us to affirm that the eigenvalues with large real parts of the matrix of (4.2) and (4.3) suffer from an ill-conditioning which increases with n.

Since the conditioning of the eigenvalues also depends on the associated eigenvectors (see e.g. [17, §4]), several matrices belonging to the family of all tridiagonal matrices similar to the one in (4.2) and (4.3) (see Remark 4.1) were tested. But, the phenomenon of ill-conditioning described above always appeared and always became more and more evident as n was increased, the only appreciable differences among the treated cases being slight variations in the seriousness of the ill-conditioning and some negligible details depending on n, on the machine used, etc. Figures 4, 5 and 6 give examples by comparing outputs related to the matrix of (4.2) and (4.3) and to the matrix defined by taking

$$\delta_1 := -1; \quad \delta_h := 0, \ h \geq 2; \quad \sigma_h = -\tau_h := -1/(4h^2 - 1)^{1/2} \quad (5.1)$$

in (4.2). This time a Personal Computer (486 DX2 66 MHz RAM 8 MB) and Mathematica Enhanced Version 2.2 for Windows were employed. Moreover, the machine precision was defined according to the conventions in [11], by $t = 16$. Note that in the case of $n = 40$, there is now no real computed eigenvalue. Also, note

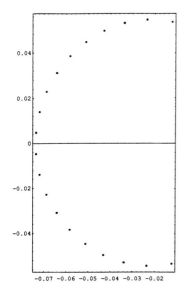

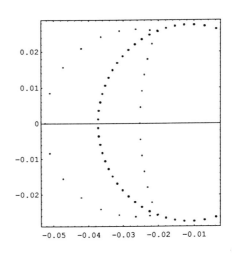

Figure 1: $n = 20$. General method (big dots). Alternative method (small dots). Matrix in (4.2) and (4.3).

Figure 2: $n = 40$. General method (big dots). Alternative method (small dots). Matrix in (4.2) and (4.3).

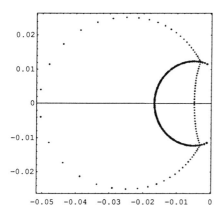

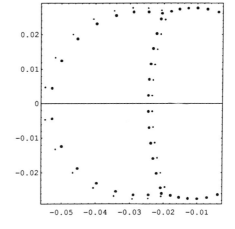

Figure 3: $n = 90$. General method (big dots). Alternative method (small dots). Matrix in (4.2) and (4.3).

Figure 4: $n = 40$. Eigenvalues of matrix in (4.2) and (4.3) (big dots). Eigenvalues of matrix in (4.2) and (5.1) (small dots).

523

that the difference between the two outputs disappears as n tends to infinity (see Figure 6).

The matrix defined by (4.2) and (5.1) will be referred to briefly as matrix (5.1). It is the one which was systematically used in all the tests discussed from here to the end of this section.

The ill-conditioning of the eigenvalues of the family of similar tridiagonal matrices mentioned above results in serious consequences when n is large. This can be effectively documented by using the theorems in §3 and by looking at Figures 7, 8 and 9, where, besides the computed eigenvalues of matrix (5.1), the zeros $\zeta_{nh}, h = 1, 2, \ldots, n$, computed by the general method, and the inclusion regions $C(n) \cap S(n) \cap D(n)$ defined by those theorems, are represented. Here, the machine precision we used was the basic one.

Finally, the remaining Figures 10, 11 and 12 and Table 1 complete the above information. They are representative of a set of tests that were carried out to find the machine precision needed by the alternative method to get outputs comparable in accuracy to those obtained with the general method and the basic machine precision. To easily repeat the tests with any small change in the machine precision, we used a PC of the above described type and the version of Mathematica quoted above to compute the eigenvalues. The machine precision is given in the figures as the value of t, used according to the conventions in [11]. It is worth stressing that the results of the general method were always obtained by using the basic machine precision.

Figures 10, 11 and 12 are related to the case $n = 125$. Table 1 sums up results related to some of the tested cases. The quoted values of t_n^* denote the values of t needed to make the graphic representation of the output obtained by the alternative method coincide with the output yielded by the general method.

The sensitivity of eigenvalues of matrix (5.1) to pertubations is rather interesting. In fact, matrix (5.1) is a rank one perturbation of a normal matrix (see Remark 5.3). A very interesting approach to the sensitivity to pertubations of the eigenvalues of nonnormal matrices has been developed recently by Trefethen and co-workers (see, e.g., [40,45]).

Remark 5.1 In all cases that we tried, the zeros computed by the general method with the basic machine precision satisfied Theorems 3.1, 3.2 and 3.3. Moreover, their graphic representation always looked like the ones in Figures 1-3, and 7-12.

Remark 5.2 The CPU times required by the two methods (t_{GM} for the general method and t_{AM} for the alternative method), when the basic machine precision is used, are shown in Table 2. The comparison is definitely in favor of the alternative method, and the difference between the times grows quickly and almost reaches four seconds for $n = 200$. It can be observed, however, that the CPU time required by the general method reduces appreciably when poor accuracy is required, and that a certain amount of CPU time is due to the generality of the supplied software and spent deliberately to get the advantages in i), ii) and iv) of Remark 2.16. In any case, the considerable difference between the accuracies of the outputs one

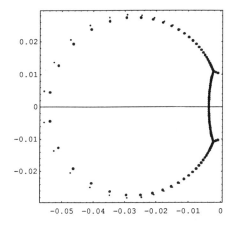

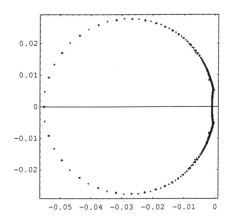

Figure 5: $n = 100$. Eigenvalues of matrix in (4.2) and (4.3) (big dots). Eigenvalues of matrix in (4.2) and (5.1) (small dots).

Figure 6: $n = 200$. Eigenvalues of matrix in (4.2) and (4.3) (big dots). Eigenvalues of matrix in (4.2) and (5.1) (small dots).

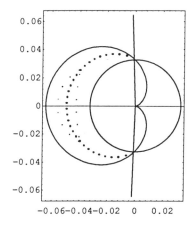

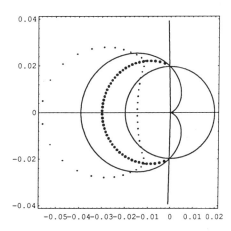

Figure 7: $n = 30$. General method (big dots). Alternative method (small dots). Matrix in (4.2) and (5.1).

Figure 8: $n = 50$. General method (big dots). Alternative method (small dots). Matrix in (4.2) and (5.1).

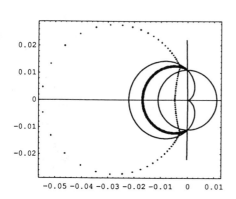

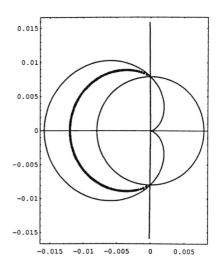

Figure 9: $n = 90$. General method (big dots). Alternative method (small dots). Matrix in (4.2) and (5.1).

Figure 10: $n = 125$. $t = 56$. General method (big dots). Alternative method (small dots). Matrix in (4.2) and (5.1).

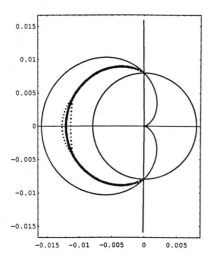

Figure 11: $n = 125$. $t = 64$. General method (big dots). Alternative method (small dots). Matrix in (4.2) and (5.1).

Figure 12: $n = 125$. $t = 72$. General method (big dots). Alternative method (small dots). Matrix in (4.2) and (5.1).

obtains by the two methods strongly lessens the importance of comparing the required CPU times.

Remark 5.3 It is possible that techniques for updating the QR factorization (see [16], §12.6), based on the observation that matrix (5.1) differs from a skew-symmetric matrix only by a rank-one matrix, can improve the performance of the alternative method [15].

n	t_n^*
30	18
50	26
90	48
125	72
160	100

n	t_{GM}	t_{AM}
5	2.09	2.09
10	2.03	2.00
15	2.20	2.05
40	3.74	2.63
105	48.51	8.42
200	427.04	39.48

Table 1: Machine precision t_n^* required by the alternative method.

Table 2: CPU times: general method (t_{GM}); alternative method (t_{AM}).

6 PROOFS

Proof of Theorem 2.1: By virtue of (2.3), the point $\vec{\zeta}_n$ belongs to the set A in (2.1b). Then it easily follows from (1.1) and (2.2) that one has

$$(p_n''/(2p_n'))(\zeta_{ni}) - R(\zeta_{ni}) = 0, \quad i = 1, 2, \ldots, n, \tag{6.1}$$

where R is the rational function introduced in (2.1e). Since one has

$$(p_n''/(2p_n'))(\zeta_{ni}) = \sum_{j=1,n; j \neq i} 1/(\zeta_{ni} - \zeta_{nj}), \tag{6.2}$$

the first part of the assertion is proved.

To prove the second part, note first that assumption $\vec{\zeta}_n \in A$ allows equality (6.2). Then use this equality to derive the n equalities (6.1) from the assumption $\mathbf{F}_n(\vec{\zeta}_n) = 0$ and, finally, rewrite these n equalities in the more convenient form

$$(a_2 p_n'' + a_1 p_n')(\zeta_{ni}) = 0, \quad i = 1, 2, \ldots, n.$$

It follows that the polynomial $a_2 p_n'' + a_1 p_n'$ vanishes at the zeros of p_n. Thus, an $a_0 \in \mathbb{P}$ must exist such that $a_2 p_n'' + a_1 p_n' = -a_0 p_n$. This concludes the proof. ∎

Preliminary notation and results are required to give the proofs of Theorems 2.2 and 2.3. We denote by $\omega_{n;i_1 i_2 \ldots i_h}$, $i_j \in \mathbb{N}^+$, $i_1 \neq i_2 \neq \cdots \neq i_h$, $1 \leq h \leq n$, the monic polynomial of degree $n - h$ defined by

$$\omega_{n;i_1 i_2 \cdots i_h}(z) := \omega_n(z)/((z - z_{i_1})(z - z_{i_2}) \cdots (z - z_{i_h})), \tag{6.3}$$

where, as usual,

$$\omega_n(z) := (z - z_1)(z - z_2) \cdots (z - z_n).$$

Also, we consider $\omega_n^{(h)}(z_i)$, the hth derivative of ω_n at the point z_i, and we shall be interested in regarding it as a function of all the coordinates $z_1, z_2, \ldots, z_n$ of $\mathbf{z}$. In this case, for the sake of clarity, we shall denote this function by Ω_{hi}:

$$\Omega_{hi} : \mathbb{C}^n \to \mathbb{C}, \qquad \Omega_{hi}(\mathbf{z}) := \omega_n^{(h)}(z_i). \tag{6.4}$$

Finally, according to (6.3) and (2.2b), we indicate by $p_{n;i_1 i_2 \cdots i_h}$ the polynomial defined by

$$p_{n;i_1 i_2 \cdots i_h}(z) := p_n(z)/((z - \zeta_{n i_1})(z - \zeta_{n i_2}) \cdots (z - \zeta_{n i_h})).$$

Lemma 6.1 *One has*

$$\omega_n^{(h)}(z_i) = h\omega_{n;i}^{(h-1)}(z_i), \tag{6.5}$$

$$(\partial \Omega_{hi}/\partial z_j)(\mathbf{z}) = -h\omega_{n;ij}^{(h-1)}(z_i), \; j \neq i. \tag{6.6}$$

Proof: The proof follows by standard arguments. Thus, it is omitted. ∎

Lemma 6.2 *Suppose that (2.2) and (2.3) are satisfied. Then*

$$(\mathbf{J_{F_n}}(\vec{\zeta_n}))_{ij} = G_i(\vec{\zeta_n}) \, (Lp_{n;j})(\zeta_{ni}), \; i = 1, 2, \ldots, n, \quad j = 1, 2, \ldots, n, \tag{6.7}$$

where G_i is the function defined by

$$G_i : A \subset \mathbb{C}^n \to \mathbb{C}, \quad G_i(\mathbf{z}) := 1/(2a_2(z_i)\Omega_{1i}(\mathbf{z})) = 1/(2a_2(z_i)\omega_n'(z_i)), \tag{6.8}$$

and A is the set defined in (2.1b).

Proof: It follows from (6.2) that

$$\sum_{j=1, n; j \neq i} 1/(z_i - z_j) = \omega_n''(z_i)/(2\omega_n'(z_i)) \tag{6.9}$$

for every $\mathbf{z} = (z_1, z_2, \ldots, z_n) \in A$. Thus, substituting this equation in (2.1d) and taking (2.1e) and (6.8) into account, yields

$$F_{ni}(\mathbf{z}) = -G_i(\mathbf{z}) \, (a_1 \omega_n' + a_2 \omega_n'')(z_i), \quad i = 1, 2, \ldots, n. \tag{6.10}$$

It is now convenient to begin by differentiating (6.10) with respect to z_j in the case of $j \neq i$. By using (6.4) and (6.6), one obtains

$$(\partial F_{ni}/\partial z_j)(\mathbf{z}) = -G_i(\mathbf{z}) \, (-a_1 \omega_{n;ij} - 2a_2 \omega_{n;ij}')(z_i) - F_{ni}(\mathbf{z}) \, (\partial G_i/\partial z_j)(\mathbf{z})/G_i(\mathbf{z}).$$

Moreover, by replacing ω_n with $\omega_{n;j}$ in (6.5) and by letting $h = 1$ and $h = 2$, it is immediately seen that $\omega_{n;ij}(z_i) = \omega'_{n;j}(z_i)$ and $2\omega'_{n;ij}(z_i) = \omega''_{n;j}(z_i)$. Thus, one finally gets

$$(\mathbf{J_{F_n}}(\mathbf{z}))_{ij} = G_i(\mathbf{z}) \, (a_1\omega'_{n;j} + a_2\omega''_{n;j})(z_i) -$$

$$F_{ni}(\mathbf{z}) \, (\partial G_i/\partial z_j)(\mathbf{z})/G_i(\mathbf{z}), \; j \neq i. \tag{6.11}$$

For $\mathbf{z} = \vec{\zeta}_n$ the last term in (6.11) vanishes and $\omega_{n;j}$ becomes $p_{n;j}$. This easily leads to assertion (6.7) in the case of $j \neq i$.

This partial result can be used to complete the proof. Actually, (2.12) shows that

$$(\mathbf{J_{F_n}}(\vec{\zeta}_n))_{ii} = R'(\zeta_{ni}) - \sum_{j=1,n;j\neq i} (\mathbf{J_{F_n}}(\vec{\zeta}_n))_{ij}.$$

Consequently, taking also (2.1e) into account, one can write

$$(\mathbf{J_{F_n}}(\vec{\zeta}_n))_{ii} = ((a_1a'_2/a_2 - a'_1)/(2a_2))(\zeta_{ni}) - G_i(\vec{\zeta}_n) \sum_{j=1,n;j\neq i} (Lp_{n;j})(\zeta_{ni}). \tag{6.12}$$

Moreover, for $\mathbf{z} = \vec{\zeta}_n$, (6.9) says that

$$\sum_{j=1,n;j\neq i} 1/(\zeta_{ni} - \zeta_{nj}) = (p''_n/(2p'_n))(\zeta_{ni}),$$

and this equality leads to

$$0 = F_{ni}(\vec{\zeta}_n) = -(p''_n/(2p'_n))(\zeta_{ni}) + R(\zeta_{ni}) = -(p''_n/(2p'_n) + a_1/(2a_2))(\zeta_{ni}),$$

i.e., $(a_1/a_2)(\zeta_{ni}) = -(p''_n/p'_n)(\zeta_{ni})$. Substituting this in (6.12) gives

$$(\mathbf{J_{F_n}}(\vec{\zeta}_n))_{ii} = G_i(\vec{\zeta}_n) \, \{(-a'_2p''_n - a'_1p'_n)(\zeta_{ni}) - \sum_{j=1,n;j\neq i} (Lp_{n;j})(\zeta_{ni})\},$$

and since $(-a'_2p''_n - a'_1p'_n)(\zeta_{ni}) = (-a'_2p''_n - a'_1p'_n + (Lp_n)')(\zeta_{ni}) = (Lp'_n)(\zeta_{ni})$, one obtains

$$(\mathbf{J_{F_n}}(\vec{\zeta}_n))_{ii} = G_i(\vec{\zeta}_n) \, \{(Lp'_n)(\zeta_{ni}) - \sum_{j=1,n;j\neq i} (Lp_{n;j})(\zeta_{ni})\} =$$

$$= G_i(\vec{\zeta}_n) \, (L(p'_n - \sum_{j=1,n;j\neq i} p_{n;j}))(\zeta_{ni}) = G_i(\vec{\zeta}_n) \, (Lp_{n;i})(\zeta_{ni}).$$

This concludes the proof. ∎

Immediate consequences of Lemma 6.2 are pointed out in the following two corollaries. Only the second one merits a brief proof.

Corollary 6.1 Let (2.2) and (2.3) be satisfied. Let $\mathbf{L}$ be the matrix defined by

$$\mathbf{L} := [\ell_{ij}], \quad \ell_{ij} := (Lp_{n;j})(\zeta_{ni}), \ i = 1, 2 \ldots, n, \ j = 1, 2 \ldots, n. \tag{6.13}$$

Then $\operatorname{rank}(\mathbf{J_{F_n}}(\vec{\zeta}_n)) = \operatorname{rank}(\mathbf{L})$.

Corollary 6.2 Suppose that (2.2) and (2.3) are satisfied, but assume that (2.4) does not hold. Then $\det \mathbf{J_{F_n}}(\vec{\zeta}_n) = 0$.

Proof: Denying (2.4) is the same as saying that the restriction of L to $\mathbb{P}^{n-1}$ is not an invertible operator. This means that the n polynomials $Lp_{n;j}$, $j = 1, 2, \ldots, n$, which belong to $L(\mathbb{P}^{n-1})$, are linearly dependent. Thus, a column in the matrix in (6.13) is a linear combination of the remaining ones. This, by virtue of Corollary 6.1, concludes the proof. ∎

Also, it is convenient to take note of the following lemma, which is a direct consequence of the definition (6.13) of the matrix $\mathbf{L}$.

Lemma 6.3 Let $\mathbf{L}$ be the matrix in (6.13). Moreover, let P belong to $L(\mathbb{P}^{n-1})$, let $w \in \mathbb{P}^{n-1}$ be a counter-image of P and let $\mathbf{c} = (c_1, c_2, \ldots, c_n)^T \in \mathbb{C}^n$ be the vector of the coefficients of w in the basis $\{p_{n;j}\}_{j=1,2,\ldots,n}$

$$P := Lw \in L(\mathbb{P}^{n-1}), \ w := \sum_{j=1,n} c_j p_{n;j} \in \mathbb{P}^{n-1}, \ \mathbf{c} := (c_1, c_2, \ldots, c_n)^T \in \mathbb{C}^n.$$

Then

$$P(\zeta_{ni}) = 0, \ i = 1, 2, \ldots, n,$$

if and only if $\mathbf{c}$ is such that

$$\mathbf{Lc} = 0.$$

Now we are in a position to give a brief proof.

Proof of Theorem 2.2: First it is convenient to introduce the notation $\mathbf{Q}_{n0} := \{0\}$. This enables us to include the first assertion in the statement of the theorem, the one related to the case $k = 0$, in the following one and, consequently, to unify the proof.

Then a simple argument shows that, no matter what $k \geq 0$ is, one can write

$$Lw = P \in \mathbf{Q}_{nk} \Leftrightarrow P(\zeta_{ni}) = 0, \quad i = 1, 2, \ldots, n.$$

Thus, by virtue of Lemma 6.3, (2.11) implies $\det \mathbf{L} \neq 0$ and vice versa. The proof follows then by invoking again Corollary 6.1. ∎

Remark 6.1 To better understand Theorem 2.2, it is worth noting that from (1.1) and (2.9), it easily follows

$$L(\mathbb{P}^{n-1}) \subseteq \mathbb{P}^{n+k-1} \tag{6.14}$$

(with equality in (6.14) only if (2.4) is satisfied and $k = 0$) and that in the statement of the theorem, condition (2.11) can be replaced by the following one

$$\mathbb{P}^{n+k-1} = L(\mathbb{P}^{n-1}) \oplus \mathbf{Q}_{nk}.$$

Proof of Theorem 2.3: Consider the matrix defined by (6.13). Then

$$\text{rank}(\mathbf{L}) = \text{rank}(\mathbf{J}_{\mathbf{F}_n}(\vec{\zeta}_n))$$

by virtue of Corollary 6.1 and

$$\dim(\ker(\mathbf{L})) = \dim(L(\mathbb{P}^{n-1}) \cap \mathbf{Q}_{nk}) + d$$

as a straightforward consequence of Lemma 6.3 and of a standard argument of linear algebra. Thus,

$$n = \text{rank}(\mathbf{L}) + \dim(\ker(\mathbf{L})) =$$

$$\text{rank}(\mathbf{J}_{\mathbf{F}_n}(\vec{\zeta}_n)) + \dim(L(\mathbb{P}^{n-1}) \cap \mathbf{Q}_{nk}) + d \leq \text{rank}(\mathbf{J}_{\mathbf{F}_n}(\vec{\zeta}_n)) + k + d,$$

and this proves the lower bound in the assertion. Because the upper bound is an obvious consequence of Theorem 2.2 in the case $d = 0$, and of Corollary 6.2, in the case $d = 1$, the proof is concluded. ∎

Acknowledgements. It is a pleasure to thank Dr. Nicola Nunzio Cafarelli of the University of L'Aquila for his various computations and graphics which were a great help to carry out the comparison in §5.

REFERENCES

[1] Bochner S. Über Sturm-Liouvillesche Polynomsysteme. *Math. Zeit.*, **29**:730–736, 1929.

[2] de Bruin M.G. Convergence in the Padé table for $_1F_1(1; c; x)$. *Nederl. Akad. Wetensch., Proc. Ser. A*, **79**. Also published in *Indag. Math.*, **38**:408–418, 1976.

[3] de Bruin M.G., Saff E.B., Varga R.S. On the zeros of the generalized Bessel polynomials I & II. *Nederl. Akad. Wetensch., Proc. Ser. A*, **84**(1). Also published in *Indag. Math.*, **43**(1), 1981.

[4] Carpenter A.J. Asymptotics for the zeros of the generalized Bessel polynomials. *Numer. Math.*, **62**:465–482, 1992.

[5] Decker D.W., Kelley C.T. Newton's method at singular points I. *SIAM J. Numer. Anal.*, **17**:66–70, 1980.

[6] Decker D.W., Kelley C.T. Newton's method at singular points II. *SIAM J. Numer. Anal.*, **17**:465–471, 1980.

[7] Decker D.W., Keller H.B., Kelley C.T. Convergence rates for Newton's method at singular points. *SIAM J. Numer. Anal.*, **20**:296–314, 1983.

[8] Dočev K. On the generalized Bessel polynomials. *Bulgar. Akad. Nauk. Izv. Mat. Inst.*, **6**:89–94, 1962.

[9] Dubrulle A. A Short Note on the implicit QL algorithm for symmetric tridiagonal matrices. *Numer. Math.*, **15**:450, 1970.

[10] Dubrulle A., Martin R.S., Wilkinson J.H. The implicit QL algorithm. In J.H.Wilkinson and C. Reinsch, *Linear Algebra, Handbook for Automatic Computation*, Vol. II, pages 241–248. Springer-Verlag, Berlin, Heidelberg, New York 1971.

[11] Forsythe G.E., Malcolm M.A., Moler C.B. *Computer Methods for Mathematical Computations*. Prentice Hall, Englewood Cliffs, N.J., 1977.

[12] Gautschi W. On the zeros of polynomials orthogonal on the semicircle. *SIAM J. Math. Anal.*, **20**:738–743, 1989.

[13] Gautschi W., Landau H.J., Milovanović G.V. Polynomials orthogonal on the semicircle, II. *Constr. Approx.*, **3**:389–404, 1987.

[14] Gautschi W., Milovanović G.V. Polynomials orthogonal on the semicircle. *J. Approx. Theory*, **46**:230–250, 1986.

[15] Golub G.H. Private communication at the Conference.

[16] Golub G.H., Van Loan C.F. *Matrix Computation*, Second edition. The Johns Hopkins University Press, Baltimore and London, 1989.

[17] Golub G.H., Wilkinson J.H. Ill-Conditioned eigensystems and the computation of the Jordan canonical form. *SIAM Review*, **18**:578–619, 1976.

[18] Griewank A. Starlike domains of convergence for Newton's method at singularities. *Numer. Math.*, **35**:95–111, 1980.

[19] Griewank A., Osborne M.R. Newton's method for singular problems when the dimension of the nullspace is > 1. *SIAM J. Numer. Anal.*, **18**:145–149, 1981.

[20] Griewank A., Osborne M.R. Analysis of Newton's method at irregular singularities. *SIAM J. Numer. Anal.*, **20**:747–773, 1983.

[21] Grosswald E. *Bessel Polynomials*. Lecture Notes in Mathematics **698**, Springer-Verlag, New York, 1978.

[22] Grosswald E. Recent applications of some old work of Laguerre. *Amer. Math. Monthly*, **86**:648–658, 1979.

[23] Krall H.L., Frink O. A New Class of orthogonal polynomials: the Bessel Polynomials. *Trans. Amer. Math. Soc.*, **65**:100–115, 1949.

[24] Lakshmikanthám V., Trigiante D. *Theory of Difference Equations: Numerical Methods and Applications.* Academic Press, New York, 1988.

[25] Littlejohn L.L., Shore S.D. Non classical orthogonal polynomials as solutions to second order differential equations. *Can. Math. Bull.*, **25**:291–295, 1982.

[26] Luke Y.L. *Special Functions and their Applications*, Vol. 2. Academic Press, Inc., 1969.

[27] Marcellán F., Ronveaux A. On a class of polynomials orthogonal with respect to a Sobolev inner product. *Indag. Math. (N.S)*, **1**:451–464, 1990.

[28] Maroni P. Prolégomènes à l'étude des polynômes orthogonaux semi-classique. *Ann. Mat. Pura Appl.*, **149**:165–184, 1987.

[29] Martinez J.R. Transfer functions of generalized Bessel polynomials. *IEEE CAS*, **24**:325–328, 1977.

[30] Martin R.S., Peters G., Wilkinson J.H. The QR algorithm for real Hessenberg matrices. *Numer. Math.*, **14**:219–231, 1970. Also published in J.H.Wilkinson and C. Reinsch, *Linear Algebra, Handbook for Automatic Computation*, Vol. II, pages 359–371. Springer-Verlag, Berlin, Heidelberg, New York 1971.

[31] Martin R.S., Wilkinson J.H. The implicit QL algorithm. *Numer. Math.*, **12**:377–383, 1968.

[32] Martin R.S., Wilkinson J.H. The modified LR algorithm for complex Hessenberg matrices. *Numer. Math.*, **12**:369–376, 1968. Also published in J.H.Wilkinson and C. Reinsch, *Linear Algebra, Handbook for Automatic Computation*, Vol. II, pages 396–403. Springer-Verlag, Berlin, Heidelberg, New York 1971.

[33] Olver F.W.J. The asymptotic expansions of Bessel functions of large order. *Phil. Trans. Roy. Soc. London*, Ser. A, **247**:338–368, 1954.

[34] Pasquini L. Polynomial solution to second order linear homogeneous ordinary differential equations. Properties and Approximation. *Calcolo*, **26**:165–183, 1989.

[35] Pasquini L. Computation of the zeros of the orthogonal polynomials. In *Orthogonal Polynomials and their applications*, pages 359–364. J.C.Bolzer AG, Scientific Publishing Co, IMACS 1991. C.Brezinski, L.Gori and A.Ronveaux, eds.

[36] Pasquini L. On the simultaneous computation of the zeros of semi-classical orthogonal polynomials. In preparation.

[37] Rall L.B. Convergence of the Newton process to multiple solutions. *Numer. Math.*, **9**:23–37, 1966.

[38] Reddien G.W. On Newton's method for singular problems. *SIAM J. Numer. Anal.*, **15**:993–996, 1978.

[39] Reddien G.W. Newton's method and high order singularities. *Comput. Math. with Appl.*, **5**:79–86, 1979.

[40] Reichel L., Trefethen L.N. Eigenvalues and pseudo-eigenvalues of Toeplitz matrices. *Linear Algebra Appl.*, **162–164**:153–185, 1992.

[41] Ronveaux A., Marcellán F. Differential equation for classical-type orthogonal polynomials. *Can. Math. Bull.*, **32**:404–411, 1989.

[42] Saff E.B., Varga R.S. Zero-free parabolic regions for sequences of polynomials. *SIAM J. Math. Anal.*, **7**:344–357, 1976.

[43] Saff E.B., Varga R.S. On the sharpness of theorems concerning zero-free regions for certain sequences of polynomials. *Numer. Math.*, **26**:345–354, 1976.

[44] Szegö G. *Orthogonal Polynomials*. Amer. Math. Soc. Colloq. Publ. **23**, American Mathematical Society, Providence, RI, 1975.

[45] Trefethen L.N. Pseudospectra of matrices. In *Numerical Analysis 1991*, pages 234–266. Longman **260**, 1992. D. F. Griffiths and G. A. Watson, eds.

[46] Underhill C. On the zeros of generalized Bessel polynomials. Internal note, Univ. of Salford, 1972.

[47] Van Iseghem J. Polynomes orthogonaux. *Pub. IRMA, Lille*, **16**, II, 1988.

International Series of Numerical Mathematics, Vol. 119, ©1994 Birkhäuser

SOME PROPERTIES OF POLYNOMIALS BOUNDED AT EQUALLY SPACED POINTS

Theodore J. Rivlin*

Dedicated affectionately to Walter Gautschi on his 65th birthday

Abstract. Suppose the absolute value of a real polynomial, $p(x)$, of degree d is bounded by 1 at k equally spaced points of the real line. For pairs (d, k) we present some results about how large are: (i) the absolute value of $p(x)$ for a given real x; (ii) the maximum norm of $p(x)$ on the span of the k points; (iii) the absolute values of the coefficients.

1   INTRODUCTION

Let $\mathcal{P}_d$ denote the real polynomials of degree at most d. Suppose $t_0 < t_1 < \cdots < t_m$ are given points of the real line. Let

$$P(d, m) = \{p \in \mathcal{P}_d : |p(t_i)| \leq 1, \ i = 0, \dots, m\}, \tag{1}$$

and for $p \in \mathcal{P}_d$

$$\|p\| := \max\{|p(x)| : \ t_0 \leq x \leq t_m\}. \tag{2}$$

For $p \in P(d, m)$ we consider the following norms of linear functionals:

$$B(d, m, x) := \max |p(x)|, \ x \text{ fixed and real} \tag{3}$$

$$B(d, m) := \max \|p\|, \tag{4}$$

and

$$A(k; d, m) := \max |a_k(p)|, \ k = 0, 1, \dots, d, \text{ where } p(x) = a_0 + a_1 x + \cdots + a_d x^d. \tag{5}$$

This work is concerned mainly with the case of *equally spaced points*, i.e., $t_{i+1} - t_i = a > 0$, $i = 0, \dots, m - 1$. When $m < d$ the functionals (3), (4) and (5) are trivial, and so we restrict our attention to $m \geq d$. What follows is a selective survey of the behavior of $B(d, m, x)$, $B(d, m)$ and $A(k; d, m)$ which is by no means comprehensive. We begin with the simplest case, $m = d$, in Section 1, as a warm-up, (or, as more appropriate to this occasion, a "finger exercise"). Section 2 is then devoted to the general case, $m \geq d$.

*IBM Research, T.J. Watson Research Center, P.O. Box 218, Yorktown Heights, NY 10598-0218, USA

2 THE CASE $m = d$

The study of $B(d, d, x)$ and $B(d, d)$ is, clearly, equivalent to the examination of the *Lebesgue function* and *Lebesgue constant* of polynomial interpolation theory. Namely, let $\{t_0, t_1, \ldots, t_d\} = \tau_d$ then

$$B(d, d, x) = \lambda_d(\tau_d, x) := \sum_{i=0}^{d} |\ell_i(x, \tau_d)|, \tag{6}$$

where, if $\omega(x) = (x - t_0) \ldots (x - t_d)$ (and we dispense with the τ_d),

$$\ell_i(x) = \frac{\omega(x)}{\omega'(t_i)(x - t_i)}, \quad i = 0, 1, \ldots, d, \tag{7}$$

and

$$B(d, d) = \Lambda_d := \max\{\lambda_d(x) : t_0 \leq x \leq t_d\}. \tag{8}$$

In the case of equally spaced points, which is our major concern, the behavior of $\lambda_d(x)$ and Λ_d has been studied extensively throughout this century. We next present a sampling of some results. It is clear that in the case of equally spaced points $B(d, d, x)$ and $B(d, d)$ are invariant under a linear transformation of the sampling points (and the variable). Thus we need only consider the case of $t_0 = -1, t_d = 1$, i.e. $t_j = -1 + (2j)/d$, $j = 0, 1, \ldots, d$.

In 1901, Runge [14] showed that the interpolating polynomial to an analytic function, such as $1/(1 + 25x^2)$, at $d + 1$ equally spaced points of $[-1, 1]$, was unbounded as $d \to \infty$, thus implying that $B(d, d)$ was unbounded as $d \to \infty$. In a much neglected work of 1914, Heinrich Tietze [16] examined $B(d, d, x)$ and showed that the single humps that form the graph of this function (which are symmetric about the origin) have maxima which are monotonically decreasing towards the origin. This result was rediscovered and improved upon by Mills and Smith [9] in 1992, who showed that $B(d, d, x) > B(d, d, x - (2/d))$, $2/d < x < 1, x \neq t_j$. As Tietze observed, $B(d, d) = \max\{B(d, d, x) : -1 \leq x \leq 1\} = B(d, d, \bar{x})$ where $1 - \frac{2}{d} < \bar{x} < 1$, and he showed that

$$B(d, d) \geq \frac{2^{d-1} d^{-3/2}}{\sqrt{\pi} \log 2} \tag{9}$$

as $d \to \infty$. This was not good news about interpolation on equally spaced points. On the other hand, Tietze also showed that if

$$x' = \begin{cases} 0; & d \text{ even} \\ \pm\frac{1}{d}; & d \text{ odd} \end{cases}$$

then as $d \to \infty$

$$B(d, d, x') \sim \sum_{j=0}^{\left[\frac{d}{2}\right]} \left(\frac{1.3 \ldots (2j - 1)}{2.4.6 \ldots (2j)}\right)^2 \sim \frac{1}{\pi} \log d. \tag{10}$$

It is interesting that Tietze points out, that the sum in (10) is precisely the sharp upper bound which Landau [7] had obtained for $|a_0 + a_1 + \cdots + a_d|$ whenever

$$f(z) = \sum_{j=0}^{\infty} a_j z^j$$

is analytic in $|z| \le 1$ and $|f(z)| \le 1$ in $|z| \le 1$, in his famous paper of 1913.

By contrast, we remark that in the same year Faber [6] showed that for *any* choice of $t_0, \ldots, t_d$

$$B(d, d) > \frac{2}{\pi^2} \log d$$

and it is known (cf. Rivlin [12]) that for the zeros of $T_{d+1}(x)$, $t_{d-j} = \cos(2j + 1)\pi/2(d+1)$, $j = 0, \ldots, d$,

$$\frac{2}{\pi} \log(d+1) + .9 < B(d, d) < \frac{2}{\pi} \log(d+1) + 1.$$

Returning to equally spaced points, we record the result of Turetskii [17]

$$B(d, d) \sim \frac{2^{d+1}}{ed \log d}, \quad d \to \infty,$$

the slight improvement on this due to Schönhage [15]

$$B(d, d) \sim \frac{2^{d+1}}{ed(\log d + \gamma)}, \quad d \to \infty,$$

(γ is Euler's constant) and a further improvement due to Mills and Smith [10]. We also wish to mention an unpublished result of 1980 of R.-Q. Jia (communicated privately by C. de Boor, cf. Rivlin [12, p. 27]).

$$\frac{2^{d-1}}{2d(d-1)} \le B(d, d) \le 2^{d-1}, \quad d \ge 2.$$

There is an improvement on (10) due to Runck [13]. If $r > 0$ and

$$-\frac{r}{\sqrt{d}} \le x \le \frac{r}{\sqrt{d}}$$

then

$$B(d, d, x) \le \begin{cases} A(r) \log d + B(r) \\ \frac{1}{\pi}(\log d + 2e^{\frac{r^2}{2}} \log_3 d) + C(r) \end{cases} \tag{11}$$

where $A(r), B(r), C(r)$ do not depend on d.

Finally, we turn to $A(j; d, d)$. Suppose $0 \le t_0 < t_1 < \cdots < t_d$, not necessarily equally spaced. Put $\omega(x) = (x - t_0) \cdots (x - t_d)$, then

$$\omega'(t_i) = (-1)^{d-i} \prod_{j=0}^{i-1}(t_i - t_j) \prod_{j=i+1}^{d} (t_j - t_i) =: (-1)^{d-i}\sigma_i,$$

where $\sigma_i > 0$, $i = 0, \ldots, d$. Thus, for $i = 0, 1, \ldots, d$,

$$\ell_i(x) = \frac{(-1)^{d-i}\omega(x)}{\sigma_i(x - t_i)} = \frac{(-1)^{d-i}}{\sigma_i} \sum_{j=0}^{d}(-1)^{d-j}b_{i,j}x^j, \tag{12}$$

with $b_{i,j} \ge 0$, $j = 0, 1, \ldots, d$.

Suppose $\alpha_0, \ldots, \alpha_d$ satisfy $|\alpha_i| \le 1$, $i = 0, \ldots, d$, and for $q \in \mathcal{P}_d$, $q(t_i) = \alpha_i$, $i = 0, \ldots, d$. Then

$$q(x) = \sum_{i=0}^{d} \alpha_i \ell_i(x) = \sum_{i=0}^{d} \frac{(-1)^{d-i}\alpha_i}{\sigma_i} \sum_{j=0}^{d}(-1)^{d-j}b_{i,j}x^j$$

$$= \sum_{j=0}^{d} q_j x^j, \tag{13}$$

and

$$q_j = (-1)^{d-j} \sum_{i=0}^{d} \frac{(-1)^{d-i}\alpha_i}{\sigma_i}b_{i,j}, \quad j = 0, \ldots, d. \tag{14}$$

Hence

$$|q_j| \le \sum_{i=0}^{d} \frac{b_{i,j}}{\sigma_i}, \quad j = 0, \ldots, d. \tag{15}$$

But if $p \in \mathcal{P}_d$ and $p(t_i) = (-1)^{d-i}$, $i = 0, \ldots, d$,

$$p(x) = \sum_{j=0}^{d} p_j x^j$$

and q is replaced by p in (13), we obtain, upon putting $\alpha_i = (-1)^{d-i}$, $i = 0, \ldots, d$,

$$|q_j| \le |p_j|, \quad j = 0, \ldots, d.$$

It is easy to see that we have equality, if, and only if, $\alpha_i = \pm(-1)^{d-i}$, $i = 0, \ldots, d$. Thus we have established the following result.

Theorem If $0 \leq t_0 < t_1 < \cdots < t_d$ (not necessarily equally spaced) and if $q(x) = q_0 + q_1 x + \cdots + q_d x^d$ is in $P(d, d)$, then we have

$$A(k; d, d) = |p_k|, \quad k = 0, \ldots, d$$

if and only if $p(x) = p_0 + p_1 x + \cdots + p_d x^d \in P(d, d)$ and satisfies $p(t_i) = (-1)^{d-i}$ (or $(-1)^{d-i-1}$) for $i = 0, \ldots, d$.

The same result holds if $t_0 < t_1 < \cdots < t_d \leq 0$. However, it may fail if the interpolation points contain the origin in the interior of their span, as the following example shows. Suppose $t_0 = -1, t_1 = 0, t_2 = 1, t_3 = 2$. Then $p(x) = (4/3)x^3 - 2x^2 - (4/3)x + 1$ yields $p(t_i) = (-1)^{3-i}$, $i = 0, 1, 2, 3$. But if $q(t_0) = 1, q(t_1) = 1, q(t_2) = -1$ and $q(t_3) = 1$, then $q(x) = x^3 - x^2 - 2x + 1$ and $|q_1| > |p_1|$.

If the points $t_0, t_1, \ldots, t_d$ are symmetrically disposed about zero, i.e. $t_i + t_{d-i} = 0$, $i = 0, \ldots, d$ then an argument (essentially involving polynomials in x^2), similar to that given in the proof of the theorem, yields the following result. With notation as above,

$$|q_{d-2j}| + |q_{d-2j-1}| \leq |p_{d-2j}|, \quad 0 \leq j < d/2. \tag{16}$$

This result has the following history. In the early seventies I noticed that if $t_j = \cos(d-j)\pi/d$, $j = 0, \ldots, d$ (so that $p(x) = T_d(x)$) and $q(x)$ is as above, then (16) holds. I submitted this as a problem to the Problems and Solutions section of the American Mathematical Monthly and it appeared [11] in 1976. David G. Cantor [1] provided a solution of the general problem, i.e. (16), the next year.

3   THE CASE $m \geq d$

Suppose $m \geq d$. Micchelli and Rivlin [8] (see also Schönhage [15]) show that if $-1 \leq t_0 < \cdots < t_m \leq 1$ (not necessarily equally spaced), $\tau_m = \{t_0, \ldots, t_m\}$ and $\beta \subseteq \tau_m$ consists of $d+1$ distinct points of τ_m, then if $m \geq d$,

$$B(d, m, x) = \min\{\lambda_d(\beta, x) : \ \beta \subseteq \tau_m, |\beta| = d + 1\}, \tag{17}$$

and

$$B(d, m) = \max\{B(d, m, x) : \ t_0 \leq x \leq t_m\}. \tag{18}$$

In some sense this result solves our problem, but it does not give the useful bounds we are seeking, to which we shall turn shortly.

But first let us record a useful fact about (17) and (18) in the case of equally spaced points. Since we may, with no loss of generality, consider $t_j = j$, $j = 0, 1, \ldots, m$ or $t_j = j$, $j = 0, 1, \ldots, m + 1$, it is clear that $B(d, m) \geq B(d, m + 1)$ and $B(d, m, x) \geq B(d, m + 1, x)$. Henceforth, $t_0, t_1, \ldots, t_m$ are equally spaced points, unless an explicit statement to the contrary is made.

3.1 $B(d,m)$ and $B(d,m,x)$

Schönhage [15] shows that $B(d, m)$ is bounded for all (m, d) if $m > d^2$. His proof is quite elaborate. Ehlich and Zeller [3], [4] give an elegant and brief proof that $B(d, m)$ is bounded if $m > (2/\pi^2)d^2$, and Ehlich [5], using an explicit construction involving Zolotarev polynomials, showed that if $m = o(d^2)$, as $d \to \infty$ then $B(d, m) \to \infty$ as $d \to \infty$. Coppersmith and Rivlin [2] recently closed the gap in these results by establishing a converse to Ehlich's result. In fact, we showed that $B(d, m)$ is bounded, as $d \to \infty$, if, and only if,

$$\lim \frac{m}{d^2} = \delta > 0, \quad d \to \infty. \tag{19}$$

Moreover, if (19) holds

$$B(d, m) < ae^{b/\delta}, \tag{20}$$

while a constructed example shows that for $m = \delta d^2$

$$B(d, m) > \alpha e^{\beta/\delta}, \tag{21}$$

where a, b, α, β are positive constants. The proofs are rather long. Vértesi [18] provides a shorter proof of (19) based on a 1969 result of Erdös.

$B(d, m, x)$ poses a more difficult problem. Coppersmith and I have several results in this area which are not yet in finished form.

3.2 $A(k; d, m)$

I now conclude with the beginning. About four years ago Richard Beigel, of the Computer Science Department at Yale, asked me about bounds for $A(k; d, m)$ when $m > d^2$. He had obtained bounds of the form $O(d^{10}/km)^k$ when $t_j = j + 1, j = 0, \ldots, m - 1$, but thought that the exponent 10 was too large. In the course of examining this problem I became interested in the properties of the polynomials in $P(d, m)$ in the case of equally spaced points.

I observed that if $t_j = j, j = 0, \ldots, m$ and $m > d^2$, then

$$\left(\frac{2}{m}\right)^k \frac{T_d^{(k)}(1)}{k!} \leq A(k; d, m) \leq 2\left(\frac{2}{m}\right)^k \frac{T_d^{(k)}(1)}{k!}. \tag{22}$$

The upper bound uses the observation that if $q \in P(d, m)$, then since $m > d^2$, the bound given by Ehlich and Zeller [3], [4] yields $B(d, m) \leq 2$. Thus if $q(x) = q_0 + q_1 x + \cdots + q_d x^d$

$$|q_k| = |\frac{q^{(k)}(0)}{k!}| \leq 2\left(\frac{2}{m}\right)^k \frac{T_d^{(k)}(1)}{k!}. \tag{23}$$

This follows from the fact that $T_d(2(x/m) - 1)$ is the Chebyshev polynomial of the interval $[0, m]$ and the Duffin-Schaeffer theorem (cf. Rivlin [12, p. 136]). The lower bound is an immediate consequence of $T_d(2(x/m) - 1) \in P(d, m)$.

Now

$$\frac{T_d^{(k)}(1)}{k!} = \frac{2^k}{(2k)!}d^2(d^2 - 1^2)\ldots(d^2 - (k-1)^2)$$

$$= \frac{2^k}{(2k)!}d\frac{(d+k-1)!}{(d-k)!}, \tag{24}$$

and a careful examination of "Stirling bounds", above and below, yields

$$A(k;d,m) \le c\frac{2^{2k}}{(2k)!}\left(\frac{d^2 - (k-1)^2}{m}\right)^k, \tag{25}$$

for a constant $c > 2$ and $0 \le k \le d$. But in view of (24)

$$\frac{2^{2k}}{(2k)!}\left(\frac{d^2 - (k-1)^2}{m}\right)^k \le \frac{T_d^{(k)}(1)}{k!}, \quad 0 \le k \le d$$

which together with (25) gives, for $0 \le k \le d$,

$$1 \le \frac{A(k;d,m)}{\frac{2^{2k}}{(2k)!}\left(\frac{d^2 - (k-1)^2}{m}\right)^k} \le c.$$

REFERENCES

[1] Cantor D.G. Solution of Problem 6084. *American Math. Monthly*, **84**:832–833, 1977.

[2] Coppersmith D., Rivlin T.J. The growth of polynomials bounded at equally spaced points. *SIAM J. Math. Anal.*, **23**:970–983, 1992.

[3] Ehlich H., Zeller K. Schwankung von Polynomen zwischen Gitterpunkten. *Math. Zeitschr.*, **86**:41–44, 1964.

[4] Ehlich H., Zeller K. Numerische Abschätzung von Polynomen. *ZAMM*, **45**: T20–T22, 1965.

[5] Ehlich H. Polynome zwischen Gitterpunkten. *Math. Zeit.*, **93**:144–153, 1966.

[6] Faber G. Über die interpolatorische Darstellung stetiger Funktionen. *Jahresber. der deutschen Math. Verein*, **23**:190–210, 1914.

[7] Landau E. Abschätzung der Koeffizientsumme einer Potenzreihe (Zweite Abhandlung). *Archiv. der Math. u. Phys*, (3) **21**:250–255, 1913.

[8] Micchelli C.A., Rivlin T.J. Optimal recovery of best approximations. *Resultate Math.*, **3**:25–32, 1980.

[9] Mills T.M., Smith S.J. On the Lebesgue function for Lagrange interpolation with equidistant nodes. *J. Austral. Math. Soc.*, (Series A) **52**:111–118, 1992.

[10] Mills T.M., Smith S.J. The Lebesgue constant for Lagrange interpolation on equidistant nodes. *Numer. Math.*, **61**:111–115, 1992.

[11] Rivlin T.J. Proposal of Problem 6084. *American Math. Monthly*, **83**:292, 1976.

[12] Rivlin T.J. *Chebyshev Polynomials*, 2nd Edition. John Wiley & Sons, New York, 1990.

[13] Runck P.O. Über Konvergenzfragen bei Polynominterpolation mit äquidistanten Knoten. I. *J. Reine Angew. Math.*, **208**:51–69, 1961.

[14] Runge C. Über empirische Funktionen und die Interpolation zwischen äquidistanten Ordinaten. *Z. f. Math. u. Phys.*, **46**:224–243, 1901.

[15] Schönhage A. Fehlerfortpflantzung bei Interpolation. *Numer. Math.*, **3**:62–71, 1961.

[16] Tietze H. Eine Bemerkung zur Interpolation. *Z. angew. Math. u. Phys.*, **64**:74–90, 1914.

[17] Turetskii A.H. The bounding of polynomials prescribed at equally distributed points (Russian). *Proc. Pedag. Inst. Vitebsk*, **3**:117–127, 1940.

[18] Vértesi P. Remark on a theorem about polynomials. To appear.

International Series of Numerical Mathematics, Vol. 119, ©1994 Birkhäuser

Numerical Methods via Transformations

Frank Stenger[*]

Dedicated to Walter Gautschi on the occasion of his 65th birthday

Abstract. This article begins with the setting of equi-spaced approximation, and then connects these methods with various other numerical methods via simple transformations. In particular, one thus traverses the classes of methods to which Gautschi made many contributions. We also comment on the multistep formulas derived by Gautschi, based on exactness for trigonometric polynomials which deserves further study, owing to its potential power.

1 TRANSFORMATION PROCEDURES

Transformation procedures may be used to link up many numerical methods. We explore such transformations in this section.

1.1 The Finite Strip and the Trapezoidal Rule

Numerical methods can be connected to one another in various ways. Suppose that we start with the midordinate rule on $(-\pi/h, \pi/h)$, i.e.,

$$\mathcal{M}_{2N}(F) \equiv \frac{\pi}{hN} \sum_{j=-N+1}^{N} F((j-1/2)h_1); \quad h_1 = \frac{\pi}{hN}, \tag{1.1}$$

as a means to approximate an integral,

$$\mathcal{I}(F) \equiv \int_{-\pi/h}^{\pi/h} F(\xi)\, d\xi \approx \mathcal{M}_{2N}(F). \tag{1.2}$$

The usual form of the error of this approximation under the assumption that $F \in \mathbf{C}^2[-\pi/h, \pi/h]$ is

$$\mathcal{I}(F) - \mathcal{M}_{2N}(F) = \frac{2\pi}{h} F''(\xi) \frac{h_1^2}{24}, \tag{1.3}$$

for some $\xi \in (-\pi/h, \pi/h)$.

There is another form for the error of this approximation; it is derived under the assumption that F belongs to the space $\mathbf{P}(\delta, 2\pi/h)$ of functions that are analytic in the infinite strip

[*]Department of Computer Science, University of Utah, Salt Lake City, Utah 84112

$$\mathcal{D}_\delta \equiv \{\xi \in \mathbb{C} : |\Im\xi| < \delta\}, \tag{1.4}$$

and which are periodic, with period $2\pi/h$, i.e., $F(\xi + 2\pi/h) = F(\xi)$, $\xi \in \mathbb{R}$. It follows that every $F \in \mathbf{P}(\delta, 2\pi/h)$ has the representation

$$F(\xi) = \sum_{n=-\infty}^{\infty} c_n e^{inh\xi}, \quad \xi \in \mathcal{D}_\delta. \tag{1.5}$$

Moreover, it then follows (see (1.15) below), that if F is bounded by M in $\mathcal{D}_\delta$, then

$$|c_n| \leq M e^{-|n|h\delta}. \tag{1.6}$$

Using this bound, we may bound the error of approximation of $\mathcal{I}(F)$ by $\mathcal{M}_{2N}(F)$ as follows:

$$|\mathcal{I}(F) - \mathcal{M}_{2N}(F)| = \left| \sum_{\sigma=-\infty, \sigma\neq0}^{\infty} (-1)^{\sigma(1+N)} c_{2N\sigma} \right| \leq \frac{2M e^{-|2N|h\delta}}{1 - e^{-|2N|h\delta}}. \tag{1.7}$$

The coefficients c_n of the expansion (1.5) may be efficiently approximated by means of the formula (see [1])

$$c_n = \frac{h}{2\pi} \int_{-\pi/h}^{\pi/h} F(\xi) e^{-inh\xi} d\xi \approx \mathcal{M}_{2N}(e^{-inh\xi}F). \tag{1.8}$$

Similar remarks may be made with regard to the approximation of $\mathcal{I}(F)$ via the trapezoidal rule, $\mathcal{T}_{2N}(F)$, i.e.,

$$\mathcal{T}_{2N}(F) \equiv \frac{\pi}{hN} \left\{ \frac{1}{2}F(-\pi/h) + \sum_{j=-N+1}^{N-1} F(jh_1) + \frac{1}{2}F(\pi/h) \right\}, \quad h_1 = \frac{\pi}{hN}, \tag{1.9}$$

for which the error of approximation under the assumption that $F \in \mathbf{C}^2[-\pi/h, \pi/h]$ is

$$\mathcal{I}(F) - \mathcal{T}_{2N}(F) = \frac{-2\pi}{h} F''(\eta) \frac{h_1^2}{12}, \tag{1.10}$$

where η is some point belonging to $(-\pi/h, \pi/h)$. This same error of approximation may also be bounded by the extreme right hand side of (1.7) under the assumption that $F \in \mathbf{P}(\delta, 2\pi/h)$.

1.2 The Annulus and DFT

Let us now make the transformation

$$w = e^{ih\xi}, \quad (\Leftrightarrow \xi = \frac{1}{ih} \log(w)) \tag{1.11}$$

to transform the region $\mathcal{D}_\delta \cap \{\xi \in \mathbb{C} : |\Re\xi| \leq \pi/h\}$ into the annulus

$$\mathcal{A}_R = \{w \in \mathbb{C} : R^{-1} \leq |w| \leq R\}, \quad R = e^{\delta h}. \tag{1.12}$$

Although (1.11) is not a conformal transformation, it is a one-to-one transformation for the periodic functions $\mathbf{P}(\delta, 2\pi/h)$. Under (1.11), a function $F \in \mathbf{P}(\delta, 2\pi/h)$ is transformed into a function $G \in \mathbf{H}(\mathcal{A}_R)$, i.e., $F(\frac{1}{ih}\log(w)) = G(w)$, with $\mathbf{H}(\mathcal{A}_R)$ denoting the set of all functions analytic in the annulus $\mathcal{A}_R$. In $\mathcal{A}_R$, these functions have the expansion

$$G(w) = \sum_{n=-\infty}^{\infty} c_n w^n, \quad w \in \mathcal{A}_R. \tag{1.13}$$

Under (1.11), formula (1.2) becomes, after multiplication by $h/(2\pi)$,

$$\frac{1}{2\pi i} \int_{|w|=1} G(w) \frac{dw}{w} \approx \frac{1}{2N} \sum_{j=-N+1}^{N} G\left(\exp\left(\frac{i\pi(j - 1/2)}{N}\right)\right), \tag{1.14}$$

and, of course, for the case of $|G(w)| \leq M$ in $\mathcal{A}_R$, we have the well-known contour integral estimate

$$|c_n| \leq \frac{M}{R^{|n|}}, \tag{1.15}$$

which is equivalent to (1.6). Because of the left side of (1.14), we also have the well-known DFT (Discrete Fourier Transform) approximation

$$c_n \approx \frac{1}{2N} \sum_{j=-N+1}^{N} e^{-in(j-1/2)\pi/N} G\left(\exp\left(\frac{i\pi(j - 1/2)}{N}\right)\right). \tag{1.16}$$

1.3 The Ellipse and Finite Interval Formulas

We next take that subset $\mathbf{H}_e(\mathcal{A}_R)$ of $\mathbf{H}(\mathcal{A}_R)$ consisting of all those functions $G \in \mathbf{H}(\mathcal{A}_R)$ such that

$$G(1/w) = G(w). \tag{1.17}$$

The expansion (1.13) then simplifies to

$$G(w) = c_0 + 2 \sum_{n=1}^{\infty} c_n \left(\frac{w^n + w^{-n}}{2}\right). \tag{1.18}$$

Note that assumption (1.17) implies $F(-\xi) = F(\xi)$, with F the function given in (1.1) and (1.2).

The expansion (1.18) motivates the transformation

$$z = \frac{w + w^{-1}}{2}, \tag{1.19}$$

which transforms the function G defined in (1.18) to the function

$$H(z) = c_0 + 2 \sum_{n=1}^{\infty} c_n T_n(z), \tag{1.20}$$

i.e., $H(z) = G(z + \sqrt{z^2 - 1})$, and where the T_n denote the Chebyshev polynomials,

$$\begin{aligned} T_0(z) &= 1 \\ T_1(z) &= z \\ T_{n+1}(z) &= 2\, z\, T_n(z) - T_{n-1}(z), \quad n = 1, 2, \dots. \end{aligned} \tag{1.21}$$

The transformation (1.19) maps the annulus $\mathcal{A}_R$ onto the ellipse $\mathcal{E}_R$, where

$$\mathcal{E}_R = \left\{ z \in \mathbb{C} : z = (\rho e^{i\theta} + \rho^{-1} e^{-i\theta})/2, \quad 1 \le \rho < R, \ 0 \le \theta \le 2\pi \right\}. \tag{1.22}$$

Equation (1.19) is not a conformal transformation, although it is in fact a one-to-one transformation for the family of functions $\mathbf{H}_e(\mathcal{A}_R)$.

Under (1.19), and the assumption that $G(1/w) = G(w)$, the formulas (1.2) and (1.14) are equivalent to

$$\int_{-1}^{1} \frac{H(z)}{\sqrt{1 - z^2}} dz \approx \frac{\pi}{N} \sum_{j=1}^{N} H\left(\cos\left\{\frac{2j-1}{2N}\pi\right\}\right), \tag{1.23}$$

while (1.16) becomes

$$\begin{aligned} c_n &= \int_{-1}^{1} \frac{H(z)\, T_n(z)}{\sqrt{1 - z^2}} dz \\ &\approx \frac{\pi}{N} \sum_{j=1}^{N} H\left(\cos\left\{\frac{2j-1}{2N}\pi\right\}\right) T_n\left(\cos\left\{\frac{2j-1}{2N}\pi\right\}\right). \end{aligned} \tag{1.24}$$

1.4 The Infinite Strip and Sinc Approximation

Let d denote a positive number. As in (1.4), we define the infinite strip

$$\mathcal{D}_d = \left\{ \zeta \in \mathbb{C} : |\Im \zeta| < d \right\}. \tag{1.25}$$

Let $\mathbf{H}(\mathcal{D}_d)$ denote the family of all functions that are analytic in $\mathcal{D}_d$. Let the sinc function be defined by

$$\operatorname{sinc}(\zeta) = \frac{\sin(\pi\zeta)}{\pi\zeta}. \tag{1.26}$$

It may then be readily shown [6] that if h is a positive number, if $f \in \mathbf{H}(\mathcal{D}_d)$, and if $f \in \mathbf{L}^1(\mathbb{R})$, then

$$\sup_{\zeta \in \mathbb{R}} \left| f(\zeta) - \sum_{k \in \mathbb{Z}} f(kh) \operatorname{sinc}\{(\zeta - kh)/h\} \right| \le C e^{-\pi d/h}, \tag{1.27}$$

where C is a constant independent of h.

Now, let δ be a positive number, and let $\mathbf{L}_\delta(\mathcal{D}_d)$ denote the family of all functions f that are analytic in $\mathcal{D}_d$ and satisfy a decay condition of the form

$$|f(\zeta)| \leq C_1 \exp\{-\delta|\zeta|\} \tag{1.28}$$

for all $\zeta \in \mathbb{R}$. If N denotes a positive integer, if $f \in \mathbf{L}_\delta(\mathcal{D}_d)$, and if we select h by the expression

$$h = \left(\frac{\pi d}{\delta N}\right)^{1/2}, \tag{1.29}$$

then

$$\sup_{\zeta \in \mathbb{R}} \left| f(\zeta) - \sum_{|k| \in \mathbb{Z}} f(kh) \operatorname{sinc}\{(\pi/h)(\zeta - kh)\} \right| \leq C_3 N^{1/2} \exp\left\{-(\pi d\delta N)^{1/2}\right\}. \tag{1.30}$$

Now, let $f \in \mathbf{L}_\delta(\mathcal{D}_d)$. Then (approximately, see [6, Thm. 1.8.3]) F, the Fourier transform of f, i.e.,

$$F(\xi) = \int_\mathbb{R} f(\zeta) \, e^{i\xi\zeta} d\zeta, \tag{1.31}$$

belongs to the space $\mathbf{L}_d(\mathcal{D}_\delta)$. That is, the roles of the numbers d and δ are reversed.

If we take the Fourier transform of the approximating Sinc approximation of (1.27) above, we get the DFT approximation

$$F(\xi) \approx \begin{cases} h \sum_{k \in \mathbb{Z}} f(kh) e^{ikh\xi}, & \xi \in \mathbb{R}, |\xi| < \pi/h, \\ 0, & \xi \in \mathbb{R}, |\xi| > \pi/h. \end{cases} \tag{1.32}$$

The error of this approximation may be shown to be of the order of $\exp(-\pi d/h)$, uniformly on $\mathbb{R}$. We may note, in particular, by taking $\xi = 0$ in both (1.31) and in the first expression of (1.32), that we get the trapezoidal rule approximation to the integral of f over $\mathbb{R}$, i.e.,

$$\int_\mathbb{R} f(\zeta) \, d\zeta \approx h \sum_{k \in \mathbb{Z}} f(kh). \tag{1.33}$$

Moreover [6, Sec. 3.3], if $f \in \mathbf{L}_\delta(\mathcal{D}_d)$, we again select h according to formula (1.29) above, and we set

$$F_N(\xi) = \begin{cases} h \sum_{|k| < N} f(kh) e^{ikh\xi}, & \xi \in \mathbb{R}, |\xi| < \pi/h, \\ 0, & \xi \in \mathbb{R}, |\xi| > \pi/h. \end{cases} \tag{1.34}$$

Then

$$\sup_{\xi \in \mathbb{R}} |F(\xi) - F_N(\xi)| \leq C_4 N^{1/2} \exp\left\{-(\pi d \delta N)^{1/2}\right\},$$

$$\left| \int_{\mathbb{R}} f(\zeta)\, d\zeta - h \sum_{k=-N}^{N} f(kh) \right| \leq C_4 N^{1/2} \exp\left\{-(\pi d \delta N)^{1/2}\right\}. \tag{1.35}$$

where C_4 is a constant that is independent of N.

Finally we remark that conformal transformations of the strip $\mathcal{D}_d$ to other regions, combined with the above Sinc and DFT series, lead to other methods of accurate approximation over finite and semi-infinite intervals [4,5,6]. For example, let H be analytic and bounded in the lens-shaped region $\mathcal{D} = \{z \in \mathbb{C} : |z| < d\}$, and let $|H(z)| \leq C_4(1 - z^2)^\delta$ on $(-1, 1)$. Then the map $\phi(z) = \log\{(1+z)/(1-z)\}$ is a conformal transformation of $\mathcal{D}$ onto the strip $\mathcal{D}_d$ defined as in (1.25) above. Then, by setting $G(z) = H(z)/(1 - z^2)$ and letting h be defined as in (1.29), we set $z_j = (\exp(jh) - 1)/(\exp(jh) + 1)$ to get the results [6]

$$\sup_{z \in (-1,1)} \left| H(z) - \sum_{j=-N}^{N} H(z_j) \mathrm{sinc}\{[\phi(z) - jh]/h\} \right| \tag{1.36}$$

$$\leq C_5 N^{1/2} \exp\left\{-(\pi \delta N/2)^{1/2}\right\},$$

and

$$\left| \int_{-1}^{1} G(z)\, dz - h \sum_{j=-N}^{N} \frac{2\, e^{jh}}{(1 + e^{jh})^2} G(z_j) \right| \tag{1.37}$$

$$\leq C_5 N^{1/2} \exp\left\{-(\pi \delta N/2)^{1/2}\right\},$$

where C_5 is a constant independent of N. We remark that (1.36) follows directly from (1.30) upon making the conformal transformation $\zeta = \phi(z)$. The inequality (1.37) can be obtained from (1.36), upon multiplying the expression in absolute values in (1.36) by $(1 - z^2)^{-1}$ and integrating over $(-1, 1)$. The expression (1.37) can also be obtained by making the conformal transformation $\zeta = \phi(z)$ in (1.33).

2 GAUTSCHI'S RESEARCH

Many of Gautschi's fine contributions to numerical analysis are intimately related to the methods represented in §1.1 to §1.3 above. The representative methods of §1.3 readily generalize to Gaussian quadrature with arbitrary weight W, that generate a sequence of orthogonal polynomials $\{P_n\}$. Any function $H \in \mathbf{H}(\mathcal{E}_R)$ may be expanded via a series of these polynomials that converges in $\mathcal{E}_R$. Moreover, the three-term recurrence relations that these polynomials satisfy can be connected with a continued fraction expansion of the Stieltjes transform of W, a beautiful expansion that involves weighted Gaussian quadrature.

We also remark that all the numerical methods represented by §1.1 to §1.3 are highly sensitive to a knowledge of the period of the periodic function F of §1.1. If this period is known exactly, then we obtain the very rapid rate of convergence of the approximations given in (1.7). On the other hand, if this period is not known exactly, then this accurate approximation suddenly reduces to one with the slower rate of convergence given in (1.3) or (1.10). The presence of singularities at end-points of intervals also drastically reduce the rate of convergence of these methods. For example, in evaluating the integral $\int_{-1}^{1} e^x (1 - x^2)^{\alpha-1} dx$ by Gauss–Legendre quadrature, we would get excellent results whenever α is a positive integer, but very poor results whenever α is a positive number, but not an integer. In the latter case, the rate of convergence changes from an $\mathcal{O}(e^{-c_1 n})$ rate to an $\mathcal{O}(n^{-2\alpha})$ rate, using n-point Gauss–Legendre quadrature, with c_1 some constant that is independent of n. Gautschi remedied this situation in the case when the singularity of the integrand was explicitly known, through publication of his Gaussian quadrature algorithm [3]. The approximating methods represented by §1.4 do not converge as rapidly as the fastest rate of convergence of the methods of §1.1–1.3. On the other hand, the rate of convergence of the methods of §1.4 is not appreciably reduced even if the period $(2\pi/h)$ is not known exactly, or (see (1.36) above) if the function to be approximated has singularities at the end-points of the interval of approximation. The rate of convergence to zero of the error in both of these situations is $\mathcal{O}(e^{-c_2 n^{1/2}})$. In addition, the methods of §1.4 give information on the error introduced in the methods of §1.1–1.3, when the period is not known exactly.

Finally, we mention the beautiful multistep methods that Gautschi derived for solving initial value problems [2]. These methods are very powerful, although not as well-known as they ought to be. Instead of demanding exactness for polynomials of certain degree, Gautschi demanded exactness for trigonometric polynomials up to a certain degree. He then showed a remarkable property of these formulas, namely, that as the step-size h_1 tends to zero, these formulas converge at the same rate (as a function of h_1) as formulas based solely on polynomial precision. In [2] Gautschi gave examples of this phenomenon, illustrating the application of the methods to differential equation problems for which the period was not known exactly. He thus demonstrated the superiority of his derived formulas over standard multistep formulas based on polynomial precision. On the other hand, he did not illustrate the power of the new formulas when applied to problems for which the period is known exactly, thus enabling the use of a bound such as (1.7). Admittedly, to take advantage of such a bound, one would require the order of the multistep method to be large, and one would then need to study the stability of the methods. Such an endeavor would be a worthwhile undertaking.

REFERENCES

[1] Davis P.J. On the numerical integration of periodic analytic functions. In *On Numerical Approximation*, pages 303–313. University of Wisconsin Press, Madison, 1953. R. Langer, ed.

[2] Gautschi W. Numerical integration of ordinary differential equations based on trigonometric functions. *Numer. Math.*, **3**:381–397, 1961.

[3] Gautschi W. Algorithm 351—Gaussian quadrature formulas. *Comm. ACM*, **11**:432–436, 1968.

[4] Kowalski M., Sikorski K., Stenger F. *Selected Topics In Approximation And Computation*. To be published by Oxford University Press, 1994.

[5] Lund J., Bowers K.L. *Sinc Methods for Quadrature and Differential Equations*. SIAM, 1992.

[6] Stenger F. *Numerical Methods Based on Sinc and Analytic Functions*. Springer–Verlag, N.Y., 1993.

International Series of Numerical Mathematics, Vol. 119, ©1994 Birkhäuser

COMPUTATIONAL ASPECTS OF INCOMPLETE GAMMA FUNCTIONS WITH LARGE COMPLEX PARAMETERS

N.M. Temme[*]

Dedicated to Walter Gautschi on the occasion of his 65th birthday

Abstract. The incomplete gamma functions are defined by the integrals

$$\gamma(a,z) = \int_0^z t^{a-1} e^{-t}\, dt, \qquad \Gamma(a,z) = \int_z^\infty t^{a-1} e^{-t}\, dt,$$

where $\Re a > 0$. We present several series representations and other relations which can be used for numerical evaluation of these functions. In particular, asymptotic representations for large complex values of the parameter a and z are considered. The results of numerical tests and the set-up of an algorithm will appear in a future paper.

1 INTRODUCTION

The incomplete gamma functions are defined by

$$\gamma(a,z) = \int_0^z t^{a-1} e^{-t}\, dt, \quad \Gamma(a,z) = \int_z^\infty t^{a-1} e^{-t}\, dt. \tag{1.1}$$

For $\gamma(a,z)$ we assume the condition $\Re a > 0$, and with respect to z we assume $|\arg z| < \pi$. Analytical continuation with respect to both parameters is discussed below. It is useful to have the normalizations

$$P(a,z) = \frac{\gamma(a,z)}{\Gamma(a)}, \quad Q(a,z) = \frac{\Gamma(a,z)}{\Gamma(a)}, \tag{1.2}$$

of which the sum equals unity. P and Q are in particular used in statistics and probability theory in connection with the gamma distribution.

The point $z = 0$ is a singularity for the incomplete gamma functions, except when $a = 0, 1, 2, 3, \ldots$. The singularity at $z = 0$ becomes apparent in the representation

$$\gamma^*(a,z) = z^{-a} P(a,z) = \frac{z^{-a}}{\Gamma(a)} \gamma(a,z); \tag{1.3}$$

$\gamma^*(a,z)$ is an entire function of z and a.

---

[*]CWI, P.O. Box 94079, 1090 GB Amsterdam, The Netherlands; nicot@cwi.nl

The following recurrence relations are easily derived from the definitions in (1.1):

$$\gamma(a+1, z) = a\gamma(a, z) - z^a e^{-z}, \quad \Gamma(a+1, z) = a\gamma(a, z) + z^a e^{-z}. \qquad (1.4)$$

Some care is needed when the relations in (1.4) are used in numerical computations. When the parameters are positive, the first relation is unstable.

When a is a non-negative integer the incomplete gamma functions are very simple:

$$\begin{aligned} \gamma(n+1, z) &= n! \left[1 - e^{-z} e_n(z)\right] \\ \Gamma(n+1, z) &= n! \, e^{-z} e_n(z). \end{aligned} \qquad n = 0, 1, 2, \ldots, \qquad (1.6)$$

in which $e_n(z)$ is the first part of the Taylor series of the exponential function:

$$e_n(z) = \sum_{m=0}^{n} \frac{z^m}{m!}, \quad n = 0, 1, 2, \ldots .$$

The relations with the Kummer functions (confluent hypergeometric functions) are as follows:

$$\begin{aligned} \gamma(a, z) &= a^{-1} z^a e^{-z} M(1, a+1, z) \\ &= a^{-1} z^a M(a, a+1, -z), \\ \gamma^*(a, z) &= \frac{e^{-z}}{\Gamma(a+1)} M(1, a+1, z) \\ &= \frac{1}{\Gamma(a+1)} M(a, a+1, -z), \\ \Gamma(a, z) &= z^a e^{-z} U(1, a+1, z) \\ &= e^{-z} U(1-a, 1-a, z). \end{aligned} \qquad (1.6)$$

Important series expansions are

$$\gamma(a, z) = e^{-z} \sum_{n=0}^{\infty} \frac{z^{a+n}}{(a)_{n+1}} = \sum_{n=0}^{\infty} \frac{(-1)^n}{n!} \frac{z^{a+n}}{a+n}. \qquad (1.7)$$

The series are convergent for each complex a and z, with the exception of $a = 0, -1, -2, \ldots$. Speed of convergence of the first series depends on the ratio $|z/a|$; when $|z/a| < 1$ the convergence is quite fast.

When $z \gg a$, we can use an asymptotic expansion of $\Gamma(a, z)$. This expansion follows from the representation

$$\Gamma(a, z) = z^a e^{-z} \int_0^{\infty} (t+1)^{a-1} e^{-zt} \, dt, \qquad (1.8)$$

and integrating by parts. We obtain for $N = 0, 1, 2, \ldots$

$$\Gamma(a, z) = z^{a-1} e^{-z} \left[\sum_{n=0}^{N-1} \frac{(-1)^n (1-a)_n}{z^n} + \theta_N \frac{(-1)^N (1-a)_N}{z^N} \right], \qquad (1.9)$$

(when $N = 0$ the sum is empty). The quantity θ_N is the remainder:

$$\theta_N = z \int_0^\infty (t+1)^{a-N-1} e^{-zt} \, dt,$$

which in fact is an incomplete gamma function (see (1.8)):

$$\theta_N = z^{N+1-a} e^z \Gamma(a - N, z). \tag{1.10}$$

This relation also follows from repeated application of the recurrence relation in (1.4).

For positive values of the parameters we can obtain an interesting estimate for θ_N. We need the condition $z > a - N$. Then we write

$$\theta_N = z^{N+1-a} e^z \int_z^\infty t^{a-N-1} e^{-t} \, dt$$

and next we integrate with respect to $u = t - (a - N) \ln t$. We obtain

$$\theta_N = z^{N+1-a} e^z \int_{z_0}^\infty \frac{e^{-u}}{t - (a - N)} \, du < \frac{z}{z - (a - N)},$$

in which $z_0 = z - (a - N) \ln z$.

Expansion (1.9) is re-considered in Olver (1991), in particular in connection with the Stokes phenomenon and the role of exponentially small terms in this expansion. Expansion (1.9) is valid in the sector $|\arg z| < \frac{3}{2}\pi$, with $a \in \mathbb{C}$ fixed. Olver investigates the expansion at the Stokes line $|\arg z| = \pm\pi$ and gives a uniform asymptotic expansion of $\Gamma(a, z)$ which is valid in a much larger z–domain. In fact Olver gives an expansion of the remainder θ_N for large $|z|$ and N. Considering (1.10) we observe that this problem is related with the uniform expansion of the incomplete gamma function with both parameters large given in (5.2) and (5.3).

Another important tool for numerical computations is Legendre's continued fraction:

$$\Gamma(a, z) = \cfrac{e^{-z} z^a}{z + \cfrac{1 - a}{1 + \cfrac{1}{z + \cfrac{2 - a}{1 + \cfrac{2}{z + \cfrac{3 - a}{1 + \cfrac{}{\ddots}}}}}}} \tag{1.11}$$

This expansion converges for all $z \neq 0$, $|\arg z| < \pi$ and any complex value of a. It is a splendid addition to the asymptotic expansion (1.9). The continued fraction converges better as the ratio $|z/a|$ increases. An extensive discussion on the use of (1.11) and other continued fractions for the incomplete gamma functions is given in Jones & Thron (1985).

In the literature many articles on the evaluation of the incomplete gamma functions can be found. It is outside the scope of the present paper to give a full overview. In the recent survey Lozier & Olver (1994) references to many papers and software are given.

Apart from the series expansions, continued fractions, Chebyshev and Padé expansions mentioned in the present paper, one can use other numerical techniques. For instance, in Allasia & Besenghi (1987) numerical quadrature is used.

The incomplete gamma functions occur in the literature also as generalized exponential integrals, which are usually defined by

$$E_\nu(z) = \int_1^\infty t^{-\nu} e^{-zt}\, dt, \quad \nu \in \mathbb{C}, \quad \Re z > 0.$$

The relation with $\Gamma(a, z)$ is $E_\nu(z) = z^{\nu-1}\Gamma(1 - \nu, z)$. In a set of papers Chiccoli et al. (1987–1990) the generalized exponential integrals are extensively discussed, with emphasis on real ν and positive z.

2 ALGORITHMS FOR REAL PARAMETERS

When a and $z = x$ are positive one usually concentrates on the normalized functions $P(a, x)$ and $Q(a, x)$. First the smaller of these two is computed; the relation $P(a, x) + Q(a, x) = 1$ gives the other value. We have the following rule (with slight corrections for small values of x and a) :

$0 \le a \le x$: first compute Q, then $P = 1 - Q$;

$0 \le x < a$: first compute P, then $Q = 1 - P$.

This splitting of the (a, x)−domain follows from the asymptotic relation

$$P(a, a), Q(a, a) \sim \frac{1}{2}, \quad \text{as} \quad a \to \infty.$$

The method for Q is usually based on Legendre's continued fraction (1.11), say when $x > 1$. When $x \gg a$ the asymptotic expansion (1.9) may be used for Q. The computation of P may be based on the first series occurring in (1.7), that is,

$$P(a, x) = x^a e^{-x} \sum_{n=0}^\infty \frac{x^n}{\Gamma(a + n + 1)}. \tag{2.1}$$

For large values of the parameters x and a, especially when the parameters are nearly equal, the expansions for computing P and Q converge slowly. In that case an algorithm based on a uniform asymptotic expansion can be used; see §5. References for software for the real case are DiDonato & Morris (1986), Gautschi (1979) and Temme (1994).

3 OTHER SERIES EXPANSIONS

All expansions in this section give excellent approximations for restricted domains of the parameters, although the expansions are all valid in large domains of the complex plane. It is quite difficult to give a priori information on the applicability of the expansions in efficient and reliable numerical algorithms. As a rule, the expansions for $\gamma(a, z)$ should be used in a z−domain around the origin, while the expansions for $\Gamma(a, z)$ are of interest in a z−domain containing the point ∞.

Chebyshev expansions

Let $T_n(x)$ denote the Chebyshev polynomial (that is, $T_n(\cos\theta) = \cos n\theta$) and let $T_n^*(x) = T_n(2x - 1)$ be the shifted Chebyshev polynomial. From expansions of the Kummer function $M(a, c, z) = {}_1F_1(a, c, z)$ we derive, see Luke (1969, II, p. 35),

$$\gamma^*(a, zx) = \frac{e^{-zx}}{\Gamma(a + 1)} \sum_{n=0}^{\infty} C_n(z) T_n^*(x), \quad 0 \le x \le 1.$$

The coefficients can be computed by a backward recurrence algorithm using the relation

$$\frac{2C_n(z)}{\varepsilon_n} = [-1 + 4(n + 1 + a)/z] \, C_{n+1}(z) + [1 + 4(n + 2 - a)/z] C_{n+2}(z) + C_{n+3}(z),$$

where $\varepsilon_0 = 1$, $\varepsilon_n = 2$ if $n \ge 1$.

From expansions of the Kummer function $U(a, c, z) = \psi(a, c, z)$ we derive, see Luke (1969, II, p. 25),

$$\Gamma(a, \omega z) = (\omega z)^{a-1} e^{-\omega z} \sum_{n=0}^{\infty} C_n(z) T_n^*(1/\omega), \quad |\arg z| < \tfrac{3}{2}\pi, \quad 1 \le \omega \le \infty,$$

provided that $z \ne 0$, $a \ne 1, 2, 3, \ldots$. When a is a positive integer the incomplete gamma functions reduce to elementary functions; see (1.4). We have the recurrence relation

$$\frac{2(n + 1 - a)}{\varepsilon_n} C_n(z) = -(3n + 4 + 4z - a)C_{n+1}(z) - (3n + 5 - 4z + a)C_{n+2}(z)$$
$$- (n + 2 + a)C_{n+3}(z),$$

$n = 0, 1, 2, \ldots$. Luke gives the following estimate of these coefficients:

$$C_n(z) = \frac{4(-1)^n \sqrt{\pi} \, n^{-2a/3} z^{(3-2a)/6}}{\sqrt{3} \, \Gamma(1 - a)} e^{-3n^{2/3} z^{1/3}} \left[1 + \mathcal{O}(n^{-1/3})\right],$$

$|\arg z| < 3\pi/2$, from which follows excellent convergence, in particular when z is large. The speed of convergence of the backward recurrence algorithm deteriorates, however, when z approaches the negative real axis. Also, large complex values of a may have a bad influence on the rate of convergence of both the series expansion and the recurrence algorithms. When comparing the Chebyshev expansion with the continued fraction we found that the continued fraction is easier to implement, since in that case a forward recurrence relation for evaluating the continued fraction can be used. See Gautschi (1979), where the continued fraction is evaluated as a series. Controlling the error term is easier then, although a reliable stopping criterion for obtaining a certain accuracy is difficult to obtain for general complex values of the parameters.

Padé approximations for $\gamma^*(a, z)$

From Luke (1969, II, p. 190) we obtain

$$\gamma^*(a, -z) = \frac{e^z}{\Gamma(a+1)} \left[\frac{A_n}{B_n} + V_n \right],$$

where

$$A_0 = B_0 = 1, \quad A_1 = 1 - \frac{z}{(a+1)(a+2)}, \quad B_1 = 1 + \frac{z}{a+2},$$

and both A_n and B_n satisfy the recurrence formula

$$B_{n+1} = \left[1 + \frac{az}{(2n+a)(2n+a+2)} \right] B_n + \frac{n(n+a)}{(2n+a-1)(2n+a)^2(2n+a+1)} B_{n-1},$$

$$V_n = \frac{(-1)^{n+1}\pi\Gamma(a+1)n!\,\Gamma(n+a+1)z^{2n+1}e^\chi}{2^{4n+2a}(2n+a+1)\{\Gamma[n+(a+1)/2]\Gamma[(n+(a+2)/2]\}^2} \left[1 + \mathcal{O}\left(n^{-3}\right) \right],$$

as $n \to \infty$ with z and a fixed, and $\chi = -z + z(z+4a)/[4(2n+a+1)]$.

Bessel function expansions

We have

$$\gamma^*(a, z) = \frac{e^{-\frac{1}{2}z} \left(\frac{1}{4}z\right)^{-a+\frac{1}{2}} \Gamma\left(a-\frac{1}{2}\right)}{\Gamma(a+1)} \sum_{n=0}^{\infty} \frac{(2a-1)_n(a-1)_n(-1)^n}{n!(a+1)_n} I_{a-\frac{1}{2}+n}\left(\tfrac{1}{2}z\right),$$

$$\gamma^*(a, -z) = \frac{2\sqrt{\pi}\, e^{\frac{1}{2}z}}{\sqrt{z}\,\Gamma(a+1)} \sum_{n=0}^{\infty} \frac{(1-a)_n(-1)^n}{(a+1)_n} I_{n+\frac{1}{2}}\left(\tfrac{1}{2}z\right),$$

where in both cases $a \neq -1, -2, -3, \ldots$. More expansions of this kind are given in Luke (1969, II, p. 48).

Tricomi's expansions

Let

$$E_\nu(z) := z^{-\frac{1}{2}\nu} J_\nu\left(2\sqrt{z}\right),$$

the Bessel function without its algebraic singularity; $E_\nu(z)$ is an entire function of both z and ν, with $E_\nu(0) = 1/\Gamma(\nu + 1)$.

In Luke (1969, I, p. 129) Tricomi's expansions are cast into the form:

$$\frac{e^{-hz}}{\Gamma(c)} F(a, c, z) = \sum_{n=0}^{\infty} C_n z^n E_{m+c-1}(\omega z), \quad |z| < \infty,$$

where C_n depends only on the parameters a, c, h and ω, and are determined by the generating function

$$e^{\omega z}(1 + hz)^{a-c}[1 + (h - 1)z]^{-a} = \sum_{n=0}^{\infty} C_n z^n.$$

That is, $C_0 = 1, C_1 = \omega + a - ch, 2C_2 = C_1^2 + ch^2 - 2ha + a$, and we have the recurrence formula

$$(n + 1)C_{n+1} = [C_1 - (2h - 1)n]C_n + [\omega(2h - 1) - h(h - 1)(n + c - 1)]C_{n-1}$$
$$+ \omega h(h - 1)C_{n-2},$$

where $n = 2, 3, \ldots$. Here ω and h are free parameters. We interpret this expansion for the case of the incomplete gamma functions:

$$\gamma^*(a, z) = e^{(h-1)z} \sum_{n=0}^{\infty} A_n z^n E_{n+a}(\omega z),$$

with generating function

$$e^{\omega z}(1 + hz)^{-a}[1 + (h - 1)z]^{-1} = \sum_{n=0}^{\infty} A_n z^n.$$

That is, $A_0 = 1, A_1 = \omega + 1 - (a + 1)h, 2A_2 = A_1^2 + (a + 1)h^2 - 2h + 1$ with recurrence formula

$$(n+1)A_{n+1} = [A_1-(2h-1)n]A_n+[\omega(2h-1)-h(h-1)(n+a)]A_{n-1}+\omega h(h-1)A_{n-2}.$$

Tricomi considered ω as function of a and z in order to get a series that has asymptotic properties as $a \to \infty$; see Luke (1969, I, p. 129) for more details.

4 UNIFORM ASYMPTOTIC EXPANSIONS

We give expansions for $\Gamma(-a, z)$ and $\gamma^*(a, -z)$ which are suitable when both parameters a and z are large. Let $\lambda = z/a$; in this section we assume that $\lambda + 1$ is bounded away from zero. From Gautschi (1959) we obtain

$$
\begin{aligned}
z^a \Gamma(-a, z) &= \int_1^\infty e^{-a(\lambda t + \ln t)} g_0(t) \frac{dt}{t} \\
&= -\frac{1}{a} \int_1^\infty \frac{g_0(t)}{1 + \lambda t} de^{-a(\lambda t + \ln t)} \\
&= \frac{e^{-z}}{a} \frac{1}{\lambda + 1} + \frac{1}{a} \int_1^\infty e^{-a(\lambda t + \ln t)} g_1(t) \frac{dt}{t},
\end{aligned}
$$

where $g_0(t) = 1, g_1(t) = t \frac{d}{dt}[g_0(t)/(1 + \lambda t)]$. Continuing this we obtain

$$
z^a \Gamma(-a, z) = \frac{e^{-z}}{a} \sum_{n=0}^{N-1} \frac{C_n(\lambda)}{a^n} + \frac{R_N}{a^N}, \tag{4.1}
$$

where, for $n = 0, 1, 2, \ldots$,

$$
R_n = \int_1^\infty e^{-zt} t^{-a-1} g_n(t)\, dt, \quad g_n(t) = t \frac{d}{dt} \frac{g_{n-1}(t)}{1 + \lambda t}, \tag{4.2}
$$

and the coefficients $C_n(\lambda)$ are given by $C_n(\lambda) = \frac{g_n(1)}{\lambda + 1}$. We have the recursion

$$
C_n(\lambda) = \frac{\lambda}{\lambda + 1} \frac{d}{d\lambda} C_{n-1}(\lambda), \quad C_0(\lambda) = \frac{1}{\lambda + 1}.
$$

The first few coefficients are

$$
C_0(\lambda) = \frac{1}{\lambda + 1}, \quad C_1(\lambda) = \frac{-\lambda}{(\lambda + 1)^3}, \quad C_2(\lambda) = \frac{\lambda(2\lambda - 1)}{(\lambda + 1)^5},
$$

$$
C_3(\lambda) = \frac{-\lambda(6\lambda^2 - 8\lambda + 1)}{(\lambda + 1)^7},
$$

$$
C_4(\lambda) = \frac{\lambda(24\lambda^3 - 58\lambda^2 + 22\lambda - 1)}{(\lambda + 1)^9}.
$$

Gautschi computed (for positive parameters) numerical bounds for the first 7 remainders $R_n(\lambda)$ of this expansion. For complex parameters bounds are unavailable. The expansion remains valid when a becomes negative; in fact it holds for $\Gamma(a, z)$ with $a < z$ and and also for complex parameters.

A corresponding expansion follows from the integral

$$
\Gamma(a)\gamma^*(a, -z) = \int_0^1 e^{zt} t^{a-1}\, dt \tag{4.3}
$$

where $z \in \mathbb{C}$ and $\Re a > 0$. Integrating by parts we obtain

$$\Gamma(a)\gamma^*(a, -z) = e^z \sum_{n=0}^{N-1} (-1)^n \frac{C_n(\lambda)}{a^{n+1}} + \frac{S_N}{a^N}, \qquad (4.4)$$

with the same $C_n(\lambda)$ as given above. Both expansions are useful additions to the power series expansions and continued fraction. Just bounding the remainders gives poor information on the applicability of the expansions for general a and z.

More analytic insight gives the following. Consider

$$z^a \Gamma(-a, z) = \int_1^\infty e^{-a(\ln t + \lambda t)} \frac{dt}{t}.$$

Transform: $u = \ln t + \lambda(t - 1)$. Then,

$$z^a e^z \Gamma(-a, z) = \int_0^\infty e^{-au} \frac{du}{1 + \lambda t}. \qquad (4.5)$$

Similarly,

$$\Gamma(a)\gamma^*(a, -z) = \int_0^1 e^{a(\ln t + \lambda t)} \frac{dt}{t}.$$

Transform: $v = -\ln t - \lambda(t - 1)$. Then,

$$e^{-z} \Gamma(a)\gamma^*(a, -z) = \int_0^\infty e^{-av} \frac{dv}{1 + \lambda t}. \qquad (4.6)$$

Now the asymptotic tool is Watson's lemma. Expand

$$\frac{1}{1 + \lambda t} = \sum_{n=0}^\infty c_n(\lambda) u^n = \sum_{n=0}^\infty (-1)^n c_n(\lambda) v^n \qquad (4.7)$$

and substitute this in (4.5) and (4.6). This gives the above expansions (4.1) and (4.4). For both Laplace integrals we need to know the singular points. Clearly, $t = -1/\lambda$ is the source of the singularities. Corresponding $u, v-$values are:

$$u_\pm = -\lambda - 1 - \ln \lambda \pm i\pi, \quad v_\pm = \lambda + 1 + \ln \lambda \pm i\pi.$$

When these points are close to the origin, Watson's lemma fails. Observe that $u_\pm, v_\pm \to 0$ when $\lambda \to -1$. The radius of convergence of the series in (4.7) equals $\min |u_\pm|$.

For numerical applications we need a condition like $|u_\pm| > 1$. Next, let $\alpha_1 := \arg u_-, \alpha_2 := \arg u_+$ and assume that $\alpha_1 < 0, \alpha_2 > 0$, as is the case when starting with positive values of a and z. Then, according to Olver (1974, p. 114), Watson's lemma applied to (4.5) gives expansion (4.1) holding for a-values in the sector

$$-\alpha_2 - \tfrac{1}{2}\pi + \delta \leq \arg a \leq -\alpha_1 + \tfrac{1}{2}\pi - \delta,$$

where δ is a small positive number. Similarly for $\gamma^*(a, -z)$. It is quite difficult to visualize both conditions when the parameters are complex. A numerical verification is not difficult, however.

A first impression of the nature of this expansion follows by considering ratios of successive terms in the expansion. Let $\psi_n = C_n(\lambda)/a^n$. Then we have, when λ is large, $\psi_{n+1}/\psi_n \sim 1/(\lambda a) = 1/z$. When $|\lambda| \ll 1$ we have $\psi_{n+1}/\psi_n \sim 1/a$. We see that the scale $\{\psi_n\}$ is a uniform asymptotic scale as both a and z are large, except when λ tends to -1; large or small values of λ do not disturb the asymptotic nature of the scale.

We have derived the expansion for $\gamma^*(a, -z)$ using (4.3), under the condition $\Re a > 0$, since otherwise representation (4.3) does not converge at the origin. However, by using

$$\gamma^*(a, z) = \frac{\Gamma(1-a)}{2\pi i} \int_{-1}^{(0^+)} t^{a-1} e^{zt} \, dt, \tag{4.8}$$

where the contour starts at $t = -1$ and encircles the origin once in positive direction, with $\arg t = 0$ when $t > 0$, it is possible to obtain the expansion (4.4) without the restriction $\Re a > 0$. The integral in (4.8) is defined for all complex values of z and a. The poles of $\Gamma(1 - a)$ at $a = 1, 2, 3, \ldots$ are removable singularities in this representation, because the integral vanishes for these integer values of a. It is easily verified that

$$\gamma^*(-n, z) = z^n, \quad n = 0, 1, 2, \ldots .$$

5 FURTHER UNIFORM ASYMPTOTIC EXPANSIONS

Let η be the real number defined by

$$\tfrac{1}{2}\eta^2 = \lambda - 1 - \ln\lambda, \quad \lambda > 0, \quad \text{sign}(\eta) = \text{sign}(\lambda - 1). \tag{5.1}$$

Extend λ and η to complex values. Then we write, with $\lambda = z/a$,

$$Q(a, z) = \frac{\Gamma(a, z)}{\Gamma(a)} = \tfrac{1}{2} \, \text{erfc}(\eta\sqrt{a/2}) + R_a(\eta),$$

$$P(a, z) = \frac{\gamma(a, z)}{\Gamma(a)} = \tfrac{1}{2} \, \text{erfc}(-\eta\sqrt{a/2}) - R_a(\eta). \tag{5.2}$$

The error functions are the dominant terms, that describe the transition at $a = z$. Observe that the property $P + Q = 1$ is reflected in the error functions, since erfc z + erfc $(-z) = 2$, and in having the same term $R_a(\eta)$. We have

$$R_a(\eta) \sim \frac{e^{-\frac{1}{2}a\eta^2}}{\sqrt{2\pi a}} \sum_{n=0}^{\infty} \frac{C_n(\eta)}{a^n}, \quad \text{as} \quad a \to \infty, \tag{5.3}$$

uniformly with respect to η; in particular, $|\eta|$ may be small. Further investigations are needed to obtain the (a, z)–domain in which the above expansion can be used for numerical algorithms. For positive parameters, see Temme (1987), (1994).

We need a representation and expansion like (5.2), (5.3) in the left half planes. From Tricomi (1950) we have, with $\lambda = z/a$,

$$\gamma^*(-a, -z) = \frac{z^a \sin \pi a\, \Gamma(a+1)}{\pi} \int_z^a e^t\, \frac{dt}{t^{a+1}} + \lambda^a \gamma^*(-a, -a). \tag{5.4}$$

This can be proved by using series expansions, or the fact that $y(z) = z^{-a}\, \gamma^*(-a, -z)$ satisfies the equation $zy'' + (a + 1 - z)y' = 0$.

The integral can be transformed into

$$G_a(\eta) = e^{-\frac{1}{2}\eta^2} \int_0^{\eta} e^{\frac{1}{2}a\zeta^2} f(\zeta)\, d\zeta,$$

where η is given in (5.1). $G_a(\eta)$ reduces to Dawson's integral (error function with purely imaginary argument) if $f = 1$; Dawson's integral is the main approximant in the uniform expansion of $G_a(\eta)$. Furthermore a series like the one in (5.3) plays a role. Representation (5.4) and the uniform asymptotic expansion containing Dawson's integral have to be investigated for numerical applications. The method developed in Temme (1987) may be applicable to this new expansion.

Acknowledgment. The author wishes to thank the referee for suggesting several new references.

REFERENCES

[1] Allasia G., Besenghi R. Numerical calculation of incomplete gamma functions by the trapezoidal rule. *Numer. Math.*, 50:419–428, 1987.

[2] Chiccoli C., Lorenzutta S., Maino G. A numerical method for generalized exponential integrals. *Comput. Math. Applic.*, 14:261–268, 1987.

[3] Chiccoli C., Lorenzutta S., Maino G. On the evaluation of generalized exponential integrals $E_\nu(x)$. *J. Comput. Phys.*, 78:278–286, 1988.

[4] Chiccoli C., Lorenzutta S., Maino G. Calculation of exponential integrals of real order. *Int. J. Comput. Math.*, 31:125–135, 1990.

[5] Chiccoli C., Lorenzutta S., Maino G. On a Tricomi series representation for the generalized exponential integral. *Int. J. Comput. Math.*, **31**:257–262, 1990.

[6] Chiccoli C., Lorenzutta S., Maino G. An algorithm for exponential integrals of real order. *Computing*, **45**:269–276, 1990.

[7] DiDonato A.R., Morris Jr A.H. Computation of the incomplete gamma functions. *ACM Trans. Math. Software*, **12**:377–393, 1986.

[8] Gautschi W. Exponential integral $\int_1^\infty e^{-xt} t^{-n}\, dt$ for large values of n. *J. Res. Nat. Bur. Standards*, **62**:123–125, 1959.

[9] Gautschi W. Efficient computation of the complex error function. *SIAM J. Numer. Anal.*, **7**:187–198, 1970.

[10] Gautschi W. ALGO 363, The complex error function. *Comm. ACM*, **12**:280, 1970.

[11] Gautschi W. A computational procedure for incomplete gamma functions. *ACM Trans. Math. Software*, **5**:466–481, 1979. Algorithm 542, 482–489.

[12] Lozier D.W., Olver F.W.J. Numerical evaluation of special functions. In *Proceedings of the Symposium on Mathematics of Computation*, August 1993, Vancouver. *AMS Proc. Symp. in Appl. Math*, 1994. W. Gautschi, ed.

[13] Luke Y.L. *The special functions and their approximations*, Vols. I, II. Academic Press, New York, 1969.

[14] Olver F.W.J. *Asymptotics and special functions*. Academic Press, New York, 1974.

[15] Olver F.W.J. Uniform, exponentially improved, asymptotic expansions for the generalized exponential integral. *SIAM J. Math. Anal.*, **22**:1460–1474, 1991.

[16] Poppe G.P.M., Wijers C.M.J. More efficient computation of the complex error function. *ACM Trans. Math. Software*, **16**:38–46, 1990.

[17] Temme N.M. On the computation of the incomplete gamma functions for large values of the parameters, pages 479–489. In *Algorithms for approximation*, Oxford, 1987. E.J.C. Mason and M.G. Cox, eds.

[18] Temme N.M. A set of algorithms for the incomplete gamma functions. To appear in *Probability in the Engineering and Informational Sciences*, 1994.

[19] Tricomi F.G. Asymptotische Eigenschaften der unvollständigen Gammafunktion. *Math. Z.*, **53**:136–148, 1950.

International Series of Numerical Mathematics, Vol. 119, ©1994 Birkhäuser

SENSITIVITY ANALYSIS FOR COMPUTING ORTHOGONAL POLYNOMIALS OF SOBOLEV TYPE

Minda Zhang*

Dedicated to Walter Gautschi on the occasion of his 65th birthday

Abstract. A modified Chebyshev algorithm for computing a sequence of monic orthogonal polynomials of Sobolev type was introduced in [6]. Numerical experiments indicated that the sensitivity of the modified moments in the algorithm was less than that of ordinary moments. In addition, the sensitivity was greater for associated measures of unbounded support than for measures with bounded support. In this article, we obtain bounds for the derivative of the recursion coefficient matrix with respect to the input moments, and show that the behavior of the algorithm can be explained satisfactorily in terms of those bounds.

1 INTRODUCTION

In [1], [6], we developed and discussed a modified Chebyshev algorithm as a computational process for computing a sequence of monic orthogonal polynomials $\{\pi_k(x)\}_{k=0}^n$ of Sobolev type for any given Sobolev space $H_1(R, d\lambda_0, d\lambda_1)$.

Numerical experiments from [1], [6] on the modified Chebyshev algorithm indicate that the error growth due to small perturbations in modified moments for the modified Chebyshev process is much slower than error growth due to small perturbations in ordinary moments. However, if the pair of measures $(d\lambda_0, d\lambda_1)$ has unbounded support, then the process has an error growth behavior much worse than for $(d\lambda_0, d\lambda_1)$ with bounded support.

Our objective in this paper is to study the sensitivity of the recursion coefficient matrix B defined in [1], [6] with respect to small perturbations in the input moments $\nu_k^{(i)}$ [1], [6], for $k = 0, 1, 2, \ldots, 2n - 1$ and $i = 0, 1$. We obtain upper bounds for $\|\frac{dB}{d\nu_j^{(i)}}\|$, for $i = 0, 1$; $j = 0, 1, \ldots, 2n - 1$. Based on these bounds, the behavior of the algorithm in the presence of small perturbations, as observed in a number of numerical experiments, can be explained satisfactorily.

2 SENSITIVITY OF THE MAP $\widetilde{K}_n$

We define $\widetilde{\nu}^{(0)} = [\widetilde{\nu}_0^{(0)}, \widetilde{\nu}_1^{(0)}, \ldots, \widetilde{\nu}_{2n-1}^{(0)}]^T$ and $\widetilde{\nu}^{(1)} = [\widetilde{\nu}_0^{(1)}, \widetilde{\nu}_1^{(1)}, \ldots, \widetilde{\nu}_{2n-1}^{(1)}]^T$, the vectors of normalized modified moments, $\widetilde{\nu}_k^{(i)} = \int_R \widetilde{p}_k(x) d\lambda_i(x)$, where $\widetilde{p}_k(x)$ is

*CH10-21, Intel Corporation, 5000 W. Chandler Boulevard, Chandler, AZ 85226-3699, USA; mzhang@arthur.intel.com

the normalized polynomial $p_k(x)$ considered in [1], [6]. We propose to investigate the map

$$\widetilde{K_n} : R^{4n} \to R^{\frac{n(n+1)}{2}} \tag{2.1}$$

from these modified moments to the recursion coefficient matrix $\widetilde{B} = (\widetilde{\beta}_j^i)$, which generates the normalized orthogonal polynomials $\{\widetilde{\pi}_k(x)\}$ of Sobolev type in a given Sobolev space $H_1(R, d\lambda_0, d\lambda_1)$ by means of

$$\widetilde{\beta}_{-1}^k \widetilde{\pi}_{k+1}(x) = x\widetilde{\pi}_k(x) - \sum_{j=0}^{k} \widetilde{\beta}_j^k \widetilde{\pi}_{k-j}(x). \tag{2.2}$$

For the purpose of studying the stability of the map $\widetilde{K_n}$, it is convenient to think of $\widetilde{K_n}$ as a composition of two maps

$$\widetilde{K_n} = \widetilde{H_n} \circ \widetilde{G_n}, \tag{2.3}$$

where $\widetilde{G_n}$ maps $(\widetilde{\nu}_k^{(0)}, \widetilde{\nu}_k^{(1)})$ to the moment matrices $(\widetilde{M}, \widetilde{M_1})$, and $\widetilde{H_n}$ maps $(\widetilde{M}, \widetilde{M_1})$ to the recursion coefficient matrix $\widetilde{B}$; i.e.,

$$\widetilde{G_n}(\widetilde{\nu}_k^{(0)}, \widetilde{\nu}_k^{(1)}) = (\widetilde{M}, \widetilde{M_1}), \tag{2.4}$$

and

$$\widetilde{H_n}(\widetilde{M}, \widetilde{M_1}) = \widetilde{B}, \tag{2.5}$$

where

$$\widetilde{M} = \begin{bmatrix} (\widetilde{p}_0, \widetilde{p}_0)_{H_1} & (\widetilde{p}_1, \widetilde{p}_0)_{H_1} & \cdots & (\widetilde{p}_{n-1}, \widetilde{p}_0)_{H_1} \\ (\widetilde{p}_0, \widetilde{p}_1)_{H_1} & (\widetilde{p}_1, \widetilde{p}_1)_{H_1} & \cdots & (\widetilde{p}_{n-1}, \widetilde{p}_1)_{H_1} \\ \vdots & \vdots & \vdots & \vdots \\ (\widetilde{p}_0, \widetilde{p}_{n-1})_{H_1} & (\widetilde{p}_1, \widetilde{p}_{n-1})_{H_1} & \cdots & (\widetilde{p}_{n-1}, \widetilde{p}_{n-1})_{H_1} \end{bmatrix} \tag{2.6}$$

and

$$\widetilde{M_1} = \begin{bmatrix} (x\widetilde{p}_0, \widetilde{p}_0)_{H_1} & (x\widetilde{p}_1, \widetilde{p}_0)_{H_1} & \cdots & (x\widetilde{p}_{n-1}, \widetilde{p}_0)_{H_1} \\ (x\widetilde{p}_0, \widetilde{p}_1)_{H_1} & (x\widetilde{p}_1, \widetilde{p}_1)_{H_1} & \cdots & (x\widetilde{p}_{n-1}, \widetilde{p}_1)_{H_1} \\ \vdots & \vdots & \vdots & \vdots \\ (x\widetilde{p}_0, \widetilde{p}_{n-1})_{H_1} & (x\widetilde{p}_1, \widetilde{p}_{n-1})_{H_1} & \cdots & (x\widetilde{p}_{n-1}, \widetilde{p}_{n-1})_{H_1} \end{bmatrix}. \tag{2.7}$$

Here the inner product $(., .)_{H_1}$ in Sobolev space $H_1(R, d\lambda_0, d\lambda_1)$ is defined by

$$(f, g)_{H_1} = \int_R f(x)g(x)d\lambda_0 + \int_R f'(x)g'(x)d\lambda_1. \tag{2.8}$$

In what follows, we assume that the pair of measures $(d\lambda_0, d\lambda_1)$ is such that the space of all polynomials P is a subspace of $H_1(R, d\lambda_0, d\lambda_1)$.

Now $\widetilde{M}$ is an $n \times n$ symmetric positive definite matrix. This fact follows readily from the expression (2.6) by calculating a quadratic form $x^T \widetilde{M} x$ for any $x = (x_1, x_2, \ldots, x_n)^T$ in R^n, producing

$$x^T \widetilde{M} x = \int_R \left(\sum_{i=1}^n x_i \widetilde{p}_{i-1}(x) \right)^2 d\lambda_0 + \int_R \left(\sum_{i=1}^n x_i \widetilde{p}'_{i-1}(x) \right)^2 d\lambda_1.$$

Hence, $x^T \widetilde{M} x > 0$, if $x \neq 0$, proving our assertion. It should be noted that $\widetilde{M}_1$ is not symmetric if $d\lambda_1 \neq 0$.

2.1 Sensitivity of the map $\widetilde{G}_n$

To get an upper bound for the sensitivity of the map $\widetilde{G}_n$ with respect to small perturbations in the moment vectors $(\widetilde{\nu}^{(0)}, \widetilde{\nu}^{(1)})$, we begin by establishing relations between the normalized modified moment vectors $(\widetilde{\nu}^{(0)}, \widetilde{\nu}^{(1)})$ and the moment matrices $(\widetilde{M}, \widetilde{M}_1)$. To do this, we denote $P(x) = (\widetilde{p}_0(x), \widetilde{p}_1(x), \ldots, \widetilde{p}_{2n-1}(x))^T$ and by e_k as the k-th coordinate vector of appropriate dimension. We assume throughout that $e_j = 0$ if $j \leq 0$ or $j \geq 2n+1$. Furthermore, we denote by T the $2n \times 2n$ symmetric tridiagonal matrix

$$T = \begin{bmatrix} a_0 & \sqrt{b_1} & & & \\ \sqrt{b_1} & a_1 & \sqrt{b_2} & & \\ & \ddots & \ddots & \ddots & \\ & & \sqrt{b_{2n-2}} & a_{2n-2} & \sqrt{b_{2n-1}} \\ & & & \sqrt{b_{2n-1}} & a_{2n-1} \end{bmatrix}, \qquad (2.1.1)$$

where the a_j's are real and the b_j's nonnegative.

We first describe a simple shift property of the operator acting on coordinate vectors.

Lemma 2.1.1 If $X_k \in \mathrm{span}\{e_{2n-k+1}, e_{2n-k+2}, \ldots, e_{2n}\}$, i.e., $X_k = \sum_{j=0}^{k-1} x_j^{(k)} \cdot e_{2n-j}$, then $X_{k+1} = TX_k$ is a vector in $\mathrm{span}\{e_{2n-k}, e_{2n-k+1}, \ldots, e_{2n}\}$, i.e

$$X_{k+1} = \sum_{j=0}^k x_j^{(k+1)} e_{2n-j}. \qquad (2.1.2)$$

Proof: By (2.1.1), we obtain

$$X_{k+1} = TX_k = \sum_{j=0}^{k-1} x_j^{(k)} T e_{2n-j}$$

$$= \sum_{j=0}^{k-1} x_j^{(k)} \left(\sqrt{b_{2n-j-1}} e_{2n-j-1} + a_{2n-j-1} e_{2n-j} + \sqrt{b_{2n-j}} e_{2n-j+1} \right)$$

$$= \sum_{j=0}^k x_j^{(k+1)} e_{2n-j},$$

where

$$\begin{cases} x_k^{(k+1)} = x_{k-1}^{(k)}\sqrt{b_{2n-k}} \\ x_{k-1}^{(k+1)} = x_{k-2}^{(k)}\sqrt{b_{2n-k+1}} + x_{k-1}^{(k)}a_{2n-k} \\ x_j^{(k+1)} = x_{j-1}^{(k)}\sqrt{b_{2n-j}} + x_j^{(k)}a_{2n-j-1} + x_{j+1}^{(k)}\sqrt{b_{2n-j-1}} \quad \text{for} \quad j = 1, 2, \ldots k-2, \\ x_0^{(k+1)} = x_0^{(k)}a_{2n-1} + x_1^{(k)}\sqrt{b_{2n-1}}. \end{cases}$$ ∎

Throughout the rest of the paper, $[I_n \ 0_n]$ is an $n \times (2n)$ matrix, where I_n stands for the $n \times n$ identity matrix and 0_n for the $n \times n$ zero matrix. Furthermore, the sequence of orthonormal polynomials $\{\widetilde{p}_k(x)\}$ satisfies a three-term recursion relation

$$\sqrt{b_{k+1}}\widetilde{p}_{k+1}(x) = (x - a_k)\widetilde{p}_k(x) - \sqrt{b_k}\widetilde{p}_{k-1}(x). \tag{2.1.3}$$

Proposition 2.1.2 *For $k = 0, 1, \ldots, n - 1$, we have*

$$[I_n \ 0_n](\widetilde{p}_k(x)P(x)) = [I_n \ 0_n](\widetilde{p}_k(T)P(x)). \tag{2.1.4}$$

Proof: Using the fact that $[I_n \ 0_n]e_{2n-k} = 0$ for $k = 0, 1, \ldots, n - 1$, one can see that (2.1.4) is a consequence of the identity

$$\widetilde{p}_k(x)P(x) = \widetilde{p}_k(T)P(x) + X_k(x), \quad \text{where} \quad X_k(x) = \sum_{j=0}^{k-1} x_j^{(k)}(x)e_{2n-j}. \tag{2.1.5}$$

We now prove (2.1.5) by induction.

The case of $k = 0$ i.e., $\widetilde{p}_0(x)P(x) = \widetilde{p}_0(T)P(x)$ is obvious, since $\widetilde{p}_0(x)$ is a constant.

Now we prove (2.1.5) in the case $k + 1$ under the assumption that (2.1.5) is true up to k. For $j = 0, 1, \ldots, 2n - 1$, using (2.1.3), we have

$$\begin{aligned} \sqrt{b_{k+1}}\widetilde{p}_{k+1}(x)\widetilde{p}_j(x) &= \Big((x - a_k)\widetilde{p}_k(x) - \sqrt{b_k}\widetilde{p}_{k-1}(x)\Big)\widetilde{p}_j(x) \\ &= \Big((x - a_j)\widetilde{p}_k(x) + (a_j - a_k)\widetilde{p}_k(x) - \sqrt{b_k}\widetilde{p}_{k-1}(x)\Big)\widetilde{p}_j(x) \\ &= \Big((x - a_j)\widetilde{p}_j(x) - \sqrt{b_j}\widetilde{p}_{j-1}(x)\Big)\widetilde{p}_k(x) \\ &\quad + (a_j - a_k)\widetilde{p}_k(x)\widetilde{p}_j(x) + \sqrt{b_j}\widetilde{p}_{j-1}(x)\widetilde{p}_k(x) - \sqrt{b_k}\widetilde{p}_{k-1}(x)\widetilde{p}_j(x) \\ &= \Big(\sqrt{b_{j+1}}\widetilde{p}_{j+1}(x) + (a_j - a_k)\widetilde{p}_j(x) + \sqrt{b_j}\widetilde{p}_{j-1}(x)\Big)\widetilde{p}_k(x) \\ &\quad - \sqrt{b_k}\widetilde{p}_{k-1}(x)\widetilde{p}_j(x). \end{aligned} \tag{2.1.6}$$

By Lemma 2.1.1 and introducing $P(x)$, we can rewrite (2.1.6) in matrix form as

$$\sqrt{b_{k+1}}\widetilde{p}_{k+1}(x)P(x)$$
$$= (T - a_k I)\left(\widetilde{p}_k(x)P(x)\right) - \sqrt{b_k}\left(\widetilde{p}_{k-1}(x)P(x)\right) + \sqrt{b_{2n}}\widetilde{p}_{2n}(x)e_{2n}$$
$$= \left((T - a_k I)\widetilde{p}_k(T) - \sqrt{b_k}\widetilde{p}_{k-1}(T)\right)P(x) + \sqrt{b_{k+1}}X_{k+1}(x)$$
$$= \sqrt{b_{k+1}}\left(\widetilde{p}_{k+1}(T)P(x) + X_{k+1}(x)\right),$$

$$(2.1.7)$$

where

$$X_{k+1}(x) = \frac{1}{\sqrt{b_{k+1}}}\left((T - a_k I)X_k(x) - \sqrt{b_k}X_{k-1}(x) + \sqrt{b_{2n}}\widetilde{p}_{2n}(x)e_{2n}\right)$$

$$(2.1.8)$$

$$= \sum_{j=0}^{k} x_j^{(k+1)}(x)e_{2n-j}.$$

This completes the induction step, and proves the proposition. ∎

It is evident that $\widetilde{p}_k'(x)$ can be expressed as a linear combination of $\{\widetilde{p}_0(x)\}_{j=0}^{k-1}$, i.e.,

$$\widetilde{p}_k'(x) = \sum_{j=1}^{k} \tau_j^k \widetilde{p}_{j-1}(x).$$

$$(2.1.9)$$

As a consequence, if we define the $(2n) \times (2n)$ matrix τ by (cf. [1] [6])

$$\tau = \begin{bmatrix} 0 & & & & & \\ \tau_0^1 & 0 & & & & \\ \vdots & \vdots & & & & \\ \tau_0^{2n-2} & \tau_1^{2n-2} & \cdots & \tau_{2n-3}^{2n-2} & 0 & \\ \tau_0^{2n-1} & \tau_1^{2n-1} & \cdots & \tau_{2n-3}^{2n-1} & \tau_{2n-2}^{2n-1} & 0 \end{bmatrix},$$

$$(2.1.10)$$

we can rewrite (2.1.9) in matrix form as

$$P'(x) = \tau P(x).$$

$$(2.1.11)$$

Corollary 2.1.3 For $k = 0, 1, \ldots, n-1$, we have

$$[I_n\ 0_n](\widetilde{p}_k'(x)P'(x)) = [I_n\ 0_n](\tau\widetilde{p}_k'(T)P(x)),$$

$$(2.1.12)$$

where τ is defined in (2.1.10).

Proof: By (2.1.9), (2.1.11) and using Proposition 2.1.2 for $k = 0, 1, \ldots, n-1$, we have

$$[I_n \ 0_n](\widetilde{p}_k'(x)P'(x)) = [I_n \ 0_n]\sum_{j=0}^{k-1}\tau_j^k\widetilde{p}_j(x)\tau P(x) = [I_n \ 0_n]\tau\sum_{j=0}^{k-1}\tau_j^k(\widetilde{p}_j(x)P(x))$$

$$= [I_n \ 0_n]\tau\sum_{j=0}^{k-1}\tau_j^k\left(\widetilde{p}_j(T)P(x) + X_j(x)\right)$$

$$= [I_n \ 0_n]\tau\widetilde{p}_k'(T)P(x) + \sum_{j=0}^{k-1}\tau_j^k[I_n \ 0_n]\tau X_j(x).$$

(2.1.13)

Bearing in mind that τ is a strict lower triangular matrix and using (2.1.5), one can easily see that $\tau X_j(x) \in \text{span}\{e_{2n-j+2}, e_{2n-j+3}, \ldots, e_{2n}\}$, for $j = 0, 1, \ldots, n-1$. Hence, $[I_n \ 0_n]\tau X_j(x) = 0$, i.e., (2.1.13) yields the corollary. ∎

To describe the map $\widetilde{G}_n$, we still need to establish expressions for $[I_n \ 0_n] \cdot (x\widetilde{p}_k(x)P(x))$ and $[I_n \ 0_n](x\widetilde{p}_k(x))'P'(x)$. To achieve this, we use a matrix form of (2.1.3), i.e, $xP(x) = TP(x) + \sqrt{b_{2n}}\widetilde{p}_{2n}(x)e_{2n}$. Thus,

$$[I_n \ 0_n](x\widetilde{p}_k(x)P(x)) = [I_n \ 0_n](\widetilde{p}_k(x)(xP(x)))$$

$$= [I_n \ 0_n](\widetilde{p}_k(x)(TP(x) + \sqrt{b_{2n}}\widetilde{p}_{2n}(x)e_{2n})$$

$$= [I_n \ 0_n](T(\widetilde{p}_k(x)P(x)))$$

$$= [I_n \ 0_n]T\widetilde{p}_k(T)P(x) + [I_n \ 0_n]TX_k(x).$$

(2.1.14)

By Lemma 2.1.1, we have $TX_k(x) \in \text{span}\{e_{2n-k}, e_{2n-k+1}, \ldots, e_{2n}\}$, for $k = 0, 1, \ldots, n-1$, producing $[I_n \ 0_n]TX_k(x) = 0$. Hence, (2.1.14) yields

$$[I_n \ 0_n](x\widetilde{p}_k(x)P(x)) = [I_n \ 0_n](x\widetilde{p}_k)(T)P(x).$$

(2.1.15)

Similarily, we have

$$[I_n \ 0_n]((x\widetilde{p}_k(x))'P'(x)) = [I_n \ 0_n]\tau(x\widetilde{p}_k)'(T)P(x).$$

(2.1.16)

Now, we are in a position to describe the map $\widetilde{G}_n$, i.e., $\widetilde{G}_n(\widetilde{\nu}^{(0)}, \widetilde{\nu}^{(1)}) = (\widetilde{M}, \widetilde{M}_1)$. By (2.6), (2.7), (2.8), (2.1.4), (2.1.12), (2.1.15) and (2.1.16), we have, for $k = 0, 1, \ldots, n-1$,

$$\widetilde{M}e_{k+1} = (\widetilde{p}_k(x), [I_n \ 0_n]P(x))_{H_1}$$

$$= \int_R [I_n \ 0_n](\widetilde{p}_k(x)P(x))d\lambda_0 + \int_R [I_n \ 0_n](\widetilde{p}_k'(x)P'(x))d\lambda_1$$

$$= [I_n \ 0_n]\int_R \widetilde{p}_k(T)P(x)d\lambda_0 + [I_n \ 0_n]\tau\int_R \widetilde{p}_k'(T)P(x)d\lambda_1$$

$$= [I_n \ 0_n](\widetilde{p}_k(T)\widetilde{\nu}^{(0)} + \tau\widetilde{p}_k'(T)\widetilde{\nu}^{(1)}).$$

(2.1.17)

Similarily,

$$\widetilde{M}_1 e_{k+1} = [I_n \ 0_n]\Big((x\widetilde{p}_k)(T)\widetilde{\nu}^{(0)} + \tau(x\widetilde{p}_k)'(T)\widetilde{\nu}^{(1)}\Big). \tag{2.1.18}$$

The relations (2.1.17) and (2.1.18) indicate that the map $\widetilde{G}_n$ can be thought of as a linear transformation. That is, once the orthonormal polynomial sequence $\{\widetilde{p}_k(x)\}$ has been chosen, the perturbations $\Delta\widetilde{M}$, $\Delta\widetilde{M}_1$ of the moment matrices $\widetilde{M}, \widetilde{M}_1$ caused by errors $\Delta\widetilde{\nu}_j^{(i)}$ in the input modified moments $\widetilde{\nu}_j^{(i)}$ are linear functions of these input errors. Based on this observation, we can obtain an upper bound for the amount of perturbation of $\widetilde{M}$ and $\widetilde{M}_1$ in terms of errors in the input modified moments.

To do this, we need the following definitions.

Definition 2.1.4 *For* $i = 0, 1; \ j = 0, 1, \ldots, 2n - 1$, *define*

$$\begin{cases} \dfrac{d\widetilde{M}}{d\widetilde{\nu}_j^{(i)}} = \lim_{|\Delta\widetilde{\nu}_j^{(i)}| \to 0} \dfrac{\Delta\widetilde{M}}{\Delta\widetilde{\nu}_j^{(i)}}, \\[2mm] \dfrac{d\widetilde{M}_1}{d\widetilde{\nu}_j^{(i)}} = \lim_{|\Delta\widetilde{\nu}_j^{(i)}| \to 0} \dfrac{\Delta\widetilde{M}_1}{\Delta\widetilde{\nu}_j^{(i)}}, \\[2mm] \dfrac{d\widetilde{B}}{d\widetilde{\nu}_j^{(i)}} = \lim_{|\Delta\widetilde{\nu}_j^{(i)}| \to 0} \dfrac{\Delta\widetilde{B}}{\Delta\widetilde{\nu}_j^{(i)}}. \end{cases}$$

Evidently, $\frac{d\widetilde{M}}{d\widetilde{\nu}_j^{(i)}}, \frac{d\widetilde{M}_1}{d\widetilde{\nu}_j^{(i)}}$ and $\frac{d\widetilde{B}}{d\widetilde{\nu}_j^{(i)}}$ are $n \times n$ matrices.

Proposition 2.1.5 *Let* $\{x_i\}_{i=1}^{2n}$ *be the zeros of the orthonormal polynomial* $\widetilde{p}_{2n}(x)$. *Then for* $j = 0, 1, \ldots, 2n - 1$,

$$\begin{cases} \left\| \dfrac{d\widetilde{M}}{d\widetilde{\nu}_j^{(0)}} \right\|_F \leq \left\{ \displaystyle\sum_{k=0}^{n-1} (\max_{1 \leq i \leq 2n} |\widetilde{p}_k(x_i)|)^2 \right\}^{\frac{1}{2}}, \\[4mm] \left\| \dfrac{d\widetilde{M}}{d\widetilde{\nu}_j^{(1)}} \right\|_F \leq \|\tau\|_2 \left\{ \displaystyle\sum_{k=0}^{n-1} (\max_{1 \leq i \leq 2n} |\widetilde{p}_k'(x_i)|)^2 \right\}^{\frac{1}{2}}, \\[4mm] \left\| \dfrac{d\widetilde{M}_1}{d\widetilde{\nu}_j^{(0)}} \right\|_F \leq \left\{ \displaystyle\sum_{k=0}^{n-1} (\max_{1 \leq i \leq 2n} |(x\widetilde{p}_k)(x_i)|)^2 \right\}^{\frac{1}{2}}, \\[4mm] \left\| \dfrac{d\widetilde{M}_1}{d\widetilde{\nu}_j^{(1)}} \right\|_F \leq \|\tau\|_2 \left\{ \displaystyle\sum_{k=0}^{n-1} (\max_{1 \leq i \leq 2n} |(x\widetilde{p}_k)'(x_i)|)^2 \right\}^{\frac{1}{2}}, \end{cases}$$

where $\|\tau\|_2$ *is the spectral norm of the matrix* τ.

569

Proof: It is sufficient to consider the case of $\|\frac{d\widetilde{M}}{d\widetilde{\nu}_j^{(0)}}\|_F$, the other cases being treated similarily. We begin with the identity $\|\widetilde{p}_k(T)\|_2 = \max_{1 \leq i \leq 2n} |\widetilde{p}_k(x_i)|$. This follows readily from the equation $TP(x) = xP(x) - \sqrt{b_{2n}}\widetilde{p}_{2n}(x)e_{2n}$, showing that the zeros of $\widetilde{p}_{2n}(x)$ are eigenvalues of the matrix T, and the fact that the spectral norm of a symmetric matrix is the largest absolute value of all its eigenvalues. Therefore, by (2.1.17) and Definition 2.1.4, we have

$$\left\|\frac{d\widetilde{M}}{d\widetilde{\nu}_j^{(0)}}\right\|_F^2 = \sum_{k=0}^{n-1} \lim_{|\Delta\widetilde{\nu}_j^{(0)}| \to 0} \left(\frac{\|\Delta(\widetilde{M}e_{k+1})\|_2}{|\Delta\widetilde{\nu}_j^{(0)}|}\right)^2 \leq \sum_{k=0}^{n-1} \left\|[I_n \ 0_n]\widetilde{p}_k(T)\right\|_2^2$$

$$\leq \sum_{k=0}^{n-1} \|\widetilde{p}_k(T)\|_2^2 = \sum_{k=0}^{n-1} \left(\max_{1 \leq i \leq 2n} |\widetilde{p}_k(x_i)|\right)^2. \qquad \blacksquare$$

It is worth pointing out that $\widetilde{M}$ and $\widetilde{M}_1$, and hence the map $\widetilde{G}_n$, are well behaved under perturbations on the modified moments , if $\{\widetilde{p}_k(x)\}_{k=0}^{2n-1}$ is a sequence of orthonormal polynomials with respect to a measure with bounded support. The simplest special case, the Chebyshev polynomials of the first kind, furnishes a good illustration of this, since both $|\widetilde{p}_k(x_i)|$ and $|\widetilde{p}'_k(x_i)|$ are relatively small in this case.

However, if the sequence $\{\widetilde{p}_k(x)\}_{k=0}^{2n-1}$ is orthonormal with respect to a measure with unbounded support, then the map $\widetilde{G}_n$ would become much more sensitive to the perturbations in the modified moments. This can be seen roughly by the special case of normalized Laguerre polynomials. In fact, it is known [2, page 115] that the largest zero x_0 of $L_{2n-1}^{(\alpha)}(x)$, i.e., the largest eigenvalue of the matrix T satisfies $x_0 > 4n + \alpha - 2$, and

$$\widetilde{p}_n(x) = \widetilde{L}_n^{(\alpha)}(x) = \left[\binom{n+\alpha}{n}\Gamma(\alpha+1)\right]^{-\frac{1}{2}} L_n^{(\alpha)}(x).$$

So, if $x \approx 4n$, then [2, Eq. (8.22.1)] $L_n^{(\alpha)}(x) \approx n^{-\frac{1}{3}}e^{\frac{x}{2}}$, and since the largest eigenvalue of T is of the order $4n$, we arrive at $\widetilde{p}_n(x_0) \approx \binom{n+\alpha}{n} n^{-\frac{1}{3}}e^{2n}$. The exponential power e^{2n} in this expression is the decisive factor for the increased sensitivity of the map $\widetilde{G}_n$.

2.2 Sensitivity of the map $\widetilde{H}_n$

J. Kautsky and G. Golub studied the relation between the Jacobi matrix and the moment matrix in [3]. Using their techniques, we describe the map $\widetilde{H}_n$, i.e., $\widetilde{H}_n(\widetilde{M}, \widetilde{M}_1) = \widetilde{B}$, and obtain upper bounds for $\|\frac{d\widetilde{B}}{d\widetilde{\nu}_j^{(i)}}\|_F$, for $i = 0, 1; j = 0, 1, \ldots,$ $2n - 1$.

Proposition 2.2.1 *The matrix $\widetilde{B}$ of recursion coefficients can be expressed as*

$$\widetilde{B} = \widetilde{L}^{-1} \widetilde{M}_1 \widetilde{L}^{-T}, \tag{2.2.1}$$

where $\widetilde{M} = \widetilde{L}\widetilde{L}^T$ is the Cholesky decomposition of $\widetilde{M}$.

Proof: Let $\widetilde{\pi}(x) = [\widetilde{\pi}_0(x), \widetilde{\pi}_1(x), \ldots, \widetilde{\pi}_{n-1}(x)]^T$ and $\widetilde{p}(x) = [I_n \ \ 0_n]P(x)$. We rewrite (2.2) in matrix form as

$$\begin{cases} x\widetilde{\pi}(x)^T = \widetilde{\pi}(x)^T\widetilde{B} + \widetilde{\beta}_{-1}^{n-1}\widetilde{\pi}_n(x)e_n^T \\ (x\widetilde{\pi}(x)^T)' = (\widetilde{\pi}(x)^T)'\widetilde{B} + \widetilde{\beta}_{-1}^{n-1}\widetilde{\pi}_n'(x)e_n^T, \end{cases}$$

pre-multiply the two equations respectively by $\widetilde{\pi}(x)$ and $\widetilde{\pi}'(x)$, and integrate with respect to the measures $d\lambda_0, d\lambda_1$. Noting that

$$\int_R \widetilde{\pi}(x)\widetilde{\pi}(x)^T d\lambda_0 + \int_R \widetilde{\pi}'(x)\widetilde{\pi}'(x)^T d\lambda_1 = I_n \tag{2.2.2}$$

and

$$\int_R \widetilde{\pi}(x)\widetilde{\pi}_n(x)d\lambda_0 + \int_R \widetilde{\pi}'(x)\widetilde{\pi}_n'(x)d\lambda_1 = 0, \tag{2.2.3}$$

we obtain

$$\int_R \widetilde{\pi}(x)(x\widetilde{\pi}(x))^T d\lambda_0 + \int_R \widetilde{\pi}'(x)(x\widetilde{\pi}(x)^T)'d\lambda_1 = \widetilde{B}. \tag{2.2.4}$$

Since both $\{\widetilde{\pi}_0(x), \widetilde{\pi}_1(x), \ldots, \widetilde{\pi}_{n-1}(x)\}$ and $\{\widetilde{p}_0(x), \widetilde{p}_1(x), \ldots, \widetilde{p}_{n-1}(x)\}$ are bases of the polynomial space P_{n-1}, there exists a unique lower triangular matrix L such that $\widetilde{\pi}(x) = L\widetilde{p}(x)$. Hence, (2.2.2) yields

$$L\widetilde{M}L^T = I_n, \tag{2.2.5}$$

i.e., $\widetilde{M} = L^{-1}L^{-T}$. But $\widetilde{M}$ is a symmetric positive definite matrix, hence has a Cholesky decomposition $\widetilde{M} = \widetilde{L}\widetilde{L}^T$. We conclude that $\widetilde{L} = L^{-1}$. Therefore, (2.2.4) yields (2.2.1). ∎

Proposition 2.2.2 *For $i = 0, 1; j = 0, 1, \ldots, 2n - 1$, we have*

$$\left\| \frac{d\widetilde{B}}{d\widetilde{\nu}_j^{(i)}} \right\|_F \le \left\| \widetilde{M}^{-1} \right\|_2 \left(\sqrt{2}\|\widetilde{B}\|_2 \left\| \frac{d\widetilde{M}}{d\widetilde{\nu}_j^{(i)}} \right\|_F + \left\| \frac{d\widetilde{M}_1}{d\widetilde{\nu}_j^{(i)}} \right\|_F \right). \tag{2.2.6}$$

Proof: By taking the derivative with respect to each component of $\widetilde{\nu}^{(i)}$ in (2.2.1), we obtain

$$\frac{d\widetilde{B}}{d\widetilde{\nu}_j^{(i)}} = \left(\frac{d\widetilde{L}^{-1}}{d\widetilde{\nu}_j^{(i)}} \widetilde{L} \right) \left(\widetilde{L}^{-1} \widetilde{M}_1 \widetilde{L}^{-T} \right) + \widetilde{L}^{-1} \frac{d\widetilde{M}_1}{d\widetilde{\nu}_j^{(i)}} \widetilde{L}^{-T} + \left(\widetilde{L}^{-1} \widetilde{M}_1 \widetilde{L}^{-T} \right) \left(\widetilde{L}^T \frac{d\widetilde{L}^{-T}}{d\widetilde{\nu}_j^{(i)}} \right)$$

$$= \left(\frac{d\widetilde{L}^{-1}}{d\widetilde{\nu}_j^{(i)}} \widetilde{L} \right) \widetilde{B} + \widetilde{B} \left(\widetilde{L}^T \frac{d\widetilde{L}^{-T}}{d\widetilde{\nu}_j^{(i)}} \right) + \widetilde{L}^{-1} \frac{d\widetilde{M}_1}{d\widetilde{\nu}_j^{(i)}} \widetilde{L}^{-T}. \tag{2.2.7}$$

Similarily by taking the derivative in the equation $\widetilde{M}^{-1} = \widetilde{L}^{-T}\widetilde{L}^{-1}$, we obtain

$$-\widetilde{M}^{-1}\frac{d\widetilde{M}}{d\widetilde{\nu}_j^{(i)}}\widetilde{M}^{-1} = \frac{d\widetilde{M}^{-1}}{d\widetilde{\nu}_j^{(i)}} = \widetilde{L}^{-T}\frac{d\widetilde{L}^{-1}}{d\widetilde{\nu}_j^{(i)}} + \left(\frac{d\widetilde{L}^{-1}}{d\widetilde{\nu}_j^{(i)}}\right)^T \widetilde{L}^{-1}. \qquad (2.2.8)$$

Pre-multiplying (2.2.8) by $\widetilde{L}^T$ and post-multiplying (2.2.8) by $\widetilde{L}$, we get

$$\frac{d\widetilde{L}^{-1}}{d\widetilde{\nu}_j^{(i)}}\widetilde{L} + \left(\frac{d\widetilde{L}^{-1}}{d\widetilde{\nu}_j^{(i)}}\widetilde{L}\right)^T = -\widetilde{L}^{-1}\frac{d\widetilde{M}}{d\widetilde{\nu}_j^{(i)}}\widetilde{L}^{-T}. \qquad (2.2.9)$$

By a result of [4] we have that $A + A^T = S$ implies $\|A\|_F \leq \frac{1}{\sqrt{2}}\|S\|_F$. Therefore,

$$\left\|\frac{d\widetilde{L}^{-1}}{d\widetilde{\nu}_j^{(i)}}\widetilde{L}\right\|_F \leq \frac{1}{\sqrt{2}}\left\|\widetilde{L}^{-1}\frac{d\widetilde{M}}{d\widetilde{\nu}_j^{(i)}}\widetilde{L}^{-T}\right\|_F. \qquad (2.2.10)$$

Combining (2.2.7) and (2.2.10) now yields

$$\left\|\frac{d\widetilde{B}}{d\widetilde{\nu}_j^{(i)}}\right\|_F \leq \sqrt{2}\|\widetilde{B}\|_2\left\|\widetilde{L}^{-1}\frac{d\widetilde{M}}{d\widetilde{\nu}_j^{(i)}}\widetilde{L}^{-T}\right\|_F + \left\|\widetilde{L}^{-1}\frac{d\widetilde{M}_1}{d\widetilde{\nu}_j^{(i)}}\widetilde{L}^{-T}\right\|_F. \qquad (2.2.11)$$

Since $\|\widetilde{L}^{-1}\frac{d\widetilde{M}}{d\widetilde{\nu}_j^{(i)}}\widetilde{L}^{-T}\|_F \leq \|\widetilde{L}^{-1}\|_2^2\|\frac{d\widetilde{M}}{d\widetilde{\nu}_j^{(i)}}\|_F$, and $\|\widetilde{L}^{-1}\|_2^2 = \|\widetilde{M}^{-1}\|_2$, the proposition follows from (2.2.11). ∎

An immediate consequence of propositions 2.1.5 and 2.2.2 is:

Corollary 2.2.3 *Let $\{x_i\}_{i=1}^{2n}$ be the zeros of the orthonormal polynomial $\widetilde{p}_{2n}(x)$. Then for $j = 0, 1, \ldots, 2n - 1$,*

$$\left\|\frac{d\widetilde{B}}{d\widetilde{\nu}_j^{(0)}}\right\|_F \leq \left\|\widetilde{M}^{-1}\right\|_2\left(\sqrt{2}\left\|\widetilde{B}\right\|_2 + 1\right)\left\{\sum_{k=0}^{n-1}\left(\max_{1\leq i\leq 2n}|\widetilde{p}_k(x_i)|^2 + \right.\right.$$
$$\left.\left.\max_{1\leq i\leq 2n}|(x\widetilde{p}_k)(x_i)|^2\right)\right\}^{\frac{1}{2}}$$

$$\left\|\frac{d\widetilde{B}}{d\widetilde{\nu}_j^{(1)}}\right\|_F \leq \left\|\widetilde{M}^{-1}\right\|_2\|\tau\|_2\left(\sqrt{2}\left\|\widetilde{B}\right\|_2 + 1\right)\left\{\sum_{k=0}^{n-1}\left(\max_{1\leq i\leq 2n}|\widetilde{p}_k'(x_i)|^2 + \right.\right.$$
$$\left.\left.\max_{1\leq i\leq 2n}|(x\widetilde{p}_k)'(x_i)|^2\right)\right\}^{\frac{1}{2}}.$$

Corollary 2.2.3 provides computable upper bounds for both $\|\frac{d\widetilde{B}}{d\widetilde{\nu}_j^{(0)}}\|_F$ and $\|\frac{d\widetilde{B}}{d\widetilde{\nu}_j^{(1)}}\|_F$ for any sequence of orthonormal polynomials $\{\widetilde{p}_k(x)\}_{k=0}^{2n-1}$. Based on these upper bounds, we can compute an upper bound for the condition number of the map $\widetilde{K_n}$, since the condition number of the map $\widetilde{K_n}$, measured in Frobenius norm, can be defined by

$$\mathrm{cond}_F(\widetilde{K_n}) = \frac{1}{\|\widetilde{B}\|_2} \sqrt{\sum_{j=0}^{2n-1} \left(\left\| \frac{d\widetilde{B}}{d\widetilde{\nu}_j^{(0)}} \right\|_F^2 + \left\| \frac{d\widetilde{B}}{d\widetilde{\nu}_j^{(1)}} \right\|_F^2 \right)} \sqrt{\|\widetilde{\nu}^{(0)}\|_2^2 + \|\widetilde{\nu}^{(1)}\|_2^2}.$$

(2.2.12)

However, in order to gain more insight into the nature of the sensitivity of the matrix $\widetilde{B}$ to the errors in the input modified moments, we need some more analysis.

Proposition 2.2.4 *Let $\{x_i\}_{i=1}^{2n}$ be the zeros of the orthonormal polynomial $\widetilde{p}_{2n}(x)$. Then for $j = 0, 1, \ldots, 2n-1$,*

$$\begin{cases} \left\| \dfrac{d\widetilde{B}}{d\widetilde{\nu}_j^{(0)}} \right\|_F \leq (1 + \sqrt{2}) \left(\|\widetilde{B}\|_2 + |\widetilde{\beta}_{-1}^{n-1}| \right) \|\widetilde{M}^{-1}\|_2^{\frac{1}{2}} \left\{ \displaystyle\sum_{k=0}^{n} (\max_{1 \leq i \leq 2n} |\widetilde{\pi}_k(x_i)|)^2 \right\}^{\frac{1}{2}}, \\[6mm] \left\| \dfrac{d\widetilde{B}}{d\widetilde{\nu}_j^{(1)}} \right\|_F \leq (1 + \sqrt{2}) \left(\|\widetilde{B}\|_2 + |\widetilde{\beta}_{-1}^{n-1}| \right) \|\widetilde{M}^{-1}\|_2^{\frac{1}{2}} \|\tau\|_2 \left\{ \displaystyle\sum_{k=0}^{n} (\max_{1 \leq i \leq 2n} |\widetilde{\pi}_k'(x_i)|)^2 \right\}^{\frac{1}{2}}, \end{cases}$$

where $\widetilde{\beta}_{-1}^{n-1}$ is a recursion coefficient defined in (2.2).

Proof: In the proof of Proposition 2.2.1, we have seen that $\widetilde{\pi}(x) = \widetilde{L}^{-1}\widetilde{P}(x)$, where $\widetilde{M} = \widetilde{L}\widetilde{L}^T$ is the Cholesky decomposition of the symmetric positive definite matrix $\widetilde{M}$. Now we write

$$\widetilde{L}^{-1} = \begin{bmatrix} \widetilde{l}_{11} & & & \\ \widetilde{l}_{21} & \widetilde{l}_{22} & & \\ \vdots & \vdots & \ddots & \\ \widetilde{l}_{n1} & \widetilde{l}_{n2} & \cdots & \widetilde{l}_{nn} \end{bmatrix}.$$

(2.2.13)

Then, for $k = 0, 1, \ldots, n-1$

$$\widetilde{\pi}_k(x) = \sum_{s=0}^{k} \widetilde{l}_{k+1,s+1} \widetilde{p}_s(x).$$

(2.2.14)

Also, for $k = 0, 1, \ldots, n-1$ and $i = 0, 1$, we use (2.1.17), (2.1.18) and (2.2.13),

(2.2.14) to derive following relation:

$$\left(\widetilde{L}^{-1}\frac{d\widetilde{M}}{d\widetilde{\nu}_j^{(i)}}\widetilde{L}^{-T}\right)e_{k+1} = \widetilde{L}^{-1}\frac{d\widetilde{M}}{d\widetilde{\nu}_j^{(i)}}\left(\widetilde{L}^{-T}e_{k+1}\right)$$

$$= \widetilde{L}^{-1}\frac{d\widetilde{M}}{d\widetilde{\nu}_j^{(i)}}\left(\sum_{s=0}^{k}\widetilde{l}_{k+1,s+1}e_{s+1}\right) = \widetilde{L}^{-1}\sum_{s=0}^{k}\widetilde{l}_{k+1,s+1}\frac{d\widetilde{M}}{d\widetilde{\nu}_j^{(i)}}e_{s+1}$$

$$= \widetilde{L}^{-1}[I_n\ 0_n]\left\{\sum_{s=0}^{k}\widetilde{l}_{k+1,s+1}\widetilde{p}_s(T)\frac{d\widetilde{\nu}^{(0)}}{d\widetilde{\nu}_j^{(i)}} + \tau\left(\sum_{s=0}^{k}\widetilde{l}_{k+1,s+1}\widetilde{p}_s\right)'(T)\frac{d\widetilde{\nu}^{(1)}}{d\widetilde{\nu}_j^{(i)}}\right\}$$

$$= \widetilde{L}^{-1}[I_n\ 0_n]\left(\widetilde{\pi}_k(T)\frac{d\widetilde{\nu}^{(0)}}{d\widetilde{\nu}_j^{(i)}} + \tau\widetilde{\pi}_k'(T)\frac{d\widetilde{\nu}^{(1)}}{d\widetilde{\nu}_j^{(i)}}\right).$$

(2.2.15)

Similarily, we have

$$\left(\widetilde{L}^{-1}\frac{d\widetilde{M_1}}{d\widetilde{\nu}_j^{(i)}}\widetilde{L}^{-T}\right)e_{k+1} = \widetilde{L}^{-1}[I_n\ 0_n]\left((x\widetilde{\pi}_k)(T)\frac{d\widetilde{\nu}^{(0)}}{d\widetilde{\nu}_j^{(i)}} + \tau(x\widetilde{\pi}_k)'(T)\frac{d\widetilde{\nu}^{(1)}}{d\widetilde{\nu}_j^{(i)}}\right).$$

(2.2.16)

A further analysis of (2.2.16), together with (2.2) and (2.2.15), reveals that for $k = 0, 1, \ldots, n-2$,

$$\left(\widetilde{L}^{-1}\frac{d\widetilde{M_1}}{d\widetilde{\nu}_j^{(i)}}\widetilde{L}^{-T}\right)e_{k+1}$$

$$= \widetilde{L}^{-1}[I_n\ 0_n]\left\{\sum_{s=-1}^{k}\widetilde{\beta}_s^k\widetilde{\pi}_{k-s}(T)\frac{d\widetilde{\nu}^{(0)}}{d\widetilde{\nu}_j^{(i)}} + \tau\left(\sum_{s=-1}^{k}\widetilde{\beta}_s^k\widetilde{\pi}_{k-s}\right)'(T)\frac{d\widetilde{\nu}^{(1)}}{d\widetilde{\nu}_j^{(i)}}\right\}$$

$$= \sum_{s=-1}^{k}\widetilde{\beta}_s^k\left\{\widetilde{L}^{-1}[I_n\ 0_n]\left(\widetilde{\pi}_{k-s}(T)\frac{d\widetilde{\nu}^{(0)}}{d\widetilde{\nu}_j^{(i)}} + \tau\widetilde{\pi}_{k-s}'(T)\frac{d\widetilde{\nu}^{(1)}}{d\widetilde{\nu}_j^{(i)}}\right)\right\}$$

$$= \sum_{s=-1}^{k}\widetilde{\beta}_s^k\left(\widetilde{L}^{-1}\frac{d\widetilde{M}}{d\widetilde{\nu}_j^{(i)}}\widetilde{L}^{-T}\right)e_{k-s+1} = \left(\widetilde{L}^{-1}\frac{d\widetilde{M}}{d\widetilde{\nu}_j^{(i)}}\widetilde{L}^{-T}\right)\sum_{s=-1}^{k}\widetilde{\beta}_s^k e_{k-s+1},$$

that is,

$$\left(\widetilde{L}^{-1}\frac{d\widetilde{M_1}}{d\widetilde{\nu}_j^{(i)}}\widetilde{L}^{-T}\right)e_{k+1} = \left(\widetilde{L}^{-1}\frac{d\widetilde{M}}{d\widetilde{\nu}_j^{(i)}}\widetilde{L}^{-T}\right)\widetilde{B}e_{k+1}.$$

(2.2.17)

In case $k = n-1$, we have

$$\left(\widetilde{L}^{-1}\frac{d\widetilde{M}_1}{d\widetilde{\nu}_j^{(i)}}\widetilde{L}^{-T}\right)e_n =$$

$$\left(\widetilde{L}^{-1}\frac{d\widetilde{M}}{d\widetilde{\nu}_j^{(i)}}\widetilde{L}^{-T}\right)\widetilde{B}e_n + \widetilde{\beta}_{-1}^{n-1}\widetilde{L}^{-1}[I_n\ 0_n]\left(\widetilde{\pi}_n(T)\frac{d\widetilde{\nu}^{(0)}}{d\widetilde{\nu}_j^{(i)}} + \widetilde{\pi}_n'(T)\frac{d\widetilde{\nu}^{(1)}}{d\widetilde{\nu}_j^{(i)}}\right).$$

$$(2.2.18)$$

It follows from (2.2.17) and (2.2.18) that

$$\widetilde{L}^{-1}\frac{d\widetilde{M}_1}{d\widetilde{\nu}_j^{(i)}}\widetilde{L}^{-T} = \widetilde{L}^{-1}\frac{d\widetilde{M}}{d\widetilde{\nu}_j^{(i)}}\widetilde{L}^{-T}\widetilde{B} + \widetilde{\beta}_{-1}^{n-1}e_n^T\widetilde{L}^{-1}[I_n\ 0_n]\left(\widetilde{\pi}_n(T)\frac{d\widetilde{\nu}^{(0)}}{d\widetilde{\nu}_j^{(i)}} + \widetilde{\pi}_n'(T)\frac{d\widetilde{\nu}^{(1)}}{d\widetilde{\nu}_j^{(i)}}\right).$$

$$(2.2.19)$$

Now the matrices $\widetilde{\pi}_k(T)$ and $\widetilde{\pi}_k'(T)$ are symmetric and $\widetilde{\pi}_k(x_i)$ for $i = 1, 2, \ldots,$ $2n$ are the eigenvalues of the matrix $\widetilde{\pi}_k(T)$. Likewise, $\widetilde{\pi}_k'(x_i)$ for $i = 1, 2, \ldots, 2n$ are the eigenvalues of the matrix $\widetilde{\pi}_k'(T)$. Hence, $\|\widetilde{\pi}_k(T)\|_2 = \max_{1 \le i \le 2n} |\widetilde{\pi}_k(x_i)|$ and $\|\widetilde{\pi}_k'(T)\|_2 = \max_{1 \le i \le 2n} |\widetilde{\pi}_k'(x_i)|$. Combining (2.2.15) and (2.2.19), (2.2.11) yields the proposition. ∎

3 CONCLUSION

Proposition 2.2.4 illustrates the factors responsible for the magnitudes of $\|\frac{d\widetilde{B}}{d\widetilde{\nu}_j^{(i)}}\|_F$, for $i = 0, 1$; $j = 0, 1, \ldots, 2n - 1$, and $\text{cond}_F(\widetilde{K}_n)$. The first factor is the norm $\|\widetilde{M}^{-1}\|_2^{\frac{1}{2}}$, which is solely determined by the properties of the sequence $\{\widetilde{p}_k(x)\}_{k=0}^{2n-1}$. The remaining factors depend on the properties of the orthogonal polynomials of Sobolev type, $\{\widetilde{\pi}_k(x)\}$.

Evidently, the Chebyshev algorithm based on ordinary moments is severe unstable, as was observed in a number of numerical experiments [5]. The reason is that the norm of the resulting inverse moment matrix $\widetilde{M}$ is very large. The same was observed when the algorithm is used with modified moments, if the underlying measures have unbounded support [5]. The reason is that $\max_i |\widetilde{\pi}_k(x_i)|$ would be a large quantity, as discussed in Section 2.1.

However, the Chebyshev algorithm usually is stable if the pair of measures $(d\lambda_0, d\lambda_1)$ has bounded support, if one chooses $\{\widetilde{p}_k(x)\}$ to be the sequence of orthonormal polynomials with respect to the measure $d\lambda_0$. Then, not only the value, and also the derivative, of $\widetilde{\pi}_k(x)$ at the point x_i in the support of $d\lambda_0$ are quite small, as discussed in Section 2.1, but also for this particular choice of the modified moments, we have $\|\widetilde{M}^{-1}\|_2 = 1$. Thus, it can be concluded from Proposition 2.2.4 that the magnitude of the sensitivity of the matrix $\widetilde{B}$ is under control.

The assertion $\|\widetilde{M}^{-1}\|_2 = 1$ in the last case follows readily from the identity $\|\widetilde{M}^{-1}\|_2 = (\min_{\|x\|_2=1} x^T \widetilde{M} x)^{-1}$. In fact, if $x = (x_0, x_1, \ldots, x_{n-1})^T$ and $\|x\|_2 = 1$, then

$$x^T \widetilde{M} x = \int_R \left(\sum_{j=0}^{n-1} x_j \widetilde{p}_j(x) \right)^2 d\lambda_0 + \int_R \left(\sum_{j=0}^{n-1} x_j \widetilde{p}_j'(x) \right)^2 d\lambda_1$$

$$\geq \int_R \left(\sum_{j=0}^{n-1} x_j \widetilde{p}_j(x) \right)^2 d\lambda_0 = \sum_{j=0}^{n-1} x_j^2 \int_R \widetilde{p}_j(x)^2 d\lambda_0 = 1,$$

i.e., $\min_{\|x\|_2=1} x^T \widetilde{M} x \geq 1$. On the other hand, we have

$$\min_{\|x\|_2=1} x^T \widetilde{M} x \leq e_1^T \widetilde{M} e_1 = \int_R \widetilde{p}_0(x)^2 d\lambda_0 = 1,$$

proving the assertion.

REFERENCES

[1] Gautschi W., Zhang M. Computing orthogonal polynomials in Sobolev spaces. In preparation.

[2] Szegö G. *Orthogonal Polynomials*. American Mathematics Society, 1939.

[3] Kautsky J., Golub G.H. On the calculation of Jacobi matrices. *Linear Algebra and Its Applications*, **52-53**, 1983.

[4] Stewart G.W. Perturbation bounds for the QR factorization of a matrix. *SIAM J. Numer. Anal.*, **64**, 1977.

[5] Gautschi W. On generating orthogonal polynomials. *SIAM J. Sci. Stat. Comput.*, **3**:289–317, 1982.

[6] Zhang M. *Orthogonal Polynomials in Sobolev Spaces: Computational Method*. Ph.D dissertation, Purdue University, West Lafayette, Indiana, USA , 1993.

SUBJECT INDEX

Author Index

Group photograph of the Purdue Conference Participants, December 3, 1993.

Front Row: Charles Micchelli, Amos Carpenter, Shikang Li, Arieh Iserles, Dirk Laurie, Daniella Calvetti, Walter Gautschi, Luigi Gatteschi, Giuseppe Mastroianni, Gradimir Milovanović, Ward Cheney, James Lyness

Second Row: Ray Zahar, Frank Olver, Lothar Reichel, Martin Gutknecht, Ed Saff, Frank Stenger, Sotiros Notaris, Dick Askey, John Butcher, Richard Varga, Jim Douglas

Third Row: Henry Landau, Roderick Wong, Bernard Flury, Jacob Korevaar, Daniel Lozier, Giovanni Monegato, Ted Rivlin, Nico Temme, Jet Wimp, Carl Rohwer

Top Row: John Rice, Sven Ehrich, David Masson, Bill Jones, Martin Muldoon, Gene Golub, Jeff Cash, Paul Van Dooren, Carl de Boor, Jörg Peters, Tom Kunkle, Lionello Pasquini